Symposia D, E and F from MRS International Materials Research Conference 2008

Journal of Physics: Conference Series Volume 152

Chongqing, China
9-12 June 2008

ISBN: 978-1-61738-386-1

Printed from e-media with permission by:

Curran Associates, Inc.
57 Morehouse Lane
Red Hook, NY 12571
www.proceedings.com

Some format issues inherent in the e-media version may also appear in this print version.

Printed by Curran Associates, Inc. (2010)

Institute of Physics
Dirac House, Temple Back
Bristol BS1 6BE UK

Phone: 44 1 17 929 7481
Fax: 44 1 17 920 0979

techtracking@iop.org

TABLE OF CONTENTS

ELECTRONIC MATERIALS

PROCESSES FOR FLAT-PANEL DISPLAYS

FUNCTIONAL CERAMICS

PREFACE

This volume contains selected papers presented at the MRS International Materials Research Conference (IMRC-2008) held in Chongqing, China, 9–12 June 2008. IMRC-2008 included 9 symposia of A. Eco/Environmental Materials, B. Sustainable Energy Materials, C. Electronic Packaging Materials, D. Electronic Materials, E. Materials and Processes for Flat-panel Displays, F. Functional Ceramics, G. Transportation Materials, H. Magnesium and I. Biomaterials for Medical Applications. Nearly 1200 participants from 33 countries attended the conference, and the conference organizers received more than 700 papers. After the peer review processes, 555 papers were selected to be published in 9 Journals or proceedings, including *J. of Materials Research* (JMR), *Rare Metal Materials and Engineering*, *J. of Univ. Science and Technology Beijing*, *Biomedical Materials: Materials for Tissue Engineering and Regenerative Medicine*, *Chinese Journal of Aeronautics*, *Materials Science Forum*, and *Journal of Physics: Conference Series*. Among the 555 selected papers, 91 papers are published in this volume, and the topics mainly cover electronic matrials, processes for flat-panel displays and functional ceramics.

The editors would like to give special thanks to the graduate students Liwu Jiang, Ming Li and Di He from Beihang University for their hard work compiling and typesetting each paper in this volume.

Zhanguo Wang, Yong Qiu and Yongxiang Li
Editors

Dates

Issue 1 (2009)

Interface Characterization of $SrTiO_3$/Sr/Si Heterostructure through X-ray Reflectivity

H. Ji[1*], X.Y. Zhou[2], W. Wang[3], Y.R.Li[1], and Y.Wang[2]

1 *State Key Laboratory of Electronic Thin Films and Integrated Devices*, University of Electronic Science and Technology of China, Chengdu 610054, China.
2 Department of Applied Physics and Material Research Centre, The Hong Kong Polytechnic University, Hong Kong, China.
3 College of Electronics Engineering, Chongqing University of Posts and Telecommunications, Chongqing, 400065, China.

[*]Corresponding author: **H. Ji**
Email: jihong@uestc.edu,cn

Abstract. The growth of hetero-epitaxial strontium titanate ($SrTiO_3$) on single-crystal silicon (Si) substrate is of considerable scientific interest because of its wide range application. Unlike other hetero-epitaxial systems, there exists a fairly large lattice mismatch between $SrTiO_3$ and Si. This leads to the possibility of strain in $SrTiO_3$ thin film. In order to prevent the formation of amorphous SiO_2 phase and interdiffusion of component atoms at the interface, an ultrathin strontium (Sr) layer is deposited on the underlying Si-substrate. The X-ray diffraction (XRD) is used for characterization of hetero-structures. In this paper, we report the characterization of SrTiO3 thin film on Si (100) substrate with Sr interlayer by a variety of X-ray measurements. Based on the data obtained from $SrTiO_3$ thin film, the growth characteristics of the sample is objectively appreciated. The sample is prepared by laser molecular beam epitaxy (L-MBE) under optimized conditions of substrate temperature and oxygen pressure. $2\theta/\omega$ scan indicates a high crystallization quality and epitaxial grown of $SrTiO_3$ thin film in the nanometre scale. Φ-scans have further revealed the in-plane orientation relationship between $SrTiO_3$ and Si. The α- and c- lattice parameters of the $SrTiO_3$ thin film are found to be 0.3898, and 0.3901 nm, respectively, which suggests a slight elastic distortion. More detailed investigations by X-ray reflectivity have therefore been carried out, in order to get better understanding to the effect of the buried heterointerface.

Keywords. strontium titanate ($SrTiO_3$), Sr termination, X-ray diffraction (XRD), X-ray reflectivity (XRR)

1. Introduction

Strontium titanate ($SrTiO_3$) is an incipient ferroelectric material in perovskite structure, which has been extensively studied for its application in tunable microwave devices for having a large and variable dielectric constant. $SrTiO_3$(STO) is also a perfect prototype to investigate phase transition, especially to understand the surface and interface structural configuration for the perovskite films grown on it. STO thin film on Si has potential application[1-2]. STO thin film on Sr terminated (100) Si substrate has been achieved.

In this paper, we report the characterization of STO thin film on Si (100) substrate with Sr interlayer by X-ray measurements. Phase determination, thin film reflectivity, orientation relationship, lattice strain and texture are discussed .The growth characteristic of the sample is objectively appreciated.

2. Experimental procedure

Approximately 10-nm-thick STO films were deposited by laser molecular beam epitaxy (laser-MBE) on (100)-oriented, silicon wafer. Ceramics of STO was used as target. In the fabrication process, the wafers were ultrasonically cleaned in acetone, followed by dipping into a 10% HF solution for 5 mins to remove the surface oxide layer. The wafer was heated by a hot plate at 850°C for 15 mins to clean the surface free of the hydrogen. In order to prevent the formation of amorphous SiO_2 phase and interdiffusion of component atoms at the interface, an ultrthin strontium (Sr) layer was deposited on the wafer by Laser-MBE at 650°C and followed by a thermal treatment. The Sr buffer layer acting as seed for STO nucleation can be considered as a "sacrificial" epitaxial template.The required minimum thickness of Sr for the epitaxial growth of STO can be less than 2 nm. The termination of the Si (100) surface by ultrathin Sr drastically improves the growth behavior of STO films.Fig.1 is the flow diagram of sample preparation mentioned above.

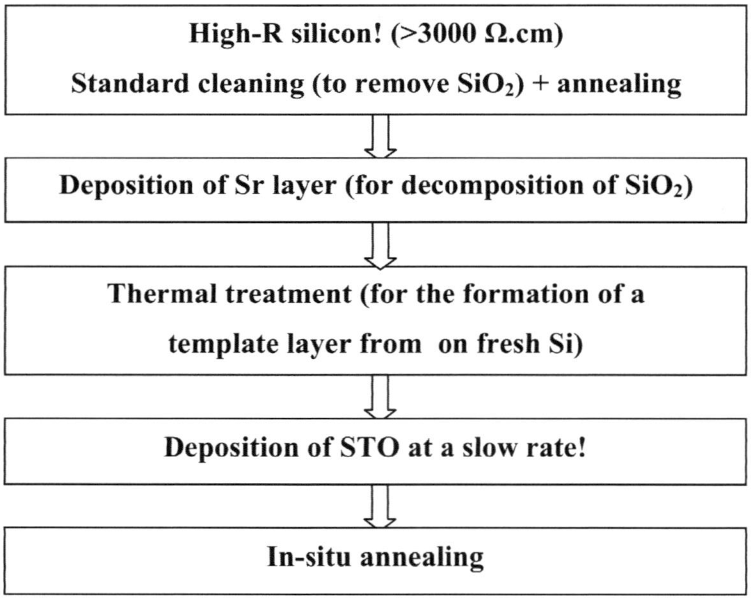

Figure 1. The flow diagram of sample preparation for the STO /Sr/ Si heterostructure.

The greatest use of XRD is for characterization of heterostructures. To check the the STO /Sr/ Si heterostructure ,the crystal structure and orientation of STO thin film is identified using the Bede D1 four-circle X-ray diffractometer (Cu Kα, λ=0.1541874nm) by wide (2θ-ω) scanning, specular reflective curve[3], pole figure and Φ-scan analysis[4].

3. Measurement of lattice constants

Strain field in STO thin film on large mismatched Si (100) substrate is determined by the spatial distribution of lattice constants and, in principle, can be deduced from X-ray diffraction measurements. Fig.2 (a) and (b) illustrate the symmetric and asymmetric diffraction patterns of STO thin film , respectively. Notice that there is only the STO (00l) reflection in addition to that from the substrate(Fig.2 (a)). The full width at half maximum (FWHM) for STO (002) rocking curve (ω scan) is 0.31° (not shown here), indicating comparable structural quality to STO grown on on Sr terminated (100) Si substrate.The profile refinement of all STO thin film peaks determines a tetragonal structure with a=b=0.3898 nm and c=0.3901 nm, revealing a slight elastic distortion. The cause of strain is primarily the larger lattice mismatch between the film and the substrate,which achieves the mismatch about 7 %.

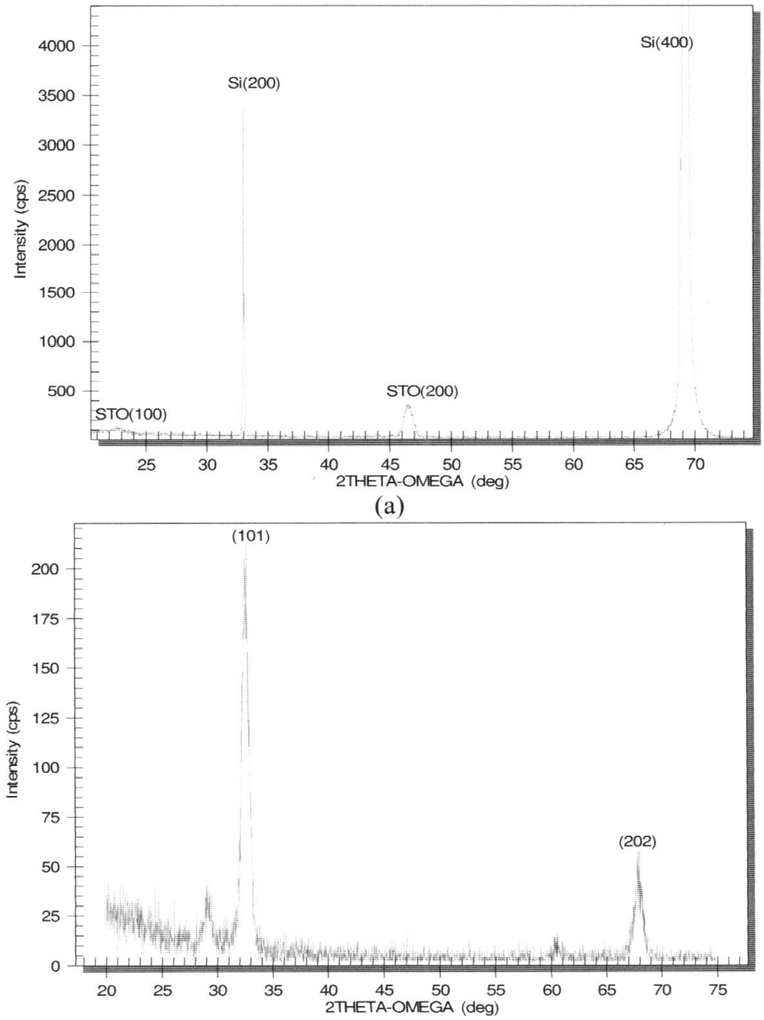

(b)

Figure 2. Symmetric 2θ/ω scan of STO (a) and asymmetric 2θ/ω scan of STO (b).

4. Measurement of Φ-scan and texture

The STO thin film is purposely grown with a lattice mismatch so that the in-plane orientation of the film and substrate is established by X-ray pole figure and Φ-scan. Fig. 3 gives the 360°Φ-scan on the STO (101) and Si (202) reflections, and it shows that the two groups of reflections are separated by 45°, consistent with the χ value (χ=45°). It implies that the STO thin film is also epitaxial, and the in-plane epitaxial relation is [100] STO // [110] Si. This plus (001) STO // (001)Si from Fig.1(a) completely determine the epitaxial relation between the film and the substrate.

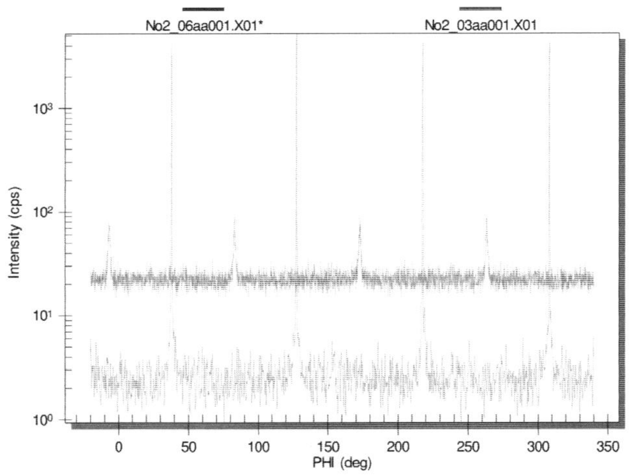

Figure 3. Φ-scan over SrTiO3 (101) (top) and Si (202) (bottom) reflections.

High qulity texture on the STO thin film is needed . Fig.4 shows the measured X-ray pole figure of STO (110). Clearly there is a strong preferred alignment of these STO grain. Especially note that the four poles corresponding to the (110) reflection in film are elongated along the Chi (χ) direction, which is related to deviation of orientation distribution . This phenomenon is normal since the competition between the elastic restore for STO unit cell and the lattice mismatch between STO film and silicon substrate may result in the elongation of the four poles corresponding to the (110) plane in film.

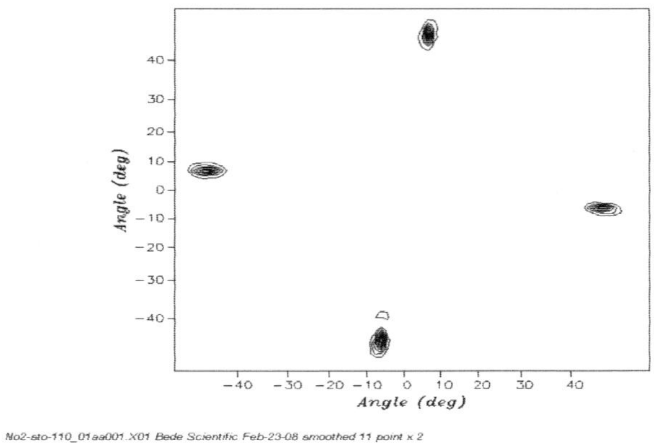

Figure 4. Pole figures of (110) STO thin film.

5. Measurement of X-ray reflectivity curves

X-ray reflectivity can provide microscopic information about the thickness, the interfacial roughness and the electron density within the utra-thin film. Here we briefly introduce our investigation carried out for the STO /Sr/ Si heterostructure. Fig.5 is the specular X-ray reflectivity curves. The black solid lines represent the experiment data and the red dotted lines represent the fitted data. Obviously, the match between experimental data and calculated data is pretty good. Here note that the interference oscillations, called "Kiessig fringes", can obviously be observed in the reflectivity profile. This means that the densities of thin films are different from that of the substrate.The thickness of STO overlayer is given as follows,

$$2t(\sin\theta_{i+1} - \sin\theta_i) = \lambda$$

where, t is the thickness of the film, λ is the wave length of X-ray, θi+1 and θi are the angles of the angles of the (i+1)th and the ith peaks of the oscillations in the reflectivity profile, respectively. By cross-sectional high-resolution transmission electron microscopy (HRTEM) characterization, we found that the film has a sharp interface (see Fig.6). In fact, the structural parameters, such as thichness of films and interface and surface roughness, are obtained by simulating the X-ray reflectivity curve.To simulate the measured reflectivity curve by dynamical diffraction theory, we have already been contiously trying to apply different structure model. According to our study, the structure model in Fig.7 is very reliable. By using a non-linear least-squares curve-fitting technique, we obtained the structural parameters of the sample,which are listed in Table 1.

Table 1. The structural parameters ascertained from XRR

Fitted thickness (Å)		Fitted density (%)		Root-mean-square surface and interface roughness (Å)		
dSTO	dSr	ρSTO	ρSr	σSTO	σSTO/Sr	σSr/Si
100.12	18.72	90.04	80.53	1.77	1.01	2.00

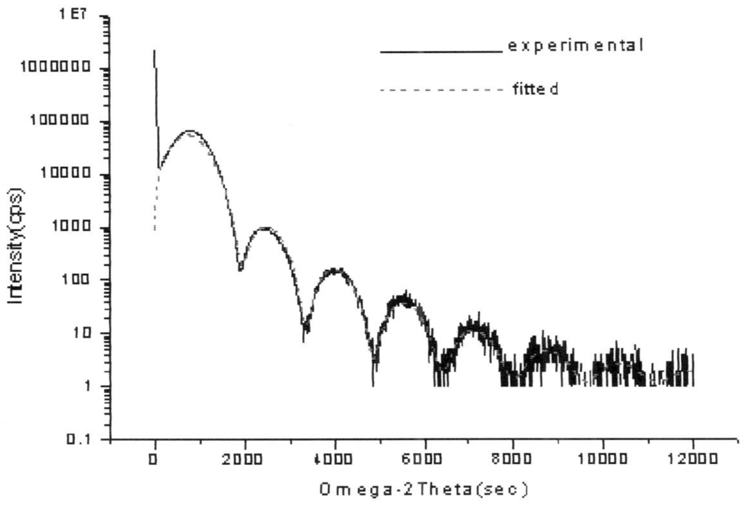

Figure 5. X-ray reflectivey spectra for STO /Sr/ Si heterostructure. The black solid lines represent the experiment data and the red dotted lines represent thefitted data.

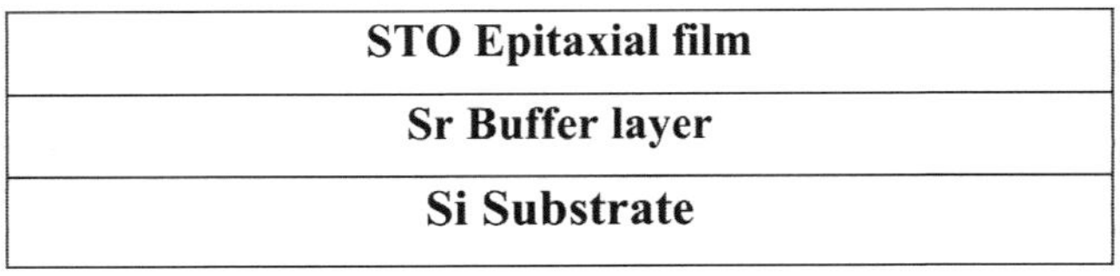

Figure 7. Schematic diagram of sample structure model.

6. Summary

Sr-terminated Si(100) surface provides a sacrificial epitaxial template for STO thin film. The (001) epitaxial growth of STO films on the template is achieved by an in-situ laser-MBE. The X-ray measurement technique used here gives us a convenient way to investigate the required microstructure information for STO /Sr/ Si heterostructure. We have demonstrated that STO film has a slightly different lattice parameter compared with bulk STO. The XRR experimental results confirm that the STO/Sr/Si interface of the sample appeared to be very abrupt and STO film was very dense. The thickness of STO film is approximately 10 nm, which is in good agreement with cross-sectional HRTEM.

We gratefully acknowledge financial support from the Hong Kong Polytechnic University internal funding (A-PG18).

References

[1] J. Robertson, J. Vacuum Sci. Tech. B 18, 1785 (2000).

[2] Z. Yu, Y. Liang, C. Overgaard, X.Hu, J. Curless, H. Li, Y. Wei, B. Craigo, D. Jordan, R. Droopad, J. Finder, K. Eisenbeiser, D. Marshall, K. Moore, J. Kulik, and P. Fejes, Thin Solid Films 462/63, 51 (2004).

[3] S. K. Sinha,E. B. Siroa and S. Graroff, H.B.Stanley, "*X-ray and neutron Scattering from rough surface*", Phys. Rev. B.,38(1988)2297

[4] L. G. Schulz, J. Appl. Phys. Vol. 20, pages1030-1035, 1949.

Impact of solvent and low k material surface characteristics on pore sealing efficiency and porosity measurement

D Rébiscoul[1], L Broussous[2], W Puyrenier[2], V Rouessac[3] and A Ayral[3]

[1] CEA-LETI, MINATEC, 17 Av. des Martyrs, 38000 Grenoble Cedex 09, France
[2] STMicroelectronics, 850, rue Jean Monnet, 38926 Crolles cedex, France
[3] Institut Européen des Membranes – CNRS/ENSCM/UM2, CC047 – 2 place Eugène Bataillon, 34095 Montpellier cedex 5, France

E-mail: diane.rebiscoul@cea.fr

Abstract. The penetration kinetics of different solvents having various properties (size, polarity, viscosity…) inside a porous low k SiOCH material modified by plasma treatments were investigated by ellipsometric porosimetry. The solvent penetration rates and durations to reach the saturation strongly depend on both the solvent properties and on the plasma post-treatments performed on the porous low k material. Pore sealing and barrier effect were quantified from the experimental determination of permeability coefficients for sample modified by a NH_3 and Fluocarbon based plasma using a permeation model. For these samples, the best barrier effects are observed with toluene and ethanol, respectively. In absence of barrier effect (unmodified material, plasmas He and NH_3/N_2), the solvent polarity and viscosity influence the time required for the complete saturation of the porous material. The liquid volume fraction f_L at saturation is also a function of the used solvent affecting the porosity estimation. All of these results highlight the complexity of the solvent penetration mechanisms.

1. Introduction

Porous low k materials ($k \leq 2.5$) are intensively studied for their integration as insulator in the Back End Of Line (BEOL) for 45 nm and beyond technology nodes. Integrate such materials requires to investigate the modifications generated during the integration processes. Our study is focused on the consequence of plasma post-treatments and more particularly on the efficiency of the resulting pore sealing and on their impact on the porosity measurement. Indeed, during plasma treatment, the structure, the morphology, the composition and the porosity of a few nanometers in material subsurface are modified. After specific plasma treatments the porosity of the material is sealed. This so-called "pore sealing" phenomenon has been previously characterized by standard adsorption/desorption isotherms measured by Ellipsometric Porosimetry (EP) [2-5].

Puyrenier et al [5] have also performed another type of EP measurement by collecting the evolution versus time of the refractive index after a sudden solvent pressure increase from residual vacuum to a relative pressure corresponding to the sample saturation. For a porous SiOCH dielectric modified by different plasmas, it was shown that solvents such as ethanol can more or less easily diffuse through the modified subsurface and fill the "buried" zone. These results also highlighted that the ethanol penetration kinetics strongly depended on the nature of the applied plasma.

It is known that vapor transport across a thin polymeric barrier (like a membrane) is governed both by sorption and diffusion [6]. The chemical nature of the barrier (organic/inorganic balance, hydrophylic/hydrophobic balance, etc) mainly influences the sorption process whereas the polymer reticulation controls the diffusion process [7]. Solvent viscosity and surface tension also affect solvent permeation in a nanoporous layer [8].

In this study, we attempted to determine if the solvent penetration is only governed by specific chemical interactions between the solvent and the modified zone or is also affected by a geometrical effect of the subsurface porosity limiting the solvent transport. Thus, we studied the penetration of several solvents exhibiting different characteristics (in term of size, polarity, viscosity, etc) in the same SiOCH porous material modified by different plasma post-treatments. The mean objective was to improve the understanding concerning the interactions between modified surface of porous material and the used solvent to quantify the low k material porosity and pore sealing efficiency.

2. Materials and Method

2.1. Experiments

Samples (20 mm x 20 mm x 1 mm sized) were cut from a 300 mm silicon wafer coated with a PECVD SiOCH porous material (k=2.5) unmodified or modified by different plasma post-treatments. The thickness and the density of the different zones in the materials were determined from X-ray reflectometry measurements. The characteristics of the samples are summarized in the table 1. More detailed information about these samples is available in [4][5].

Table 1. Samples characteristics.

Plasma	Sample label	thickness (nm)		density ($g.cm^{-3}$)		water contact angle (°)
		Zone 1	Zone 2	Zone 1	Zone 2	
no plasma	ref	158	-	1.23	-	97
He	He	140	10	1.23	1.36	92
NH_3	NH_3	149	9	1.23	1.76	62
NH_3/N_2	NH_3/N_2	153	5	1.23	1.28	52
$C_4F_8/Ar/N_2$	Etch	98	9	1.23	1.61	89

After annealing at 250°C during 1 hour in order to remove the residual moisture, the samples were characterised at 25°C in a closed EP cell connected to a solvent source and to a vacuum pump. The evolution versus time of the refractive index was recorded with a single wavelength ellipsometer (Plasmos SD 2300, λ = 632.8 nm) at fixed incidence angle (70°), after a sudden pressure increase from residual vacuum to a relative solvent pressure corresponding to complete filling of the pores, later named the sample saturation. The thickness variations (less than 5%) were neglected. Several solvents presented in table 2 were used as penetrants.

Table 2. Characteristics of the solvents used as penetrants.

	water	methanol	ethanol	isopropanol	toluene	cyclohexane
Formula	H_2O	—OH	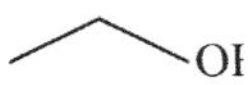	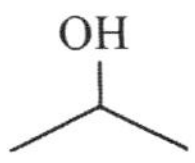	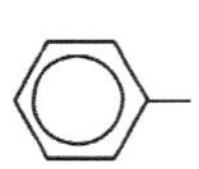	
Vapour pressure at 25°C (mbar)	31.67	169.9	75.9	58.01	37.05	130.7
Refractive index	1.333	1.329	1.361	1.377	1.496	1.426
Dipole moment (Debye)	1.85	1.71	1.69	1.66	0.38-0.43	0.04
kinetic diameter (nm)	0.27	0.38-0.41	0.44	0.47	0.59	0.60
Viscosity at 25°C (mPa.s)	1.00	0.56	1.04	1.98	0.55	0.88
Surface tension against air at 20°C ($mN{\cdot}m^{-1}$)	72.9	22.5	22.4	23	28.5	25

The solvent volume fraction in the material f_L was determined using the Lorentz-Lorenz formula (1):

$$f_L = \frac{\frac{n^2-1}{n^2+2} - \frac{n_0{}^2-1}{n_0{}^2+2}}{\frac{n_{sol}{}^2-1}{n_{sol}{}^2+2}} \tag{1}$$

with n the low k material refractive index, n_0 the initial refractive index and n_{sol}, the solvent refractive index.

2.2. Permeation model

The solvent penetration rate in the mesoporous thin layer through the top zone acting as a barrier layer/membrane was investigated from the evolution versus time of the refractive index. Gas transport in a polymeric membrane can be decomposed into three consecutive steps: sorption in the membrane, diffusion through the membrane and desorption on its opposite side. Sorption usually occurs very fast and is difficult to observe [5]. In our study, permeation which appeared as the time-limiting process was more precisely investigated. Solvent transport in the porous layer with thickness d and across a barrier with a thickness e (figure 1) was quantified using a permeation model. The samples were considered as bilayer systems (table 1) corresponding to a dead-end cell setup: the modified top layer and the "buried" porous material act as membrane and closed reception tank, respectively. The permeability coefficient Pe was determined from the permeance P (2) and permeance P was evaluated using (3), derived from Fick and Crank equation [5].

$$\text{Pe} = \text{P x e} \tag{2}$$

$$P = \frac{d}{V_{ML}} \frac{\Delta f_L}{\Delta t} \frac{1}{\Delta P} \tag{3}$$

with V_{ML}, the solvent molar volume; ΔP, the transmenbrane pressure and, $\Delta f_L/\Delta t$, the initial slope of the curve giving the evolution versus time of the solvent volume fraction in the material, f_L.

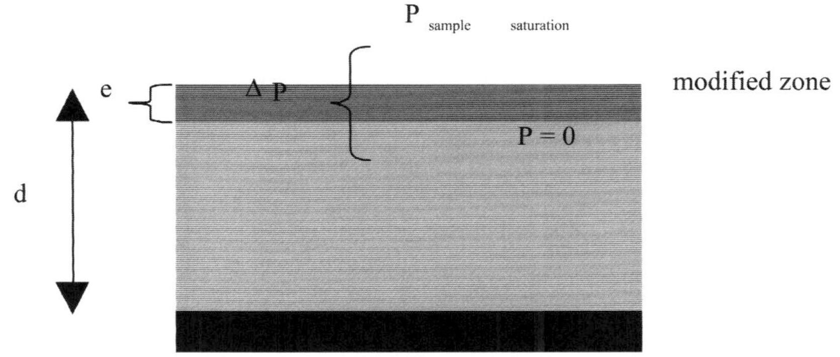

Figure 1. Bilayer model

3. Results and discussion

The evolutions versus time of the refractive index for the different samples and the different solvents are shown in figure 2. The origin of the abscissa axis (t = 0) corresponds to the sudden increase of solvent pressure in the EP cell.

Whatever the solvent, these results highlight two types of sample behaviors. For the unmodified material (sample ref) and for samples He and NH_3/N_2, a fast solvent penetration (less than 2 minutes) is observed, excepted for water. For samples Etch and NH_3, the penetration kinetics are significantly longer and the barrier effect is thus potentially measurable.

3.1. Barrier effect of the zone modified by plasma

The barrier effect for samples NH_3 and Etch samples was quantified by calculating the permeability coefficient of the modified top layer (table 3).

Table 3. Permeability coefficients Pe ($mol.m^{-1}.s^{-1}.Pa^{-1}$) of the modified top layer.

Sample	methanol	ethanol	isopropanol	toluene	cyclohexane
NH_3	$1.9.10^{-19}$	$7.4.10^{-20}$	$4.3.10^{-19}$	$5.1.10^{-19}$	$3.9.10^{-18}$
Etch	$5.3.10^{-18}$	$2.1.10^{-18}$	$1.3.10^{-19}$	$3.7.10^{-20}$	$3.5.10^{-19}$

The measured values of Pe are ranging in the same order of magnitude as that of vitreous polymers with ethanol [10] and strongly depend on the used solvent. For instance, the barrier effect of the sample NH_3 with the ethanol is from one to two orders of magnitude higher than that with the other solvents. For sample Etch, the permeability of the modified layer is the lowest with toluene.

Attempts were made for explaining these experimental results by the chemical affinity of the solvent with the modified subsurface and also by a plasma densification of this top zone (geometrical restriction). In order to estimate each contribution, the initial rates of solvent volume fraction variation, $\Delta f_L/\Delta t$, were plotted as a function of two parameters: A, defined as the "affinity parameter" (eq. 4) and S, defined as the "size parameter" (5) :

$$A = \frac{\theta_{H2O}}{P_{sol}} \qquad \text{(eq. 4)}$$

$$S = \frac{1}{\phi_{sol} \cdot \rho_{surf}} \qquad (5)$$

with θ_{H2O}, the water contact angle (°); P_{sol}, the dipole moment of the solvent (Debye); ϕ_{sol} (nm), the solvent kinetic diameter; ρ_{surf} ($g.cm^{-3}$), the mass density of the plasma-modified top zone [5].

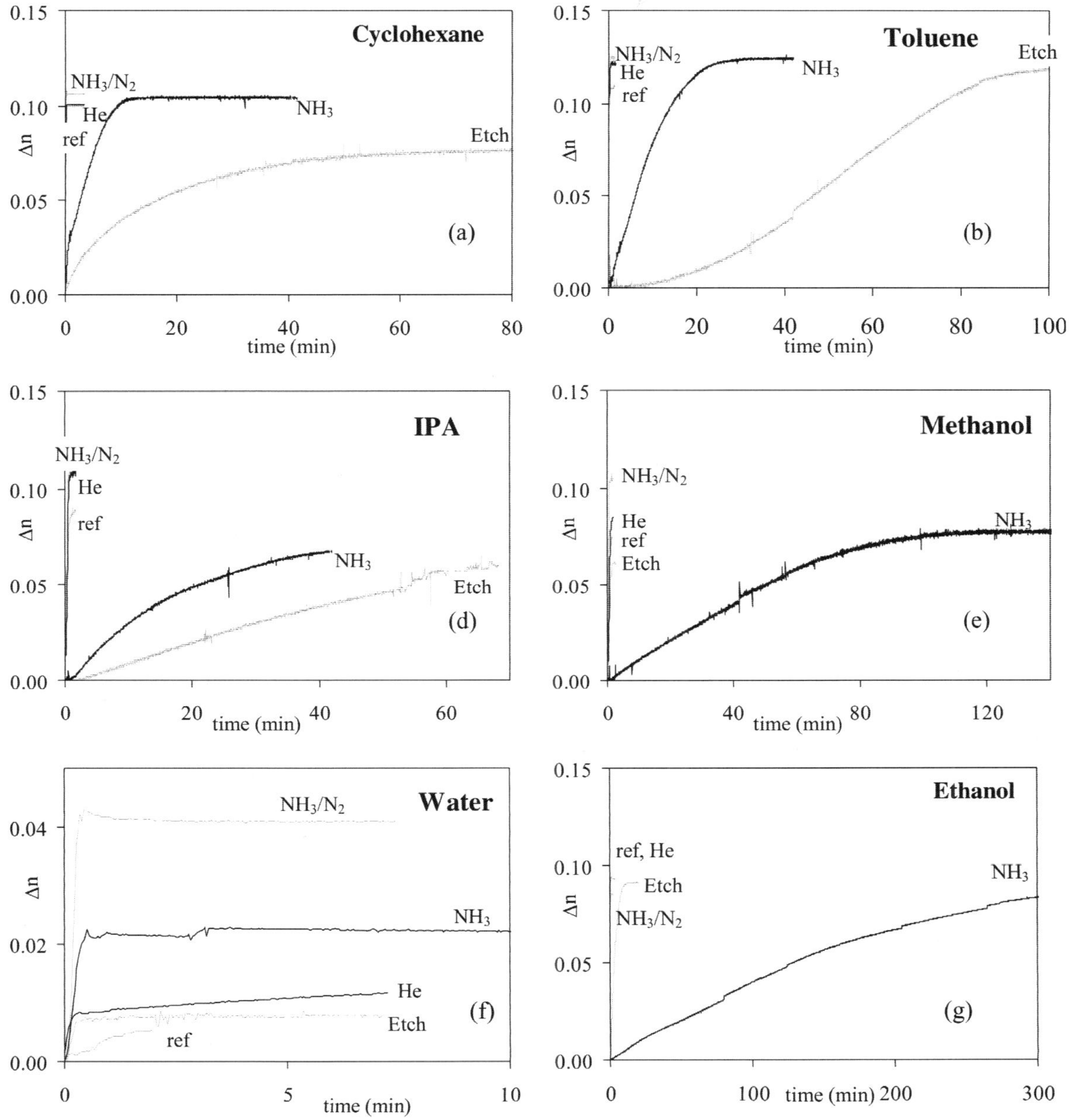

Figure 2. Evolution versus time of the refractive index for the different samples using: (a), cyclohexane; (b), toluene; (c), isopropanol; (d), methanol; (e), water; (f), ethanol.

The evolution of $\Delta f_L/\Delta t$ as a function of these two parameters is presented in figure 3. From figure 3, no overall trend can be extracted between penetration rate, subsurface properties and solvent characteristics. Indeed, the most favourable cases, i.e. low density surface coupled with small size solvent, hydrophobic surface coupled with apolar solvent, hydrophilic surface coupled with polar solvent do not present the highest solvent penetration rates. Meanwhile, figure 3 highlights specific trends for each sample. For Etch sample the increase of size parameter S is in good agreement with the increase of the solvent penetration rate, meaning that the barrier effect of the modified zone is mainly due to a geometrical restriction. For the NH_3 sample, the penetration rate of the solvent increases with the increase of the A parameter, i.e with the decrease of the solvent polarity. This phenomenon can demonstrate that, for this sample, the penetration rate of the solvent is mainly driven by the affinity of the "buried" zone.

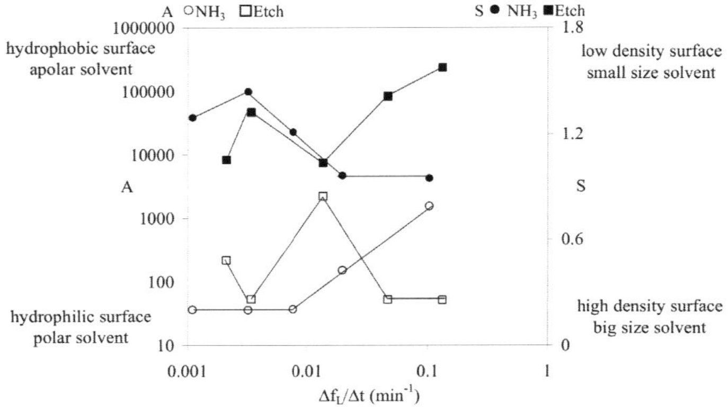

Figure 3. Evolutions of $\Delta f_L/\Delta t$ as a function of A and S.

3.2. Impact of solvent properties in absence of barrier effect

For the unmodified material (sample ref) and for samples He and NH_3/N_2, the fast solvent penetration is due to the absence of significant plasma densification of the subsurface. Consequently parameters which could influence the solvent flow inside the mesoporous low k material should be the solvent polarity and the solvent viscosity as previously discussed by Yang et al [8].

The time required to reach saturation equilibrium, later named "solvent penetration duration", was determined from the different curves shown in figure 2. The evolution of the solvent penetration duration as a function of the solvent polarity and viscosity is presented in figure 4. These data highlight that the solvent transport inside the material is mainly driven by affinity of the "buried" zone located underneath the modified zone: apolar solvents penetrate faster than polar and/or small-size solvents. Viscosity influences also the solvent penetration duration: solvent such IPA presents some stabilization duration higher than one minute.

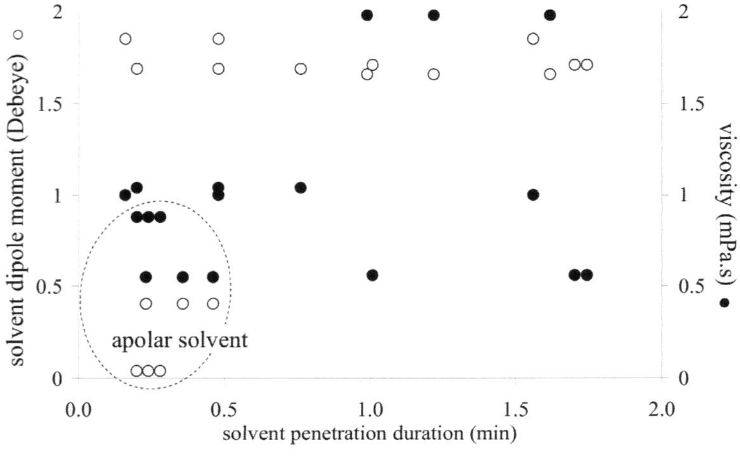

Figure 4. Evolution of the stabilization duration to reach the sample saturation as a function of the solvent polarity and viscosity for samples ref, He and NH_3/N_2.

3.3. Impact on porosity determination

Usually the porosity is determined by EP from the calculation of the liquid fraction f_L at saturation. However for a given sample, this value also depends on the solvent nature as shown in figure 5. Water is also a specific case. The water amount inside the material at saturation increases with the material hydrophilicity induced by some plasma treatments: NH_3/N_2 and NH_3. This phenomenon can be explained by the filling with water of the chemically modified zone. This phenomenon has been already characterized using neutron reflectometry and deuterated water [11].

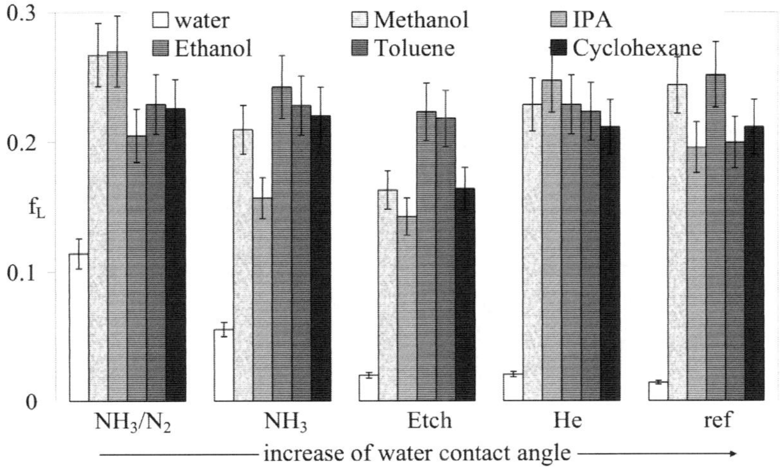

Figure 5. Solvent volume fraction at saturation for the different samples.

4. Conclusion

The penetration kinetics of different solvents inside a porous low k material modified by plasma treatments were investigated by ellipsometric porosimetry. The results shown that the solvent penetration rates and durations to reach the saturation strongly depend on both the solvent properties and on the plasma post-treatments performed on the porous low k material. NH_3 and Etch plasmas induce significant pore sealing and barrier effect which was quantified from the experimental determination of permeability coefficients for the different used solvents. For samples Etch and NH_3, the best barrier effects are observed with toluene and ethanol, respectively. In absence of barrier effect (unmodified material, plasmas He and NH_3/N_2), the solvent polarity and the solvent viscosity influence the time required for the complete saturation of the porous material. The liquid volume fraction f_L at saturation is also a function of the used solvent affecting the porosity estimation from EP measurements.

All of these results highlight the complexity of the solvent penetration mechanisms. Profilometric analyses during solvent penetration should be very useful in order to model the solvent penetration as already shown by promising preliminary results obtained using neutron reflectometry coupled with solvent adsorption [11].

References

[1] Arnal V, Farcy A, Aimadeddine M, Icard B, Guedj C, Maitrejean S, Todeschini J, Besling W, Brun P, Ollier E, Jacquemin J P, Delsol R, Vannier P, Mellier M, Richard E, Fox R, Imbert G, Lefriec Y, Toffoli A, Torres J, 2006 *Interconnect Technology Conference, 2006 International* 213

[2] Posseme N. et al. 2005 *Solid-State Phenomena* **103-104** 337

[3] Assous M, Simon J, Broussous L, Bourlot C, Fayolle M, Louveau O, Roman, A, Tabouret E,

Feldis H, Louis D, Torres J 2003 *Interconnect Technology Conference, 2003. Proceedings of the IEEE 2003 International* 97
[4] Puyrenier W, Rouessac V, Broussous L, Rébiscoul D, Ayral A 2006 *Microelec. Eng.* **83** 2314
[5] Puyrenier W, Rouessac V, Broussous L, Rébiscoul D, Ayral A 2007 *Microporous Mesoporous Mat.* **106** 40
[6] Favre E, Schaetzel P, Nguygen Q T, Clément R, Néel J 1992 *J. Membr. Sci.* **190** 169
[7] Roualdes S, Durand J, Field R W 2003 *J. Membr. Sci.* **211** 113
[8] Yang X J, Livingston A G, Freitas dos Santos L 2001 *J. Membr. Sci.* **190** 45
[9] Revol P, Perret D, Bertin F, Fusalba F, Rouessac V, Chabli A, Passemard G, Ayral A 2005 *J. Porous Mat.* **12** 113
[10] Barrer R M, Barrie J A, Rogers M G 1965 *J. Polym. Sci.* **10** 111
[11] Rébiscoul D, Broussous L, Puyrenier W, Rouessac V, Ayral A, Cubitt R 2008 *Microelec. Eng.*, **85** 2089

Epitaxy Growth and Electrical Properties of $La_2Hf_2O_7$ Thin Film on Si(001) substrate by Pulsed Laser Deposition

Feng Wei, Hailing Tu[1], Yi Wang, Shoujing Yue, Jun Du

Advanced Electronic Materials Institute, General Research Institute for Nonferrous Metals, Beijing, 100088, China

[1]E-mail: tuhl@grinm.com

Abstract. $La_2Hf_2O_7$ (LHO) thin films have been epitaxially grown on Si(001) substrates using an ultrahigh vacuum pulsed laser deposition (PLD) system. The high-resolution transmission electron microscopy (HRTEM) results show the predominant orientation of LHO grown on Si(001) is $(001)_{LHO}//(001)_{Si}$ and $[110]_{LHO}//[110]_{Si}$, indicating a remarkable tendency of cube-on-cube epitaxy. The composition of the sample was investigated by X-ray photoelectric spectroscopy (XPS). High-frequency C-V characteristics for n-type MIS capacitors with epitaxial LHO film show that the dielectric constant (k value) of the LHO dielectrics is estimated approximately to be 22 from the accumulation area of 1M frequency C-V curves. I-V characteristic of the LHO dielectrics shows that the leakage current density is in an acceptable magnitude, ranging from 10^{-7} to 10^{-2} A/cm^2 in the voltage range from 0 to 3 V.

1. Introduction

The aggressive scaling of CMOS devices is driving SiO_2-based gate dielectrics to their physical limits [1-2]. Recently, complex rare earth (RE) and transition metal (TM) oxides (such as $LaAlO_3$, $Dy_{0.1}Hf_{0.9}O_x$, and $La_2(Zr,Hf)_2O_7$) attract considerable interests as candidate high-κ gate dielectrics for silicon MOS devices [3-5]. $La_2Hf_2O_7$(LHO) is one of the promising high-k dielectrics for its significant advantages: (i) the crystallization temperature can be raised without lowing κ; (ii) La-based oxides have a larger conduction band offset and lower leakage current; (iii) LHO exhibits better adjustable work function ability with metal gate.

Epitaxial oxide thin films have been intensively investigated as high-k dielectrics because of their particular interfacial properties on Si substrates [6]. However, most of these oxides that have been epitaxially grown on Si(001) substrate are adopting the unfavorable orientation (such as $(110)_{Re2O3}//(001)_{Si}$) which may cause twinning and other defects [7]. Therefore, to fabricate cube-on-cube epitaxy of metal oxides on Si(001) is a vital issue on high-κ oxide preparation. Crystalline LHO is in pyrochlore phase with space group symmetry Fd3m and has a little lattice mismatch of -0.74 % with respect to two times of Si(001) spacing at room temperature [8]. Therefore, it is quite suitable for epitaxial growth on Si substrate. Up until now, both molecular beam epitaxy (MBE) and pulsed laser deposition (PLD) processes can be employed to grow the LHO films [9, 10]. In order to achieve further application, studies on the structural and electrical properties of the LHO film on Si(001) are significant.

In this paper, the epitaxial LHO films have been grown on Si(001) substrates using an ultrahigh vacuum PLD system. The interfacial composition and bonding states of the LHO/Si structure were investigated by high-resolution transmission electron microscopy (HRTEM) and x-ray photoelectron spectroscopy (XPS). In addition, the epitaxial LHO thin films show promising electrical properties which are suitable for high-k dielectric application in the future.

2. Experimental

The LHO ceramic target with purity of 99.99% for PLD deposition was obtained by conventional sintering method. LHO thin films were grown by a ultra-high vacuum PLD system. Prior to the growth, the n-type Si(001) substrates with a resistivity of 2-5 Ω·cm were cleaned via a standard RCA cleaning process to obtain a H-terminated Si surface. The Si substrates were immediately transferred into the vacuum chamber. When the hydrogen atoms were completely removed from Si substrate surface at ~500 °C, a (2×1) reconstruction of reflection high-energy electron diffraction (RHEED) pattern indicated that a clean Si(100) surface was formed. The vacuum chamber was under 1×10^{-7} Torr during the deposition. Ultrathin LHO films were deposited using a KrF excimer laser running at 2 Hz repetition rate with an energy density of ~2 J/cm^2 focused at a growth rate of ~0.04 Å per pulse. The substrate temperature ranged form 700 to 850 °C during the film growth.

HRTEM was used to determine the crystallinity and structure of the LHO films. The chemical composition and bonding states of the films were investigated by XPS. The measurements were carried out in VG MK II spectrometer using monochromated Al Kα irradiation and Ar^+ ions etching was performed for 5 mins. The electrical properties, such as capacitance-voltage (*C-V*) and current-voltage (*I-V*) characteristics, were measured by an HP 4284 *LCR* meter and a Keithley 2400 sourcemeter at room temperature respectively.

3. Results and discussion

HRTEM demonstrates the interface and crystallinity information of the epitaxial LHO films deposited at 780 °C. As shown in Fig. 1, LHO film and Si substrate reveals a good lattice matching and there is a slight interface in the HRTEM image. It is clearly observed that the LHO sample with uniform atomic pattern shows high crystallinity and has no distinguished grain boundary within all TEM observation. The orientation relationship between LHO and Si can be absolutely determined as [110]LHO//[110]Si and (001)LHO //(001)Si which is in agreement with the RHEED and Φ-scan results (not shown). The in-plane epitaxy of [001] LHO that is parallel with Si [001] suggests a super-cell lattice match, that is, the spacing of LHO [001] lattice (10.78 Å) matches that of twice of Si [001] (10.86 Å). The electron diffraction pattern of LHO layer in the inset further indicates a cubic structure in the LHO film.

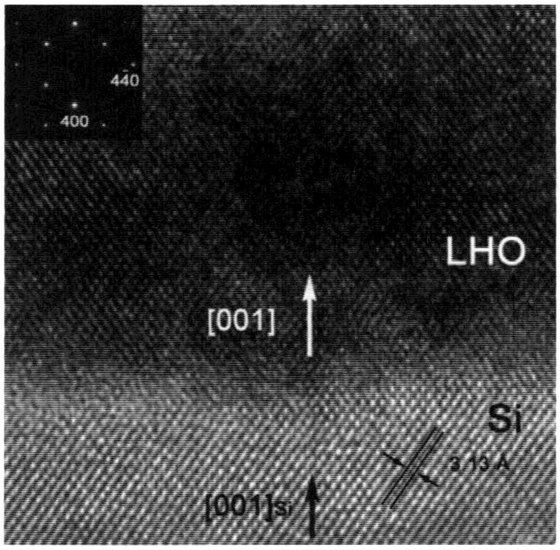

Figure 1. HRTEM image of the epitaxial LHO film on Si substrate.

The chemical properties of the thin films were measured by XPS. Fig. 2 shows XPS Hf 4f, La 4d spectra for LHO thin films. The binding energies of the core level were calibrated by setting the adventitious C 1s peak at 284.6 eV. Fig. 2(a) is the Hf 4f spectrum of LHO film. Detailed curve fitting of the Hf 4f spectrum has been carried out with Shirley background and Gaussian-Lorentzian lines. In the fitting curves, there are two typical peaks at (i)~19 eV, (ii)~17.4 eV, corresponding 4f spin-orbit energy splitting which was calculated to be ~1.6 eV. It implies that Hf in the film presents in Hf 4+ state. Fig. 2(b) shows the XPS spectrum of La 3d portion of LHO. The binding energy and spin-orbit splitting energy of La 3d for LHO are similar to those for La_2O_3. According to the XPS data, the element composition has been calculated. The element ratio of La/Hf/O is 1.74:2:7.5 which is in agreement with the expected theoretical ratio. The XPS results indicate that the as-grown LHO film is in stable chemical state.

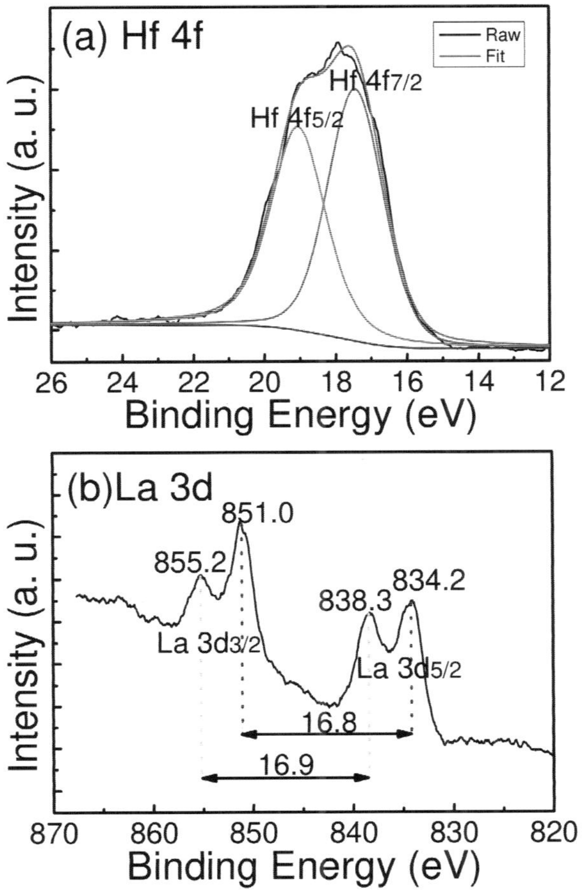

Figure 2. Core-level XPS spectra of (a) Hf 4f and (b) La 3d of epitaxial LHO thin film on Si substrate.

Fig. 3 shows high-frequency C-V characteristics for n-type MIS capacitors with 25 nm-thick epitaxial LHO film with Pt top electrodes. The circular Pt electrodes with diameter of 100 μm were sputtered through a shadow mask and the back contacts were made by evaporating Al. In the typical C-V traces, a transition from accumulation to depletion mode occurs at ~ 1.6 V. The capacitive effective thickness (CET) ~4.3 nm are extracted from the C-V curve without the quantum mechanical correction, and the dielectric constant (k value) of the LHO dielectrics is estimated approximately to be 22 from the accumulation. The forward and backward measurements show a very small hysteresis below 0.1 V(~ 20 mV) indicating a very low density of defects causing hysteresis. The calculated flat-band voltage (VFB) is estimated to be ~0.6 V implying a negative flat band voltage shift, which can be attributed to positive charges arising from fixed charges or slow interface traps [11]. In addition, an obvious frequency dispersion phenomenon occurs in the accumulation area, indicating a larger interfacial state density.

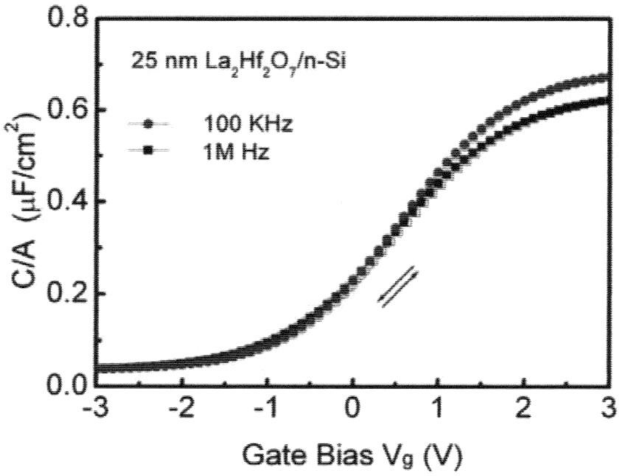

Figure 3. C–V characteristic of a Pt/LHO/Si(001) MOS structure measured at high frequency.

I-V characteristic of the LHO dielectric is plotted in Fig. 4. The leakage current density is in an acceptable magnitude for the LHO film, ranging from 10-7 to 10-2 A/cm2 in the voltage range from 0 to 3 V and the leakage current at 1 V is 10−4 A/cm2. This result is relatively higher than that in other reports and need further investigation.

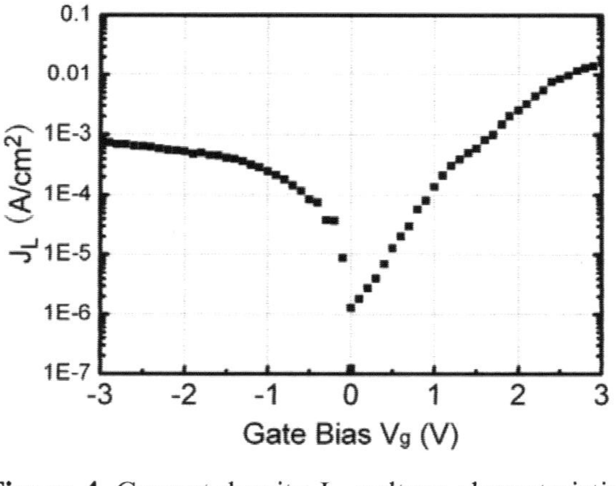

Figure 4. Current density J - voltage characteristic of a Pt/LHO/Si(001) MOS structure.

4. Conclusion

To conclude, we have epitaxially grown LHO film on Si(001) substrate using PLD method. The LHO film and Si substrate exhibit a cube-on-cube relationship (001)LHO//(001)Si and [110]LHO//[110]Si. The element composition of as-deposited LHO film is near stoichiometry. The capacitors show acceptable dielectric properties at room temperature; such as a dielectric constant of ~22, leakage current of 10^{-4} A/cm^2 at 1 V for layers with 25 nm thickness.

References

[1] Wilk G D, Wallace R M and Anthony J M 2001 *J. Appl. Phys.* **89** 5243
[2] Wong H, Iwai H 2006 *Microelectronic Engineering* **83** 1867
[3] Busani T and Devine R A B 2005 *J. Appl. Phys.* **98** 044102
[4] Adelmann C, Sriramkumar V, Elshocht S. Van, Lehnen P, Conard T and Gendt S De 2007 *Appl. Phys. Lett.* **91** 162902
[5] Seo J W, Fompeyrine J, Guiller A, Norga G, Marchiori C, Siegwart H and Locquet J-P 2003 *Appl. Phys. Lett.* **83** 5211
[6] Dimoulas A, Vellianitis G, Travlos A, Ioannou-Sougleridis V and Nassiopoulou A G 2002 *J. Appl. Phys.* **92** 426
[7] Apostolopoulos G, Vellianitis G, Dimoulas A, Alexe M, Scholz R, Fanciulli M, Dekadjevi D T and Wiemer C 2002 *Appl. Phys. Lett.* **81** 3549
[8] Rittersma Z M, Hooker J C, Vellianities G, Locquet J-P, Marchiori C, et. al. 2006 *J. Appl. Phys.* **99** 024508
[9] Dimoulas A, Vellianities G, Mavrou G, Apostolopoulos G, Travols A, Wiemer C, Franciulli M, and Rittersma Z M 2004 *Appl. Phys. Lett.* **85** 3502.
[10] Wei F, Tu H, Wang Y, Yue S and Du J 2008 *Appl. Phys. Lett.* **92** 012901
[11] Fissel A, Czernohorskyb M, Osten H J 2006 *Superlattices and Microstructures* **40** 551.

Journal of Physics: Conference Series **152** (2009) 012004 doi:10.1088/1742-6596/152/1/012004

Gd_2O_3 High-*K* Gate Dielectrics Deposited by Magnetron Sputtering

Shoujing Yue, Feng Wei, Yi Wang, Zhimin Yang, Jun Du

Advanced Electronic Materials Institute, General Research Institute for Nonferrous Metals, Beijing 100088, China

E-mail: dujun@grinm.com

Abstract: Gd_2O_3 thin films as high-*K* gate dielectrics were deposited on Si substrates by magnetron sputtering at 550 °C and 13 Pa. X-ray diffraction and cross sectional high-resolution transmission electron microscopy studies revealed that the Gd_2O_3 film was crystalline and it was consisted of a mixture of cubic and monoclinic phases. And the film show preferential orientation growth: $(111)_{cubic}//(001)_{Si}$ and $(401)_{monoclinic}//(001)_{Si}$. The Gd_2O_3 thin film presented acceptable electrical properties such as low leakage current density of 3.6×10^{-5} A/cm^2 at +1 V with capacitance equivalent SiO_2 thickness of 2.54 nm. Rapid thermal annealing in N_2 atmosphere can passivate a large amount of oxygen vacancy and improve the electrical properties.

Key words: Gd_2O_3 thin film; high-*K* gate dielectric; magnetron sputtering

1. Introduction

Aggressive scaling of device dimensions in the silicon based complementary metal-oxide-semiconductor (CMOS) technology demands the replacement of conventional silicon dioxide (SiO_2) gate dielectric by a higher permittivity (high-*K*) material [1, 2]. In recent years, rare earth oxides have formed an interesting group as high-*K* candidates based on their thermodynamic stability with Si and high conduction band offset over 2 eV.

Among the rare earth oxides, gadolinium oxide (Gd_2O_3) has attractive features for high-*K* applications. It exhibits a suitable permittivity, a wide band gap (E_g = 5.4 eV) and high band offsets with respect to Si (ΔE_C = 3.2 eV, ΔE_V = 3.9 eV) [3]. Thus, Gd_2O_3 has been widely investigated and showed promising results [4-10]. It is interesting to note that crystalline Gd_2O_3 thin films are usually formed with cubic structure. This observation may be understood on the basis that cubic phase is stable at room temperature while monoclinic phase is stable at higher temperature over 1200 °C [11]. However, A. Fissel et al. deposited Gd_2O_3 thin films on 6H-SiC(0001) substrates at 670 °C by MBE technique and observed the formation of Gd_2O_3 thin films in a mixture of [111]-oriented cubic bixbyite and monoclinic structure [12]. According to our literature survey, such orientation textures in Gd_2O_3/Si system by magnetron sputtering are truly seldom observed.

In present work, Gd_2O_3 thin films were deposited directly on Si(001) substrates by magnetron sputtering at 550 °C and a structure similar to reference [12] was observed. Pt/Gd_2O_3/Si MOS capacitors were fabricated and measured. The structure of Gd_2O_3 thin films was studied using X-ray diffraction (XRD) and transmission electron microscopy (TEM). The electrical properties of Pt/Gd_2O_3/Si MOS capacitors were investigated by basic MOS electrical characterization.

2. Experimental

The Gd_2O_3 ceramic target with a purity of 99.95% was obtained by conventional sintering method. The Si substrate with a resistivity of 2-5 Ω·cm is n-type and (001)-oriented. After standard RCA (Radio Corporation of America) cleaning and a buffered HF strip for 60 seconds to remove the native surface oxide, the Si substrates were immediately moved into the vacuum chamber and heated in situ to 550 °C to acquire a clean Si surface. Then Gd_2O_3 thin films (~ 15 nm) were deposited by radio frequency (rf) magnetron sputtering at a low pressure of 13 Pa in Ar ambient with target-substrate distance of 50 mm and sputtering power of 30 W. After deposition, the films were cooled down to room temperature. Subsequent postdeposition annealing (PDA) process was carried out at 600 °C for 30 seconds in N_2 (or O_2) atmosphere by rapid thermal annealing (RTA). For electrical measurements, circular platinum (Pt) dots of 300 μm in diameter as electrodes were deposited on the surfaces of Gd_2O_3 thin films through a shadow mask.

The Gd_2O_3 film thickness was determined by ellipsometry. The crystal structure was characterized by XRD and cross sectional high-resolution transmission electron microscopy (HRTEM). For electrical measurement of the Pt/Gd_2O_3/Si MOS capacitors, *I-V* and *C-V* curves were taken by Keithley 2400 source meter and Agilent 4284A *LCR* analyzer respectively in a probe station. The two probes were placed on the top of Pt electrode and the back of Si substrate respectively for both *I-V* and *C-V* measurements.

3. Results

Fig.1 shows a typical XRD (θ-2θ scan) pattern of a Gd_2O_3 thin film grown on Si(001). Besides the Si diffraction peaks, there were four broad peaks indicating that the Gd_2O_3 thin film was crystalline. The polycrystalline Gd_2O_3 film was composed of two phases as indicated by the marks of C for cubic phase and M for monoclinic phase. Diffraction peaks were identified according to PDF card 88-2165 and 43-1015. The peak around 28° corresponded to C(222), and the ones around 59°, 30°, 62° corresponding to C(444), M(401), M(802) respectively. And we noted that the intensities of cubic peak and monoclinic peak were comparable.

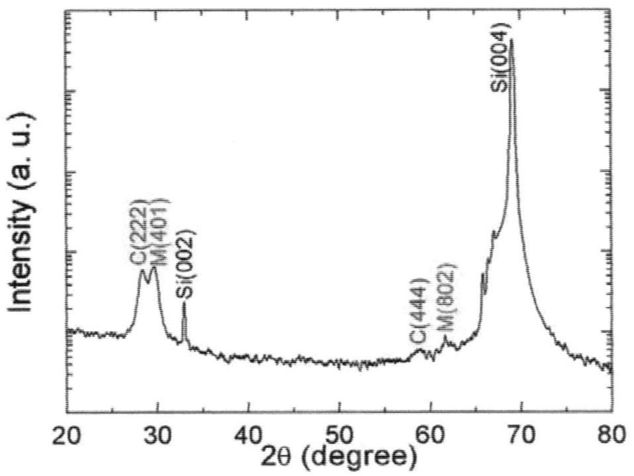

Fig.1 XRD pattern of the 15 nm Gd_2O_3 film grown on Si(001) by magnetron sputtering. C and M denote cubic and monoclinic, respectively.

Another structural feature is that the film showed preferential orientation growth since the C(222), C(444), M(401) and M(802) were the only diffraction peaks, other peaks were not evident or relatively too week. This suggested the following growth orientation relationships: $(111)_C//(001)_{Si}$ and $(401)_M//(001)_{Si}$. Similar relationships have also been found in the Gd_2O_3 films grown on 6H-SiC(0001) substrates by MBE. It is believed that the C(111) plane with the highest atomic density is usually the surface plane for some other oxide films, because such an oriented growth could minimize the total surface energy.

The interface and structure of the Gd_2O_3/Si heterostructure was further investigated by cross sectional TEM. The TEM images of Gd_2O_3/Si are shown in Fig.2 and 3. Fig.2 is a low resolution TEM image showing that the as-grown Gd_2O_3 film is 15 nm thick. HRTEM image of the Gd_2O_3 film (Fig.3) indicated that an amorphous oxide layer between Gd_2O_3 and Si. The image clearly shows that both cubic and monoclinic phases existed in the film and they are well oriented.

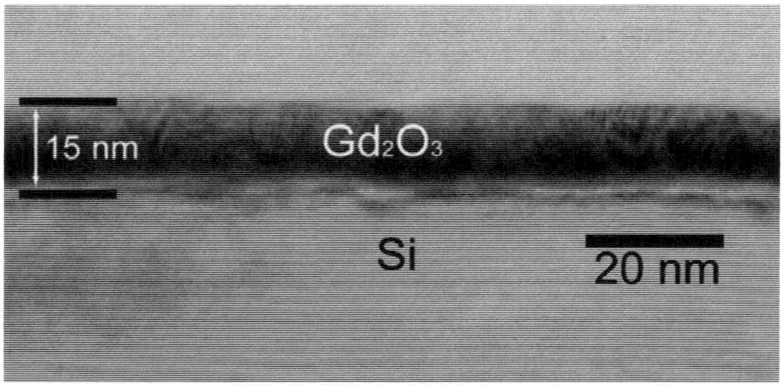

Fig.2 Cross-sectional TEM image of the Gd_2O_3 thin film grown on Si(001). The bar scale corresponds to 20 nm.

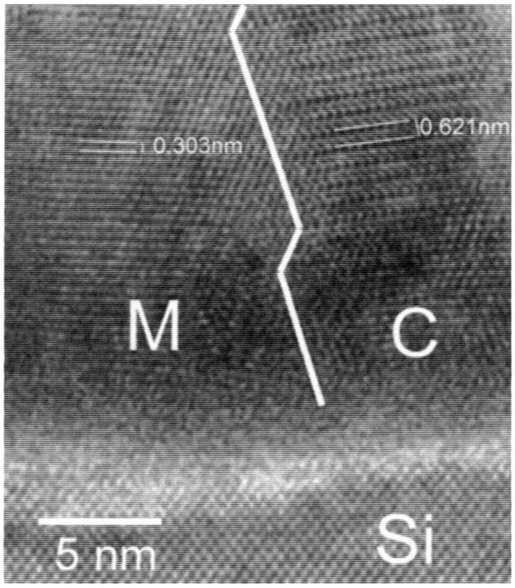

Fig.3 Cross-sectional HRTEM image of the Gd_2O_3 thin film grown on Si(001). The bar scale corresponds to 5 nm.

The electrical properties of Gd_2O_3 films as gate bias were also investigated. The dependence of leakage current density (J_L) on bias for Gd_2O_3 thin films is shown in Fig.4. It was noted that the as grown film showed a relatively low J_L of about 10^{-6} A/cm^2 at zero bias and 3.6×10^{-5} A/cm^2 at +1 V. The breakdown field was 3.5 MV/cm. The J_L result was acceptable in the bias range below 5 V in microelectronic industry. After PDA process, the leakage current density of Gd_2O_3 film decreased for RTA in N_2 but increased for RTA in O_2. This indicated that nitrogen incorporation can passivate a large amount of oxygen vacancy and improve the electrical properties. The asymmetry noted in J_L-V data is possibly due to the effects of air exposure before the electrode was deposited and interacted with the electrode interface [4, 5].

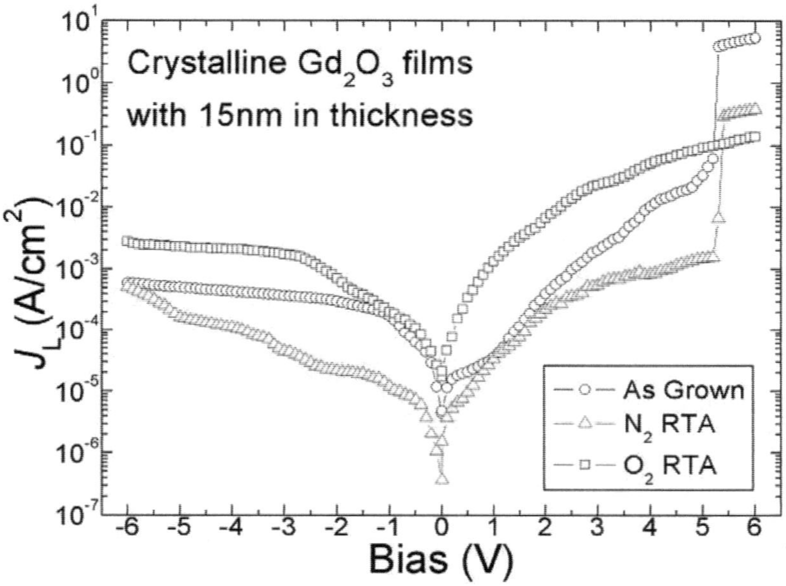

Fig.4 *I-V* characteristic of the 15 nm Gd_2O_3 films including the one as grown (○), the one after N_2 RTA at 600 °C for 30 s (△) and the one after O_2 RTA at 600 °C for 30 s (□). The positive bias means that the top Pt electrode was positive with respect to the Si substrate.

The dependence of capacitance on bias at a frequency of 500 kHz for Pt/Gd_2O_3/Si MOS capacitors is shown in Fig.5. The transition from accumulation to depletion was evident for the MOS capacitor with as grown Gd_2O_3 film. And the *C-V* curve didn't present a noticeable shift relative to the ideal one. The capacitance equivalent SiO_2 thickness (CET) estimated directly from the accumulated capacitance was 2.54 nm and the effective permittivity of the overall capacitor was 23. However, the accumulation was not saturated implying that a large amount of oxygen vacancy existed in the film. After PDA process in N_2, the accumulation became saturated and the effective permittivity of the overall capacitor became 16. However, the *C-V* curve for the Gd_2O_3 film after PDA process in O_2 presented a shift of about +1 V implying a large amount of negative fixed charge existed in the film [1]. This is consistent with the results of *I-V* curves.

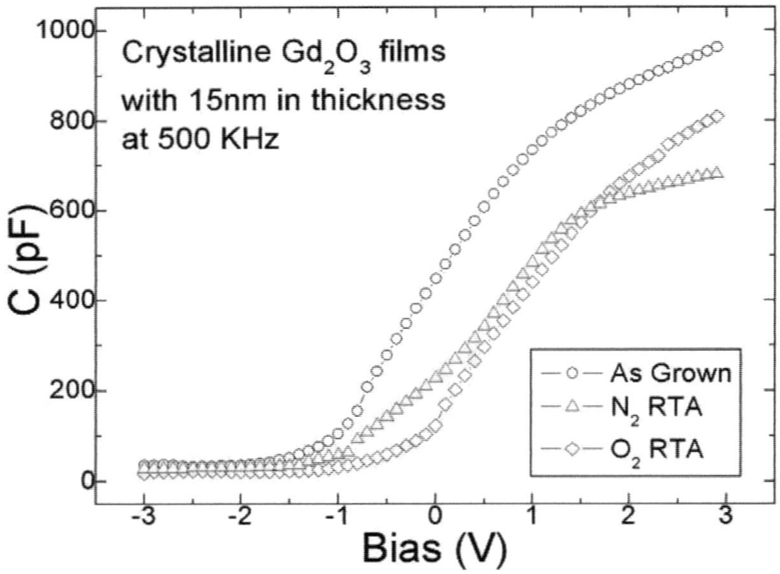

Fig.5 High frequency (500 kHz) *C-V* characteristic of the Pt/Gd_2O_3/Si MOS capacitors with Gd_2O_3 film as grown (○), after N_2 RTA at 600 °C for 30 s (△) and after O_2 RTA at 600 °C for 30 s (□). The area of top Pt electrodes was 7.1×10^{-4} cm^2.

4. Summary

In summary, both cubic and monoclinic phases co-existed in Gd_2O_3 thin film grown by magnetron sputtering at 550 °C. These two phases presented preferential orientation growth with regard to Si(001) substrate. The Gd_2O_3 thin film presented acceptable electrical results and RTA treatment in N_2 atmosphere can passivate a large amount of oxygen vacancy and improve the electrical properties. Considering the commercial application, the low-cost Gd_2O_3 thin film grown by magnetron sputtering could be a promising candidate as high-*K* gate dielectric in CMOS devices. Further work is in process to improve the electrical properties of the Gd_2O_3 thin films.

References:

[1] G. D. Wilk, R. M. Wallace, and J. M. Anthony, J. Appl. Phys. **89**, 5243 (2001).

[2] H. Wong, and H. Iwai, Microelec. Eng. **83**, 1867 (2006).

[3] M. Badylevich, S. Shamuilia, V. V. Afanas'ev, A. Stesmans, A. Laha, H. J. Osten, and A. Fissel, Appl. Phys. Lett. **90**, 252101 (2007).

[4] J. Kwo, M. Hong, A. R. Kortan, K. T. Queeney, Y. J. Chabal, J. P. Mannaerts, T. Boone, J. J. Krajewski, A. M. Sergent, and J. M. Rosamilia, Appl. Phys. Lett. **77**, 130 (2000).

[5] J. Kwo, M. Hong, A. R. Kortan, K. L. Queeney, Y. J. Chabal, R. L. Opila, Jr., D. A. Muller, S. N. G. Chu, B. J. Sapjeta, T. S. Lay, J. P. Mannaerts, T. Boone, H. W. Krautter, J. J. Krajewski, A. M. Sergent, and J. M. Rosamilia, J. Appl. Phys. **89**, 3920 (2001).

[6] T. M. Pan, C. S. Liao, H. H. Hsu, C. L. Chen, J. D. Lee, K. T. Wang, and J. C. Wang, Appl. Phys. Lett. **87**, 262908 (2005).

[7] Jaakko Niinistö, Nikolina Petrova, Matti Putkonen, Lauri Niinistö, Kai Aratila, and Timo Sajavaara, J. Cryst. Growth **285**, 191 (2005).

[8] M. Czemohorsky, E. Bugiel, H. J. Osten, A. Fissel, and O. Kirfel, Appl. Phys. Lett. **88**, 152905

(2006).
[9] A. Laha, H. J. Osten, and A. Fissel, Appl. Phys. Lett. **89**, 143514 (2006).
[10] A. Laha, H. J. Osten, and A. Fissel, Appl. Phys. Lett. **90**, 113508 (2007).
[11] G. Y. Adachi, and N. Imanaka, Chem. Rev. **98**, 1479 (1998).
[12] A. Fissel, M. Czemohorsky, and H. J. Osten, Superlattices and Microstructures **40**, 551 (2006).

Journal of Physics: Conference Series **152** (2009) 012005 doi:10.1088/1742-6596/152/1/012005

Preparation and characterization of Gd_2O_3-doped HfO_2 high-k gate dielectric thin films by RF sputtering

Mei Ji[1], Lei Wang, Jun Du[2]

Advanced Electronic Materials Institute, General Research Institute for Nonferrous Metals, Beijing, 100088, China

E-mail: [1]jimei@griaem.com, [2]dujun@grinm.com

Abstract. Research was carried out on the growth and characterization of Gd_2O_3-doped HfO_2 thin films as gate dielectrics for Si CMOS technology prepared by RF magnetron co-sputtering. The crystal morphology, composition and electrical properties of both Gd_2O_3-doped and as grown HfO_2 gate dielectrics have been comparatively investigated. Cross-section high-resolution transmission electron microscopy (HRTEM) reveals that all the deposited high-k gate dielectrics display amorphous structures with a thickness of ~10nm. C-V and I-V measurements indicate the pronounced reductions of dielectric loss, equivalent oxide thickness (EOT), as well as the leakage current for Gd_2O_3-doped HfO_2 thin films. Moreover, the calculated dielectric constants significantly increase from 20.2 to 31.6 and less hysteresis in the reverse sweeping could also be observed as the doping content of Gd_2O_3 increases.

1. Introduction

The aggressive down-scaling of complementary metal-oxide-semiconductor (CMOS) device dimension has led to the introduction of high-k materials to replace SiO_2 as gate dielectrics [1]. Many binary metal oxides for high-k gate dielectric applications, such as Ta_2O_5, ZnO_2, ZrO_2, HfO_2, CeO_2, Y_2O_3 and Gd_2O_3 have already been widely researched [2-5]. Among them, HfO_2 has been considered to be the most attractive competitor for advanced CMOS devices due to its relatively higher dielectric constant (~25), larger band gap (~6.0eV), higher electron offset (1.5eV) and better thermodynamic stability on Si wafers [6]. However, HfO_2 shows poor electrical properties especially the leakage current, which is brought down by the large concentration of charge traps mostly caused by oxygen vacancies [7]. Recently doping transitional and rare earth metals or metal-oxides into HfO_2 such as Y and Al-doped HfO_2 are considered promising methods to resolve the problems for future microelectronic devices [8-9]. In this paper, particular attention was paid to the possibility of further reducing the leakage current of HfO_2 by Gd_2O_3 doping.

2. Experimental details

The Gd_2O_3-doped HfO_2 high-k gate dielectric thin films were deposited by radio frequency (RF) magnetron co-sputtering using separate HfO_2 and Gd_2O_3 targets with the purity of 99.99%. All the films were grown on n-type Si(100) wafers with a resistivity of 2-5 Ω cm. The Si wafers were previously cleaned by RCA washing process. After cleaning, they were etched into the diluted hydrofluoric acid (~2%) for 30 s in order to remove the native oxides and form H-terminated silicon surfaces. After quickly blowing with nitrogen gas, they were immediately transferred into the vaccum chamber. The base pressure of the deposition chamber was 5.0×10^{-7}Torr. The sputtering ambient was a mixture of Ar/O_2 (4:1) and the chamber pressure during deposition was kept at 18.75mTorr. Depositions were performed at room temperature for two purposes. One is to prevent the formation of low k interfacial oxide layers between HfO_2 and Si. The other is to form amorphous films which can have lower leakage current than crystal thin films. The sputtering power of Gd_2O_3 was varied from 0-60W to control the doping content while the power of HfO_2 was kept at 120W. In addition, a pure HfO_2 thin film was also deposited to serve as a parallel comparison to the Gd_2O_3-doped films. Post deposition annealing (PDA) process was carried out in Ar atmosphere at 450℃ for 20 minutes to improve the qualities of the thin films. To fabricate MOS structures, circular Pt electrodes of 200μm in diameter were deposited over the grown dielectric thin films through a shadow mask.

Electrical properties for capacitance-voltage (C-V) and current-voltage (I-V) measurements were performed in a probe station using Agilent 4284A*LCR* analyzer and Keithley 2400 source meter. The film thickness was determined by Sopra GES5 spectra ellipsometry. Crystal morphology and cross sectional TEM observation were made using JEOL 2010 high-resolution transmission electron microscope. The surface compositions and bonding states of the Gd_2O_3-doped HfO_2 thin films were determined by X-ray photoelectron spectroscopy (XPS) measurements using ESCALAB MkⅡ.

2. Results and details

Fig.1 is the HRTEM cross-section image of the Gd_2O_3-doped HfO_2 thin film at 40W. It is seemed that the physical thickness of the films is about 10 nm. The Gd_2O_3-doped HfO2 thin film shows an amorphous structure. An obscure interfacial layer can be observed with a thickness of about 1nm. This image can indicate that the deposited Gd_2O_3-doped HfO_2 thin films at room temperature are amorphous.

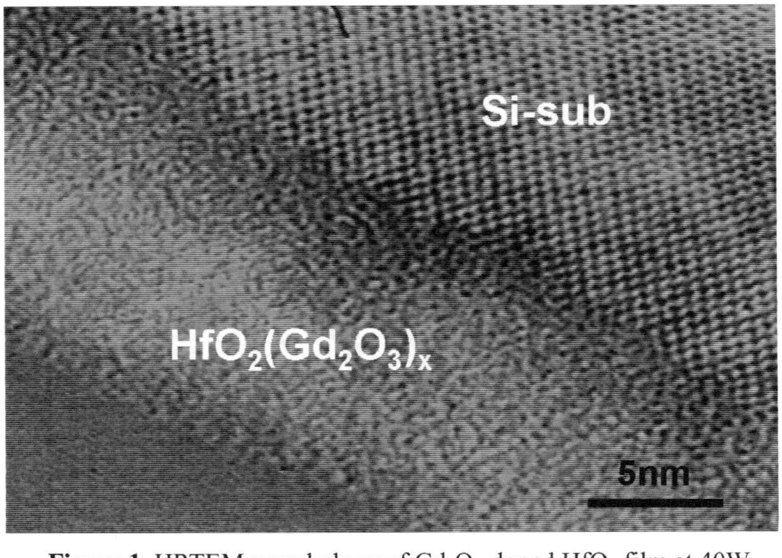

Figure 1. HRTEM morphology of Gd_2O_3-doped HfO_2 film at 40W

In order to investigate the film composition and chemical bonding states, XPS analysis has been performed for both doped and as grown HfO_2 thin films. The binding energy (BE) has been calibrated by centering the C1s at 284.8eV. As the Gd_2O_3 doping power varies from 0W to 60W, the Gd/(Gd+Hf) atomic ratio increases from 0 to 13.4% according to the XPS measurement. In figure 2 (a), a succession of peaks from Hf4f, Hf4d, Hf4p, Gd4d, Gd3d and O1s are clearly present and the signal from C1s is very weak. Figure 2(b) shows the XPS spectra of Hf4f core levels for HfO_2 gate dielectrics deposited on Si wafers. The spectra collected for Hf4f of as grown HfO_2 can be fitted into two peaks at 19.1eV and 17.5eV. The $Hf4f_{7/2}$ at 17.5eV is separated by 1.6eV from $Hf4f_{5/2}$, which attributes to Hf bounded to oxygen atom [9-11]. While the Hf4f binding energies of Gd_2O_3 doped HfO_2 thin films at 60W were at 19.8eV and 18.6eV corresponding to $Hf4f_{5/2}$ and $Hf4f_{7/2}$, respectively. The peak positions of Hf4f in Gd_2O_3 doped HfO_2 samples shift to higher binding energies compared to that of as grown HfO_2 sample. The increase of Hf4f binding energies is caused by the presence of gadolinium in HfO_2 films [12]. As when mixing two oxides, the cation of the more ionic metal oxide is expected to become even more ionic after formation of the complex oxide [13]. In the Gd-doped HfO_2 films, Hf is more ionic than Gd. Consequently, as the Gd atomic ratio increases, Hf becomes more ionic and loses more charges. Therefore, the corresponding binding energies of the Hf4f peaks shift to higher energies as the Gd/(Gd+Hf) atomic ratio increases.

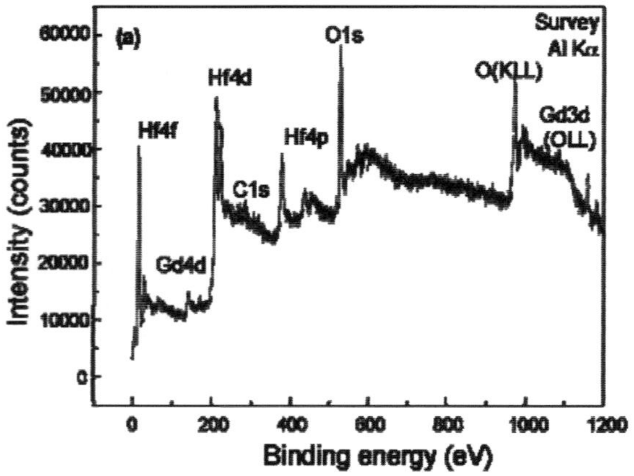

Figure 2. (**a**)XPS Spectra for amorphous Gd_2O_3 doped HfO_2 thin films deposited on n-Si (100) wafers and annealed at 450℃ for 20 min in Ar ambient. (a)Typical X-ray photoelectron spectrum of Gd_2O_3 doped HfO_2 thin film at 60W.

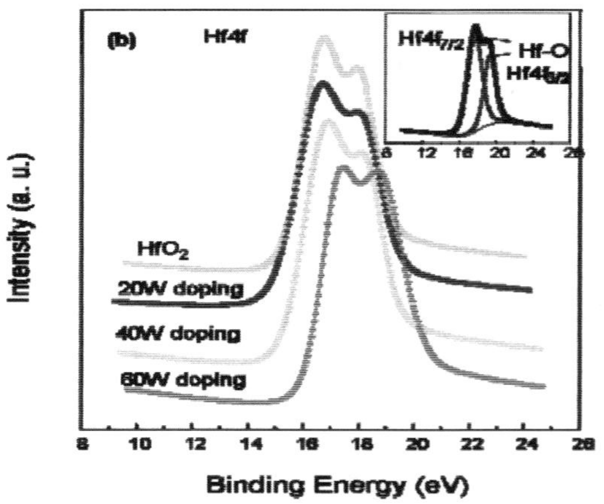

(**b**)Hf4f core-level XPS spectra of both doped and as grown films annealed at the same conditions. The inset shows the Hf4f core-level XPS spectra of pure HfO_2 thin film in details. Spectras were recorded using Al Kα radiation.

The I-V characteristics of both as grown and Gd_2O_3-doped HfO_2 thin films are shown in Fig 3. It is seen that the leakage current densities of all the samples are in the rang from $1\times10^{-3}A/cm^2$ to $5\times10^{-2}A/cm^2$ at gate bias voltage 1V. The lowest leakage current density is found for films with Gd_2O_3 doping at 60W. The leakage current is evidently reduced as Gd_2O_3 deposited power increases. All the Gd_2O_3 doped HfO_2 films show lower leakage current than the as grown HfO_2 films. This phenomenon elucidates that the leakage current in HfO_2/Si heterostructure could be effectively suppressed by Gd_2O_3 doping.

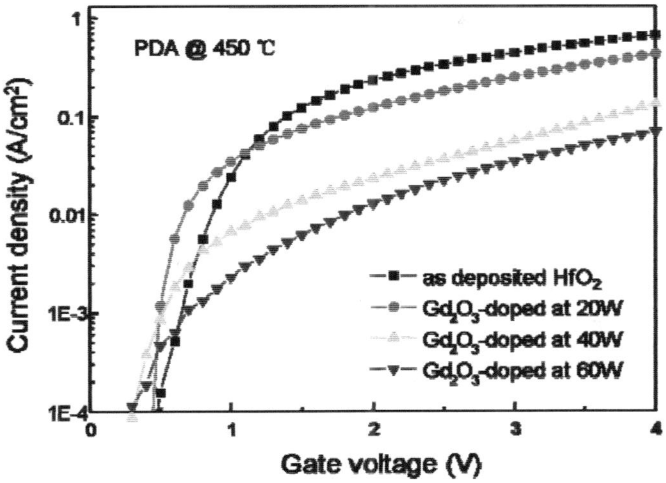

Figure 3. Gate leakage current densities for MOS capacitors with 10nm thick gate dielectrics

Different hysteresis obtained by sweeping the voltage from inversion to accumulation and back again can be observed from figure 4. When Gd_2O_3 doping power is as low as 20W, the film shows less hysteresis than the as grown HfO_2 film. When the doping power increases to 40W and 60W, the hysteresis was further reduced. The dielectric constant, equivalent oxide thickness, dielectric loss as well as the hysteresis ΔV_{FB} extracted from the C-V curve are shown in table 1. All the doped films show lower dielectric loss. The calculated EOT from accumulation capacitances ranges from 1.9 nm to 1.2 nm. The relative k value ranges from 20.2 to 31.6 which can meet the aggressive down scaling of complementary metal-oxide-semiconductor (CMOS) dimensions.

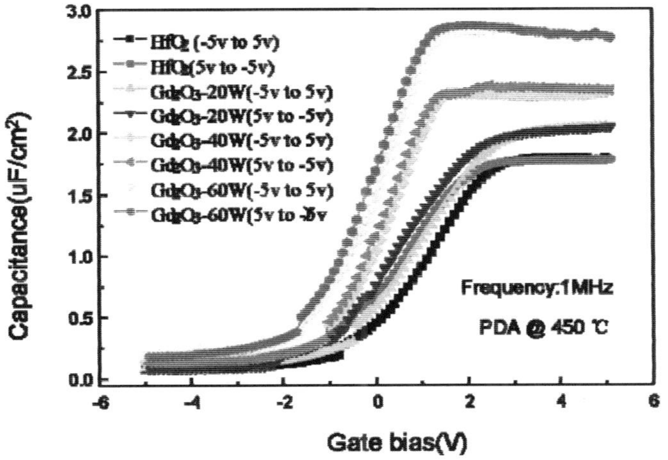

Figure 4. C-V curve of MOS capacitors with 10nm dielectric layers for both doped and pure HfO_2 thin films.

Table 1. Dielectric constant (k value), equivalent oxide thickness (EOT), dielectric loss and ΔV_{FB} of both pure HfO_2 and Gd_2O_3-doped HfO_2 gate dielectrics

Samples	k value	EOT(nm)	Dielectric loss	ΔV_{FB}(V)
As-grown	20.2	1.9	0.35	-0.46
20W doping	23.1	1.7	0.199	-0.31
40W doping	26.1	1.5	0.076	-0.15
60W doping	31.6	1.2	0.02	-0.12

4. Conclusion

In this work, Gd_2O_3-doped HfO_2 high-k gate dielectric thin films have been prepared and characterized. The doping effects are compared with those as grown HfO_2 dielectrics. The results indicate that by doping rare earth oxide Gd_2O_3 in HfO_2 thin films for CMOS gate dielectrics, the leakage current, dielectric loss and equivalent oxide thickness would be evidently reduced. Moreover, the dielectric constant was obviously increased and showing an increasing trend as adding more Gd_2O_3. Less hysteresis in the reverse sweeping can also be observed as the content of Gd_2O_3 increases. These results suggest that the Gd_2O_3-doped HfO_2 thin film is a promising candidate for high k gate dielectric application.

References

[1] Wong H and Iwai H 2006 *Microelectronic Engineering*. **83** 1867-1904
[2] Kwo J, Hong M, Kortan A. R 2000 *Appl. Phy. .Lett.* **77** 130-132
[3] Wilk G D, Wallace R M and Anthony J. M 2001 *J. Appl. Phys.* **89** 5243-5275
[4] Robertson J 2006 *Rep. Prog. Phys.* **69** 327-396
[5] Locquet J P, Marchiori C and Sousa M 2006 *J. Appl. Phys.* **100** 051610-1-13
[6] Moon T H, Lee J W, Ham M. H 2006 *Microelectronic. Engineering* **83** 2452-2457
[7] Wong H, Zhan N, Ng K L 2004 *Thin. Solid. Films* **462** 96-100
[8] Zhu W J, Tamagawa T 2002 *Electron. Device. Letters, IEEE.* **0741** 649-651
[9] Dai J Y, Lee P F, Wong K H 2003 *J. Appl. Phys.* **94** 912-915
[10] Yamamoto K, Hayashi S, Niwa M 2003 *Appl. Phys. Lett.* **83** 2229-31
[11] Sahiner M A, Woicik J C 2007 *Thin. Solid. Films 515* 6548-51
[12] Yan Kai, Chiou and Chang Che-Hao 2007 *J. Eletrochem. Soc.* **154(4)** G99-G102
[13] Guittet M J, Crocombette J P, Gautier-Soyer M 2007 *Phy. Rev. B* . **63** 125117-125117.7

Morphology and microstructure control of sputter-deposited copper films by addition of Cr

Xinjian Wang*[1], Xianping Dong[1], Bo Zhang[1], Xiaofeng Xu[2]

[1] School of Materials Science and Engineering, Shanghai Jiao Tong University, Shanghai, 200240, P. R. China
[2] Applied Physics Department, Dong Hua University, Shanghai, 200030, P. R. China

*Xinjian_Wang@sjtu.edu.cn

Abstract. Pure Cu and Cu with 1 at.% and 2.6 at.% Cr films were sputtered on (100) single crystal Si wafers in the presence of native suboxide (SiO_x). Samples were vacuum-annealed to investigate effects of Cr on the microstructural characteristics of Cu films. Cr refines the columnar structure and decreases the surface roughness of Cu films. Upon annealing, most films undergo the process of crystallite growth and coalescence. Cross-sectional images show that Cr stabilizes the columnar structure and inhibits the agglomeration of annealed Cu(Cr) films at high temperatures. The fine grains of Cu(Cr) films suggest that the extensive Cu crystallite/grain growth at 400 °C is effectively impeded by the presence of Cr. The noticeable grain growth occurs around 450 °C with Cr precipitates in Cu(Cr) films. After annealing, the final resistivity of the annealed Cu(Cr) film is approached to that of the Cu film. Adding Cr to films has obtained significant changes in morphology and microstructure of Cu films before and after annealing.

1. Introduction

For the 250 nm technology node and beyond, Cu interconnections have replaced Al alloy ones in the advanced metallization because of their reduced *RC* delay [1]. The shift from Al alloy to Cu has made a significant difference in interconnect fabrication methods.

The method being developed for Cu metallization was referred to as the damascene method [2]. In multilevel metallization schemes, preserving a smooth planar morphology of the copper layers is needed to build upper levels. Nevertheless, copper undergoes significant grain growth at 400 °C, an interesting annealing temperature in Si-LSIs, and this can lead to roughening of copper films [3]. Alloying has been suggested to address this problem. This alloying scheme is intended to dissociate the dopant out of the alloy film upon annealing to change the microstructure of films [4]. Cu films containing alloy elements such as Ti, Ag, Mo, Zr et al., respectively, have been investigated to have enhanced thermal stability than pure Cu films [5-8].

However, for a potential application, resistivities of Cu alloy films after annealing around 400 °C must be lower than those of Al alloy films (< 3 μΩ·cm). Solutes satisfying the requirement are Zr, Cr, Ag [9-11]. Among these, Cu-Cr alloy films have been studied to have the low electrical resistivity, high corrosion resistance and improved adhesion to SiO_2 owing to the segregation of Cr at the surface and interface [12-15]. Carl et al. have reported the effect of microstructural changes in Cu-1 at.%Cr films on the final resistivity after annealing [12]. Yet, the morphology and microstructure of low-alloyed Cu(Cr) films, as an important characteristic of thermal stability, have not been received enough recognition and research, although they play a significant role in affecting the properties of Cu(Cr) films.

The present work is directed toward Cu(Cr) films in order to investigate effects of Cr dopant on the morphology and microstructure of the Cu/SiO_x/Si(100) system before and after annealing, which will provide an experimental reference for the application of Cu(Cr) in the microelectronic field as a new interconnect material or a seed layer.

2. Experimental

Cu(Cr) films were deposited on Si(100) substrates by magnetron sputtering a pure Cu (99.98%, purity) target inlaid with Cr sticks(99.99%, purity). Different numbers of Cr sticks were inlaid to control the Cr content. Before loading in the sputtering system, the silicon wafers were successively cleaned in acetone and alcohol, followed each time by deionized water rinsing, in an ultrasonic tank. Because the substrates were not etched with hydrofluoric acid, there was a native suboxide (SiO) layer about 15 Å thick on the surface [16].

Prior to the deposition process, the base pressure in the chamber was lower than 0.5×10^{-3} Pa. All films were sputtered in pure Ar at a working pressure of 0.67 Pa with a power of 100 W. The substrates were unheated, resulting in a deposition temperature of ~ 80 °C. The Cr content in Cu-Cr films was determined with energy dispersive X-ray spectrum (EDX) to be 1 at.% (referred to as Cu-1%Cr) and 2.6 at.% (referred to as Cu-2.6%Cr), respectively. After deposition, all samples were vacuum-annealed in the 300 °C-600 °C temperature range for 1 h. Film morphology was observed by Sirion 200 field emission scanning electron microscopy (FESEM) and the surface roughness was obtained by Nanoscope III atomic force microscopy (AFM). Microstructural analysis was carried out in a JEM-2100 transmission electron microscopy (TEM) operating at 200 kV. The resistivity as a function of annealing temperature was measured by the four-point probe method.

3. Results and discussion

3.1. The feature of film morphology

Figure.1 shows the cross-sectional FESEM images of Cu/SiO_x/Si and Cu(Cr)/SiO_x/Si samples before and annealing. A columnar structure could be observed in the as-deposited Cu films, which was textured in (111) orientation [7,17]. However, the poorly textured Cu film had a significant fraction of films ((a), (d), (g) as-deposited, (b), (e), (h) annealed at 400 °C and (c), (f), (i) annealed at 500 °C). regions with equiaxed grains near the surface of film. Distinctly, the typical columnar structure of Cu(Cr) films was much finer than that of copper films. The formation of equiaxed grains was attributed to the *in-situ* annealing on the growing film during the sputtering process [18]. In such an *in-situ* annealing process, the impurity effect of Cr atoms is prominent. The film thickness decreased noticeably with the increasing Cr content as a result of decreases in the deposition rate.

After annealing at 400 °C, with continuous, extensive growth and coalescence of crystallites taking place, the columnar structure was invisible and the film became coarse for pure Cu and Cu-1%Cr films. At 500 °C, the Cu film agglomerated and cavities formed in the film and the substrate. The coalescence of Cu films, perpendicular to sample surface, resulted in the great roughness of Cu/SiO_x/Si samples. At even higher temperatures (> 650 °C), the formation of Cu-Si compound crystallites would aggravate the surface roughness [19].

In contrast, although the aggregated crystallites grew, the annealed Cu-2.6%Cr film was much more continuous and compact than the annealed Cu and Cu-1%Cr films. Because the wettablity of film on the substrate can be defined by the agglomeration of films [20], the investigation of alloying effect on Cu agglomeration suggested that the addition of Cr increased the Cu film wettability on substrate and enhanced the thermal stability of Cu films. Therefore, the Cu-Cr alloy film is rather stable and keeps a smooth surface at the processing temperature of Si-LSIs.

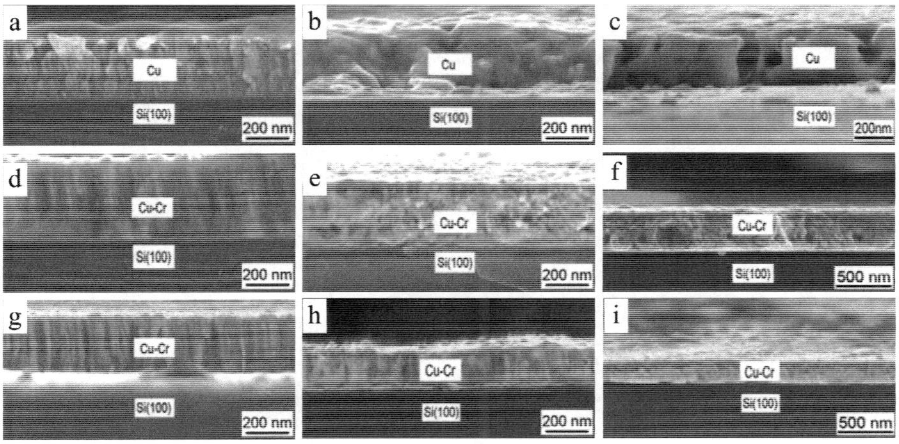

Figure 1. Cross-sectional FESEM images of Cu (a, b, c), Cu-1%Cr (d, e, f), and Cu-2.6%Cr (g, h, i)

AFM has been employed to obtain the corresponding surface roughness RMS (root mean square) of samples, which is shown in Figure 2. The measurement was performed in tapping mode with 256×256 data collected over an area of 5×5 μm and a scan speed of 2 Hz. The RMS value is determined by the following formula:

$$RMS = \frac{1}{(256)^2} \sum_{i,j=1}^{256} \left| z(i, j) - \bar{z} \right|$$

where $z(i, j)$ represents the height measured at position (i, j) , and $\bar{z}$ is the height averaged over the entire area. It was clear that the as deposited Cu(Cr) films had smaller RMS values than Cu films. The RMS values decreased from 14.128 nm of Cu films to 3.837 nm of Cu-1%Cr films. This suggested that Cr atoms mixed with Cu atoms and prevented from the growth of Cu grains on the growing film surface by reducing the diffusion rate of Cu atoms during the deposition process. However, the RMS value did not change much as the Cr content was increased to 2.6%. After annealing, RMS values of Cu films increased owing to the grain growth and agglomeration at above 500 °C. Exactly, annealed Cu(Cr) films had much smaller surface roughness, indicating the high thermal stability upon annealing. In fact, the Cu-2.6%Cr film was heated to a maximum temperature of 700 °C in this research and no agglomeration had taken place, whereas the pure Cu film was ruptured at all so that the RMS value could not be measured accurately.

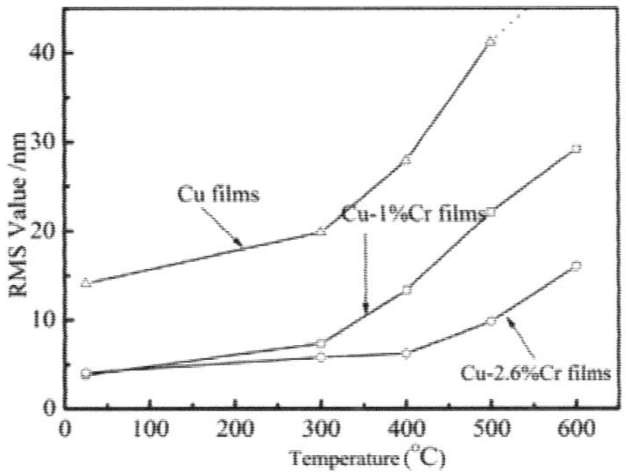

Figure 2. RMS values of Cu, Cu-1%Cr and Cu-2.6% films at various annealing temperatures.

3.2. TEM analysis

Transmission electron microscopy was performed on the sputtered Cu and Cu-2.6%Cr films before and after annealing (Figure 3). After removal from the substrate, the films were glued onto 3 mm Cu double cirques and thinned to electron transparency in a Gatan 691 precision ion polishing system (PIPS). The as-deposited Cu film had a fine microstructure with the average grain size around 50 nm. After Cr was added, the crystallinity of films became incomplete, with some little crystallites agglomerated or dispersed in the film. After annealing at 400 °C, Cu and Cu(Cr) films showed evidence of grain growth. The selected area diffraction (SAD) showed that the film was mainly the face-centered-cubic (fcc) Cu phase. The increase in diffracted intensity of Cu was taken as an indicator of grain nucleation and growth. The grains had grown to a few hundred nanometers in diameter and were frequently twinned in the Cu film. However, the average grains didn't grow much in size for the Cu(Cr) film. At 500 °C, the Cu(Cr) film was composed entirely of large grains. The smallest Cu grains were about 40 nm and the largest grains were 200-300 nm, indicating the abnormal grain growth. In contrast, the extensive grain growth resulted in Cu grains about 1 μm in the Cu film, which led to the agglomeration of copper films at high temperatures.

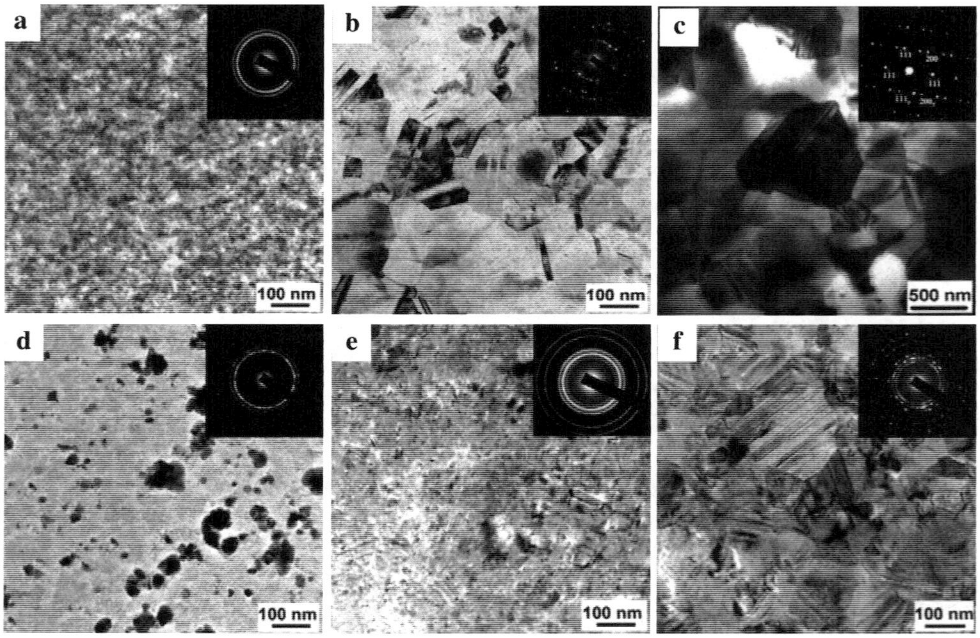

Figure 3. Bright-field transmission electron micrograph and selected area diffraction patterns of Cu (a, b, c) and Cu(Cr) films (d, e, f) ((a, d) as-deposited, (b, e) annealed at 400 °C, (c, f) at 500 °C.

To understand the microstructural evolution of Cu(Cr) films upon annealing more clearly, we examined the microstructure of Cu(Cr) films annealed at 450 °C for 1 h (Figure 4). The smallest Cu grains were about 40 nm in size and the largest grains with typical twins were about 150 nm, exhibiting a bimodal grain size distribution.

It was reported that the dissociation of insoluble alloy elements, such as Co, Ag, B, et al, out of Cu alloy films resulted in an evident grain growth and coalescence at 400 °C [21]. However, the large Cu grains with equiaxed shape in annealed Cu-2.6%Cr films formed at above 450 °C, although Cr had the most approximate atomic size with Cu. K. Barmak suggested that this might be due to the effect of oxygen impurity [12]. In addition, that the insolubility of Cr in Cu and the delocalization of Cr 3*d* electron owing to the transition character of Cr contributes to the higher transition temperature. The delocalization resulting from the interaction between Cr 3*d* and Cu 4*s* electrons increases the Cr atomic radius in solution, leading to the unexpectedly large grain boundary (GB) solute binding energy in the Cu-Cr alloy system [22,23]. Cr precipitating in the film and between grains inhibited the extensive Cu grain growth below 500 °C [12,17,23]. We think that microstructural changes will bring about great effects on the mechanical property and electromigration resistance of Cu(Cr) films, which needs further work to be confirmed.

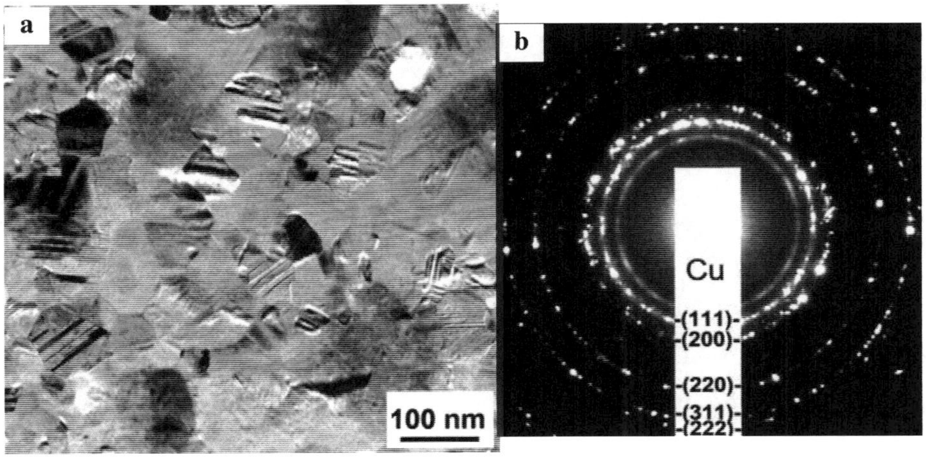

Figure 4. TEM results of Cu(Cr) films annealed at 450 °C. (a) bright-field micrograph, (b) SAD pattern. Rings corresponding to $Cu_2O(111)$ and $Cu_2O(220)$ could also be seen.

Evidently, the presence of insoluble Cr dopant has appeared to play significant roles in refining the microstructure, affecting the crystallinity and improving the thermal stability of Cu films. Firstly, the impurity effect of Cr atoms is prominent in inhibiting the crystallinity of as-deposited Cu films. Since Cr has essentially no solid solubility with Cu at room temperature, the addition of Cr atoms reduces the rapid diffusion rate of Cu adatoms, which in turn alters the growing structure of as-deposited Cu films [24].

Similarly, the beneficial effect of Cr atoms on the morphology and microstructure of Cu films upon annealing can be revealed. It is known that the grooving process occurs preferentially at the groove root (eg. the intersection of the GB and the film surface) upon annealing [25]. The local energy equilibrium in the as-deposited thin film has not been reached at GB triple joints on both sides of films. When the thin film is annealed at an elevated temperature, adjacent grains will groove at GB triple joints in order to reach an equilibrium state by the movement of boundaries, which results in the agglomeration finally. Based on a simplifying model for polycrystalline thin films with columnar grains, Nolan T P et.al suggested a maximum grain size *Lc* for which agglomeration didn't occur, and that addition of thin layer of high-melting materials or adding contaminants to the film could inhibit the agglomeration [26].

Upon annealing at low temperatures (~ 400 °C), Cr atoms inhibited the prevalent grain growth effectively and prohibited from the agglomeration of Cu(Cr) films. The substantial grain growth took place at 450 °C, where the Cr pinning effect on the Cu grain growth appeared to be overcome. The distribution of grain size became bimodal. Moreover, further annealing at a higher temperature resulted in the abnormal grain growth, which was often thought to arise from the growth of favorably oriented grains [21]. The AES result (not displayed here) has shown that most of Cr segregated at the surface after annealing, in the form of Cr oxide, and formed a protective cap layer on the film surface [12,17]. In any case, the segregation of Cr at grain boundaries, the film surface as well as the interface between the film and the substrate, contributes to the thermal stability of the Cu film by inhibiting the diffusion of Cu atoms.

3.3. Resistivity Data

Resistivities of Cu/SiO $_x$ /Si and Cu(Cr)/SiO $_x$ /Si samples at various temperatures are shown in Figure 5. Cr increased the resistivity of as-deposited Cu films markedly. As the annealing temperature increased, the resistivity of all films decreased until a certain temperature range was reached. This was most likely due to the Cu grain growth and the annihilation of point defects. The dominant grain growth mechanism in this temperature range was the growth and coalescence of Cu grains. For the Cu-Cr alloy film, the grain growth was accompanied with the segregation of Cr [12,17]. An abrupt increase in resistivities of the pure Cu and Cu-1%Cr films occurred at 500 °C and 600 °C respectively, owing to the agglomeration of films. The observation of morphology after annealing was in agreement with the resistivity result. The minimum of resistivity was about 2.55 μΩ·cm and 2.76 μΩ·cm for Cu-1%Cr and Cu-2.6%Cr films respectively, which was approached to that of Cu films (2.1 μΩ·cm). Notably, a low resistivity (< 3 μΩ·cm) can be obtained for annealed Cu-2.6%Cr films till 600 °C. Therefore, Cu films doped with an optimal amount of Cr are suitable for the potential interconnect material resistant to high temperatures.

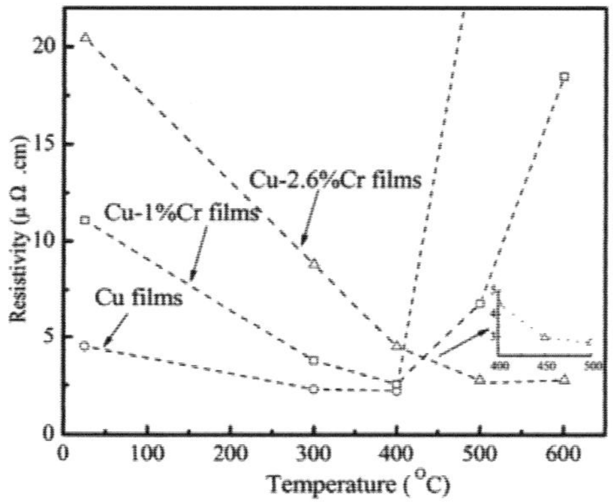

Figure 5. Resistivity variation of Cu/SiO $_x$ /Si, Cu(1% Cr)/SiO $_x$ /Si and Cu(2.6% Cr)/SiO $_x$ /Si samples upon annealing at various temperatures.

4. Conclusions

As-deposited Cu and Cu(Cr) alloy films had a nanocrystalline microstructure. Cr dopant inhibited the crystallinity and refined the columnar structure, decreasing the surface roughness of as-deposited Cu films markedly. Upon annealing, the extensive Cu crystallite/grain growth was effectively impeded by the presence of Cr. For the annealed Cu-2.6%Cr film, noticeable grain growth occurred at 450 °C, resulting in a bimodal grain size distribution. The fine-grained microstructure inhibited the agglomeration of Cu(Cr) films at higher temperatures. A low resistivity (< 3 μΩ·cm) can be obtained for annealed Cu(Cr) films till 600 °C.

Low alloyed Cu(Cr) films with the improved surface morphology and refined microstructure have potential superiority to be used as the interconnecting material or the seed layer in the Damascence technology of Cu interconnection.

Acknowledgements

The authors gratefully acknowledge the Instrumental Analysis Centre of SJTU and Dong Hua University. This work was supported from Applied Materials of Shanghai Funds (AM.0525).

References

[1] Yousuke Koie, Toshio Inase and Shinji Takayama 2007 *Solid State Phenomena* **127** 147
[2] Dalal H M, Joshi R V, Rathore H S and Fillipl R 1993 *Technical digest-international electron devices meeting* 273
[3] Lanford W A, Ding P, Wang J W, Hymes S and Murarka S P 1995 *Mater. Chem. Phys.* **41** 192
[4] Harper J M E and Rodbell K P 1997 *J. Vac. Sci. Technol.* B **15** 763
[5] Liu C L, Jeng J S and Chen J S 2002 *J. Vac. Sci. Technol.* B **20** 2361
[6] Strehle S, Menzel S, Wendrock H, Acker J, Gemming T. and Wetzig K. 2004 *Microelectronic Engineering* **76** 205
[7] Lee W H, Ko Y K, Jang J H, Kim C S, Reucroft P J and Lee J G 2001 *J. Elect. Mater.* **30** 1042
[8] Liu B, Song Z X, Xu K W 2007 *Surface and Coatings Technology* **201** 5419
[9] Ko Y K, Jang J H, Lee S, Yang H J, Lee W H, Reucroft P J, Lee J G, 2003 *J. Mater. Sci.* **38** 217
[10] Igarashi Y, Ito T 1998 *J. Vac. Sci. Technol.* B **16** 2745
[11] Barmak K, Lucadamo G A, Cabral C Jr, Lavoie C and Harper J M E 2000 *J. Appl. Phys.* **87** 2204
[12] Cabral C Jr, Harper J M E, Holloway K, Smith D A and Schad R G 1992 *J. Vac. Sci. Technol.* A **10** 1706
[13] Kim J, S H Wen and D Yee 1988 *J. Vac. Sci. Technol.* A **6** 2366
[14] Russell S W, Rafalski S A, Spreitzer R L, Li J, Moinpour M, Moghadam F, Alford T L 1995 Thin Solid Films **262** 154
[15] Li J, Mayer J W and Colgan E G 1991 *J. App. Phys.* **70** 2820
[16] Grundner M, Jacob H 1986 *Appl. Phys.* A **39** 73
[17] Wang X J, Dong X P, Liu J C, Jiang C H 2007 *Trans Nonferrous Met. Soc. China* **17** *s*730
[18] Detavernier C, Deduytsche D, Meirhaeghe R l Van, Baerdemaeker J De and Dauwe C 2003 ***Appl. Phys. Lett.,*** 82 1863
[19] Benouattas N, Mosser A, Bouabellou A 2006 Appl. Sur. Sci. **252** 7572
[20] Hoon Kim, Toshihiko Koseki, Takayuki Ohba, Tomohiro Ohta, Yasuhiko Kojima, Hiroshi Sato, Yukihiro Shimogaki 2005 Thin Solid Films **491** 221
[21] Barrack K, Gungor A, Cabral C Jr, Harper J M E 2003 *J. Appl. Phys.* **94** 1605
[22] Gupta D 2003 Interface Science **11** 7
[23] Mclntyre D, Sundgren J E and Greene J E 1988 *J. Appl. Phys.* **64** 3689
[24] Reed-hill R E and Abbaschian R 1991 *Physical Metallurgy Principles*, 3 ed (PWS, Boston, MA) PP.227-271.
[25] Mullins W W and Shewmon P G 1959 *Acta Metall.* **7** 163
[26] Nolan T P, Sinclair R and Beyers R 1992 *J. Appl. Phys.* **71** 720

Growth of InGaN-based laser diode structure on silicon (111) substrate

Bin Abu Bakar Ahmad Shuhaimi[1,2], **Hiroyuki Kawato**[1,2], **Youhua Zhu**[1,2] **and Takashi Egawa**[1,2]

[1] Research Center for Nano-Device and System, Nagoya Institute of Technology, Nagoya 466-8555, Aichi, Japan.
[2] Dep. of Eng. Physics, Electronics and Mechanics, Graduate School of Engineering, Nagoya Institute of Technology, Nagoya 466-8555, Aichi, Japan.

E-mail: shuhaimi@msn.com

Abstract. This paper reports growth and material characterization of InGaN-based laser diode (LD) structure grown on Si(111) substrate by metal organic chemical vapor deposition (MOCVD) with a high-temperature-grown thin AlN buffer layer and an AlN/GaN multi-layer (ML) intermediate layers. Two types of LD structures on Si were prepared, (a) with bulk n- and p-type $Al_{0.02}Ga_{0.98}N$ cladding layer, and (b) with SLS n- and p-type $Al_{0.06}Ga_{0.94}N$/GaN cladding layer. Active layer with 3 periods of $In_{0.10}Ga_{0.90}N/In_{0.03}Ga_{0.97}N$ multi-quantum-well (MQW) is sandwiched between the cladding layers. A standard LD structure (c) with identical MQW grown on GaN(0001) free-standing substrate was also prepared for comparison. Mirror-like and meltback-etching free surfaces have been achieved for both of the samples grown on Si. High-resolution x-ray diffraction (HRXRD) $\omega/2\theta$ scan reveals that a good quality of MQW has been achieved for both of the samples grown on Si. Temperature-dependence PL measurement reveals that LD structure grown on Si with SLS cladding layer has higher internal quantum efficiency η_{int} than that of the sample with bulk cladding layer.

1. Introduction

GaN and its tertiary and quaternary alloys are very promising materials for light-emitting optoelectronic devices such as LEDs and LDs, with emission range from deep-ultra-violet (DUV) to infra-red (IR) region. A tremendous research has been performed for optoelectronic and electronic devices based on GaN. However, most of the growth of GaN and its related alloys have been performed on sapphire, SiC and GaN free-standing substrates which have the advantages of small thermal and small lattice mismatch with GaN epilayer. On the other hand, Si is also a promising substrate for GaN growth, which offers many advantages such as low cost, large diameter, good thermal conductivity for high-performance devices, vertical electrode capability and the possibility for integration with Si-based electronics to create a monolithic optical and electronics integrated circuit (OEIC).

Despite its large thermal and lattice mismatch with GaN, many groups, including ours, have reported good quality GaN epitaxial growth and devices on Si(111) substrate [1-5]. For light emitting devices, however, growth of GaN-based devices on Si are more focused for LEDs [1-6]. On the other hand, growth of GaN-based LDs are performed on sapphire [7-9], SiC [10,11] and GaN free-standing [12,13] substrates. To our knowledge, successful growth of GaN-based LD structure on Si substrate for current pumping has never been published, due to the complexity during layer growth and necessity for thick structure, which may result to cracks in the epitaxial layer. AlN is widely used as buffer layer for GaN growth on Si. Several reports use thick AlN buffer layer to prevent the formation of meltback-etching and cracks [3,14]. However, thickening the AlN buffer layer will produce higher current-resistance to the final device due to the insulating nature of AlN, which is not practical for LD with vertical electrode. This study reports the growth and material characterization of LD structures on Si(111) substrate by utilizing thin AlN buffer layer and AlN/GaN multilayer (ML) intermediate layer. A standard LD structure on GaN(0001) free-standing substrate is also prepared for comparison.

2. Experiment

The growth of LD structures in this report was carried out in a commercially available MOCVD with horizontal reactor system. Trimethylgallium (TMG), trimethylindium (TMI), trimethylaluminum (TMA) and ammonia (NH_3) were used as source materials, and hydrogen (H_2) was used as the carrier gas. Two type of LD structures grown on Si(111) substrate, denoted as sample A and sample B, and one LD structure grown on GaN free-standing substrate, denoted as sample C were prepared for comparison. In this study, sample C is a standard LD structure, and the growth for sample A and B is optimized to reach or exceed the performance of the standard sample.

Prior to the growth of both LD structures on Si(111) substrate, a thin high-temperature-grown AlN buffer layer and an AlN/GaN multilayer (ML) intermediate layer were grown to improve the structural and optical quality of the subsequent layers [1-5]. Sample A consists of a 500 nm thick n-$Al_{0.02}Ga_{0.98}N$ cladding layer, n-GaN waveguide layer, active layer with 3 periods of 4 nm thick undoped $In_{0.10}Ga_{0.90}N$ multi-quantum-wells and 8.0 nm thick Si-doped $In_{0.03}Ga_{0.97}N$ barrier layers, GaN waveguide layer, p-AlGaN electron-block layer, 500 nm thick p-$Al_{0.02}Ga_{0.98}N$ cladding layer and p-GaN contact layer. Sample B consists of almost identical structure with sample A, except n- and p-type cladding layers which were replaced with n- and p-type $Al_{0.06}Ga_{0.94}N$/GaN strained-layer-superlattices (SLS) layers. SLS layers are generally used to reduce threading dislocation density in the active layer [9,15] and to enhance carrier concentration in the cladding layer [15,17]. In sample C, a thick n-GaN was used as buffer layer prior to the growth of LD layers. All subsequent layers in this sample are identical to sample A. Full structure for each sample is shown in figure 1.

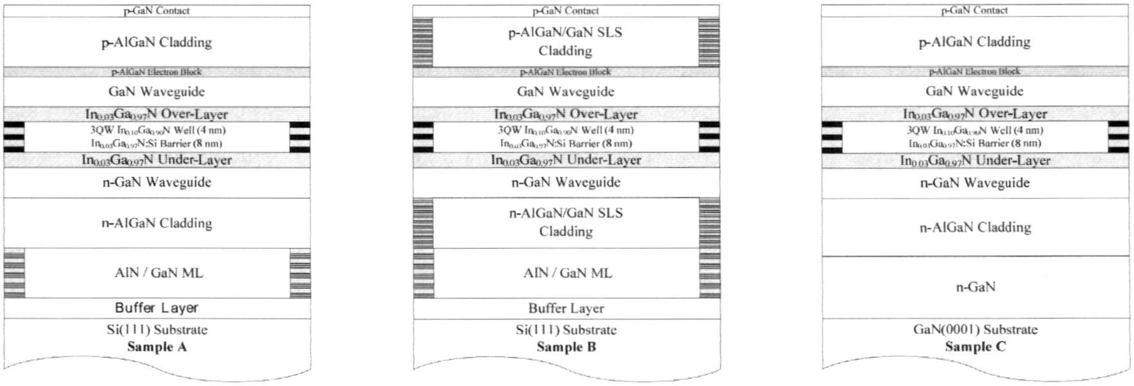

Figure 1. Layer structure for sample A (left), B (center) and C (right).

Surface morphology of the samples was characterized using atomic force microscopy (AFM). Crystal quality and structural property were characterized by high-resolution x-ray diffractometer (HRXRD). Temperature-dependence photoluminescence (PL) measurement from 10 K to 300 K was performed using a 325 nm line of He-Cd laser as excitation source, to characterize the MQW performance. Emission from the MQW was collected using CCD positioned normal to the sample surface. The MQW indium composition x was estimated from PL peak energy at 300 K on sample C using the relation $E_g(x)=3.412(1-x)+0.65x-2.6x(1-x)$ eV, where E_g is 3.412 eV for GaN [21] and 0.65 eV for InN [22]. E_g is the effective excitonic band gap energy and the value -2.6 is the bowing parameter for strained *c*-plane $In_xGa_{1-x}N$ layer on GaN for $0<x<0.2$ [16,23]. The MQW layer thickness was determined by software simulation based on HRXRD $\omega/2\theta$ characterization and TEM cross-sectional image (not shown).

3. Results and Discussion

AFM, HRXRD and PL properties of sample A, B and C are summarized in table 1. Specular and meltback-etching free surfaces have been achieved for all samples in our study. AFM surface morphology image in figure 2 shows that terraces can be clearly observed in samples grown on Si substrate. Screw component of dislocations and pure edge dislocations are clearly visible in the samples [9,19]. In contrast, the sample grown on GaN substrate shows straight steps [19]. The RMS roughness measured on 3×3 祄 area is 0.44 nm and 0.48 nm for sample A and B, respectively, while the peak-to-valley (P-V) is 3.58 nm and 3.59 nm for sample A and B, respectively. Despite a slight increase of roughness and P-V value for sample B compared to sample A, these values are more than 3 times larger than the structure on GaN substrate indicated by sample C.

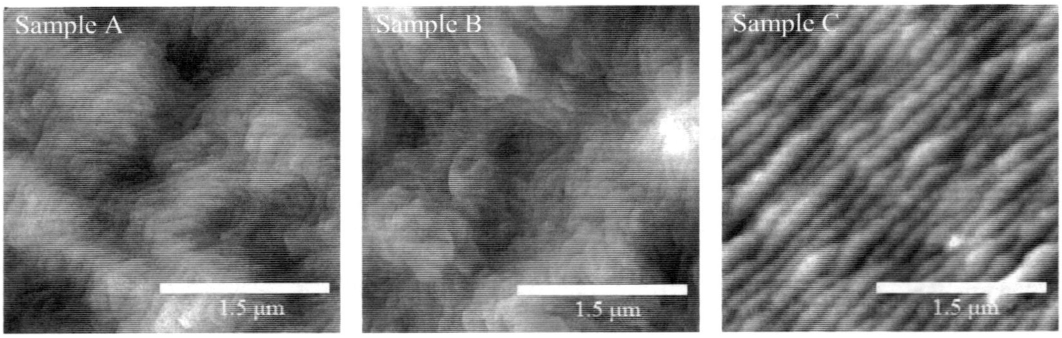

Figure 2. AFM images of the sample surface in 3×3 祄 area.

As shown in table 1, FWHM value of GaN (0004) x-ray rocking curve (XRC) is 990 arcsec, 1153 arcsec and 247 arcsec for sample A, B and C, respectively. The increase of the FWHM value in sample B, compared to that of sample A is consistent with AFM roughness and P-V for each of the

Table 1. Sample properties from each characterization method.

Characterization Method		Sample A	Sample B	Sample C
AFM 3×3 祄	RMS (nm)	0.44	0.48	0.14
	P-V (nm)	3.58	3.59	1.13
XRD	FWHM (0004) (arcsec)	990	1153	247
	FWHM $(20\bar{2}4)$ (arcsec)	1967	1899	257
	Mean Period Thickness (nm)	12.1±0.3	12.4±0.3	11.8±0.3
PL at 10 K	Intensity (a. u.)	93	33	100
	MQW Peak (eV)	2.84	2.88	2.89
	FWHM (meV)	173.3	~139.7 (at 0.6)	106.4
at 300 K	MQW Peak (eV)	2.81	2.86	2.88
	FWHM (meV)	93.5	97.2	196.6

samples. The FWHM value of GaN $(20\bar{2}4)$ XRC is 1967 arcsec, 1899 arcsec and 257 arcsec for sample A, B and C, respectively. These results suggest that despite a higher screw component of dislocations generated in the sample with SLS cladding layers, edge and mixed dislocations can be suppressed by the SLS cladding layers.

Figure 3 shows the HRXRD GaN (0004) $\omega/2\theta$ scan for sample A, B and C. The result shows strong $In_{0.10}Ga_{0.90}N/In_{0.03}Ga_{0.97}N$ MQW fringes of 0th, -1st, -2nd and -3rd orders in all samples. The $+1^{st}$ and $+2^{nd}$ orders of the fringes, however, are weak in all samples. Note that the $+1^{st}$ and $+2^{nd}$ order fringes of the MQW in samples grown on Si overlap with fringes from AlN/GaN ML, resulting a higher intensity fringes at 36.7 deg. and 37.2 deg. relative to other AlN/GaN ML fringes. The MQW mean period thickness is 12.1±0.3 nm, 12.4±0.3 nm, and 11.8±0.3 nm for sample A, B and C, respectively. The higher-order fringes indicate that a good quality MQW layer with abrupt interfaces has been achieved in all of the samples. The consistent thickness in all samples indicates that MQW growth on Si substrate does not show any enormous change in the period thickness compared to growth on GaN substrate.

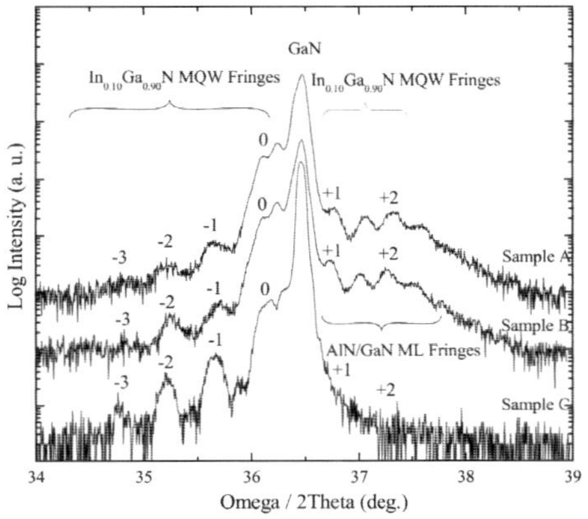

Figure 3. HRXRD (0004) $\omega/2\theta$ scan for sample A, B and C.

PL characteristics at 10 K and 300 K for sample A, B and C is shown in figure 4. The band-to-band photon energy emitted from the $In_{0.10}Ga_{0.90}N/In_{0.03}Ga_{0.97}N$ MQW in the LD structures measured at 10 K is 2.84 eV, 2.88 eV and 2.89 eV for sample A, B and C, respectively. The strong peaks at ~3.06 eV, ~3.16 eV and ~3.24 eV are from Mg-dopant in the 500 nm thick p-type cladding layer and contact layer in the samples. The appearance of MQW fringes modulated by Fabry-Perot interference at ~2.7 eV confirms that a smooth and abrupt layer interfaces has been achieved in our samples. Measurements at 300 K reveals a small red-shift in all of the samples, with MQW band-to-band photon energy of 2.81 eV, 2.86 eV and 2.88 eV in sample A, B and C, respectively. Note than wider FWHM in sample C at 300 K is attributed by the combination of slightly-lower-energy fringe peaks (~2.8 eV) into the main peak, which is commonly seen in very high quality interfacial layer.

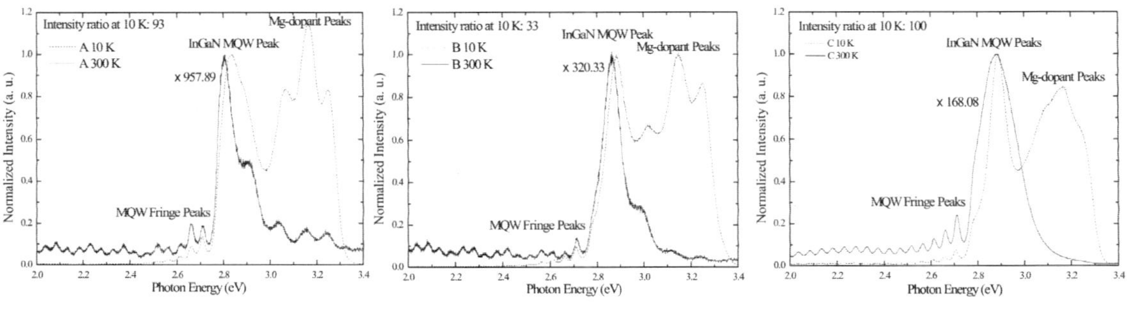

Figure 4. PL characteristics at 10 K and 300 K of sample A (left), B (center) and C (right).

Light emission intensity ratio from the samples at 10 K is 93, 33 and 100 for sample A, B and C, respectively. The lower emission intensity in sample B is attributed by the higher Al composition in the p-type cladding layer which limits He-Ne laser light from reaching the MQW, and limits the extraction of photon from MQW. Interestingly, sample B shows smaller photon emission intensity difference when the temperature is increased to 300 K, when compared to that of sample A.

In order to investigate the peak behavior of the MQW, the MQW peak from temperature-dependence PL measurement is plotted in figure 5, with 10 K interval for temperature from 10 K to 100 K, and 20 K interval for the next temperature increment until 300 K. A clear "S-shaped" peak behavior can be observed in all samples. The MQW emission peak red-shifts when the temperature is increased from 10 K to 70 K, blue-shifts when the temperature is increased from 70 K to 120-140 K, and returns to red-shifts when the temperature is increased from 120-140 K to 300 K. This result is consistent with report by Cho *et al.* [20]. However, with the increase of temperature, a stronger red-shift trend is observed in both samples grown on Si substrate compared to that of GaN substrate. This trend is influenced by quantum-confined-stark-effect (QCSE), due to the strong tensile stress for GaN structure grown on Si substrate [6]. Larger QCSE can be confirmed in Sample A, as the whole emission peak was red-shifted, compared to that of sample B and C.

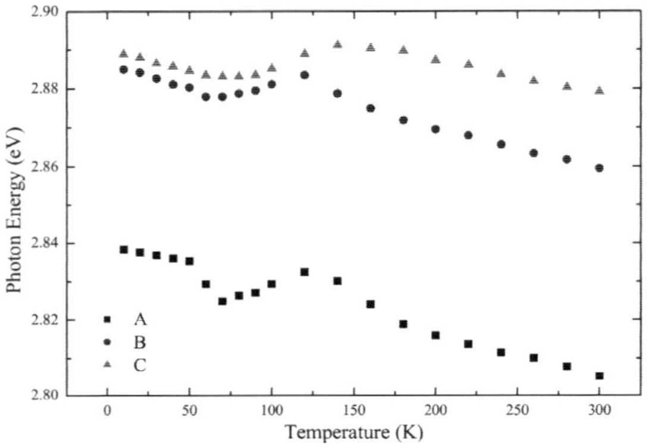

Figure 5. Temperature-dependence MQW peak behavior.

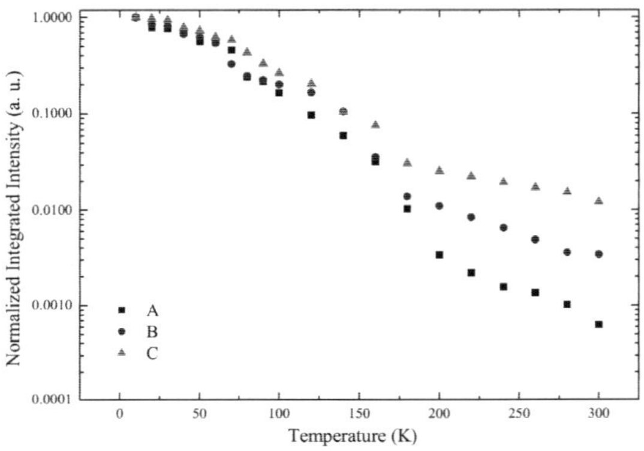

Figure 6. Temperature-dependence normalized integrated intensity.

Gaussian fitting has been performed to the temperature-dependence PL using a software called Origin to obtain the MQW normalized integrated intensity, as shown in figure 6. Assuming that the internal quantum efficiency η_{int} of the samples at 10 K as 100%, the η_{int} at 300 K is estimated to be 0.06%, 0.34% and 1.21% for sample A, B and C, respectively. The Sample B shows one order of magnitude higher η_{int} than that of sample A, probably due to lower QCSE in the MQW produced by the SLS cladding layer. The η_{int} values in this study are generally low due to the thick p-type AlGaN cladding layer above the active layer in the LD structure. The significant decrease of luminescence at ~70 K until ~180-200 K in our normalized integrated intensity result can be speculated by the activation of non-radiative process which dominates the carrier-recombination at this temperature [20]. The temperature-dependence peak behavior and temperature-dependence normalized integrated intensity reveal that $Al_{0.06}Ga_{0.94}N$/GaN SLS layer is effective to reduce QCSE and improve η_{int} in InGaN-based MQW grown on Si(111) substrate.

4. Conclusion

In summary, we have successfully grown good quality $In_{0.10}Ga_{0.90}N/In_{0.03}Ga_{0.97}N$ LD structures on Si(111) substrate, using thin high-temperature-grown AlN buffer layer and AlN/GaN ML intermediate layers. The samples grown on Si(111) substrate in this study are mirror-like and free from meltback-etching. Our results reveal that the replacement of $Al_{0.02}Ga_{0.98}N$ bulk layer with SLS layer consisted of $Al_{0.06}Ga_{0.94}N$/GaN into the n- and p-type cladding layer could reduce QCSE in the MQW, and increase the internal quantum efficiency η_{int} of the sample.

Acknowledgement
The author would like to thank Dr. H. Ishikawa and Y. Maki for their help in PL measurements, Y. Nomura for sample growth, and Dr. J. Zhang for helpful discussions.

References

[1] Zhang B, Egawa T, Ishikawa H, Liu Y, and Jimbo T 2003 *Jpn. J. Appl. Phys.* **42** L226
[2] Egawa T, Moku T, Ishikawa H, Ohtsuka K and Jimbo T 2002 *Jpn. J. Appl. Phys.* **41** L663
[3] Egawa T, Zhang B, Nishikawa N, Ishikawa H, Jimbo T and Umeno M 2002 *J. Appl. Phys.* **91** 528
[4] Ishikawa H, Zhao G Y, Nakada N, Egawa T, Jimbo T and Umeno M 1999 *Jpn. J. Appl. Phys.* **38** L492
[5] Egawa T, Zhang B and Ishikawa H 2005 *IEEE Elect. Dev. Lett.* **26** 169
[6] Dadgar A, Christen J, Riemann T, Richter S, Bläsing J, Diez A, Krost A, Alam A and Heuken M 2001 *Appl. Phys. Lett.* **78** 2211
[7] Nakamura S, Senoh M, Nagahama S, Iwasa N, Yamada T, Matsushita T, Sugimoto Y and Kiyoku H 1996 *Appl. Phys. Lett.* **69** 1477
[8] Nakamura S, Senoh M, Nagahama S, Iwasa N, Yamada T, Matsushita T, Sugimoto Y and Kiyoku H 1996 *Appl. Phys. Lett.* **69** 3034
[9] Nakamura S, Senoh M, Nagahama S, Iwasa N, Yamada T, Matsushita T, Kiyoku H, Sugimoto Y, Kozaki T, Umemoto H, Sano M and Chocho K 1998 *Appl. Phys. Lett.* **72** 211
[10] Kuramata A, Kubota S, Soejima R, Domen K, Horino K, Hacke P and Tanahashi T 1999 *Jpn. J. Appl. Phys.* **38** L481
[11] Edmond J, Abare A, Bergman M, Bharathan J, Bunker K L, Emerson D, Haberern K, Ibbetson J, Leung M, Russel P and Slater D 2004 *J. Cryst. Growth* **272** 242
[12] Kuramoto M, Sasaoka C, Hisanaga Y, Kimura A, Yamaguchi A, Sunakawa H, Kuroda N, Nido M, Usui A and Mizuta M 1999 *Jpn. J. Appl. Phys.* **38** L184
[13] Nakamura S, Senoh M, Nagahama S, Iwasa N, Yamada T, Matsushita T, Kiyoku H, Sugimoto Y, Kozaki T, Umemoto H, Sano M and Chocho K 1998 *Jpn. J. Appl. Phys.* **37** L309
[14] Kim D K 2007 *Solid-State Electronics* **51** 1005
[15] Shiori M, Čeh M, Šturm S, Chuo C C, Hsu J T, Yang J R and Saijo H 2006 *J. Appl. Phys.* **100** 013110
[16] Kubota M, Okamoto K, Tanaka T and Ohta H 2008 *Appl. Phys. Lett.* **92** 011920
[17] Kumakura K, Makimoto T and Kobayashi N 2000 *Jpn. J. Appl. Phys Part 1* **39** 2428
[18] Kozodoy P, Hansen M, DenBaars S P, Mishra U K and Kauffman J 1999 *Appl. Phys. Lett.* **74** 3681
[19] Heying B, Tarsa E J, Elsass C R, Fini P, DenBaars S P and Speck J S 1999 *J. Appl. Phys.* **85** 6470
[20] Cho Y H, Gainer G H, Fischer A J, Song J J, Keller S, Mishra U K and DenBaars S P 1998 *Appl. Phys. Lett.* **73b** 1370
[21] Chichibu S F, Torii K, Deguchi T, Sota T, Setoguchi A, Nakanishi H, Azuhata T and Nakamura S 2000 *Appl. Phys. Lett.* **76** 1576
[22] Inushima T, Mamutin V V, Vekshim V A, Ivanov S V, Sakon T, Motokawa M and Ohoya S 2001 *J. Cryst. Growth* **227-228** 481
[23] Wetzel C, Takeuchi T, Yamaguchi S, Katoh H, Amano H and Akasaki I 1998 *Appl. Phys. Lett.* **73** 1994
[24] Narukawa Y, Kawakami Y, Funato M, Fujita S, Fujita S and Nakamura S 1997 *Appl. Phys. Lett.* **70** 981
[25] Narukawa Y, Kawakami Y, Fujita S, Fujita S and Nakamura S 1997 *Phys. Rev. B* **55** R1938

Journal of Physics: Conference Series **152** (2009) 012008 doi:10.1088/1742-6596/152/1/012008

AlGaN/GaN heterostructures for hot electron and quantum effects

S A Vitusevich[1], V A Sydoruk[1], A M Kurakin[1], N Klein[1], M V Petrychuk[2], A V Naumov[3] and A E Belyaev[3]

[1]Institut für Bio- und Nanosysteme and CNI – Center of Nanoelectronic Systems for Information Technology, Forschungszentrum Jülich, Jülich D-52425, Germany
[2]Taras Shevchenko National University, Kiev 01033, Ukraine
[3]Institute of Semiconductor Physics, NASU, 03028 Kiev, Ukraine

E-mail: s.vitusevich@fz-juelich.de

Abstract. This paper reports on the transport and material characterization of AlGaN/GaN heterostructures. We compared the overheating temperatures in the transmission line model patterned devices during normal operation and analyze their dependence on the buffer thickness in order to optimize the thermal budget of the structures. It is demonstrated that noise spectra can be used to monitor the transport mechanisms and determine the activation energy of the traps. Small-dose gamma irradiation is used as an effective treatment for improving the structural properties. Based on cathodoluminescence and X-ray diffraction spectra shape, we explain the improvement as a result of the relaxation of elastic strains and structural-impurity ordering in the AlGaN barrier layer under irradiation. An irreversible improvement in mobility of the electrons in the channel, controlled by a proper dose of gamma radiation treatment, at considerably reduced self-heating is promising to achieve novel quantum transport regimes on the nanoscale.

1. Introduction

Advanced electronic systems require ultra-high miniaturization and integration. Therefore novel materials such as wide band gap III-nitrides, which can be scaled to sizes three times smaller than middle gap silicon, are attracting significant attention. Barker and co-workers [1] showed that the high field velocity characteristics of the GaN-based material appear to be advantageous for high-speed applications compared to other materials. Overshoot phenomena were registered in vacuum by Collazo et al. in intrinsic AlN film using an electron spectrometer at increased electrical field in a pulse regime [2]. An optical time-of-flight technique with femtosecond resolution allowed Wraback et al. to detect an electron velocity overshoot effect [3] in an AlGaN/GaN heterojunction p-i-n structure. At high electrical field the increase in transit time was explained as the onset of negative differential resistance, which is promising for the development of high-frequency generators. In AlGaN/GaN heterostructures with high mobility electron gas at the interface of the structure scaled to nanosizes, the integer quantum Hall effect is registered [4] at 300 mK. The authors of Ref.[5] demonstrated theoretically that controlled modification of AlGaN/GaN properties may lead to an increase in electron mobility and the realization of faster transport devices. There are many factors that influence the performance of two-dimensional gas (2DEG) at the AlGaN/GaN interface. It is well known that spontaneous and piezoelectric polarization effects in the nitride heterostructures result in ultra-high 2DEG in the channel. Interface roughness scattering and strain relaxation are identified as the factors that have a significant impact [6] on the transport properties of 2DEG at present. In addition, electron mobility dependence on scattering caused by the strain variation of the AlGaN barrier layer in AlGaN/GaN heterostructures has been established [7]. Gamma radiation treatment allows a controlled modification of the arrangement in native defects of the structure. The introduced defects may rearrange the native defect structure and stimulate strain relaxation [8], which in the case of optimized conditions results in increased mobility. Additionally, in order to realize the excellent potential of GaN-based materials on the nanoscale and utilize novel hot electron and quantum effects, the role of self-heating effects has to be investigated, which considerably influence material transport properties and screen high-field effects caused by down-scaling.

In this paper, the influence of the thickness of the buffer layer on the 2DEG transport properties in AlGaN/GaN heterostructures was investigated. Noise spectroscopy was used to monitor the self-heating effect. Cathodoluminescence (CL) spectra were measured at different acceleration voltages in order to study the depth-resolved material properties. It is found that mobility increases when the AlGaN barrier becomes partially relaxed after a small dose of gamma radiation treatment.

2. Experimental details

The investigated structures were grown by MOCVD and patterned to form transmission line model (TLM) devices. Ungated high electron mobility transistor (HEMT) structures were studied in order to avoid the non-uniform electric field redistribution introduced by the gate metal layer. The dimensions of the conducting channel were chosen in such a way as to insure a negligibly small contribution of contact resistance. The set of the samples tested included a single interface structure of $Al_{0.30}Ga_{0.70}N$/GaN heterostructure with a 30-nm barrier and a 3-μm-thick GaN buffer on a sapphire substrate (HEMT A), and samples with interfaces which differ from structure A by an additional 1-nm AlN spacer grown in between the 30-nm $Al_{0.25}Ga_{0.75}N$ barrier and the 7.5-μm GaN buffer (HEMT B). Structures A and B have GaN cap layers of 4 nm and 1.3 nm, respectively. The electrical and structural performance of the samples was analyzed. Cathodoluminescence was excited using a LEO 1550 scanning electron microscope with a 3, 5 and 20 keV electron beam. Emission spectra were analyzed with a Zeiss VIS grating monochromator using a 1221 lines/mm grating blazed at 463 nm and a photomultiplier coupled to a phase-sensitive Zeiss VIS detection system. Spectra were obtained at room temperature. The I-V characteristics were measured in the steady state. Spectral noise measurements were performed simultaneously in the frequency range from 1Hz to 100KHz using a low-noise preamplifier and spectrum analyzer HP 35670A.

3. Results and discussion

In order to understand the influence that wafer design apparently has on TLM device performance, we compared the simulation results of the overheating temperature in the active region in the TLM device with a common geometry (two-layer structure: first layer – substrate, second layer - buffer) with heat sources uniformly distributed over a rectangular area of the conducting channel on top of the second layer. The temperature rise $\Delta T = T - T_0$ in the channel is related to dissipated power, P_{dis} through the thermal impedance $\theta = \Delta T/P_{dis}$. The thermal impedances were obtained for L =25 μm using the substrate thermal conductivity, $\lambda = 0.35$ W/cmK for sapphire, and $\lambda = 1.2$ W/cmK for the GaN buffer, respectively. Our results show (figure 1a) that the temperature increase in the channel of AlGaN/GaN heterostructures has a strong dependence on buffer thickness. It should be emphasized that the overheating temperature decreases rapidly with increasing thickness of the GaN buffer layer. This result is in good agreement with measured I-V characteristics for both samples (figure 1b).

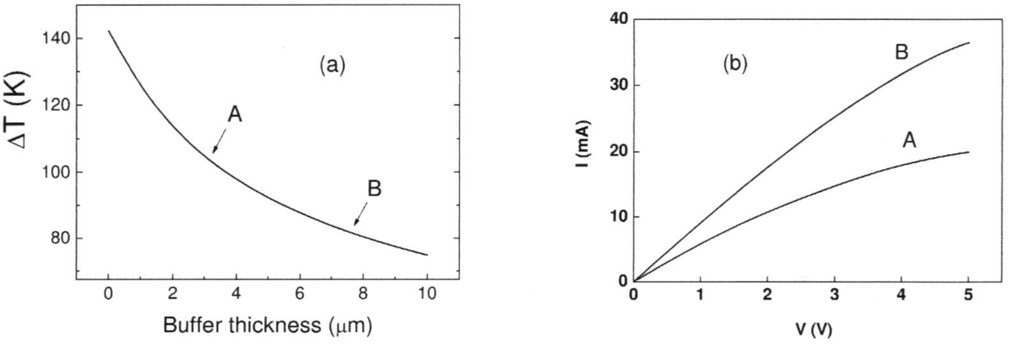

Figure 1. (a) The calculated temperature rise in the channel of the TLM device processed on AlGaN/GaN heterostructure grown on 300 μm thick substrate as a function of buffer thickness. The arrows show the thickness of HEMT A and HEMT B, respectively, (b) Measured I-V characteristics for AlGaN/GaN HEMT structures with a channel L=25 μm: (A) – for HEMT A, (B) – for HEMT B.

Noise spectroscopy is used to monitor the self-heating effect. Deviation from conventional *1/f* flicker noise was observed at some transient frequency (f_T) with a tendency towards noise level suppression in a wide frequency range below f_T (figure 2a). It was found that the transient frequency strongly depends on dissipated power. In view of the fact that self-heating occurs, the analysis of the noise spectra in both low-frequency and transition intervals requires that the temperature increase in the channel in the AlGaN/GaN heterostructures be estimated. We estimated the temperature rise ΔT versus dissipated power Pdis on the basis of the theoretical model with (i) heat-dissipation and heat-transfer modeling in the device and (ii) the self-consistent solution of coupled nonlinear equations for the channel current I and the channel temperature rise ΔT. The latter result allowed us to plot the time parameter τ, which corresponds to the frequency separating the low-frequency 1/f range and the transition region, as a function of ΔT. An exponential dependence of τ can be clearly seen in figure 2b.

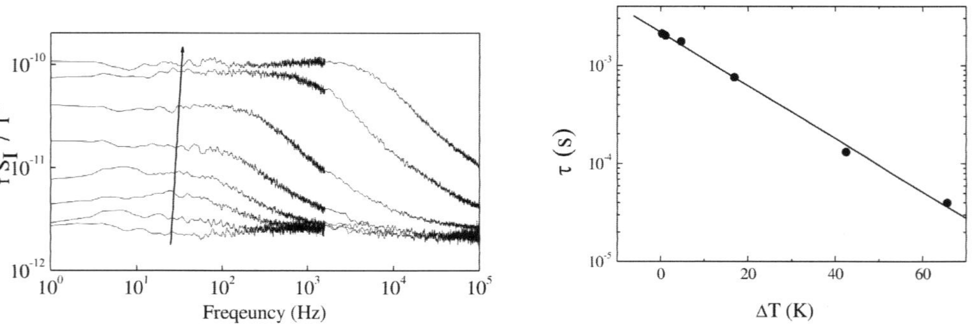

(a) (b)

Figure 2. (a) Spectra of the current noise of AlGaN/GaN structure (HEMT A) with a channel L=25 μm, measured for different electric fields, E(kV/cm): 0.04,0.09,0.18,0.35,0.75,1.52, 2.73,3.78; (b) Dependence of the characteristic time on the channel overheating.

This reflects the activation process of hopping conductivity, which results in an increase in the low-frequency component of the noise. The estimated value of the trap energy is about 0.6 eV. This energy is associated with a nitrogen vacancy, which is in good agreement with literature data [9].

Transport and noise characteristics improved after a small dose of gamma radiation treatment of 10^6rad. Our results demonstrate that the carrier concentration remains practically unchanged after an irradiation dose of 10^6 rad, while the 2DEG mobility exhibits a considerable increase. For example, room temperature mobility of HEMT A sample is 1260 cm^2/Vs before gamma irradiation and is 1350 cm^2/Vs after the irradiation of the sample at a dose of 10^6 rad. Furthermore, after such a dose of gamma radiation the reduced dispersion of 1/f noise spectral density in the set of devices is revealed. This indicates an improvement of performance and reliability of the HEMT structures.
Similar changes in mobility were also registered for three other structures. For HEMT B with the highest mobility of 1859 cm^2/Vs before gamma irradiation, a mobility of 2000 cm^2/Vs was registered after its gamma irradiation at a dose of 10^6 rad. The latter is consistent with a decrease in strain accompanied by the structural ordering of defects [10].

Depth-resolved cathodoluminescence spectra were measured in order to investigate structural changes caused by gamma radiation. The depth excitation spectra for samples A and B at different electron-beam energies were calculated using the Monte Carlo simulation CASINO [11], and are shown in figure 3a,b. It can be seen that for a beam voltage of 3kV most of the energy dissipates in the AlGaN barrier region in both structures. At higher beam voltages the excitation of the buffer is dominant.

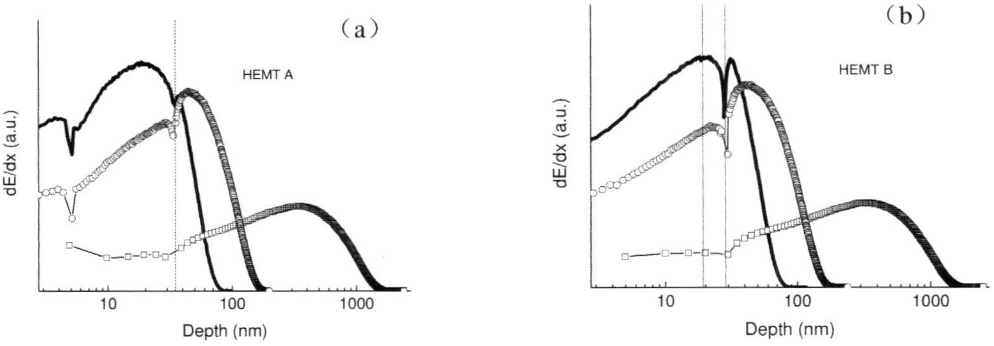

Figure 3. Monte Carlo simulation of energy loss per unit length dE/dx versus electron-beam penetration depth in the HEMT structures for different acceleration voltages: (solid) – 3 kV, (circle) – 5 kV, (square) – 20 kV: (a) for HEMT A, (b) for HEMT B.

The CL spectra measured before and after radiation of samples in the investigation contain two main emission peaks (see figure 4 – figure 6): a high energy peak near the band edge (BE) at 3.4 eV and an intense yellow luminescence (YL) broad band between 2.0 and 2.8 eV. In figure 4.a and 4.b, the CL spectra of samples A and B excited by a 3keV electron beam are shown. The higher intensity of the BE peak in the case of sample A is the direct consequence of the thicker GaN cap layer used in this sample. It should be noted that yellow luminescence demonstrates stronger intensity with respect to BE emission at small penetration depth, due to a significant contribution by luminescent sources of the barrier layer. Elsner et al. showed [12] that yellow luminescence in GaN can be explained by a gallium vacancy and its complex with oxygen positioned at the edge dislocation core. The complex is electrically active and increases the intensity of the yellow luminescence in GaN. These deep acceptors are the radiative centers and recombination is limited by the product of its whole capture cross-section and thermal velocity. According to a DLTS study [13], deep levels in GaN can have capture cross sections as large as $10^{-14}cm^2$, and, in this case, the capture product can become larger than the radiative constant of the band edge. Therefore a broad YL peak can be higher than the band edge emission peak of GaN at 3.4 eV.

At the energy of 5 keV (figure 5), the magnitude of the BE peak becomes greater and the amplitude ratio of the first and second peak increases. These changes confirm that the barrier region of the structures contributes significantly to wide band emission in the energy range from 2.0 to 2.8 eV.

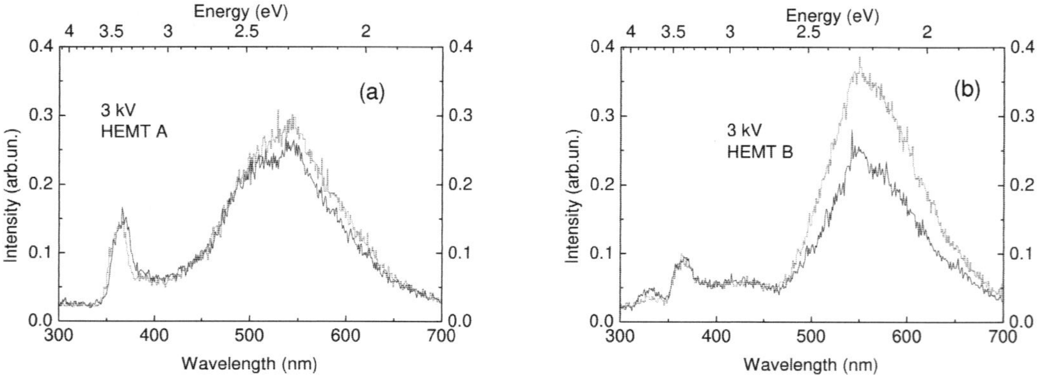

Figure 4. CL spectra of HEMT structures before (black) and after (gray) irradiation obtained at an acceleration voltage of 3 kV: (a) for HEMT A, (b) for HEMT B.

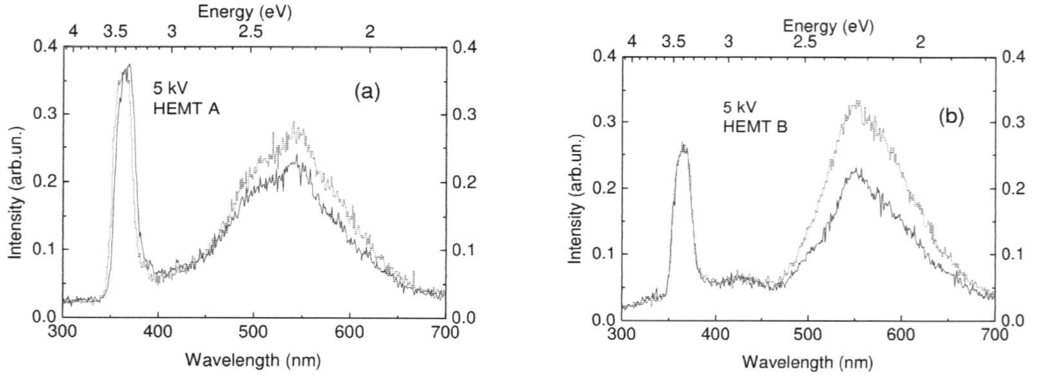

Figure 5. CL spectra of HEMT structures before (black) and after (gray) irradiation obtained at an acceleration voltage of 5 kV: (a) for HEMT A, (b) for HEMT B.

When the energy of the beam is 20 keV and most beam energy is dissipated in the GaN buffer layer the blue luminescence (BL) peak appears at an energy of 2.85 eV in both samples (figure 6). BL is termed the donor-acceptor pair in GaN, which demonstrates direct correlation with material transport properties [14]. After irradiation, an increase of CL intensity from BL and YL is observed on sample B. The uniform increase of CL intensity over the wide energy range from 1.8 to 3.2 eV can be explained by a gamma-radiation-stimulated reduction of the concentration of non-irradiative centers in the AlGaN barrier as well as in the GaN buffer layer.

The increase of BL intensity after gamma irradiation provides evidence of the reduced density of non-radiative centers. This conclusion is supported by the fact that the generated point defects are highly mobile in GaN and can easily rearrange the native defect structure. It should be noted that gamma radiation in HEMT structures can result in the generation of extra defects or the compensation and structural ordering of native defects.

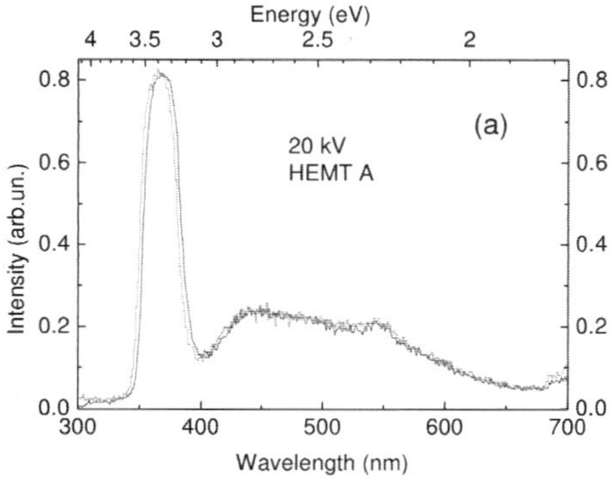

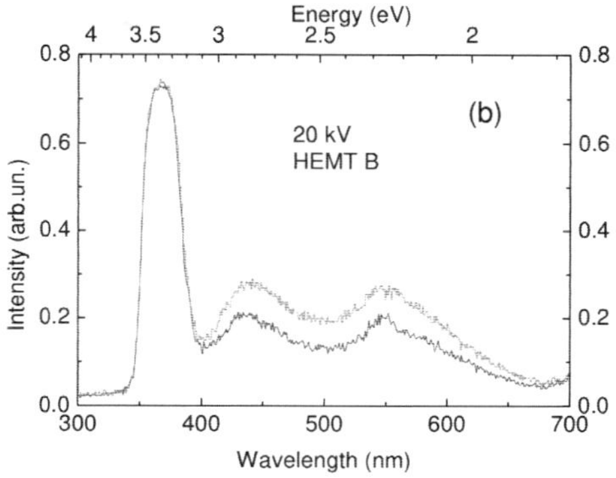

Figure 6. CL spectra of HEMT structures before (black) and after (gray) irradiation obtained at an acceleration voltage of 20 kV: (a) for HEMT A, (b) for HEMT B.

The observed room temperature mobility improvement (up to 10% at a dose of 10^6 rad) indicates that the latter is dominant during radiation-stimulated relaxation in the investigated samples.

4. Conclusions

We investigated the 2DEG transport properties in gateless AlGaN/GaN (TLM) heterostructures grown on sapphire with different thicknesses of the GaN buffer layer. The strong dependence of overheating temperature in the channel on buffer thickness was shown. The analysis of noise spectra allowed us to evaluate the activation energy (0.6 eV) of the trap influencing the transport of AlGaN/GaN 2DEG. Additionally, depth-resolved CL spectra revealed that this trap can be associated with defect states in the GaN buffer layer. CL spectra transformation of AlGaN/GaN heterostructures after gamma irradiation provides evidence of the reduced density of non-radiative centers. Furthermore, the gamma-treated devices exhibited more stable operation due to stabilization in the 2DEG transport at the AlGaN/GaN interface. The results suggest that the transport characteristics can be managed in a controlled way by using a careful design of the layer structure and active treatment in processing technology. 2DEG with improved characteristics may demonstrate novel features and can be used in devices based on the quantum effects identified.

Acknowledgments

This work was supported by the Ukrainian Ministry of Education and Science, the Deutsche Forschungsgemeinshaft under Contract No.KL1342 and in part by the Office of Naval Research. The authors are grateful to H.Hardtdegen from Forschungszentrum Jülich, Germany and Z.Bougrioua from CRHEA, Valbonne, France, for providing them with epitaxial material.

References

[1] Barker J M, Akis R, Thornton T J, Ferry D K and Goodnick S M 2002 High field transport studies of GaN. Phys. Stat. Sol. (a) 190 263

[2] Collazo R, Schlesser R, Roskowski A, Davis R F and Sitar Z 2003 Observations of electron velocity overshoot during high-field transport in AlN Mat. Res. Soc. Symp. Proc. 743 L10.2.1

[3] Wraback M, Shen H, Rudin S and Bellotti E 2002 Experimental and theoretical studies of transient electron velocity overshoot in GaN Phys. Stat. Sol. (b) 234 810

[4] Schmult S, Manfra M J, Sergent A M, Punnoose A, Chou H T, Goldhaber-Gordon D and Molnar R J 2006 Quantum transport in high mobility AlGaN/GaN 2DEGs and nanostructures. Phys. Stat. Sol. (b) 243 1706

[5] Pokatilov E P, Nika D L and Balandin A A 2006 Built-in field effect on the electron mobility in

AlN/GaN/AlN quantum wells Appl. Phys. Lett. 89 113508

[6] Cao Y and Jena D 2007 High-mobility window for two-dimensional electron gases at ultrathin AlN/GaN heterojunction Appl. Phys. Lett. 90 182112

[7] Zhao J, Lin Z, Corrigan T D, Wang Z, You Z and Wang Z 2007 Electron mobility related to scattering caused by the strain variation of AlGaN barrier layer in strained AlGaN/GaN heterostructures Appl. Phys. Lett. 91 173507

[8] Belyaev A E, Breza J, Venger E F, Vesely M, Il'in I Yu, Konakova R V, Linday J, Lyapin V G, Milenin V V, Prokopenko I V and Thorik Yu A 1998 Radiation resistance of GaAs-based microwave Schottky-barrier devices (Kiev: Interpress Ltd)

[9] Pernot J and Muret P 2008 Electronic properties of the Ec-0.6 eV electron trap in n-type GaN J. Appl. Phys. 103 023704

[10] Vitusevich S A, Petrychuk M P, Klein N, Danylyuk S V, Belyaev A E, Konakova R V, Avksentyev A Yu, Kurakin A M, Lytvyn P M, Danilchenko B A, Tilak V, Smart J, Vertiatchikh A and Eastman L F 2003 Barrier material improvement in AlGaN/GaN microwave transistors under gamma irradiation treatment Mat. Res. Soc. Symp. Proc. 764 C3.31

[11] Drouin D, Couture A R and Gauvin R 2001 CASINO V2.0: An advance simulation tool for scanning electron microscope users. Micro.Microanal. 7 684

[12] Elsner J, Jones R, Heggie M I, Sitch P K, Haugk M, Frauenheim Th, Oberg S, and Briddon P R 2008 Deep acceptors trapped at threading-edge dislocations in GaN. Phys. Rev. B 58 12571

[13] Cho H K, Kim K S, Hong C H and Lee H J 2001 Electron traps and growth rate of buffer layers in unintentionally doped GaN J. Cryst. Growth 223 38

[14] Walker D E, Fitch R C, Gillespie J K, Jessen H, Cassity P D, Breedlove J R and Brillson L J 2006 Controlled gate surface processing of AlGaN/GaN high electron mobility transistors Appl. Phys. Lett. 89 183523

Journal of Physics: Conference Series **152** (2009) 012009 doi:10.1088/1742-6596/152/1/012009

Preparation and Photoluminescence of SiO_2-CdTe multilayer film by layer-by-layer self-assembly

Ping Yang, Norio Murase

Photonics Research Institute, National Institute of Advanced Industrial Science and Technology (AIST), Midorigaoka, Ikeda-city, Osaka, 563-8577, Japan

E-mail: n-murase@aist.go.jp

Abstract. Preparation parameters of layer-by-layer self-assembled films consisting of functional SiO_2 layers and emitting CdTe quantum dots (QDs) layers were investigated. A surface-treated plate was successively dipped into three different solutions: surfactant (thioglycolic acid, TGA), colloidal QDs, and functional silicon alkoxide. Since the free TGA molecules in the first solution became linkers between the two kinds of layers, the condition of the first solution (TGA and Cd^{2+} ion concentrations and pH) affected the properties of the film. Under optimal conditions, the QDs were dispersed at a high concentration (0.01 M) in a closely packed manner without agglomeration. This resulted in a narrowing of the photoluminescence (PL) spectrum and a red shift of the peak wavelength due to reabsorption of the emission. The PL efficiency of films thus prepared with ten double layers (ca. ~50 nm thick) was 25%. This high efficiency together with the high concentration resulted in a brightness much higher than that of a commercial phosphor when their thickness was same. However, when there were excess free TGA molecules in the first solution, the film had a higher concentration (0.02 M) and lower PL efficiency due to agglomeration of the QDs.

1. Introduction

Layer-by-layer (LbL) assembly techniques have been extensively studied using various driving forces such as electrostatic, hydrogen bonding, charge transfer interaction, and coordination bonding [1, 2]. LbL assembly has been achieved for a wide variety of materials including polymers, inorganic nanoparticles, nanowires, organic components, carbon nanotubes, and biological macromolecules such as proteins, viruses, and DNA [3–14]. Such assembly requires control of the molecular orientation and organization at the nanometer level, so it is essential to investigate and develop methods for the controlled assembly of multi-component nanostructures [15]. LbL self-assembly, based on sequential adsorption of oppositely charged components, is one of the most established methods for the preparation of thin films with controlled properties [16]. The potential applications of such LbL-assembled films range from drug delivery and cell resistant coatings to antireflection coatings, batteries, and electrochromic devices [17].

Thin films of quantum dots (QDs) are emerging as an important class of materials for electronic and optoelectronic devices such as field-effect transistors, photodetectors, light-emitting diodes, and solar cells [18]. The incorporation of QDs into the films using LbL assembly has been accomplished using different approaches. For example, a novel type of composite thin film consisting of gold QDs and polyimide was prepared through binding based on the combined effect of metal-ligand coordination and electrostatic interaction [19]. Kotov's group reported free-standing LbL-assembled films of magnetite QDs [20]. They also reported LbL films of semiconductor QDs and a polymer with multicolor capability [21]. Hybrid thin films of a conjugated polymer and CdSe QDs have been fabricated using a LbL assembly method based on covalent coupling reactions [22]. A graded semiconductor film with rainbow-like luminescence was prepared using CdTe QDs [23]. In short, LbL assembly is a promising method for fabricating structured and functional films for the construction of nano-devices.

We previously developed a novel method for preparing multilayer films of CdTe QDs by LbL self-assembly using very thin functional SiO_2 layers for linkage [24]. The films have intense photoluminescence (PL) and a high QDs concentration. They have a PL efficiency of 24% and are 40 times brighter than a conventional rare-earth phosphor of the same thickness when excited at the same intensity at 365 nm [25]. This high brightness should lead to a new variety of applications with a functional SiO_2 layer composed of amino or mercapto groups.

We have now investigated the relationship between the preparation conditions and properties of SiO_2-CdTe multilayer films prepared by LbL self-assembly. We found that free thioglycolic acid (TGA) molecules (without connected Cd^{2+} ions) in the TGA and Cd^{2+} ion solution play an important role in the arrangement of the CdTe QDs in the film. This arrangement affects drastically the PL properties of the film.

2. Experimental section

A colloidal solution containing TGA-capped red-emitting CdTe QDs was prepared using a procedure described in detail elsewhere [26]. The QDs had a diameter of 3.9 nm and a PL efficiency of 70%. SiO_2-CdTe multilayer films were prepared using LbL self-assembly [24]. We reduced the dipping time at each step in accordance with the optimal preparation procedure. Briefly, a glass slide (Matsunami, No. S8111) coated with a 3-aminopropyltrimethoxysilane (APS) layer was first dipped in the solution of TGA and Cd^{2+} ($Cd(ClO_4)_2$) for 0.5−1 min. The sample was then dipped in the colloidal solution of CdTe QDs for 1−2 min. Finally, the sample was immersed in the solution of APS and toluene (volume ratio of APS to toluene of 0.33) for 1−2 min. The cycle was repeated until the desired number of layers was obtained. After each cycle, the sample was rinsed with Milli-Q water and dried under N_2. Scheme 1 illustrates the steps in the formation process.

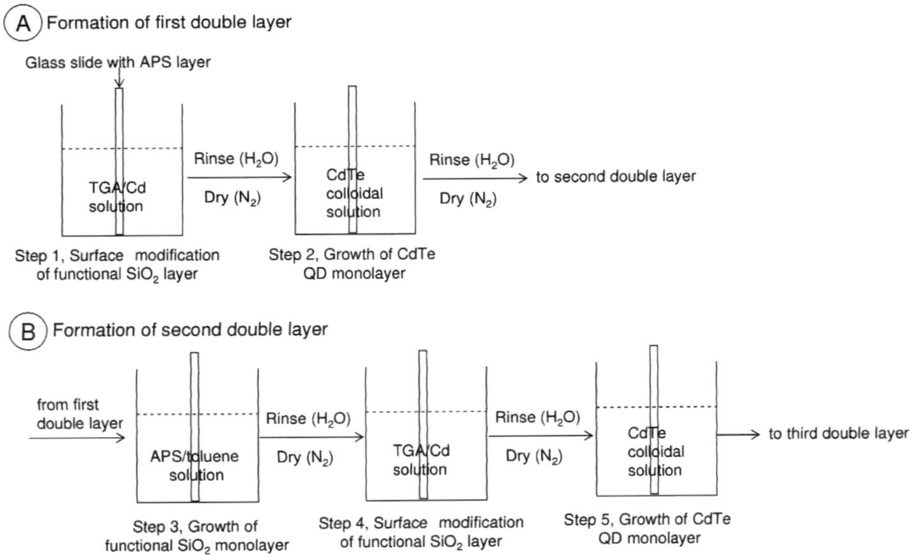

Scheme 1. Preparation of SiO_2-CdTe multilayer film by LbL self-assembly.

Because the free TGA molecules act as linkers between the QDs and SiO_2 layers, we investigated the relationship between the first solution conditions (TGA and Cd^{2+} ion concentrations, pH) and the PL properties. The molar ratio of TGA to Cd^{2+} ranged from 2 to 8. The concentration of TGA ranged from 0.01 to 0.20 M. The pH was changed from 7 to 11 by adding a NaOH solution.

The absorption and PL spectra were measured with conventional spectrometers (Hitachi U-4000 and F-4500). Transmission electron microscopy (TEM) analysis was performed using a Hitachi H-9000 microscope. Optical images of the films under visible light and UV irradiation (365 nm) were taken by a digital camera. The film thickness was measured with a surface roughness measuring instrument (Mitutoyo Surftest SV-624) or an atomic force microscope (Digital Instruments Nanoscope). The PL efficiencies of the colloidal solution were estimated by comparison with a solution of quinine sulfate in 0.5 M H_2SO_4. Because the prepared films were transparent, their PL efficiencies were estimated by comparison with solutions of standard dye having a similar optical path length and optical density. The details are described elsewhere [27].

3. Results and discussion

Figure 1 shows the absorption and PL spectra of a SiO_2-CdTe multilayer film with five double layers. It was prepared using a TGA and Cd^{2+} solution with a molar ratio of TGA to Cd^{2+} of ~5, a pH of ~10, and a TGA concentration of ~0.1 M. Comparing with the CdTe colloidal solution, the film revealed a decreasing (~10 nm) of full-width at half maximum (FWHM) of PL spectrum and a red-shift (~17 nm) of PL peak. This phenomenon is ascribed to the reabsorption of the emissions from the smaller QDs by the larger QDs in the matrix. The PL efficiency of the film was 25% while that of the QDs in the colloidal solution was 70%. The brightness of the film was estimated to be more than 40 times that of a commercial rare-earth phosphor (P22-RE3) on the basis of previous quantitative analysis [25]. This much higher brightness means that the film is attractive for various potential applications.

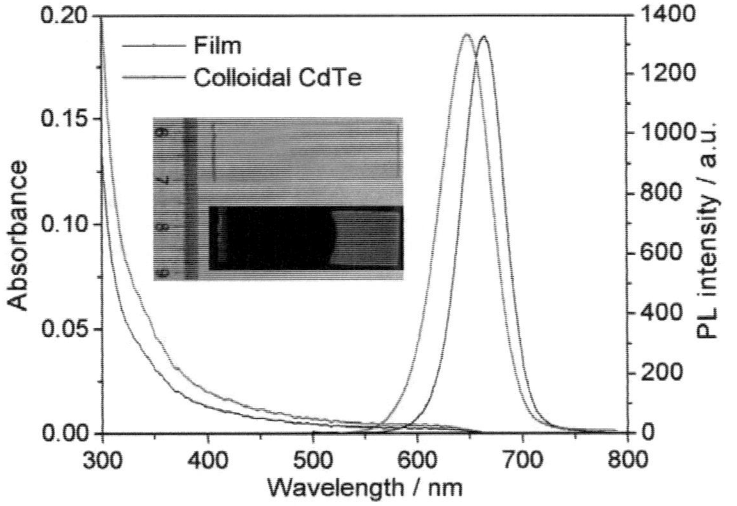

Figure 1. Absorption and PL spectra of SiO_2-CdTe multilayer film with five double layers. Inset shows color image of film (lower) under UV light and optical picture of it (upper) under visible light. Absorption and PL spectra of CdTe colloidal solution are shown for comparison.

Figure 2 shows the TEM photograph of a SiO_2-CdTe multilayer film with ten double layers. The QDs, with a mean diameter of ~ 4 nm, were uniformly dispersed in the film. Their concentration was estimated to be 0.01 M on the basis of the absorbance at the first absorption peak and the thickness

(~50 nm) of the film. Furthermore, calculation using the thickness of the film and the absorbance at the first absorption peak revealed that the immobilized QDs (3.9 nm in diameter as estimated from the first absorption peak) were arranged in a closely packed manner with a thin functional sol-gel SiO_2 layer.

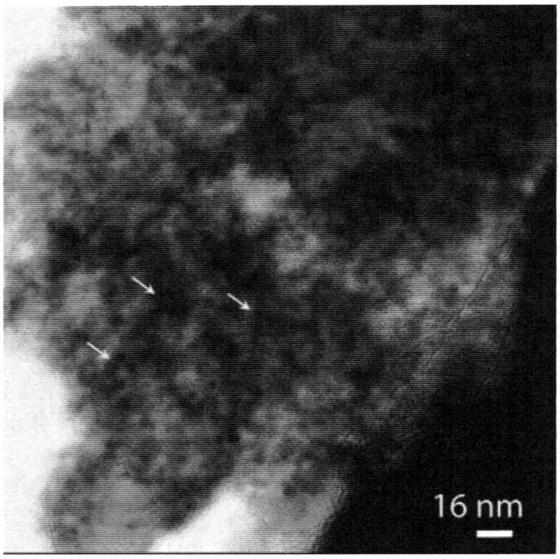

Figure 2. TEM photograph of SiO_2-CdTe multilayer film with ten double layers. Specimen was prepared by scratching the film. White arrows show locations of CdTe QDs with mean diameter of ~ 4 nm uniformly dispersed in film.

The formation of a Cd-thiol layer on the surface of aqueous CdTe QDs effectively passivates them, resulting in high PL efficiency. The Cd^{2+} ions were added to the TGA solution to maintain the initial surface state of the QDs during film preparation. Because the carboxyl and mercapto groups in free TGA (without connection to Cd^{2+} ions) act as interaction mediums between the QDs and the functional SiO_2 layers consisting of condensed APS, the form of the TGA molecules in the TGA and Cd^{2+} solution affected the arrangement of the QDs in the film. According to previous reports on CdTe QDs, a variety of Cd-TGA complexes formed depending on the molar ratio of TGA molecules to Cd^{2+} ions in the solution. When the molar ratio is less than 2, a dithio-complex is formed in two consecutive steps, as described by eqs. 1 and 2, through the reaction between Cd^{2+} ions and TGA molecules [28].

$$Cd^{2+} + {}^{-}SR \leftrightarrow (Cd\text{-}SR)^{+} \quad (1)$$

$$(Cd\text{-}SR)^{+} + {}^{-}SR \leftrightarrow Cd(SR)_2 \quad (2)$$

Additionally, the dithio-Cd complex (a molar ratio of TGA molecules to Cd^{2+} ions of more than 2) has better stabilization than that of a monothio-Cd complex (a molar ratio close to 1). In our experiments (with an alkaline condition), the molar ratio ranged from 2 to 8. Therefore, there were two main forms of TGA molecules in the solution. One was the thio-Cd complex form, and the other was the ionization state form.

Scheme 2 illustrates the morphology of the SiO_2-CdTe multilayer film prepared by LbL self-assembly for three different concentrations of free TGA in the solution. The optimal concentration resulted in the formation of layers of CdTe QDs in a closely packed manner (Scheme 2, middle). This means that the average distance between QDs is ~5.5 nm fored-emitting CdTe QDs with a thickness of the functional SiO_2 layer of 1–2 nm [24]. With the low concentration (Scheme 2, left), there were fewer QDs in each layer. With the high concentration (Scheme 2, right), the number of free TGA molecules connected with a functional SiO_2 layer was higher. This resulted in an increased number of CdTe QDs assembled into each layer. The results of additional experiments support these findings, as we explain in detail below. All the films had five double layers (CdTe QDs layer and functional SiO_2 layer) unless otherwise noted.

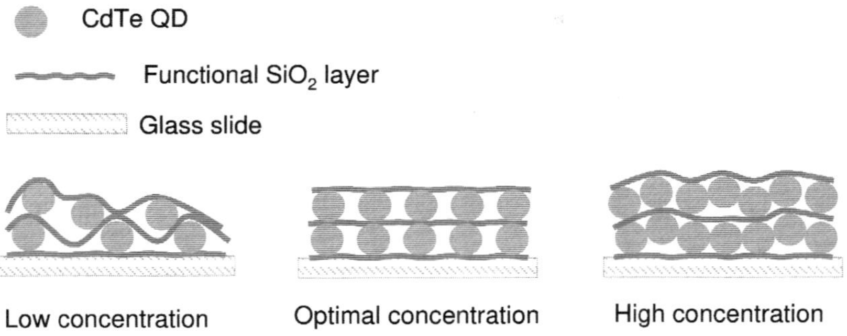

Scheme 2. Morphologies of SiO_2-CdTe multilayer films prepared with different concentrations of free TGA in solution. Number of CdTe QDs in each monolayer increased with the concentration.

Figure 3 shows the effect of the molar ratio of TGA to Cd^{2+} in the solution on the PL efficiency and the peak wavelength of the PL spectra of SiO_2-CdTe multilayer films. The inset shows the absorption and PL spectra for three molar ratios. The absorbance at the first absorption peak and the red shift of the PL peak increased with increasing the ratio. This means the number of QDs assembled in the film increased. Additionally, the films had relatively high PL efficiency when the ratio was 5–6. Beyond a ratio of 7, the efficiency decreased quickly. This phenomenon was ascribed to the number of free TGA molecules in the solution. With an increasing molar ratio of TGA to Cd^{2+} in the solution, the amount of thio-Cd complex decreased and the number of free TGA molecules increased because the concentration of TGA remained unchanged and the concentration of Cd^{2+} decreased. Therefore, many free TGA molecules connected with a functional SiO_2 layer due to the formation of a hydrogen bond between the amino group in APS and the carboxyl group in TGA. The large number of mercapto groups in each layer resulted in an increased number of QDs. This high QD concentration led to an increased red shift of the PL peak and lower PL efficiency. In addition, when the ratio was more than 7, the number of Cd^{2+} ions in the solution was less because the ions on the surface of the QDs escaped into the solution. This change in the surface state resulted in reduced PL efficiency.

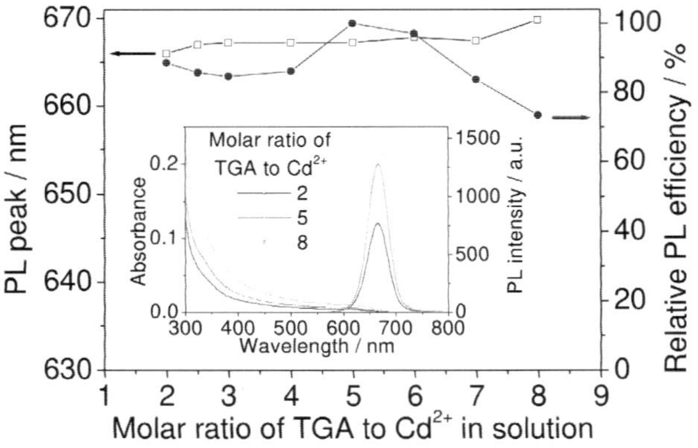

Figure 3. Effect of molar ratio of TGA to Cd^{2+} in solution (TGA concentration ~0.01 M, pH ~10) on PL efficiency and peak wavelength of PL spectra of SiO_2-CdTe multilayer films. Inset shows absorption and PL spectra of multilayer films for three molar ratios

Figure 4 shows the effect of the concentration of TGA in the solution on PL efficiency and on absorbance at the first absorption peak for SiO_2-CdTe multilayer films. When the concentration was less than 0.1 M, the films had relatively high efficiency. Above 0.1 M, it decreased dramatically with an increasing concentration. The absorbance at the first absorption peak increased linearly with the concentration. When the concentration was 0.1 and 0.2 M, the concentration of QDs in the film was 0.01 and 0.02 M, respectively, estimated from the thickness and absorbance at the first absorption peak. This increase in QDs concentration was associated with the increase in the number of free TGA molecules in the solution. The QDs were arranged in a closely packed manner when the concentration was 0.01 M. At the higher concentration (0.2 M), the QDs aggregated. The resulting concentration quenching caused the PL efficiency of the film to decrease remarkably.

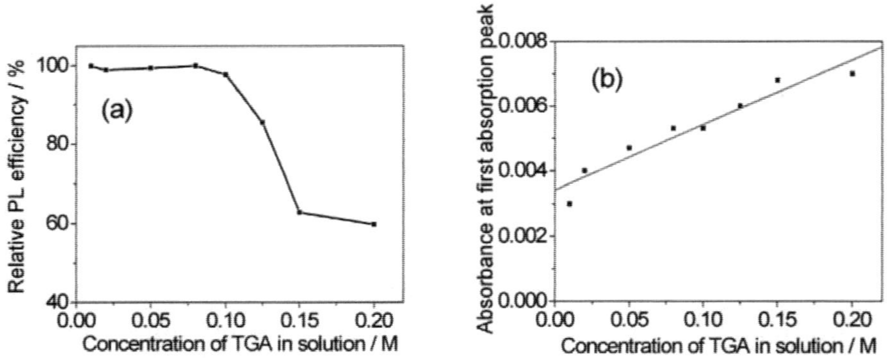

Figure 4. Effect of concentration of TGA in solution (molar ratio of TGA to Cd^{2+} ~5, pH ~10) on PL efficiency and absorbance at first absorption peak of SiO_2-CdTe multilayer films: (a) PL efficiency against concentration of TGA; (b) absorbance at first absorption peak against concentration of TGA.

Figure 5 shows the effect of the pH of the TGA and Cd^{2+} ion solution on the PL efficiency and on the absorbance at the first absorption peak for SiO_2-CdTe multilayer films. The PL efficiency was relatively high when the pH was from 7 to 10. Furthermore, the absorbance at the first absorption peak was relatively high in this range. When the pH was about 11, both the efficiency and absorbance at the first absorption peak were lower. These results are related to the pH dependence of the formation of sol-gel derived SiO_2 layers because such layers are partly dissolved when the pH is high [29].

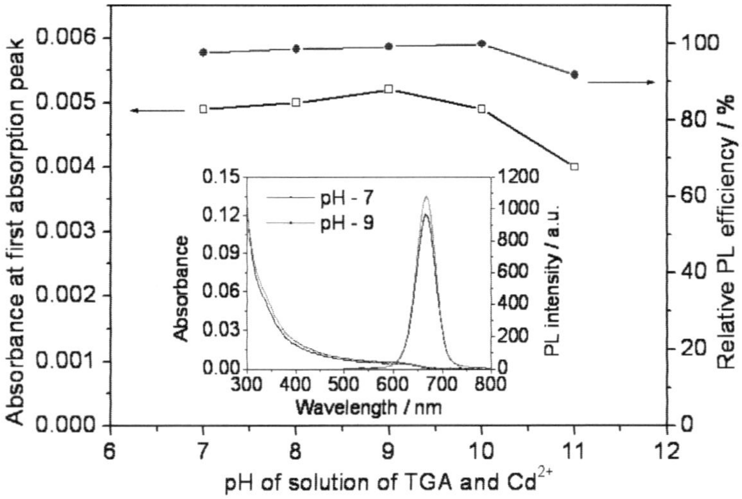

Figure 5. Effect of pH of solution (molar ratio of TGA to Cd^{2+} ~5, concentration of TGA ~0.05 M) on PL efficiency and absorbance at first absorption peak of SiO_2

4. Conclusion

To clarify the effect of the preparation conditions on the properties of LbL self-assembled films, we prepared SiO_2-CdTe multilayer films by LbL self-assembly. Because the mercapto and carboxyl groups in the TGA were linkers between the CdTe QDs and functional SiO_2 layers, the free TGA molecules (not connected with Cd^{2+}) in the aqueous solution of TGA and Cd^{2+} ions used for a successive dipping during preparation played an important role in determining the properties of these films, such as the concentration of CdTe QDs, the efficiency of the PL, and the peak wavelength of the PL spectra. When the molar ratio of TGA to Cd was 5–6, the pH was 7–10, and the concentration of TGA was 0.01–0.1 M, the PL efficiency was relatively high. The optimal QD concentration (0.01 M) was obtained by optimizing the preparation conditions. The resulting PL efficiency was 25%, and the QDs were arranged in a closely packed manner. This close-packed arrangement resulted in a red-shifted PL peak and a narrowed PL spectrum. Further work will address the use of these films as biodetectors and optical devices, making use of their higher PL brightness compared with that of a commercial phosphor.

Acknowledgements

This study was supported in part by the Creation and Support Program for Start-ups from Universities, sponsored by the Japan Science and Technology Agency (JST). P.Y. is grateful for the Research Fellowship received from the Japan Society for the Promotion of Science (JSPS).

References

[1] Jiang C and Tsukruk V V 2006 Adv. Mater. 18 829
[2] Tang Z and Kotov N A 2005 Adv. Mater. 17 951
[3] Zucolotto V, Daghastanli K R P, Hayasaka C O, Riul Jr A, Ciancaglini P and Oliveira Jr O N 2007 Anal. Chem. 79 2163
[4] Xing Q, Eadula S R and Lvov Y M 2007 Biomacromolecules 8 1987
[5] Brennecka G L, Parish C M, Tuttle B A, Brewer L N and Rodriguez M A 2007 Adv. Mater. 20 1407
[6] Kim H, Lee J, Kim T, Lee T S and Kim J 2008 Adv. Mater. 20 1117
[7] Peng H, Zhu Y, Peterson D E and Lu Y 2008 Adv. Mater. 20 1199
[8] Hartgerink J D, Beniash E and Stupp S I 2001 Science 294 1684
[9] Percec V, Ungar G and Peterca M 2006 Science 313 55
[10] Yamaguchi A, Uejo F, Yoda T, Uchida T, Tanamura Y, Yamashita T and Teramae N 2004 Nature Materials 3 337
[11] Zhao H, Yang P, Deng J, Liu L, Zhu J, Sui Y, Lu J and Yang W 2007 Langmuir 23 1810
[12] Isaacs S R, Choo H, Ko W and Shon Y 2006 Chem. Mater. 18 107
[13] Ho J C, Yerushalmi R, Jacobson Z A, Fan Z, Alley R L and Javey A A 2008 Nature Materials 7 62
[14] Calhoun M F, Sanchez J, Olaya, D, Gershenson M E and Podzorov V 2008 Nature Materials 7 84
[15] Crespo-Biel O, Dordi B, Reinhoudt D N and Huskens J 2005 J. Am. Chem. Soc. 127 7594
[16] Srivastava S, Ball V, Podsiadlo P, Lee J, Ho P and Kotov N A 2008 J. Am. Chem. Soc. 130 3748
[17] Zacharia N S, DeLongchamp D M, Modestino M and Hammond P T 2007 Macromelecules 40 1598
[18] Luther J M, Law M, Song Q, Perkins C L, Beard M C and Nozik A J 2008 ACS Nano 2 271
[19] Zhang F and Srinivasan M P 2007 Langmuir 23 10102
[20] Mamedov A A and Kotov N A 2000 Langmuir 16 5530
[21] Wang Y, Tang Z, Correa-Duarte M A, Liz-Marzán L M and Kotov N A 2003 J. Am. Chem. Soc. 125 2830
[22] Liang Z, Dzienis K L, Xu J and Wang Q 2006 Adv. Funct. Mater. 16 542

[23] Mamedov A A, Belov A, Giersig M, Mamedova N N and Kotov N A 2001 J. Am. Chem. Soc. 123 7738
[24] Yang P, Li C and Murase N 2005 Langmuir 21 8913
[25] Murase N, Li C, Yang P and Ando M 2007 Jpn. J. Appl. Phys. 46 7545
[26] Li C L and Murase N 2005 Chem. Lett. 34 92
[27] Murase N and Li C 2008 J. Lumin. 128 1896
[28] Guo J, Yang W and Wang C 2005 J. Phys. Chem. B, 109 17467
[29] Brinker C J and Scherer G W 1990 Sol-gel science: the physics and chemistry of sol-gel processing (San Diego: Academic press) p 104

Preparation and properties of a novel near infrared photovoltaic detector single-crystal material

Linghang Wang[1], Wanqi Jie, and Yang Yang

State Key Laboratory of Solidification Processing, Northwestern Polytechnical University, Xi'an 710072, China

[1]E-mail: uswlh@yahoo.com.cn

Abstract. A new photoelectronic semiconductor crystal, mercury indium telluride (MIT), with dimensions of 15mm in diameter and 175mm in length has been successfully grown by using the vertical Bridgman method (VB) at the optimized growth conditions. The crystal was determined through the power X-ray diffraction to be defect zinc-blende structure with the space group $F\bar{4}3m$. The transmittance spectra from 2.5 to 25μm shows high middle and far-infrared transmittance of 50~55%, which decreases gradually with the increase of wavenumber due to the lattice absorption and free carriers absorption. Hall measurements at room temperature show that the resistivity, carrier density and mobility of the MIT crystal are 4.79×10^2 Ωcm, 2.83×10^{13} cm^{-3} and 4.60×10^2 $cm^2V^{-1}s^{-1}$, respectively. The reduction of carrier mobility and the increase of the resistivity are related to the adding of In_2Te_3 into HgTe, which changes the energy band structure of the crystal.

1. Introduction

$Hg_{(3-3x)}In_{2x}Te_3$ (x=0.5) is a novel photoelectronic semiconductor material for photovoltaic devices, which is an equimolar solid solution of Hg_3Te_3 and In_2Te_3 and crystallizes in the defect zincblende structure [1-3], where the sites of the anion are occupied by Te ions and the sites of the cation by Hg ions, In ions and structural vacancies. Because of the valence difference of Hg^{2+} and In^{3+} ions, high concentration of neutral vacancies (about $10^{21}cm^{-3}$) is introduced into the structure of Mercury indium telluride (MIT) material. The number of structural vacancies is proportional to the content of In_2Te_3, which has large effects on the properties of the crystals. In In_2Te_3-HgTe system MIT is a compound of chemical inertness that shows wide spectral photosensitivity, electrical inertia of introduced impurities, and high resistance to gamma and neutron radiation [4-6]. MIT photodiode remains operating even up to temperature of $T \approx 85$ ℃. This makes MIT especially attractive for the application of fast and efficient photodetectors, optical filters, visible and infrared radiation sensors, X-ray detectors and temperature sensors, etc [4]. There are a few previous papers concerning the reaction kinetics in view of material chemistry and elastic behavior from mechanical properties and so on [3, 7]. So far, few works on single crystals growth, optical properties and electrical properties of MIT is reported. In this paper we report on the growth and characterization of MIT crystals as well as the optical and electrical properties.

2. Crystal growth and measurements

The single crystals of MIT (x=0.5) were grown by vertical Bridgman method with a quartz crucible (15mm in diameter). The starting materials of the high purity elements (of at least 7N purity) of Te, In, Hg were charged into the dried quartz crucible with a stoichiometric ratio corresponding to the compound to be grown, then evacuated to about 10^{-4} torr and sealed. The crucible was designed to be sharp-pointed cone tip to form a seed crystal in the initial region and had been well treated with the aqua regia and acetone to acquire clean inner surface and to decrease the chance of side wall nucleation. The starting materials in the crucible were synthesized in a rocking furnace. The crystal growth was carried out in the same crucible in a vertical Bridgman furnace to avoid extraneous impurity pollution. Mixtures to be grown were fused in the VB furnace and annealed at 1040K for 24h to obtain homogeneous melts, then gradually lowered the crucible from the high temperature zone to the cold zone in the temperature gradient of about 10K/cm. The crucible translation speed was less than 1mm/h. The as-grown MIT crystal of Ø 15mm×175mm with good quality and free crack is shown in figure 1.

Parts of the as-grown crystal of MIT were crashed and finely ground into powders for X-ray diffraction analysis, which were carried out in the 2θ range of 20°~100° using an X'Pert MPO Pro X-ray diffractometer with the step of 0.0330°. The X-ray source was CoKα, which has the wavelength of 1.79021Å. The accelerating voltage was 35KV and the current was 40mA. The X-ray diffraction data were indexed using the DICVOL programs.

The MIT wafer with 1mm thick was cut from the as-grown crystal. The FT-IR transmission spectrum was analyzed using NICOLET NEXUS FT-IR spectrometer in the region from 2.5μm to 25μm.

The conduction type as well as resistivity, carrier density and mobility of the MIT crystal was examined by a H7000 type Hall measurement apparatus in the Van der Pauw configuration. The magnetic field Bz applied was 2k Gauss and the measurement temperature was 300K. The good ohmic contacts were obtained by indium painting on the surface of the MIT sample. The wafers for the measurement were chemo-mechanically polished and etched with 5% Br-MeOH solution to eliminate the damaged layer.

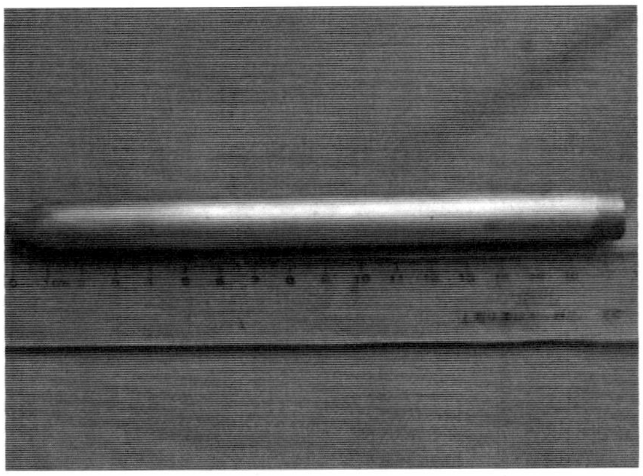

Figure 1. MIT ingot grown by vertical Bridgman method.

3. Results and discussions

3.1. The structure of MIT crystals

The MIT compound can be considered as being derived from HgTe by replacing every three mercury atoms by two In ions plus one vacancy. The zincblende structure of HgTe is composed of two equivalent face-center sublattice of Te ions and Hg ions. When In_2Te_3 is added into HgTe, neutral vacancies are produced:

$$In_2Te_3 \xrightarrow{HgTe} 3Te_{Te} + 2In_{Hg} + V_{Hg} \tag{1}$$

The structural vacancies locate in the cation sublattice inhabiting the lattice sites and are proportional to the content of In_2Te_3. When In atoms are incorporated into the HgTe lattice, vacancies of Hg are formed in the cation sublattice. If MIT crystallizes in zincblende structure, there predicatively exist structural vacancies in the cation sublattice of the defect zincblende structure of MIT material.

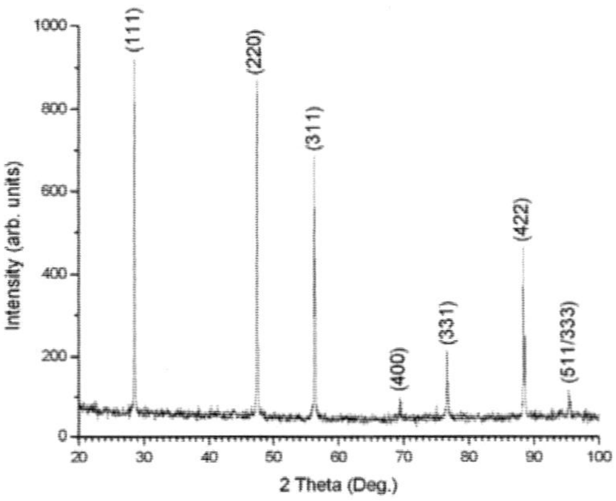

Figure 2. Powder X-ray diffraction spectrum of MIT crystals.

The structural investigation of the as-grown MIT crystals was performed by means of powder X-ray diffraction. Figure 2 shows the X-ray diffraction pattern (XRD) of the as-grown crystal measured at room temperature. The diffraction indices of the crystal are also shown in figure. 2. The pattern measured on MIT compound can be interpreted as that of a face-centered cubic lattice. The data indexed by the DICVOL programs indicate that the structure of MIT crystal is similar to that of sphalerite ZnS. The pattern shows that the as-grown crystal is a single phase of zinc-blende structure with high cation vacancies, which belongs to space group $F\bar{4}3m$. The unit cell parameters was determined to be a = b = c = 6.2930Å, $\alpha = \beta = \gamma = 90°$. The volume of the unit cell is 249.21Å^3.

3.2. Optical properties of MIT crystals

FT-IR was used to analyze the infrared transmittance of the crystals. The result is shown in figure 3. The transmittance in the value of 50~55% indicates that the as-grown crystal has good quality and low defect density. Because the MIT crystal is a direct band-gap semiconductor compound, the electrons in the valence band cannot intrinsically absorb the photons with wavelength in the range of middle and far infrared band. Therefore, the transmittance is higher in this range. In figure 3, the transmittance curve is descending gradually in the region from 400cm^{-1} to 4000cm^{-1}, which is likely to be caused by the absorption of the lattice and free carrier. On one hand, with the increase of the frequency, the stronger absorption of the lattice leads to the decrease of the transmittance of the MIT crystal.

On the other hand, the free carriers are scattered by the phonons at the infrared radiation and transfer the energy to the lattice, which will reduce the intensity of the transmittance of infrared light. The absorption coefficient, α, of the free carriers can be denoted by:

$$\alpha = \left\{ \frac{Nq^2 v}{4\pi^2 nm^* \varepsilon_0 c^3} \right\} \lambda^2 \tag{2}$$

where, N is the free carriers concentration, $N = 4.96\times10^{13}\text{cm}^{-3}$; q is the electric charge of the free carriers; v is the scattering frequency; n is refractive index, $n = 3.17$; m^* is the effective mass of the free carriers, $m^* = 0.8m_0$ (m_0 is the electron mass); ε_0 is vacuum dielectric coefficient; c is vacuum propagation velocity of the light and λ is the wavelength of the infrared light. It is found that the absorption coefficient of the free carriers is proportional to the square of wavelength (λ). With the decrease of the wavelength, the transmittance of the crystal increases.

In conclusion, the transmittance is higher at low wavenumber than at high wavenumber due to the stronger lattice absorption at high frequency and less free carrier absorption at low frequency.Some text.

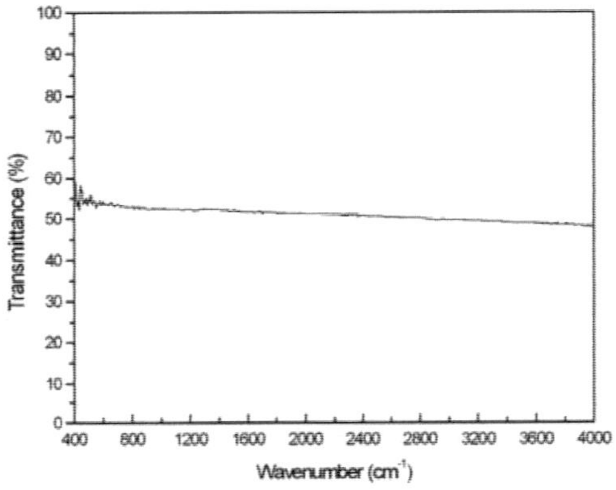

Figure 3. Infrared transmission spectrum of MIT crystals.

3.3. Electrical properties of MIT crystals

The typical I-V curve for the MIT crystal is shown in figure 4. From the curve, we can see that the I-V curve agrees well with the Ohmic Law and can be used to calculate the resistivity of the MIT wafer. Hall Effect measurements on the electrical properties at room temperature in the Van der Pauw configuration were used to find out resistivity, carrier density and mobility of the MIT crystal.

In n-type semicondutors, Hall coefficient can be expressed as

$$R_H = -\gamma \frac{1}{ne} \tag{3}$$

where, R_H is Hall coefficient, " —" represents n-type conductivity, γ is scattering parameter, which is related to energy band-gap, lattice scattering, ionized and impurities scattering, $\gamma \approx 1$, n is the carrier electron density, and e is the electron charge. In general, R_H is the summation of low-resistivity grains and high-resistivity boundaries while the measured Hall coefficient R_H is approximately equal to that of low-resistivity grains due to the extremely small dimensions of boundaries compared with that of grains.

The electron mobility μ_e and the resistivity ρ are, respectively, defined as

$$\mu_e = |R_H|\sigma \tag{4}$$

Journal of Physics: Conference Series **152** (2009) 012010 doi:10.1088/1742-6596/152/1/012010

$$\rho = \frac{1}{ne\mu_e} \tag{5}$$

where, σ is conductivity. Hall measurement at room temperature revealed that the MIT crystal was n-type conduction with $R_H = -5.58\times10^5 cm^3C^{-1}$. The resistivity, carrier density and mobility of the MIT crystal at room temperature were 4.79×10^2 Ωcm, 2.83×10^{13} cm^{-3} and 4.60×10^2 $cm^2V^{-1}s^{-1}$. As comparison, the main electrical properties of HgTe are also listed in table 1.

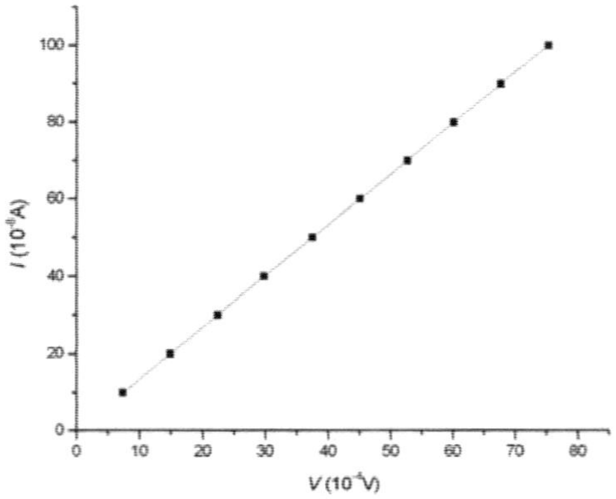

Figure 4. The typical I-V curve for the MIT crystal.

Table 1. The main electrical properties of MIT crystals.

Parameters	Energy band-gap (eV)	Electrons Mobility ($cm^2V^{-1}s^{-1}$)	Intrinsic carrier concentration (cm^{-3})	Resistivity (Ωcm)
MIT (x=0.5)	0.73	460	2.83×10^{13}	478.5
HgTe [8]	-0.11	30000	3×10^{17}	6.94×10^{-4}

From table 1, we can conclude that after adding 50mol% In_2Te_3, the resistivity increases six orders of magnitude while carrier density and electrons mobility reduces. It is obvious that the energy band-gap become larger, which may be the main reason that results in the increase of resistivity and the decrease of mobility. HgTe itself is a semimetal in consequences of a small band overlap with an extremely high electron mobility and low resistivity, whose energy band-gap is negative. Even the conduction band is much lower than the valence band. The addition of In_2Te_3 into HgTe causes the energy band overlap to decrease until it goes through zero and become positive, which means that the band edges become parted. Therefore, the resistivity increases and the carrier mobility and carrier density become smaller. In the mean time, the addition of In_2Te_3 introduces high concentration of vacancies, which act as acceptors in the crystal and lead to the reduction of carrier density and the increase of resistivity.

4. Conclusions

We have successfully grown mercury indium telluride (MIT) crystals with dimensions of 15mm in diameter and 175mm in length by the vertical Bridgman method (VB) at the optimized growth conditions. The results of power X-ray diffraction show that MIT crystals have defect zinc-blende structure with the space group F43m. The infrared transmittance of MIT crystal in the region from

$400cm^{-1}$ to $4000cm^{-1}$ was determined by FT-IR to be 50~55%, which decreases gradually with the increase of wavenumber due to the lattice absorption and free carriers absorption. As compared to semimetal HgTe, alloying In_2Te_3 improves the semiconducting properties of MIT crystal. The resistivity, carrier density and mobility of the MIT crystal determined by Hall measurements at room temperature are 4.79×10^2 Ωcm, 2.83×10^{13} cm^{-3} and 4.60×10^2 $cm^2V^{-1}s^{-1}$, respectively.

Acknowledgements

The authors sincerely acknowledge the financial support provided by the National High Technology Research and Development Program of China (863 Program), graduate starting seed fund of Northwestern Polytechnical University (Z200612) and the Doctorate Foundation of Northwestern Polytechnical University (CX200606).

References

[1] Spencer P M and Ray B 1968 Brit. J. Appl. Phys. (J. Phys. D) **1** 299
[2] Weitze D and Leute V 1996 J. Alloys Comp. **236** 229
[3] Leute V and Schmidtke H M 1988 J. Phys. Chem. Solids **49** 409
[4] Grushka G G, Bakhtinov A G and Grushka Z M 1997 J. Adv. Mater. **4** 36
[5] Grushka G G, Grushka Z M and Gavaleshko N P 1985 Ukrainian Phys. J. **30** 304
[6] Bakumenko V L, Bonakov A K and Grushka G G 1983 Elektronnaya Technika: Materialy **2** 75
[7] Halling T, Sauders G A and Lambson W A 1982 Phys. Rev. B **26** 5786
[8] Lewis J E and Wright D A 1966 Brit. J. Appl. Phys. **17** 783

Journal of Physics: Conference Series **152** (2009) 012011 doi:10.1088/1742-6596/152/1/012011

Microstructual and Optical properties of novel $Fe_{1-x}Os_xSi_2$ semiconducting alloys

Y Gao[1], R S Chen[1], H W Liu[1], Y X Xia[1], Q R Deng[1], G Shao[2]

[1] The faculty of physics and electronic technology, Hubei University, Wuhan, P. R. China

[2] Centre for Materials Research and Innovation, University of Bolton, Bolton BL3 5AB, U. K.

[1]E-mail: gaoyun@hubu.edu.cn

Abstract. Fe1-xOsxSi2 is a novel semiconducting opto-electronic material designed by energy-band engineering. This material has potential application in the Silicon-based light emitting device and thin film photovoltaic devices. In this work, novel Fe1-xOsxSi2 thin films were fabricated by magnetic sputtering deposition. The microstructures were studied by X-ray diffraction and transmission electron microscopy, and compared with FeSi2 and OsSi2 thin films. The atomic ratios between Fe and Os were determined by energy dispersion X-ray spectrometry. Optical absorption was measured to study the band gap properties. Experimental results show that all the Fe1-xOsxSi2 films have direct band gap property. The strongest optical absorption was obtained when x is equal to 0.5 and the results are valuable for the practical applications.

1. Introduction

Metal silicides with semiconducting properties have attracted great research interests in the past tens years for its potential application in silicon-based technology as monolithic/single chip optoelectronic devices.[1] Among them, binary metal silicides such as $FeSi_2$, $RuSi_2$, $MnSi_{1.7}$ films on silicon substrate have been studied extensively. The most promising material used as interconnection component is β-$FeSi_2$, which has a band gap at about 0.87 eV corresponding to the optical communication window at a wavelength of 1.55 μm.[2] However, the pseudo-direct band gap property of β-$FeSi_2$ limits its further application.[1] This problem has been behind our research effort to synthesize a ternary alloy system with direct band gap property by using band-gap engineering. One possible choice is to combine isostructural semiconducting β-$FeSi_2$ and β-$OsSi_2$ materials to form new ternary alloys. Both β-$FeSi_2$ and β-$OsSi_2$ have orthorhombic structure with the Cmca space group. The lattice parameters are a=9.863, b=7.791 and c=7.833 Å for β-Fesi2 and a=10.1496, b=8.1168 and c=8.223 Å for β-$OsSi_2$.[1] The similar lattice parameters with a mismatch less than 5% made them idea candidates to form a band engineered alloy system with tunable energy band structures. Miglio's group calculated the electronic band structure of $Fe_{1-x}Os_xSi_2$ system and found that ternary alloys with direct band gaps could be obtain by replacing certain Fe atom positions with Os atoms.[3] However, no group reported the synthesis of such ternary alloys. The most possible reason is that Os has very low vapor pressure and poor adhesion on Si substrate, which limits the β-$OsSi_2$ film growth on Si substrate and therefore makes the growth of $Fe_{1-x}Os_xSi_2$ film more difficult.[4] Here we report the growth of $Fe_{1-x}Os_xSi_2$ films on Si substrate by magnetron sputtering, with efforts towards correlation of experimentally measured band gaps with theoretical justifications.

2. Experimental details

$Fe_{1-x}Os_xSi_2$ thin films were deposited on a Si (100) substrate by RF sputtering at room temperature in a high vacuum chamber with a base pressure of about $2x10^{-4}$ Pa. The substrate was biased at -100 V and the argon partial pressure was fixed at 140 mTorr. Si, Fe and Os single targets were used in sequence with different deposition time for each target during a sequential circle, in order to ensure the required compositions. About 10 nm Si layer was pre-deposited in order to get good adhesion between the alloy layer and the Si substrate. After deposition, a 30 nm Si layer was deposited on the alloy layer to protect it from oxidation. Post-annealing was performed in a furnace at 700°C and 850°C for 2 hours with N_2 as protect gas.

Cross-sectional TEM samples were prepared by ion beam thinning at 4.5 kV. Energy dispersive X-ray spectroscopy (EDXs) was performed under a CM20 transmission electron microscope operated at 120 kV. A Bruker D8 X-ray diffractometry (XRD) was used to determine the crystal structures. UV-vis-infra spectrometry was carried out to study the optical absorption properties of these $Fe_{1-x}Os_xSi_2$ thin films in the wavelength range from 1000 to 1800 nm.

3. Results and discussion

The thickness and the Fe:Os ratio of the samples after annealing at 700°C were determined using cross-section TEM (XTEM) and EDX, respectively. Figure 1a, 1b and 1c show the XTEM pictures of $Fe_{1-x}Os_xSi_2$ alloys with x equal to 0, 0.5 and 1. The corresponding selective area diffraction (SAD) patterns were obtained and presented in Fig. 1d, 1e, and 1f. Similar SAD patterns are observed from the three samples. The diffused rings presented in the diffraction patterns indicated that deposition layers in the three samples are of extremely fine nano-crystalline structures. The first and second diffraction rings in the Fig. 1d and 1f could be indexed as (220)/(202) and (040)/(040) plane of β-$FeSi_2$ or β-$OsSi_2$, which are of the same crystal structure but slightly different lattice parameters. The same SAD pattern from the $Fe_{0.5}Os_{0.5}Si_2$ sample indicated the same crystal structure as well.

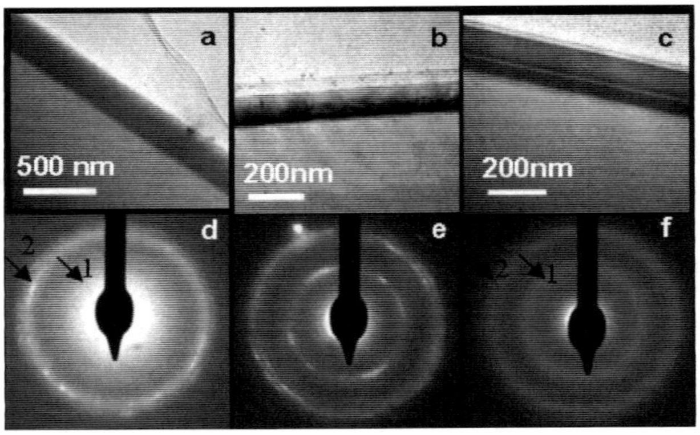

Figure 1. (a), (b) and (c) show the XTEM pictures of the $FeSi_2$, $Fe_{0.5}Os_{0.5}Si_2$ and $OsSi_2$ thin films, respectively. (d), (e), and (f) are the corresponding SAD patterns.

Figure 2 presents the XRD spectra of the FeSi2, $Fe_{0.5}Os_{0.5}Si_2$ and $OsSi_2$ samples after annealing at 850°C for 2 hours and compared with the standard $FeSi_2$ and $OsSi_2$ spectra. The peaks of the $FeSi_2$ and $OsSi_2$ films match very well with the standard spectra. The similar peak positions are also observed in the $Fe_{0.5}Os_{0.5}Si_2$ sample but the peak positions are in between the peaks of $FeSi_2$ and the $OsSi_2$. These results indicated that $Fe_{0.5}Os_{0.5}Si_2$ is isostructual with β-$FeSi_2$ and β-$OsSi_2$, but with slight difference in lattice parameters. The simulated lattice parameters are summarized in Table 1. The XRD observation was thus consistent with the results from electron diffraction.

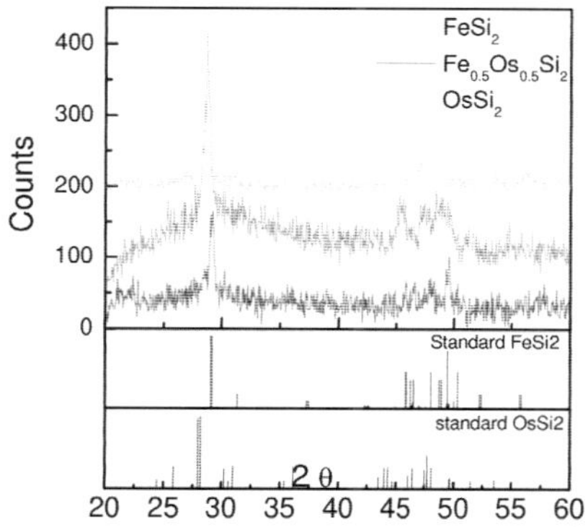

Figure 2. The XRD spectra of the FeSi2, $Fe_{0.5}Os_{0.5}Si_2$ and $OsSi_2$ samples after annealing at 850°C for 2 hours

Table 1. The simulated lattice parameters of $Fe_{1-x}Os_xSi_2$ thin films.

	$FeSi_2$			$Fe_{0.5}Os_{0.5}Si_2$			$OsSi_2$		
	a	b	c	A	b	c	a	b	c
Previous reports[1]	9.863	7.791	7.833	--	--	--	10.15	8.117	8.223
Experimental results	9.83	7.711	7.805	10.14	8.118	8.22	10.2	8.25	8.39

Figure 3a shows the optical absorption spectra of the $Fe_{1-x}Os_xSi_2$ films after annealing at 700°C for 2 hours. The linear dependence of α^2 on the photon energy over an adequate energy range indicates that all these films have a direct transition property. The strongest optical absorption was obtained in the sample with x = 0.5. According to the theoretical calculation, both $FeSi_2$ and $OsSi_2$ are indirect band semiconductors, but $Fe_{0.5}Os_{0.5}Si_2$ shows obvious direct transition property.[3][5] Therefore, it is easy to deduce that $Fe_{0.5}Os_{0.5}Si_2$ should have a larger optical absorption coefficient, which is consistent with our experimental results. The energy band gap value is plotted in Fig. 3b against the atomic percentage of Fe in the metal sublattice sites. It is interesting to note that the band gap decreases as the Fe percentage decreases. According to the theoretical calculation, -$OsSi_2$ has an indirect band semiconductor with a band gap energy of 0.92 eV. However, our experimental results showed that $OsSi_2$ is of a direct band property with an energy gap of 0.74 eV in this experiment. This discrepancy can be attributed to the ultrafine nano structure in our materials where considerable strain/stress can exist,[6][7] while the theoretical prediction was made for bulk strain-free materials. To clarify the strain effect, the annealing temperature was further increased to 850°C for 1 hour. Experimental results showed that the optical absorption coefficient has little change for $Fe_{0.5}Os_{0.5}Si_2$ sample, but is reduced substantially for $OsSi_2$. Hence, it is reasonable to conclude that the stress relaxation leads $OsSi_2$ to show indirect band property, but has little effect on $Fe_{0.5}Os_{0.5}Si_2$ which has a direct band nature. It is worth pointing out that density functional theory is often inadequate in accurate prediction of energy gaps, largely due to the approximation in the exchange functional. Further work is in progress for a more accurate theoretical prediction to account for the stress and size effects.

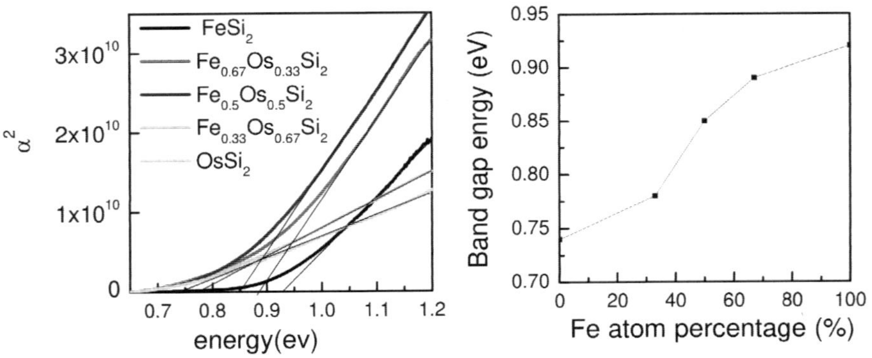

Figure 3. (a) the optical absorption spectra of the $Fe_{1-x}Os_xSi_2$ films after annealing at 700°C for 2 hours with x varied from 0 to 1. (b) The band gap energy value as a function of atomic fraction

4. Conclusion

$Fe_{1-x}Os_xSi_2$ thin films with direct band transition properties were for the first time synthesized by magnetron-sputtering deposition. TEM and XRD analysis find that all these films have the same lattice structure as β-$FeSi_2$. Optical absorption results showed that the band energy decreases with the x increases. All these results are consistent with the prediction of the band-engineering theory.

Acknowledgments

This work is partially supported by the by the NSFC (Ref. No: 50702019) and the Cheng Guang grant of Wuhan City (Ref. No: 200750731274).

References:

[1] V. E. Borisenko 2000 *Semiconducting silicides* (New York: Springer-Verlag Berlin Heidelberg)
[2] D. Leong, M. A. Harry, K. J. Reeson and K. P. Homewood 1997 *Nature* (London) **387** 686
[3] D. B. Migas, L. Miglio 2001 *Appl. Phys. Lett.*.**79** 2175
[4] F. Z. Amir, R. J. Cottir, T. D. Golding, W. Donner, N. Anibou, D. W. Stokes 2006 *J. Cryst. Growth* **294** 174
[5] D. B. Migas, L. Miglio, W. Henrion, M. Bebien, F. Marabelli, B. A. Cook, V. L. Shaposhnikov, V. E. Borisenko 2001 *Phys. Rev. B.* **64** 075208
[6] G Shao, S Ledain, YL Chen, JS Sharpe, RM Gwilliam, KP Homewood, K Reeson Kirkby and MJ Goringe 2000 *Appl. Phys. Lett.* **76** 2529
[7] G. Shao and K.P. Homewood 2000 *Intermetallics* **8** 1405

Journal of Physics: Conference Series **152** (2009) 012012 doi:10.1088/1742-6596/152/1/012012

Self-assembled SnO_2 film with low density

Hongjun Ji [1], Xiaoheng Liu [1], Xin Wang *[,1], Xujie Yang [1], and Lude Lu [1]
Xiutao Ge [2] and Yonghong Li [2]

[1] Key Laboratory for Soft Chemistry and Functional Materials of Ministry of Education, Nanjing University of Science and Technology, Nanjing 210094, China;
[2] Department of Chemistry and Life Science, Chuzhou University, Chuzhou 239012, China

E-mail address: wangx@public1.ptt.js.cn

Abstract. A new kind of metallic oxide film consisted of nanodisks with porous structure has been prepared, which is different from that of targetlike multirings previously reported by our group. Herein, the SnO_2 film has been prepared by using tin(IV) tert-butoxide as precursor, sodium dodecyl sulfonate (SDS) as template and gelatin as stabilizer. Results show that the SnO_2 film is usually composed of series nanodisks with the diameters ranging from 0.1μm to 1μm and each disk displays worm-like porous structure in the transmission electron microscope (TEM) observation. These SnO_2 nanodisks are suggested to born inside the solution and gradually rise up to the air-water interface due to low density of the disk or disk cluster. After further treatment, the well ordered SnO_2 thin film is considered to have a potential application in gas-sensing materials. Remarkably, this may be a distinctive method to get SnO_2 film and may be widely applied to produce other metallic oxide films with low density.

1. Introduction

SnO_2 is a typical n-type semiconductor material and has been proved to be sensitive to many reducing gases (e.g. H_2, CO, C_2H_5OH) and oxidizing gases (e.g. NO_x) in multiple produced pure [1-4] or doped gas sensors [5-10]. Many past researches paid attention to the good morphologies, mechanisms [11], excellent properties and etc. Remarkably, a new self-assembly technology has been created with many advantages in preparing nano metallic oxide films, which can obtain well ordered structures and excellent properties through simple apparatus, few operations and mild reaction conditions etc. White J W, Elder K J and our group et al. did some important work on TiO_2 [12-14], ZrO_2 [15, 16], ZnO_2 [17, 18], SiO_2 [19, 20], which showed good potential applications in optics and electrics etc. Herein, we will put attention into a self-assembly process inside the reaction solution, to obtain air-water interfacial SnO_2 film with low density and investigate the gas sensitivity of the samples.

2. Experimental

2.1. Chemicals
Tin(IV) tert-butoxide ($Sn(OC_4H_9)_4$, Sigma Aldrich, 99.99%), sodium dodecyl sulfonate (SDS, Alfa Aesar, 99%), chlorhydric acid (Nanjing chemical reagent Co.Ltd, 36–38%) and gelatin (G, AR, Tianjin Kermel chemical reagent Co.Ltd) were used as purchases.

2.2. Preparations
Preparations for SnO_2 thin film were as follows. First, 0.65mmol SDS and 0.15g gelatin were mixed with 18.4ml deionized water in a 25ml beaker by stirring magnetically at 40℃ for 10mins, beyond which the uniform surfactant solution was kept and cooled at room temperature. Meanwhile, 0.5mmol tin(IV) tert-butoxide was added into 6.5mmol HCl in a 10ml beaker, after which the solution was stirred for 5mins at room temperature until it became clear as the precursor. Then, the precursor was transferred to a Petri dish with a diameter of 90mm and a depth of 10mm and subsequently, the surfactant solution was poured into the Petri dish. Last, the Petri dish was kept standing at room temperature (21℃) for 12h, and then it could be found that the film has floated at the air-water interface.

2.3. Characterizations
SnO_2 thin film was transferred to a glass substrate and dried for 24h, and then the substrate was directly detected by X-ray diffraction (XRD) with a Bruker D8 diffractometer, using monochromatic Cu Kα radiation operated at 40kV and 30mA. Scanning electron microscope (SEM) of JEOL-6380LV and EDS with EDAX Genesis2000 were applied to detect the dried film. The fresh sample was dropped to a copper grid after ultrasonic dispersion for 10mins, which was observed by a JEOL-2100 transmission electron microscopy (TEM) at 200 kV and a Gatan 794 charge-coupled device (CCD) camera.

2.4. Gas sensitivity
Gas sensitivity measurements of raw SnO_2 product from direct hydrolysis of $Sn(OC_4H_9)_4$ and modified SnO_2 product from the assembled system were carried out on Hanwei intelligent testing equipment (Hanwei Company, Henan, China). The apparatus contained a closed container with a volume 30L, which was used to store the target gases for testing. The sample was calcined at 500℃ for 6h and cooled in the air, following which was ground in an agate motor and deposited to gold electrodes with alumina ceramic tubes as supports. Then the assembled gas sensors were tested by different kinds and concentrations of gases with a constant heat voltage (5V) and circle voltage (5V), such as liquid petroleum gas (LPG) and C_2H_5OH et al, which was quickly heated, evaporated and filled the container. Sensitivity (Rair/Rgas) was defined as the ratio of the gas sensor resistance in air (Rair) to that in target gas (Rgas).

3. Results and discussion

3.1. Morphology and structure
Figure 1 displays the intensity versus angle pattern of SnO_2 film with 2θ degree in the range 2°-10°. Note that there are no obvious diffractions detected from 10° to 80°. The strongest diffraction peak at 2θ=3.10° with d=2.85nm indicates that a high degree of ordered structure has been retained in the film. Additionally, there are some other peaks in figure 1, which may be caused by the surfactant SDS.

Many SnO_2 disks with different diameters (the maximum diameter is ~1μm) in the film could be observed by SEM in figure 2. This SEM image is similar to such a TEM result in figure 4(a). EDS analysis of the film is displayed in figure 3. Importantly, the element percents of N (12.44%) and S (1.56%) in the SnO_2 film suggest that substantial G and SDS have been transmitted into the self-assembled structure, which provides essential information for following analysis of SnO_2 film structure.

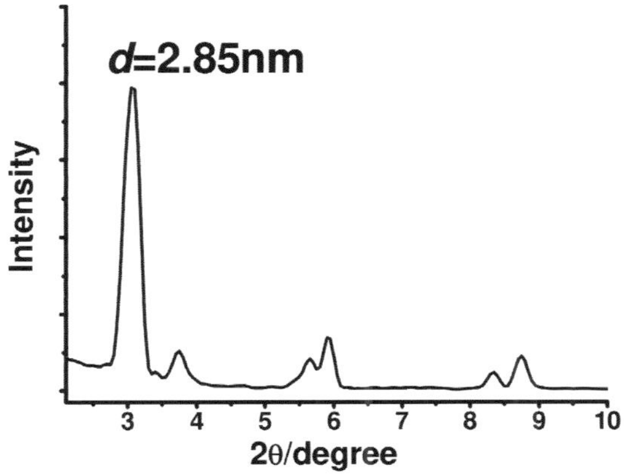

Figure 1. XRD pattern of SnO_2 film.

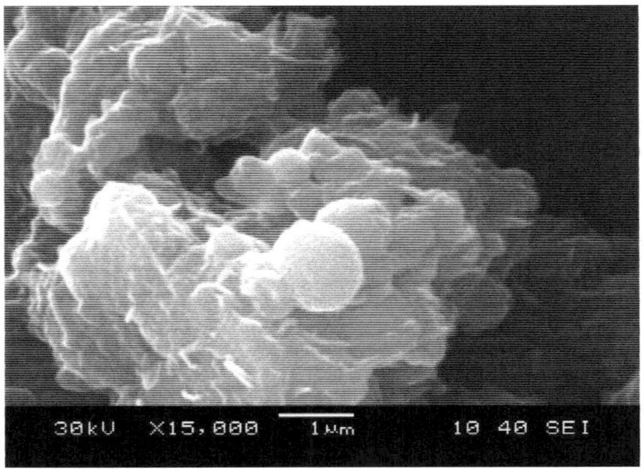

Figure 2. SEM observation of SnO_2 film

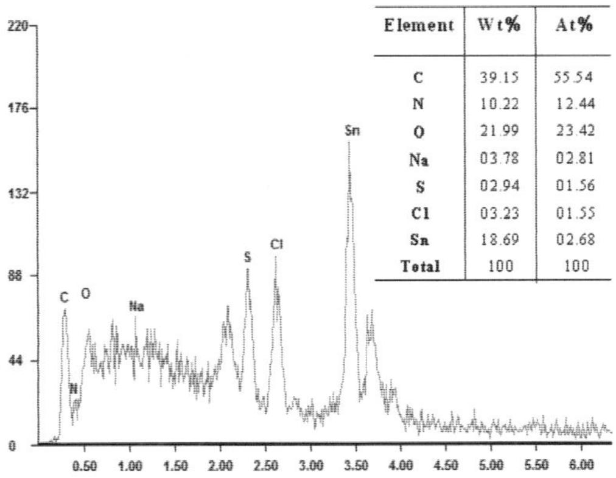

Element	Wt%	At%
C	39.15	55.54
N	10.22	12.44
O	21.99	23.42
Na	03.78	02.81
S	02.94	01.56
Cl	03.23	01.55
Sn	18.69	02.68
Total	100	100

Figure 3. EDS pattern of SnO_2 film

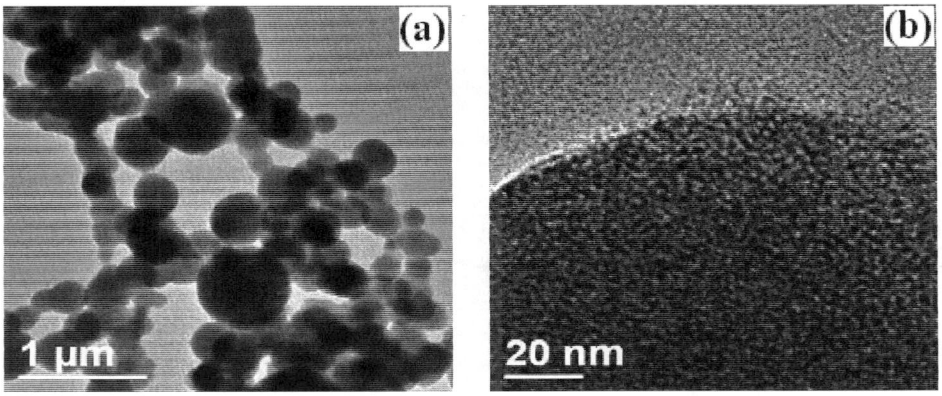

Figure 4. TEM observations of SnO_2 film with different scale bars
(a) with 1μm; (b) with 20nm

TEM observations of SnO_2 film are displayed in figure 4. The disk cluster consisted of many disks and containing many spatial gaps could be observed in figure 4(a). The disk diameters in the cluster range from 0.1μm to 0.8μm. The HRTEM in figure 4(b) shows a worm-porous structure in the scale bar of 20nm, where the diameters of the pores are measured ~3nm, which is consistent with such an ordered structure in the term of the XRD date in figure 1.

3.2. Formation of porous SnO_2 disk

Self-assembly mechanism of ZrO_2 target-like nanodisk from our previous report [16] could be employed to explain the formation of SnO_2 nanodisk. figure 5 shows the probable formation mechanism of self-assembly process of a SnO_2 disk. Due to stable interactions between SDS and G [21, 22], SDS as template [16] and G as stabilizer can be transmitted into self-assembly structure. However, we have observed an obvious porous structure in figure 4(b), instead of target-like multirings of ZrO_2 nanodisk, which may be attributed to the collapses of SnO_2 lamellar structure with less toughness than ZrO_2 lamellas [16].

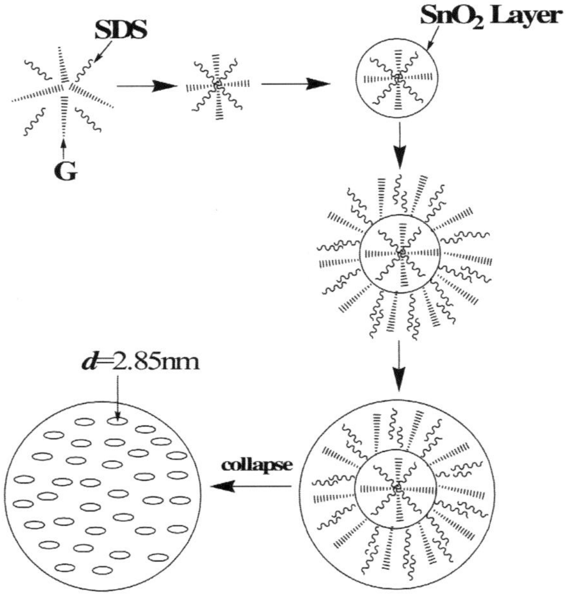

Figure 5. Schematic diagram of the self-assembly process of SnO_2 film

3.3. Structure of SnO_2 film with low density

The suggestion to explain the formation of SnO_2 film with low density is shown in figure 6. There may be two reasons resulting in the generation of this low-density material. First, there were many micropores in each isolated disk. Second, there existed some large gaps inside the disk cluster. The disk cluster and relative gaps have been found in figure 2 and figure 4(a). Hence, the film was suggested to born inside the solution and rise up gradually to the air-water interface.

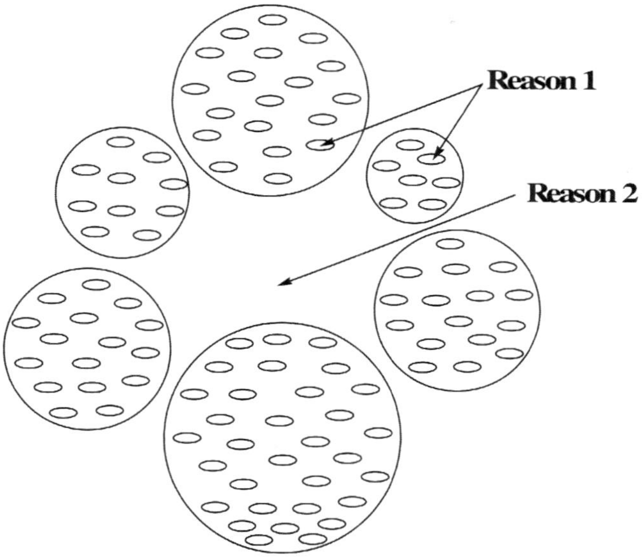

Figure 6. Structure of SnO_2 film with low density

3.4. Gas sensitivity testing

Sensitivities of two SnO_2 products to different concentrations of gases are displayed in figure 7. The gas sensor of raw SnO_2 product with little sensitivity showed no obvious differences to C_2H_5OH and LPG (L_1). Simultaneously, the gas sensor of modified SnO_2 product from the assembled system was a little more sensitive to LPG, and the sensitivity was 4.8 when the concentration was 600ppm. Note that this low density material performed much more sensitive to C_2H_5OH, which had an average sensitivity above 8 in all the concentrations and achieved 11 at 800ppm. This phenomenon could be attributed to the low density of modified SnO_2 product and well dispersed SnO_2 nanodisks or disk clusters. Accordingly, the more C_2H_5OH adsorbed and the more electrons exchanged, the more sensitivity was attached.

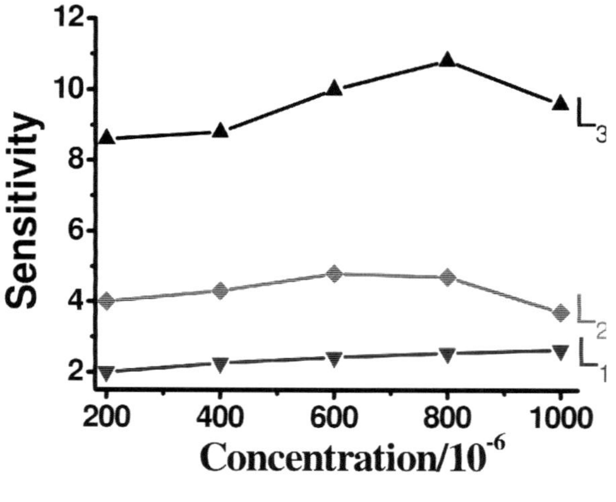

Figure 7. Sensitivities of SnO_2 to various target gases L_1: the raw product to C_2H_5OH or LPG; L_2: the modified product to LPG; L_3: the modified product to C_2H_5OH

4. Conclusions

It was found that a new kind of SnO_2 film with low density could be prepared successfully by a self-assembly process with SDS as template and gelatin as both dispersant and stabilizer. According to the characterizations of XRD, SEM and TEM, the formation mechanism of nanodisks with porous structure in the film had been reasonably conjectured. Furthermore, SnO_2 product obtained from the self-assembly process performed more sensitive to C_2H_5OH or LPG than the raw SnO_2 product. Our next work will concern on more gas sensitivity of this SnO_2 product, including its response to hydrogen, carbon monoxide.

Acknowledgement
This work was supported by the National Natural Science Foundation of China (No. 50772048) and two R&D programs (No.BK2006199 and No.JHB06-04) from Jiangsu Province, China.

References

[1] Xi L J, Qian D, Tang X C, Chen C J 2008 *Materials Chemistry and Physics* **108** 232
[2] Velasquez C, Rojas F, Esparza J M, Ortiz A and Campero A 2006 *J. Phys. Chem. B* **110** 11832
[3] Luyo C, Fábregas I, Reyes L, Solís J L, Rodríguez J, Estrada W and Candal R J 2007 *Thin Solid Films* **516** 25
[4] Wang H Z, Liang J B, Fan H, Xi B J, Zhang M F, Xiong S L, Zhu Y C and Qian Y T 2008 *Journal of Solid State Chemistry* **181** 122
[5] Majumder S, Hussainb S, Das S N, Bhar R B, Pal A K 2008 *Vacuum* **82** 760
[6] McCue J T and Ying J Y 2007 *Chem. Mater.* **19** 1009
[7] Pourfayaz F, Mortazavi Y, Khodadadi A and Ajami S 2008 *Sensors and Actuators B* **130** 625
[8] Gong J W, Sun J R and Chen Q F 2008 *Sensors and Actuators B* **130** 829
[9] Carotta M C, Gherardi S, Guidi V, Malagu C, Martinelli G, Vendemiati B, Sacerdoti M, Ghiotti G, Morandi S, Bismuto A, Maddalena P and Setaro A 2008 *Sensors and Actuators B* **130** 38
[10] Srivastava A, Rashmi and Jain K 2007 *Materials Chemistry and Physics* **105** 385
[11] Habgood M and Harrison N 2008 *Surface Science* **602** 1072
[12] Zhao L L, Yu Y, Song L X, Hua X F, Larbot A 2005 *Applied Surface Science* **239** 285
[13] Kan C, Liu X H, Wang X, Ji H J, Yang X J and Lu L D 2007 *Colloids and Surfaces A: Physicochem. Eng. Aspects* **311** 93
[14] Brennan T, Roser S J, Mann S and Edler K J 2003 *Langmuir* **19** 2639
[15] Henderson M J, Gibaud A, Bardeau J –F, Rennie A R, White J W 2005 *Physica B* **357** 27
[16] Liu X H, Kan C, Wang X, Yang X J and Lu L D 2006 *J. Am. Chem. Soc.* **128** 430
[17] Ju S Y, Lee K and Janes D B 2005 *Nano Lett.* **5** 2281
[18] Lu J G, Ye Z Z and Zhang Y Z 2006 *Appl. Phys. Lett.* **89** 023122
[19] Brown A S, Holt S A, Reynolds P A, Penfold J and White J W 1998 *Langmuir* **14** 5532
[20] Suzuki K, Ikari K and Imai H 2004 *J. Am. Chem. Soc.* **126** 462
[21] Saxena A, Antony T and Bohidar H B 1998 *J. Phys. Chem. B* **102** 5063
[22] Cooke D J, Dong C C and Thomas R K 2000 *Langmuir* **16** 6546

Direct synthesis of straight SiO_2 nanorods

Guang Zhu [1 2], Xiaoping Zou[1], Yi Su[1], Jin Cheng[1], Maofa Wang[1]

[1] Beijing Information Science & Technology University,Beijing Key Laboratory for Sensor, Beijing 100101, China
[2] Suzhou College, Anhui 234000, China

[*]E-mail address: xpzou2005@gmail.com

Abstract. The straight SiO_2 nanorods with a diameter of about 200nm and smooth surface have been directly synthesized by high temperature vapor deposition method at 1300℃. The as-synthesized samples were characterized by means of scanning electron microscopy, energy dispersive x-ray, and transmission electron microscopy. The results show that as-synthesized silica nanorods have a uniform size, well-defined shape, and smooth surface. However, the morphologies and microstructures of silica nanorods are affected by synthesis conditions, such as the concentration of the SiO_x and the deposition temperature. On the basis of these experimental results, a possible growth mechanism of silica nanorods in this process is proposed.

1. Introduction

Since silica nanotubes have been first reported by Nakamura in 1995 [1], there has been an increasing interest to the synthesis of various quasi one-dimensional nanometer-sized materials such as nanotubes, nanowires, nanorods etc.. These one-dimensional nanomaterials show some novel physical and chemical properties due to their peculiar structure and size effect, and they are one of the most promising elements for the fabrication of nanoelectronic devices [2, 3] and integrated optical device [4]. Among them, as an important member in the one-dimensional nanomaterials family, silica nanomaterials have been actively studied for a long time. Silica nanomaterials with various microstructures have been synthesized utilizing a variety of methods, such as, laser ablation [5], vapor–liquid–solid (VLS) [6], vapor–solid (VS) [7], and sol–gel [8] methods.

The well straight silica nanorods are important to the future micro-optoelectronic devices [9]. Generally speaking, aligned silica nanostructures can be obtained by using metallic catalysts, such as Ga–In [10], In–Ni [11], Ga [12, 13], Sn [14], and Ni [15]. Whereas, silica nanowires are not very straight if no catalyst is used [16]. However, this communication introduces a simple and effective method to synthesize large quantities of straight silica nanorods without using any catalysts. This method is base on high temperature solid state reactions of the mixed powders of silica and graphite, via the VS mechanism [7]. Further more, the possible growth mode corresponding to these structures was proposed in this paper.

2. Experimental procedures

Silica nanorods were prepared in a horizontal electronic resistance tube furnace with a gas supply and a control system. The procedure is depicted as follows: at first, the mixed powders of silica (99.00%, Shanghai Chemical Co.) and carbon (Analytically pure, Beijing Chemical Reagents Co.) was placed in a ceramic boat, and then the boat was covered with pretreated Si wafer. Subsequently the boat was transferred into the centre of a ceramic tube mounted in the horizontal tube furnace. 3% H_2/Ar gas mixture and N_2 gas were introduced into the tube at the rate of 50 sccm at atmosphere pressure, respectively. 10 min later, the furnace temperature was heated to 1300℃ and held for 2 hours. Then the furnace was cooled to about 700℃ at 10℃/min with protection gas, and then cooled naturally to room temperature. The white products were observed on the surface of substrates.

The morphology of as-synthesized products were examined by scanning electron microscopy (SEM, JEOL 6500F), structural analysis was carried out using transmission electron microscopy (TEM, JEM-200CX) and selected area electron diffraction (SAD). Compositional analysis was performed by an energy dispersive x-ray spectrometer (EDS) attached to the SEM. The specimens for TEM analysis were prepared by dispersing the samples in ethanol followed by sonication for 10 min. A few drops of the suspension were dripped onto a micro-grid covered with a holey graphite thin film.

3. Results and discussion

Figure 1(a) is a typical SEM micrograph of the as-synthesized products, it is clearly found the substrate is covered with a large amount of silica nano-rods. Figure 1(b) is the closer view image of silica nanorods in Figure 1(a), it shows silica nanorods with a smooth surface and a good uniformity. The diameter of silica nanorods is about 200 nm, and the lengths are up to several micrometers. We can also find these silica nanorods have a straight and cross structure. In order to clarify the composition of deposition, the sample was examined by EDS spectrometer. The inset is a typical EDS spectrum of individual nanorod. It clearly shows that the nanorods consist of oxygen and silicon. The spectrum shows the as-synthesized products consisting of elements Si and O with an atomic ratio about 1:2.

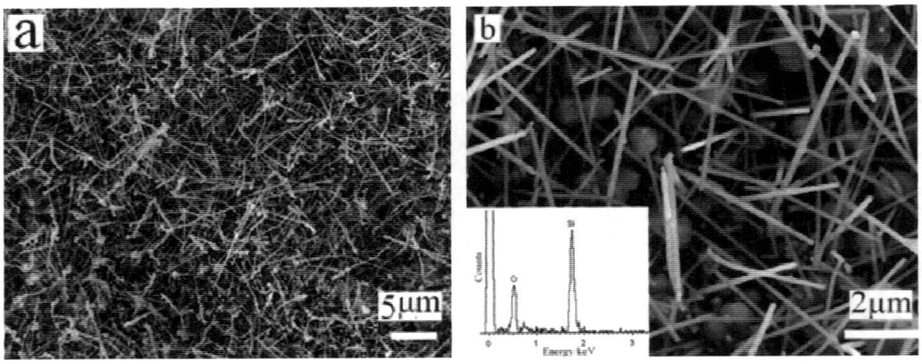

Figure 1. (a) Top–down SEM image of silica nanorods (b) Closer view showing silica nanorods in (a).

Figure 2 indicates SEM images of the silica nanorods at different growth zones. Figure 2(a) shows schematically different growth zones of the silica nanorods on the substrate. From Figure 2(b), it clearly shows some silicon oxide nanoparticles have deposited on the zone Ⅰ (Figure 2(a) light zone) and some nuclei of silica nanorods have also formed. Figure 2(c) shows a few of silica nanorods are synthesized on the zone **II** (Figure 2(a) grey zones), and Figure 2(d) shows a large amount of silica nanorods are synthesized on the zone **III** (Figure 2(a) black zone). Well then what leads to the formation of the phenomenon here?

Although the detailed mechanism for the formation of different morphology of silica nanorods structures is not well understood, we still believe that the different local concentration of the SiO_x may be responsible for the different morphologies of rods structures. A possible explanation is depicted as follows: when the furnace is heated to high temperature, the growth temperature is high enough to produce vapor SiO_x in the ceramic boat. As the process proceeds, a continuous accumulation of the vapor SiO_x in the ceramic boat leads to a continuous increase of vapor pressure and vapor concentration of SiO_x. However, their concentrations would increase gradually with the increasing distance from the centre position of growth substrate. These vapors SiO_x will diffuse along in all directions at the growth zone III (black arrows in Figure 2(a)), where is the joint between substrate and boat. So the concentration gradient of SiO_x is formed and distributed along the radial in the ceramic boat. At the deposition area near the Zone III, the concentration of the SiO_x is relatively high, and the silica nanorods would grow fast and long. The presence of a large amount of straight nanorods should be attributed to the strong limiting effect of the high SiO_x concentration to the silica nanostructures surface (see Figure 2(d)). If the SiO_x concentration was reduced a little (zone II), a small quantity of silica nanorods are synthesized on the growth substrate. In the mean time, the second nucleation perhaps could not achieve and finally only leaving the straight silica nanorods(see Figure 2(c)). If the SiO_x concentration reduced more (zone I), few silica nanorods are achieved and only leaving the nuclei of silica nanorods(see Figure 2 (b)).

Figure 2. (a) Schematic diagram depicting different growth zones
(b) SEM image of silica nanorods grown on the zone I in (a)
(c) SEM image of silica nanorods grown on the zone II in (a)
(d) SEM image of silica nanorods grown on the zone III in (a)

The morphology and structure of the nanorods have been characterized in further detail using TEM. Figure 3 is a typical TEM image of silica nanorods. An individual straight silica nano-rod (Figure 3(a)) has smooth surfaces with diameter of about 200 nm, in agreement with SEM observations, and highly dispersed selected-area electron diffraction (SAED) pattern taken from one rod (inset) with only diffusive rings (without diffraction spots), revealing the amorphous nature of the silica nano-rods. Furthermore, Figure 3(b) is the HRTEM image of the black rectangular area marked in Figure 3(a), revealing that no fringes exist in the silica nanorods.

On the basis of the experimental results described above and the conventional VS growth mechanism, we have proposed a possible mechanism for the growth of the straight silica nanorods. at first, when the furnace is heated to high temperature, the reaction of the mixed powders of carbon and silica in the boat should be occurred, and forming a large amount of SiO_x molecule have deposited on the surface of the substrate and some nuclei of silica nanorods have also formed. Then, the newly formed SiOx will stack on the SiO_x nuclei, forming silica nanorods on the substrate. Well then what affects the straight growth direction of rods here?

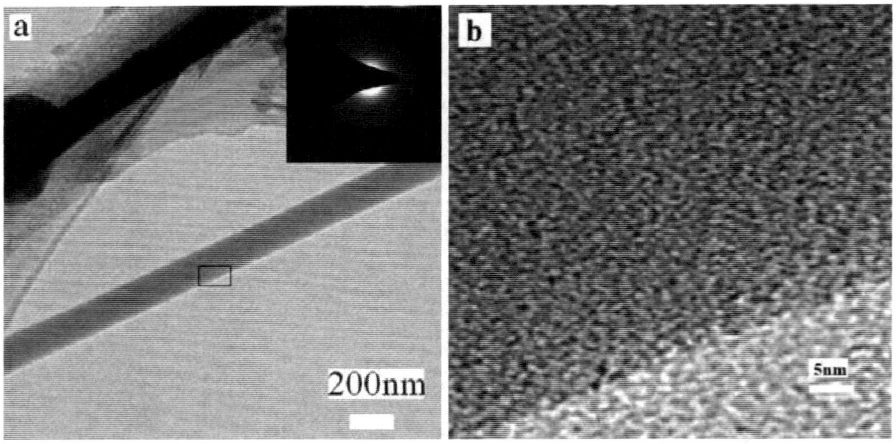

Figure 3. TEM images of (a) a straight silica nanorods and the inset is the corresponding SAED pattern. (b) HRTEM image of Figure 3(a).

We supposed that the existence of a lot of SiO_x molecules results in fast growth of silica nanorods, but there is not enough time for SiO_x to stack themselves into tortuous shape. On the other hand, there is no change of the external conditions during the growth of silica nanorods. They are also capable of inducing a subsequent growth in some proper conditions and promoting the growth of straight structures. At last, the straight nanorods would be formed by depositing SiO_x continuously.

According to observations and analysis of SEM and TEM images, the results indicate that silica nanorods lack any catalyst at nanorods tips and the VS model could be accepted. Further experiments are clearly needed in order to explain the exact process for the growth of the straight silica nanorods not being assisted by catalyst, and to research optical properties of these nanorods.

4. Conclusion

Straight silica nanorods have been synthesized through high temperature solid state reactions process without using any catalysts. These silica nanorods have uniform diameters of about 200 nm, neat smooth surface, and lengths of up to several micrometers. The morphologies and microstructures of silica nanorods are affected by synthesis conditions, such as the concentration of the SiO_x. On the basis of these experimental results, a possible growth mechanism was proposed for the silica nanorods. The straight silica nanorods offer potential application possibilities in nanodevices design and fabrication.

Acknowledgments

The authors thank Prof. X. M. Meng of Technical Institute of Physics and Chemistry of CAS for the aid in SEM analysis, Prof. X. D. Xu of Beijing University of Technology for the assistance in TEM analysis, and Prof. W. Xu, Mr. Su, Sun, and Miss Sun of Institute of Biophysics of CAS for the assistance in TEM characterization. One of Authors (X.P. Zou) thanks the partial financial support from Founding Program of Science & Technology Activity for Chinese Homecoming Fellow Abroad and Research Program of Beijing Key Lab for Sensor ((No.KM200810772009).).

References

[1] Nakamura H and Matsui Y, 1995 *J. Am. Chem. Soc.* **117** 2651-2653
[2] Dai H, Wong E W and Lieber C M, 1996 *Science* **272** 523-526
[3] Ebbesen T W, Lezec H J, Hiura H, Bennett J W, Ghaemi H F and Thio T, 1996 *Nature* **382** 54 – 56
[4] Yu D P, Hang Q L, Ding Y, Zhang H Z, Bai Z G, Wang J J, Zou Y H, Qian W and Feng S Q, 1998 *Appl. Phys. Lett.* **73** 3076-3079
[5] Morales A M and Lieber C M, 1998 *Science* **279** 208-211
[6] Liang C H, Zhang L D, Meng G W, Wang Y W and Chu Z Q, 2000 *J. Non-Cryst. Solids* **277** 63-67
[7] Peng X S, Wang X F, Zhang J, Wang Y W, Sun S H, Meng G W and Zhang L D, 2002 *Appl. Phys.* A, 74 831-835
[8] Zhang M, Bando Y, Wada L and Kurashima K, 1999 *J. Mater. Sci. Lett.* **18** 1911–1914
[9] Zheng B, Wu Y Y, Yang P D and Liu J, 2002 *Adv. Mater.* **14** 122-124
[10] Wang J C, Zhan C Z and Li F G, 2003 *Solid State Commun.* **125** 629-631
[11] Ma R Z and Bando Y, 2003 *Chem. Phys. Lett.* **377** 177-183
[12] Pan Z W, Dai Z R, Ma C and Wang Z L, 2002 *J. Am. Chem. Soc.* **124** 1817-1822
[13] Pan Z W, Dai S, Beach D B and Lowndes D H, 2003 *Nano Lett.* **3** 1279-1284
[14] Sun S H, Meng G W, Zhang M G, Tian Y T, Xie T and Zhang L D, 2003 *Solid State Commun.* **128** 287-290
[15] Yu D P, Xing Y J, Hang Q L, Yan H F, Xu J, Xi Z H and Feng S Q, 2001 *Physica* E **9** 305-309
[16] Wei Q, Meng G W, An X H, Hao Y F and Zhang L D, 2006 *Solid State Commun.* **138** 325-330

Journal of Physics: Conference Series **152** (2009) 012014 doi:10.1088/1742-6596/152/1/012014

Synthesis of flower-like silica microstructures by a simple thermal evaporation method

Guang Zhu [1 2], Xiaoping Zou[1], Yi Su[1], Jin Cheng[1], Maofa Wang[1]

[1] Beijing Information Science & Technology University,Beijing Key Laboratory for Sensor, Beijing 100101, China
[2] Suzhou College, Anhui 234000, China

[*]E-mail address: xpzou2005@gmail.com

Abstract. Novel flower-like silica microstructures have been synthesized through heating silicon, silicon monoxide and active carbon mixed powders under an H2/Ar (3%) atmosphere at 1050℃ without assistance of any metal catalyst. Scanning electron microscopy (SEM), transmission electron microscopy (TEM), high resolution transmission electron microscopy (HRTEM) and energy dispersive x-ray spectrometer (EDS) analyses reveal that the wires are amorphous and consist only of silicon oxide, generated from the reaction of CO with SiC. A possible growth model based on both of vapor-solid (VS) and the Oxide-assisted growth (OAG) mechanism has been proposed to explain this phenomenon.

1. Introduction

In recent years, the silicon nanostructures materials with special size and morphology have attracted a lot of attention due to the potential applications in various fields, such as nanoscale electronic and optoelectronics devices [1–3], and sensor[4, 5]. Up to now, various nanostructures of silica including nanowires, nanorods, and nanotubes have been prepared by using various approaches, for instance, laser ablation [6], vapor phase transport[7], bio-mimetic strategies [8,9], and sol–gel [10] methods.

The flower-like silica nanostructures is an unusual three dimensions consisting of silicon oxide nanofibers. Liu et al. [11] synthesized flower-like silicon nanostructure through combining the thermal evaporation of silicon monoxide powders in an evacuated quartz tube with the VLS growth mechanism at 1100℃. Zhu et al. [12] synthesized flower-like silica nanofibers through heating SiC, deposited on a silica substrate, in the presence of Co powder. However, these methods suffer from metal catalyst existing in the as-synthesized products.

In this paper, we report the generation of flower-like microstructures by using thermal evaporation of the mixed powders of silicon, silicon monoxide and active carbon, deposited on a quartz substrate, in the absence of metal catalyst. The resulting flower-like microstructures are composed of silica microwires, and appear to have grown via both of VS process and the OAG process. These microstructures may prove to be useful as three dimensional composite materials and as novel catalytic supports.

2. Experimental procedures

Flower-like silica microstructures were prepared in a horizontal electronic resistance tube furnace with a gas supply and a control system. Figure 1 shows a schematic diagram of the experimental setup employed during the course of the present work. Two quartz plate ($9\times2.5\times0.1cm^3$) plates were used as substrates for the growth of the novel flower-like nanostructures. These plates were ultrasonically washed in acetone for several minutes to clean the surface of plate, and then rinsed with de-ionized water. The raw material was a mixture of silicon (99.00%, Shanghai Chemical Co.), silicon monoxide (99.00%, Beijing Chemical Reagents Co.) and active carbon (Analytically pure, Beijing Chemical Reagents Co.) powders at a weight ratio of 6:3:4. It was placed in a ceramic boat that covered with cleaned quartz plate, and then the boat was transferred into the centre of a ceramic tube mounted in the horizontal tube furnace. The other cleaned quartz plate in another ceramic boat was placed next to the first boat on the downstream side of the flowing high-purity Ar. Prior to heating the tube furnace was purged with a mixture gas of Ar (97%) and H_2 (3%) for 10 min to eliminate air in the furnace. Under the ambient pressure and a constant flow of high-purity Ar (60 sccm), the furnace temperature was heated to 1050℃ and held for 1 hour. After furnace was cooled down to room temperature, gray products were observed on the surface of substrates.

The morphology of the as-synthesized products was examined by scanning electron microscopy (SEM, JEOL 6500F), structural analysis was carried out using transmission electron microscopy (TEM, JEM-200CX) and selected area electron diffraction (SAD). Compositional analysis was performed by an energy dispersive x-ray spectrometer (EDS) attached to the SEM. The specimens for TEM analysis were prepared by dispersing the samples in ethanol followed by sonication for 10 min. A few drops of the suspension were dropped onto a microgrid covered with a holey carbon thin film.

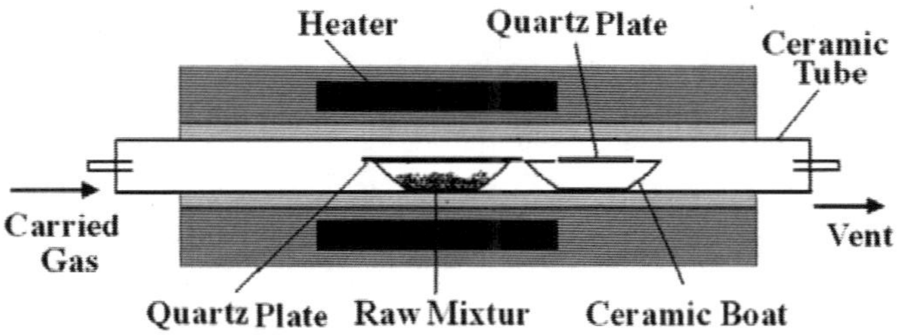

Figure 1. A schematic diagram of the experimental setup

3. Results and discussion

The morphology of the products was examined using SEM. Figure 2a is a low-magnification SEM image and it shows that the as-prepared flower-like microstructures products grown on the quartz substrate. It clearly shows that the flower-like microstructures is made up of many microwires. All the wires have cone-shaped tips at their ends. In addition, these microwires are rooted in one center and have a length of several micrometers, and the diameter of wires decrease along with the growth of wires increase. Figure 2b is the close view image in Figure 2a, it shows that the surface of microwires is neat and smooth. Figure 2c is a SEM image of as-prepared flower-like nanostructures grown on substrate at downstream position. Figure 2d is the close view image in Figure 2c, it shows that the surface of the wires is severely roughened. We speculate that the different local concentration of the SiO_x may be responsible for the different surface morphologies of silica microwires. In our system, their concentrations would decrease gradually with the increasing distance from the source. Figures 3a and b are the corresponding energy dispersive X-ray spectroscopy (EDX) data of Figure 2b and Figure 2d, it indicates that they consist of elements Si and O with an atomic ratio about 1:2.

The morphology and structure of flower-like silica microstructures have been characterized in further detail using TEM and selected area electron diffraction (SAED). Some short microrwires with length of about several μm are shown in Figure 4a.They may be broken from a whole flower-like silica microstructure because of the ultrasonic vibrations. The selected-area electron diffraction (SAED) pattern from white rotundity area marked in one arm (inset in Figure 4a) with only diffusive rings (without diffraction spots), revealing the amorphous nature of the silica wires. Figure 4b is the HRTEM image of the white pane area marked in Figure 4a, revealing that no fringes exist in the silica microwires.

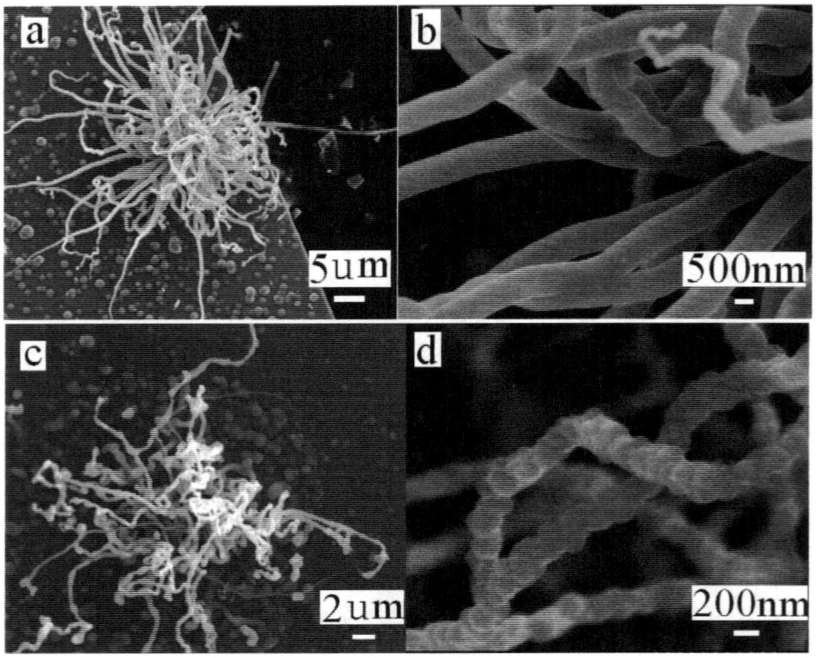

Figure 2. (a) SEM image of as-synthesized flower-like silica microstructures grown on the quartz substrate; (b) closer view showing flower-like silica microstructures in (a); (c) as-prepared flower-like silica microstructures grown on Si substrate at downstream position; (d) closer view showing flower-like silica microstructures in (c).

In order to understand the growth formation of flower-like silica microstructures, many questions about it formation must be clarified. For example, (1) How do silicon powders supply? (2) How do carbon powders supply? (3) How about the reactions of the mixed powers should be occurred in the chamber? In the next sections, detailed analysis and explanation are made to answer above questions.

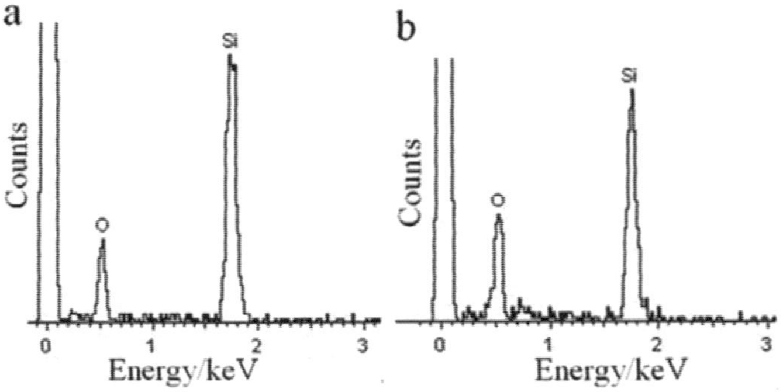

Figure 3. EDX spectrums of (a) silica microstructures in Figure2 b; (b) silica microstructures in Figure2 d.

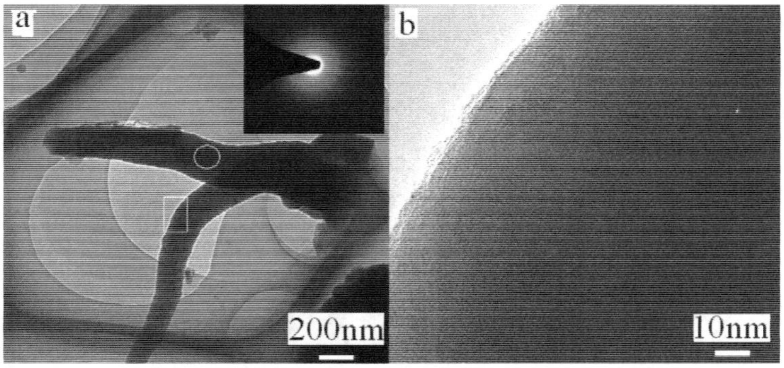

Figure 4. (a) TEM image of the as-synthesized silica microstructures (broken from the flower-like microstructures) and the inset is the corresponding SAED pattern; (b) HRTEM image of (a).

In our experiment, at first, when the furnace is heated to high temperature, silicon and silicon monoxide powders could be oxidized in the furnace, the chemical reaction equation can be written as below:

$$Si(s) + O_2(g) \rightarrow SiO_2(s) \tag{1}$$

$$2SiO(s) + O_2(g) \rightarrow 2SiO_2(s) \tag{2}$$

In our system, the carrier gas (Ar) was filled into the tube furnace, and most of oxygen can be eliminated. However, due to the poor obturation of ceramic tube system, there was still leaked or residual oxygen. The concentration of the residual oxygen met the requirement on O_2 partial pressure, thus the above reaction could occur. Well then how do active carbon powders supply in the process of flower-like nanostructures formation?

In previous reports, carbon powders often played an important role during the growth of silica nanostructures [13]. Here, we propose a possible reaction between the carbon powders and as-obtained silicon oxide to explain its role:

$$SiO_2(s) + 3C(s) \rightarrow SiC(s) + 2CO(g) \tag{3}$$

leading to the formation of solid SiC and in situ liberation of CO. We have found SiC nanowires in our products, which provide a strong support for Equation 2. Details of these results will be reported separately. The CO then immediately reacts with the SiC, at an unspecified temperature [12]:

$$SiC(s) + CO(g) \rightarrow SiO(g) + 2C(g) \tag{4}$$

The resulting silicon oxide have deposited and aggregated in one centre lead to the formation of flowers. Carbon (in Equation 3) may have been oxidized and vented with Ar gas after reaction. So we speculate that the formation of flower-like silica microstructures involves solid-solid reaction (Equation 3) and gas-solid reaction (Equations 1, 2 and 4) processes.

In order to provide support for the effects of the adding active carbon powder on the formation of the flower-like silica microstructures, we performed the additional experiment by using raw sources without adding carbon powders under the same experiment parameters. However, we have not found any products on the substrates through observing SEM. This result indicated that reaction (3) did not occur. It provides strong evidence for adding active carbon powder play a key role during the growth of flower-like silica microstructures.

On the basis of the results of above analysis, it can be considered that the growth mechanism of those flower-like silica microstructures is via both of VS process and the OAG [14] process. In the initial stage, a large amount of silicon oxide nanoparticles (in Equation 3) have deposited on the surface of the substrate and some nuclei of silica wires have also formed (see black arrow in Figure 5a), which is described as the VS process. Then, the newly formed silicon oxide SiO will stack on the silicon oxide nuclei, forming shorter silica wires on the substrate (see white arrow Figure5 b), which is described as the OAG process. In the mean time, a lot of silica microwires grow and aggregate in one center. As a result, flower-like microstructures are formed (see Figure 5c). At last, the novel flower-like silica microstructures would be formed by growing continuously silica wires in one centre (see Figure 5d).

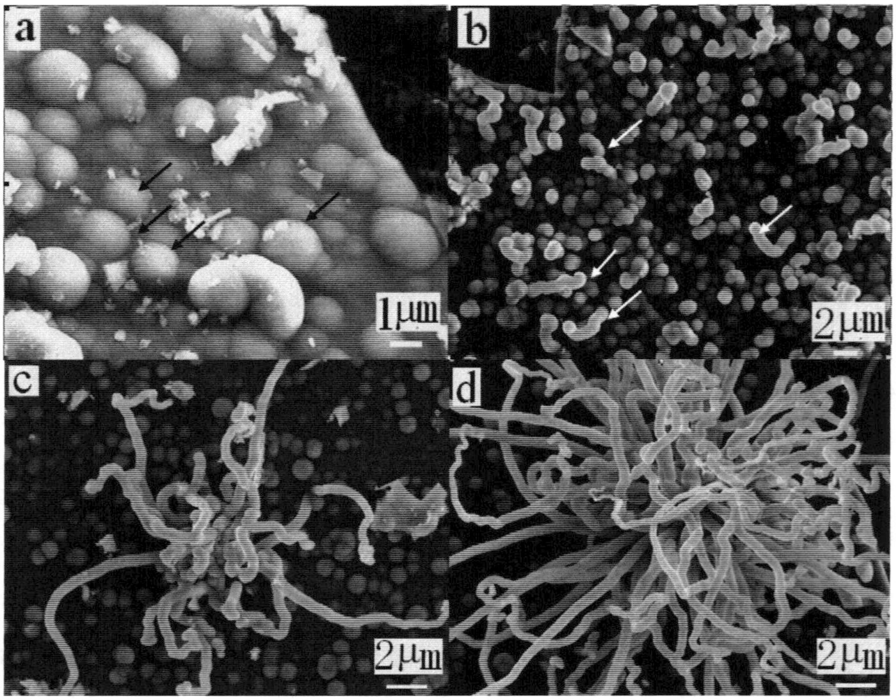

Figure 5. SEM images (a) some nucleis of silica wires grown on the quartz substrate;(b) short silica wires on the substrate; (c) forming flower-like microstructures;(d) formed flower-like silica microstructures.

4. Conclusion

The novel flower-like silica microwires have been fabricated by heating the mixed powders of silicon, silicon monoxide and active carbon powders. Investigation results demonstrate that these microflowers are made up of silica microwires. The surface morphology of these wires is affected by different local concentration of the silicon oxide. The formation of flower-like silica microstructures is via both of VS process and the OAG process. These microflowers may prove useful in reinforcing composite materials or in further modifying other nanostructures.

Acknowledgments

The authors thank Prof. X. M. Meng of Technical Institute of Physics and Chemistry of CAS for the aid in SEM analysis, Prof. X. D. Xu of Beijing University of Technology for the assistance in TEM analysis, and Prof. W. Xu, Mr. Su, Sun, and Miss Sun of Institute of Biophysics of CAS for the assistance in TEM characterization. One of Authors (X.P. Zou) thanks the partial financial support from Founding Program of Science & Technology Activity for Chinese Homecoming Fellow Abroad and Research Program of Beijing Key Lab for Sensor ((No.KM200810772009).).

References

[1] Dai H, Wong E W and Lieber C M, 1996 *Science* **272** 523-526
[2] Ebbesen T W, Lezec H J, Hiura H, Bennett J W, Ghaemi H F and Thio T, 1996 *Nature* **382** 54-56
[3] Cui Y, Zhong Z, Wang D, Wang W U, and Lieber C M, 2003 *Nano Lett.* **3** 149-152
[4] Patolsky F and Lieber C M, 2005 *Mater. Today* **8** 20-28
[5] Cui Y, Wei Q, Park H and Lieber C M, 2001 *Science* **293** 1289-1292
[6] Morales A M and Lieber C M, 1998 *Science* **279** 208-211
[7] Xu C X, Sun X W, Yu M B, Xiong Y Z, Dong Z L and Chen J S, 2004 *Appl. Phys. Lett.* **85** 5364-5367
[8] Davis S A, Burkett S L, Mendelson N H and Mann S, 1997 *Nature* **385** 420-423
[9] Huo Q S, Zhao D Y, Feng J L, Weston K, Buratto S K, Stucky G D, Schacht S and Schuth F, 1997 *Adv. Mater.* **9** 974-978
[10] Zhang M, Bando Y, Wada L and Kurashima K, 1999 *J. Mater. Sci. Lett.* **18** 1911-1914
[11] Liu Z H, Sha J, Yang Q, Su Z X, Zhang H and Yang D, 2007 *Physica* E **38** 27-30
[12] Zhu Y Q, Hsu W K and Terrones M, 1998 *J. Mater. Chem.* **8** 1859-1864
[13] Du X W, Zhao X, Jia S L, Lu Y W, Li J J and Zhao N Q, 2007 *Mater. Sci. Eng.* B **136** 72-77
[14] Pan Z W, Dai Z R, Xu L, Lee S T and Wang Z L, 2001 *J. Phys. Chem.* B **105** 2507-2514

Journal of Physics: Conference Series **152** (2009) 012015 doi:10.1088/1742-6596/152/1/012015

Diamond Film Ultraviolet Detectors with Different Diamond Grain Size

K Tang, L J Wang[1], J Huang, Y Ma, G Hu, X F Zhu, Y B Xia

School of Materials Science and Engineering, Shanghai University, Shanghai 200072, China

[1]E-mail: ljwang@shu.edu.cn

Abstract. Diamond films with different grain sizes from 1.8 μm to 3.1 μm were prepared using the bias-enhanced hot-filament chemical vapor deposition (HFCVD) technique. The film structure and morphology were characterized using the optical microscope and Raman spectrometer. Ultra-violet (UV) detectors with interdigital patterns were fabricated using above diamond films. I-V measurement results showed that the performance of the detector was closely related with the grain size of the diamond film. With increasing the grain size, the photo-current, the ratio of photo-current to dark-current and the net photo-current all increased. The above results indicate that diamond films with larger grain size are more suitable for being fabricated as UV detectors.

1. Introduction

Ultra-violet (UV) detector, a special converter for UV signal into electrical signal, has many important applications, including satellite-based missile plume detection, air quality monitoring, environment monitoring, space research and high temperature flame detection. Military purpose, especially, such as UV countermeasures and anti-countermeasures technique, UV navigation and UV communication has aroused the great attention [1-3]. Si, SiC, GaN, ZnO and diamond are the semiconductors most commonly used for UV photodetectors [4]. Diamond, however, is especially suitable for the fabrication of UV photodetectors, due to its wide direct bandgap of 5.5eV (λ=225nm), small or even negative electron affinity, which results in highly responsive and solar-blind [5]. In addition, diamond could profitably be used for realizing of electronic devices where harsh operating conditions are required, for its highest thermal conductivity, good chemical and temperature stability.

However, some problems that should be solved in CVD diamond UV detection, such as the repeatability of electrical properties [6], the persistent photocurrent [7]. Consequently, growth technique, device structure, surface treatment and, especially the quality of CVD diamond, which was of great importance for diamond photodetectors, were deeply studied presently. In general, photoconductivity was restricted by the intra-grain defect, impurity and density of grain boundary which were arbitrarily distributed in the polycrystalline film [8]. Moreover, the factors mentioned above were directly affected by the grain size. Accordingly, in this work, the different grain sizes of the diamond films were prepared using the hot-filament chemical vapor deposition (HFCVD) technique. The fabrication and electrical characterization of UV detectors with different diamond grain size were studied. The aim of this work is to evaluate the relationship between the detector performance and the diamond grain size for the performance improvement of diamond film UV photodetectors.

2. Experimental

In our experiment, diamond films were deposited on mirror-polished (100) silicon wafers (1.5×1.5 cm^2) by HFCVD technique using acetone carried by hydrogen as described in detail elsewhere [9]. Prior to depositing diamond film, the substrates were firstly etched in hydrofluoric acid (HF) aqueous solution, then mechanically scratched by hand with 100 nm diamond powder in water for about 5 min to promote the nucleation of diamond. After that the substrates were cleaned with deionized water and acetone finally. The detailed growth parameters were listed in Table 1, and diamond films with grain size (A, B, C) were acquired by controlling different deposition time, which were 10h (A), 15h (B), 20h (C) respectively. After diamond deposition, the films were chemically oxidized in H_2SO_4+H_2O_2 solution, rinsed and organically cleaned, in order to remove non-diamond carbon impurities.

Table 1 Process parameters for diamond films by HFCVD method

	Gas pressure /kPa	Ratio of gas flow rate (CH_3COCH_3: H_2)	Substrate temperature /°C	Bias voltage/ bias current of substrate (V/A)
Nucleation period	1	50:100	660	+20V/4A
Growth period	4~5	20:160	680~700	0

Related characteristics of the films were analyzed by VXV digital metallographic microscope and Raman spectroscopy (HR800UV) with an Ar+ laser (514 nm). Co-planar MSM UV detectors were realized using Au layer deposited on the above diamond films with different grain size (TA, TB, TC) by thermal evaporation. An interdigitated contact structure with a spacing of 25 μm and a pitch of 50 μm was defined by standard photolithography. By using a Xe-lamp (180-800 nm, 150 W) connected with a monochromator as UV light source, the current of the detector, which was placed on the monochromator's exit slit in air at room temperature, was measured by a KEITHLEY 4200SCS semiconductor characterization system. The wavelength of incident light used in our experiments is 225 nm.

3. Results and discussion

Figure 1 presents the optical microscopy images of diamond films with different grain size with 5000 times magnification. We can find that all samples are typical polycrystalline with a lot of grain-boundaries, and present triangular grains corresponding to (111)-orientated films. The average grain sizes of sample A, B, C are 1.8 μm, 2.5 μm and 3.1 μm, respectively. The thickness of all films is about 15 μm by observing the cross-sectional morphologies of the films (not shown here).

The Raman spectra are shown as figure 2. As evidenced in figure 2, a strong Raman scattering peak of diamond exists at about 1334 cm^{-1} and a weak peak of non-diamond carbon exists at about 1550 cm^{-1}. By considering that the sensitivity of the Raman signal for the non-diamond carbon phase is about 50 times that for diamond, Raman spectroscopy is also used to estimate the quality of diamond film [10]. As mentioned above, it is concluded that all films have high quality with low-density non-diamond. Especially, the peak intensity ratio of 1334 cm^{-1} to 1550 cm^{-1} of sample C is 2.56, and the FWHM of 1334 cm^{-1} is 4.995 cm^{-1} which is the smallest compared with sample A（5.650 cm^{-1}）and B（5.005 cm^{-1}）. These results indicate that the sample C has a better quality.

Figure 1. Optical microscopy images of diamond films with different grain size by HFCVD method

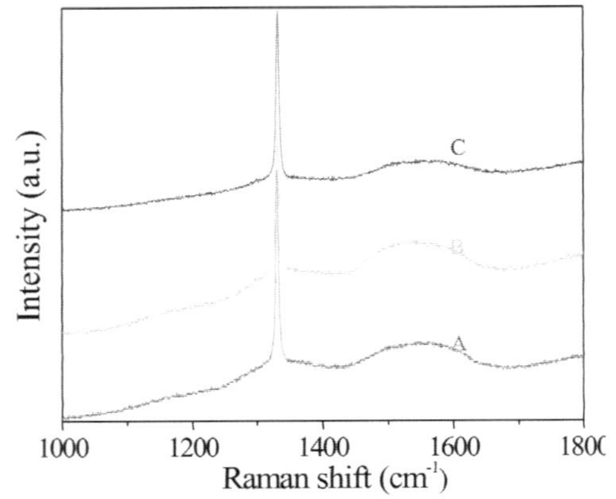

Figure 2. Raman spectra for diamond films grown by HFCVD method.

Figure 3. The optical micrograph of the electrode structure for diamond film UV detectors.

Figure 3 presents the electrode structure of the fabricated diamond film UV photodetectors amplified at 50 times. 15 pairs of interdigitated electrodes are successfully fabricated, which realize the negative electrode and positive electrode of the detector. The electrode width of interdigitated regions is 25 μm and the spacing between two electrodes is 50 μm.

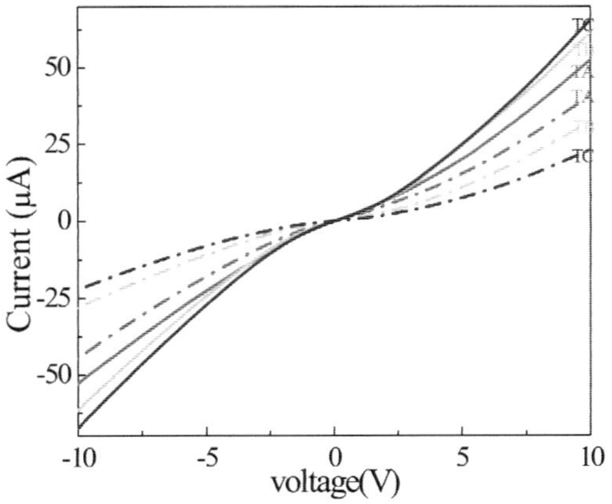

Figure 4 The I-V characteristics for the diamond UV detectors (the dark-current and photo-current are marked dashed line and solid line, respectively).

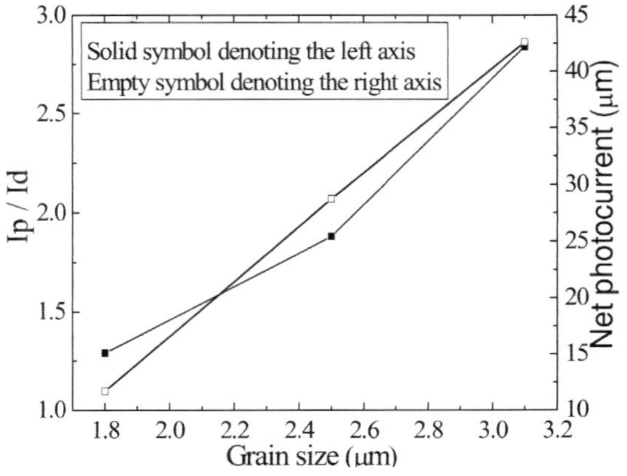

Figure 5 The relationship between Ip/Id, net photocurrent and grain size of diamond film UV photodetectors.

Figure 4 shows the I–V characteristics of the diamond film UV detectors (TA, TB, TC), in which the dark-current (Id) and photo-current (Ip) are marked dashed line and solid line, respectively. The photo-current (Ip) was obtained under the irradiation of UV light with 225 nm at room temperature. The I-V characteristic appears nearly symmetric and linear, suggesting that a fine ohmic contact is formed. From comparing with the dark-current of the photodetectors, when induced by UV light, the photo-currents of TA, TB and TC all increased, indicating that all of the three UV detectors have good response to UV light.

Table 2 Dark-current and photo-current of diamond film UV photodetectors operated at 10 V

Electrical parameter	TA	TB	TC
Dark-current Id / μA	40.9	32.8	23.2
Photo-current Ip / μA	52.6	61.5	65.8
Ip/Id	1.29	1.88	2.84
Net photo-current /μA	11.7	28.7	42.6

Table 2 exhibits the dark-current (Id), photo-current (Ip), the ratio of photo-current to dark-current (Ip/Id) and the net photo-current (Ip-Id). TA has the largest Id, and the Id of TC is the smallest. This means for the diamond films with same grain-orientation, the larger the grain size, the smaller the dark-current. As showed in figure 5, with the increase of the gain size, the ratio photo-current, dark-current and the net photo-current become larger. Thus the grain size has a great influence upon the performance of the UV detectors.

Usually, when diamond photodetectors are irradiated by UV light, free carriers (electron–hole pairs) are generated, when in an external electric field, move towards electrodes. Owing to the surface interdigitated contact structure of the diamond detector we fabricated, the motion of the carriers on the film surface and the surface quality of the diamond films are key factors affecting the performance of diamond detectors. The diamond films are polycrystalline structure with a lot of grain-boundaries. Because grain-boundaries are full of traps just like defects or impurities, which would capture free carriers, this will lead to the shorting of the carriers lifetime. There is a close relationship between the collection efficiency and the carrier lifetime. The longer lifetime carrier has, the less probabilities of being trapped will be. Then there will be a higher collection efficiency and a better device performance. Figure 5 exhibits that with the increase of grain size, the increase of the photocurrent becomes stronger. This suggests that the grain size will directly affect the detector performance.

Therefore, high performance diamond film UV detector can be obtained by using diamond film with larger gain size.

4. Conclusions

(111) diamond thin films with different grain sizes are prepared by using the hot-filament chemical vapor deposition (HFCVD) technique. The UV detectors are fabricated from the above films by the microelectronic processing with Au as the electrode material. I–V characteristics of diamond UV photodetectors are studied and a significant photoresponse is observed under UV light illumination. The dark-current and the photo-current of the UV detectors are relative to the diamond grain size. For the photodetector with a bigger grain size, a weaker dark-current and a stronger photo-current are obtained. Diamond with larger gain size is a key factor to making high performance UV photodetectors.

Acknowledgments

This work was supported by the National Natural Science Foundation of China under Grant No. 60577040, Program for Changjiang Scholars and Innovative Research Team in University (No:IRT0739), and Shanghai Leading Academic Disciplines (T0101).

References

[1] Teraji T, Yoshizaki S, Wada H, Hamada M and T Ito 2004 *Diamond and Related Materials* **13** 858
[2] Rossi M C , Spaziani F, Conte G and Ralchenko V 2005 *Diamond and Related Materials* **14** 552
[3] Askari S J, Akhtar F, Chen G C. He Q, Wang F Y, Meng X M and Lu F X 2007 *Materials Letters* **61** 2139
[4] Moon T H, Jeong M C, Lee W and Myoung J M 2005 *Applied Surface Science* **240** 280
[5] Polyakov V I, Rukovishnikov A I, Avdeeva L A, Kun'kova Z E, Varnin V P, Teremetskaya I G and Ralchenko V G 2006 *Diamond and Related Materials* **15** 1972
[6] Kagan H 2005 *Nuclear Instruments and Methods in Physics Research Section A: Accelerators, Spectrometers, Detectors and Associated Equipment* **541** 221
[7] Bergonzo P, Tromson D, Descamps C, Hamrita H, Mer C, Tranchant N and Nesladek M 2007 *Diamond and Related Materials* **16** 1038
[8] Zhang F and Geng D S 1997 *Materials Science and Technology* **5** 30
[9] Wang L J, Lou Y Y, Su Q F, Shi W M and Xia Y B 2005 *Optics Express* **13** 8612
[10] Angus J C 1992 Thin Solid Films **216** 126

Journal of Physics: Conference Series **152** (2009) 012016 doi:10.1088/1742-6596/152/1/012016

The influence of annealing on mechanical properties of hydrogenated nanocrystalline silicon thin films

H Y Peng, J L Wang, L Wang and B Zhou

Department of Physics, School of Science, Beihang University, Beijing, China 100191

E-mail: wangjinliang@buaa.edu.cn

Abstract: The hydrogenated nanocrystalline silicon (nc-Si:H) thin films were deposited by plasma enhanced chemical vapor deposition(PECVD). The as-deposited films were annealed, and the microstructure and mechanical properties were investigated by XRD, Raman spectra and nanoindenter. The results show that the crystallinity, hardness and elastic modulus have been improved when annealing at 400°C, but the mechanical properties became worse when annealing at 600°C.

1. Introduction

Recently, hydrogenated nanocrystalline silicon (nc-Si:H) thin film, which is compatible with semiconductor technology, has been attractive for their potential applications in light-emitting diodes[1], solar cells[2,3] and thin film transistors[4,5] due to its particular photoelectric properties. Considerable effects including optical and electric properties of nc-Si:H thin films are explored, but little work was on its micro-mechanical properties. The piezo-resistance effect of nc-Si:H thin films have been reported in the previous work[6], and the pressure-sensitive coefficient of the films is 6-8 times than that in monocrystalline silicon. So it is necessary to research and develop high sensitivity micro-pressure sensor, displacement pickup and other micro-nano devices. Moreover, the materials of silicon films have been applied widely in MEMS. Therefore, it is significant for devices application to study the micro-mechanical properties of nc-Si:H thin films. Nanoindentation technique can investigate the micro-mechanical properties such as hardness, elastic modulus and creep and so on[7]. Nanoindentation has pretty high force resolution and motion resolution. With this technique, the changes of burdening and motion can be continuously recorded, making this method a fine way for film mechanical property measurement.

In our previous woks, the micromechanical properties of nc-Si:H thin films deposited on different substrates[8] and P-doped[9] had been investigated. The present work mainly studies the influence of annealing on mechanical properties of nc-Si:H thin films. The microstructure of the film was characterized by XRD and Raman scatter spectra. Nanoindenter system was used to measure the mechanical properties of nc-Si:H thin films, e.g., load-depth curve, hardness and Young's modulus, the effect of annealing process on mechanical properties were researched and the relation between microstructure and micromechanical properties were also probed.

2. Experiments

Nc-Si:H thin films were fabricated on monocrystalline silicon at a radio frequency(13.56 MHz and power 30 W) capacitive coupled PECVD system at a temperature of 260°C and chamber pressure of

0.5 Torr for 9 hours. The working gases are silane(highly diluted by hydrogen, silane 5% and hydrogen 95%) and hydrogen(H_2), the gas flow rate(sccm) of silane and H_2 are 5 and 45 respectively. Nc-Si:H thin films were annealed in nitrogen atmosphere for an hour at 400°C and 600°C.

The microstructure and crystallinity of the films was measured by Raman spectroscopy and XRD(Cu-Kα_1, X-ray wavelength λ=0.15405nm). The Raman spectra were obtained by micro-Raman equipment using He-Ca laser(325nm) and with a power of 4.5mW, in order to avoid any beam-induced crystallization. Measurements of mechanical properties were performed by nanoindentation with MTS Nano Indenter XP system, using the XP indentation units and accutip indenter; the burdening measuring range is fixed from 0 to 500mN with a resolution of 50nN; the stroke length was 2mm with a resolution 0.01nm. Controlling parameter: strain ratio 0.05/s, heat drift ratio 0.05nm/s, indentation depth 1000nm.

3. Results and discussion

3.1. Microstructure Analysis

Figure 3 shows the XRD diagram of the films. It is found that the diffraction peak (111) of the unannealed film exhibits a weak intensity and an obvious split, indicating that the growth of crystalline grains have a bad orientation. After annealed at 400°C, the intensity of (111) diffraction have been enhanced and the peak is symmetrical, all which mean that crystal lattice orientation becomes finer and more ordered than unannealed film. Crystallite grain-sizes D are calculated using the classical Sherrer formula:

$$D = \frac{0.89\lambda}{\beta \cos\theta} \tag{1}$$

where λ is the X-ray(Cu-Ka1) wavelength (0.15405 nm) and β is is the half-width peak integral corrected by subtracting the instrumental broadening deduced from the powder reference signal. The average grain-sizes calculated are approximately 5nm at 260°C, 4nm at 400°C and 6.5nm at 600°C, respectively. The results show that the average grain-sizes in film annealed at 400°C is smaller than that in unannealed film. As the diffraction peak intensities increases further, Crystallographic orientation for the film annealed at 600°C get a better order. And that the full width at half maximum(FWHM) decrease sequentially, indicates that the grain-size becomes larger.

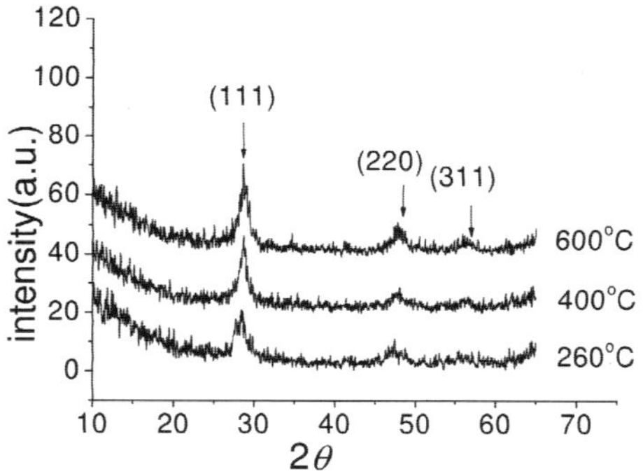

Figure 1. XRD spectra of films annealed at different temperature.

The diffraction peak (311) appeared in the annealed films, because the large numbers of hydrogen atoms escaped from the films, resulting in reconstruction of grains and boundaries with the T_a increasing. The fact that there is no the diffraction peak (311) in the XRD spectra, means that the hydrogen atoms don't have enough energy to overcome potential barrier to readjust interfaces at deposition temperature.

The Raman spectra of the films are shown in figure 2. Analyzing the transverse optical (TO) peak shift and width of the Raman spectra, the grains size D and crystalline volume fractions X_c were estimated from empirical formulas[10]:

$$D = 2\pi\sqrt{\frac{B}{\Delta\omega}}\text{, and } X_c = \frac{I_c}{I_c + \eta I_a} \qquad (2)$$

where B is a constant 2.21cm^{-1}nm^2, $\Delta\omega$ is the wave-number difference between crystalline (520cm^{-1}) and amorphous(480cm^{-1}) TO peak location, I_c and I_a are integrated intensities of the crystalline and amorphous, respectively, and the Raman scatter factor η is 1.0 approximately.

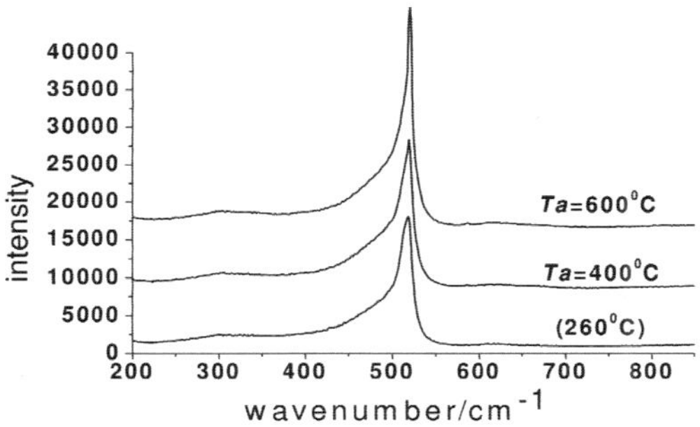

Figure 2. Raman spectra of samples with different T_a

We obtained the crystalline volume fraction X_c and grains size listed in Table 1. It can be found that X_c nearly change almost while D decreases after annealing at 400°C. While the T_a increasing to 600°C, the film's crystalline volume fraction X_c increases from 25% to 35.1% and the average grains size from 4nm to 6.5nm. This is in agreement with the results of X-ray diffraction.

Table 1. Films' X_c and D calculated from Raman spectra

Sample	T_a/°C	X_c	D
1#	260[a]	24.8%	5nm
2#	400	25%	4nm
3#	600	35.1%	6.5nm

[a]260°C is the substrate temperature.

3.2. Load-Depth Curve Analysis

Load-depth curves performed on five random points on 1#, 2# and 3# samples at a maximum indentation depth 1000nm are shown in figure 3. There is a distinct "terrace" in one of five curves illustrated in figure 3(a). It is thought that the terrace would be resulted from a crackle or gap existing[11] in the 1# film. However no crackle or gap appears at the points in annealed 2# and 3# films because the obvious terrace doesn't emerge. It implies that the crackle or gap has been eliminated possibly after annealing process.

Five load-depth curves of 2# sample repeat very well, as shown in Figure 3(b). Although there are some differences, the repeatability has been promoted evidently and the maximum loads are close to each other and the residual indentation depths are the same. One can say that the mechanical properties of nc-Si:H thin film has been enhanced by annealing at 400°C. Grains with different size, random holes and higher surface roughness in unannealed film lead to disperse the load-depth curves. Surface features and internal structure have been improved by annealing at 400°C, such as smaller grain size well-distributed and higher surface smoothness. The average applied load of 2# sample at maximum indentation depths is 160nN and the residual indentation depth is 640nm. The above two mentioned physical quantities hardly changed in comparison with that of unannealed film. All the results state that the resistance to load and elastic recovery are still preserved, and the compatibility of film/substrate deformation has been strengthened at an appropriate T_a, for example, 400°C.

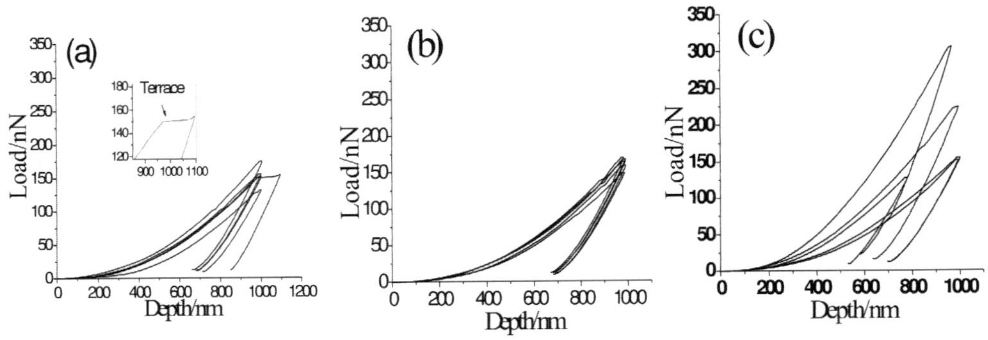

Figure 3. Load-depth curves of samples with different T_a: (a) 260°C, (b) 400°C, (c) 600°C.

Load-depth curves of 3# sample were illustrated in Figure 3(c), five curves even more scatter each other. The applied load at maximum indentation depths has a wide range of 125-305nN, which means that the resistance to load is increased at local area and decreased at another local area in the films. Accordingly, the residual indentation depth also has the same tendency, i.e., a great discrepancy comes into being among the elastic recoveries at different points.

3.3. Hardness and Elastic Modulus Analysis

The hardness and elastic modulus versus the nanoindentation depth curves are given in figure 4 and figure 5, respectively. The measured hardness and elastic modulus can't reliably reflect samples' properties due to the effect of surface roughness and of the radius of the indenter. It is considered that when the indentation depth is about 10-30% of the film's thickness[12,13], substrate's influence could be neglected. So the measured hardness and elastic modulus belong to the nc-Si:H films in the depth range of 80-170nm. The average of hardness and elastic modulus estimated from the curves are given in Table 2.

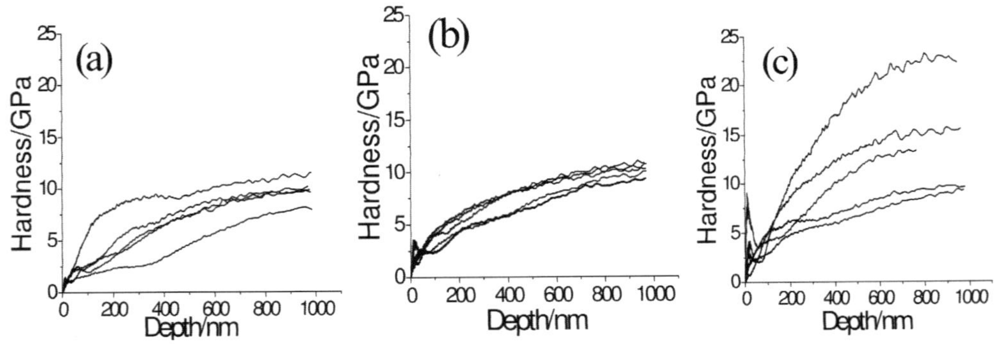

Figure 4. Hardness-depth curves of samples with different T_a: (a) 260°C, (b) 400°C, (c) 600°C.

We can find that the hardness and elastic modulus increase with the T_a increasing from 260°C to 400°C and 600°C. It indicates that the hardness and elastic would become larger with the T_a raising. Nc-Si:H film is composed of crystalline and amorphous phase. Plentiful hydrogen atoms exist in the crackles and grain boundaries. Some H atoms escaped from the boundaries and hydrogen content minishes somewhat at T_a=400°C, but most H atoms escaped from the boundaries and hydrogen content continues to decrease. Because H atom can soften the stress of films[14], hardness raised with T_a increasing. Therefore hydrogen content is an important factor influencing micromechanical properties of nc-Si:H thin films.

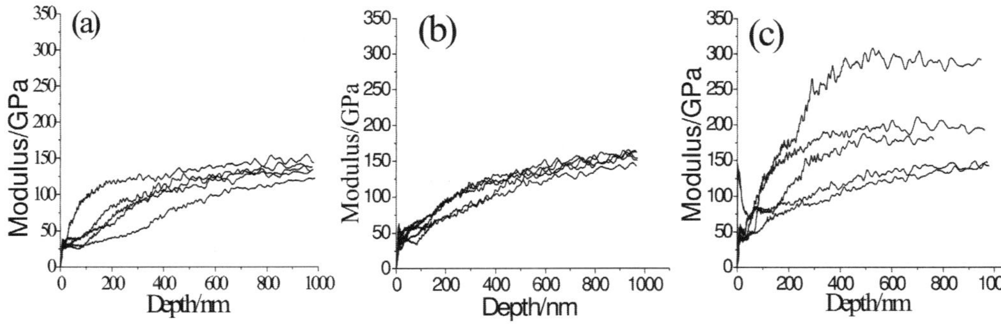

Figure 5. Elastic modulus of samples with different T_a: (a) 260°C, (b) 400°C, (c) 600°C.

The hardness-depth curves shown in Figure 4b are more repeatable than shown in figure 4(a) and figure 4(c), and the elastic modulus curves shown in figure 5(b) have the same change.

Table 2. Hardness and elastic modulus of samples with different *Ta*

Sample	T_a/°C	*H*/GPa	*E*/GP
1#	260	1.7	45
2#	400	3.0	50
3#	600	5.5	75

The uniformity of micromechanical properties were enhanced for the films annealed at 400°C, but below or above T_a is adverse to micromechanical properties. The basic cause to the change of micromechanical properties depends on the microstructure of films. Hydrogen existing in the form of Si-H hinders Si-Si bond from compact network structure in the films [15]. New Si-Si bond generated through the anneal process, but several grains unite to be a greater grain at a higher T_a and more gaps and crackles produced, thereby, the micromechanical properties was influenced. In a word, the changes of films' microstructure can explain the differences of micromechanical properties with varying T_a.

4. Conclusion

The surface feature and microstructure of nanocrystalline silicon (nc-Si:H) such as grain size, H content and distribution of defects are improved by annealing,. The mechanical properties are in a finer uniformity for the films annealed at 400°C, but crystal grains grow largely in size and films' defects in number when annealing at 600°C., resulting in the decreasing the uniformity of the mechanical properties of the film

Acknowledge
This research was supported by the National Natural Science Foundation of China (Grant Nos.10432050 and 50772006).

References

[1] Cho K S *et al* 2005 *Appl. Phys. Lett.* **86** 071909.
[2] Guha S and Yang J. 2006 *Journal of Non-Crystalline Solids.* **352** 1917–1921.
[3] Conibeer G *et al* 2006 *Thin Solid Films.* **511-512** 654–62.
[4] Lee C H, Sazonov A and Nathan A 2005 *Appl. Phys. Lett.* **86** 222106.
[5] Hatzopoulos A T *et al* 2006 *Appl. Phys. Lett.* **89** 193504.
[6] He Y L, Liu H, Yu M b and Yu X M 1996 *NanoStructured Materials.* **7** 769-777.
[7] Zhang T H and Yang Y M 2002 *Advances in Mechanics.* **32**(3):3492364(in Chinese).
[8] Lin J Z *et al* 2006 *Chinese Journal of Aeronautics.***19** S202-S205.
[9] Wang L *et al* 2008 *Proc. of SPIE.* **6984** 698411.
[10] Han D X *et al* 2003 *J. Appl. Phys.* **94** 2930-2936.
[11] Kulisch W *et al* 2004 *Diamond & Related Materials.***13** 1997–2002.
[12] Peggs G N and Leigh I C 1983 *Report MOM* UK National Physical Laboratory **62**.
[13] Panich N and Sun Y 2004 *Surface and Coatings Technology.***182** 342–350.
[14] Wei W S *et al* 2003 *Journal of Functional Materials and Devices.* **9** 3:285-290. (in Chinese).
[15] Yang H, Ma Q S, Ge M Z 1998 *Journal of Ceramics.* **19** 2:91-96 (in Chinese).

Journal of Physics: Conference Series **152** (2009) 012017 doi:10.1088/1742-6596/152/1/012017

Fabrication and Electric Characterization of the n-ZnO/ p-Nanocrystalline Diamond Film Heterojunction

J Huang, L J Wang[1], R Xu, K Tang, J M Lai, J Wang, Y Y Lou, W M Shi, Y B Xia

School of Materials Science and Engineering, Shanghai University, Shanghai, 200072, China

[1]E-mail: ljwang@shu.edu.cn

Abstract. A heterojunction of n-ZnO / p-nanocrystalline diamond (NCD) film was fabricated, where the undoped p-type NCD film and n-type ZnO film were grown by microwave plasma chemical vapor deposition (MPCVD) method and radio-frequency (RF) reactive magnetron sputtering method, respectively. The structure and morphology of the NCD film and ZnO film were analyzed by Raman spectroscopy, X-ray diffraction (XRD) and atomic force microscopy (AFM). The carrier sheet densities of p-type NCD film and n-type ZnO film measured by Hall Effect method were 1.3×10^{12} cm^{-2} and 8.9×10^{14} cm^{-2}, respectively. I-V characterization of the n-ZnO / p-NCD heterojunction indicated that this structure was rectifying in nature with a turn-on voltage of about 0.6V.

1. Introduction

Due to several unique properties including extreme high hardness, very high thermal conductivity, large band gap, chemical inertness [1-2], etc., diamond film is expected to be an important wide-band-gap semiconductor material for high temperature and high power electronics devices [3-5]. However, the development of diamond film electronics has been hampered by several problems, in which a big issue is that the search for n-type doped diamond fitted for the preparation of homojunction diode. Besides seeking an effective method to prepare n-type doped diamond films with relatively low resistivity, another route is to fabricate high quality heterojunction of diamond films. N-type doped cubic boron nitride (c-BN) has been employed to substitute n-type diamond to form the heterojunction [6]. But single crystal cubic boron nitride films are difficulty to prepare them.

However, ZnO, an n-type wide-band-gap semiconductor naturally generated due to oxygen deficiencies and interstitial Zn ions acting as donors in the ZnO lattice, can be a potential material for the fabrication of wide-band-gap p-n heterojunction when integrated with diamond film. For the fabrication of the heterojunction, the deposition of n-type ZnO films on the boron-doped p-type diamond substrate has been prepared successfully [7], while few studies involved in n-ZnO/ p-undoped NCD heterojunction have been reported. The NCD film was very attractive for many industrial applications due to the similar properties to the bulk diamond film [8-9]. In this work, an undoped p-type NCD film was deposited on a single crystalline silicon substrate by MPCVD method, and an n-ZnO film was then grown on this NCD film by RF sputtering method to fabricate an n-ZnO/ p-NCD heterojunction. The I-V characteristic of the heterojunction was reported.

2. Experimental detail

In this work, a microwave plasma chemical vapor deposition (MPCVD) technique at 2.45 GHz using a gaseous mixture of methane and hydrogen was applied to deposit p-type NCD films on mirror-polished (100) single crystalline silicon substrates (17×11×0.5 mm^3). The silicon substrates were mechanically scratched with 100nm diamond powder in water for about 10min before deposition, and then were cleaned by ultrasonic vibration in deionized water. The typical conditions for the deposition of p-type NCD films were as follows: the rate of hydrogen and methane was 100:3, the growth pressure was 0.5~ 1.5 Kpa, the substrate temperature was 650~ 700 ℃, the negative bias voltage was 0~ 150 V, the time of deposition was 3 h, the thickness of the NCD film was about 4 μ m, H plasma etching time after deposition was 0 ~ 1h. The thickness of the p-type NCD layer was about 30~50 nm The n-type ZnO films were deposited on these NCD films by radio-frequency (RF) reactive magnetron sputtering method with a ZnO target (purity 99.99%). The substrates were not heated. The sputtering chamber was evacuated to 3×10^{-3}Pa before sputtering. During the ZnO films deposition, the total working pressure was 0.3 Pa, the sputtering power was 200W. The time of sputtering was 45 min, the thickness of the ZnO was about 250 nm. Ohmic contacts were fabricated on the surface of both the NCD layer and the ZnO layer by sputtering Au layer (~ 100 nm thicknesses, ~1mm diameter and separated by a distance of ~5 mm) with proper shadow mask. The structure of n-ZnO/ p-NCD heterojunction was shown in figure 1.

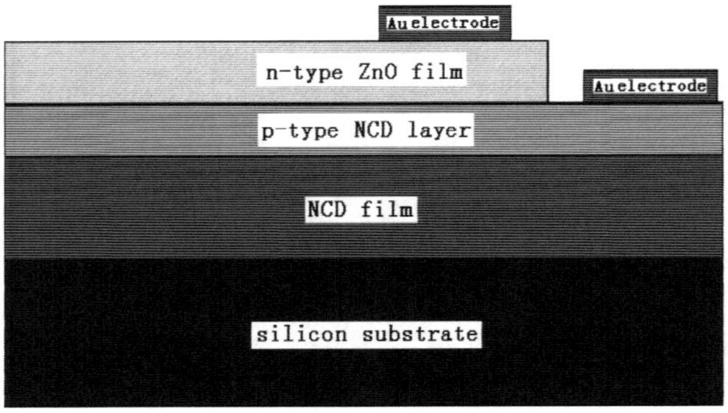

Figure 1. Structure of n-ZnO/ p-NCD heterostructure.

The morphology and microstructure of the films were analyzed by atomic force microscopy (AP-0190), X-ray diffraction (XRD D/MAX-3C) and Raman spectroscopy (Horiba Jobin Yvon HR800UV). Current-voltage (I-V) measurements of the n-ZnO/p-NCD heterojunction, the Au/NCD interface and the Au/ZnO interface were carried out using a Keithley 4200/SCS digital semiconductor characterization system. An Accent model HL5500PC Hall effect measurement system was used to characterize the film's conductivity.

3. Results and discussion

Figure 2 shows a typical AFM image for the NCD film prepared by MPCVD method. The mean surface roughness (RMS) of the film determined by AFM measurement is about 23 nm in a scan range of 2 × 2 μm^2.

The Raman spectrum of the NCD film given in figure 3 reveals several features that may be linked to NCD characters. First, the diamond peak at 1334 cm^{-1} is very weak and significantly broadened, which is well known to be caused by the decrease of the grain size to the nanometer scale [10]. Second, the bands appear at approximately 1145 cm^{-1} and 1480 cm^{-1} are assigned to trans-polyacetylene lying at grain boundaries in recent works and always been observed in chemical vapor deposition (CVD) NCD films as well as in hydrogenated amorphous carbon (a-C:H) containing nanocrystalline diamond [11]. They are often regarded as a marker of NCD phase. Third, the scattering intensity at approximately 1350 cm^{-1} and 1565 cm^{-1},corresponding to sp^2- carbon (graphite D and G bands, amorphous carbon), is obviously enhanced [12].

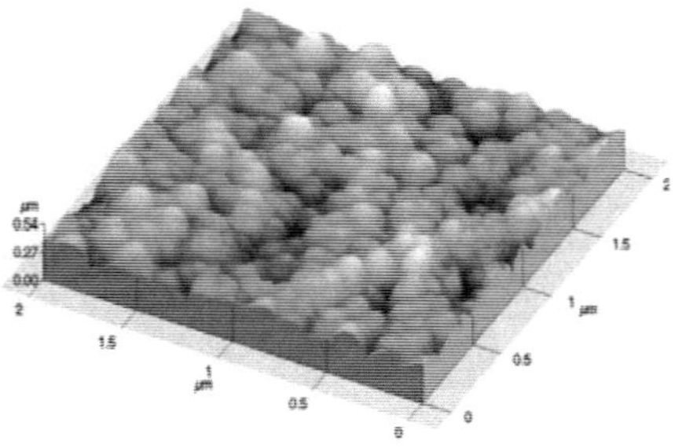

Figure 2. The AFM image of NCD film.

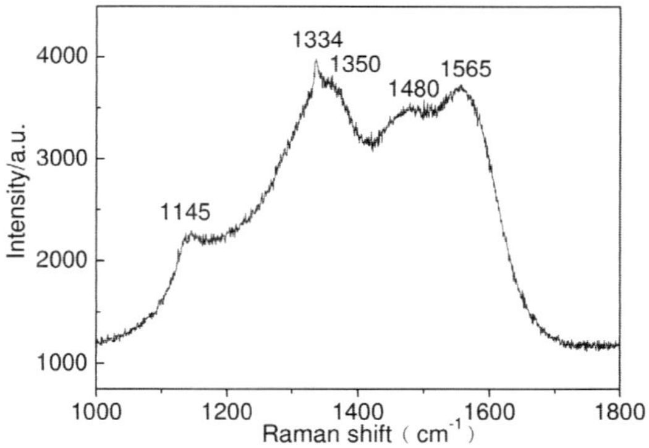

Figure 3. The Raman spectrum of NCD film.

Hall Effect measurement shows that the undoped NCD film prepared in this work is p-type conductive. The carrier sheet densities of the NCD film and the ZnO film are 1.3×10^{12} cm^{-2} and 8.9×10^{14} cm^{-2}, respectively. It is well known that undoped natural diamond with a gap of 5.5 eV is an excellent insulator. Yet in 1989 Landstrass et al. reported that as-deposited CVD diamond films have relatively high-conductivity layer (p-type semiconducting layer) near the surface [13]. This observation has been confirmed over the years [14–16]. It is proposed that the high-conductivity layer may be due to the passivation of deep levels by terminated hydrogen incorporated in the near surface [13]. This surface conductivity is unique among semiconductors and can be utilized to realize several type of diamond based devices (e.g., field effect transistors [17], heterojunction).

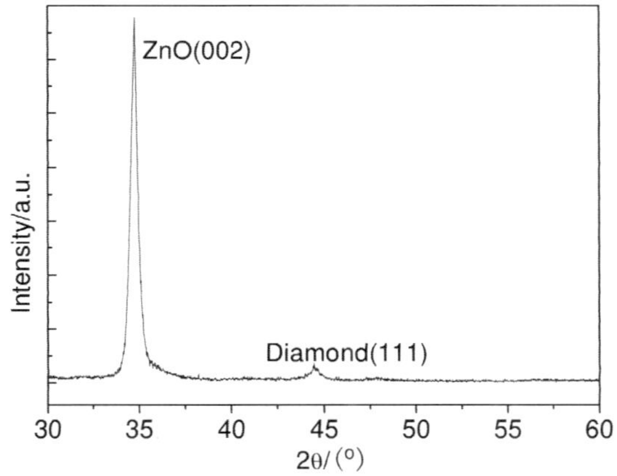

Figure 4. The XRD pattern of ZnO film deposited on NCD substrate.

XRD pattern for ZnO/NCD diamond structure is shown in figure 4, which exhibits ZnO (0 0 2) peak at 34.20° and diamond (1 1 1) peak at 44.40° . There is no other preferential orientation for the ZnO besides the (0 0 2) reflex, which indicates the ZnO layers were highly oriented with the c-axis perpendicular to the substrate surface.

In order to evaluate the ohmic nature of the Au/NCD film and Au/ZnO film interface, the interface contact is measured between two adjacent Au electrodes. The typical I-V characteristics of the contact between Au electrode and ZnO film, and the contact between Au electrode and NCD film are shown in figure 5(a) and figure 5(b), respectively. The linear current-voltage behavior indicates that the omhic contacts are formed between Au electrodes and the films.

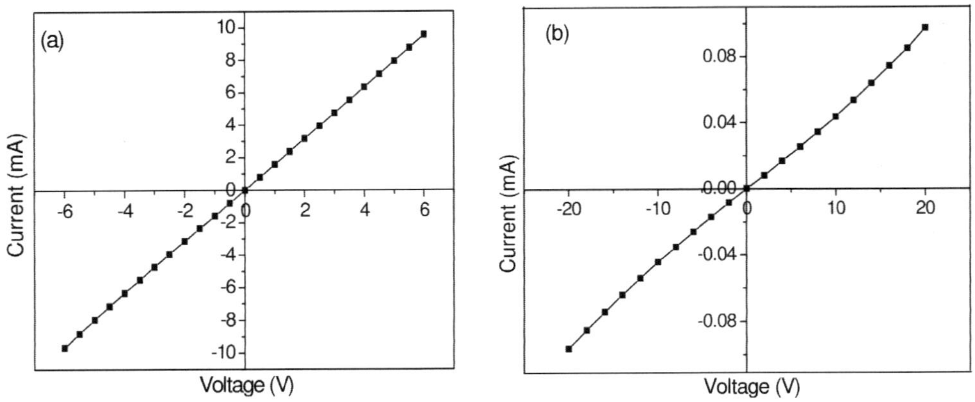

Figure 5. (a) The I–V characteristics of as-deposited Al on the ZnO film (b) The I–V characteristics of as-deposited Au on the NCD film.

The I-V characteristic of n-ZnO/p-NCD heterojunction (shown in figure 6) is measured in the dark. It is obvious that the heterojunction is rectifying in nature with a turn-on voltage of ~0.6V. A small leakage current is observed when the reverse bias is applied to -5 V, but the ratio of forward current to the reverse current is larger than 50 in the applied voltage range of -5V to 4V.

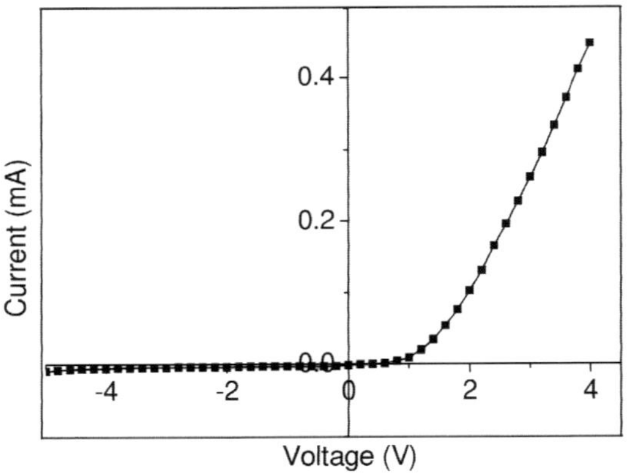

Figure 6. I-V characteristics for a ZnO/NCD heterojunction photodiode in the visible light.

4. Conclusions

The un-doped p-type NCD films were grown on (100) single-crystalline Si substrate by MPCVD method. Highly c-axis-oriented ZnO thin films were deposited on these NCD films by RF reactive magnetron sputtering and then, an n-ZnO/p-NCD heterojunction was fabricated. The carrier sheet densities of the NCD film and the ZnO film are 1.3×10^{12} cm^{-2} and 8.9×10^{14} cm^{-2}, respectively. The I-V characteristics of n-ZnO/p-NCD heterojunction were measured in the dark. A good rectification characteristic with a turn-on voltage of ~0.6V was obtained in the device.

Acknowledgements
This work was supported by the National Natural Science Foundation of China (60577040), Innovation Program of Shanghai Municipal Education Commission (08YZ04) , Program for Changjiang Scholars and Innovative Research Team in University (No:IRT0739), Shanghai Education Committee for Excellent Youth Scholars and Shanghai Leading Academic Disciplines (T0101).

References
[1] Tankala K, Debroy T and Alam M 1990 *J. Mater. Res.* **5** 2483
[2] Wong M S, Meilunas R, Ong T P and Chang R P H 1989 *Appl. Phys. Lett.* **54** 2006
[3] Lee S T, Lin Z and Jiang X 1999 *Mater. Sci. Eng.* **25** 123
[4] Vescan A, Daumiller I, Gluche P and Ebert W 1998 *Diamond Relat. Mater.* **7** 581
[5] Noda H, Hokazono A and Kawarada H 1997 *Diamond Relat. Mater.* **6** 685
[6] Wang C X, Yang G W, Zhang T C, Liu H W, Han Y H, Luo J F, Gao C X and Zou G T 2003 *Appl. Phys. Lett.* **83** 4854
[7] Wang C X, Yang G W, Zhang T C, Liu H W, Han Y H, Luo J F, Gao C X and Zou G T 2003 *Diamond Relat. Mater.* **12** 1548
[8] Kock F A M, Garguilo J M and Nemanich R J 2005 *Diamond Relat. Mater.* **14** 704
[9] Achatz P, Garrido J A and Stutzmann M 2006 *Appl. Phys. Lett.* **88** 101908-1
[10] Zhou D, Gruen D M, Qin L C, McCauley T G and Krauss A R 1998 *J. Appl. Phys.* **84** 1981
[11] Ferrari A C and Robertson J 2001 *Phys.Rev. B* **63** 121405
[12] K.E.Spear 1989 *J.Am.Ceram.Soc.* **72** 171
[13] Landstrass M I and Ravi KV 1989 *Appl. Phys. Lett.* **55** 1391
[14] Grot S A, Hatfifeld C W, Wronski C R, Badzian A R and Messier R 1990 *IEEE Electron Device Lett.* **11** 100
[15] Albin S and Watkins L 1990 *Appl. Phys. Lett.* **56** 1454
[16] Looi H J, Foord J S and Jackman R B 1998 *Appl. Phys. Lett.* **72** 353
[17] Tsugawa K, Kitatani K, Noda H, Hokazono A, Hirose K, Tajima M and Kawarada H 1999 *Diamond Relat. Mater.* **8** 927

Journal of Physics: Conference Series **152** (2009) 012018 doi:10.1088/1742-6596/152/1/012018

Growth and properties of ZnO/ZnS core/shell nanostructures

Yinshu Wang[1],Qihua Guo[1], Sanjin Lin[1], Binkun Chen[1], Dong Zheng[2]

[1]Department of Physics, Beijing Normal University, Beijing 100875
[2]Analytic and Testing Center, Beijing Normal University, Beijing 100875

[1]yswang@bnu.edu.cn

Abstract. ZnO nanorods were prepared in aqueous solution of zinc acetate dihydrate and hexamethylenetetramine with two-step methods. ZnO/ZnS core/shell structures were prepared by immerging ZnO nanorods in Na_2S and nitrate hexahydrate aqueous solution alternately for different cycles. The structures were characterized by x-ray diffraction, scanning electron microscopy, high resolution transmission electron microscopy and spectroflurometer. The results show that ZnO/ZnS nanocables formed by one cycle and then hollows appeard in the core region With increasing shell growth cycle, ZnO core dissolved completely in some region, and quasiperiodic distribution of ZnO and hollows in ZnS tubes were observed. Photoluminescence of ZnO/ZnS core/shell structures was found intense than that of pure ZnO nanorods.

1. Introduction

Surface modification has been recognized as one of the most advanced and intriguing methods to build tailored nanostructure materials effectively and reliably [1]. Core–shell composite materials have attracted a great deal of interest due to their versatile structures and properties for many significant applications in optics, biology and catalysts. ZnO is a wide gap semiconductor with bandgap of 3.3 eV at room temperature and its nanostructure arrays have been studied extensively in the lasted decades due to their unique nanoscale electronic and optoelectronic properties used in electronic and optoelectronic devices and gas sensors [2]. Surface modification of a semiconductor with semiconductors of wider band-gap has been recognized as one of the most advanced and intriguing methods to improve the luminescence properties. ZnS is a wide band gap semiconductor with band gap of 3.7eV and it is a well-known luminescence material having prominent applications in sensors and lasers [3]. Many researchers have studied ZnO/ZnS core/shell nanostructures. Square nanocables, rod nanocables of ZnO/ZnS core/shell structures synthesized by converting ZnO to ZnS have been reported [4-9]. ZnS nanotubes and ZnO/ZnS nanotubes were observed during sulfuration of ZnO nanorods. However, there are few reports on quasiperiodic structure formation during sulfuratiion of ZnO nanorods.

The In this work, ZnO nanorods were growth in aqueous solution of zinc acetate dihydrate and hexamethylenetetramine with two-step method. ZnO/ZnS core/shell structures were prepared by immerging ZnO nanorods in Na_2S and nitrate hexahydrate aqueous solution alternately. Quasiperiodic distribution of ZnO and hollows in ZnS tube was reported for the first time. Nanostructure properties were measure and the formation of the order structure was discussed.

2. Experiments

Zinc acetate dihydrate, zinc nitrate hexahydrate, hexamethylenetetramine (HMT) and sodium sulfide were selected to synthesize ZnO/ZnS core/shell nanostructures. Glass plates were used as substrates. The substrates were cleaned ultrasonically in acetone, ethanol, and deionized water in order. Growth of ZnO nanorods were performed by two-steps method. First ZnO seeds were introduced by annealing methods [10]. Then ZnO nanorods were grown at 90°C in aqueous solution of 0.005mol/L zinc acetate dihydrate and hexamethylenetetramine (HMT) for 2h. ZnO/ZnS core/shell structures were prepared by a self-assembling route [5]. Typically ZnO nanorods were immerged at 60°C in 0.16 mol/L Na_2S aqueous solution for 2h and then in 0.16 mol/L nitrate hexahydrate aqueous solution for 1h for ZnS growth. To modulate the shell thickness, growth of ZnS was prepared by repeating ZnS shell growth for 1, 2 and 3 cycles and the samples were denoted as ZnO/ZnS1, ZnO/ZnS2, ZnO/ZnS3 relatively.

The as-prepared nanostructures were analyzed using x-ray diffraction (SHIMADZU XRD-6000) with a Cu Ka radiation. The morphology and dimension of the nanocrystal were observed by scanning electron microscopy (SEM, HITACHI S-4800) and high resolution transmission electron microscopy (HRTEM, JEOL Jem-2010). Room temperature photoluminescence spectra (PL) were recorded by a spectroflurometer (Fluorolog-Tar-3) excited with 320nm light.

3. Results and discussions

Figure 1 shows the XRD pattern of the ZnO and ZnO/ZnS core/shell nanostructures. There are dominant peaks at 34.5° which belongs to the (002) diffraction of wurtzite structured ZnO. The strong (002) diffraction indicates that nanostructures grow preferential along *c*-axis orientation. For ZnO/ZnS core/shell samples, additional diffraction peaks at about 28.8° can be observed, which corresponds to diffraction from (111) planes of zinc-blend structured ZnS. This indicates that ZnS layer formed. The morphology of ZnO and ZnO/ZnS core/shell nanostructures observed by SEM is shown in Figure 2. ZnO nanorods are inclined to be vertical on the substrates. Obvious core/shell junction can be seen in the samples of ZnO/ZnS2 and ZnO/ZnS3. Core/shell junctions can not be seen clearly in ZnO/ZnS1sample, but some fingers on the top surface of the rods can be observed. TEM images of ZnO and ZnO/ZnS nanostrutures are shown Figure 3. No cavities are observed in ZnO nanorods (Figure 3(a)). Nanocables with solid cores form in the sample with the shell grown for one cycle and fingers stand on the top of the side wall can also be seen(Figure 3(b)). With an increase in growth cycle of ZnS shell, hollows form in the core region (Figure3(c)). With the further increase in growth cycle of ZnS shell, ZnO core dissolved completely in some region. ZnO and hollows distribute alternatively in ZnS tube (fig3(d)) and show a quasiperiodical distribution (fig.3(d)).

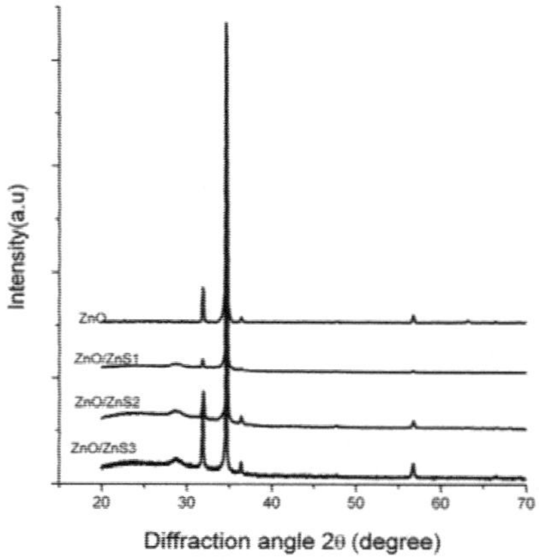

Figure 1. XRD of ZnO and ZnO/ZnS nanostruture

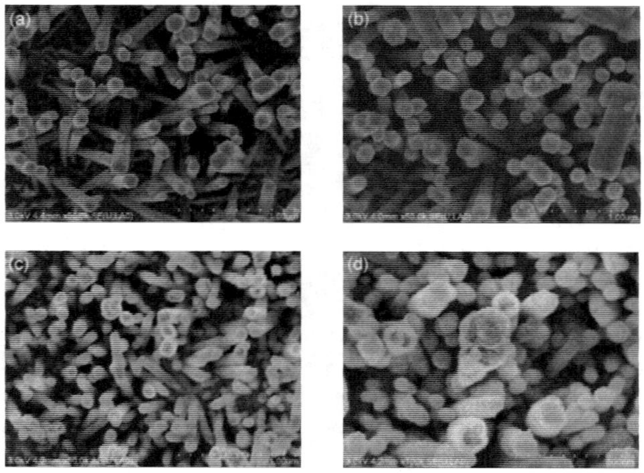

Figure 2. SEM images of (a) ZnO and (b)ZnO/ZnS1 (c) ZnO/ZnS2 (d) ZnO/ZnS3 grown on glass

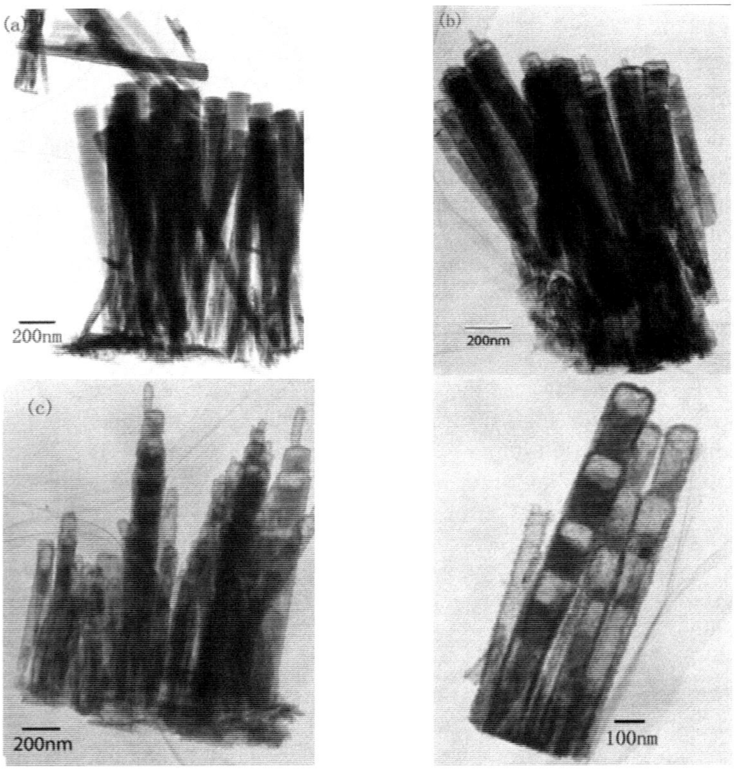

Figure 3 TEM images of (a)ZnO and (b)ZnO/ZnS1 (c) ZnO/ZnS2 (d) ZnO/ZnS3 grown on glass.

To study the structures of the fingers and the core/shell structure with periodically distributed hollows, the high-resolution TEM image and a selected area electron diffraction pattern taken from the fingers and from a cable region (denoted as S) are shown in fig.4 and fig.5 separately. The finger is tube like and composed of nanoparticles. The nanoparticles are inclined to arrange orientationally. The spacing between adjacent lattice planes in fig.4 is about 0.31 nm, which equals to the distance between two (111) planes in zinc-blend structured ZnS. The electron select area diffraction pattern of the finger is a set of rings, which is consistent very well with that of zinc-blend structured ZnS. All this indicate that the particles have zinc-blend structure.

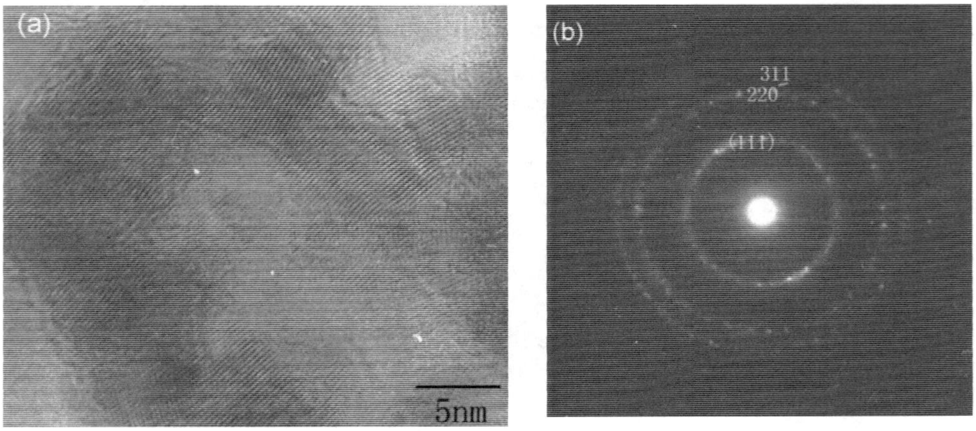

Figure 4 (a) HRTEM images and (b) SAD pattern of the finger observed in figure 3b

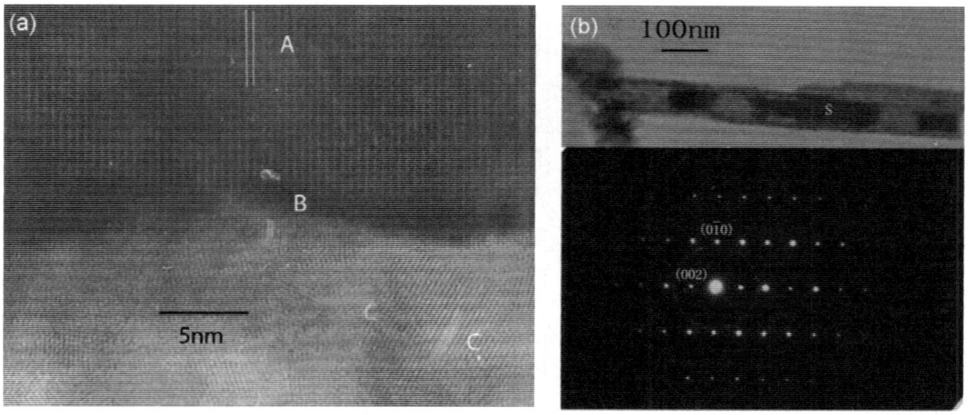

Figure 5 (a) HRTEM and (b) selected area diffraction taken from S region of the nanocable with hollows

It can be seen from fig.5(a) that the conjunction of the core and shell is compact. The nanocable has a single crystal ZnO core. The spacing between adjacent lattice planes (denoted as A) is about 0.52 nm, which corresponds to the distance between two (001) planes in wurtzite structured ZnO. At the interface of core and shell (B region), TRTEM image is different from that of ZnO core. The space is about 2.2 nm which corresponds to the distance between two (102) planes in wurtzite structured ZnS. Nanoparticles are observed at the region farer from the interface. The spacing between adjacent lattice planes (denoted as C) is about 0.31 nm which corresponds to the distance between two (111) planes in zinc-blend structured ZnS. The select area diffraction of the cable shown in fig.5(b) displays a spotted pattern and a set of ring diffraction pattern. The spots corresponds to diffraction of ZnO with $[2\bar{1}\bar{1}0]$ zone axis .The rings are diffraction of zinc-blend structured ZnS which is similar to that observed in fig.4(b). From fingure 4 and 5 we can deduce that a transitional thin layer of wurtzite ZnS forms between ZnO core and zinc-blend structured ZnS nanoparticle shell. Wang et al. [4] and Yan et al. [12] also observed zinc-blend structured ZnS formation in ZnO/ZnS nanocables. The formation of zinc-blend structured ZnS is due to the low growth temperature.

Hollows are observed in ZnO/ZnS2 and ZnO/ZnS3 samples shown in fig.3 (c) and (d). Ge et al.[13] observed the formation of tube-like ZnS crystals during conversion from ZnO crystals to ZnS. They attributed tube-like structure to the ion exchange reaction and the diffusion between zinc and sulfide ions. Wang et al. [9] studied the conversion from ZnO crystals to ZnS. They proposed that the conversion from ZnO to ZnS took place on the out surface of ZnO columns and continued across the side wall. Then the top surface was broken and ZnO dissolved in solution and reacted with S precursor inside and outside the arrayed structures. The newly formed zinc sulfide inside the surface of arrayed columns prevented the diffusion of zinc ions and the remained zinc oxide crystals were sulfurated in situ into zinc sulfide. Tube-like fingers stand on the surface wall are observed in fig.3(b) and fig.3(c). This suggests that formation mode of ZnS in this work is different from that proposed by Wang et al. Formation of the fingers indicates that ZnS grows preferentially along the side wall direction. Zn ions come from ZnO core during sulfuration. ZnS forms at the surface and wall initially. With an increase in growth time or growth cycle, the formed ZnS layer on the side wall would prevent Zn ions from diffusing out the wall. Then Zn ions diffuse to the top surface along the interface of the core and shell and supply Zn ions for finger growth. Then fingers joint together and supply as additional wall. Hollows would form due to the out diffusion of Zn ions from the rods. Self-organization of the quasiperiodic distribution hollows may be due to the compensation of Zn ions diffusion from different regions along the interface.

The PL spectra of ZnO nanorods and ZnO/ZnS nanostructures are shown fig.6. Both ZnO and ZnO/ZnS nanostructures have strong emission around 400 nm. ZnO bandgap is 3.3 eV and ZnS badgap is 3.7 eV. The emission centered at 400 nm is corresponding to ZnO band edge emission. The stronger emission of ZnO/ZnS nanostructures indicates that growth of ZnS on ZnO nanorods will enhance ZnO nanostructure band edge emission.

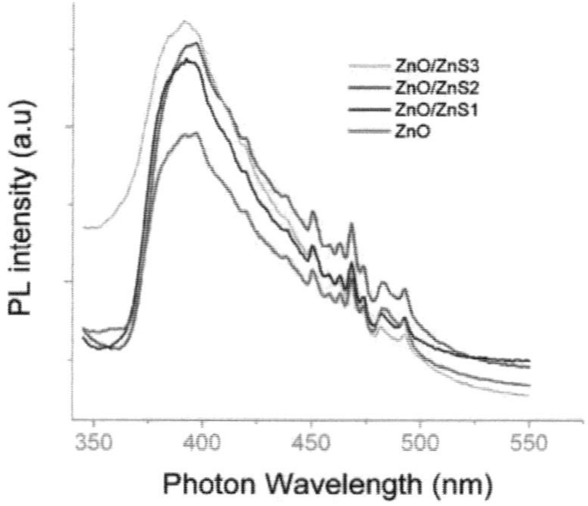

Figure 6 PL spectra of ZnO and Zno/ZnS nanostructures

4. Conclusions

ZnO/ZnS core/shell nanostructures have been synthesized from ZnO nanorods by a chemical route. ZnO core is wurtzite structured single crystal and ZnS shell consists of nanoparticles with zinc-blend structure. Thin layer of wurtzite structured ZnS forms at the interface. Nanocables form when ZnS shell is grown for one cycle. With an increase in shell growth cycle, hollows form in the core region. As the growth cycle of ZnS shell increases, ZnO core dissolved completely in some region, and ZnO and hollow alternately distribute in ZnS nanotubes. Photoluminescence of ZnO/ZnS core/shell is intense than that of pure ZnO nanorods

Acknowledgments

Authors wishing to acknowledge financial support from NSFC (Project:10574011) and from Beijing Normal University, Beijing, China

References

[1] Gao T, Li Q H and Wang T. H. 2005 *Chem. Mater.* **17** 887
[2] Wang Z L, 2007 *Materialtoday* 10, 20
[3] Zhu Y, Bando Y, Xue D and Golberg D 2003 *J. Am. Chem. Soc.* **125** , 16196.
[4] Wang D Y, Gao P X, Li J, Summers C J and Wang Z L 2002 *Adv.Mater.* **14** 1732.
[5] Li J H, Zhao D X, Meng X Q , Zhang Z Z, Zhang J Y, Shen D Z, Lu YM and Fan X W 2006 *J. Phys. Chem. B* **110** 14685
[6] Shao H F, Qian X F and Huang B C 2007 *Mater. Lett.* **61** 3639
[7] Zhang H, Yang D R, Ma X Y and Que D L 2005 *Nanotechnology* **16** 2721.
[8] Sun C W, Jeong J S and Lee J Y 2006 *J Crys. Growth* **294** 162
[9] Wang Z, Qian X F, Li Y, Yin J and Zhu Z K 2005 *J Solid State Chem.* **178** ,1589
[10] Greene L E, Law M, Tan D H, Montano M, Goldberger J, Somorjai G and Yang P D 2005 *Nano Lett.* **5** 1231
[11] Li Q C, Kumar V, Li Y, Zhang H, Marks T J, and Chang R. P H 2005 *Chem. Mater.* **17** 1001
[12] Yan C L and Xue D F 2006 *J. Phys. Chem. B* **110** 25850
[13] Ge J P and Li Y D 2004 *Adv. Funct. Mater.* **14** 157.
[14] Strite S and Morkoc H 1992 *J. Vac. Sci. Technol.* B **10** 1237

Effect of citric acid to carbon nanotube growth

Q. Jiang, Z. W. He, B. Du, X. F. Zhao, Y. Zhao

Key Laboratory of Advanced Technologies of Materials (Ministry of Education of China) and Superconductivity R&D Center, Southwest Jiaotong University, Chengdu, China, 610031

E-mail address: jiangqi66@hotmail.com, jiangqi66@163.com (Q. Jiang).

Abstract. Carbon nanotube (CNT) was synthesized by catalytic chemical vapor deposition using $LaNiO_3$ as the catalyst precursor, where $LaNiO_3$ was prepared by citric acid complexometry. The results show that the catalyst precursors prepared with different citric acid dosages could all grow CNT, but the CNT yield and patterns obtained with these catalyst precursors were different. Different citric acid dosages could result in different (111) crystal face contents of Ni particle in the catalysts (the products obtained from catalyst precursors reduced by the H_2) and thus affected the CNT yield and patterns, which is the result of the complexation between the carboxyl (came from citric acid) and metal ions in lower temperature (80℃) during the catalyst precursor preparation.

1. Introduction

Carbon nanotube (CNT) has come to be one of the most actively studied materials for their novel nanometer-hollow-carbon-tube structure and extraordinary physical properties since it was rediscovered by Iijima in 1991 [1, 2]. Based on its excellent properties, CNT could use in many field. Many methods were found to prepare CNT, amongst which the catalytic chemical vapor deposition (CCVD) appeared as the most promising technique for scaling up the CNT preparation at a relatively low cost [3]. An efficient catalyst with perovskite structure made by citric acid complexometry (for example, $LaNiO_3$) was developed to prepare CNT with this method [4]. The adding of the citric acid during the catalyst preparation was order to obtain the effective size catalyst particles (for example, Ni metal particles) to grow CNT by the complex reactions between the citric acid and catalyst metal ions [5]. And the citric acid did not exist in the obtained catalyst precursors (for example, $LaNiO_3$). But in our experiments, the experimental data showed that the change of citric acid dosage could greatly affect the CNT yield and patterns. The research results showed that the citric acid dosage affected the CNT yield and patterns via affecting the crystal face contents of the active element (Ni) in the catalysts.

2. Experimental

$LaNiO_3$ catalyst precursors were synthesized by a citric acid complexometry. AR grade Ni_2O_3 and La_2O_3 were dissolved in nitric acid at the 1:1 of Ni and La atomic ratio, and then respectively added in AR grade citric acid, the complexing agent, as different atomic ratios of the citric acid and metal ions (total of Ni and La) (0∶1, 1∶1, 2∶1, 3∶1, 4∶1, 5∶1, 6∶1, 7∶1 and 8∶1). Then, the obtained powders were named as catalyst precursor Ⅰ, Ⅱ, Ⅲ, Ⅳ, Ⅴ, Ⅵ, Ⅶ, Ⅷ and Ⅸ and used as the catalyst precursors to grow CNT [4]. The CNT was synthesized in a horizontal-chamber reactor with an electron auto control temperature system as the CCVD [4]. Transmission Electron Microscopy (TEM), X-ray Diffraction (XRD) and Infrared spectrum (IR) were used to characterize the obtained CNT and the produced catalysts.

3. Results and discussion

Table 1 is the relationship between the different $LaNiO_3$ catalyst precursors obtained with different citric acid dosages and the obtained CNT yield. From the Table, we can see that the CNT can also be prepared with the catalyst precursor Ⅰ obtained without citric acid and its CNT yield is about 5.15 (g CNT / g Catalyst precursor. h). With the adding of the citric acid, the CNT yield obtained using the catalyst precursor greatly increases to 8.47 (g CNT / g Catalyst precursor. h), indicating the adding of citric acid can obtain more effective catalysts by the complex reaction between the citric acid and metal ions. With increasing the citric acid dosage, the CNT yield further increases. And the CNT yield increases up to 13.46 (g CNT / g Catalyst precursor. h) at the 2：1 atomic ratio of the citric acid and the metal ions (Ni and La). And then, the CNT yield begins to decrease with the increase of citric acid dosage. The CNT yield is only 4.07 (g CNT / g Catalyst precursor. h) at the 3：1 atomic ratio of the citric acid and the metal ions (Ni and La). After that, the CNT yield begins to increase and drives to a relative stabilized value (about 7 g CNT / g Catalyst precursor. h) with further increasing the citric acid dosage. From the above experimental data, we see that citric acid dosage can greatly affect the CNT yield.

Table 1. Correlative data with the catalyst precursors obtained with different citric acid dosage and metal ions (Ni and La).

Citric acid：Ni and La (Atomic ratio)	Catalyst precursor	Yield of CNT (g_{CNT}/g_{LaNiO3}. h)	Ni particle Crystal face	Ni particle size / nm
0：1	Ⅰ	5.15		
1：1	Ⅱ	8.47		
2：1	Ⅲ	13.46	(111) (200)	27.5 13.7
3：1	Ⅳ	4.07	(111) (200)	22.5 68.6
4：1	Ⅴ	6.70		
5：1	Ⅵ	8.21		
6：1	Ⅶ	7.60		
7：1	Ⅷ	6.48		
8：1	Ⅸ	6.81		

Fig. 1 is the XRD patterns of the catalyst precursors Ⅲ, Ⅳ (a) and their reduced products Ⅲ' (b), Ⅳ' (c) by H_2 at high temperature. From Fig. 1a, we can see that catalyst precursors Ⅲ and Ⅳ have very similar XRD patterns. Analyzing the catalyst precursors crystal structure by the XRD PDF database with the XRD scanning program, we can find that catalyst precursors Ⅲ and Ⅳ both are $LaNiO_3$ (perovskite-type crystal structure, hexagonal system). Observing the curves Ⅲ and Ⅳ in Fig. 1a in detail, we can find that there are some fine differences between the curve Ⅲ and Ⅳ though the two curves are very similar in the trend, that is the intensity of the peaks of $LaNiO_3$ is different. In order to study the difference, we compare the intensity of the main peak ((110) crystal face peak), the secondary main peak ((202) crystal face peak) and the tertiary main peak ((122) crystal face peak) in curves III and IV. From these intensity ratio values, we can find that the value of the IIV(202) / IIV (110) is greater than that of the IIII(202) / IIII(110) and the value of the IIV(122) / IIV(110) is greater than that of IIII(122) / IIII(110). These data show that though the catalyst precursor III and catalyst precursor IV both are $LaNiO_3$, they have different crystal face contents. These differences may result in the difference of their reduction products (catalysts). Fig.1b and Fig. 1c are the XRD patterns of the reduction products from catalyst precursor' s III (catalyst III') and IV (catalyst IV'). Researching by the XRD analyzing software, we can find the catalysts III' and IV' both are the mixture of La_2O_3 and Ni, where the metal Ni peaks are marked and the others are the La_2O_3 peaks. But if we observe Fig. 1b and c in detail, we can find that the La_2O_3 peaks in Fig. 1b are different from that in Fig. 1c. La_2O_3 peaks in Fig. 1b are the mixture of orthorhombic system and tetragonal system crystal structure La_2O_3 peaks. But in Fig. 1c, there is only one crystal structure that is orthorhombic system. Of course, the multiplicity of La_2O_3 in the obtained catalyst III' can also have the contribution on dispersing the active element (Ni) and avoiding the agglomeration of the Ni particle, which can also improve the effectivity of the catalyst growing CNT. Moreover, further researching the Ni peaks in Fig. 1b and Fig. 1c, we can find that the intensity ratios of the two Ni peaks are different. The value of IIII' (111) / IIII' (200) in Fig. 1b is greater than that of IIV' (111) / IIV' (200), indicating that the two crystal face contents of Ni element in the two catalysts are different. The (111) crystal face content of Ni element in catalyst III' is higher than that in catalyst IV' . The content of Ni (111) crystal face has great effect on the CNT growth [6]. The (111) crystal face sizes of Ni element in catalyst III' and IV' are calculated by the SCHERRER equation and also listed in Table 1.

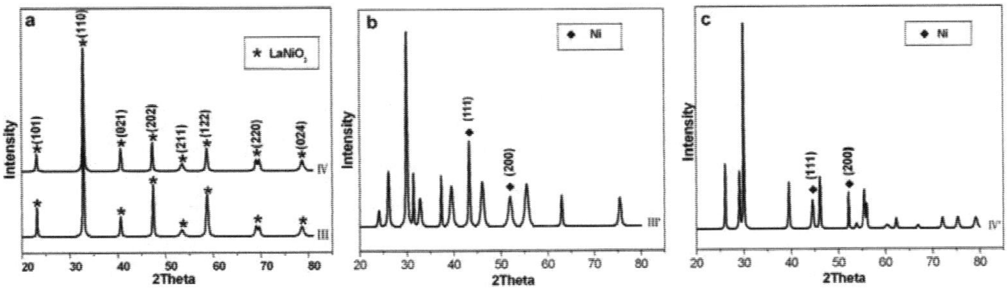

Fig. 1 XRD patterns of the catalyst precursors III, IV (a) and their reduced products III' (b), IV' (c) by H_2 at high temperature.

Fig. 2 is the TEM images of the obtained CNT with the catalyst precursor' s I, II, III, IV, V and VI (named as Fig. 2a, b, c, d, e and f). Considering the CNT yields from the catalyst precursors VII, VIII and IX are very similar to that from catalyst precursors V and VI, so we only study the CNT TEM images obtained from the frontal five catalyst precursors. From Fig. 2, we can see that the obtained CNT all have well nanometer-hollow-tube structure, and thus prove $LaNiO_3$ is a type of excellent catalyst precursor to grow CNT. Researching Fig. 2a to f in detail, we can find that the CNT in Fig. 2a have a wider tube-diameter-distribution than that in Fig. 2b, about from 24 to 70 nm; the CNT in Fig. 2b have a smaller tube-diameter-distribution than that in Fig. 2a, about from 26 to 50 nm, indicating that the adding of citric acid can contribute to obtain uniform CNT by affecting the Ni particle. The CNT in Fig. 2c have the smallest tube-diameter-distribution in all the obtained CNT, about from 30 to 40 nm; the CNT in Fig. 2d have a wider tube-diameter-distribution than that of Fig. 2c, about from 28 to 50 nm; the CNT in Fig. 2e and f both have a wider tube-diameter-distribution than that in Fig. 2d, about from 12 to 50 nm and from 26 to 50 nm, respectively. But the tube walls of the obtained CNT change to thinner and smoother with increasing the citric acid dosage. So, we can see that the modification of citric acid dosage during the catalyst precursor preparation can affect not only the CNT yields but also the CNT patterns.

Comparing Fig. 2c and d, we can find that the inner diameters of the CNT in Fig. 2c range from 25 to 30 nm and that in Fig. 2d from 20 to 30 nm, which is just right the (111) crystal face sizes of the Ni particles listed in Table 1. Based on the research results that the Ni (111) crystal face is the key factor to the CNT growth, we can deduce that the modification of citric acid dosage affect the CNT yield and patterns by affecting the Ni (111) crystal face content in the catalyst particles by the complexation reactions between the citric acid and metal ions (Ni and La).

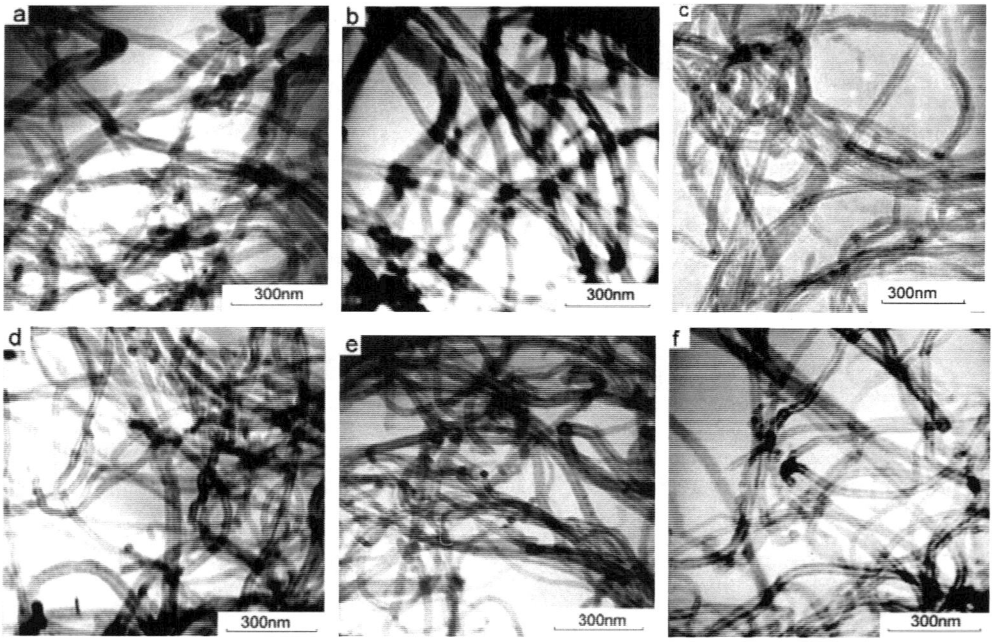

Fig. 2 TEM images of the obtained CNT with the catalyst precursors Ⅰ (a), Ⅱ (b), Ⅲ (c), Ⅳ(d), Ⅴ (e) and Ⅵ (f).

Fig. 3 is the IR graphs of the complex intermediate Ⅲ (a, dried in vacuum at 100 ℃ for 48 hours), Ⅳ (b, dried in vacuum at 100 ℃ for 48 hours) and complex intermediate Ⅲ' (c, dried in vacuum at 200 ℃ for 48 hours). There are both clear adsorption peak and adsorption bands at 1380 cm-1 and from 1410 to 1630 cm-1 in Fig. 3a and b, which are the characteristic adsorption peaks of the complex samples between the carboxyl and metal ions. And there are both broad adsorption peaks above 3000 cm-1 in Fig. 3a and b, which are the characteristic adsorption peaks of carboxyl. Comparing Fig. 3a to b, we can find that there is nice difference between Fig. 3a and b, that is the different ratio of peaks intensity between 1620 and 1380 cm-1 (i.e. I1620 cm-1 / I 1380 cm-1) in Fig. 3a and b. I1620 cm-1 / I 1380 cm-1 in Fig. 3a is greatly smaller than that in Fig. 3b, indicating that complex intermediate Ⅲ in Fig. 3a is different from complex intermediate Ⅳ in Fig. 3b, maybe by different complexation mode or maybe by different complexation degree. But we can sure that the two complex intermediates are different only by modifying the dosage of citric acid. However, there is no peak in Fig. 3c, indicating that there is no organic radical groups in the complex intermediate Ⅲ' obtained with being dried in vacuum at 200 ℃ for 48 hours, which also proves that the citric acid does not exist in the obtained catalyst precursors.

The adding of the citric acid during the catalyst preparation is order to obtain the effective size catalyst particles (Ni metal particles in our experiments) to grow CNT by the complex reactions between the citric acid and catalyst metal ions [6]. After that, the citric acid was decomposed and removed as gases during the high-heat-treatment in the latter. The citric acid does not exist in the obtained catalyst precursors and the catalysts (products reduced by H_2). The modification of the citric acid dosage during the catalyst precursor preparation can modify the crystal face content of the obtained catalyst precursors, and then modify the crystal face content of the obtained catalysts (Ni), and thus affect the CNT growth. Considering the research results in Table 1, we have the result that the modification of the citric acid dosage affect the CNT growth by affecting the (111) crystal face content of the obtained catalysts (Ni).

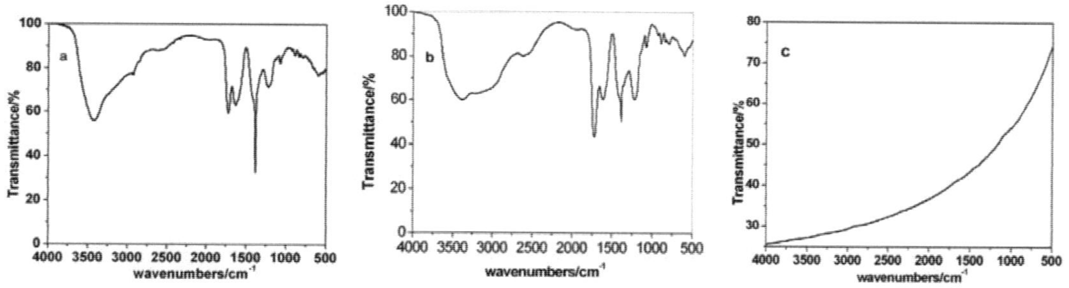

Fig. 3 IR graphs of the complex intermediate III (a), IV (b) and complex intermediate III' (c).

4. Conclusions

All the catalyst precursors obtained with different citric acid dosages could prepare CNT, but the catalyst precursor with a 2 : 1 atomic ratio of citric acid and metal ion (La and Ni) had the highest yield, about 13.46 (g CNT / g Catalyst precursor. h). Different citric acid dosage during the catalyst precursor preparation result in different Ni (111) crystal face content in the obtained catalyst (Ni), and thus resulted in different CNT yield and patterns.

Acknowledgements

The authors acknowledge the supports of Science and Technology Research Funds of Sichuan Province (05GG009-003, 2006Z02-006-1), Application Foundation Research Funds of Sichuan Province (2008JY0061) and Fundamental Science Funds of Southwest Jiaotong University (2007B20).

References

[1] Iijima S 1991 Nature. 354 56
[2] Monthioux M and Kuznetsov V L 2006 Carbon. 44 1621
[3] Mukhopadhyay K, Koshio A, Sugai T, Tanaka N, Shinohara H, Konya Z and Nagy J B 1999 Chem. Phys. Lett. 303 117
[4] Jiang Q, Song L J, Zhao Y, Lu X Y, Zhu T Y, Qian L, Ren X M and Cai Y D 2007 Mater. Lett. 61 2749
[5] Liang Q, Gao L Z, Li Q, Tang S H, Liu B C and Yu Z L 2001 Carbon. 39 897
[6] Helveg S, López-Cartes C, Sehested J, Hansen P L, Clausen B S, Rostrup-Nielsen J R, Abild-Pedersen F and Nørskov J K 2004 Nature. 427 426

Memory Properties of Ge quantum dots and rings MOS structure prepared by Pulsed Laser Deposition

Xiying Ma

Institute of Photoelectrical Materials, Shaoxing University, Shaoxing 312000, Zhejiang Province, P. R. China.

E-mail:maxy@zscas.edu.cn

Abstract. New memory structures using threshold shifting from charge stored in quantum dots (QDs) and quantum rings (QRs) of Germanium (Ge) are described. The QDs and QRs of Ge were prepared on a p-Si (100) matrix by means of pulsed laser deposition (PLD) using the droplet technique combined with rapid annealing. The perfect planar nanorings with well-defined sharp inner and outer edges were formed via an elastic self-transformation of droplet process, which is probably driven by the lateral strain of the Ge/Si layers under annealing treatment. A significant shift of threshold-voltage of 1.2V and 2.5 V were observed in Ge QDs and QRs memory devices, respectively, when an operating voltage of 8 V was implemented on the device. It is attributed to the effect of Coulomb blockade and better interface properties of the Ge nanostructures.

1. Introduction

These years, nanocrystals based memory devices have attracted much research interest because of their high speed and small scale for the next generation as compared with the conventional floating gate (FG) structures [1-3]. In particular, Ge nanoscrystals including quantum dots memory devices possess high speed for the high-mobility channel material and a charge-retention time [4,5]. Additionally, owing to the low carrier effective mass, the quasi-ballistic transport in Ge nanocrystal, the MOSFETs of Ge may lead to a better device performance than that of Si MOSFETs. In this paper, two kinds of memory nanostructures, Ge quantum dots and rings, are proposed, in which the core layer is composed either by Ge quantum dots (QDs) or by planar Ge quantum rings (QRs) that were grown by the pulsed laser deposition (PLD) technique based on the elastic strain of Ge droplets. Typically, the synthesis of Ge quantum dots and rings nanostructures relies on the self-assembly and self-organization processes of molecular beam epitaxy (MBE) and chemical vapor deposition (CVD) using the large strain property of Ge/Si heterolayers [6-8]. In the present experiments, the QDs were obtained by droplet technology of PLD, while the planar QRs were obtained by the self-transformation process of Ge quantum dots into nanorings under annealing at 600°C in the presence of Ar gas. The as-grown Ge QDs and QRs were observed through field-emission scanning-electron microscopy (FESEM). The memory devices have a significant threshold-voltage shift at a low operating voltage of 8 V due to charge trapping in the Ge microstructures.

2. Experimental details

Ge QDs and QRs thin films were deposited on p-Si (100) substrates by pulsed laser deposition at room temperature. An irradiation with a wavelength of 248 nm, an energy density of 1.2 J/cm^2 and a pulse duration of 25 ns from a pulsed KrF excimer laser (Lambda Physik LPX 305) was used to ablate the Ge single crystal (100) target (99.99%) placed on a rotating holder inside a vacuum chamber. The laser beam was focused with three pieces of lens, introduced into the deposition chamber through a ultra-violet window, and finally ripped into the target at an angle of 45 °. Two cleaned Si substrates were placed on a holder facing the target in the vacuum chamber, one was served as a reference sample. A tunneled oxide layer about 5 nm thick was formed by thermal oxidation of the substrates at 800°C for 30 min in the presence of oxygen gas after the experimental chamber being pumped down to 10^{-6} Torr. Two Ge QDs films were deposited for 30 min in Ar gas that was used as both a working and buffer gas at 100 Pa background pressure. The as-grown Ge reference uncapped sample was removed for characterization by FESEM. Then a 10 nm thick silicon oxide was further grown on the QDs layer of the other sample as a control oxide layer by means of ablating a Si single-crystal (100) target in the presence of an oxygen atmosphere at 800°C. Two Ge QRs samples are prepared almost the same procedure as the QDs samples except the as grown QRs samples are an in-situ annealing at 600°C for 5 min, then the reference QRs sample was took out for EFSEM observation, and the other QRs sample for continually thermally grown a SiO_2 caper layer. After the Al electrodes and back contact were patterned and sintered on both the QDs and QRs nanostructure, current-voltage (*I-V*) and capacitance-voltage (*C-V*) measurements were performed using a semiconductor parameter analyzer model HP4156 at room temperature to investigate the leakage behavior of the Ge nanosacle memory devices.

3. Results and discussion

Figure 1 shows the FESEM images of Ge quantum dots prepared by PLD droplet technique. As can be seen, some quantum dots are evenly scattered on the Si substrate, the average size of them is about 900 nm, and the height is about 80 nm. The QDs take almost an identical hemisphere shape except two of them are partly overlapped. An enlarged image is shown in Fig.1b, it shows a spherical dot. The top surface takes a shape of a structure as a spire that tapers to a point at the top. From this, we can deduce the process of Ge nanoring formation: first, under the powerful pulsed laser the solid Ge was melted and transformed into microdrolets in the atmosphere of cool Ar. Then Ge QDs were formed when the droplets fall on the Si substrate, and finally transformated into nanorings in situ of 600 °C annealing in the presence of Ar.

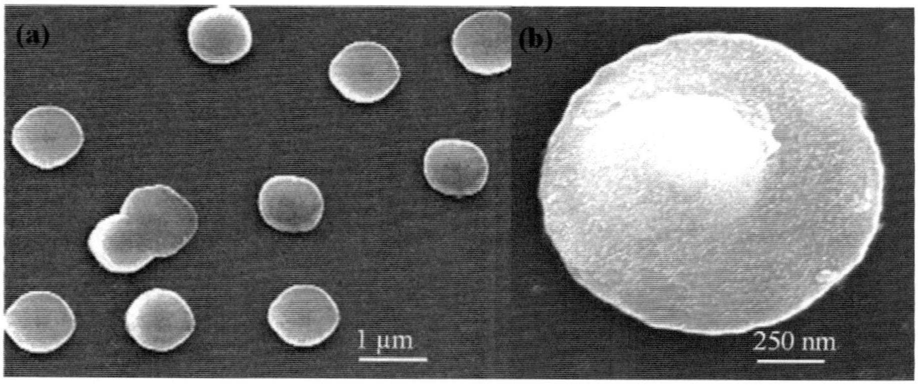

Figure 1. (a) SEM image of Ge quantum dots with diameters of 0.90 μm. (b) The high magnification shows the perfect dot with spherical symmetry.

The Ge QRs samples are prepared as the same condition as the Ge QDs in addition to following annealing at 600C for 5 mins, the obtained Ge nanoring images are presented in Fig.2a. The pattern of

Ge QRs is multiform, there are separated single nanorings, poly nanorings that some nanorings are linked together, and nanocoil rings that a few rings jointed. The Ge nanorings with nearly homogeneous radius of curvature, ring shell and thickness. Clearly, the rings are characteristic of a two-dimensional structure, the thickness is much smaller than the diameter, and the width of the ring shell does not change. They have a typical a shell width of 155-158 nm, and a thickness of 25 nm measured by means of a DEKTAK profilometer, and the density is determined to be 2.3×10^7 cm^2. The image recorded at high magnification (Fig. 2b) shows the perfect circular shape of a complete ring, with uniform shell and flat surface. The separated single nanoring is formed via a self-deformation process of a droplet during the growth under thermal annealing treatment, accordingly, no dislocations are introduced into the QRs.

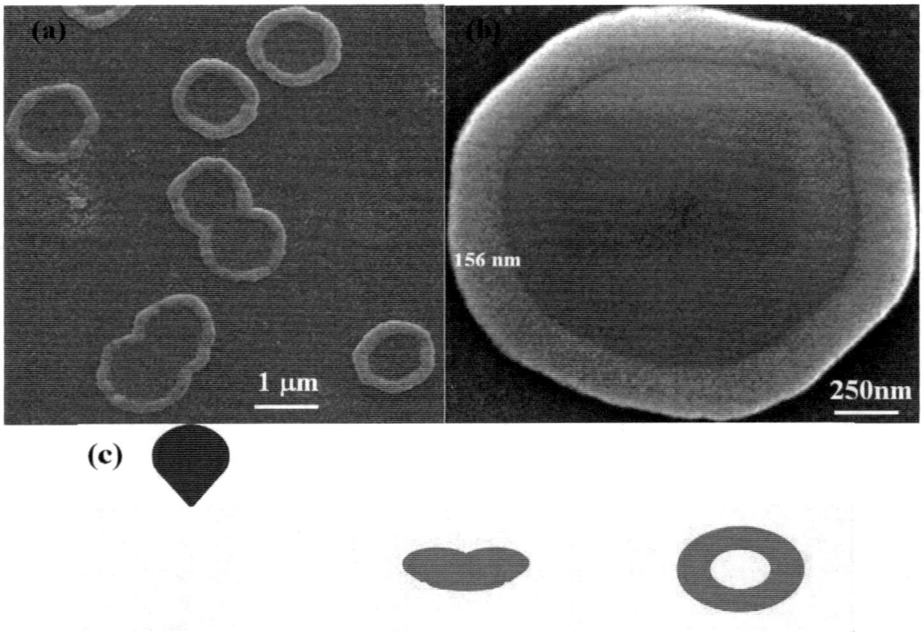

Figure 2. SEM image of Fig.2 (a) shows the Ge QDs nanorings with the diameters of 1.2-1.4 μm, and the ring shell of 155-158 nm. (b) The high magnification shows the perfect ring with uniform shell and flat surface. (c) The schematic of illustrates the nucleation and growth of a nanoring.

J. Sormunen et al [9] self-Assembled InAs/InP quantum dots into quantum rings by annealing at 560°C. They found that the effect of temperature makes the InAs/InP dot-to-ring transformation. The formation mechanism of QRs is related to As/P exchange and the elastic strain relaxation of the QD/QR system. This diffusion, together with As/P exchange, may be the driving force for the mass redistribution from QD to QR, thus result in a modification of

the QD morphology, even holes. Cui, J et.al [10] thought that the formation of SiGe/Si QRs is related to the QD strain distribution. By means of the surface segregation and diffusion, Ge atoms are released from the QD. They mix and alloy with Si on the QD side surface where the lateral lattice constant matches SiGe with varying composition. These reports verify the significance of strain-controlled surface diffusion effects. Based on these results, we developed the formation model of our Ge rings, showing in Fig.2 c. In the first step, a droplet of Ge produced by the laser moves on the substrate with a powerfully kinetic energy. In the second step, a small dip initially appears upon the droplet when it touches on the substrate because the upside of the droplet will continually go forward under the inertia force. Additionally, a small part of the residual impact force under Ar gas buffering transfers into a lateral force, which drives the material of the droplet extending that has been demonstrated by the document of [11-12]. As a result, a cone-shaped structure is formed on the substrate. In the third step, that is, in the thermal annealing process, with the jointed forces of the large strain of Ge/Si layers and the lateral thermal diffusion force, a nanoring of Ge with a plane structure is developed on the substrate. Similarly, the poly-rings are formed by the same transformation process of Ge droplets by partly overlapped falling on the substrate. Finally, Ge planar nanorings are prepared by the droplets technique of PLD without fracture in the depositions.

The capacitance-voltage (C-V) characteristic of the Ge QDs and QRs uncapped samples measured under the bias -0.5 to 0.3 V is shown in Fig. 3. There is a platform in the QDs C-V curve in the bias region of -0.2 to 0.1 V, but two platforms appear in the QRs curve, which indicates that only a boundary of space charge region is located in the dots and two boundaries in the rings because of having an inner and an outer faces. The Coulomb blocking effects occur due to large holes confined in the QDs and QRs blocking the further extension of space charge region. As the reverse bias voltage increases above 0.15 V, the space charge region extends over the dot layer so the holes in the dot are completely pumped out.

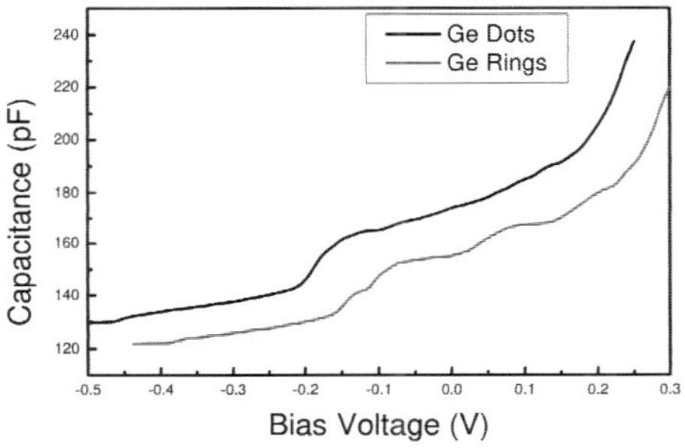

Figure 3. Capacitance-voltage traces of self-organized dots and rings. There is only a plat in QDs curve while two flats in nanorings.

Figure 4 shows the hysteresis curves of the *C-V* measurements for the capered Ge QDs and QRs memory devices in a bi-directional voltage sweep. Figure 4(a) gives a schematic diagram of the device structure used in this work, in which the core layer is composed either of QDs or QRs. The Ge nanostructure sandwiched between the oxide layers is utilized as a charge-storage element for a memory device. The voltage is swept between 5 and (-8) V or (-8) and 5 V counterclockwise, and the erasing voltage is fixed at (-8) V. It is found that it causes a threshold-voltage shift of 1.2 V in the QDs devices, a significant shift up to 2.5 V occurs in QRs sample at the low operating voltage of 8 V, which is large enough to be defined as 1 or 0 for circuit design. The large threshold-voltage shift may arise from the oxide traps, the Coulomb blockade effect, and most important, from the Ge/Si composite tunneling potential. The blockade effect of the discrete distribution of Ge nanorings leads to lower capacitance coupling of the nanoring memory device than that of a conventional FG memory device.

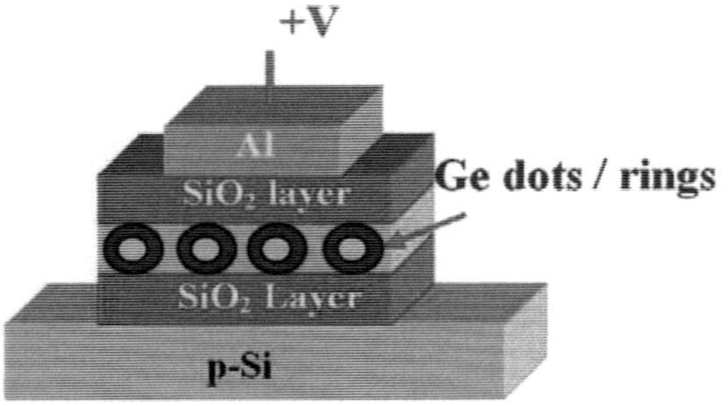

(a)

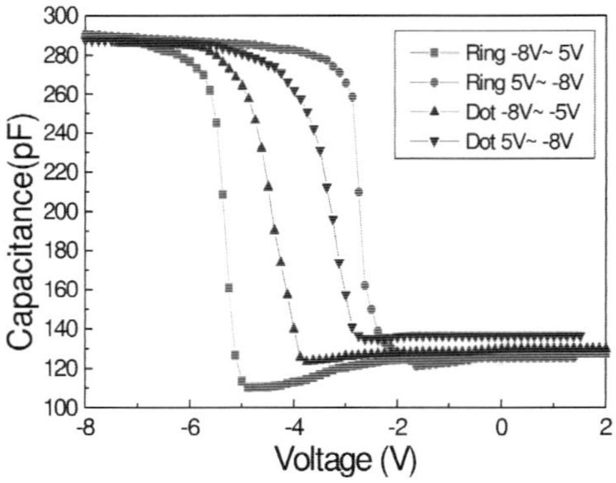

(b)

Figure 4. The hysteresis curves of the *C-V* measurements for the capered Ge QDs and QRs memory devices in a bi-directional voltage sweep. It causes a threshold-voltage shift of 1.2 V in the QDs devices, a significant shift up to 2.5 V aoouring in QRs sample at the low operating voltage of 8 V.

4. Conclusions

Ge QDs and QRs based memory devices were fabricated by pulsed laser deposition. The Ge nanorings were formed by the transformation of Ge droplets into nanorings caused by the large strain and stress of Ge/Si layers. A significant shift of threshold-voltage of 1.2V and 2.5 V was observed in QDs and QRs memory devices, respectively, when an operating voltage of 8 V was implemented on the device, which attributed to the effect of Coulomb blockade and better interface properties of the Ge nanostructure.

Acknowledgement

This research work was supported in part by the National Natural Science Foundation (No. 60776004) and the Natural Science Foundation of Zhejiang Province (No. Y405043).

Reference

[1] Guo L G, Leobandung E and Chou S Y 1997 *Science* **275** 649
[2] Shi Y, Saito K, Ishikuro H and Hiramoto T 1998 *J.Appl.Phys.* **84** 2358
[3] Welser J J, Tiwari S, Rishton S, Lee K Y, and Lee Y 1997 *IEEE Electron Device Lett.* **18** 278
[4] Chang T C, Yan S T, Liu P T, Chen C W, Lin S H and Sze S M 2004 *Electrochem. Solid State Lett.* **7** G17
[5] Chang T C, Yan S T, Hsu C H, Tang M T, Lee J, Tai Y H, Liu P T and Sze S M 2004 *Appl. Phys. Lett.* **84** 2581
[6] Zhang L, Deckardt E, Weber A, Schönenberger C and Grützmacher D 2005 *Nanotechnology* **16** 655
[7] Prinz V Y, Seleznev V A, Gutakovsky A K, Chehovsky A V, Preobrazhenskii V V, Putyato M A and Gavrilova T A 2000 *Physica E* **6** 828
[8] Dong L X, Zhang L, Bell D J, Grützmacher D and Nelson1 B J 2007 *J. Phys.: Conference Series* **61**, 257
[9] Sormunen J, Riikonen J, Mattila M, Tiilikainen J, Sopanen M, and Lipsanen H 2005 *Nano Lett.* **5** 1541
[10] Cui J, He Q, Jiang X M, Fan Y L, Yang X J, Xue F, Jiang Z M 2003 *Appl. Phys. Lett.*, **83** 2907
[11] Lorke A, Luyken R J, Garcia J M, and Petroff P M 2001 *Jpn. J. Appl. Phys.* **40** 1857
[12] Lorke A, Blossey R, Garcia J M, Bichler M and Abstreiter G 2002 *Mat.Sci.Eng.B* **88**, 225

Journal of Physics: Conference Series **152** (2009) 012021 doi:10.1088/1742-6596/152/1/012021

Synthesis and photoluminescence of well aligned ZnO nanotube arrays by a simple chemical solution method

Yuewen Wang Zuolin Cui*

Key Laboratory of Nanomaterials, Qingdao University of Science and Technology, Zhengzhou Road 53, Qingdao 266042, PR China

cuizl@public.qd.sd.cn

Abstract. The oriented ZnO nanotube arrays have been synthesized by a chemical solution method on a silicon wafer coated with ZnO thin films via thermally decomposing zinc acetate at 280°C. The morphologies, phase structure and the photoluminescence (PL) properties were investigated by field-emission scanning electron microscopy (FE-SEM), X-ray diffractometer (XRD), transmission electron microscope (TEM) and photoluminescence (PL) spectrum. It was found that the outer diameters of the ZnO nanotube ranges from 200nm to 300 nm, and the wall thickness is in the range of 30nm to 40 nm. ZnO nanotubes prefer to grow along the c axis. Photoluminescence (PL) spectrum shows a strong UV emission peaks at ~380 nm and a broad green-yellow emission at ~ 580 nm.

1. Introduction

Zinc oxide (ZnO), with a direct band gap of 3.37 eV and a large exciton binding energy of 60 meV[1-3] at room temperature, has received considerable interest due to its many important applications including blue light-emitting diodes (LEDs), field-effect transistors(FET), ultraviolet laser diodes (LD)[4] ,and so on. Since the first report of ultraviolet lasting from arrays of ZnO nanorods[5], vast interest has been devoted to the synthesis of vertically aligned one-dimensional (1D) ZnO nanostructure materials (nanorods (NRs), nanowires (NWs), nanotubes (NTs), etc.)[6]. Several methods of synthesizing ZnO nanostructure arrays have been employed, including vapor phase transport (VPT) , template-assisted growth , electrochemical deposition, pulsed laser deposition (PLD), motel-organic chemical vapor deposition, thermal evaporation.[7-9] However , this vapor phase synthesis methods are very expensive and require high temperature and low press. Therefore, a simple and fast route for the synthesis of well aligned 1D ZnO nanostructure materials on the substrate under ambient conditions has remained a great challenge until now. In this paper, we present a simple method for synthesis of vertically aligned ZnO nanotubes with diameters in the range 200–300 nm and wall thickness in the range of 30–40 nm on silicon substrates that had been pre-covered by a thin ZnO film produced via thermally decomposing zinc acetate at 280°C by a chemical solution method.

2. Experiment·

2.1. Materials

The raw materials included zinc nitrate hexahydrate ($Zn(NO_3)_2{\cdot}6H_2O$), zinc acetate dihydrate ($Zn(CHCOO)_2{\cdot}2H_2O$), hexamethylenetetramine($(CH_2)_6N_4$). All reagents and solvent used in this experiment were chemical reagent grade. Silicon wafers as substrates were rinsed ultrasonically successively in anhydrous ethanol and distilled water, then dried in the air.

2.2. Preparation

In this work, the preparation process of ZnO nanotube arrays need two steps: in the first step (seed layer deposition) ZnO seed layers were firstly deposited on silicon substrates. In brief, zinc acetate dihydrate (0.005M) was dissolved in anhydrous ethanol with continuous stirring for 30 min. Then, the clean silicon substrates were dipped into this clear solution 10min and drawn out at 2 cm/min and dry in the air, then the second time dipping was carried out. The process was repeated for six times. Subsequently, the substrates were heat-treated at 280°C for 2 h to obtain seed layers. in the second step (nanotube arrays growth), zinc nitrate hydrate ($Zn(NO_3)_2 \cdot 6H_2O$) (0.025M) and hexamethylenetetramine (HMT; $(CH_2)_6N_4$) solution (0.025M) were mixed and stirred to get a uniformly growth solution, then the substrate with ZnO seed layers was horizontally suspended in aforementioned growth solution at 85 °C for 6 h, then the growth solution was cooled down to room temperature naturally. The substrate was pulled out and thoroughly washed with deionized water to remove any residual salt or amino complex on the surface and allowed to dry in air at room temperature.

2.3. Characterization

The morphologies and sizes of the resulting products were determined by field-emission scanning electron microscopy (FE-SEM, JSM 6700F) and transmission electron microscopy (TEM, JEM 2000EX). The crystal structures of the resulting products were characterized by X-ray diffractometer (XRD, Rigaku D-max-γA XRD with Cu Kα radiation, λ=1.54178 Å). Photoluminescence (PL) at room temperature was measured using a He–Cd laser with a wavelength of 325 nm as the excitation source.

3. Results and discussion

X-ray diffraction analysis was carried out to investigate the phases of ZnO nanostructures. A typical XRD pattern of the as synthesized products is shown in figure 1. All of the diffraction peaks in figure 1 can be exactly indexed to the hexagonal ZnO with lattice constants a=0.3249 nm and c=0.5206 nm, which is in good agreement with the literature values (JCPDS 36-1451). No impurity peaks are observed, indicating a high purity of the final products. The much stronger (002) peak indicates that these ZnO nanostructures prefer to grow along the c-axis.

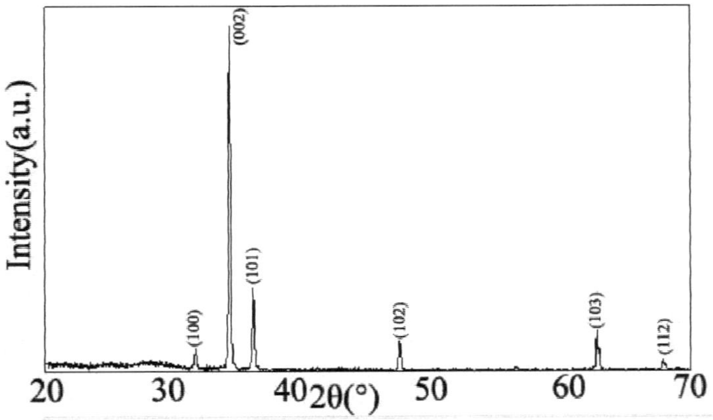

Figure1. XRD pattern of ZnO nanotube array on the silicon substrate coated with ZnO thin films

The surface morphologies of the ZnO nanotube arrays were characterized by scanning electron micrograph (SEM), as shown in figure 2.

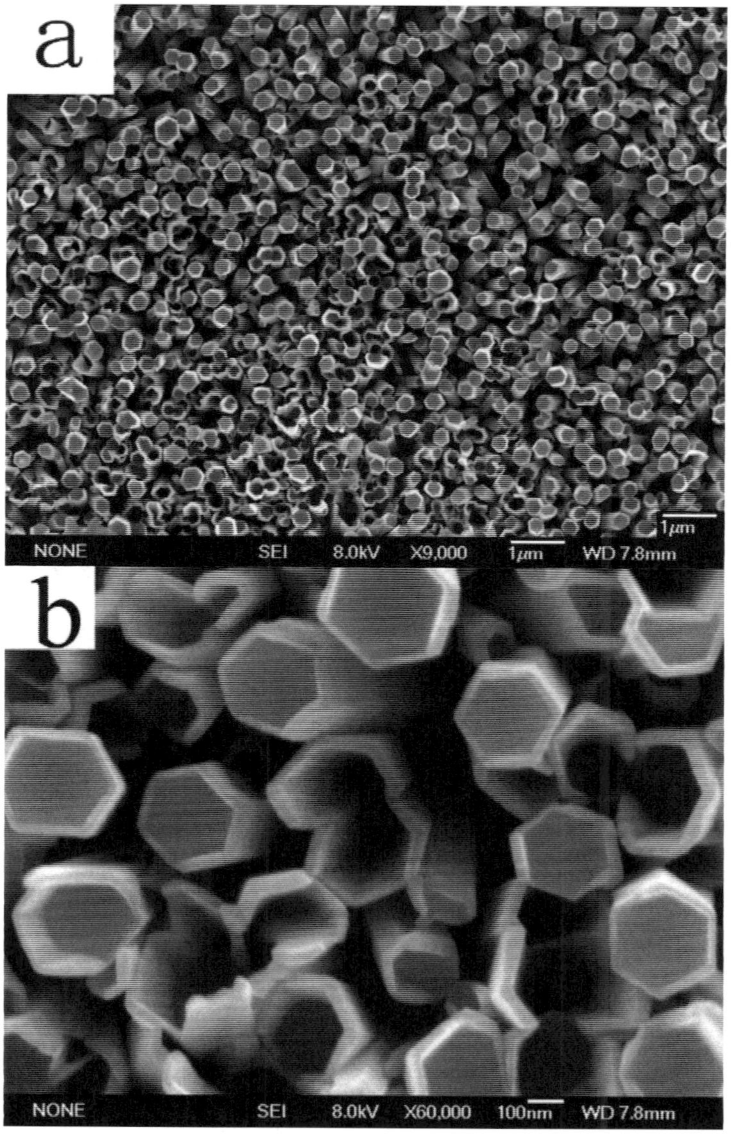

Figure. 2 Scanning electron micrographs of ZnO nanotube array on silicon substrates. (a) Top view at low magnification (b) top view at relatively high magnification

Figure 2a is the low magnification image of the ZnO nanotube arrays grown on a silicon substrate coated with a thin film of ZnO, showing the array consists of the nanotubes with diameters of 200–250 nm and most of the nanotubes are perpendicular to the substrate.

The figure 2b reveals a high-magnification SEM image of one ZnO nanotube, showing a perfect hexagonally faceted morphology with a diameter of 200-300nm, and uniform wall thickness of about 30-40 nm

Further structural characterization of the ZnO nanotube was performed by TEM. Figure 3 shows a typical TEM image of an individual ZnO nanotube, the inset of Figure 3 shows the accompanying selected-area electron diffraction pattern (SAED) pattern and reveals that the obtained ZnO nanotube exhibits crystal structure.

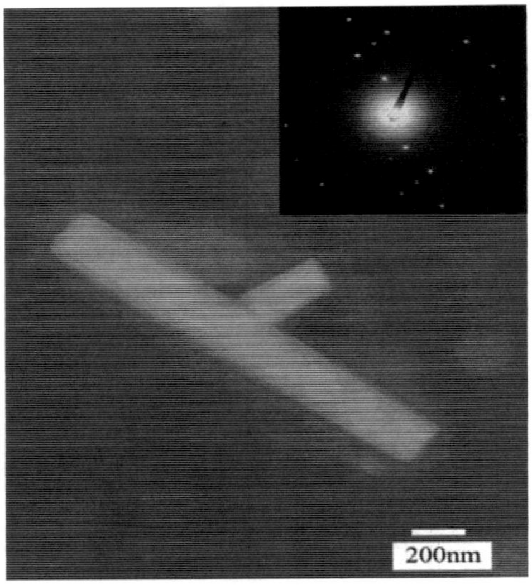

Figure.3 a TEM image of an individual the nanotube and related SAED patterns

To characterize the optical properties of ZnO nanotube arrays, we performed photoluminescence (PL) measurements. As an excitation source for the PL experiment, a He–Cd laser (325 nm) was employed. Figure 4 shows the room-temperature PL spectrum of ZnO nanotube arrays fabricated on the silicon substrate. Two typical emission peaks at ~380 nm and~ 580 nm were observed, which were assigned to UV emission and green-yellow emission, respectively. According to the literature, the UV emission with a peak at~ 380 nm has been attributed to the near band edge emission of ZnO. The green-yellow emission is a deep-level transition, originating from the recombination of the holes with the electrons occupying the singly ionized oxygen vacancy. The intensity of the deep-level emission is determined by the concentration of the oxygen vacancies in the ZnO crystal.

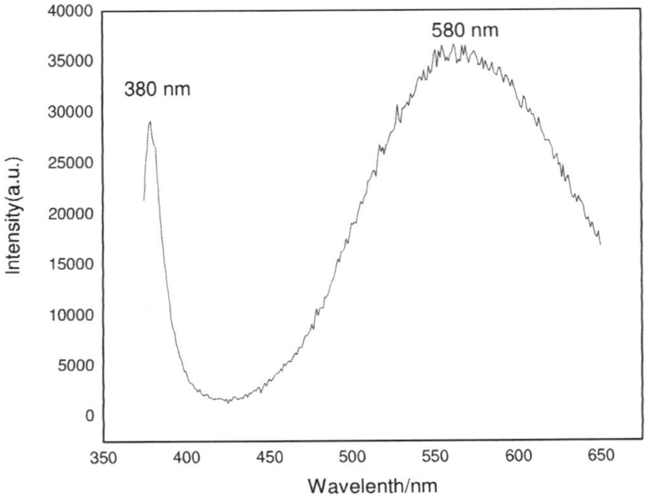

Figure.4 Photoluminescence spectrum of the ZnO nanotube arrays at room temperature

4. Conclusion

Well-aligned ZnO nanotube arrays were grown on silicon wafer substrates coated with ZnO film by a simple chemical solution method under mild conditions. The outer diameters of the nanotube range from 200nm to 300 nm, and the wall thickness is in the range of 30–40 nm. The ZnO nanotube is wurtzite structure, which prefers to grow along the c axis. The photoluminescence (PL) spectrum of the ZnO nanotube has a broad and stronger visible emission peak showing the potential application as a visible phosphor.

Acknowledgment

The Natural Science Foundation of Shandong Province supported this work financially.

References

[1] M. H. Huang, S. Mao, H. Feick, H. Yan, Y. Wu, H. Kind, E.Weber, R. Russo, P. Yang, Science 2001, 292, 1897
[2] S.Fujita, S.-W.Kim, M.Ueda, and S.Fujita, J. Cryst. Growth 2004, 272,138
[3] H.Zhou and Z.Li, Mater. Chem. Phys. 2005,89, 326
[4] D.W .Zeng, C.S.Xie, M.Dong, R.Jiang, X.Chen, A.H.W ang, J.B.W ang, and J.Shi, Appl. Phys. A 2004,79, 1865
[5] J.Zhang, L.-D.Sun, X.-C.Jiang, C.-S.Lian, and C.-H.Y an, Cryst.Growth Des. 2004, 4, 309
[6] H.Zhang, D.Y ang, X.Ma, Y.Ji, J.Xu, and D.Que, Nanotechnology, 2004 15, 622
[7] P.X.Gao and Z.L.W ang, J. Phys. Chem. B 2004,108, 7534
[8] Y.H.Leung, A.B.Djuriši 'c, J.Gao, M.H.Xie, Z.F .W ei, S.J. Xu, and W.K.Chan, Chem. Phys. Lett.2004, 394, 452
[9] F.Liu, P.J.Cao, H.R.Zhang, J.Q.Li, and H.J.Gao, Nanotechnology 2004,15,949.

Synthesis of Cu Nanowires via Solventhermal Reduction in Reverse Microemulsion System

Xiaojuan Zhang and Zuolin Cui*

Key Laboratory of Nanomaterials, Qingdao University of Science & Technology, Zhengzhou Road 53, Qingdao, 266042, PR China

E-mail: *cuizl@public.qd.sd.cn*

Abstract: Copper nanowires with high aspect ratios were synthesized by solventhermal reduction of Cu^{2+} in H_2O/CTAB/n-butanol/n-hexane reverse microemulsion system with a water to surfactant molar ratio (ω) = 40. The synthesized products were studied by XRD, SEM, and TEM respectively. The phase structure of the products was cubic face-centered Cu as straight nanowires with an average diameter of ~20 nm and lengths up to 2 μm. It was found that higher water content (ω=40) favored the formation of nanowires with high aspect ratios. The evolution of the nanowires was unidirectional with constant diameters. In aqueous solution experiments, flower-like microspheres consisted of short nanorods were obtained, which indicated that the microemulsion system was a crucial prerequisite for the formation of nanowires with high aspect ratios. A tentative formation mechanism of the nanowires is proposed, which involves unidirectional crystal growth as a consequence of irreversible micellar fusion, micellar longitudinal exchange and coalescence of micelles.

1. Introduction

In recent years researches on one-dimensional (1D) nanomaterials with unique magnetic, optical, electrical and catalytic properties in nanometer scale [1-5], have been the subject of much interest. Tremendous efforts have been made to tailor the morphology and aspect ratio of 1D nanostructure, which are highly attractive building blocks for electronic, optoelectronic, electrochemical and electromechanical devices [6-8] due to their inherent anisotropies and efficient transport of electrons and excitons within dimensions of nano-scale.

Among all the 1D nano-materials under research, special attentions have been paid on copper and related materials in the past few years. Metallic copper with a high electrical conductivity and cheap price has been used in a wide variety of commercial applications. Various nanostructures of copper with desirable properties can be used as a wide variety of potential applications such as chemical sensors [9], catalyst [10], interconnects [11], solid lubricant and etc. A number of approaches have been actively explored to synthesize 1D copper nanostructures [12, 13]. Up to now, the fabrication of copper nanostructures is mainly achieved by template-directed methods. A variety of templates have been successfully demonstrated for use by this method, and they can be classified into hard templates and soft templates. Typical hard templates include alumina membrane [14], carbon nanotubes [15, 16], $CaCO_3$ [17], silicas [18] etc. Although hard templates are effective in preparing nanowires with uniform and controllable dimensions, they usually lead to complicated processes and furthermore they are hard to remove. Whereas soft templates such as organic surfactants [19], block copolymers [20], and biomacromolecules including DNA [21]and viruses [22] seem to be better choice for synthesizing 1D nanostructures without the problems of hard templates.

As effective colloidal templates or nanostructure reaction media, reverse micelles and microemulsions have been widely used for the controlled synthesis of inorganic nanocrystals with specific size and shape [23, 24].A variety of one dimensional nanosized materials including metals [25], oxides [26], sulfides [27], sulfate [28], tungstate [29] and etc [30-34] have been successfully synthesized in the reverse micelle or reverse microemlsion system. Especially Pileni [24] synthesized cylindrical Cu nanoparticles in AOT/water/isooctane reverse miclles. However the aspect ratios of the cylindrical nanoparticles synthesized by him were low, and there were a lot of spherical nanoparticles in the products.

Very recently, solventhermal method has been widely used to synthesize nano-scale materials because it can provide a unique reaction environment, which facilitates the formation of products that are hard to obtain in normal conditions. In this paper, we demonstrated that Cu nanowires with aspect ratios as high as 100 could be synthesized by solventhermal reaction in the reverse microemulsion system. The diameters and lengths of the nanowires can be tuned by adjusting water to surfactant molar ratio (ω). With the reduction of ω (from 40 to 10), the nanowires turned thicker and shorter and finally transformed into crinkly nanoflakes. Based on the results of time-controlled experiments at a series of growth time and the experiment in aqueous solution, a tentative formation mechanism of the nanowires is proposed, which involves unidirectional crystal growth as a result of irreversible micellar fusion, micellar longitudinal exchange and coalescence of micelles.

2. Experimental Section

All the chemicals were analytic grade reagents used without further purification. A typical synthesis of Cu nanowires was as follows: Two types of water /cetyltrimethyl ammonium bromide (CTAB)/1-butanol /n-hexane microemulsion solution were prepared by solubilizing 1 ml of 0.6 mol/L $CuCl_2 \cdot 2H_2O$ solution (1ml of 1.2mol/L $NaBH_4$ solution in another microemulsion solution), 1ml 1-butanol into 30 ml CTAB n-hexane solution with a surfactant to oil concentration of 0.087 M .The final $[H_2O]/[surfactant]$ molar ratio of the two microemulsion solution was 40. The two different microemulsion solutions were vigorously stirred until they were optically transparent. And then they were mixed rapidly and magnetically stirred for another 15 min. Afterwards the mixture was transferred into an 80 ml Teflon-lined autoclave, sealed, heated and reacted at 180 °C for 12 h. Then the Teflon-lined autoclave was cooled to room temperature naturally. The obtained precipitate was separated, washed several times with distilled water and absolute ethanol, and then dried in a vacuum at 45 °C for 5 h.

The phase structure of the as-prepared products was characterized by X-ray powder diffraction (XRD) using a D/max-2500 X-ray diffractometer with Cu Kα radiation (λ = 1.5418 Å). The morphology and microstructure of the products were characterized using a JEOL JSM-6700F field-emission scanning electron microscopy (FESEM) operated at 5.0 kV, and a JEM-2000EX transmission electron microscope (TEM) an acceleration voltage of 160 kV.

3. Result and Discussion

3.1. The phase structure of the products.

The phase structure of the as-synthesized products was determined from the XRD pattern shown in Figure 1. The main reflection peaks can be indexed to cubic face-centered Cu which is very close to the reported data (JCPDS 04-0836) with a lattice parameter a=3.615 Å. Additional peaks marked with triangle can be indexed into a cubic structure of Cu_2O (JCPDS 5-667) .As is known, metallic copper can be slowly oxidized in the atmosphere. When its size reduces into nanometers, the surface energy is enhanced with the increase of specific surface area and oxidization is accelerated. Liu et al [35] have studied the oxidation process by exposing copper nanowires without oxides in atmosphere and after 24h Cu_2O appeared. In our experiment, the products had been exposed to air after vacuum drying before XRD test, so we can speculate that the as-obtained nanowires were copper nanowires with surface oxidized. Further methods to prevent the products from oxidizing need to be developed.

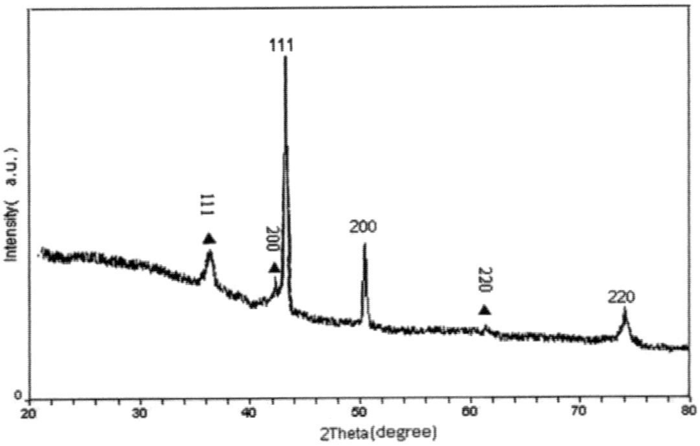

Figure 1. XRD pattern of the nanowires.

3.2. The morphology of the products.
The morphology of the products synthesized above was examined by FESEM. SEM images (Figure 2a) demonstrate that the final products were straight nanowires with diameters of 20 nm±10nm and typical lengths of 1-2μm. The diameter of a typical nanowire was uniform along the axial direction (Figure 2b) and no bent of large radian was observed.

Figure 2. FESEM images(a)and TEM image(b) of as-prepared Cu nanowires.

3.3. The effect of water content on the morphology of the products.
Keeping other reaction conditions constant, controlled experiments were carried our at four different ω values (30, 20, 10, 5). Nanowires with diameters about 20 nm were still obtained when ω = 30, but the aspect ratios turned low since their lengths reduced to less than 500nm (Figure 3a).Only short nanorods with diameters about 40-50nm and aspect ratios of about 3 were obtained when ω was reduced to 20 (Figure 3b). Crinkled nanoflakes were obtained when ω=10 (Figure 3c). And when ω was reduced to 5, the morphology of the products was polydisperse without uniformity. This controlled experiment demonstrated that the selected water content (ω=40) is beneficial to the formation of nanowires with high aspect ratios.

Figure 3.FESEM images of the products obtained at different water content: (a)30; (b)20; (c)10; (d)5.

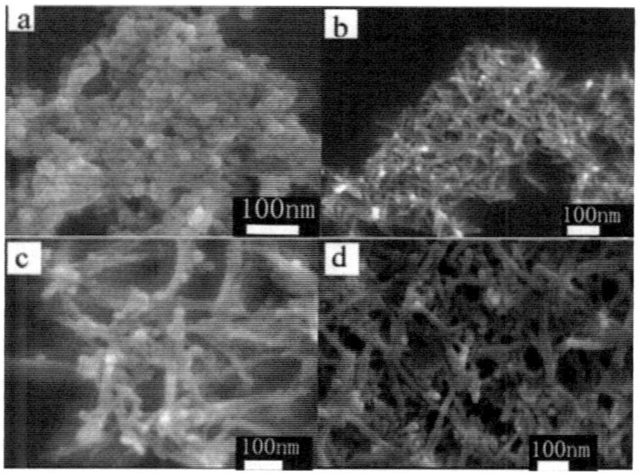

Figure 4. FESEM of the products at different growth times:(a)0.5h;(b)1h;(c)2h;(d)4h.

3.4. Investigation on the formation mechanism of the nanowires.
To investigate the formation mechanism o f Cu nanowires, samples at different growth times were studied by FESEM. Figure 4 shows FESEM images of the products obtained sloventhermal at 180℃ for 0.5h, 1h, 2h and 4h respectively, which clearly exhibits the evolution process form nanoparticles to high aspect ratio nanowires. Subsphaeroidal nanoparticles with average diameter of 20 nm were obtained after sloventhermal growth for 0.5h (Figure 4a). Nanorods with diameters about 20nm and lengths about 100nm appeared after sloventhermal treating for 1h (Figure 4b). With diameters unchanged, the lengths of the nanorods were grown to about 200nm when the sloventhermal time reached 2h (Figure 4c). The nanorods turned into nanowires at 4h of sloventhermal growth (Figure 4d) since the aspect ratio exceeded 20 at this time. To sum up, the growth process of the nanowires can be included as a longitudinal equant growth process with the aspect ratio increases with time.

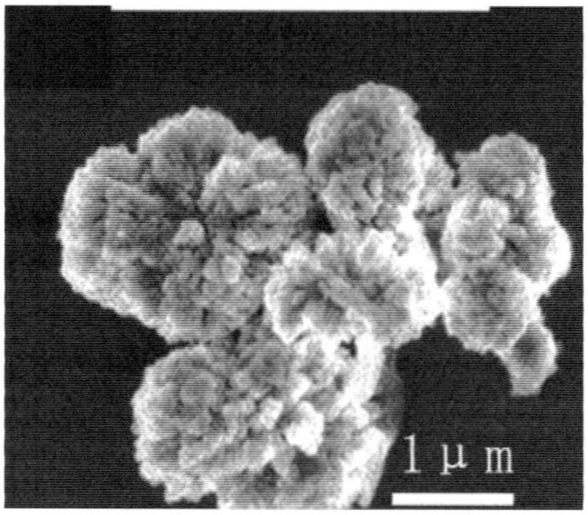

Figure 5. FESEM of the products obtained in aqueous solution at 180℃ for 12h.

Figure 6. FESEM images of the products formed at room temperature in reverse microemulsion.

It is found that the microemulsion system is a crucial prerequisite for the formation of high aspect ratio nanowires. If the reaction is conducted in aqueous solution keeping other experiment parameter unchanged (reactant concentration based on the total volume, the amount of surfactant used, reaction temperature and time). Flower-like microspheres consisted of nanorods with diameters of about 100nm were obtained (Figure 5).

3.5. The formation mechanism of Cu nanowires.

On the basis of experimental results discussed above, a tentative formation mechanism of the Cu nanowires is proposed as follows: When two microemulsion solutions containing $Cu^{2+}$and $BH_4^-$are mixed in the previous stage, the dimers formed by miceller droplet collision break down again as the Cu nucleation rate is comparatively low at room temperature . The contents from one micelle can be transferred into the other uniformly and the products obtained after 12h were spherical nanoparticles (Figure 6).

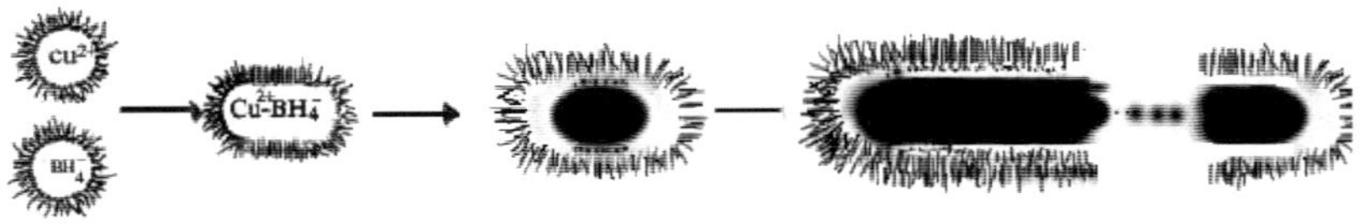

Figure7. Schematic illustration of the proposed mechanism for the formation of Cu namowires in reverse microemulsions.

When the mixed microemulsions were transferred into solventhermal condition in which the Cu nucleation was very fast, irreversible micellar fusion would occur because surfactants could be adsorbed onto the surface of previously formed rodlike nucleus in the center of a dimer, which would form a cylindrical micelle. The surfactants in the central region of the dimmer became fixed and immobile while at the end of the cylindrical micelle, where water was enriched, the surfactant molecules were relatively free. Thus, the cylindrical microemulsion micelle may dynamically exchange content with other micells at both ends forming a equant cylindrical micelle which grew simultaneously with the rodlike nucleus along the axial direction .This process (shown in Figure 7) may result in the formation of one-dimensional Cu nanowires whose diameters are constant and lengths are the function of growth time .

4. Conclusion

Cu nanowires with diameters of ~20nm and lengths up to 2μm have been synthesized in a cationic surfactant CTAB microemulsion system under solvothermal conditions. Higher water content (ω=40) favored the formation of nanowires with high aspect ratios. The evolution of the nanowires was unidirectional with constant diameters. The microemulsion system is a crucial prerequisite for the formation of nanowires with high aspect ratios. The formation mechanism of the nanowires involves unidirectional crystal growth as a result of irreversible micellar fusion, micellar longitudinal exchange and coalescence of micelles.

References

[1] Wang B, Zhu L F, Yang Y H, Xu N S and Yang G 2008 *J. Phys. Chem. C* **112** 6643-47.
[2] Jung H S, Hong Y J, Li Y, Cho J, Kim Y and Yi G 2008 *ACS Nano* **2** 637-42.
[3] Tan Y, Xue X, Peng Q, Zhao H, Wang T and Li Y 2007 *Nano Lett.***7** 3723-28.
[4] Cox A J, Louderback J G and Bloomfield L A 1993 *Phys. Rev. Lett.***71** 923.
[5] Alivisatos A P 1996 *Science* **271** 933.
[6] Liao L, Liu K, Wang W, Bai X, Wang E, Liu Y, Li, J and Liu C 2007 *J. Am. Chem. Soc.***129** 9562-63.

[7] Kauffman D R and Star A 2008 *J. Phys. Chem. C* **112** 4430-34.
[8] Gardner C E, Ghanem M A, Wilson J W, Smith D C 2006 *Anal. Chem.* **78** 951-55.
[9] Li C Z, He H X, Bogozi A, Bunch J S and Tao N J 2000 *Appl. Phys. Lett.* **76** 1333-35.
[10] Mandal S, Bose R N, Reed J W and Gould E S 1996 *Inorg. Chem.* **35** 3159-62.
[11] Lin Q and Corbett J D 2005 *Inorg. Chem.* **44** 512-518.
[12] Liu Z W and Yoshio B 2003 *Adv. Mater.* **15** 303-05.
[13] Yen M Y, Chiu C W, Hsia C H, Chen F R, Kai J J, Lee C Y and Chiu H T 2003 *Adv. Mater.* **15** 235-37.
[14] Genaro A G, Uttandaraman S and Joel A H 2005 *Macromol. Rapid Commun.* **26** 1677-81.
[15] Deng B, Xu A W, Chen G Y, Song R Q and Chen L 2006 *J. Phys. Chem. B.* **110** 11711-16.
[16] Hwang G L, Hwang K C, ShiehY T and Lin S 2003 *J.Chem. Mater.* **15** 1353-57.
[17] Zhu H, Liu Z, Wang Y, Kong D, Yuan X and Xie Z 2008 *Chem. Mater.***20** 1134-39.
[18] Delahaye E, Escax V, El Hassan N, Davidson A, Aquino R, Dupuis V, Perzynski R and Raikher Y L 2006 *J. Phys. Chem. B* **110** 26001-11.
[19] Wang W, Wang G, Wang X, Zhan Y, Liu Y and Zheng C 2002 *AdV.Mater.* **14** 67.
[20] Chai J and Buriak J M 2008 *ACS Nano* **2** 489-501.
[21] Rolland J P, Maynor B W, Euliss L E, Exner A E, Denison G and DeSimone J M 2005 *J. Am. Chem. Soc.* **127** 10096-00.
[22] Nam K T, Kim D W, Yoo P J, Chiang C Y, Meethong N, Hammond P T, Chiang Y M and Angela M 2006 *Science* **32** 885-88.
[23] Li M, Schnablegger H and Mann S. 1999 *Nature* ***402*** 393.
[24] Pileni M P 2003 *Nat. Mater.* **2** 145.
[25] Pileni M P 2001 *Langmuir* **17** 7476-86.
[26] Xiong Y, Li Z, Zhang R, Xie Y, Yang J and Wu C 2003 *J. Phys. Chem. B* **107** 3697-02.
[27] Simmons B A, Li S, John V T, McPherson G L, Bose A, Zhou W and He J 2002 *Nano Lett.* **2** 263-68.
[28] Jeremy D H and Stephen M 1997 *Chem. Mater.* **9** 1819-28.
[29] Shi H T, Qi L, Ma J M and Cheng H M 2002 *CHEM. COMMUN.* **16** 1704-05.
[30] Rees G D, Evans-Gowing R, Hammond S J and Robinson B H 1999 *Langmuir***15** 1993-02.
[31] Qi L, Ma J, Cheng H and Zhao Z 1997 *J. Phys. Chem. B* **101** 3460.
[32] Li M and Mann S 2000 *Langmuir* **16** 7088-94.
[33] Shi H T, Wang X H, Zhao NN, Qi L and Ma J M 2006 *J. Phys. Chem.* B **110** 748-53.
[34] Zhang H W, Zhang X, Li H Y, Qu Z K, Fan S and Ji M Y 2007 *Cryst. Growth Des.***7** 820 -25.
[35] Liu Z P, Yang Y, Liang J B, Hu Z K, Li S, Peng S and Qian Y T 2003 *J. Phys. Chem. B* ***107*** 12658-61.

Research progress on a novel conductive polymer —poly(3,4-ethylenedioxythiophene) (PEDOT)

Y Wang

Department of Polymer Materials and Engineering, College of Materials Science and Engineering, Shandong University of Science and Technology, 579 Qianwangang Road, Qingdao Economic & Technical Development Zone, Qingdao, Shandong Province, P. R. China 266510

Email:holylilyee@163.com

Abstract In this paper the progress in study on a new organic conducting polymer---poly(3,4-ethylenedioxythiophene) (PEDOT) is reviewed. Its ground-state structure, excellent advantages and morphology are presented. Then the modification ways such as its copolymers and composites are introduced in detail. Moreover, the applications of PEDOT are also summarized.

Keywords: poly(3,4-ethylenedioxythiophene) (PEDOT), structure, advantages, composites, copolymers, applications

1. Introduction

Although polythiophene is relatively stable compared with other conducting polymers, the decrease of its conductivity over a long period limits its industrial applications. The stability of polythiophene may be improved by decreasing its oxidising potential. This can be realised by introducing an alkoxyl group (such as ether group), which has electron-donating character. Substituted thiophene can be polymerised at a considerably lower potential than thiophene and the resulting polymers are more stable in the anion-doped state especially when the side chain is not too long. Ether substituted thiophene mainly includes 3-alkoxy thiophenes, 3,4-alkoxy thiophenes and 3,4-(crown ether) thiophene or pseudocrown ether thiophene. However, study of polythiophenes with long side chain substituents shows that the polymer chain is distorted thereby giving lower conductivities. Therefore, a shorter chain ether substitutented thiophene, 3,4-(crown ether) thiophene, has attracted many scientists' attention and many results about PEDOT have been achieved and reported. Therefore, the research progress on PEDOT is reviewed in many aspects in this paper.

2. Ground-state structures of PEDOT

In many publications, the structure of PEDOT is usually described as Figure 1. However, two possible structures for neutral PEDOT: aromatic-like (Figure 2 (B)) and quinoid-like (Figure 2 (C)). In the earlier study, it was proposed by Lapkowski and Pron[1] that the ground state of PEDOT was quinoid, based on the spectroscopic data. However, a theoretical investigation carried out by A.

Dkhissi et al.[2] along two lines suggested that the ground state of neutral PEDOT was aromatic-like. They evaluated the energy per repeat unit (E_{pru}) of the aromatic and quinoid forms from quantum-chemical calculation by both RHF (restricted Hartree-Fock) and DFT (density functional theory) techniques. On the other hand, to confirm the conclusion, the vibration spectra of the two forms were calculated and compared to the experimental spectra recorded by Kvarnström et al.[3-4]. Finally, a conclusion was drawn that PEDOT had aromatic character in the ground state, which was also the case for unsubstituted PT (Figure 2 (A)) but in contrast to polyisothianaphthene (PITN) (Figure 2 (D)).

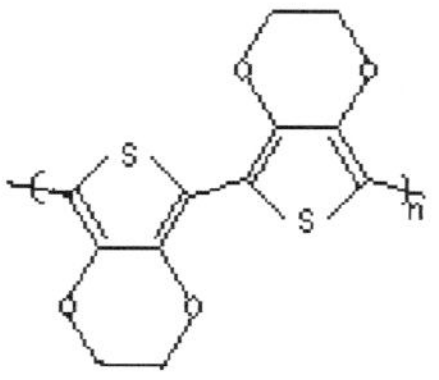

Figure 1. Structure of PEDOT.

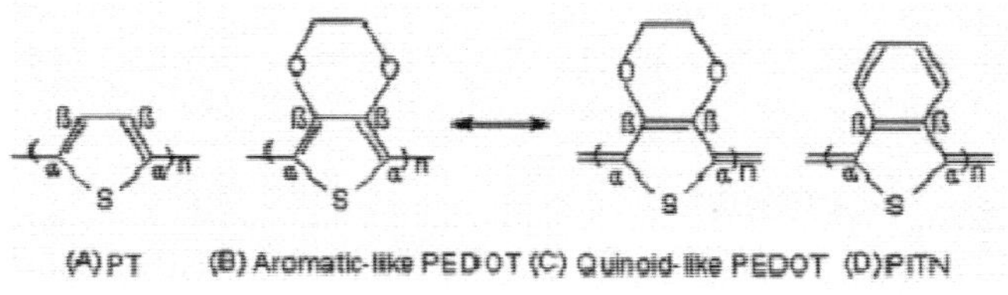

Figure 2. Ground-state structures of PEDOT.

3. Excellent characteristics of PEDOT

PEDOT is one of the most promising materials for practical applications due to its following characteristics:

3.1. Reversible doping state

PEDOT can be repeatedly doped and undoped. PEDOT is almost transparent and light blue in the oxidized state and can be easily changed into opaque and dark blue appearance in the neutral state. Thus its color changes visibly when its doped state changes and may be suitable for optical applications, such as electrochromic displays[5].

3.2. Excellent stability

PEDOT has improved chemical and thermal stability. Thermal studies show that a continuous degradation occurs above 150 °C and complete decomposition above 390 °C[6]. Electrical conducting properties appear to remain almost unaltered after aging in environmental conditions. Its high stability is attributed to favorable ring geometry and the electron-donating effect of the oxygen atoms at the 3,4-positions stabilizing the positive charge in the polymer backbone[7].

3.3. Regular structure

Due to the structure of the monomer, competing polymerizations through 3- and 4- positions as in thiophene are avoided. Thus, only the 2,5-couplings of the 3,4-ethylenedioxythiophene are expected. Therefore, PEDOT is expected to have fewer defects than the thiophene analogues.

3.4. Low band-gap (High conductivity)

PEDOT has a low band gap of 1.5-1.6 eV[8]. The lower band-gap relative to polythiophene is thought to originate from the influence of the electron-donor ethylene dioxy groups on the energies of the frontier levels of the π system[9]. Experimental results show that after doping, PEDOT exhibits reduced absorption in the visible: the oscillator strength shifts from around 1.5 eV (lowest π-π^* transition) to below 1 eV in the metallic state[10]. Thus it shows a high electrical conductivity (up to 550 S/cm) in the doped state.

3.5. Effect of temperature on conductivity

DC conductivity measurements[11] show that PEDOT has a resistivity ratio $\rho_r<2$. But for the samples with $\rho_r = \rho(1.4K)/\rho(291K) = 1.5\text{-}2.8$, the temperature coefficient of resistivity, TCR, changes signs below 10K from negative to positive which is characteristic of normal metals. For negative and positive TCR, the δ(T) is isotropic and proportional to $T^{1/2}$. The increase in conductivity below the transition temperature is approximately 2-3 %.

3.6. Electrochemical properties
Compared to other conducting polymers, electrochemically synthesized films of PEDOT have a low redox potential and excellent stability in their doped state[12]. Studied by cyclic voltammetry, it is found that the redox peaks at approximately 0 mV (oxidation) and –400 mV (reduction) remain almost unaffected during cycling. However, only under an applied negative potential of –700 mV were the neutral films found to be stable. Open circuit potential measurements show that the neutral films are rapidly oxidized[13].

4. Morphology of PEDOT
The high electrical conductivity measured for PEDOT agrees well with metallic state behaviour in the surface plane of the film. Determined by variable-angle spectroscopic ellipsometry together with intensity reflectance and transmission spectrophotometry, the optical anisotropy of PEDOT films are of uniaxial character with the optic axis normal to the film surface. The (ordinary) index of refraction in the plane shows a metallic state behaviour while the out-of-plane (extraordinary) index of refraction was character similar to that of a dielectric[14].

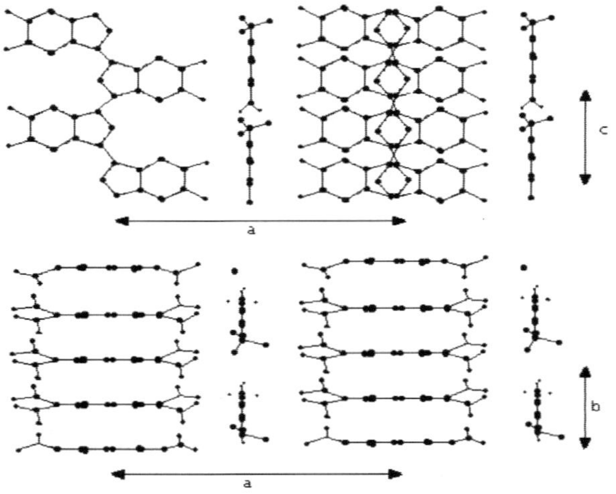

Figure 3. Structural model of tosylate-doped PEDOT.

Based on diffraction peak positions and intensities, K. E. Aasmundtveit et al.[15-16] presented a structural model of thin-film tosylate-doped PEDOT which corroborated with the optical anisotropy. As shown in Figure 3, the structural model has a pseudo-orthorhombic unit cell with four monomers and one tosylate ion per cell and lattice parameters a=14.0 Å, b=6.8 Å and c=7.8 Å. The thiophene backbone of the polymer chain along the c-axis is flat and the polymer repeat distance is assumed to be 7.8 Å. The chains stack on top of each other with a stacking distance of b/2 = 3.4 Å. These bc-layers are separated along the a-axis by layers of tosylate ions. Therefore, the orientation of the material is such that the a-axis is preferably oriented along the surface normal so that the thin films are highly anisotropic, which is in good agreement with an observed optical anisotropy of the materials[17]. In the plane of the film, the b- and c-axes are randomly distributed, ensuring cylindrical symmetry, giving a picture where the bc-layers (alternate polymer and dopant layers) lie flat down on the substrate. The structure is described as a paracrystalline one, where the lattice parameters represent only mean values of probability distributions, implying also that the crystalline order decays with increasing distance. The width of the a-axis distribution is estimated to 2 %, and the size of individual paracrystalline domains in this direction to 50 Å. It is noteworthy that hardly any amorphous scattering was observed, so the paracrystalline state accounts for the entire volume of the material. This fact may be important for explaining the high conductivity of PEDOT, because insulating barriers formed by amorphous regions have been proposed as the limiting factor for the conductivity of conducting polymers.

5. PEDOT composites

5.1. PEDOT nanocomposites

A step forward with respect to 'simple' modification with pristine CPs consists in the inclusion of metal functionalities inside the polymeric matrix, in order to further increase the performances of the resulting composite material. In particular, inclusion of noble metal nanoparticles (NPs), whose catalytic properties are well known, constitutes one of the most interesting possibilities.

Au_{nano}-PEDOT nanocomposites were successfully prepared using chemical and electrochemical routes[18] and the AuNPs were found to attach stably to the polymer nanofibrillar matrix. The advantages of these films were demonstrated for sensing biologically important compounds such as dopamine (DA) and uric acid (UA) in presence of excess ascorbic acid, with superior selectivity and sensitivity when compared to the polymer film alone. The PEDOT matrix was recognized to be responsible for the peak separation (selectivity) while also favoring catalytic oxidation of the above compounds and the nanometer-sized gold particles allowed nanomolar sensing of DA and UA (sensitivity). Thus, it is possible to detect nanomolar levels of DA and UA in presence of excess of AA.

Composite materials consisting of PEDOT including Au nanoparticles encapsulated by bulky anionic species could also be electrogenerated[19]. Notable chemical and morphological differences of the CP/NP composites were obtained with respect to the pure organic materials and reflected on the electrochemical performances of the various PEDOT-based coatings with respect to the oxidation of suitably chosen electroactive species. The sensitivity and selectivity of the electrochemical responses depended on the presence and density of the nanoparticles and the nature of the relevant encapsulating agent. The selectivity exhibited by the PEDOT/Au-Tann electrode system constitutes quite an interesting feature to exploit in electroanalysis. And this electrode coating is stable under subsequent DPV scans in the presence of DP, which is of key importance for possible analytical applications.

A new type of layered PEDOT/VS_2 nanocomposite were prepared by in situ oxidative polymerization of EDOT with VS_2 as a host material in the presence of an external oxidizing agent[20]. The polymerization proceeded concomitantly with intercalation so that there was considerable bonding interaction between the organic and inorganic components, probably due to hydrogen bonding. Compared with pristine VS_2, the improvement of electrochemical performance with reversible specific capacities up to ~130 mAh/g was attributed to higher electric conductivity and enhanced bidimensionality. The influence of intercalants on Li^+ diffusion rates and charge capacity in the PEDOT/VS_2 nanocomposite was increased relative to that for VS_2, therefore, the polymer nanocomposite acts as a better cathode material than the pristine VS_2 material by enhancing lithium diffusion.

5.2. PEDOT composites including carbon materials

Composite films of carbon nanotubes (CNTs)/PEDOT were prepared via electrochemical co-deposition from solutions containing acid treated CNTs and the corresponding monomer[21]. CNTs served as the charge carriers during electro-deposition, and also acted as both the backbone of a three-dimensional micro-and nano-porous structure and the effective charge-balancing dopant within the polymer. All the composites showed improved mechanical integrity, higher electronic and ionic conductivity (even when the polymer was reduced), and exhibited larger electrode specific capacitance and greater stability in charge–discharge cycling compared with the pure conducting polymers.

Moreover, transparent conductive films were fabricated using a composite of PEDOT and the acid-treated various CNTs by the bar coating method[22]. The dispersion stability of acid-treated CNTs is very stable in PEDOT solution. Compared with the raw PEDOT film with a transmission of 87.8% and a sheet resistance of 1182 V/sq, electrical and optical properties of CNTs/PEDOT film was improved so that a sheet resistance of 249.2 V/sq and a transmission of 82.7% were obtained when the concentration of the t-MWNTs was 0.01 wt.%. In addition, these films showed robust flexibility and did not break down upon bending and folding in contrast with commercial films such as ITO. A further improvement in the various properties of the transparent conductive film, such as its electrical conductivity and optical transmission, should be possible by enhancing the degree of dispersion of the CNTs in the PEDOT solution. These CNT/PEDOT thin films may be obtained, permitting one to fabricate transparent electronic devices such as a flexible display, solar cell and sensors, etc. It will be able to lead to important advances in developments of flexible optoelectronic devices and especially to alternatives to approach for transparent and conductive coatings such as commercial ITO films.

The PEDOT/$LiCoO_2$–KB(Ketjenblack)cathodes were made by electrochemical deposition of 3,4-ethylenedioxythiophene (EDOT) monomer on the pre-formed $LiCoO_2$–KB electrodes with 1 and 5 wt% of KB content[23]. The PEDOT layer was successfully coated on the $LiCoO_2$ and KB components. Homogeneity of the KB component and the presence of conductive PEDOT layer were found to be important factors in controlling the cathode properties such as the involved capacity, cycle ability, intercalation/de-intercalation rate and rate capability. Electrochemical characterizations on the $LiCoO_2$–KB and PEDOT/$LiCoO_2$–KB cathodes suggested that incorporation of the conductive PEDOT into cathode enhanced the intercalation/deintercalation of Li ion and also the rate capability.

5.3. PEDOT composites with other polymers

PEDOT/polypyrrole(Ppy) composite electrodes were prepared by electropolymerization of EDOT on the surface of Ppy modified tantalum electrodes to obtain the PEDOT/Ppy composite with its rapid response like PEDOT while its specific capacitance and cost close to those of Ppy [24]. The specific capacitance of composite electrodes, due to the synergic effect of PEDOT and Ppy, was much higher than the values of either pure PEDOT or pure Ppy electrodes. Moreover, the composites prepared on the surface of Ppy with horn-like structure allowed the specific capacitance up to more than 200 F g^{-1} and had a good cycleability. This implies that PEDOT/Ppy composites are promising to be used as electrode material of supercapacitors.

In order to improve the repeatability of the responses on the PEDOT modified electrode, different kinds of surfactants were included into the polymer during the polymerization. The fouling effect was strongly alleviated when the polymerization of EDOT was carried out in the presence of poly(sodium 4-styrenesulphonate) (NaPSS). PEDOT:PSS composite films were successfully prepared by LB inducing deposition method[25]. There is an intense electrostatic force between SO_3^- functional group of PSS and NH_3^+ functional group of aliphatic amines. ODA–SA/PEDOT:PSS composite film had higher film-forming ability than ODA/PEDOT–PSS composite film. From the electrical conductivity investigation, this kind of composite LB film exhibited a very potential application for organic electronic devices and the research focusing on this aspect is underway now.

PEDOT/silk fibroin (SF) composite fibers were fabricated successfully and expediently by in situ polymerization without any modification of silk fibroin surface[26]. SEM observation and elements analysis confirmed that PEDOT molecules had been coated successfully by such polymerization process without destroying SF in nature and the composite fibers still possessed their former fibrillar morphology and strength properties. Structural analysis (IR, XRD and DSC) implied that PEDOT molecules had different interactions to certain extent with SF chains. These composite fibers exhibited better electrical and thermal properties and may have potential applications in textile, biological and other novel functional materials

Composite materials containing nanoparticulate PEDOT in a polymer electrolyte matrix (PEO) containing either a Cu–Cu^{2+} or Fe^{2+}–Fe^{3+} redox couples showed rapid and reversible decreases of up to 500-fold in their electrical and microwave impedances when small DC or AC electric fields were applied across coaxial line and strip samples from their edges, which were much larger than microparticulate PEDOT composites and made signification progress in this field[27]. The composites showed large field-dependent resistances at low applied fields and good electrochemical stability. Changes in microwave impedance from 350 $\Omega\square^{-1}$ at 0.01 V to 24$\Omega\square^{-1}$ at 12 V, both measured at 1 GHz proved unequivocally that the changes observed occurred in the bulk samples and not just in a region within the immediate vicinity of the electrode–composite interface.

6. PEDOT copolymers and their electrochromic devices

Copolymerization is a powerful tool to enhance the electrochemical, optical, and electrochromic properties of the parent homopolymers. To reduce its band gap, shorten switching time, enhance its capacity stability, many efforts were dedicated to copolymerize EDOT and other monomers and the copolymers were assembled into electrochromic devices and measured in many espects as follows.

Copolymer of 4-(2,5-di(thiophen-2-yl)-1H-pyrrol-1-yl)benzenamine (SNS-NH_2) and EDOT was achieved in acetonitrile $NaClO_4$/$LiClO_4$ (0.1 M:0.1 M) solvent–electrolyte couple[28]. Copolymer revealed multichromic property with five different colors at different applied potentials. Colorimetry studies for P(SNS-NH_2-co-EDOT) proved that it is possible to provide fine tuning of these colors by varying applied potential during synthesis. The band gap of the copolymer was calculated as 1.7 eV and 0.6 s was found as the time required to attain 95% of the total transmittance difference. A dual-type complementary ECD was assembled with a configuration of ITO/P(SNS-NH_2-co-EDOT)/gel electrolyte/PEDOT/ITO and characterized for its performance. Suitable potential range for operating the device between purple to blue was found to be between -1.5 and +1.5 V. Electrochromic switching study results revealed that optical contrast (%DT) and switching time were 11% and 0.8 s at 491 nm, respectively. In addition, the device showed good open circuit memory and stability. Copolymer of SNS-NH_2 with EDOT can be thought as a feasible nominee for anodically coloring electrochromic layers in ECDs.

Electropolymerizations (EPs) of the mixed monomers of Th and EDOT in 0.05M Et_4NClO_4/propylene carbonate (PC) solution were performed to prepare polymer films[29]. The incorporation of EDOT units into pure polythiophene (PTh) chain leads to large alternations on the experimental conditions of EPs and the properties of the resulting polymer films. The resulting copolymer films further served as cathode materials to test their applicability in lithium ion battery. With incorporation of EDOT units, the corresponding polymers showed the enhancement of reversibility compared to PTh, i.e. during the repeatedly positive and negative sweeps, no considerable overoxidation was observed. With varying scanning rate, PTh–EDOT copolymers also exhibited faster responses than PTh. Advantages of EDOT incorporation also reflects in the better capacity stability toward successive cycling. Almost 100% recovery on the capacity was observed for PTh–EDOT (1/1) copolymer, which is different from the 40% loss on PTh after 25 times of cycling.

The synthesis of a new copolymer utilizing 2,5-di(thiophen-2-yl)-1-*p*-tolyl-1*H*-pyrrole (DTTP) and EDOT was successfully achieved in ACN/$LiClO_4$/$NaClO_4$ (0.1 M) solvent–electrolyte couple[30]. The copolymerization with EDOT not only decreases the band gap Eg, but also enhanced the electrochromic properties such as optical contrast and switching time. The copolymer has an electronic band gap of 1.65 eV. The contrast was measured as the difference between %T in the reduced and oxidized forms and noted as 20% at 487 nm. Then dual-type complementary colored polymer ECD were assembled with a configuration of ITO/P(NTP-*co*-EDOT)||gel electrolyte||PEDOT/ITO and electrochromic switching study results showed that its optical contrast (%T) and switching time were 18% and 1.5 s at 615 nm. In addition, the device has good environmental and redox stability.

1-benzyl-2,5-di(thiophen-2-yl)-1H-pyrrole (SNBS) and EDOT was electrochemically copolymerized[31]. The resulting copolymer displayed exclusive color changes between claret red, yellow, green, and blue upon applied potential and a full switch in 1 s with an optical contrast of 40%.

As an application, absorption/transmission type electrochromic device with indium tin oxide (ITO)/copolymer/gel electrolyte PEDOT/ITO configuration was constructed, where copolymer and PEDOT functioned as the anodically and the cathodically coloring layers, respectively. Results implied the successive use of this copolymer in electrochromic device applications, since the device exhibited short switching times with a wide color variation upon applied potential.

7. Applications of PEDOT

Since PEDOT has so many excellent properties, it inspires much great interest of research and a wide range of potential applications. Its high conductivity and stability in the doped state cause it to be of major interest for applications such as antistatic and electrostatic coatings, for metallization of insulators, and as electrodes for capacitors or photodiodes[32-34]. If the durability of the capacitors using PEDOT is enhanced, they will have a wider range of applications such as in electronic circuits for cars and man-made satellites[35]. Due to its property of being repeatedly doped and de-doped, it is very attractive for electrochromic applications[36-37]. With proper choice of counter-ions, PEDOT is also of interest for applications as an electrochemical actuator[38]. PEDOT has been a successful commercial product due to the accessibility of the monomer and the possibility of producing the conducting material as a water dispersion by using a water-soluble polymeric counterion poly(styrene sulphonate). It has been extensively used as an antistatic coating on photographic films and as an electrode layer in flexible displays and organic light-emitting devices[39]. Further progress has been achieved through recent studies on the application of PEDOT in the electrochemical field such as actuators, capacitors, OLED, photovoltaic cells, sensors etc.

7.1. Actuators

PEDOT has very high conductivity, particularly high stability in the doped state and the reversibility of the doping process. The expulsion/inclusion of ions during its doping/de-doping processes results in the dimensional change properties, leading to interesting potential applications in the design of actuator with large mechanical stress and strain values for robocics, prosthetics, microvalves. For trilayer actuators with the configuration of Ppy(DBS)/PEDOT(DBS)/Ppy(DBS), the addition of the PEDOT(DBS) layer produced significant improvements in the strain and force difference between reduced and oxidized states at faster scan rates without any negative effects or delamination[40]. Additionally, the dispersion of PEDOT in the interpenetrating polymer networks (IPNs) of poly(ethylene oxide) and polybutadiene with a room temperature ionic liquid, 1-ethyl-3-methylimidazolium bis-(trifluoromethylsulfonyl)imide (EMITFSI) could enhance the actuation result of the actuator based on them. By adjusting the PEDOT mass quantity loaded in the conducting IPN, the electrical resistance along the surface was improved at least 1000 times compared with that in the bulk[41]. Moreover, solid actuators were prepared based on PEDOT/NBR films with the absorption of room temperature ionic liquids (RTIL) including imidazolium salts, e.g. 1-butyl-3-methyl imidazolium X [where X=BF_4^-, PF_6^-, $(CF_3SO_2)_2N^-$] into the composite films. The PEDOT content was gradually decreased from the outside towards the center of the matrix film in order to yield a similar structure to that of an interpenetrating polymer network (IPN). The conformational effects of the polymer chains influenced their actuation and the displacement of the actuator increased with increasing the anion-size of the ionic liquids. The $(CF_3SO_2)_2N^-$-doped PEDOT actuator showed a large displacement with a maximum displacement generated electrically of 5.49 mm, which was much higher than that of the films doped with BF_4^-(1.10 mm) or PF_6^- (3.09 mm)[42].

7.2. Capacitors

The highly stable polymer electrode material PEDOT is of particular interest for being used to develop high-performance supercapacitor owing to the combination of high energy density, high specific power density, lower cost, easy to be prepared and fabricated on needs. PEDOT electropolymerized in neat [bmim][BF_4] for actuators possessed nearly ideal capacitive characteristic of 130 F/g and especially excellent level of stability with a long time of redox cycling up to 70,000 cycles, close to that of carbon electrode material with a general lifetime of 100,000 cycles[43]. The developed Ta/PEDOT capacitor showed the advantages of high capacitance per unit volume, excellent frequency characteristics greatly improved in high-frequency range in comparison with Ta/MnO_2 one, as well as superior durability without any deterioration for more than 1000 h at 125°C in air and at 85C/85%RH due to the environmental stability of PEDOT. Therefore, the commercialization of the Ta/PEDOT capacitor will give rise to further wide-ranging applications[44]. Introduction of crosslinked HBPSi onto the aluminum oxide/PEDOT counter electrode for capacitor resulted in a large number of vinyl groups to improve the interfacial affinity between aluminum oxide and PEDOT so that it is effective in improving the capacitance and equivalent series resistance of the capacitor as compared with linear, vinyl terminated polydimethylsiloxanes[45]. The specific capacitance of the PEDOT-PSS-$RuO_2{\cdot}xH_2O$ for supercapacitor increased with increasing loading of $RuO_2{\cdot}xH_2O$ particles embedded into the porous PEDOT-PSS matrix with a uniform dispersion and excellent incorporation and a maximum specific capacitance of 653 F/g was achieved[46].

7.3. Photovoltaic cell

Compared with the conventional electrodes such as expensive Pt or C with poor quality and relatively low conductivity, PEDOT:PSS has high room temperature conductivity, much better film-forming ability, relatively low cost, remarkable stabilities and the electronically catalytic activity, expected to have improved performances as the counter electrodes for dye-sensitized solar cells (DSSCs). PEDOT:PSS counter electrodes for highly transparent DSSCs with PMMA/PC gel polymer electrolyte showed similar I^{3-}/I^- catalytic activities to Pt based counter electrodes. The best performance obtained gave a short circuit current of $2mAcm^{-2}$ and a open circuit voltage of 625mV in quasi-solid-state construction assemblies and had a a longer lifetime, which is a prerequisite for any future application[47]. The transparent thin graphene/PEDOT:PSS film as a counter electrode in DSSC possessed high transmittance (>80%) at visible wavelengths and the high electrocatalytic activity of graphene and PEDOT:PSS. As a result, the DSSC with the counter electrode of graphene/PEDOT:PSS composite film containing 1 wt% or more graphene showed a high energy conversion efficiency of 4.5% under irradiation of 100mWAM 1.5 white light. Thus the high transmittance of the films provides them with potential applications in power-producing windows or metal-foil-supported DSSC[48]. The conductivity of the PEDOT:PSS films treated with DMSO exhibited high conductivity of 85±15 S/cm due to the increased surface roughness. The DSSC using carbon black (0.1 wt%)-modified DMSO-PEDOT:PSS conductive coating as a counter electrode reached a cell efficiency of 5.81% under 100 mW/cm^2, which is attributed to the low chargetransfer resistance and the low Nernstian diffusion resistance[49].

7.4. OLED

PEDOT anodes for OLED substantially reduce the turn-on voltage and made current injection almost linear after turn-on, attributed to the reduced barrier at the organic-organic interface in PEDOT, the 'ladder' effect of stepping the band offset over several interfaces, and the favorable PEDOT film morphology. The benefit of the PEDOT anode is clearly seen in the improvement in device brightness and the high external quantum efficiency obtained. The best performance achieved for OLED based on Alq3/TPD active regions with PEDOT:PSS as an anode layer yielded a brightness of 1720 cd/m^2 at 25 V, a turn-on voltage of 3 V, and electroluminescence (EL) efficiency and external quantum efficiency of 8.2 cd/A and 2%, respectively, at a brightness of 100 cd/m^2 and 5 V[50]. The metal or semimetal properties of the MWCNTs might contribute to the improved electric properties of the nanocomposites of PEDOT:PSS and MWCNTs, which leads to the improvement of the OLEDs such as the improved luminance intensity and decrease of turn-on voltage[51]. The appropriate mild oxygen plasma treatment of PEDOT:PSS layers could form uniformly distributed nanoislands on the surface of PEDOT:PSS layers, which may improve the contact between PEDOT:PSS anode buffer layer and hole transport layer, leading to improvement of device performance. The OLEDs with PEDOT:PSS after mild oxygen plasma treatment as anode buffer layer exhibited significantly enhanced lifetime and decreased driving voltage[52]. An electrophilic addition reaction involving the PPV vinylene bond and protons from PEDOT:PSS occurred between PEDOT:PSS layer and PPV layer for OLED and PEDOT:PSS has a considerable effect on molecular dopants dispersed in the polymeric host. Defect states in an interface region between them were created by the interaction and resulted in a considerable quenching of the PPV photoluminescence so that the photoluminescence intensity of PPV in a working device under operating conditions can be quenched by at least 20% with respect to the intensity at zero applied volt. Chemical interactions between layers in the light-emitting device can result in the occurrence of additional quenching mechanisms that influence both the host polymer and molecular dopants dispersed in the host[53].

7.5. Sensors

PEDOT is an attractive candidate for sensing material in ion selective electrodes because it shows high selectivity toward the dopant ion in its oxidized state, a selectivity which depends more on the size of the ion than its lypophilicity. And PEDOT-based ion selective sensors are expected to deteriorated more slowly than other CP-based electrodes due to it environmental stability in the oxidized state. PEDOT was successfully used for sensing perchlorate in a new solid-state potentiometric sensor for perchlorate ion. Over the 8 month period of this study, the PEDOT(ClO_4^-) sensors exhibited a stable, linear response spanning at least five orders of magnitude in concentration (1Mto 1×10^{-5}M perchlorate) with near-Nernstian slopes approaching -59 mV/decade of ClO_4^- concentration and a limit of detection of 5×10^{-6} M[54]. The GC/PEDOT($CB_{11}H_6Br_6^-$) electrode for sensors is sensitive to Ag^+ in the concentration range of 10^{-1}-10^{-5} M $AgNO_3$ solutions with sub-Nernstian slope[55]. A significant enhancement of the potentiometric response was achieved after the electrodes were pretreated maybe due to a decrease in the amount of elemental silver (Ag^0/Ag^+ ratio) present in the polymer film, thereby facilitating ion transfer at the polymer/solution interface. The electrode showed high selectivity to Ag^+ in comparison to many alkali-, alkaline-earth, and transition-metal cations except for Hg^{2+}. For simple pressure sensors based on polyimide membrane and a PEDOT/PSS strain sensor, an approximate value of 0.8 for the strain gauge factor could be derived from membrane mechanics, clearly demonstrating the potential of PEDOT/PSS as a mechanical sensor material and might lead to completely organic MEMS[56].

References

[1] Lapkowski M and Pron A, 2000 *Synth. Met.* **110** 79
[2] Dkhissi A, Louwet F, Groenendaal L, Beljonne D, Lazzaroni R and Brédas J L 2002 *Chem. Phys. Lett.* **359** 466
[3] Kvarnström C, Neugebauer H, Blomquist S, Ahonen H J, Kankare J and Ivaska A 1999 *Electrochim. Acta* **44** 2739
[4] Kvarnström C, Neugebauer H, Ivaska A and Sariciftci N S 2000 *J. Mol. Struct.* **521** 271
[5] Reynolds J R, Kumar A, Reddinger J L, Sankaran B, Sapp S A and Sotzing G A 1997 *Synth. Met.* **85** 1295
[6] Kiebooms R, Aleshin A, Hutchison K and Wudl F 1997 *J. Phys. Chem. B.* **101** 11037
[7] Ahonen H J, Lukkari J and Kankare J 2000 *Macromolecules* **33** 6787
[8] Pei Q, Zuccarello G, Ahskog M and Inganäs O 1994 *Polymer* **35** 1347
[9] Cornil J, dos Santos D A, Beljonne D and Brédas J L 1995 *J. Phys. Chem.* **99** 5604
[10] Heywang G and Jonas F 1992 *Adv. Mater.* **4** 116
[11] Aleshin A A, Kiebooms R and Heeger A J 1999 *Synth. Met.* **101** 369~370
[12] Dietrich M, Heinze J, Heywang G and Jonas F 1994 *J. Electroanal. Chem.* **369** 87
[13] Kvarnström C, Neugebauer H, Blomquist S, Ahonen H J, Kankare J, Ivaska A and Sariciftci N S 1999 *Synth. Met.* **101** 66
[14] Pettersson L A A, Johansson T and Carlsson F 1999 *Synth. Met.* **101** 198
[15] Aasmundtveit K E, Samuelsen E J, Pettersson L A A, Inganäs O, Johansson T and Feidenhans R 1999 *Synth. Met.* **101** 561

[16] Aasmundtveit K E, Samuelsen E J, Inganäs O, Pettersson L A A, Johansson T and Ferrer S 2000 *Synth. Met.* **113** 93
[17] Travers J P, Sixou B, Berner D, Wolter A, Rannou P, Beau B, Pépin-Donat B, Barthet C, Guglielmi M, Mermilliod N, Gilles B, Djurado D, Attias A J and Vautrin M 1999 *Synth. Met.* **101** 359
[18] Mathiyarasu J, Senthilkumar S, Phani K L N and Yegnaraman V 2008 *Mater. Lett.* **62** 571
[19] Terzi F, Zanardi C, Martina V, Pigani L and Seeber R 2008 *J. Electroanal. Chem.* **619-620** 75
[20] Murugan A V, Gopinath C S and Vijayamohanan K 2005 *Electrochem. Commun.* **7** 213
[21] Peng C, Jin J and Chen G Z 2007 *Electrochim. Acta* **53** 525
[22] Moon J S, Park J H, Lee T Y, Kim Y W, Yoo J B, Park C Y, Kim J M and Jin K W 2005 *Diamond Relat. Mate.* **14** 1882
[23] Her L J, Hong J L and Chang C C 2006 *J. Power Sources* **161** 1247
[24] Xu Y, Wang J, Sun W and Wang S 2006 *J. Power Sources* **159** 370
[25] Yang Y, Jiang Y, Xu J and Yu J 2007 *Colloids Surf., A : Physicochem. Eng. Aspects* **302** 157
[26] Xia Y and Lu Y 2008 *Compos. Sci. Technol.* **68** 1471
[27] Wang Y, Barnes A, Wright P and Chambers B 2008 *Eur. Polym. J.* **44** 257
[28] Yildiz E, Camurlu P, Tanyeli C, Akhmedov I and Toppare L 2008 *J. Electroanal. Chem.* **612** 247
[29] Chang C C, Her L J and Hong J L 2005 *Electrochim. Acta* **50** 4461
[30] Yigitsoy B, Varis S, Tanyeli C, Akhmedov I M and Toppare L 2007 *Electrochim. Acta* **52** 6561
[31] Camurlu P, Tarkuç S, Şahmetlioğlu E, Akhmedov İ M, C Tanyeli and L Toppare, 2008 *Sol. Energy Mater. Sol. Cells* **92** 154
[32] Joans F, Krafft W and Muys B 1995 *Macromol. Symp.* **100** 169
[33] de Leeuw D M, Kraakman P A, Bongaerts P F G, Mutsaers C M J and Klaasen D B M 1994 *Syth. Met.* **66** 263
[34] Carlberg J C and Inganäs O 1997 *J. Electrochem. Soc.* **144** L61
[35] Kudoh Y, Akami K, Kusayanagi H and Matsuya Y 2001 *Synth. Met* **123** 541
[36] Pei Q, Zucarello G, Ahlskog M and Inganäs O 1994 *Polymer* **35** 1347
[37] Sotzing G A, Reddinger J L, Reynolds J R and Steel P J 1997 *Synth. Met.* **84** 199
[38] Chen X, Xing K and Inganäs O 1996 *Chem. Mater.* **8** 2439
[39] Dkhissi A, Louwet F, Groenendaal L, Bljonne D, Lazzaroni R and Brédas J L 2000 *Chem. Phys. Lett.* **359** 466
[40] Plesse C, Vidal F, Randriamahazaka H, Teyssié D and Chevrot C 2005 *Polymer* **46** 7771
[41] Zainudeen U L, Careem M A and Skaarup S 2008 *Sens. Actuators B* **134** 467
[42] Cho M S, Seo H J, Nam J D, Choi H R, Koo J C, Song K G and Lee Y 2006 *Sens. Actuators B* **119** 621
[43] Liu K, Hu Z, Xue R, Zhang J and Zhu J 2008 *J. Power Sources* **179** 858
[44] Kudoh Y, Akami K and Matsuya Y 1999 *Synth. Met.* **102** 973
[45] Nogamia K, Sakamoto K, Hayakawa T and Kakimoto M 2007 *J. Power Sources* **166** 584
[46] Huang L-M, Lin H-Z, Wen T-C and Gopalan A 2006 *Electrochim. Acta* **52** 1058
[47] Biancardo M, West K and Krebs F C 2007 *J. Photochem. Photobiol., A* **187** 395
[48] Hong W, Xu Y, Lu G, Li C and Shi G 2008 *Electrochem. Commun.***10** 1555
[49] Chen J-G, Wei H-Y and Ho K-C 2007 *Sol. Energy Mater. Sol. Cells* **91** 1472
[50] Mu H, Li W, Jones R, Steckl A and Klotzkin D 2007 *J. Lumin.* **126** 225
[51] Wang G-F, Tao X-M and Wang R-X 2008 *Compos. Sci. Technol.* **68** 2837
[52] Zhou Y, Yuan Y, Lian J, Zhang J, Pang H, Cao L and Zhou X 2006 *Chem. Phys. Lett.* **427** 394
[53] van Dijken A, Perro A, Meulenkamp E A and Brunner K 2003 *Org. Electron.* **4** 131
[54] Bendikov T A and Harmon T C 2005 *Anal. Chim. Acta* **551** 30
[55] Mousavi Z, Bobacka J, Lewenstam A and Ivaska A 2006 *J. Electroanal. Chem.* **593** 219
[56] Lang U, Rust P and Dual J 2008 *Microelectron. Eng.* **85** 1050

Design and fabrication of a GaAs/$Al_{0.4}Ga_{0.6}As$ micro-accelerometer based on piezoresistive effect

Guowen Liu[1,2], Binzhen Zhang[1,2] and Kairui Zhang[1,2]

[1]National Key Laboratory for Electronic Measurement Technology, North University of China Taiyuan, Shanxi, 030051, P.R.China
[2]Key Lab on Instrumentation Science & Dynamic Measurement of the Ministry Education, North University of China, Taiyuan 030051, P.R.China

E-mail: jacky.mucklow@iop.org

Abstract. In this paper, a novel piezoresistive accelerometer based on the piezoresistive effect of GaAs/$Al_{0.4}Ga_{0.6}As$ thin films was designed. The piezoresistive accelerometer contains four suspended flexural beams and a central proof mass configuration. The piezoresistive effect of a piezoresistor or thin film was used to make a resistor changing the output that is proportional to applied acceleration. The GaAs-based piezoresistive accelerometer was prepared with advanced surface micromachining processes, and bulk micromachining processes. Finally, the static pressure experiments were conducted on the sensing element. The experimental results showed that the combined semiconductor heterostructures and mechanical cantilevers have a good stress sensitive characteristic. The integration of these technologies promises to bring about a revolution in the applications of the semiconductor fine-structure devices.

1. Introduction

Although, the Si piezoresistor accelerometers have a good performance in practice, the temperature effect is very vital to it, the measurement result will shift with the variety of temperature. However, the GaAs material can make up the limitation of Si. Epitaxially grown III–V materials are widely used for the production of high-speed VLSI circuits, monolithic microwave integrated circuits, and laser based optical communication systems. More recently, there have also been growing interests in III–V materials for the fabrication of MEMS devices. For example, III–V surface micromachining has been demonstrated for applications including wavelength division multiplexing [1],optical telecommunication [2], and far infra-red photodetection [3], and GaAs has been widely used for the fabrication of MEMS components such as thermoconvertors [4-5] , infrared detectors [6], and mechanical sensors [7]. MEMS based accelerometer is one of the most important types of the mechanical sensors, since there have been large demands for accelerometers in automotive applications, where they are used for crash detection, and for vehicle stability systems. In addition, due to small size and light weight, they are also used in biomedical and robotics applications for active motion monitoring, and in consumer for stabilization of pictures in camera, head-mounted displays.

Piezoresistive accelerometers have simplicity of the structure and batch-fabrication process, as well as a dc response, simple readout circuits, ability to meet the requirement of high reliability and low cost in addition to the potential for mass production [8-11]. But, the sensitivity of piezoresistive accelerometers is not as high as that of the capacitive and tunneling accelerometers. But these problems can be improved by using novel materials. In contrast with Si, III–V compounds offer a number of material-related and technology advantages. There are some physical intrinsic properties of the direct band gap, low thermal conductivity, piezoelectricity effect, and its high saturation velocity of electrons [12-14].

In this study, a novel micromachined piezoresistive accelerometer based on the piezoresistive effect of $GaAs/Al_{0.4}Ga_{0.6}As$ thin films is designed. And the accelerometer is fabricated using the lithography, wet etching, ion implantation and evaporated metal liftoff pattering technique. At last, Static pressure experiments on piezoresistor have been conducted. It is desirable that the application of ingenious materials of the sensor may improve the sensitivity.

2. Design of the accelerometer

We designed a piezoresistive microaccelerometer configuration with four suspended symmetric beams and a seismic mass based on the $GaAs/Al_{0.4}Ga_{0.6}As$. The piezoresistive thin films on each flexure beam is patterned into a transducer element, thus four piezoresistive transducers are arranged on the four beams symmetrically to form the sensing devices in the structures. Thin films are formed by airbridge technique at suitable places on the surface of the GaAs sensing beam, which themselves are fixed to the rigid frame at the ends as Figure 1. When an external acceleration is applied to the accelerometer, the seismic block will be displaced due to the inertial force. This movement of the seismic mass makes the beams deformed. As a result, the resistance of $GaAs/Al_{0.4}Ga_{0.6}As$ piezoresistors will be changed. When there is incentive direct current, the change of resistance will be converted to an output voltage change by an external circuit. So we can get the external pressure by measuring the shift of current-voltage characteristics.

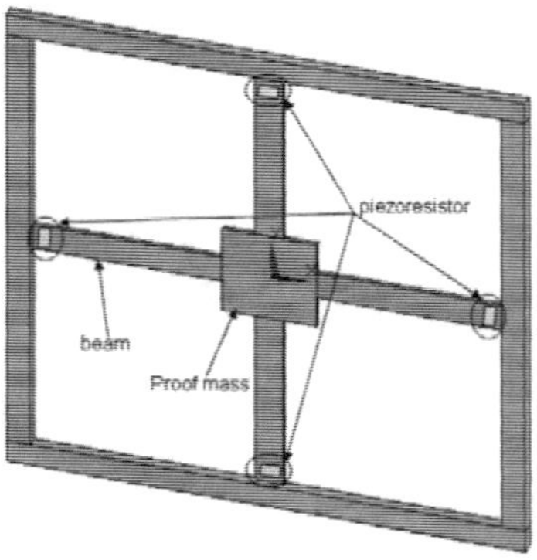

Fig 1. Illustration of the top view of the GaAs piezoresistor accelerometer

3. Fabrication of the accelerometer

After the $GaAs/Al_{0.4}Ga_{0.6}As/AlAs$ piezoresistive thin films are grown, piezoresistor microstructures are fabricated, using the lithography, wet etching, ion implantation and evaporated metal liftoff pattering technique. In step 1, the inductively coupled plasma (ICP) is introduced to etch two slots down to the piezoresistive layer, And Ti/Pt/Au metal is deposited and patterned on the top of the piezoresistive thin films mesas to form the ohmic contacts. In step 2, Using ICP etching forms the piezoresistor. In step 3, the ion implantatation mask defines the area to be isolated, and the implantation of B^+ with a dose of 2×10^{17} cm^2 produces an isolated layer among the piezoresistive device. In step 4, a 1240 Å PECVD Si_3N_4 film is deposited at low temperature (200℃) to form the insulating layer, then the Si_3N_4 film is etched by reactive ion etching (RIE) and the collector contact layer is exposed. In step 5, the polyimide sacrifice layer is coated and etched to fabricate air-bridge; down-lead holes are etched in the C and E electrodes, and a Ti/Au alloy metal is vaporized and etched

Journal of Physics: Conference Series **152** (2009) 012024 doi:10.1088/1742-6596/152/1/012024

to form the deck of the air-bridge. In step 6, Au/Ge/Ni metal is deposited on the deck to strength the contact layers and air-bridge, and then alloyed at 410 ℃ to form the electrodes. In step 7, the emitter electrode and collector electrode are fabricated through the metal layer by a shallow etch, using wet etching. In step 8, the sacrifice layer is wiped off and the double air-bridge structure of piezoresistive structure is formed as Figure 2.

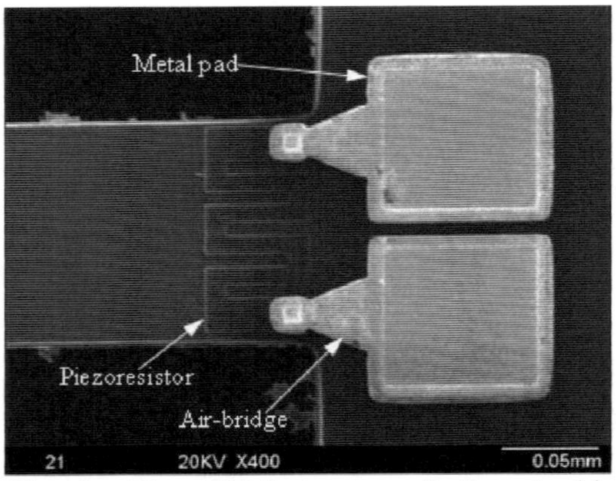

Fig 2. The SEM image of the double air-bridge Heterostructure

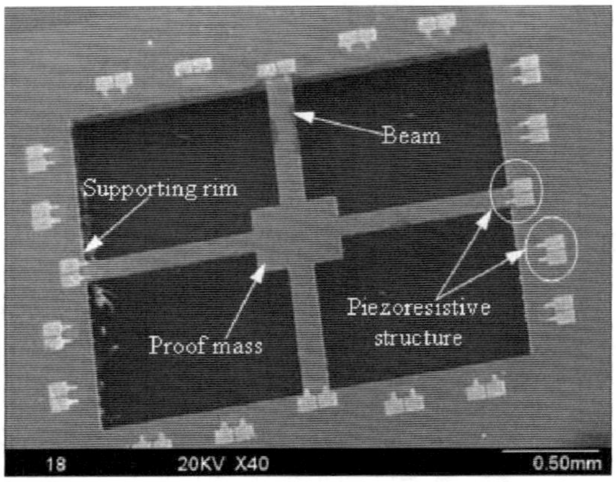

Fig 3. The SEM image of GaAs microaccelerometer

Figure 3 describes the fabricating steps of the cantilever proof mass structure. The control hole etching technology is introduced to control the thickness of the beams accuracy.15 The three dimensional patterns of four independent thin GaAs cantilever beams are defined by a double side aligned selective of GaAs material through the opening in the masks to the membrane. After thinning the substrate down to 130 um by mechanically rubbing, the photolithography is used for masking the chip of the frontside of the membrane, and then 20 um cantilever beams are defined and etched initially from the front side by deep frontside vertical etching. Next, the cantilever beams and proof mass are formed successfully by etching 110 um from the backside of membrane again, when the frontside holes are visible. To make the sensing elements get a much lager stress signal, the sensing elements of piezoresistor are placed at the edge of the support rim and proof-mass where the stress variation is maximum.

4. Experimental and discussion

To study the piezoresistive effect of uniaxial pressure on sensitive element, the static pressure experiments have been conducted on Piezoresistor. The static pressure measuring system is shown in Figure 4.In this experiment, the Raman spectrum system was introduced as the quantitative analysis of the external pressure. The piezoresistor sensing element is fixed on the probe station and lighted by Raman laser. When compressive uniaxial pressures are applied to the sensing element along [110] direction using probes, the strain of the piezoresistor can be reflected by the Raman dispersion spectrum. So the Raman shifts acquire new values when the resistance is subjected to external mechanical stresses. The change in the Raman shifts can be calculated as [15]:

$$\Delta\omega_j = \varpi_j - \varpi_0 \cong \frac{\lambda_i}{2\varpi_0} \quad j = 1,2,3 \tag{1}$$

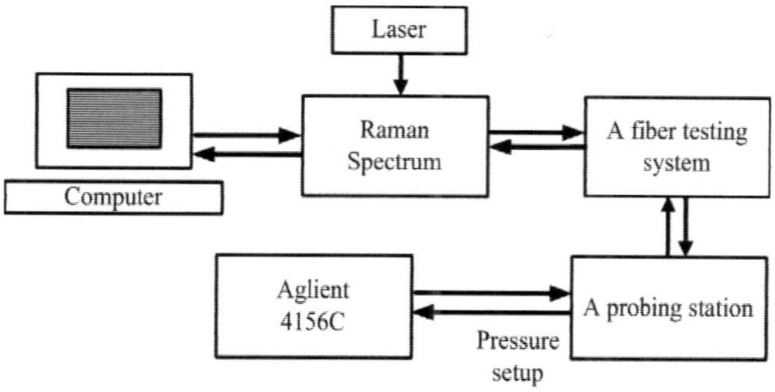

Fig 4. The schematic diagram of the pressure experiment.

Where, λ_j (j = 1, 2, 3) are the eigenvalues of the matrix shown in (2).

$$\varphi_{ij} = \begin{bmatrix} p\varepsilon_{11} + q(\varepsilon_{22} + \varepsilon_{33}) & 2r\varepsilon_{12} & 2r\varepsilon_{13} \\ 2r\varepsilon_{12} & p\varepsilon_{22} + q(\varepsilon_{33} + \varepsilon_{11}) & 2r\varepsilon_{23} \\ 2r\varepsilon_{13} & 2r\varepsilon_{23} & p\varepsilon_{33} + q(\varepsilon_{11} + \varepsilon_{22}) \end{bmatrix} \quad (2)$$

The following stress (σ_{xx}) and strain (ε_{ij}) tensors of the crystal coordinate system is resulted from A simple coordinate transformation [16].

$$\varepsilon_{ij} = \begin{bmatrix} \frac{(s_{11} + s_{12})}{2}\sigma_{xx} & \frac{s_{44}}{2}\sigma_{xx} & 0 \\ \frac{s_{44}}{2}\sigma_{xx} & \frac{(s_{11} + s_{12})}{2}\sigma_{xx} & 0 \\ 0 & 0 & s_{12}\sigma_{xx} \end{bmatrix} \quad (3)$$

Here, s_{11}, s_{12} and s_{44} are elements of the compliance tensor of crystal. The p, q and r are the phonon deformation potentials and ϵ_{ij} are elements of the strain tensor referred to the crystal coordinate system. From (1), (2), (3) and the basic physical quantities of GaAs, we can obtain:

$$\sigma = -576\Delta\varpi \quad (4)$$

Where, "+" means the stress is a compression one, and "–" means the stress is a tensile stress. $\Delta\varpi$ is the Raman shift, and σ is the external pressure. Corresponding to the Raman shift, it is a uniquely determined value of the local uniaxial stress in the GaAs piezoresistor. At the same time, the current–voltage (I–V) characteristics are tested by using Agilent 4156C semiconductor characteristic analyzer. The measured I–V curves of piezoresistor at room temperature are presented in Figure 5, and the resistance obtained by linear fitting in Origin 7.0. The shift of resistance under different pressures is shown in Figure 6. The resistance linearly increases with the strengthening of pressure.

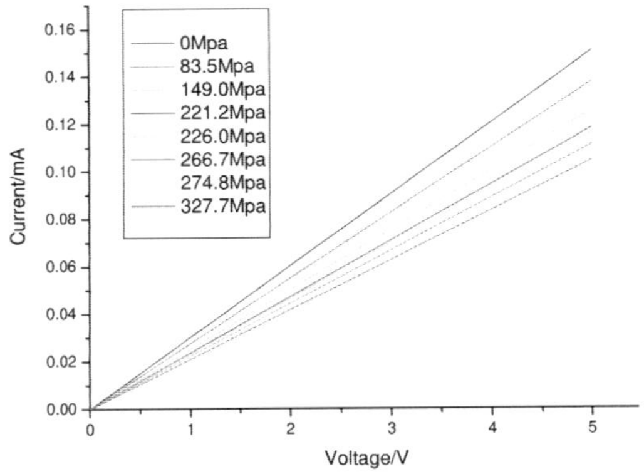

Fig 5. Pressure dependent *I–V* characteristics.

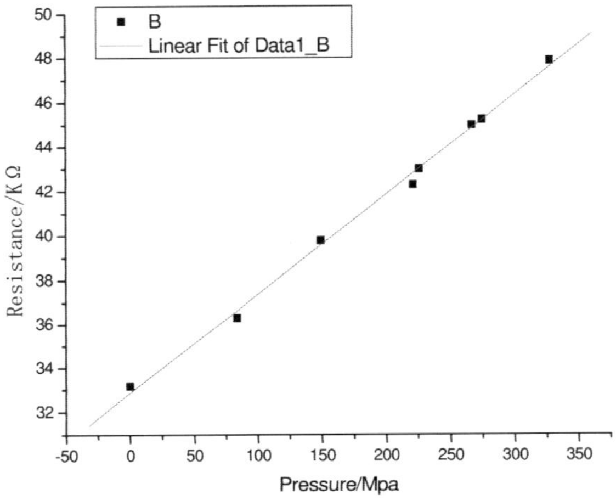

Fig 6. Resistance change under different pressure.

5. Conclusions

A micromachined piezoresistive accelerometer with a high sensitivity is designed. The technology for the fabrication of the piezoresistive accelerometer based on $GaAs/Al_{0.4}Ga_{0.6}As$ thin films has been presented. The static pressure experiments show that the GaAs exhibits an excellent piezoresistive response, which may benefit to the future integration of the microsensors and the actuators.

References and notes

[1] G. Jacquemod, K. Vuorinenk, F. Gaffiot, A. Spisser, C. Seassal, J.-L. Leclereq, P. Rojo-Romeo, P. Viktorvitch, MOEMS modeling for optoeletromechanical co-simulations, J. Model. Simulat. Microsyst.1 (1) (1999) 39.

[2] C. Seassal, J.L. Leclercq, P. Viktorvitch, Fabrication of InP-based free standing micro structures by selective surface micromachining, J. Micromech. Microeng. 6 (2) (1996) 261.

[3] P. Viktorovitch, III–V semiconductor-based MOEMS, in: Proceedings of the SPIE Symposium on Design, Test, and Microfabrication of MEMS and MOEMS, vol. 3680, Paris, France, 1999, p. 30.

[4] J. Miao, H.L. Hartnagel, B.L. Weiss, R.J. Wilson, Improved freestanding semi insulating GaAs membranes for sensor applications, Electron. Lett. 31 (13) (1995) 1047.

[5] K. Hjort, Sacrificial etching of III–V compounds for micromechanical devices, J. Micromech.

Microeng. 6 (4) (1996) 370.
[6] H. Ukita, Y. Uenishi, H. Tanaka, A photomicrodynamic system with a mechanical resonator monolithically integrated with laser diode on GaAs, Science 260 (5) (1993) 786.
[7] Jie Hu, Chenyang Xue, Wendong Zhang, Jijun Xiong, Binzhen Zhang and Hui Qiao,'A Piezoresistive Microaccelerometer Based on Resonant Tunneling Heterostructures', Sensor Letters Vol 6, No.1, p. 197-203 (2008).
[8] N.Y azdi, F.A yazi, and K.Najafi, Proceeding of the IEEE 86, 8 (1998).
[9] P.W .Barth, F.Pourahmadi, R.Mayer , J.Po ydock, and K.Peterson, Technical Digest of IEEE Solid State Sensor and Actuator Workshop, USA, June (1998).
[10] H.Sandmaier , K.K uhl and E.Obermeier, Technical Digest of the 4th Intl. Conf. Solid State Sensors and Actuators (Transducer87), Japan (1987).
[11] H.Chen, S.Shen, and M.Bao, Sensor and Actuators, A 58, 197 (1997).
[12] J. L. Leclercq, R. P .Ribas, J. M. Karam, and P. Viktorovitch, Microelectron. J 29, 613 (1998).
[13] K. Hjort, J. Soderkvist, and Schweitz, J. Micromech. Microeng. 4, 13 (1994).
[14] S.C.Jian, M.W ilander, and H.Maes, Semicond. Sci. Technol. 11, 641 (1996).
[15] DeWolf, H. E. Maes, and S. K. Jones, J. Appl. Phys. 79, 7148 (1996).
[16] J. F. Nye, Physical Properties of Crystals Clarendon, Oxford, U.K (1985).

Journal of Physics: Conference Series **152** (2009) 012025 doi:10.1088/1742-6596/152/1/012025

Theoretical study of composition fluctuation in InGaN films on various substrates

K Mukose[1] and M Sano[1]

[1] Faculty of Engineering, Tokyo University of Science, 1-3 Kagurazaka, Shinjuku-ku, Tokyo 162-8601, Japan

E-mail: mukose@ee.kagu.tus.ac.jp

Abstract. The residual strain due to composition fluctuation in InGaN films grown epitaxially on various InGaN/GaN multiple quantum well (MQW) structures was simulated numerically by a novel method based on first principles. It was found that the strain energy due to the interlayer shear stress caused by the composition fluctuation in InGaN/GaN MQW could be controlled by the InGaN/GaN layer thickness ratio. The critical temperature, which is the maximum substrate temperature at which composition fluctuation can occur, was also investigated as well, and was also found to be dependent on the structure of the MQW.

1. Introduction

$In_xGa_{1-x}N$ with a wurtzite structure is a semiconductor, and its direct band gap is considered to be potentially variable from 3.510 to 0.78 eV [1] depending on the composition x. This compound has been used to successfully fabricate some optical devices such as light emitting diodes (LEDs) [2] and laser diodes [3]; in the future, it is expected that this compound will be used to develop tunable optical devices with a wide wavelength range from near-infrared to ultraviolet. The objective of this study is to understand the mechanisms of epitaxial growth and control properties of InGaN films on various substrates.

In the ternary alloy InGaN, the indium composition has been known to exhibit spatial inhomogeneity under different growth conditions [4−6]. Some theoretical studies conducted in the past have revealed that a ternary system is not thermodynamically stable [7, 8]. The influence of strain on phase separation has also been studied theoretically [9]. It is generally believed that the fluctuation in the composition of indium in InGaN is responsible for the formation of quantum disk structures in InGaN quantum wells; these quantum disks localize excitons and influence the spontaneous emission rate [5, 6, 10].

Some long-range atomic orderings have been observed to coexist with the compositional phase separation in wurtzite InGaN [11−13]. These chemical ordering have been explained theoretically with a model of adsorption of indium atoms at certain positions on the surface of the alloy during the crystal growth process [14].

In this study, the structure dependency of the interlayer shear strain caused by fluctuation in the composition of an InGaN film grown epitaxially on an InGaN/GaN multiple quantum well (MQW) structure was investigated quantitatively by using the first-principles-based approach.

2. Simulation method

Our simulation model for InGaN films contains triangular pillar-shaped cells, where the composition ratio, strain, and stress in each cell follow an equation of state, which has been determined by *ab initio* electronic structure calculations, as shown in figure 1. The volume and shape of the cells were optimized to minimize the free energy in the system. In this study, the residual strain due to spatial fluctuation in the composition of indium in InGaN films grown epitaxially on the InGaN/GaN MQW was simulated theoretically by using this novel model.

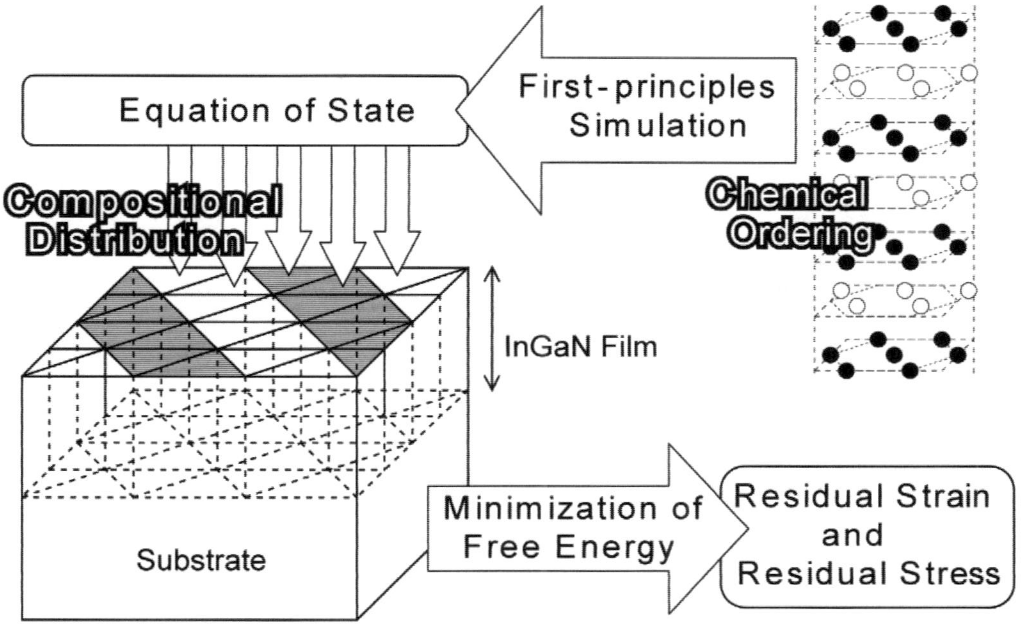

Figure 1. Flow chart of the simulation model for an InGaN film on a substrate.

The equation of state, which is the relation between the lattice constants and the internal energy in InGaN of various composition ratios, was evaluated using a first-principles electronic state calculation. The calculations were performed using the "PWscf" code [15], which is a package of programs within the density functional theory (DFT), and by using *ab initio* pseudopotentials and a plane-wave basis set.

In this study, it is assumed that phase separation coexists with the *c*-plane atomic ordering in InGaN thin films on substrates. The thin films are assumed to have one of five structures: two of these are the GaN and InN type while the other three are the *c*-plane atomic ordered structures of $In_{0.25}Ga_{0.75}N$, $In_{0.5}Ga_{0.5}N$, and $In_{0.75}Ga_{0.25}N$, as shown in figure 2. The In atoms and Ga atoms are stacked alternately at equilibrium positions near the (1/3, 2/3, 0; 2/3, 1/3, 1/2) sites, and the N atoms are at the equilibrium positions near the (1/3, 2/3, u; 2/3, 1/3, 1/2+u) sites of the pseudo-hexagonal structure. The stacking sequences of the layers of the group-III atoms are In-Ga-Ga-Ga-In-Ga-Ga-Ga-... for $In_{0.25}Ga_{0.75}N$, In-Ga-In-Ga-... for $In_{0.5}Ga_{0.5}N$, and In-In-In-Ga-In-In-In-Ga-... for $In_{0.75}Ga_{0.25}N$. These structures of $In_{0.25}Ga_{0.75}N$ and $In_{0.5}Ga_{0.5}N$ are consistent with the former observation by Behbehani *et al* [12].

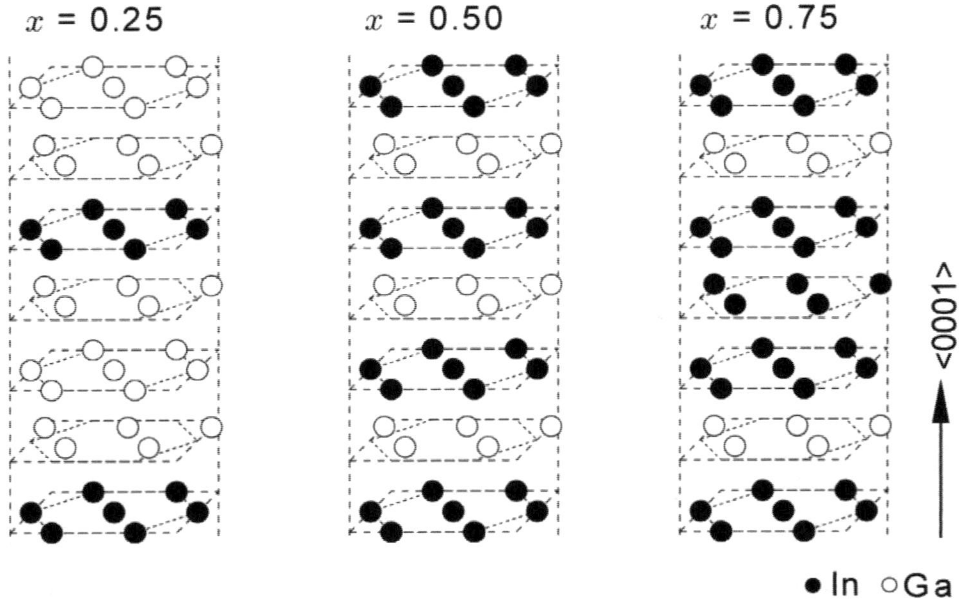

Figure 2. c-plane atomic orderings in $In_xGa_{1-x}N$. Nitrogen atoms are not shown in the figure.

The model of the InGaN/GaN MQW is shown in figure 3. The InGaN layers and GaN layers are alternately stacked in the direction of the c-axis. Here the N_W is the number of atomic layers in the each well layer of InGaN, and the N_B is that in the each barrier layer of GaN. There has been assumed to be no lattice defect in the model. The average linear atomic densities of in-layer directions were fixed to the values of bulk GaN, under an assumption that the MQW was grown on a thick enough GaN substrate, while that of the out-of-layer direction was optimized. A phase separation in one InGaN layer is assumed to influence the strain in the next InGaN layer and cause a similar separation in that InGaN layer.

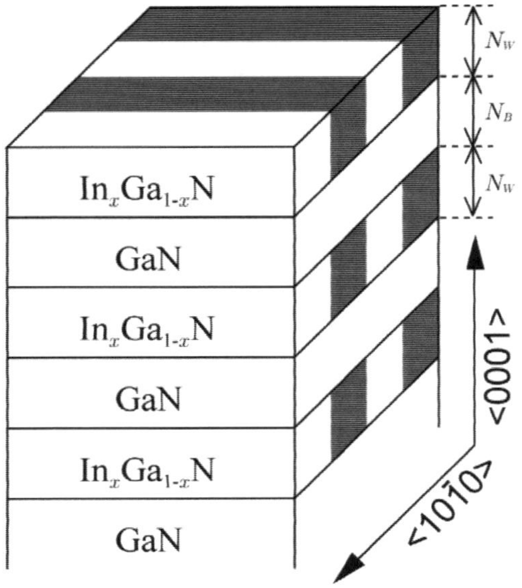

Figure 3. Model of the InGaN/GaN MQW.

One can investigate various spatial patterns of phase separations by using triangular pillar-shaped cells models. In this study, simple stripe patterns were investigated systematically for different average compositions, as shown in figure 4.

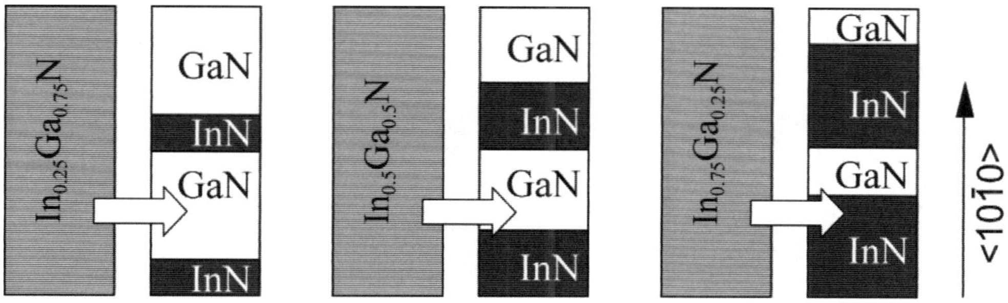

Figure 4. Simple stripe models of compositional phase separation of $In_xGa_{1-x}N$.

3. Results and discussions

The critical temperature, which is the maximum substrate temperature at which compositional phase separation in InGaN can occur, was estimated by dividing the free energy difference from the homogeneous state to the separated state by the corresponding entropy. The average composition dependence of the critical temperature of an InGaN film on an InGaN/GaN MQW is shown in figure 5. It was observed that the larger the InGaN/GaN layer thickness ratio, the higher is the critical temperature, as shown in figure 6. In the other words, the small InGaN/GaN layer thickness ratio suppresses the composition fluctuation.

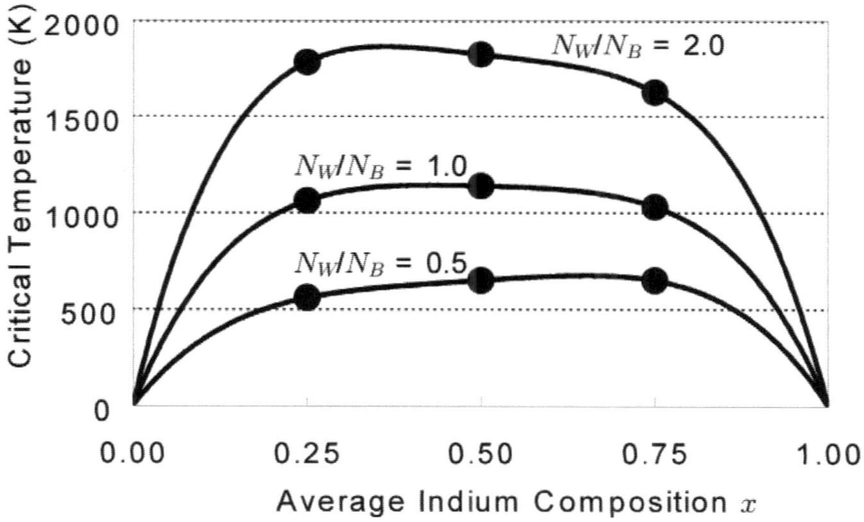

Figure 5. Average composition dependence of the critical temperature of $In_xGa_{1-x}N$ film on $In_xGa_{1-x}N$/GaN MQW for different layer thickness ratios N_W/N_B.

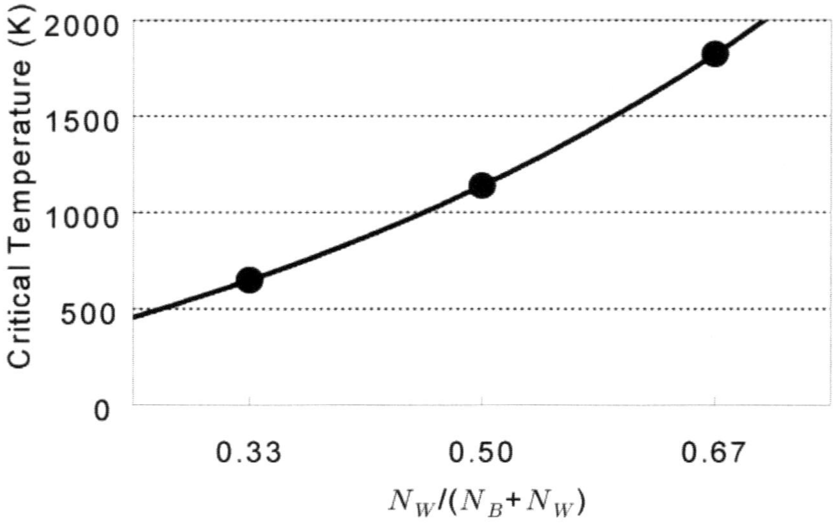

Figure 6. Structure dependence of the critical temperature of an $In_{0.5}Ga_{0.5}N$ film on $In_{0.5}Ga_{0.5}N$/GaN MQW structures.

For the separated state in each $In_xGa_{1-x}N$/GaN MQW structure, the strain energy due to the interlayer shear stress along the $<10\bar{1}0>$ direction of the pseudohexagonal heterostructure was evaluated, as shown in figures 7 and 8. It was found that the strain energy can be widely controlled by the structure, as shown in figure 10. The residual strain energy is maximized when the average indium composition x is about 0.5 and the InGaN/GaN layer thickness ratio is approximately 1.

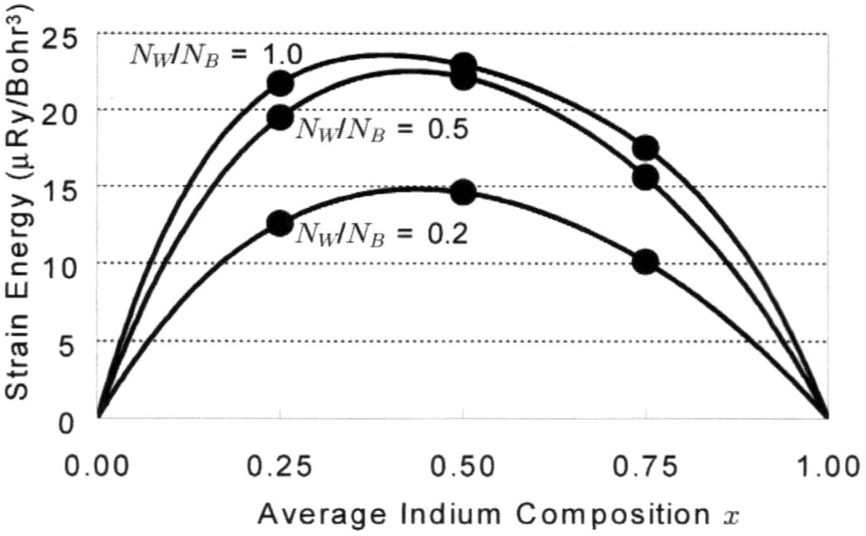

Figure 7. Average composition dependence of the strain energy due to interlayer shear stress along the $<10\bar{1}0>$ direction in the phase-separated $In_xGa_{1-x}N$/GaN MQW with layer thickness ratio $N_W/N_B \leq 1$.

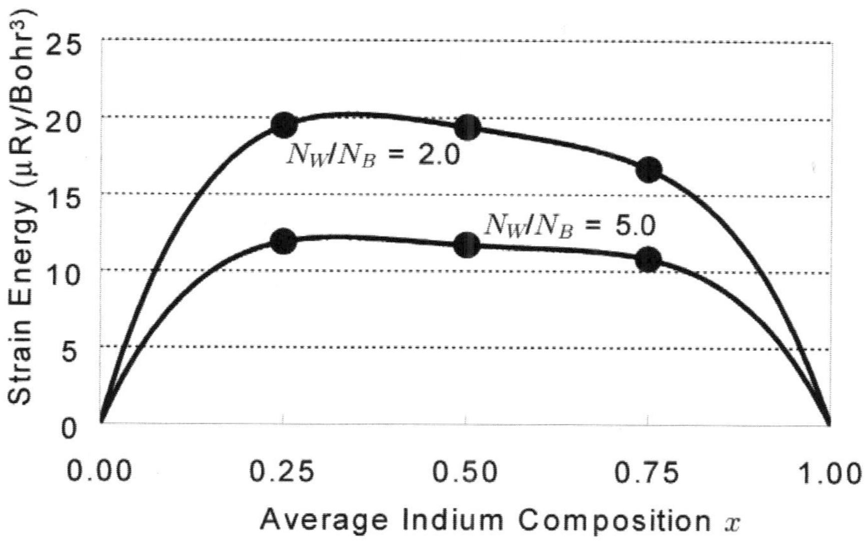

Figure 8. Average composition dependence of the strain energy due to interlayer shear stress along the $<10\bar{1}0>$ direction in the phase-separated $In_xGa_{1-x}N$/GaN MQW with layer thickness ratio $N_W/N_B > 1$.

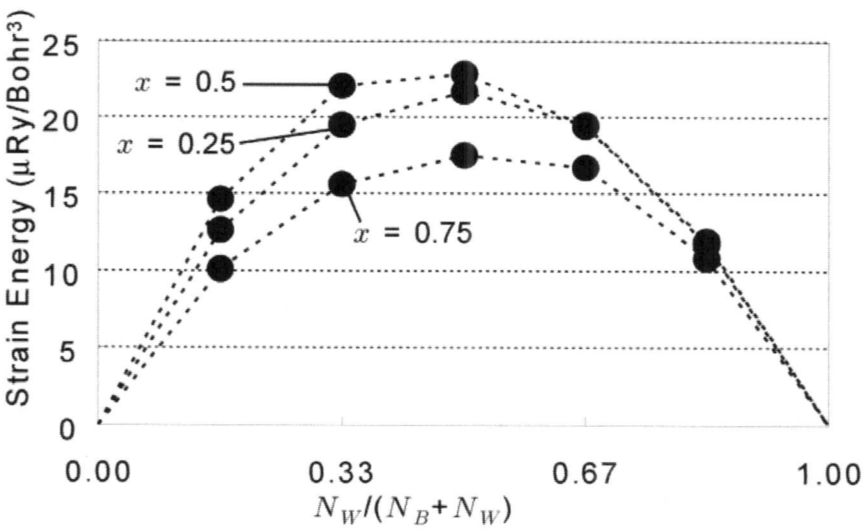

Figure 9. Structure dependence of the strain energy due to interlayer shear stress along the $<10\bar{1}0>$ direction in the phase-separated $In_xGa_{1-x}N$/GaN MQW with layer thickness ratio N_W/N_B.

A rough estimate of the minimum average length between defects that can be caused by this interlayer shear strain in $In_{0.5}Ga_{0.5}N$/GaN MQW with $N_W/N_B = 1$ is approximately 70 nm, as dividing the length of the corresponding Burgers vector by the strain. That is, it seems that a quantum disk structure of a diameter of that is shorter than about 140 nm can be formed with no defect in the $In_{0.5}Ga_{0.5}N$ layers of the MQW structure, if no other mechanism is present.

4. Conclusions

We used a novel simulation method based on first principles to numerically investigate the structure dependency of the interlayer shear strain caused by composition fluctuations in InGaN films grown epitaxially on an InGaN/GaN MQW structure. The residual strain energy was found to be strongly dependent on the structure. The critical temperature, which is the maximum temperature at which the composition fluctuation can occur, was found to reduce significantly when the InGaN/GaN layer thickness ratio in the MQW was reduced.

References

[1] Vurgaftman I and Meyer J R 2003 *J. Appl. Phys.* **94** 3675
[2] Nakamura S, Mukai T and Senoh M 1994 *Appl. Phys. Lett.* **64** 1687
[3] Nakamura S, Senoh M, Nagahama S, Iwase N, Yamada T, Matsushita T, Kiyoku H and Sugimoto Y 1996 *Japan. J. Appl. Phys.* **35** L74
[4] Singh R, Doppalapudi D, Moustakas T D and Romano L T 1997 *Appl. Phys. Lett.* **70** 1089
[5] Narukawa Y, Kawakami Y, Fujita S, Fujita S and Nakamura S 1997 *Phys. Rev. B* **55** R1938
[6] Narukawa Y, Kawakami Y, Funato M, Fujita S, Fujita S and Nakamura S 1997 *Appl. Phys. Lett.* **70** 981
[7] Ho I H and Stringfellow G B 1996 *Appl. Phys. Lett.* **69** 2701
[8] Takayama T, Yuri M, Itoh K and Harris J S 2001 *J. Appl. Phys.* **90** 2358
[9] Karpov S Y 1998 *MRS Internet J. Nitride Semicond. Res.* **3** 16
[10] Chichibu S, Sota T, Wada K and Nakamura S 1998 *J. Vac. Sci. Technol. B* **16** 2204
[11] Doppalapudi D, Basu S N, Ludwig K F and Moustakas T D 1998 *J. Appl. Phys.* **84** 1389
[12] Ruterana P and Deniel F 1999 *Mater. Sci. Eng. B* **59** 186
[13] Behbehani M K, Piner E L, Liu S X, El-Masry N A and Bedair S M 1999 *Appl. Phys. Lett.* **75** 2202
[14] Northrup J E, Romano L T and Neugebauer J 1999 *Appl. Phys. Lett.* **74** 2319
[15] Baroni S, deGironcoli S, DalCorso A and Giannozzi P *PWscf (Plane-Wave Self-Consistent Field) codes* http://www.pwscf.org/

Journal of Physics: Conference Series **152** (2009) 012026 doi:10.1088/1742-6596/152/1/012026

Reactively sputtered Ti-Si-N films for application as heating layers for low-current phase-change memory

You Yin, Tomoyuki Noguchi, Kazuhiro Ota, Naoya Higano, Hayato Sone, and Sumio Hosaka

Department of Production Science and Technology, Gunma University, 1-5-1 Tenjin, Kiryu, Gunma 376-8515, Japan

E-mail: yinyou@el.gunma-u.ac.jp

Abstract. In this study, we investigate the properties of Ti-Si-N films for the application as the heating layers in phase-change memory (PCM). The experimental results show that the resistivity of Ti-Si-N films can be varied by over six orders of magnitude from 2.18×10^{-4} to 3.9×10^{2} $\Omega\cdot$cm by increasing the flow rate ratio [$N_2/(N_2+Ar)$] from 0 to 10%. The controllability of resistivity might result from the concentration change from Ti-Si to mixture of TiN and Si_3N_4. Reversible switching was also successfully demonstrated by using a lateral PCM with these heating layers. The stability of the Ti-Si-N films at high temperatures implies that they can be used as the heating layers in the conventional vertical PCMs for current reduction.

1. Introduction

These guidelines, Phase-change memory (PCM) is widely regarded as the most promising non-volatile memory (NVM) substitute for the current flash memory [1-5]. PCM is based on the large resistivity difference between amorphous and crystalline phases, which can be reversibly transformed by the application of pulses of the order of ns [1, 4]. Although PCM as an NVM technology has almost perfect properties such as excellent endurance, direct overwriting, huge read dynamic range, fast speed, high performance, multi-state storage [7-8], and so on, its high programming current is the most important issue for its commercialization since its invention in 1960s [9-10].

Solutions to the issue of the high programming current mainly include selection or change of phase-change (PC) materials based on material properties and structure optimization [11-12]. Particularly, the programming current is expected to be greatly reduced due to the thermal confinement by introducing a heating layer such as GeN, SiGe, SiGeN and TiON into the conventional bottom contact PCM structure [13-15].

Furthermore, a lot of PC films except the intensively researched $Ge_2Sb_2Te_5$ were proposed and exhibited good performance in recent years [16-18]. These materials in poly-crytalline phase have a resistivity of $10^{-4}-1\Omega\cdot$cm , one-to-two orders of magnitude either higher or lower than that of $Ge_2Sb_2Te_5$, i.e., around 10^{-2} $\Omega\cdot$cm . Further reducing the programming current of the PCM with these PC materials must optimize device structure, for example, inserting a heating layer close to PC layers. Correspondingly, heating materials with a very wide range of resistivity, for example,

$10^{-3} - 10^{2}\ \Omega\cdot\text{cm}$, should be developed for these PC films. As far as we know, there is lack of such a universal heating material for PCM application.

In this study, we report the controllability of resistivity of TiSiN films by N_2 partial flow rate during reactive sputtering and how great the effect of the resistivity of the heating layer in PCM has on the programming current.

2. Experimental

Ti-Si-N films with a SiO_2 capping layer on glass substrates were prepared by simultaneously introducing N_2 and Ar into the chamber when sputtering. The radio-frequency sputtering equipment (MNS-3000-RF, ULVAC, Inc.) was adopted. The schematic diagram is shown in Fig. 1. Six Si chips (around 10×10 mm^2) were mounted on the Ti target. The sputtering conditions are as follows: a background pressure below 5×10^{-5} Pa, a sputtering pressure of 0.2 Pa and a power of 100 W. Flow rate ratio of $N_2/(N_2+Ar)$ increased from 0 to 10%. The total gas flow rate of (N_2+Ar) was set to be 10 sccm.

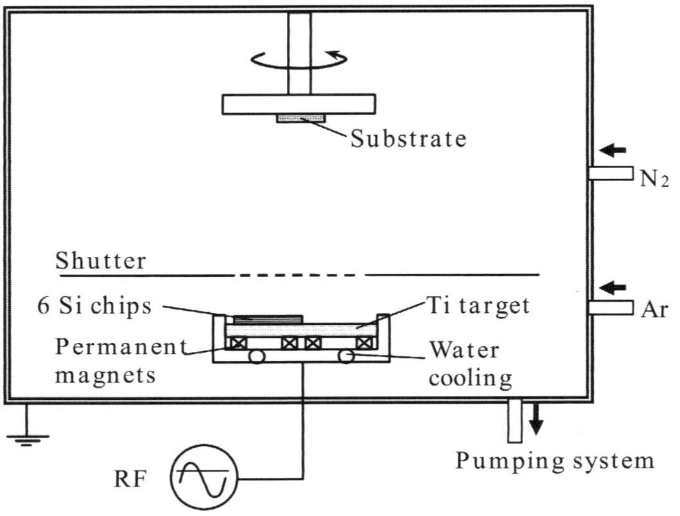

Fig 1. Schematic diagram of reactive RF magnetron co-sputtering apparatus.

The crystal structures of these films were characterized by means of an X-ray diffractometer (RINT 2000, Rigaku Co.) after the films were annealed on a hot plate up to 400°C for 5 min. The resistivity, as a function of the annealing temperature of the films annealed in the same way, was measured using square-shaped film samples defined by Ti electrodes.

The current-voltage (*I-V*) characteristics of the device samples were measured by a semiconductor parameter analyzer (4155B, Agilent Technologies, Ltd.). A waveform generator (Model 2571, Tabor Electronics, Ltd.) was used to apply single pulses to the devices.

3. Results and discussion

Sputtering rate of the Ti-Si-N films was determined by atomic force microscope (AFM) in the profile mode. It slightly decreases with increasing N_2 percentage in the mixed gas as shown in Fig. 2. Typical scanning electronic microscope (SEM) images are shown in Fig. 3. All of these reactively sputtered films were very smooth.

Resistivity of Ti-Si-N films with increasing N_2 percentage in the mixed gas as shown in Fig. 4 was investigated. The resistivity of the Ti-Si film (N_2: 0%) exhibits a very low resistivity of $2.18\times10^{-4}\ \Omega\cdot\text{cm}$, which is very close to the reported data. It can be seen that increasing N_2 percentage from around 5% to 10% in the mixed gas can effectively control the resistivity of the Ti-Si-

N films. A resistivity with a very wide range of $5.7\times10^{-3}-3.9\times10^{2}\ \Omega\cdot\text{cm}$ is available by simply adjusting the N_2 partial flow rate in the range of 5-10%. The resistivity of Ti-Si-N films increases with a strong dependence on flow percentage ($N_2/(N_2+Ar)$). One percentage of the increase in flow percentage ($N_2/(N_2+Ar)$) from 5% to 10% causes around one order of magnitude of resistivity increase.

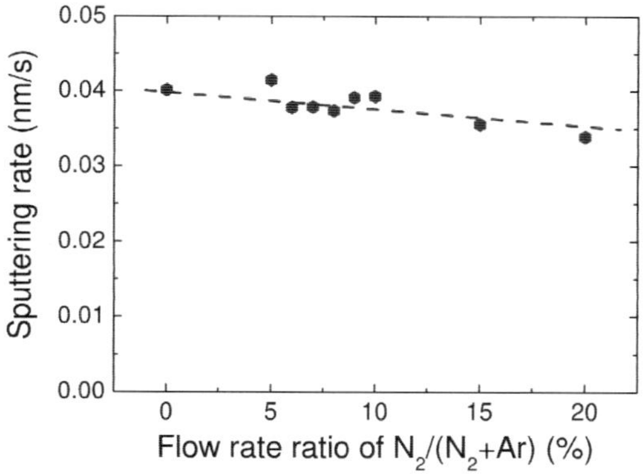

Fig 2. Sputtering rate as a function of the flow rate ratio of $N_2/(N_2+Ar)$.

Fig 3. SEM images of Ti-Si-N film with the flow rate ratio ($N_2/(N_2+Ar)$) of 9%.

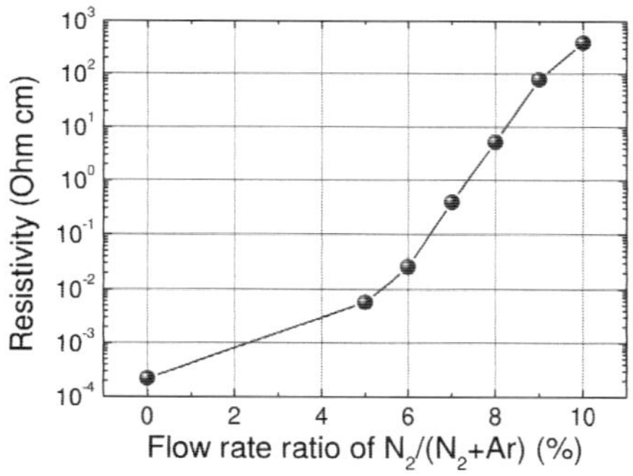

Fig 4. Resistivity of Ti-Si-N films as a function of the flow rate ratio of $N_2/(N_2+Ar)$.

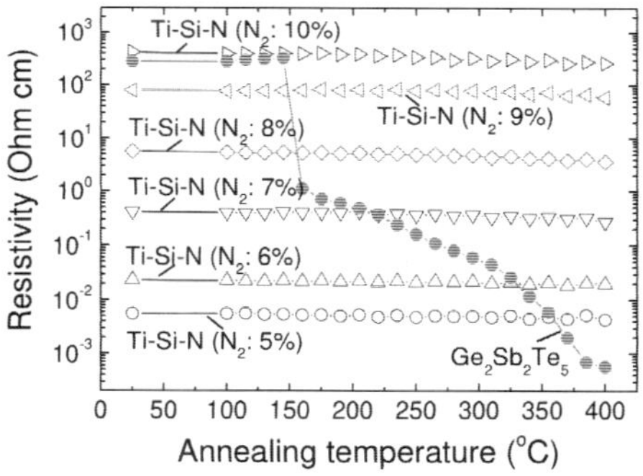

Fig 5. Resistivity of TiSiN and $Ge_2Sb_2Te_5$ films as a function of annealing temperature.

The effect of annealing on the resistivity was investigated and is shown in Fig. 5. The resistivity of $Ge_2Sb_2Te_5$ (GST, most widely used in PCM) films is also plotted for reference. Compared with the marked change in resistivity of $Ge_2Sb_2Te_5$ due to phase transformations, all of these Ti-Si-N films showed relative stability with temperature. Ti-Si film (N_2: 0%) has a very low resistivity of around $2.18\times10^{-4}\ \Omega\cdot\text{cm}$, which is approximately the resistivity of hexagonal GST. When the flow percentage ($N_2/(N_2+Ar)$) increases to 6%, the resistivity of Ti-Si-N films increases up to the order of $10^{-2}\ \Omega\cdot\text{cm}$, corresponding to that of the crystalline GST with a face-centered cubic structure. The resistivity of Ti-Si-N films (N_2: 10%) reaches up to $390\ \Omega\cdot\text{cm}$, which is a little higher than that of amorphous GST.

Furthermore, thickness as a function of annealing temperature of the Ti-Si-N films was studied and showed weak temperature dependence as shown in Fig. 6. This implies that the properties of Ti-Si-N films are thermally stable.

M-Si-N (M= Mo, Ta, Ti, or W) films were investigated in the past years for application to the barrier materials for Al as well as Cu metallizations [19-21]. Although some papers revealed that the resistivity variation of the films is mostly effected by the film composition, a maximum resistivity is usually the order of 1 mΩ cm, which can not meet the requirements of the heater in PCM. These Ti-Si-N films are composed of TiN nanocrystalline grains and Si_3N_4 amorphous phase according to analysis. However, the X-ray diffraction patterns of our sputtered Ti-Si-N films after annealing show no peaks of TiN as shown in Fig. 7. This means that the Ti-Si-N films in this study might be composed of amorphous TiN and Si_3N_4. The composition and resistivity of the Ti-Si-N films are expected to be relatively stable even above the melting of PC materials (500-635°C) because of the very high crystallization temperature of TiN and Si_3N_4. The controllability of resistivity of the Ti-Si-N as shown in Fig. 4 is assumed to be based on the change of volume percentage of Si_3N_4 in the films.

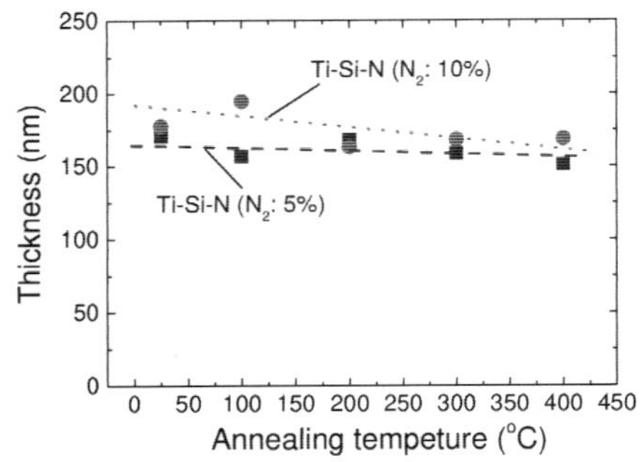

Fig 6. Thickness of TiSiN films as a function of annealing temperature.

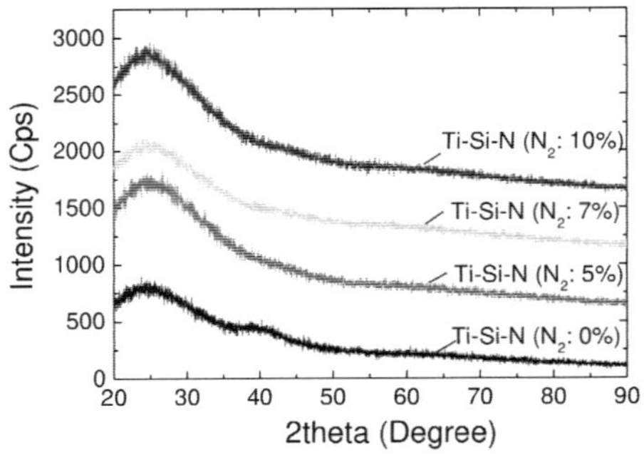

Fig 7. XRD patterns of Ti-Si and Ti-Si-N films.

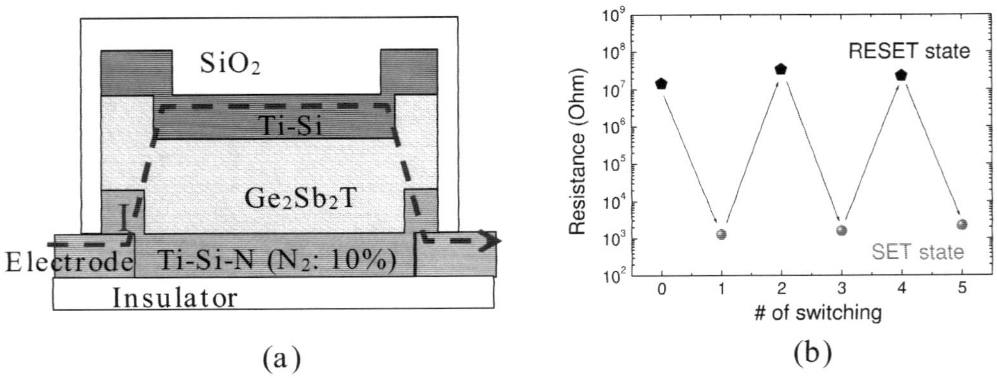

Fig 8. (a) Schematic diagram of a lateral PCM with Ti-Si-N layers. (b) Resistance change as a function of the switching number.

To know if the Ti-Si-N films are possible to apply to PCM, a lateral PCM as shown in Fig. 8(a) was adopted for the testing of the stability of Ti-Si-N films at high temperature. In the channel between two electrodes, the GST layer was sandwiched with the top Ti-Si and the underlying Ti-Si-N

(N_2: 10%) layers. The current flows as shown by the dashed line. From Fig. 8(b), the device with Ti-Si-N layers can be reversibly switched between high-resistance amorphous set and low-resistance crystalline reset states. This means that the Ti-Si-N layers are able to be used as a heating layer in PCM because they can remain stable even at the high temperature above the melting point of GST (632 °C) during switching.

Next, we will show the effect of the resistivity of the heating layer on the reset current of the today's widely researched vertical PCM cell, which was simulated by finite element method. The PCM cell is assumed to have a Si substrate/W/TiN contact/additional heating layer/PC/W-layered structure as shown in the inset of Fig. 9. The thicknesses of all the layers from the bottom W layer are 50, 200, 5, 80, and 50 nm. The reset current can dramatically drop from 1.4 mA to several tens of μA by adopting a resistive heater for the resistivity of crystalline GST of $10^{-2}\ \Omega\cdot\text{cm}$ as shown in Fig. 9.

The simulation results suggest that a heater with a resistivity of one-to-two orders of magnitude is very effective to further reduce the reset current without a significant drop in the difference in device resistance between set and reset states. As a result, the resistivity of the material used for a resistive heater is expected to have a wide range of around $10^{-3}-10^{2}\ \Omega\cdot\text{cm}$ if we take the practical PC materials with a resistivity of $10^{-4}-1\,\Omega\cdot\text{cm}$ into account. The resistivity of Ti-Si-N films can enter into the required range.

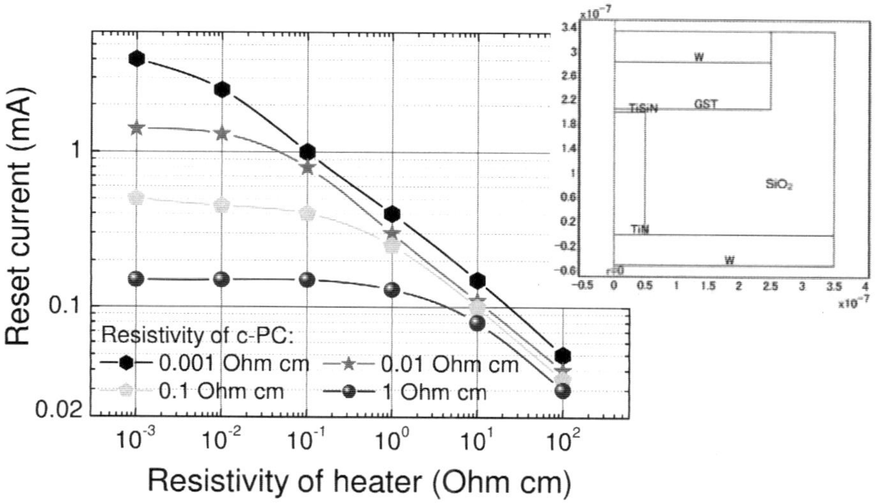

Fig 9. Reduction in reset current with increasing the resistivity of the Ti-Si-N heating layer.

4. Conclusions

In summery, Ti-Si-N films showed a controllable resistivity window from 2.18×10^{-4} to 390 $\Omega\cdot\text{cm}$ and its good thermal stability. It is promising for Ti-Si-N films to be applied to the future's phase-change memory for reducing its operation current.

Acknowledgements

The authors would like to thank for financial support by Semiconductor Technology Academic Research Center (STARC) of Japan.

Reference

[1] S. Lai: Tech. Dig. IEDM, 2003, p. 255.
[2] Y. Yin, D. Niida, K. Ota, H. Sone, and S. Hosaka: Rev. Sci. Instrum. **78** (2007) 126101.
[3] H. Y. Cheng, Y. C. Chen, R. J. Chung, and T. S Chin: Semicond. Sci. Technol. **21**(2006) 1196.
[4] Y. Yin, A. Miyachi, D. Niida, H. Sone and S. Hosaka: Jpn. J. Appl. Phys. **45** (2006) L726.
[5] K. Nakayama, K. Kojima, F. Hayakawa, Y. Imai, A. Kitagawa, and M. Suzuki: Jpn. J. Appl. Phys. **39** (2000) 6157.
[6] Pirovano, A. L. Lacaita, A. Benvenuti, F. Pellizzer, and R. Bez: IEEE Trans. Electron Devices **51** (2004) 452.
[7] Y. Yin, H. Sone, and S. Hosaka: Microelectron. Eng. **84** (2007) 2901.
[8] F. Rao, Z. Song, M. Zhong, L. Wu, G. Feng, B. Liu, S. Feng, and B. Chen: Jpn. J. Appl. Phys. **46** (2007) L25.
[9] D. H. Kang, D. H. Ahn, K. B. Kim, J. F. Webb, and K. W. Yi: J. Appl. Phys. **94** (2003) 3536.
[10] Y. Yin, H. Sone, and S. Hosaka: Jpn. J. Appl. Phys. **45** (2006) 8600.
[11] Y. H. Ha, J. H. Yi, H. Horii, J. H. Park, S. H. Joo, S. O. Park, U. I. Chung, J. T. Moon: Symp. VLSI Tech. Dig. Tech. Pap., 2003, p. 175.
[12] Y. Yin, H. Sone and S. Hosaka: Jpn. J. Appl. Phys. **45** (2006) 6177.
[13] J. Shen, B. Liu, Z. Song, C. Xu, F. Rao, S. Liang, S. Feng, and B. Chen: Applied Physics Express, **1** (2008) 011201.
[14] D. H. Kang, D. H. Ahn, M. H. Kwon, H. S. Kwon, K. B. Kim, K. S. Lee, and B. Cheong: Jpn. J. Appl. Phys. **43** (2004) 5243.
[15] S. Y. Lee, K. J. Choi, S. O. Ryu, S. M. Yoon, N. Y. Lee, Y. S. Park, S. H. Kim, S. H. Lee, and B. G. Yu: Appl. Phys. Lett. **89** (2006) 053517.
[16] Y. Ling, Y. Lin, B. Qiao, Y. Lai, J. Feng, T. Tang, B. Cai and B. Chen: Jpn. J. Appl. Phys. **45** (2006) L349.
[17] K. Nakayama, K. Kojima, Y. Imai, T. Kasai, S. Fukushima, A. Kitagawa, M. Kumeda, Y. Kakimoto and M. Suzuki: Jpn. J. Appl. Phys. **42** (2003) 404.
[18] Y. Yin, H. Sone, and S. Hosaka: J. Appl. Phys., **102** (2007) 064503.
[19] J. S. Reid, E. Kolawa, C. M. Garland, M. A. Nicolet, F. cardone, D. Gupta and R. P. Ruiz: J. Appl. Phys., **79** (1996) 1109.
[20] W. H. Lee, S. K. Park, B. J. Kang, P. J. Reucroft, and J. G. Lee: J. Electronic Materials, 30 (2001) 84.
[21] X. Sun, J. S. Reid, E. Kolawa, and M. A. Nicolet: J. Appl. Phys., **81** (1997) 656.

Journal of Physics: Conference Series **152** (2009) 012027 doi:10.1088/1742-6596/152/1/012027

Research on Phase-Change of Disc Recording Media GaSbBi

Degui ZHU[1, a], Shuangquan GUO[1, b], Xuexiang Yuan[1] and Hongliang SUN[1]

[1]Key Laboratory of Advanced Technologies of Materials, Ministry of Education, School of Materials Sci. & Eng., Southwest Jiaotong University, Chengdu 610031, P. R. of China

[a]dgzhu@home.swjtu.edu.cn, [b]shuangquan2006@126.com

Keywords: Phase-change disc; GaSbBi; Hot isostatic pressing (HIP); Crystallization property; Optical property

Abstract: A new phase-change recording materials, Bi-doped eutectic composition GaSb, was researched experimentally. The films were prepared by vacuum evaporation. The crystallization temperature decreases as the Bi concentration increases. The crystallization activation energy, calculated by Kissinger method, is more than 2.0eV and meets the requirement of the films stability. The crystallization rate enhances with the increase of the Bi concentration. The $Ga_{12}Sb_{79.2}Bi_{8.8}$ satisfies the demand of optical property by means of spectrophotometer. The absorption of the $Ga_{12}Sb_{79.2}Bi_{8.8}$ crystalline films is more than that of the $Ga_{12}Sb_{79.2}Bi_{8.8}$ amorphous films, which can reduce the change when reads information. The least reflectivity of $Ga_{12}Sb_{79.2}Bi_{8.8}$ is 28.78%, corresponding with the reflection demand.

1. Introduction

An important aspect of all information transmission is the storage and detection of encoded information [1]. Compact disc (CD) storage technology is a high and new technology, and initiated in the early 1970s. In comparison with other recording media, it combines excellent properties, such as higher density packing, large capacity, random access available, long lifetime of information storage. To date phase-change disc is the most important compact disc-rewritable (CD-RW). Three systems of phase-change disc recording media materials, containing binary, ternary, and quaternary, have been researching (as shown in Table 1) through decades of development [2].

Table1. Common phase-change disc recording media materials [2]

Binary	Ternary	Quaternary
GaSb	$Ge_2Sb_2Te_5$	AgInSbTe
InSb	InSbTe	(Ge,Sn)SbTe
InSe	GaSeTe	GeSb(SeTe)
Sb_2Te_3	$SnSbTe_4$	$Te_{81}Ge_{15}Sb_2S_2$
GeTe	InSbGe	

Although Ge–Sb–Te alloys has been widely researched [3,4], non-tellurium (Te)-based phase-change disc materials have been mostly studied nowadays, for Sb-Te alloys have the disadvantage, such as amorphous state stability and bigger noise. Non-Te based phase-change disc materials have a very high stibium (Sb) content, for instance, germanium stibium (GeSb) and gallium stibium (GaSb), GaSb was first used for amorphous state and crystal condition transformation in 1987 [5]. These non-Te-based materials have the advantage of thermal stability of amorphous recording point and fast crystalline growth rate. Crystalline growth velocity of $Ga_{12}Sb_{88}$ in GaSb series can reach 23.3m/s, besides a good thermal stability. However, the melting point of $Ga_{12}Sb_{88}$ is high. The melting point of bismuth (Bi) is lower than that of $Ga_{12}Sb_{88}$. Moreover, Sb and Bi belong to a same family. Based on the reasons above, this article attempts to gain a new phase-change disc recording materials GaSbBi by Bi-doped eutectic composition GaSb. The following research indicates optical property of new GaSbBi is consistent with that of phase-change disc materials.

2. Experimental procedure

Ingredients of GaSbBi targets prepared according as literature [6] were shown in Table 2. The films were prepared by vacuum evaporation. The substrate was quartz glass with dimensions of 25.4mm×76.2mm×1.0mm. The degree of vacuum was 5×10^{-5} Torr. The annealing temperature of films was 350℃, maintained for 90 minute in the quartz tube by argon shielding. The amorphous powders were got by lightly scraping the films on the substrate without heat treatment. Differential scanning calorimeter (DSC: STA 449C) was used to analyze the crystallization temperature dynamics of the films. The optical property of the films was analyzed using spectrophotometer (UV-22101 and UV-22100).

Table2. Target ingredients

Sample ID	Recipe of targets	Sample ID	Recipe of targets
1#	$Ga_{12}Sb_{88}$	4#	$Ga_{12}Sb_{81.4}Bi_{6.6}$
2#	$Ga_{12}Sb_{85.8}Bi_{2.2}$	5#	$Ga_{12}Sb_{79.2}Bi_{8.8}$
3#	$Ga_{12}Sb_{83.6}Bi_{4.4}$		

3. Results and discussion

3.1. Thermodynamic properties of amorphous powders

Write and stability of the disc are closely related to thermodynamic properties of phase-change materials. Phase-change materials have to be with suitable melting and phase-change temperature because the write of recording layer corresponded with formation of amorphous state. Generally, the crystallization temperature of amorphous state exceeds 150℃, and the crystallization activation energy was over 2.0eV [7].

In this study, the crystallization temperatures of amorphous powders are got through DSC curves at different heating rate. The crystallization activation energy E_a and frequency factor K_0 are acquired with Kissinger method [8] (as shown in formula 1). The integral constant n which is associated with crystal mechanism was obtained by Ozawa method [9] (as indicated in formula 2).

$$\ln(\frac{\alpha}{T_p^{\ 2}}) = \ln A + \ln(\frac{K_0 R}{E_a}) - \frac{E_a}{RT_p} \approx \ln(\frac{K_0 R}{E_a}) - \frac{E_a}{RT_p} \tag{1}$$

Where R is Avogadro constant, α is rate of temperature rise, E_a is crystallization activation energy, A is constant, T_p is crystallization temperature.

$$-n = \frac{d\{\ln\ln[1/(1-\chi)]\}}{d\ln\alpha} \tag{2}$$

Where χ is nucleation, α is rate of temperature rise.

The thermodynamics coefficients calculated in the experiment are shown in Table 5. The results indicated that the ratio of T_p / T_m is about 0.46. Moreover, the value of the ratio increased with the increase of Bi content. Thus the increase of Bi content is in favor of the amorphization. The crystallization activation energy depends on the mobility of the atoms in the order-disorder transition [10]. The weak bond easily makes the phase-change occurred, as a result the crystallization temperature is declined. The E_a from Table 3 are all over 2.0eV, these values of the E_a can be content the requirement of the stability comes up with by Zhong Chao'an [7]. The mode of the crystal growth is bi-dimensional growth when n fall in between two and three, while the mode is one-dimensional growth when n fall in between one and two, according to nucleus theory presented by Avrami [11]. The n calculated in this study fall in between one and two, thus crystal growth of the GaSbBi is reckoned as one-dimensional growth.

Table3. The thermodynamics coefficients of GaSbBi

Sample ID	T_m (℃)	T_p (℃)	Ea(eV)	K_o(min^{-1})	n
1#	601	267.436	2.807	1.57E+26	1.04295
3#	587	266.01	2.229	5.12E+20	1.01762
5#	573.4	264.484	2.1046	4.14E+19	1.00534

3.2. Heat treatment of films

Sample 1# and sample 2# have more diffraction maximum, compared with other three different samples before annealing, according to Figure 1. It is indicated that Bi is good for the amorphization of the films. In contrast with the as-deposited films, the crystallization of the annealed sample 3#, 4# and 5# has obviously occurred.

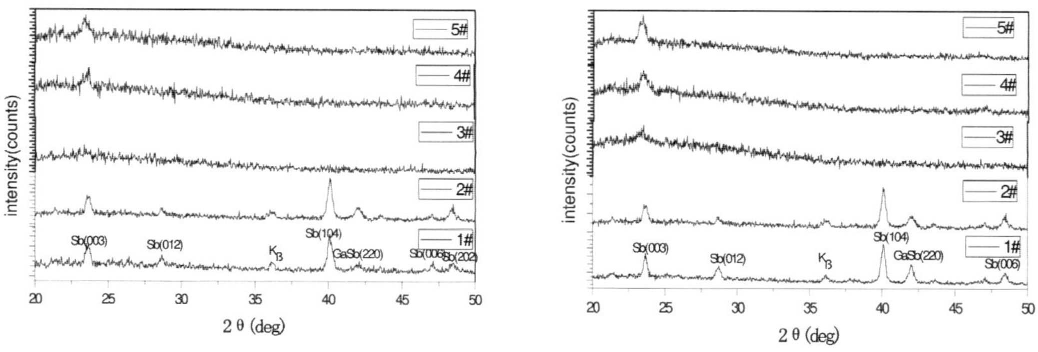

(a) as-deposited films (b) annealed films

Fig 1. X-ray diffraction patterns of films

3.3. Optical properties of films

The optical maser wavelength used in the DVD disc is 650 nm nowadays, while blue-ray disc have a good developing potential. So 450nm and 650nm is used for comparison in this experiment. We only investigate optical properties of sample 1#, 3#, and 5# according to their other good properties. The reflectance, transmittance, reflectivity and absorption of films with sample 1#, 3#, and 5# are followed in Figure 2. The results indicate that the light stability of eutectic composition GaSb is minimally affected by Bi-doped. It is reported that reflectivity of the films have to be over 20% [12]. As far as the reflectivity, both sample 1# and sample 5# are over 20%, which could corresponded with the reflection demand, and the least reflectivity of sample 5# (Ga12Sb79.2Bi8.8) is 28.78%. While sample 3# is not up to the reflection demand. The high value of sample 3# is caused by the experimental error.

The absorption of sample 5# crystalline films is more than that of amorphous films, which could reduce the change when reading information, particularly. When a disc is designed, design practice demands for a higher monolithic reflectivity of the disc. In comparison with sample 1# and 5#, sample 3# has the highest reflectivity. It comes to the conclusion that not only crystalline films but also amorphous films have the same variation tend caused by Bi-doped.

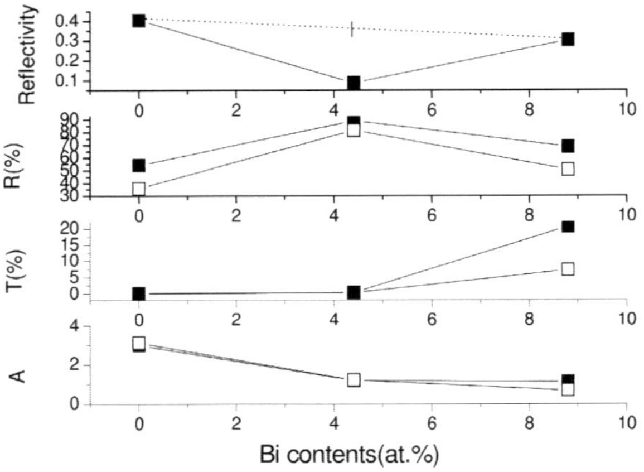

(a) Wavelength of 650nm

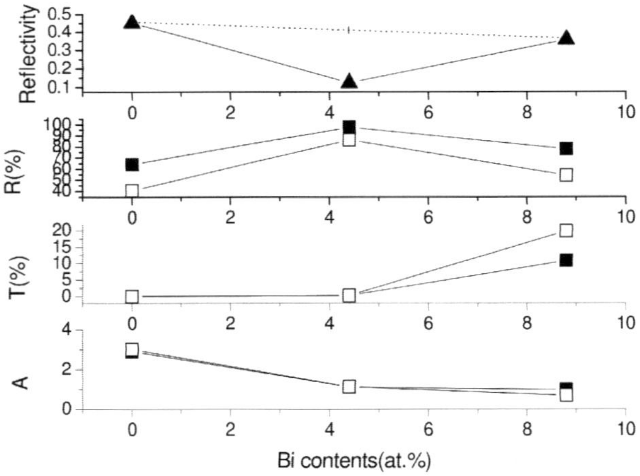

(b) Wavelength of 450nm

Fig 2. Optical properties of films (The solid symbols represent the annealed state, the open ones correspond to the as-deposited state)

The energy gap of films was calculated by Tauc equation [13] (as indicated in formula 3). The results of sample 1#, 3# and 5# are shown in Figure 3. The values of energy gap raised by Bi-doped have marginal discrepancies. So it is reasonable to consider that a handful of Bi contents have little effect of the energy gap.

$$\alpha h\nu = A(h\nu - E_g)^2 \tag{3}$$

Where α is absorption coefficient, $h\nu$ is photon energy, E_g is energy gap, A is constant.

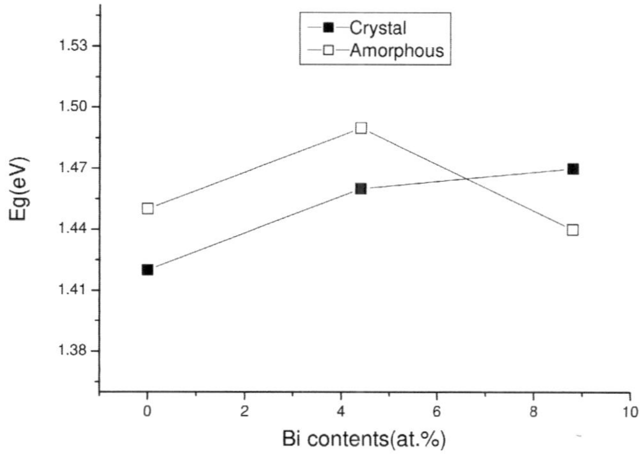

Fig 3. The energy gap of different compositions

4. Conclusion

1. The crystallization temperature decreases as the Bi concentration increases;
2. The crystallization activation energy is more than 2.0eV, which meets the requirement of the films stability;
3. The crystallization rate enhances with the increase of the Bi concentration;
4. The least reflectivity of $Ga_{12}Sb_{79.2}Bi_{8.8}$ is 28.78%, which corresponds with the reflection demand.

References

[1] Ernst Binz, Walter Schempp. Computational and Applied Mathematics 144 (2002) 85-103
[2] Information on http://www.ovonyx.com/technology.pdf
[3] Alexander V. Kolobov, Paul Fons, Junji Tominaga, T. Uruga. Non-Crystalline Solids 352 (2006) 1612-1615
[4] A.V. Kolobov, P. Fons, J. Tominaga. Thin Solid Films 515 (2007) 7534-7537
[5] M.H.R. Lankhorst et al. Jpn. Appl. Phys. 42 (2003) 863
[6] YUAN Xuexiang, ZHU Degui. Information Recording Materials. 2006, 7 (6) 18-21
[7] Zhong Chao'an. Industrial Ttechnology Rresearch Institute 2002 (12)11-19, Taiwan
[8] H. E. Kissinger, Anal. Chem. 1957 (29) 1702
[9] T.Ozawa. Polymer 1971 (12) 150
[10] Qianmin Li. Research on structure, kinetics of crystallization, optical property of phase-change recording materials for Ga-Sb-Te series. Taiwan Tsinghua University, Minguo 92
[11] Schneider, T. and Stoll, E. Phys. Rev. B13, 1216 (1976)
[12] J. Li, F. Gan. Thin Solid Films. 402 (2002) 232-236
[13] Yuzeng Li. Thermal analysis. Tsinghua University Press. 1987, 220-246

Synthesis and Electrical Characterization of a MOS Memory Containing Si Nanoparticles Embedded in High-k HfO_2 Thin Film

Lei Wang[1], Mei Ji, Shiwei Zhu, Hailin Tu, Jun Du[2]

Advanced Electronic Materials Institute, General Research Institute of Nonferrous Metals, No.2 Xinjiekou Wai Street, Beijing, P.R.China

[1]E-mail:wanglei@griaem.com
[2]E-mail:dujun@griaem.com

Abstract. The size-controlled, non-agglomerated, 5nm Si nanoparticles were fabricated in the middle of HfO_2 layers using a pulsed laser ablation (PLA) system at a relatively low pressure and formed trilayer structure (HfO_2/Si-nps/HfO_2) on n-type Si. These high density nanoparticles ($>1x10^{11}$ cm^{-2}) were imaged with SEM. The crystallinity of HfO_2 and Si nanoparticles were characterized by HRTEM. The memory effect of a Si nanoparticles floating gate memory structure consisting of HfO_2 tunneling and control oxides have been investigated by means of high-frequency (1 MHz) capacitance–voltage (C–V) measurements. HRTEM study revealed that the Si nanoparticles are single crystals with average size of about 5 nm in diameter and are well distributed within the amorphous HfO_2 matrix. The memory effect was observed by the hysteresis loop in the C-V curves and a high storage charge density of about $5\times10^{11}cm^{-2}$ and a large flat-band voltage shift of 4.8V have been achieved due to charging and discharging to the nanoparticles.

1. Introduction

Recently, a novel memory cell structure consisting of nanocrystals embedded in dielectrics has received considerable attention because of its high endurance, low operating voltage, reduced lateral discharge paths, promise larger retention time, lower power consumption, and faster operation [1]. Many materials such as Si, Ge, SiGe, silicide and Pt have been considered as promising candidates for the charge storage nodes in nanocrystal floating gate memories [1–15]. Compared with other nanocrystals, Si nanoparticles are easy to nucleate on the high-k substrate and the benefits of both high density and high crystalline can be derived at relatively low temperature, and the Si nanocrystal is considered to be an ideal memory node[3,5,14,15]. Several groups of researchers have demonstrated that Si nanocrystals have very promising properties to act as a storage electrical charge for future memory devices [3,12,16-17]. Si nanocrystals can be synthesized by various methods, including Si ion implantation [18], laser ablated Si target [10], and the thermal annealing of Si grown by electron-beam evaporation [18,19] or cosputtering [2] with the dielectric mixture layers. On the other hand, being physically much thicker than SiO_2, high-k dielectric materials such as HfO_2 with leakage current several orders of magnitude smaller than SiO_2 for the same equivalent oxide thickness (EOT), results in superior data retention property. Mantey et al [3] have also fabricated monodisperse Si nanoparticles on HfO_2 matrix by low pressure chemical vapor deposition.

In this paper, we reported on the formation of silicon nanoparticles at room temperature using a pulsed laser ablation, and the memory effects of the memory structure. The majority of current experimental studies utilized 5~10nm Si nanocrystals embedded in high-k HfO_2 matrix, instead of SiO_2, formed trilayer structure. Compared with the pure HfO_2 system, the charge storage property of HfO_2/Si-nps/HfO_2 system has been studies.

2. Experimental

HfO_2 high-k gate dielectric thin films were deposited by RF magnetron co-sputtering using HfO_2 target (99.9%) in Ar/O_2 (4:1) mixture ambient. By varying the sputtering pressure and sputtering power, amorphous HfO_2 thin film with a thickness of 50Å was directly deposited on n-type Si(100) wafer with a resistance of 2-5Ωcm, which was cleaned by a standard RCA cleaning process, finally immerged into the diluted hydrofluoric acid($HF:H_2O$=1:30) for 30 seconds to prevent the wafers from oxide. After the deposition of Si nanoparticles, a same HfO_2 dielectric thin film was deposited form a trilayer structure consisting of an embedment of Si nanoparticles in HfO_2 matrix. A KrF excimer laser (λ=248nm) with laser fluence of

2 Jcm^{-2} was used for the deposition of size-controlled and non-agglomerated Si nanoparticles in the middle of HfO_2 layers using a pulsed laser ablation (PLA) system at a relatively low Ar pressure (30Pa).

In order to reduce the defects in the dielectric layers and keep the structure stability under high temperature, the as-deposited trilayer structure was annealed at 400°C for 10min in N_2 ambient. Pt dot electrodes with a diameter of 0.2mm were subsequently deposited by PLD. The memory effect of Si nanocrystals floating gate capacitors was characterized by capacitance-voltage (C-V) measurement using an Agilent HP4284 impedance analyzer. The leakage current density of the capacipors was characterized by means of Keithley 2410 source meter. The structure of Si nanoparticles embedded in HfO_2 high-k dielectric were investigated by high-resolution transmission electron microscopy (HRTEM) using a JEOL 2010 electron microscope. Morphology of the Si nanoparticles on the HfO_2 films was observed by Hitachi S4800 SEM.

3. Results and discussion

HfO_2 film with the thickness of 5nm was deposited on n-type Si(100) wafer by rf sputtering and then Si nanoparticles were directly deposited on the HfO_2 film by PLA. The surface morphology of HfO_2 film and the Si nanoparticles were shown in Fig 1, respectively. It can be seen that the surface morphology of HfO_2 film was smooth and free of porosity in Fig 1a. A Quartz Resonators for monitoring thickness of the film was used to form HfO_2 film with the thickness of 5nm. At the same time, Si nanoparticles were characterized by SEM and the nonagglomerated Si particles dispersed on the HfO_2 film with high area density were shown in Fig 1b.

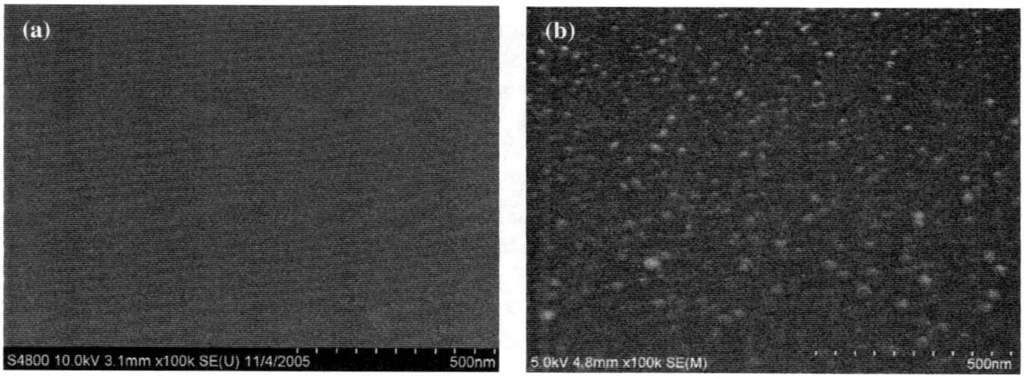

Figure 1. (a)A typical SEM image of HfO_2 film on the Si wafer; (b) SEM image of Si nanoparticles with high area density on HfO_2 film

In order to character structural properties of Si nanoparticles, the TEM and HRTEM were also involved to measure the Si nanoparticles on the HfO_2 film with, which were directly deposited on carbon grid to expediently detect by TEM. In Fig 2a, it can be seen that Si nanoparticles with a diameter of ~5nm are randomly distributed without agglomeration. Most of these sphere-like nanoparticles are single crystals as revealed by the inset SAD image. The corresponding amorphous ring may be come from the carbon grid or thin HfO_2 layer. Further, the structural properties of Si nanocrystal were characterized by HRTEM. A region is multiplied in Fig 2.b. It was showed that the Si-NPs were evidently single crystals with a clear lattice plane spacing of 0.31nm. This value corresponds to (111) plane spacing of the bulk Si (0.312nm)。The amorphous HfO_2 matrix was also revealed in Fig 2b. The corresponding FFT was showed in figure 2.c, and diffraction spots and amorphous diffraction ring confirmed the coexistence of a higher crystallinity of the Si-NPs and amorphous HfO_2 films.

Figure 2. (a) TEM image showing the general morphology of the Si nanoparticles on the HfO_2 film. The inset is SAD (selected area diffraction) pattern (b) HRTEM image of the Si nanoparticle as marked by arrow A in figure 2 a. (c) FFT image for Si nanoparticle

Generally, C-V measurements were very useful for studying the charging storage of the Si nanoparticles by measuring the flat-band voltage of the MOS capacitors. Fig 3 showed the 1 MHz C-V curves measured at room temperature for the MOS capacitor without Si nanoparticles. The HfO_2 film with the thickness of 10nm is used to act as the high-k dielectric, and the structure of MOS is illustrated on the top left cornet in the Fig 3. The sample without Si nanoparticles shows a typical C-V curve without obvious hysteresis loop, which means that the HfO_2 layer and the Si interface fabricated under these condition did not contain any defect states. At the same time, it is worth noting that the effect of mobile ions on the charge trapping can be ruled out, since no significant hysteresis was observed in the sample without Si nanoparticles. The obvious advantage of HfO_2 high-k dielectric layer was low leakage current and relatively thin tunneling layer. Compared with reported leakage current results based on SiO_2 tunneling layer [20,21], a low leakage current can be seen in the right corner of the Fig 3. It is seen that the leakage current density of the sample without Si nanoparticles is 1×10^{-5} A/cm^2 at gate bias voltage 1V.

Storage charge density of the Si nanoparticles was concluded from the trilayer MOS based on the formula given by Tiwari[1] and Kim et al [21]. The schematic of MOS was shown on the top left corner of Fig 4. Nonagglomerated Si nanoparticles with size of 5nm were embedded in the middle of the HfO_2 high-k dielectric films. The density of Si nanoparticles was estimated 5×10^{11}cm^{-2}, which was calculated using surface areas of the ellipsoidal and spherical nanoparticles. High-frequency (1MHz) C-V measurement of the MOS capacitors incorporating the Si nanoparticles revealed significant hysteresis in the range of the gate voltage from -10V to 10V, indicating successful charge storage in the Si nanoparticles in Fig 4. Actually, in this curve, the center of the capacitance hysteresis at 0V bias voltage implies a possibility of low-power device applications.. The insulator capacitance was taken as the maximum capacitance when the MOS structure was in the accumulation region. Then, the flat-band voltage was defined by the absence of a DC potential across the Si nanoparticles [22], which means that the Si nanoparticles were charged electrically. At the same time, the maximum flat-band voltage shift ($\triangle V_{FB}$) for the 5nm Si nanoparticles sample is about 4.8V from the width of the hysteresis, and the calculated storage charge density corresponding to the hysteresis loop is up to $7.5\times10^{12}cm^{-2}$, which was related to charging of carriers into Si nanoparticles [23]. On the other hand, with the introduction of Si nanoparticles, a decrease of the leakage current density of 5×10^{-6} A/cm^2 at gate bias voltage 1V was obtained compared with that of the sample without Si nanoparticles.

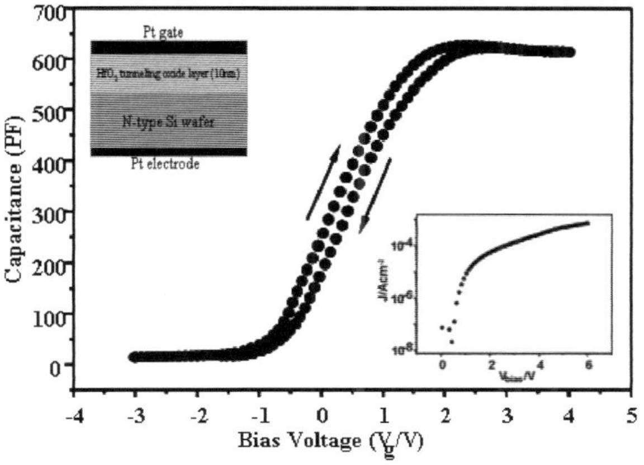

Figure 3. high –frequency (1MHz) C-V characteristics of the MOS capacitor without Si nanopartilces. The inset shows a schematic of the MOS capacitor system on the top left corner. The inset shows an I-V characteristic of the sample on the right corner.

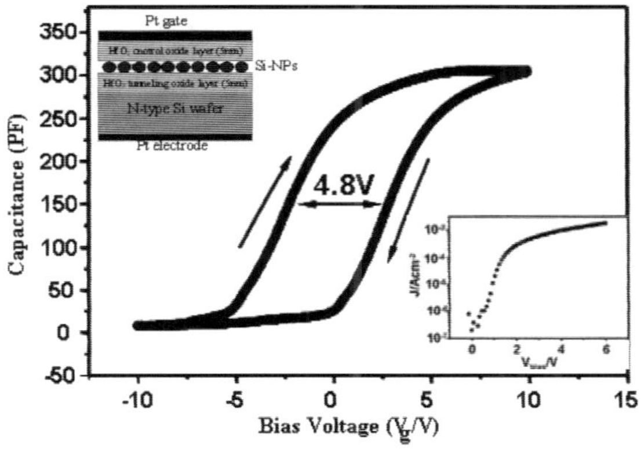

Figure 4. high –frequency (1MHz) C-V characteristics of the MOS capacitor with Si nanopartilces. The inset shows a schematic of the trilayer structure MOS capacitor system on the top left corner. The inset shows an I-V characteristic of the sample on the right corner.

4. Conclusion

In conclusion, single crystals Si nanoparticles with diameters of ~ 5 nm embedded in HfO_2 thin films were made by using the pulsed laser ablation method. According to measurements of leakage current of the capacitors, low leakage current of 5×10^{-6} A/cm^2 has been obtained because of the introduction of the HfO_2 high-k dielectric material. The memory effects of the capacitors with and without Si nanoparticles have been characterized by using C-V measurement. The sample without Si nanoparticles showed a typical C-V curve without obvious hysteresis loop. By contrast, counter-clockwise hysteresis loop are present in the sample with Si nanoparticles. The maximum flat-band voltage shift ($\triangle V_{FB}$) for the 5nm Si nanoparticles sample is about 4.8V from the width of the hysteresis, and the calculated storage charge density corresponding to the hysteresis loop is up to 7.5×10^{12}cm^{-2}, which was contributed to charging of carriers into Si nanoparticles.

Acknowledges
The authors would like to acknowledge financial support from the Dow-Coring Co. We are also grateful to Professor. An-sheng Liu for assistance during the measurement of the HRTEM.

Reference

[1] S. Tiwari, F. Rana, H. Hanafi, A. Hartstein, E. F. Crabbe´, and K. Chan 1996 Appl. Phys. Lett. 68 1377.
[2] N. Nedev, D. Nesheva,F.Manolov, R. Bruggemann, S.Meier, Z.Levi, Microelectronics, 2008. MIEL 2008. 26th International Conference, 117.
[3] O.M.Nayfeh, D.A.Antoniadis, K.Mantey andM.H.Mayfen 2007 Appl. Phys. Lett. 90 153105-1.
[4] M. She and T.J. King 2003 IEEE Trans. Electron Devices 50 1934.
[5] A. Dima, F. G. Della Corte, C. J. Williams, K. G. Watkins, et al. 2008 Microelectronics Journal, Vo39(5) 768.
[6] O.M. Nayfeh, D.A. Antoniadis, K. Mantey, and M.H. Nayfeh, "Towards MOS memory devices containing 1 nm silicon nanoparticles," 2006 IEEE Silicon Nanoelectronics Workshop., Honolulu, HI.
[7] T. Feng, H. Yu, M. Dicken, J. Heath, and H. Atwater 2005 Appl. Phys. Lett. 86 033103.
[8] O. Akcakir, J. Therrien, G. Belomoin, N. Barry, E. Gratton, and M. Nayfeh 2000 Appl. Phys. Lett. 76 1857.
[9] G. Belomoin, J. Therrien, A. Smith, S. Rao, S. Chaieb, and M. H. Nayfeh 2002 Appl. Phys. Lett. 80 841.
[10] P F Lee, X B Lu, J Y Dai, H LWChan, Emil Jelenkovic and K Y Tong 2006 Nanotechnology 17 1202.
[11] S T, Zela V, Gustafsson A, Pietzonka I and Seifert W 2002 Appl. Phys. Lett. 81 3455.
[12] Y-C King, T-J King and Hu C M 2001 IEEE Trans. Electron Devices 48 696.
[13] W L Liu, P F Lee, J Y Dai, J Wang, H L W Chan, C L Choy, Z T Song and S L Femg 2005 Appl. Phys. Lett 86 013110.
[14] Y D Kim, E K Kim, S Lee and W J Cho 2006 Journal of the Korean Physical Society Vol 49(3) 1192.
[15] J J Lee, X Wang, W Bai, N Lu, and D-Lee Kwong 2003 IEEE TRANSACTIONS ON ELECTRON DEVICES Vol50(10) 2067.
[16] A Kanjilal, J L Hansen, P Gaiduk, A N Larsen, N Cherkashin, A Claverie, P Normand, E Kapelanakis, D Skarlatos and D Tsoukalas 2003 Appl. Phys. Lett. 82 1212
[17] M Kanoun, A Souif, T Baron and F Mazen 2004 Appl Phys Lett 84 5079.
[18] H H I Hanafi, S Tiwari and I Khan 1996 IEEE Trans. Electron Devices 43 1553.
[19] Q Wan, T H Wang, M Zhu and C L Lin 2003 Appl. Phys. Lett.82 4708.
[20] H.Wong, H. Iwai. Microelectronic Engineering 2006 83 1867.
[21] J K Kim, H J Cheong, Yong Kim, Jae-Yel Yi, and Hong Jun Bark 2003 Applied Physics Letters Vol 82 2527.
[22] M.H.An, J.Korean 2005 Physics.Soc. 47,847.
[23] J.H.Kim, E.K.Kim, M.S.Song, Y.H.Kim and J.Kim 2004 J.Korean Physics.Soc. 45 795.

Journal of Physics: Conference Series **152** (2009) 012029 doi:10.1088/1742-6596/152/1/012029

Growth and properties of self-assembled monolayers on metals

Y X Zhuang[1, 2], O Hansen[2] and J C He[1]

[1]Key laboratory of Electromagnetic Processing of Materials, Ministry of Education, Northeastern University, Shenyang, 110004, P.R. China
[2] Department of Micro- and Nanotechnology, DTU Nanotech and CINF, Technical University of Denmark, Building 345 east, DK-2800, Kgs. Lyngby, Denmark

[1]Email: yxzhuang@epm.neu.edu.cn

Abstract. Self-assembled monolayers (SAMs) grown from organosilane are promising candidate for anti-stiction coatings. It is well know that the application of hydrocarbon- and fluorocarbon-based SAMs can significantly reduce stiction and adhesion in silicon micro/nanostructures in Micro-Electro-Mechancial System (MEMS). There are often various metals involved in MEMS, such as Au, Al, Ti, etc. In the process of growing anti-stiction SAMs on the silicon microstructures, SAM will be unavoidably deposited on the metals, which will significantly change surface properties of the materials. Therefore, it is necessary to investigate how the organosilane SAMs grow on the metals, and affect the surface properties of the materials. In this paper, $CF_3(CF_2)_5(CH_2)_2SiCl_3$ (FOTS) SAMs were grown on various metals using a vapour phase process in a home-made setup. The metals investigated are thin film Pd, Au, Ni, Al, Ti, Cr with thickness of 100 nm, which are deposited on monocrystal-silicon substrates using e-beam evaporation. The SAMs were characterized by static contact angle, surface energy, roughness, nano-scale adhesion force and friction force.

1. Introduction

Self-assembled monolayer (SAM) coatings have attracted considerable scientific and industrial interest due to their good bonding strength, low surface energy, low friction forces, and good thermal stability [1-6]. Applications range from functionalization of surfaces for bio-technology application to control of adhesion, friction and wear in microdevices and in Nanoimprint Lithography (NIL). Self-assembled monolayers (SAMs) can be deposited either in a liquid phase process or in a vapor phase process. Liquid phase deposition of SAM coatings has significant disadvantages, such as complicated process control, generation of large amounts of contaminated effluents, insufficient stiction prevention and high production costs. Vapor phase SAM deposition can eliminate some drawbacks of liquid-based SAM deposition, and thereby attract strong attention [7-8].

It is well know that the application of hydrocarbon- and fluorocarbon-based SAMs can significantly reduce stiction and adhesion in silicon micro/nanostructures. Meanwhile, there are often various metals involved in MicroElectroMechanical System (MEMS), such as Au, Al, Ti, Pt, Ni, etc. In the process of growing anti-stiction self-assembled monolayers on the silicon microstructures, SAMs will be unavoidably deposited on the metals. This significantly changes surface properties of the materials. Therefore, it was decided to investigate the growth and properties of organosilane SAMs on the metals. A self-assembled monolayer was deposited onto the various metal surfaces using the same process as that used for silicon microstructures. The water contact angles, surface energy, nanoscale adhesion force and friction force of the SAM coatings were characterized.

2. Experimental techniques

The metals investigated are Au, Ni, Al, Pd, Cr, and Ti. The various metals with the thickness of 100 nm were deposited onto monocrystal silicon substrates using e-beam evaporation, where a 10 nm Ti layer was used as an adhesion layer for the deposition of Au, Ni, Al and Pd. The wafers with metal and monocrystal silicon (as a reference) were produced using the processes of photolithography, metal deposition, and lift-off. Afterwards, self-assembled monolayers were grown on silicon and various metals using a vapor phase process with $CF_3(CF_2)_5(CH_2)_2SiCl_3$ (FOTS) as a precursor. The vapor phase coating process was carried out in a home-made setup, which comprises a chamber, precursor injection system, vacuum system etc. Before the actual coating process, an O_2-plasma pretreatment step was applied in order to terminate the wafer surface with -OH groups. The substrates were left in the air (clean-room environment) for 3-4 hours to get enough water adsorbed on the surfaces, and then were loaded into the chamber. After pumping down the system, a certain amount (around 25 μL) FOTS was injected into the system, and kept 15 minutes for reaction. Then, the reaction chamber was pumped down and vented to take out the samples with FOTS self-assembled monolayers.

The contact angle measurements were performed using a contact angle meter DSA10 from Krüss GmbH equipped with an automatic dispensing system for four liquids and a frame grabber. Static contact angles were used to evaluate the SAM coatings. The static contact angle values were taken 5 seconds after deposition of the droplets on the surface to allow droplets relaxation. At least ten measurements were performed for each droplet. The static contact angle values reported in the work are the average value of measurements on at least ten droplets. The surface energy of coatings was calculated from static contact angles of deionised water and diiodomethane according to the Owens–Wendt–Rabel–Kaelble method, as described in ref [9].

The microtopography and roughness of the FOTS coatings were investigated using a commercial AFM system (NanoMan, Digital Instrument, Santa Barbara, CA, USA) in tapping mode with commercial silicon tips. The images were analyzed using the software Nanoscope 6.13 (Digital Instrument, Santa Barbara, CA, USA). The average roughness R_a, root-mean-square roughness RMS, and maximum peak roughness R_{max} were extracted from the images. All roughness data were obtained from a 2×2 μm^2 scanning area in order to eliminate the scanning length effect. AFM observations were performed at different positions to characterize the coating homogeneity.

Nanoscale adhesion and friction force measurements were carried out using a commercial AFM system (NanoMan, Digital Instrument, Santa Barbara, CA, USA) that is operated under ambient conditions at 22°C and 45%–55% relative humidity. Square pyramidal Si_3N_4 tips with a nominal tip radius in the range of 20–60 nm that were mounted on gold-coated Si_3N_4 cantilevers with a nominal spring constant of 0.12 N/m (Digital Instrument, Santa Barbara, CA) were used in this work. The real spring constant of the AFM cantilever used in this work was 0.07 N/m which was determined using "cantilever on cantilever" method. The adhesion force measurements were carried out in the force calibration mode. The normal force calibration constant was obtained from the force-distance curve of the same AFM tip on a nature diamond [6]. For each sample, the adhesion forces were measured in more than 15 different locations. AFM friction measurements were done in lateral force microscopy mode. All scans were performed in a direction that is perpendicular to the long axis of the cantilever beam with a tip velocity of 4 μm/s and a scan length of 2 μm. For each sample, the friction tests were performed at at least three different locations. For easy comparison, all data presented in one plot were obtained with the same AFM tip. A detailed explanation of adhesion and friction measurement can be found in ref [6].

3. Results and Discussions

The static contact angles of water on the surfaces were measured at different process steps, such as after lift off, after O_2-plasma treatment, immediately after the FOTS deposition, and 2 weeks after the

FOTS deposition. Figure 1 gives the static water contact angle of the surfaces at each step. It can be seen that all the surfaces have similar wettability (water contact angle is around 80 degrees) after lift-off process. After 2 min O_2 plasma treatment, the water can completely spread on the silicon surface, indicating that the pre-treatment is good enough for growing FOTS self-assembled monolayer on silicon substrate. Even though the water contact angle on metal surfaces decreases after the oxygen plasma treatment, the wettability is not as good as that of silicon. After FOTS depositing on the surfaces, the water contact angle increases for all the surfaces. The water contact angles are larger than 110 degrees for freshly deposited FOTS on the Ti, Al, and Si substrates, indicating that the FOTS has high coverage and is well grown on those substrates. Meanwhile, the water contact angles remain same after 14 days. FOTS on other metals behave differently though. For example, the water static contact angles of freshly deposited FOTS on Au, Pd, and Ni substrates are only about 46, 84 and 62 degrees, respectively, suggesting that the FOTS coating has very poor coverage, and the FOTS molecules might be physically adsorbed onto those substrates. And the water contact angle increased after 14 days for the FOTS coatings on Au and Ni, which might be induced by contamination.

Figure 2 shows the AFM images of FOTS grown on the various metals and silicon. It should be mentioned that the z-scale used in the AFM images are different. The z-scale is 60 nm for FOTS on Au, Pd, Ni, Ti, and Al, 20 nm for FOTS on Cr, and 5 nm for the FOTS on Si, respectively. It can be seen that the roughness of FOTS on various metals is significantly larger than that of FOTS on monocrystal-silicon, which is caused by larger roughness of metal substrates. Meanwhile, there are aggregations observed on some silicon reference samples (not shown in this paper), which is due to larger amount of water adsorbed on the substrates before the sample being loaded into the reaction chamber. Figure 3 summarizes the RMS roughness, nano-scale adhesion force, and static contact angle of freshly deposited FOTS. The contact angles are strongly dependent on the surface energy of the surface materials and the topography of the surfaces. Larger roughness of Ti and Al explains that the water contact angle of FOTS on Ti and Al substrate is larger than that of FOTS on silicon. The surface energy of FOTS on Au, Pd, Ni, Al, Ti, Cr, and Si substrates (after 14 days storing) are 37.8, 26.2, 14.9, 10.2, 9.9, 11.5, and 11.2 mJ/m^2, respectively. The static water contact angle of freshly deposited FOTS on Cr, Si, Al and Ti are larger than 100°, while the contact angles of other surfaces are less than 90°. The roughness of FOTS on various metals is larger than that of silicon. It can be seen that FOTS on monocrystal silicon has largest nano-scale adhesion force even though it has higher static contact angle than others. It is well know that the nano-scale adhesion force is determined by the surface energy and the roughness of surfaces. The smaller the surface energy, the smaller the adhesion force is. The larger the roughness, the smaller the adhesion force is. This explains why FOTS on Au, Pd and Ni has smaller nano-scale adhesion force.

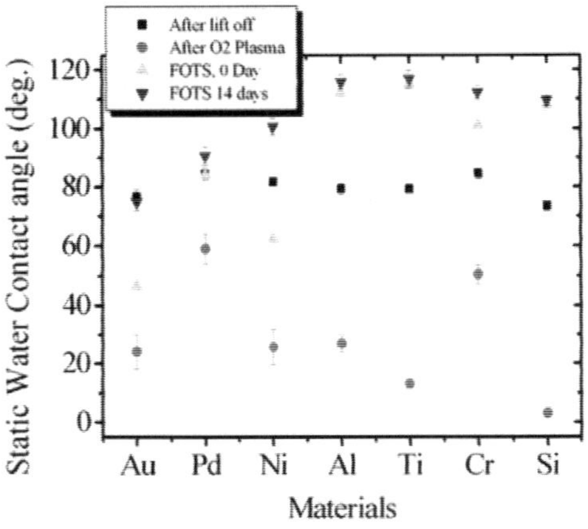

Figure 1. Static water contact angle of the various surfaces (Au, Pd, Ni, Al, Ti, Cr, and Si) measured after individual step, such as after lift off, after O_2 plasma pretreatment, on freshly deposited FOTS, and on FOTS 14 days after deposition.

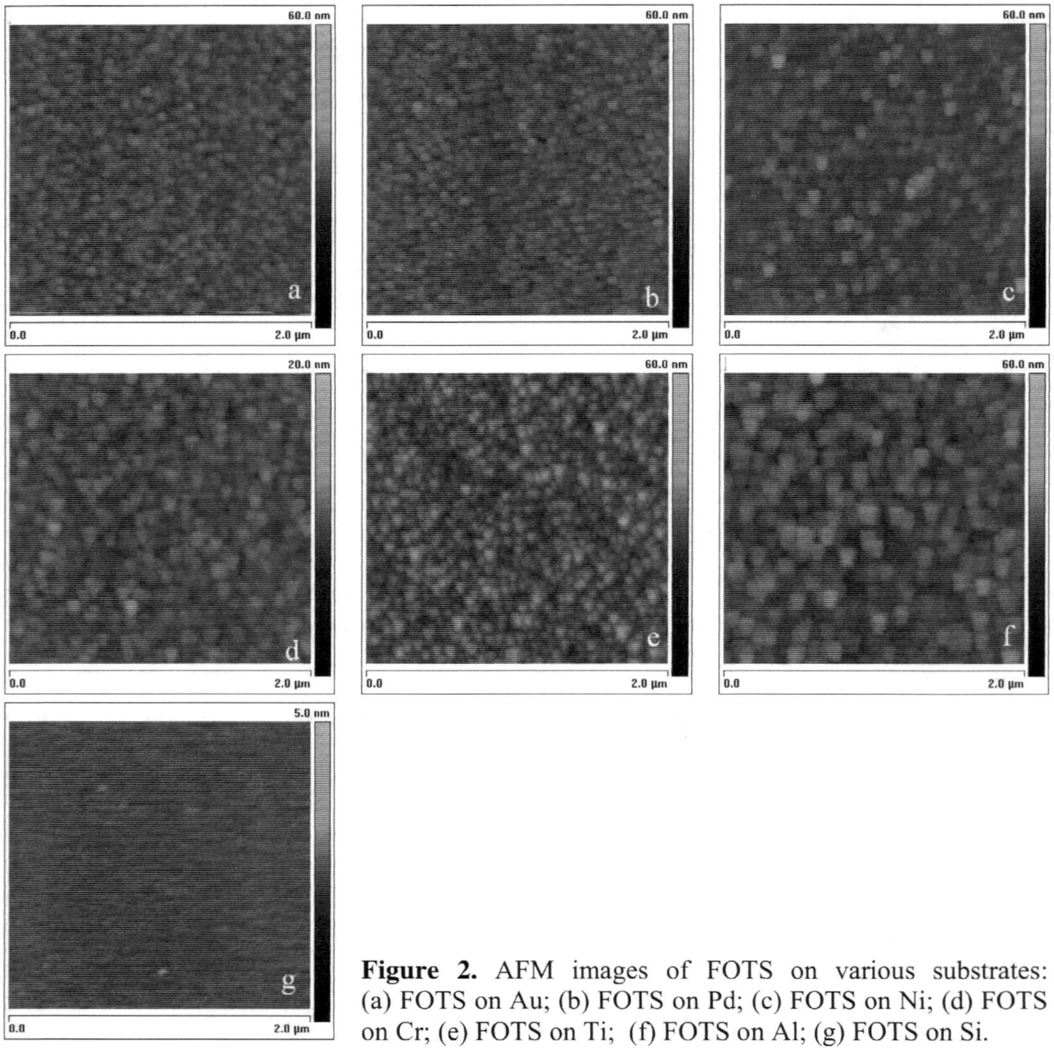

Figure 2. AFM images of FOTS on various substrates: (a) FOTS on Au; (b) FOTS on Pd; (c) FOTS on Ni; (d) FOTS on Cr; (e) FOTS on Ti; (f) FOTS on Al; (g) FOTS on Si.

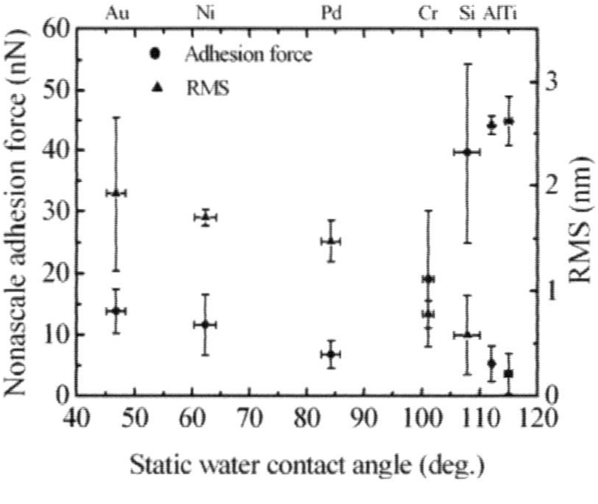

Figure 3. Nano-scale adhesion force, static water contact angle and roughness of FOTS on various metals and monocrystal-silicon.

Figure 4a gives variation of friction signal with normal load for FOTS on Si, Pd and Al. The FOTS coatings on other substrates have similar behavior. An approximately linear response is observed in the plot of friction signal versus normal load, suggesting that no substantial plastic deformation occurs in the testing range. Furthermore, the curve of FOTS on Pd is well above that of FOTS on Al and Si. The relative friction coefficient is defined as the slope of the friction signal versus normal load and is used as a parameter in comparing the friction behavior of FOTS on various substrates. It is well known that the following relationship exists between the friction force F and external normal load W [10]

$$F = \mu(W + W_a) \tag{1}$$

where μ is the friction coefficient and W_a is the adhesion force between the surface and AFM tip. Based on this equation and the data in Figure 4a, the μ and W_a values can be calculated. Figure 4b illustrates the relative friction coefficient of FOTS on various substrates, and the inset in Figure 4b shows the adhesion force determined from equation (1) and data in Figure 4a. It can be seen that the substrate can be considered as two groups. One group has larger static water contact angle, and low friction force, such as FOTS on Cr, Al, Ti and Si, which normally has a native oxide on the metal surface. The other group is Au, Pd, and Ni, which has smaller water contact angles and larger friction force. It has been suggested [10] that a chemical adsorbed self-assembled molecule on a surface can be regarded as an assembled molecular spring anchored to the surface. Therefore, a Si_3N_4 tip sliding on the surface of SAMs is similar to the tip sliding on the top of molecular springs or a molecular brush, which consists of a number of molecular springs. The monolayer can be elastically compressed under certain conditions, and the monolayer becomes disorder at the higher load. Based on these assumptions, a molecular spring model was proposed [10]. The molecular spring assembly has compliant features and can experience change in orientation and compression under normal load. The tilt angle of the molecules relative to the surfaces changes under the normal load applied by the AFM tip. The orientation of the "molecular springs or brush" under normal load reduces the shearing force at the interface, which in turn reduces the friction forces. In elasticity deformation range, when the tip slips away, the orientated and compressed molecules can recover to their initial state. The possibility of orientation is determined by the spring constant of a single molecule, as well as the interaction between the neighboring molecules, which is reflected by the packing density. The more molecules anchored on the unit surface, the higher coverage is, and the higher packing density is. The different friction forces of FOTS on various substrates can be explained by different coverage of the coatings (or packing density). The lower relative friction force of FOTS on Al, Cr, and Si might indicate that higher coverage of FOTS on those surfaces. However, it can not explain why the FOTS on Ti has larger friction force. The reason for that is still unknown.

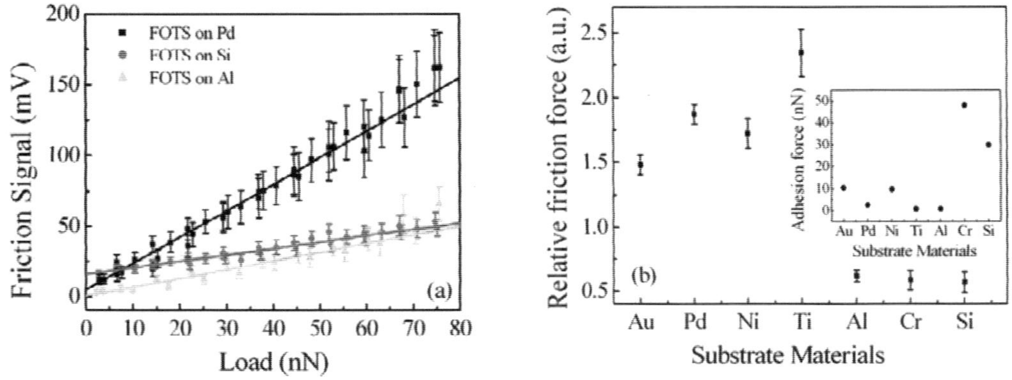

Figure 4 Friction signal vs. normal load applied by AFM tip (a), and the relative friction coefficient of FOTS on various substrates (b). The inset in (b) displays the adhesion force deduced from friction signals.

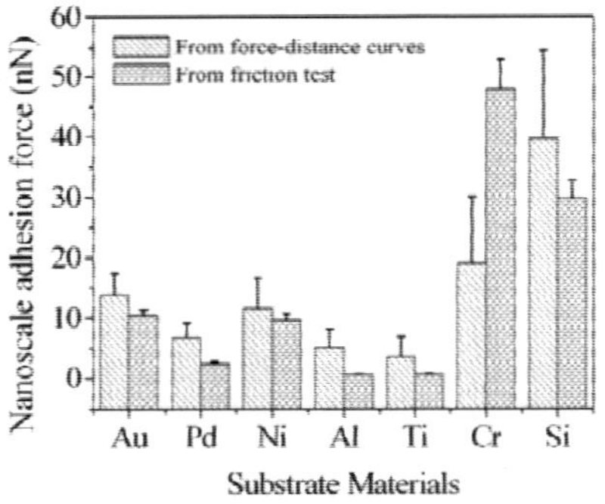

Figure 5. Comparison of the adhesive forces of FOTS on various substrates measured using force-distance curves and friction test in ambient air.

Figure 5 compares the adhesion forces between AFM tip and FOTS on various substrates measured by force-distance curves and friction test, respectively. The results measured by these two methods are in good agreement. The adhesion force between AFM tip and a surface is determined by the surface energy and the roughness of the surface. The competence of these two factors induces that the FOTS on Au, Pd, Ni, Al, and Ti has lower nano-scale adhesion force due to its larger roughness and FOTS on monocrystal silicon has largest nano-scale adhesion force.

4. Conclusions

$CF_3(CF_2)_5(CH_2)_2SiCl_3$ (FOTS) self-assembled monolayers have been grown on thin films of Au, Pd, Ni, Ti, Cr, and Al using a same process as that used for silicon substrates. The contact angle, surface energy, nano-scale adhesion force, and nano-scale friction behavior have been characterized. The metals can be ranked into two groups. One group has a behavior similar to silicon, for example, Ti, Al, and Cr. FOTS on those substrates displays an overall good anti-stiction performance, including high water contact angle, low surface energy, low adhesion and friction force. On the other hand, FOTS on other group of metals exhibit very low contact angle, high adhesion and friction force, indicating that FOTS molecule might be physically adsorbed onto those surface and have low coverage.

Acknowledgement

Financial support from Danish Research Council for Technology and Production Science is acknowledged, the Center of Individual Nanoparticle Functionality (CINF) is sponsored by the Danish National Research Foundation. Y X Zhuang would like to thank support from Northeastern Univeristy.

Reference

[1] Schreiber F 2000 *Prog. Surf. Sci.* **65** 151-256
[2] Baker M A and Li J 2006 *Surf. Interface Anal.* **38** 863-7
[3] Cox J D, Curry M S, Skirboll S K, Gourley P L and Sasaki D Y 2002 *Biomaterials* **23**: 929-35
[4] Lee N, Choi S and Kang S 2006 *Appl. Phys. Lett.* **88** 073101
[5] Zhuang Y X, Hansen O, Knieling T, Wang C, Rombach P, Lang W, Benecke W, Kehlenbeck M and Koblitz J 2006 *J. Micromech Microeng* **16** 2259-64
[6] Zhuang Y X, Hansen O, Knieling T, Wang C, Rombach P, Lang W, Benecke W, Kehlenbeck M and Koblitz J 2007 *J. Microelctromech. S.***16** 1451-60
[7] Mayer T M, de Boer M P, Shinn N D, Clews P J and Michalske T A 2000 *J. Vac. Sci. Technol. B* **18** 2433-39
[8] Ashurst W R, Carraro C and Maboudian R 2003 *IEEE Trans. Devices Mater. Reliab.* **3** 173-8
[9] Zhuang Y X and Menon A 2005 *J. Vac. Sci. Technol. A* **23** 434-9
[10] Liu H and Bhushan B 2003 *Ultramicroscopy* **97** 321-40
[11] Bhushan B and Liu H 2001 *Phys. Rev. B* **63** 245412

Journal of Physics: Conference Series **152** (2009) 012030 doi:10.1088/1742-6596/152/1/012030

Influence of the gas flow of Argon and the distance between substrate and plasma on properties of Al-doped zinc oxide films

Y j Jiang, D x Zhang, H k Cai, K Tao , Y Xue, Y p Sui, L s Wang, J f Zhao and J Wang

Department of Electronic Science and Technology, Nankai University, Tianjin, 300071, P.R. China

Email: Dexian Zhang@nankai.edu.cn

Abstract. Al-doped ZnO(ZAO) films were deposited by DC magnetron sputtering using facing zinc oxide targets at room temperature and in argon atmosphere. The effects of the gas flow of Argon and the distance between substrate and plasma on the properties of the ZAO thin films were characterized by several techniques. By optimizing the craft of preparation, the electrical resistivity as low as 3.3×10^{-4} Ω·cm and the optical transmittance over 80% in the visible range were obtained for these thin ZAO films. Therefore, the ZAO thin films were suitable for the window layers of n-i-p thin film solar cells or transparent conductive films.

1. Introduction

The group III metal-doped ZnO materials, particularly the Al-doped ZnO (ZAO), have been studied with a focus on application to sensors, transparent electrode display, heat mirrors and transparent conducting oxide (TCO) coatings for solar cells [1–5]. ZnO:Al films have been prepared by several deposition techniques such as metal organic chemical vapor deposition, photo-MOCVD, reactive evaporation, spray pyrolysis, sol-gel process, dc and rf sputtering [6–16]. The latter method is the most widely used technique for preparing ZAO films. Facing targets magnetron sputtering was used because of their desirable features, such as higher deposition rate and film uniformity. Naturally, this technology was applied rarely and the investigation was not detailed. Therefore, it is meaningful to study the influence of the gas flow of Argon and the distance between substrate and plasma in order to obtain ZAO films with optimum properties using facing targets magnetron sputtering.

In this work, the ZAO films were grown by DC magnetron sputtering system using two facing Al-doped ZnO (with Al atom concentration of 2.5%) ceramic targets in an argon atmosphere at room temperature. We investigated the gas flow of Argon and the distance between substrate and plasma dependence of the properties of the ZAO thin films.

2. Experimental procedures

The ZAO films were deposited on 45mm×92mm×2mm soda–lime glass substrates by a magnetron sputtering system in which two targets were situated facing each other. Each target was sintered from zinc oxide (99.99%) and aluminum oxide (99.99%) mixture with aluminum content of 2.5 wt%. The substrate was placed outside the plasma to reduce the direct bombardment and thus to improve film uniformity [17, 18]. When the base pressure was 3×10^{-3}Pa, Ar (99.999%) was introduced into the chamber with a pressure of 0.6 Pa. The DC sputtering current applied on both targets was kept at 400 mA. The thickness of all the ZAO films was maintained at 800nm by controlling the deposition time. Film thickness was measured using a surface profiler(AMBIOS XP-2). The structure of the films was characterized by X-ray diffractometer (XRD, Philips PANalytical X'Pert, CuKa). Electrical resistivity, mobility and carrier concentration were measured using a Hall automatic measuring system (ACCENT HL5550 LN2,USA).Because the ZAO films are one kind of n-type high degenerate semiconducting material with high conductivity, there is no need to plate metal electrodes on the samples measured by the machine above. The optical transmission spectra of films were recorded by the UV–VIS–IR pectrophotometer (VARIAN CARY5000).

3. Results and discussion

3.1. The distance between substrate and plasma dependence of structural、electrical and optical properties of the ZAO thin films

The distance between substrate and plasma is from 0mm to 35mm. The gas flow of Argon is 40 sccm.

3.1.1. Structure analysis.

XRD patterns of ZAO films deposited on different distances are shown in Fig. 1. All ZAO films show (002) and (004) diffraction peaks, indicating a polycrystalline structure with a preferential orientation of c-axis perpendicular to the substrate. Grain pitch and (002) peak's angle of ZAO as a function of the distance between substrate and plasma are shown in Fig. 2. As the distance between substrate and plasma increases, (002) peak's angle of ZAO films decreases gradually, and depart from (002) peak 's angle of ZnO films(34.45°) ; Grain pitch of ZAO films increases gradually, and departs from that of ZnO film(0.2603nm).

This means that: as the distance between substrate and plasma increases, Lattice distortion intensifies and residual stress of the films increases.

The crystal size of ZAO films with different distance can be evaluated from the full-width at halfmaximum (FWHM) of (002) peak using Scherrer's formula [19]:

$$D = \frac{0.9\lambda}{\beta COS\theta} \qquad (1)$$

Where λ , θ and β are the X-ray wavelength, Bragg diffraction angle and FWHM, respectively.

The calculated values of crystal sizes of ZAO films are about 33 nm.

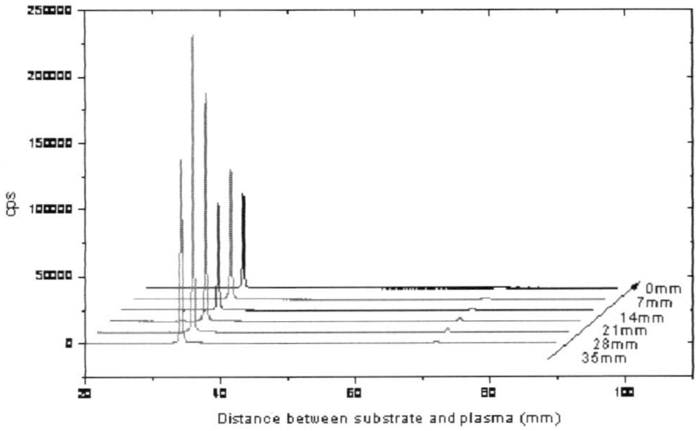

Figure 1. XRD patterns of ZAO films deposited at different distance

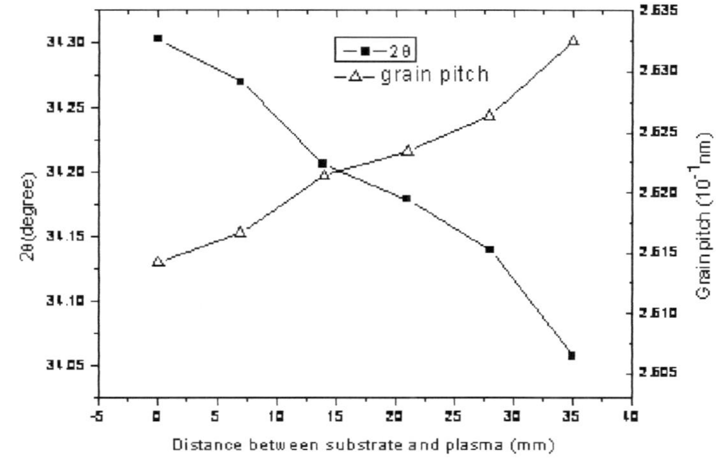

Figure 2. Grain pitch and (002) peak's angle of ZAO as a function of the distance between substrate and plasma

Deposition rate of ZAO films as a function of the distance between substrate and plasma is shown in Fig.3. Deposition rate of ZAO films decreases linearly with the increase of the distance.

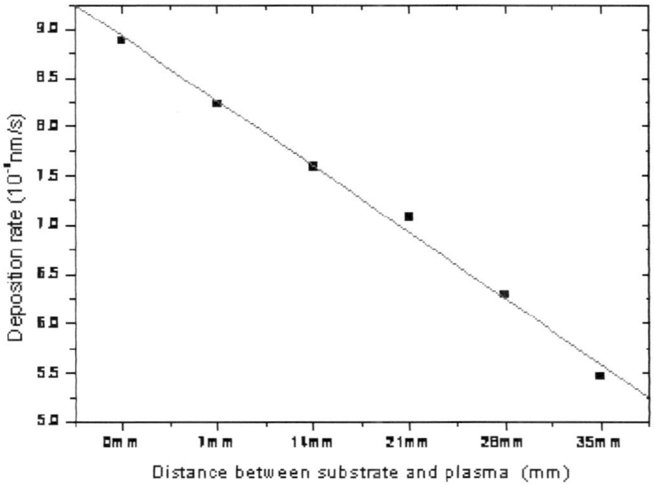

Figure 3. Deposition rate of ZAO films as a function of the distance between substrate and plasma

3.1.2. Electrical properties.

The resistivity and hall mobility of the ZAO films deposited at various distances between substrate and plasma are shown in Fig. 4. The resistivity decreases to a minimum of $3.3\times10^{-4}\Omega$cm at 14mm.

The hall mobility increases with the distance between substrate and plasma, and reaches a maximum of 24.2 cm2/Vs at 14mm, and then decreases as the distance increases to 35mm. The carrier concentration has the same trend with the carrier mobility. The improvement of conductivity is caused

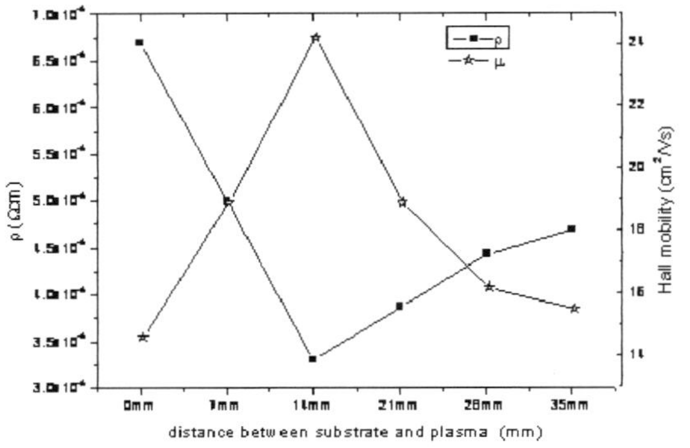

Figure 4. Resistivity and hall mobility of the ZAO films as a function of the distances between substrate and plasma.

by the improved crystallinity and decreases amount of interstitial atoms. When the distance between substrate and plasma is short, the deposition rate is so fast that the ions sputtered are covered by the latter ones before they move to the lowest position of energy at the substrate and trapped in the non-equilibrium position. So the film's crystallinity is poor. As the distance increases, the deposition rate decreases gradually, the crystallinity improves, many carriers are released from the poorly crystallized areas, thus causing the increase in carrier concentration and mobility. However, when the distance is too long, energies of the ions sputtered are so low that they aren't able to shift to the lowest position of energy at the substrate. That makes the increase of amount of interstitial atoms and defects.

3.1.3. Optical properties

The effects of the distances between substrate and plasma on optical properties are depicted in Fig. 5. A transmission of all films reached nearly 80% in visible region. In this work, the transmission of the ZAO film prepared at 0mm was lower than that at 14mm at near-UV spectral region but higher at infrared spectral region. Furthermore, there is a "blue-shift" phenomena found, which is caused by effect of BursteinMoss[20]. The transmission of films deposited at 14mm is over 80% in visible region, which was mainly due to the good crystallinity. The fluctuation in Fig. 5 is caused by interference effect of the incident light, it means that the uniformity of the film deposited is good.

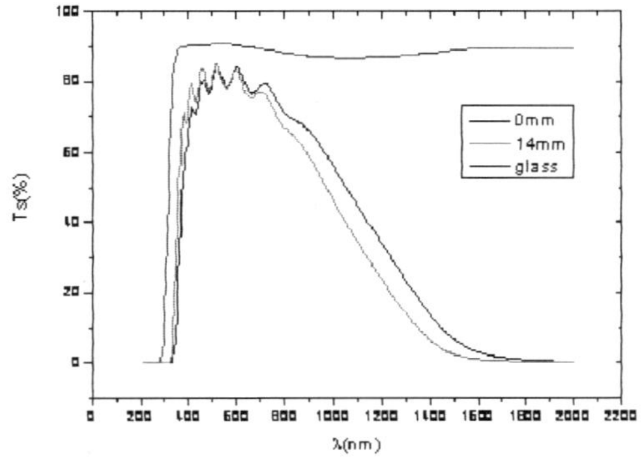

Figure 5. Transmission of the ZAO as a function of the wavelengths at the different distance between substrate and plasma.

3.2. The effects of the gas flow of Argon and the distance between substrate and plasma on the properties of the ZAO thin films

The gas flow of Argon is from 10sccm to 50 sccm. The distance between substrate and plasma is 35mm.

3.2.1. Structure analysis.

XRD patterns of ZAO films deposited on different gas flow of Argon are shown in Fig. 6. All ZAO films show (002) and (004) diffraction peaks, indicating a polycrystalline structure with a preferential orientation of c-axis perpendicular to the substrate. As the gas flow of Argon increases, (002) peak's angle of ZAO films departs from (002) peak's angle of ZnO films(34.45°); Grain pitch of ZAO films departs from that of ZnO film(0.2603nm).Diffraction angle's change origins from the substitute Al^{3+} for Zn^{2+}.The atomic radius of Al^{3+} is smaller than that of Zn^{2+},this causes the Lattice distortion and residual stress of the films. If we anneal the film in Argon, that situation will improve. Grain size is about 33nm.

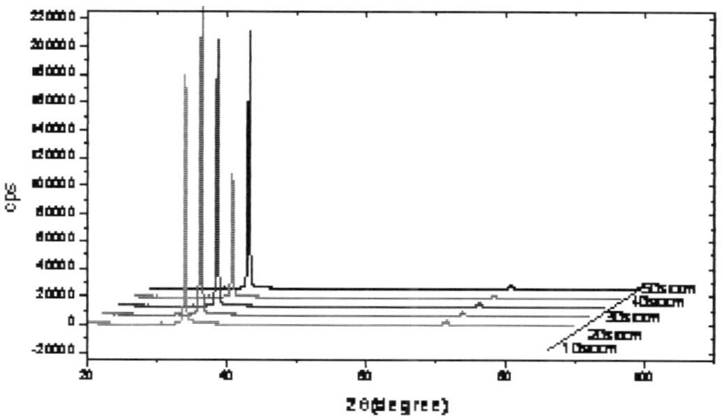

Figure 6. XRD patterns of ZAO films deposited on different gas flow of Argon

3.2.2. Electrical properties

ZAO is one kind of n-type and wide band gap semiconducting oxide. Carriers in the film are mostly from the absence of oxygen and the substitute Al^{3+} for Zn^{2+}.The latter one is the main source of carriers. So the carrier concentration is dependent on Al concentration in the film of ZAO. The gas flow of Argon has important influence on the absence of Oxygen and the substitute Al^{3+} for Zn^{2+}.

The resistivity and carrier concentration, deposition rate of the ZAO films deposited at various gas flows of Argon is shown in Fig. 7, Fig. 8 respectively. The resistivity decreases with the gas flow of Argon and reaches a minimum of 4.26×10^{-4} Ωcm at 20sccm, and then increases as the gas flow increases to 50sccm. The carrier concentration has the opposite trend with the resistivity. It reaches a maximum of $1.8\times10^{21}cm^{-3}$ at 20 sccm. However, the hall mobility has the opposite trend with the carrier concentration, it reaches a minimum of 8.16 $cm^2/(Vs)$ at 20sccm. It is probably because the scattering between the carriers which cause the decrease of mobility. The improvement of conductivity is caused by the improved crystallinity and decreases amount of interstitial atoms. When the gas flow of Argon is small, the deposition rate is so low that energies of the ions sputtered are too low to make them shift to the lowest position of energy at the substrate. So the crystallinity of film is poor. As the gas flow increases, the deposition rate increases gradually, the crystallinity improves, then carrier concentration increases and the scattering between the carriers causes the decrease of mobility. However, when the gas flow is too big , the deposition rate is too fast , the ions sputtered were covered by the latter ones before they moved to the lowest position of energy at the substrate . So the film's crystallinity is poor .That makes the increase of amount of interstitial atoms and defects. So the changes mentioned above exist.

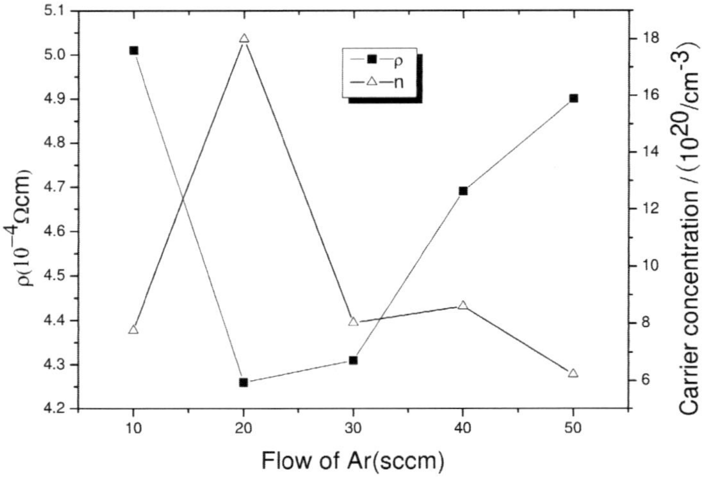

Figure 7. The resistivity and carrier concentration of the ZAO films deposited at various gas flows of Argon

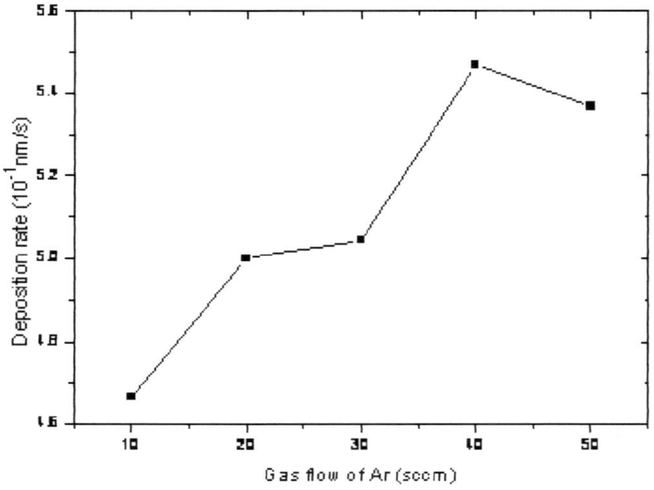

Figure 8. Deposition rate of ZAO films as a function of the gas flow of Argon

4. Conclusion

Through extensive research on the performance of ZAO films, high-quality ZAO films by controlling the distance between substrate and plasma and the gas flow of Argon in fabricated process were obtained. All ZAO films exhibit a strong (002) c-axis oriented normal to the substrate. When the distance between substrate and plasma increased from 0 to 35 mm, the lattice distortion intensifies and the residual stress of the films increase; the deposition rate of ZAO films decreases linearly, and the resistivity decreases to a minimum and then increases, while the hall mobility has an opposite tendency with the resistivity. At 14mm, the ZAO film was obtained with a minimum resistivity of

$3.3\times10^{-4}\Omega$cm, a maximum hall mobility of 24.2 cm^2/Vs and a transmission over 80% in the visible range. When the gas flow of Argon increased from 10 to 50 sccm, the deposition rate of ZAO films increases, the resistivity decreases to a minimum and then increases, however, the carrier concentration has the opposite trend with the resistivity. At 20 sccm, the ZAO film was obtained with a minimum resistivity of $4.26\times10^{-4}\Omega$cm, a maximum carrier concentration of $1.8\times10^{21}cm^{-3}$ and a transmission over 80% in the visible range. Therefore, the ZAO thin films are suitable for the window layers of n-i-p thin film solar cells or transparent conductive films.

References

[1] Martinez M A., Herero J, and Gutierez M T 1997 *Sol. Energy Mater. Sol. Cells* **45**, 75.
[2] Chen M, Pei Z L, Sun C, Gong J, Huang R F, and Wen L S 2001 *Mater. Sci. Eng.* B **85**, 212.
[3] Rye Y R, Zhe S, Look D C, Wrobe J M l, Amjeony, and White H W 2000 *J. Cryst. Growth* **216**, 330.
[4] Stambolova I, Konstantinov K, Vassilev S, Peshev P, andTsacheva Ts 2000 *Mater. Chem. Phys.* **63**, 104 .
[5] Reddy K T R., Gopalaswamy H, Reddy P J, and Miles R W 2000 *J. Cryst. Growth* **210**, 516.
[6] Shen W, Duan Y, Wang J, et al 2006 *Proc. SPIE* 6029 60290G.
[7] Li X, Asher S E, Limpijumnong S, et al., 2006 *J Vac Sci. Technol.* A **24** 1213.
[8] Mina mi T, Nanto H, Takata S 1985 *Thin Solid Films* **43** 124.
[9] Jinsoo Song, I-Jun Park, Kyung-Hoon Yoon, Woo-Yeong Cho, and Koeng-Su Lim, J 1996 *Korean Phys. Soc.* **29**,219.
[10] Lee J H, Ko K H, and Park B O 2003 *J. Cryst. Growth* **247**, 119.
[11] Lee J H and Park B O 2003*Thin Solid Films* **426**, 94.
[12] Nunes P, Costa D, Fortunato E, and Martins R 2002 Vacuum **64**, 293 .
[13] Kelly P J, Zhou Y, and Postill A 2003 *Thin Solid Films* **426**, 111.
[14] Zhou Y, Kelly P J, Postill A, Abu-Zeid O, and Alnajjar A A 2004 *Thin Solid Films* **447/448**, 33 .
[15] Muller J, Schope G, Kluth O, Rech B, Ruske M, Trube J, Szyszka B, Jiang X, and Briluer G 2001*Thin Solid Films* **392**, 327 .
[16] Hosono E, Fijihara S, and Kimura T 2004 *J. Mater. Chem.* **14**, 881.
[17] Hanabusa T, Hosoda H, Kusaka K, Tominaga K 1999 *Thin Solid Films* **343–344** 164.
[18] Zafar S, Ferekides C S, Morel D L 1995 *J. Vac. Sci. Technol.* A 13 (4) 2177.
[19] Sanon G, Rip R, Mansingh A 1990 *Thin Solid Films* **190** 287.
[20] Chang J F, Hon M H 2001 The effect of deposition temperature on the properties of Al-doped zinc oxide thin films [J].*Thin Solid Films*,386(1):**79-86**.

Effects of Ta_2O_5 addition on properties of $Ba_{0.6}Sr_{0.4}TiO_3$-$Mg_{0.9}Zn_{0.1}O$ ceramics for phase shifter applications

Shuwang Ma Guixia Dong, Zhimin Yang, Jun Du

Advanced Electronic Materials Institute, General Research Institute for Nonferrous Metals, No.2 Xinjiekou Wai Street, Beijing 100088, PR China

E-mail: shuwangma@163.com

Abstract. Influence of Ta_2O_5 addition on the dielectric properties of $Ba_{0.6}Sr_{0.4}TiO_3$ (BST) mixed with 65 wt% $Mg_{0.9}Zn_{0.1}O$ (MZO) composite was investigated. The XRD results indicate that the main phases in all Ta_2O_5 doped samples are BST and MZO. With increasing the amount of Ta_2O_5, the lattice constant of BST increases. The SEM morphology shows that the grain size of BST-MZO decreases with a certain amount of Ta_2O_5 addition. The dielectric constant decreases while loss tangent decreases first and then increases with increasing frequency. The dielectric constant, the tenability and loss tangent decrease with increasing addition of Ta_2O_5. Optimum doping amount of Ta_2O_5 reduces the loss tangent of BST-MZO compound which can also ensure the moderate dielectric constant and tenability. When Ta substituted 1.0wt%, the dielectric constant, the dielectric loss and the tenability are 85.2, 0.0005 (at 1MHZ) and 17% (under electric field 7kv/mm), respectively. At microwave frequency (4.025GHz), the dielectric constant decreases to 81.03 but the dielectric loss increases to 0.0049. This material shows good dielectric properties, which is one promising candidate for phase shifter applications.

1. Introduction

Barium strontium titanate (BST) ferroelectric system with high dielectric constant is one of the attractive candidates for tunable high-frequency devices such as phase shifters, frequency agile filters and tunable capacitors [1].In phase shifters, it is desirable to have a moderate permittivity for good impedance matching, a low dielectric loss to minimize the insertion losses of devices, and a high dielectric tunability, which is the change of permittivity with dc field, for less power driving and high speed of phase shift. BST-MgO bulk ceramics were reported to possess good dielectric properties [2-4]. However, this BST-MgO system has a relatively high dielectric loss even though having high tunability. Many studies focus on the effects of various rare earth oxide dopants to reduce the dielectric loss [5-6]. Moreover, the sintered temperature of the material is relatively high. In this study, $Mg_{0.9}Zn_{0.1}O$ was involved as the substitute of MgO to reduce the sintered temperature. Influence of Ta_2O_5 addition on the dielectric properties of $Ba_{0.6}Sr_{0.4}TiO_3$ (BST) mixed with 65 wt%$Mg_{0.9}Zn_{0.1}O$ (MZO) composite (BST-MZO) was investigated.

2. Experimental procedure

The material was synthesized by conventional solid-state reaction method with starting materials $SrCO_3$ (99%), TiO_2 (98.5%), $BaCO_3$ (99%), MgO (98.5%), ZnO (99%) and Ta_2O_5 (99.9%). The starting materials $SrCO_3$, TiO_2 and $BaCO_3$ were mixed using alcohol and agate milling media for 24 hour. After drying, the mixture was calcined in an alumina crucible at 1100℃ for 4 h in air to produce BST powder. MgO and ZnO were processed using the same method to produce MZO powder. Then 65wt% MZO and different amounts of Ta_2O_5 were added to the calcined BST. The mixture was re-milled and then dried. After that, the mixed powder was pressed using uniaxial pressing. These disks with 20mm diameter and 5mm thick were sintered at 1450℃ for 4 h in air. Some sintered pellets were polished down to 0.5mm in thickness and then terminated with silver electrodes for dielectric property testing at low frequency. Some others were polished down to 5 mm in thickness and 10 mm in diameter for dielectric property testing at microwave frequency.

X-ray diffraction (Cu K_ radiation MSAL XD2) was carried out for phase identification and lattice parameter measurements. The microstructure of the ceramics was investigated using a scanning electron microscope (SEM, Model S-4200, Hitachi, Japan). The capacitance and the loss tangent of the samples were measured using Agilent 4284A LCR meter (in the frequency range of 1 kHz to 1 MHz) and Agilent E8363 (at microwave frequency).

3. Results and discussions

Investigation of the microstructure of x wt% Ta_2O_5 doped ceramics sintered at 1450℃ (Fig.1) revealed the existence of two phases. The main phase is MZO with lager grain size. The BST distributes in the grain boundary. Without Ta_2O_5 addition, the BST continuously distributes in the MZO grain boundary while discontinuous grain growth is observed the other way round. The grain size of BST-MZO decreased with small addition of Ta_2O_5 (Fig.1a-c), while at larger concentration the grain size becomes bigger (Fig. 1d).

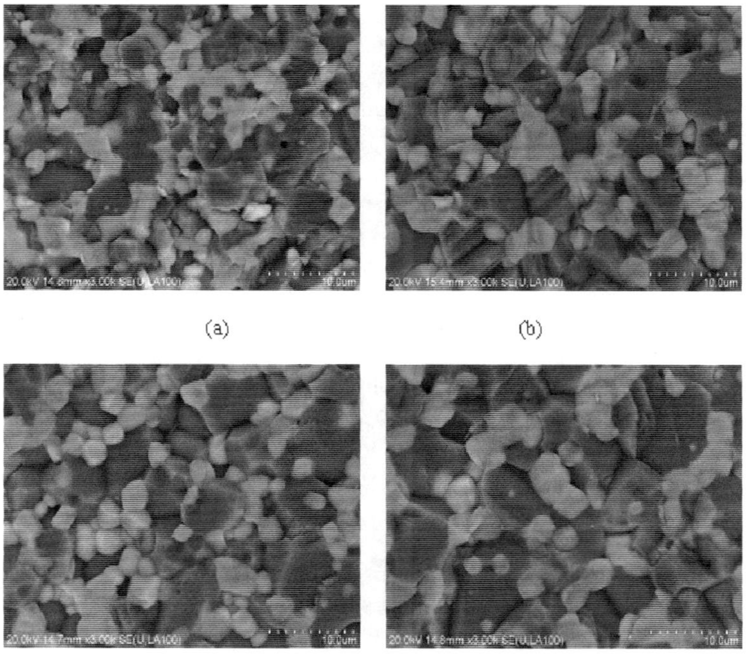

Figure 1. SEM morphology of BST-MZO sintered at 1450℃ with Ta_2O_5 added where (a) 0wt%, (b)0.5wt% (c) 1.0wt%, (d) 2.0wt%.

XRD patterns of BST-MZO-x wt% Ta_2O_5 ($0 \le x \le 2.0$) ceramics are shown in Fig.2. The XRD results clearly show that the main phases in all Ta_2O_5 doped samples are BST and MZO and both the crystal structures of them are cubic phase. The (110) peak of BST is shifted low angle with increasing Ta concentration, which indicates that B-site substitution of the Ti ions by the larger Ta^{5+} cations will cause the increase of the lattice parameter.

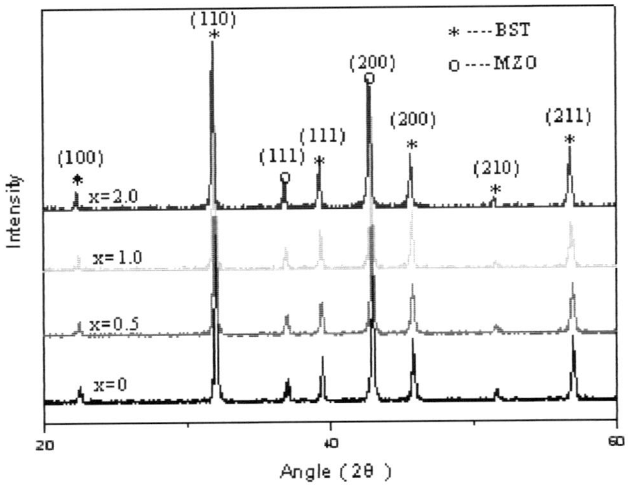

Figure 2. XRD patterns of x wt% Ta_2O_5 doped BST-MZO.

Table 1 shows the lattice parameter of BST and dielectric properties of the ceramics at 1 MHz. The lattice parameter of BST increases from 3.964 Å to 3.975 Å with increasing of Ta_2O_5 concentration x. This is because Ta ions are substituted into the B site of ABO_3-type perovskite BST and the ionic radii of Ta^{5+} (0.78 Å) are larger than that of Ti^{4+} (0.745 Å). In doped $BaTiO_3$, large trivalent cations that substitute for Ba^{2+}, such as La^{3+}, or small pentavalent or hexavalent cations that substitute for Ti^{4+}, such as Nb^{5+}, Ta^{5+}, or W^{6+}, act as donors. The sintered ceramic is light-colored, which means that the excess oxygen contained in the donor oxide is retained, and compensation appears to be primarily by Ti vacancies [7]:

$$2Ta_2O_5 + BaO \xrightarrow{(4TiO_2)} 4Ta_{Ti}^{\cdot} + Ba_{Ba} + V_{Ti}^{4'} + 11O_o \qquad (1)$$

Where $Ta_{Ti}^{\cdot}$ means Ta^{5+} substituted for Ti^{4+}, $V_{Ti}^{4'}$ means Ti vacancy. A dot is used for an extra positive charge, and a slash denotes an extra negative charge.

This gives a light-colored insulating material. The positively charged donor centers suppress the concentration of positively charged oxygen vacancies, the mobile species that leads to degradation of the leakage resistance. When Ta ions substitute for Ti ions, the single-phase compositions which compensated by Ti vacancies under oxidizing conditions should be $BaTi_{1-5x/4}Ta_xO_3$ [8]. Unless the composition corresponded to the proper concentration of compensation titanium vacancies, the material self-adjusted by splitting out the necessary amount of second phase [9].

Table 2 shows the dielectric properties of the samples at microwave frequency. The dielectric constant decreases and loss increases at X-Band compared with the properties at 1 MHz. It is mainly because of the relaxation of polarization mechanisms at higher frequencies.

From table 1 and table 2 it can be seen that the addition of non-ferroelectric oxide Ta_2O_5 will cause a dramatic decrease in the relative permittivity of BST-MZO. It is expected that the decrease of dielectric constant by substitution of Ta occurs because a slightly large Ta^{5+} ion is incorporated to a B-site Ti^{4+} with slightly smaller ionic radius hence, it is uneasily displaced under an electric field. On the other hand, the loss tangent decreases as Ta concentration increases. The result is accompanied by a drastic change in microstructure. Moreover, from the defect chemistry interview, when Ta^{5+} substitutes for Ti^{4+}, the excess oxygen is retained, which suppresses the Ti ions change from $Ti^{4+}$to Ti^{3+}.

Table 1. The lattice parameter of BST and dielectric properties of the BST-MZO with addition of x wt% Ta_2O_5 at 1 MHz.

Ta_2O_5 content (wt %)	Lattice parameter (Å)	Relatively dielectric constant	Loss tangent
X=0	3.964	110.15	0.00103
X=0.5	3.965	99.96	0.00092
X=1.0	3.966	85.21	0.00050
X=2.0	3.975	65.32	0.00047

Table 2. Dielectric properties of the BST-MZO with addition of x wt% Ta_2O_5 at microwave frequency

Ta_2O_5 content (wt %)	Relatively dielectric constant	Loss tangent	Measure frequency(GHz)
X=0	106.35	0.0060	4.005
X=0.5	95.06	0.0055	4.012
X=1.0	81.03	0.0049	4.025
X=2.0	62.42	0.0042	4.059

Fig.3. shows the dielectric tunability of the ceramics (under an electric field of 7kv/mm). The dielectric tunability of the ceramic is defined as $\{(\varepsilon(0)-\varepsilon(E))\}/\varepsilon(0)$, where $\varepsilon(0)$ and $\varepsilon(E)$ represent the dielectric constant at zero and applied electric field, respectively. E is the applied dc electric field. The calculated dielectric tunability of x wt% Ta_2O_5 doped BST-MZO are 20%, 19%, 17% and 15% for x value 0, 0.5, 1.0 and 2.0, respectively. With more Ta ions substitute for Ti ions, the ferroelectricity and interactions of Ti^{4+} ions weaken, which causes the decrease of tunability.

Fig.4. shows frequency dependence of the loss tangent of x wt% Ta_2O_5 doped BST-MZO. With increasing frequency, the loss tangent decreases first (below 100 kHz around) and then increases. For all samples, the loss tangent at high frequency (around 1MHz) is higher than that at low frequency. As Ta_2O_5 addition increases, the slope of the curve becomes flat, which benefit for obtaining lower loss tangent at microwave frequency.

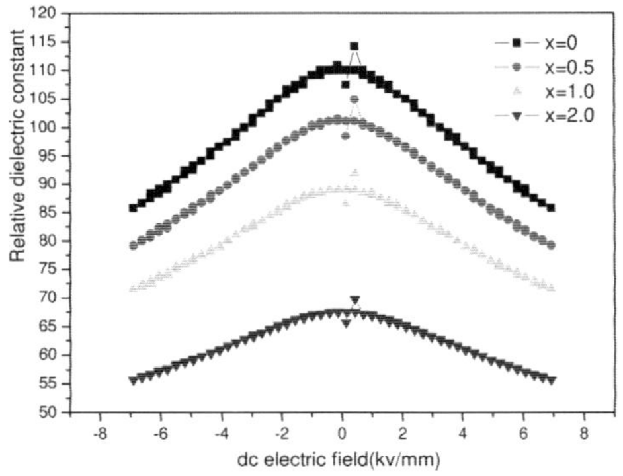

Figure 3. Dielectric tunability of x wt% Ta_2O_5 doped BST-MZO.

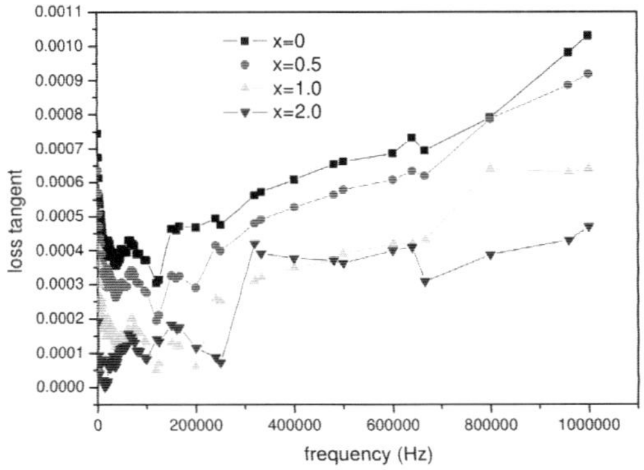

Figure 4. Frequency dependence of the loss tangent of BST-MZO with addition of x wt% Ta_2O_5.

Fig.5 shows temperature dependence of the relative dielectric constants of BST-MZO with different addition of Ta_2O_5. The Curie temperature shifted negative direction with increasing Ta concentration. It indicated that the Curie temperature should be below -40℃ for all the samples. Above the Curie temperature, the system is at non-ferroelectric phase. The Ta doped bulk ceramics show a broad dielectric constant–temperature curve, which is caused by an inhomogeneous distribution of Ta ions in the Ti sites and mechanical stress in the grain [10]. The characteristic of broad dielectric constant-temperature will benefits the application of BST-MZO in phase shifter.

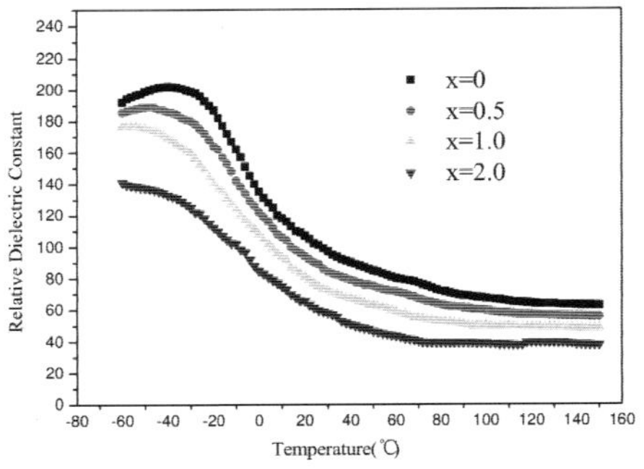

Figure 5. Temperature dependence of the relative dielectric constant of BST-MZO with addition of x wt% Ta_2O_5.

4. Conclusions.

In Ta_2O_5 doped 35wt%$Ba_{0.6}Sr_{0.4}TiO_3$-65wt%$M_{0.9}Z_{0.1}O$ ceramics, the main phases are BST and MZO. With increasing Ta_2O_5 amount, the lattice constant of BST increases and the dielectric constant, the tenability and the loss tangent decrease. The grain size of BST-MZO decreases at a certain amount of Ta_2O_5 addition. With increasing frequency, the loss tangent decreases first and then increases. Optimum doping amount of Ta_2O_5 can reduce the loss tangent of the BST-MZO composite which can also ensure the moderate dielectric constant and tenability. With a 1.0 wt %Ta substitution, the dielectric constant, the dielectric loss and the tenability are 85.2, 0.0005 (at 1MHz), 17% (under an electric field of 7kv/mm), respectively. At microwave frequency (4.025GHz), the dielectric constant decreases to 81.03 but the dielectric loss increases to 0.0049.

References

[1] Tagantsev A K, Sherman V O, Astafiev K F, Venkatesh, J and Setter, N 2003 *J. Electroceram.* **11** 5-66
[2] Chang W and Sengupta L C 2002 *J. Appl. Phys.* **10** 3941-3946
[3] Sengupta L C, Ngo E, O'Day M E, Stowell, S and Lancto, R In ISAF '94, Proceedings of the Ninth IEEE International Symposium on Applications of Ferroelectrics, 7 August 1991, PP. 622-625
[4] Wang X H, Lu W Z, Liu J, Zhou Y L and Zhou D X 2006 *J. Eur. Ceram. Soc.* **26** 1981-1985
[5] Chiu L H, Sengupta L C and Sengupta S US patent 6074971, 13 Jun 2000
[6] Sengupta L C and Drach W C 1994 *Ferroelectr.* **153** 359-364
[7] Smyth D M, The defect chemistry of metal oxides. (Oxford University Press, New York, 2000), pp. 276-277
[8] Chan H M, Harmer M P and Smyth D M 1986 *J. Am. Ceram. Soc.* **69**, 507-510
[9] Eror N G and Smyth D M, In the Chemistry of Extended Defects in Non-Metallic Solids, Eds. By Eyring L and O'Keeffe M (Amsterdam, North-Holland, 1970), p. 62-74
[10] Jong Y H, Ji W C, Chong Y K, Jin S K, Seok J Y, Doo J C and Hyun J K 2007 *J. Eur. Ceram. Soc.* **27**, 2747-2751

Preparation of $Ca_2Ti_2O_6$ Pyrochlore Films by MOCVD

Yuanwei Zhang[1], Jie Yu[1], Rong Tu[2], Takashi Goto[2]

[1]School of Physics and Electronic Technology and Key Laboratory of Ferroelectric and Piezoelectric Materials and Devices of Hubei Province, Hubei University, Wuhan 430062, PR China

[2]. Institute for Materials Research, Tohoku University, Sendai, Japan.

E-mail address: hdzhyw@hotmail.com (Yuanwei Zhang)

Abstract. Ca-Ti-O films were prepared on quartz substrates by MOCVD using the precursors of $Ti(O\text{-}i\text{-}Pr)_2(dpm)_2$ and $Ca(dpm)_2$. The effects of deposition conditions on the crystal structure and microstructure of Ca-Ti-O films were investigated. The deposition temperatures (T_{dep}) were changed from 787 to 1187 K and the total pressures in the reaction chamber (P_{tot}) was kept at 0.8 kPa. The molar ratio of Ca to Ti precursors ($R_{Ca/Ti}$) was changed from 0.34 to 0.95. At temperatures higher than 1073 K, $CaTiO_3$ perovskite films in a single phase were obtained. Mixtures of $Ca_2Ti_2O_6$ in a pyrochlore structure and $CaTiO_3$ in a perovskite structure were obtained at 787 to 1073 K and the content of $Ca_2Ti_2O_6$ phase increased with decreasing $R_{Ca/Ti}$. $Ca_2Ti_2O_6$ pyrochlore films in a single phase were obtained at T_{dep} = 1073K and $R_{Ca/Ti}$ = 0.38. $Ca_2Ti_2O_6$ is a face-centered cubic structure with a=0.990 nm. Pyrochlore showed a granular microstructure about 500 nm in grain size. The transmittance of the $Ca_2Ti_2O_6$ films in a single phase was 50% to 70% in the wavelength range of 500 to 2500 nm. The band-gap of pyrochlore was about 3.3 eV that calculated by the transmittance. The permittivity of $Ca_2Ti_2O_6$ films was about 100 from room temperature to 700K.

1. Introduction

Among the ternary metallic oxides, compounds of general formula, $A_2B_2O_7$ (A and B are metals), represents of phases isostructural to the mineral pyrochlore. *A* is nominally a 3+ valence cation and *B* is a 4+ valence cation [1-3]. The pyrochlore structure is closely related to fluorite and can be considered as an ordered defective fluorite $A_2B_2O_6O'$. The pyrochlore structures have many special properties. They are being considered for use in applications in many aspects, such as solid oxide fuel cell, catalysts to potential phases for the immobilisation of actinides in nuclear waste and the giant magnetoresistance effect in $Ti_2Mn_2O_7$. The $Ca_2Ti_2O_6$ pyrochlore structure can be considered as an ordered vacancy fluorite structure and has the ideal composition of $A_2B_2O_6O'$ [4-5]. In 1979 Savenko first prepared a stable $Ca_2Ti_5O_{12}$ phase by hydroxide thermal decomposition. However, in 1986, Ball precisely analyzed that the pyrochlore $Ca_2Ti_2O_6$ was existed through the XRD data of that Savenko reported [6-9]. Furthermore, they reported that $Ca_2Ti_2O_6$ might be ionic conductivity. However, no study on the $Ca_2Ti_2O_6$ films has been reported so far. In the present study, we tried to prepare $Ca_2Ti_2O_6$ films by the method of chemical vapor deposition (CVD) and investigated their properties, such as structure, transmittance, bandgap and permittivity.

2. Experimental Procedures

Figure.1 shows a sketch of the vertical cold-wall type CVD apparatus applied in the present study [10]. Source precursors of $Ti(O\text{-}i\text{-}Pr)_2(dpm)_2$ and $Ca(dpm)_2$ were vaporized at 787 to 1187 K, and carried by Ar gas into the CVD reactor. O_2 gas was separately introduced by using a double tube nozzle, and mixed with the precursor vapor in a mixing chamber placed above a substrate holder. Fused quartz glass plats were used as substrates. The total gas flow rate was fixed at $0.17 \sim 1.5\times10^{-6}$ m^3 s^{-1}. The total pressure in the CVD reactor was controlled at 0.8 kPa. The crystal structure was studied by X-ray diffraction (XRD). The microstructure and thickness of films were examined by Scanning Electron Microscopy (SEM) and Transmission Electron Microscopy(TEM). The transmittance was measured by a UV-VIS-NIR scanning spectrophotometer (Shimazu UV-3101PC) in the range of 200 to 2500nm. The permittivity was measured by a AC impedance analyzer (Hewllet Packard,HP-4194) in the frequency range of 100 Hz to 10 MHz at room temperature to 773 K.

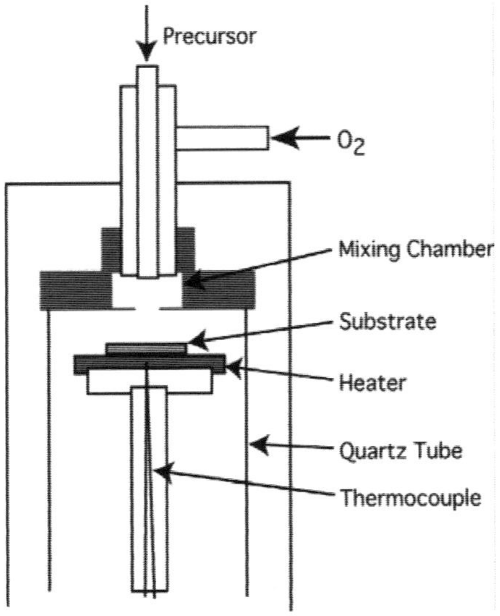

Fig1. CVD apparatus

3. Results and Discussion

Figure.2 shows XRD patterns of Ca-Ti-O films with different $R_{Ca/Ti}$ at 1073 K. At $R_{Ca/Ti}$=0.95, main phase of perovskite $CaTiO_3$ with a small amount of $Ca_2Ti_2O_6$ was prepared. With the decrease of $R_{Ca/Ti}$, the content of perovskite $CaTiO_3$ decreased and that of pyrochlore $Ca_2Ti_2O_6$ increased. At $R_{Ca/Ti}$ =0.38, $Ca_2Ti_2O_6$ in a single phase was obtained. At $R_{Ca/Ti}$ less than 0.38, anatase TiO_2 and $CaTiO_3$ was appeared.

Figure.3 shows the XRD patterns at $R_{Ca/Ti}$=0.38 and T_{sub}=1013 to 1113 K. At T=1013 K, mixed phases of pyrochlore $Ca_2Ti_2O_6$, perovskite $CaTiO_3$ and anatase TiO_2 were identified. With increasing T_{sub}, content of $Ca_2Ti_2O_6$ in the film increased and $Ca_2Ti_2O_6$ film in a single phase was obtained. At T_{sub} =1073 K, $R_{Ca/Ti}$=0.38, T=1073 K may be the best condition to obtain the pyrochlore $Ca_2Ti_2O_6$ film by CVD.

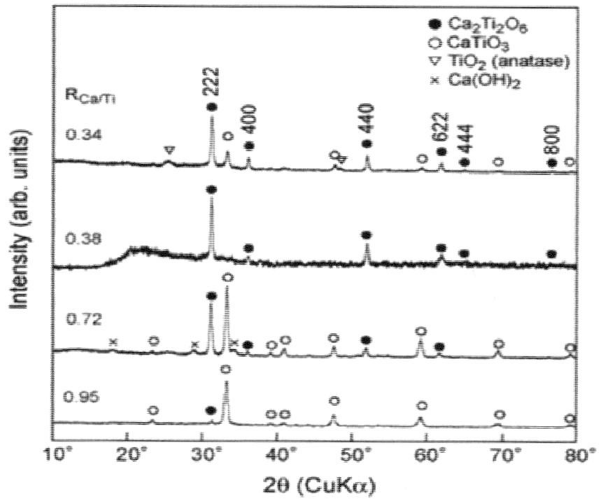

Fig 2. XRD patterns at 1073 K, P_{tot}=0.8 kPa and different $R_{Ca/Ti}$

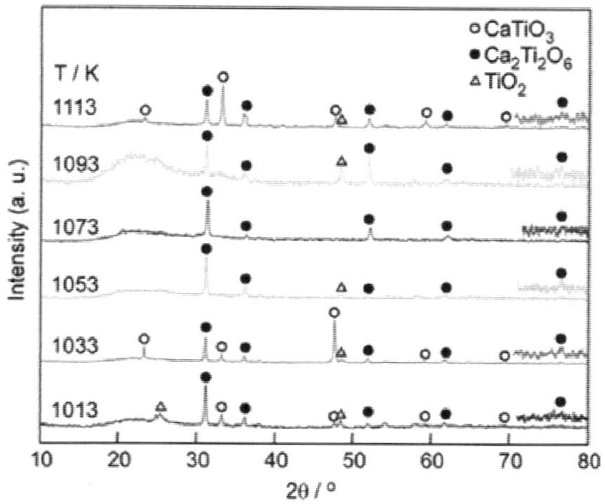

Fig 3. XRD patterns at $R_{Ca/Ti}$=0.38, P_{tot}=0.8 kPa and different T_{sub}

Figure 4 depicts the SEM of $Ca_2Ti_2O_6$ films prepared at different $R_{Ca/Ti}$ and T_{sub}=1073 K. The thickness of films was about 1 micrometer. The film prepared at $R_{Ca/Ti}$ =0.34 had a granular surface of 300 nm in diameter. Furthermore the agglomerated grains consisted of smaller grains about 50 nm in grain size. The the grain size of pyrochlore $Ca_2Ti_2O_6$ ($R_{Ca/Ti}$=0.38) was about 50 nm.

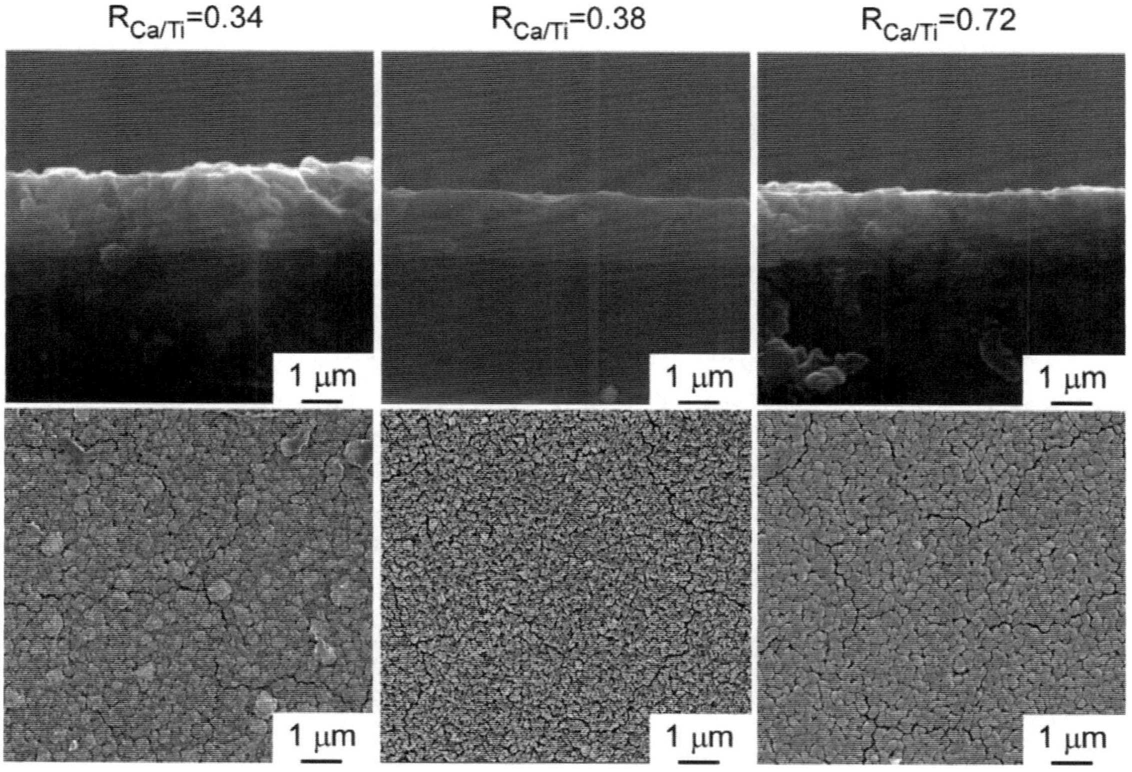

Fig 4. SEM images of the Ca-Ti-O films at $R_{ca/ti}$=0.34, 0.38, and 0.72

Journal of Physics: Conference Series **152** (2009) 012032 doi:10.1088/1742-6596/152/1/012032

Figure 5 depicts the electron diffraction pattern of $Ca_2Ti_2O_6$ prepared at $R_{Ca/Ti}$ =0.38. The incidence zone axis was (001). From the electron diffraction pattern, the lattice constant of $Ca_2Ti_2O_6$ was calculated as a= 0.990 nm. Compared with the ICDD, which shows a= 0.9953 nm, and all electron diffraction spots were indexed as pyrochlore $Ca_2Ti_2O_6$.

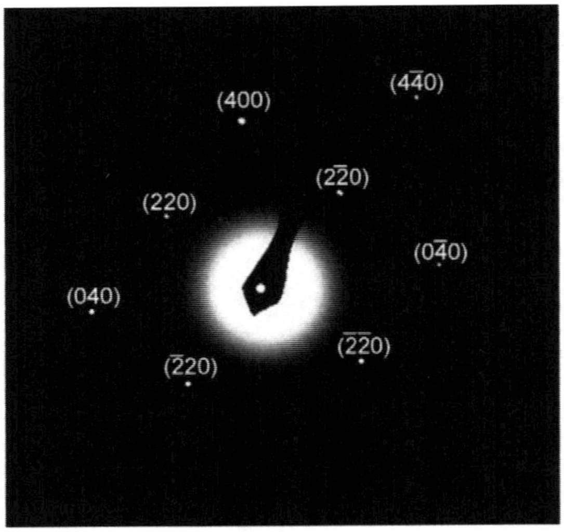

Fig 5. TEM of the Pyrochlore

Figure 6 depicts the transmittance of $Ca_2Ti_2O_6$ film. The transparency of the films exhibited a sharp decrease in the UV region. The absorption edge of the transmittance shifted, for different contents of $Ca_2Ti_2O_6$ phases indicated the pyrochlore $Ca_2Ti_2O_6$ phase has a similar bandgap with $CaTiO_3$. In the visible region, the absorption coefficient is influenced by the scattering of light on the surface roughness. Being a transparent material, the surface roughness dominates over the absorption, and it can be obtained from the approximate relation [11].

$$T = \frac{(1 - R)^2 \exp(-\alpha(\lambda)d)}{1 - R^2 \exp(-2\alpha(\lambda)d}$$

Where T is the transmittance, R is the reflectance, d is the thickness of the film and λ is the wavelength. However, at shorter wavelengths close to the optical bandgap, the scattering losses are dominated by the fundamental absorption and the following relation is often used:

$$\alpha = d^{-1} \ln(1/T)$$

Above the threshold of fundamental absorption, the dependence of α on incident light energy is given by the expression

$$\alpha E = \alpha_0 (E - E_g)^m$$

Where $E=h\gamma$ is the photon energy, E_g is the optical bandgap and α_0 is a constant which does not depend on E. The value of m may be taken m=2, a characteristic value for the indirect allowed transition which dominates over the optical absorption, according to the theoretical and experimental results. For photon energies $h\gamma>E_g$, the material can absorb these photons. For $h\gamma<E_g$, α=0 and the g photons cannot be absorbed anymore. Figure 7 shows the relationship between coefficient ($\alpha^{1/2}$) and photon energy (E) of $Ca_2Ti_2O_6$ thin films. The optical gap for $Ca_2Ti_2O_6$ films was about 3.3 eV.

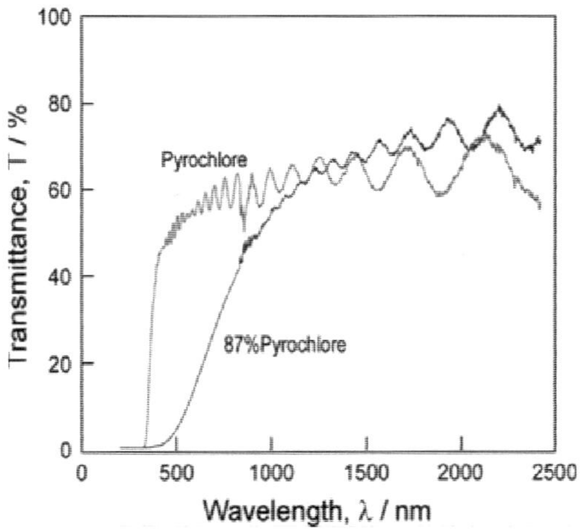

Fig 6. Transmittance spectra of $Ca_2Ti_2O_6$ thin films deposited on glass substractes

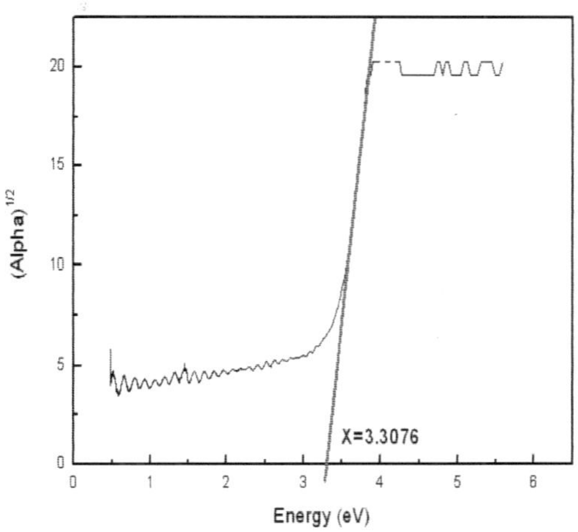

Fig 7. Optical absorption spectra plotted as $\alpha^{1/2} = f(E)$

Figure.8, shows the cole-cole plot of the Ca-Ti-O film at 773 K, the electrical resistance of the film was approximate 150 kΩm at 773 k. Figure 9 shows the permittivity of the $Ca_2Ti_2O_6$ film at room temperature to 773 K and f=100 kHz. The permittivity was almost 100 to 110, showing independent of temperatures.

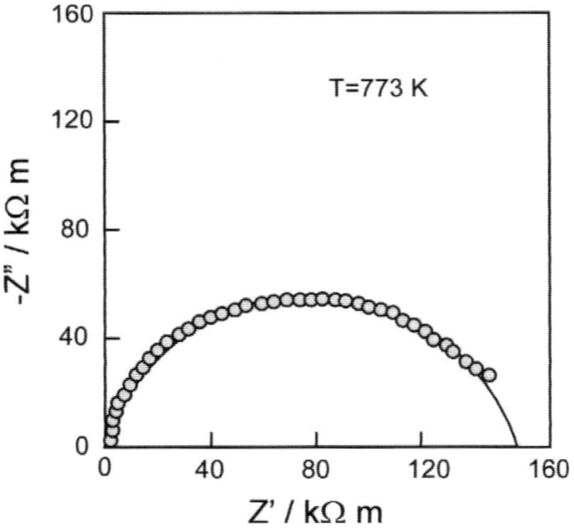

Fig 8. Impedance of $Ca_2Ti_2O_6$ film at 773 K

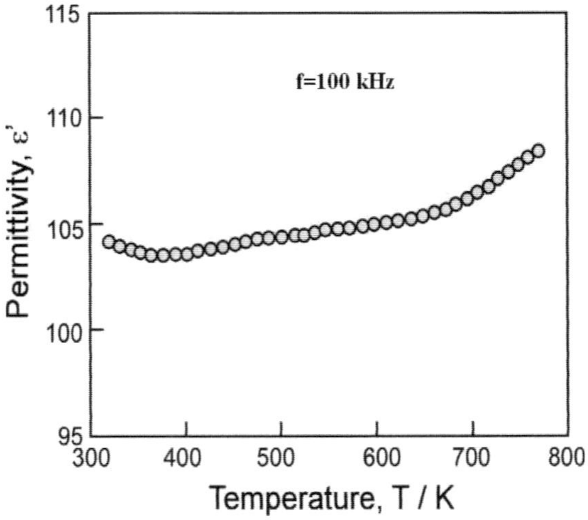

Fig 9. The permittivity of $Ca_2Ti_2O_6$ film at f=100 kHz

4. Conclusion

$Ca_2Ti_2O_6$ films were prepared by MOCVD using $Ca(dpm)_2$ and $Ti(O\text{-}i\text{-}Pr)_2(dpm)_2$ precursors. Single-phase pyrochlore $Ca_2Ti_2O_6$ was obtained at T_{sub}=1073 K, P_{tot}=0.8 kPa and $R_{Ca/Ti}$=0.38. The lattice constant was a=0.990 nm. The content of $Ca_2Ti_2O_6$ phase increased with decreased of $R_{Ca/Ti}$. The Ca-Ti-O film mainly containing the pyrochlore $Ca_2Ti_2O_6$ showed a grain size of about 50 nm. The band-gap of $Ca_2Ti_2O_6$ was about 3.3 eV. The permittivity of $Ca_2Ti_2O_6$ film was about 100 at room temperature, and up to 773 k.

References

[1] Diana Mardare, M. Tasca, M. Delibas, Applied Surface Science Vol.156 (2000).200–206.
[2] Brendan J.Kennedy, Physica B Vol.241-243 (1998) 303-31.
[3] Mohsin Pirzada, Robin W. Grimes, Licia Minervini, John F. Maguire, Solid State Ionics. Vol.140 (2001) 201–208.
[4] P.J. Wilde, C.R.A.Catlow, Solid State Ionics. Vol.112 (1998) 173–183.
[5] Rong Tu, Takashi Goto, Materials Science Forum. Vol.475-479 (2005) 1219-1222.
[6] Ball C J, White T J, Japan Solid State Chem. Vol.65, No.1 (1986) 148-150.
[7] Shimikawa Y , Kubo Y , Manako T, Nature. Vol.379 (1996) 53-55.
[8] Xie Yahong, Liu Ruiquan, Wang Jide, Progress In Chemistry. Vol.17, No .4 (2005) 673-677.
[9] T.Bak, J.Nowotny, C.C.Sorrell, M.F.Zhou, Journal of materials sciences: materials in electronics. Vol.15 (2004) 635-644.
[10] Rong Tu, Teiichi Kimura, Takashi Goto, Materiacls Transactions. Vol.43, No. 9 (2002) 2354-2356.
[11] Diana Mardare, M. Tasca, M. Delibas, G.I. Rusu, Applied Surface Science. vol.156 (2000) 200-206.

Synthesis and gas-sensing properties of Ce-doped SnO_2 materials

Chao Li[1,2] *, Zhishuo Yu[1], Shaoming Fang[1], Shide Wu[1], Yanghai Gui[1] and Rongfeng Chen[2]

[1]Henan Province Key Laboratory of Surface & Interface Science, Zhengzhou University of Light Industry, Zhengzhou 450002, China.
[2] Henan Academy of Science, Zhengzhou 450002, China.

* The corresponding author's e-mail address: zzulilc@zzuli.edu.cn (C. Li)

Abstract. Ce-doped SnO_2（2, 5, 10%Ce） nanopowders were successfully synthesized by a sol-gel method and their gas sensing properties were investigated. The samples were characterized by X-ray diffraction (XRD), transmission electron microscopy (TEM), and thermogravimetric-differential thermal analyzer (TG-DTA). The results indicated that well-crystallized Ce-doped SnO_2 particles in size of about 30nm were obtained at annealing temperature 800℃. Ce-doped SnO_2 sensors showed a high response to H_2S, whose sensitivity is about sextuple of the undoped SnO_2 sensors to H_2S, and to ethanol, about triplication of the undoped SnO_2 sensors. These results were thought to be correlated with Ce distribution on the SnO_2 surface and its effective contribution towards the conductivity and crystallite size of the grains.

1. Introduction

Tin oxide semiconductor sensors are extensively used for the detection of various pollutant and combustible gases. The advantages of the sensors are high response, simple design, and low weight and cost. However, the major problems associated with SnO_2 sensors are their low selectivity and stability. Many attempts has been done to solve these problems by using suitable additives such as noble metals (Pd, Pt) and transition metal oxides [1]. On the other hand, rare earth doping in semiconductor materials such as TiO_2, Al_2O_3, etc [2-3], have obtained widespread interest, since the addition of rare earth ions can control morphology and size, which can contribute for technological innovation and development of new optoelectronic devices [4]. In this work, Ce-doped SnO_2 were successfully synthesized by a sol–gel method and succedent annealing procedure. The gas-sensing properties of Ce-doped SnO_2 are also investigated.

2. Experimental

2.1. Synthesis of materials

In the typical process, $SnCl_4{\cdot}5H_2O$ and citric acid in a 1:1 molar ratio were added to the glycol solution to produce a tin citrate solution. The concentration of tin ion in the solution is 0.15 M. The $Ce(NO_3)_3{\cdot}6H_2O$ (Ce:Sn molar ratio is 0.02,0.05 and 0.10, respectively) mixed with an appropriate volume of citric acid was added to the tin citrate solution to form a turbid mixture. $NH_3{\cdot}H_2O$ was added until pH reached 1.5. The mixture was stirred for 4h at 55℃ until the gel was formed, and then dried at 100 ℃. The dried gel precursor was heated at 800℃ for 3h, and the samples were obtained after ground into fine powders. For comparison, pure SnO_2 powders were also prepared with the same procedures described previously. All the chemicals used in the process are analytical grade.

The thermal analysis of the dried gel was obtained by thermogravmetric-differential thermal analysis (TG-DTA, ZRY-1) in ambient air with increasing the temperature at a speed of 15□C/min. The structure and morphology of the products were characterized using XRD (Bruker D8, Cu Kα radiation, λ=1.5418Å) and TEM (FEI TECNALG^2).

2.2. Gas sensitivity test

The final powders were mixed and ground with an adhesive in an agate mortar to form a gas-sensing paste. The paste was coated on a ceramic tube on which a pair of Au electrodes was previously printed, dried under IR light for several minutes in air, and then sintered at 400°C for 1h. At last, a Ni–Cr heating wire was inserted to form a side-heated gas sensor. Its structure was shown in figure 1. The prepared gas sensors were aged at 300℃ for 240h in order to improve their stability.

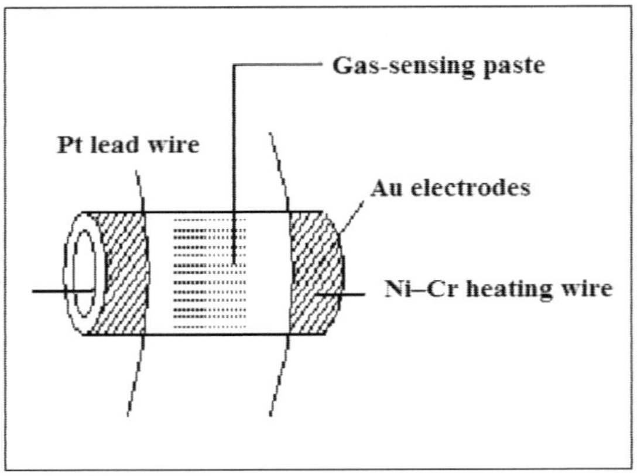

Figure 1. The structure of the gas sensor.

The test was operated in a measuring system of HW-30A (Hanwei Electronics Co. Ltd., Henan, China). The concentration of all gases was 50 ppm. The gas sensitivity was measured in static state. The circuit voltage was 10V, and outputs V_{out} were the terminal voltage of the load resistor. The working temperature of a sensor was adjusted through varying the heating voltage. The resistance of a sensor in air or test gas was measured by monitoring V_{out}. The gas sensitivity (response magnitude) in this work was defined as $S = R_{air}/R_{gas}$, where R_{air} and R_{gas} were the resistances of a sensor in air and in a test gas, respectively. The response time was defined as the time required for the variation in conductance to reach 90% of the equilibrium.

3. Results and discussion

Figure 2 shows the TG-DTA curves of the dry gel precursor powder for the 2%Ce-doped SnO_2. In the DTA curve, there were broad exothermal peaks from 300℃ to 750℃ and no peak at other temperatures. The exothermic effects may correspond to the oxidation and decomposition of organic materials. Besides the weight loss at the temperature stage from 300℃ to 750℃, a obvious weight loss can also be observed in TG curve from room temperatue to 300℃. The weight change can be thought from the releasing of the adsorbed water, crystal water and other gases. From 750℃ to 900℃, there is no obvious weight loss. In order to obtain the pure CeO_2-doped SnO_2, we chose 800℃ as a appropriate calcination temperature.

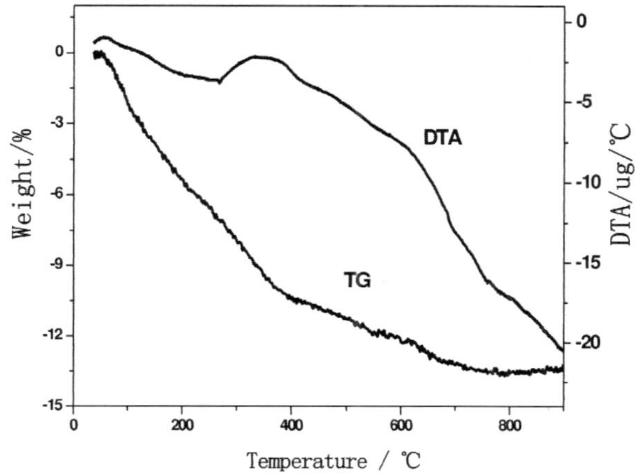

Figure 2. TG-DTA curves of the dry gel precursor

Figure 3 showed XRD patterns of Ce-doped SnO_2 and undoped SnO_2 samples calcinated at 800℃. From these patterns, it was observed that the strong diffraction peaks of the SnO_2 with tetragonal rutile structure became sharper with the increasing amount of Ce, which indicated doping Ce can impove the crystallization of the SnO_2 particles. When the Ce content was 5% and 10%, the reflections of CeO_2 appeared. As already reported [5], the second phase (CeO_2) of the tetragonal structure for 5%Ce-doped SnO_2 nanoparticles were also observed at temperature 1000℃. Due to the large difference of the radius and charge between Ce^{4+} ions(0.092nm) and Sn^{4+} ions(0.071nm), it is difficult for Ce^{4+} ions to enter the lattice of SnO_2.

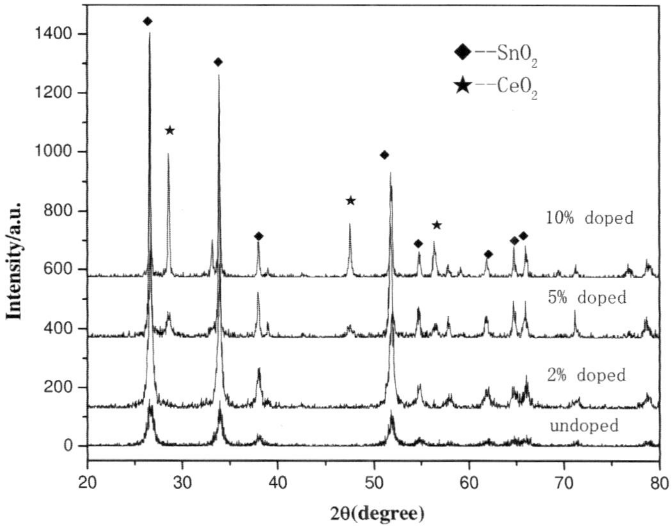

Figure 3.XRD patterns of Ce-doped SnO_2 and undoped SnO_2

Figure 4 shows TEM image of 2% Ce-doped SnO_2 nanoparticles. Sol-gel technology is an efficient synthetic method for SnO_2 nanocrystals with uniform shape and size. The image reveals that the doped materials present small particles with sizes about 30nm, which is in accord with the XRD analysis by

Scherrer equation. Ce ion has a bigger radius than Sn ion (0.071nm), so it has a smaller solubility in the SnO_2 lattice, and the excess content simply acted as a stabilizer of the SnO_2 crystallites [6].

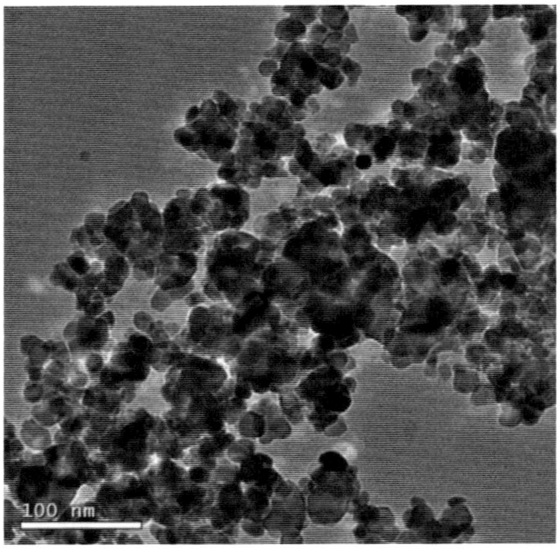

Figure 4. TEM micrograph of Ce-doped SnO_2

We selected 11 kinds of gases as detecting gases, including ethanol, formaldehyde, dimethylbenzene, acetonitrile, toluene, sulfureted hydrogen, hydrogen, ethene, methane and carbon monoxide. The gas sensing properties of samples were tested under heating voltage 5V (working temperature about 332℃). Figure 5 gives the sensitivities to gases of sensors based on SnO_2 and 10%Ce-doped SnO_2. As for pure SnO_2, it showed a certain response to almost all the gases, but the sensitivities are very low. But Ce-doped SnO_2 sensors showed a high response to H_2S, whose sensitivity is about sextuple of the undoped SnO_2 sensors to H_2S, and to ethanol, about triplication of the undoped SnO_2 sensors to ethanol. So that it may have excellent potential applications to detect sulfureted hydrogen and ethanol.

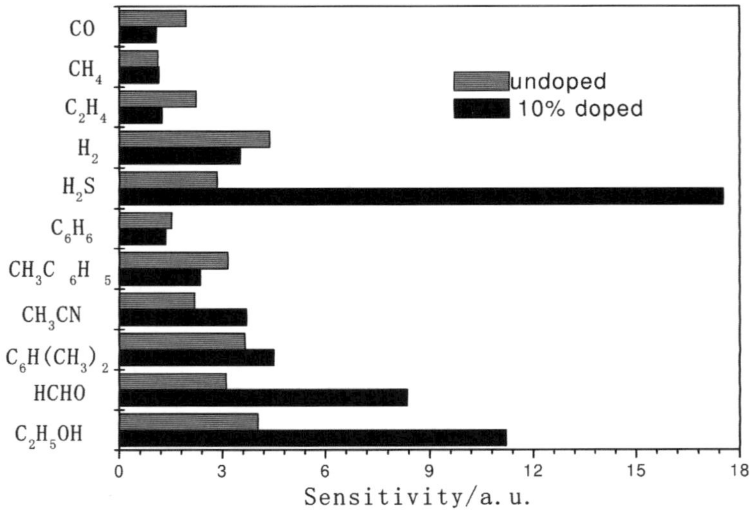

Figure 5. Sensitivity of gas sensors based on SnO_2 and Ce-doped SnO_2

Rare earth metal has a high alkalescence and good catalysis [7-8], which may be in favor of improving the sensitivities to H_2S and ethanol while did not favor that of orther tested gases. Based on the spillover mechanism proposed by Morrison [9], the surface catalyst (in our case, Ce) dissociates the gas molecule, and reactive atoms formed are spilled over on the semiconductor surface and influence the conductivity. In the case of H_2S detection, the sensing mechanism was presumed as follows [10]:

$$2H_2S\ (g) + 3O_2^{x-}\ (ad) \rightarrow 2H_2O\ (g) + 2SO_2\ (g) + 3xe^- \quad (1)$$

When oxygen is adsorbed on the surface of Ce-dopoed SnO_2 grains, the high alkalescence formed by doping Ce species with strong localization can be avail to dissociate H_2S gas molecules. The effects may be the reason that the materials had the high sensitivity to H_2S gas.

The reason of increasing response with the addition of Ce to ethanol should be similar to the above description to H_2S. The increasing of alkalescence of the surface of SnO_2 particles is good for enhancing the ethanol sensitivity [11-12]. The sensitivity of acetaldehyde by dehydrogenation and oxygenation from the surface of alkalescence is higher than that of ethylene by dehydrogenation and oxygenation from the surface of acidity.

On the orther hand, the addition of Ce to SnO_2 can impove the crystallization of the SnO_2 particles, modified the particles size distribution and electrical conductivity. These changes also affected gas-sensing properties to some extent.

4. Conclusions

Tin dioxide (SnO_2) and Ce-doped SnO_2 were successfully synthesized by a sol–gel method using inorganic salts as raw materials. The Ce-SnO_2 powders were characterized by XRD, TEM, and so on. The results showed that well-crystallized nano-sized Ce-doped SnO_2 in size of about 30nm was obtained at annealing temperature 800℃. The gas-sensing properties were characterized. It was found that ceria doped SnO_2 showed the high sensitivity to ethanol and sulfureted hydrogen. The increase of alkalescence of the surface of SnO_2 particles is good for enhancing the response to ethanol and sulfureted hydrogen. It may have excellent potential applications to detect sulfureted hydrogen and ethanol.

Acknowledgements

The authors gratefully acknowledge the financial supports of He'nan Outstanding Youth Science Fund (No. 0612002700) and the National Natural Science Foundation of China (No. 20871107).

References

[1] Gaidi M, Chenevier B and Labeau M 2000 *Sens.Actuators B*, **62** 43
[2] Wincenty T, Mieczyslaw L, Agnieszka K, Helena T and Roman K 2007 *Appl. Surf. Sci.* **253** 5920
[3] Yu Q C , Zhang S C , Wang X D , Zhang J and Lu Z M 2007 *J. Rare Earths* **25** 184
[4] Ge C Q, Xie C S and Cai S Z 2007 *Mater. Sci. Eng. B* **137** 53.
[5] Maciel A P, Lisboa-Filho P N, Leitea E R, Paiva-Santos C O, Schreiner W H, Maniette Y and Longo E 2003 *J. Eur. Cer. Soc.* **23** 707
[6] Gong J, Chen Q, Lian M R., Liu N C, Stevenson R.G and Adami F, 2006 *Sens. Actuators B* **114** 32
[7] Geng L N, Zhao Y Q, Huang X L, Wang S R, Zhang S M and Wu S H. 2007 *Sens. Actuators B* **120** 568
[8] Pan T M and Liao K M. 2007 *Sens. Actuators B* **128** 245
[9] Morrison S.R. 1987 *Sens. Actuators B* **12** 425
[10] Ionescu R, Hoel A, Granqvist C G, Liobet E and Heszler P 2005 *Sens. Actuators B* **104(1)** 132.
[11] Maekawa T andYamazoe 1992 *Sens. Actuators B* **139** (1) 63
[12] Yu J H and Choi G M. 2001 *Sens. Actuators B* **75** 56

Journal of Physics: Conference Series **152** (2009) 012034 doi:10.1088/1742-6596/152/1/012034

Influences of deposition parameters on microstructure and dielectric properties of BST thin film deposited by R.F. magnetron sputtering

Chao Xiang, Tianjin Zhang * , Ruikun Pan, Jinzhao Wang, Zhijun Ma, Min Xia

Department of Materials Science and Engineering, Hubei University, Wuhan,430062 China

zhangtj@hubu.edu.cn(T.Zhang)

Abstract. Down graded barium strontium tianate(BST) thin films were prepared on Pt/Ti/SiO_2/Si(100) substrate by radio frequency magnetron sputtering. The BST thin films were deposited at various substrate temperatures,then annealed at 700 °c for an hour. The influences of the deposition parameters on the crystallization behavior, microstructure and dielectric properties of BST thin films were investigated by X-ray diffraction, field emission scanning electron microscopy and dielectric frequency spectra. XRD results indicate that the BST thin films deposited at higher temperature have improved crystallization structure.The SEM observations show that the surface of the films is smooth with homogeneous grains.The dielectric properties of the films have been examined and discussed. The experiments show that BST thin films deposited at 650°c, 3.0Pa working pressure exhibits superior dielectric properties: the highest dielectric constant is 448 and lowest dielectric loss is 0.013 at 100 kHz,respectively.These results make BST thin films be a promising candidate for microelectronic device application.

1. Introduction

$Ba_xSr_{1-x}TiO_3$ is thought to be the most promising candidate material of sensitive elements for uncooled infrared focal plane arrays (UFPA).The properties of $Ba_xSr_{1-x}TiO_3$ can be tailored for special applications including piezoelectric transducers and optical signal processing[1–4]. Recently, there have been some studies on the preparation of BST thin films. A variety of techniques has been used to prepare BST thin films[5]. Among these methods, radio frequency magnetron sputtering has obtained much interest because of its many advantages such as deposition from a single target, high deposition rate, low cost, large homogeneity and large area smooth film. However, very limited information of the influences of sputtering parameters on BST thin films has been reported, such as substrate temperature and sputtering pressure.

In this work, the influences of substrate temperature,sputtering pressure on the crystallization behavior,microstructure and dielectric property of compositionally graded BST thin films were studied

2. Experimental

The n-type(100) silicon substrates were rinsed in acetone,ethanol,distilled water, and dried by nitrogen gas sequentially, then the substrate was thermally grown SiO_2 about 100 nm , BST thin films with different compositions were prepared on Pt/Ti/SiO_2/Si substrate by R.F.magnetron sputtering at substrate temperatures of 550℃, 600℃, 650℃.The sputtering was carried out at 0.5Pa,3.0Pa,6.5Pa, using a gas mixture of oxygen and argon.The compositionally graded BST thin films were achieved by continuously sputtering four different kinds of ceramic targets layer by layer with compositions of $BaTiO_3$, $Ba_{0.9}Sr_{0.1}TiO_3$, $Ba_{0.8}Sr_{0.2}TiO_3$ and $Ba_{0.7}Sr_{0.3}TiO_3$, respectively. All the films were annealed at 700 °c for an hour,then they were cooled down to room temperature naturally.X-ray diffraction (XRD) technique was carried out to identify the crystal structure of the compositionally graded BST thin films by using a D/max-IIIC diffractometer (Cu Kα radiation with a working voltage of 35 kV and current of 25 mA, wave-length λ = 0.154176 nm, scan speed of 10°/min, and scan range 20–67°).The microstructure of the graded BST thin films was examined by scanning electron microscopy (JOEL JSM 6700F).To measure the dielectric property, Pt top electrode with diameter of 300 μm was deposited on the top surface of the graded BST thin films by DC sputtering. The dielectric constant and dielectric loss were measured by an Agilent 4294A impedance analyzer.

3. Results and discussion

3.1. The effect of deposition parameters on the crystallization of BST thin films

Fig.1.was XRD patterns of graded BST thin films deposited at different temperature and sputtering pressure.It can be seen that the intensity of (100), (110), and (111) peaks of the BST thin films enhance with the increase of the deposition temperature. (100) and (200) peaks become sharper and more intense at 600℃,which reveal that the better crystalinity of the BST thin films can be obtained at higher substrate temperature.

The temperature dependence of the film quality can be interpreted mainly by the mobility of the particles at different temperature, the particles with low surface mobility will prevent the crystallization of the thin films at low temperature [6]. The dominant diffraction peaks of BST thin films deposited at 3.0Pa are (110) and (111).(110) peak intensity decreases apparently when the sputtering pressure increases up to 6.5 Pa,which indicates that the crystallizabillity is not very well at 6.5pa.

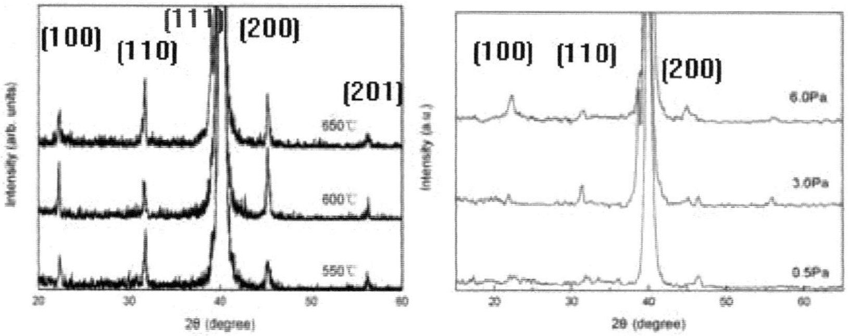

Fig 1. XRD patterns of graded thin films deposited at different temperature(a) and sputtering pressure(b)

3.2. The effect of deposition parameters on Microstructure of BST thin films

The microstructures of the graded BST thin films as a function of substrate temperature and sputtering pressure were examined by SEM images. It can be observed from Fig.2 that the surface morphologies vary with the change of the substrate temperature and the sputtering pressure.While deposited at 600℃, sputtering pressure was 3.0pa, the films have smooth and densely packed structure with homogeneous grains,the grain size is about 70nm,which is showed in Fig.2.(b),(e).This is due to that the mobility of species becomes higher and induces a higher speed of the coalescence of grain islands as the substrate temperature increases. Further increasing the pressure to 6.5pa, the graded BST grains do not grow remarkably and the roughness of graded BST thin film increases to 6.32nm.

Journal of Physics: Conference Series **152** (2009) 012034 doi:10.1088/1742-6596/152/1/012034

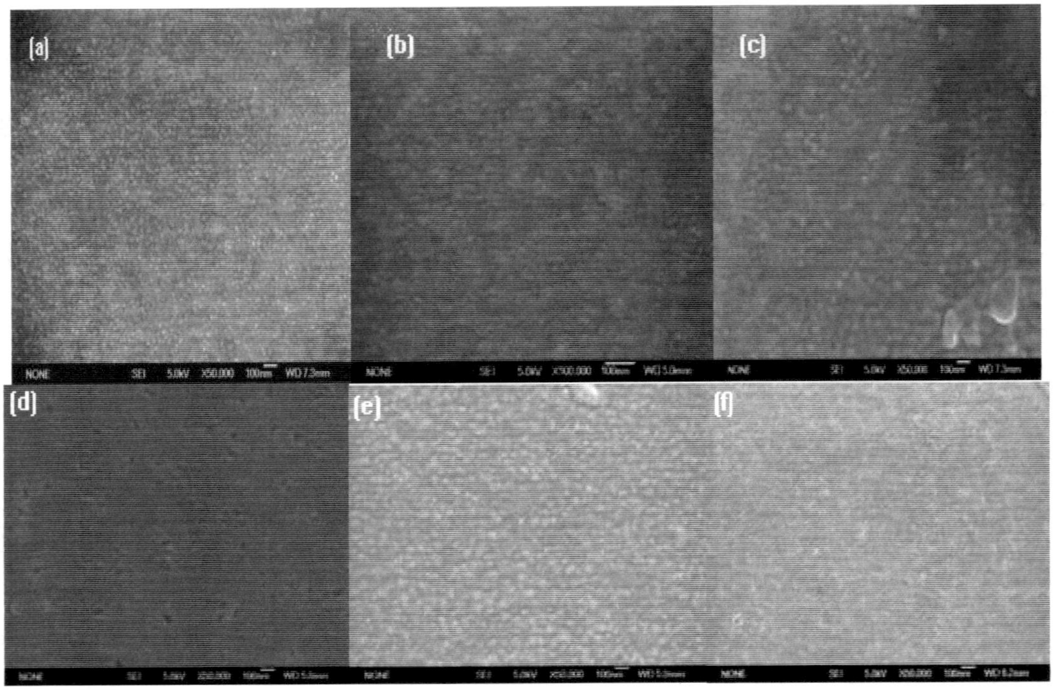

Fig 2. SEM images of compositionally graded BST thin films deposited at different substrate temperatures: (a) 550 °c;(b) 600 °c; (c) 650 °c; (d)0. 5pa; (e)3. 0pa; (f)6. 5pa

3.3. The effect of deposition parameters on the dielectric properties of BST thin films

The dielectric properties of the graded BST thin films was measured by a metal-ferroelectric-metal (MFM) configuration, in which the graded films were sandwiched between bottom and top Pt electrodes. The frequency dependence of dielectric constant of graded BST thin films deposited at different substrate temperature and different sputtering pressure is shown in Fig.3(a),(b). It's found that the dielectric constant of all the samples decrease with the increase of the frequency. The decrease of the dielectric constant may be caused by the accumulation of space charge that involves lots of oxygen vacancies in the graded BST thin films[7]. The experiments show that the highest dielectric constant and the lowest dielectric loss at 100KHz of the graded BST thin films deposited at 650℃ and 3.0 Pa are 448,0.013,respectively,which exibit excellent dielectric property of the graded BST thin films.

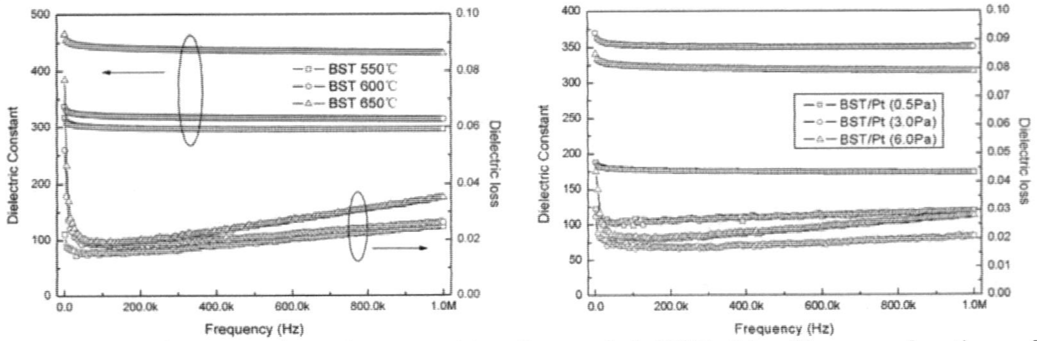

Fig 3.dielectric properties of compositionally graded BST thin films as function of frequency,(a) deposited at different temperature; (b) deposited at different sputtering pressure.

Journal of Physics: Conference Series **152** (2009) 012034 doi:10.1088/1742-6596/152/1/012034

4. Conclusions

Compositionally graded BST thin films with good crystallization structure were successfully fabricated by RF magnetron sputtering technique at a higher temperature of 600℃ and in the sputtering pressure of 3.0Pa, the dominant X-ray diffraction peaks (100),(200) became sharper and more intense.The graded BST thin films deposited at 650℃ and 3.0Pa possess the highest dielectric constant 448 and the lowest dielectric loss 0.013 at 100kHz, which exhibit excellent dielectric property of the BST thin films.These results showed that the deposition parameters have important effect on the crystallization,microstructure and dielectric properties of the BST thin films.

References

[1] PADMINI P, TAYLOR T R, LEFEVRE M J, NAGRA A S, YORK RA, SPECK J S 1999 Realization of high tenability barium strontium titanate thin films by rf magnetron sputtering (California:Appl Phys Lett) pp 3186−3188

[2] PARK W Y, AHN K H, HWANG C S 2003 Effects f in-plane compressive stress on electrical properties of (Ba,Sr)TiO_3 thin film capacitors prepared by on- and off-axis rf magnetron sputtering (California: Appl Phys Lett) pp 4387−4389

[3] TSAI M S, SUN S C, TSENG T Y 1997 Effect of oxygen to argon ratio on properties of (Ba,Sr)TiO_3 thin films prepared by ratio-frequency magnetron sputtering (California: J Appl Phys)pp 3482−3487

[4] JANG J W, CHUNG S J 1997 Thickness dependence of room temperature permittivity of polycrystalline $BaTiO_3$ thin films by ratio-frequency magnetron sputtering (California: J Appl Phys)pp 6322−6327

[5] CHANG S T, LEE J Y 2002 Electrical conduction mechanism in high-dielectric-constant ($Ba_{0.5}$,$Sr_{0.5}$)TiO_3 thin films (California: Appl Phys Lett) pp 655−657

[6] ZAFAR S, JONES R E, JIANG B, WHITE B, KAUSHIKV, GILLESPIE S 1998 The electronic conduction mechanism in barium strontium titanate thin films (California: Appl Phys Lett)pp 3533−3535

[7] S. Lahiry, V. Gupta, K. Sreenivas, et al 2000 Dielectric properties of sol–gel derived barium–strontium–titanate thin films, (California:IEEE Trans.Ultrason. Ferroelec) pp 854–859

Journal of Physics: Conference Series **152** (2009) 012035 doi:10.1088/1742-6596/152/1/012035

Electrodeposition route to synthesize SnO_2 microparticles as an anode material for Li-ion batteries

Chao Li[1,2] * , Wei Wei[1], Huanxin Wang[1], Shide Wu[1], Shaoming Fang[1] and Rongfeng Chen[2]

[1] Henan Province Key Laboratory of Surface & Interface Science, Zhengzhou University of Light Industry, Zhengzhou 450002, China.
[2] Henan Academy of Science, Zhengzhou 450002, China.

* The corresponding author's e-mail address: zzulilc@zzuli.edu.cn (C. Li)

Abstract. Tin dioxide (SnO_2) samples were prepared by direct current electrodeposition method and their electrochemical properties as anode materials for Li-ion battery were investigated with Li as counter electrode. The as-prepared SnO_2 powders were characterized by X-ray diffraction (XRD), scanning electron microscopy (SEM) and N_2-adsorption-desorption experiments. It was found that SnO_2 can be synthesized by one-step electrodeposition and the crystallinity improved markedly after annealed. The samples consisted of uniform spherical SnO_2 crystal particles with size in the order of 100~200nm. After dried at 120℃ and annealed at 400℃, 600℃, 800℃ respectively, the samples were used as anode materials for Li-ion battery. The results showed that the samples annealed at 600℃ had a high coulombic efficiency and had the best electrochemical performance.

1. Introduction

Although carbonaceous materials are most commonly used as anode materials for commercial Li-ion batteries, new anode materials are needed to improve the safety of Li-ion batteries as well as their capacities and rate capabilities.Tin dioxide is considered to be one of the most promising anode materials for Li-ion battery because it has high energy density (781mAh/g). However, Tin oxides had the drastic volume expansion occurring during the lithiation, which generally resulted in poor cycle performance [1]. Some different approaches have been used to improve the drawback of tin oxides, including introducing nanostructured electrodes [2] , doping other elements or compouds [3], and so on. In fact, the preparetion method and process also plays an important role in improving the property of many materials. And so, many methods such as sol-gel, co-precipition etc, were used to prepare SnO_2 as an anode material for Li-ion battery. In this work, we prepared submicron-SnO_2 powders by one-step electrodeposition method and the electrochemical proformance of materials treated in different temperatures was also investigated.

2. Experimental

The traditional electrolytic cell with two electrodes was used in the process. A sheet of copper (99.99%) in 200mm × 500 mm × 1 mm size was used as the working electrode. It was cleaned with acetone, diluted hydrochloric acid and deionized water. The tin tablet (99.99%) was the counter electrode. The distance between the working electrode and the counter electrode was 2cm. Electrolyte solutions contained $SnCl_2$ (c=0.02mol/L), $NaNO_3$ (c=0.1mol/L) and HNO_3 (c=0.075mol/L). All the used reagents were AR purity. WYJ-30 DC was used as the power supply. The current density was 8.0mA/cm^2, the time of the deposition was 10h, and the electrolyte temperature was controled at around 50℃. The product was collected and cleaned with deionized water, and then dried at 120℃. The sample was recorded as Dried at 120℃. The parallel samples were treated at 400℃, 600℃, 800℃ for 2h and recorded as Annealed at 400℃, 600℃, 800℃ respectively.

The crystal structure of the as-prepared samples were characterized by XRD (Bruker D8), using CuK α (λ =0.15418nm) radiation at 40kV and 60mA in a 2 θ range from 20° to 80° at room temperature. The morphology and the microstructure were observed by SEM (JEM-5600 scanning electron microscopes) with an accelerating voltage of 200kV. N_2-adsorption-desorption experiment instrument (3H-2000, Beijing Haihong) was employed to test the specific surface area of samples by the Brunauer-Emmett-Teller (BET) method.

Electrochemical performance of samples versus Li metal was also evaluated for possible use in lithium-ion batteries. The electrochemical reactions of samples with lithium were investigated using a simple two-electrode cell: Li/LiPF6(EC+DMC)/SnO_2. The working electrode consists of 80wt.% as-prepared powders, 10wt.% Super P carbon as the conducting agent, 10wt.% polyvinylidene fluoride (PVDF) as the binder, and Cu foils as the substrates (current collector). The cells were assembled in an Ar-filled glove box using polypropylene (PP) micro-porous film as the separator, a solution of 1M $LiPF_6$ in ethylene carbonate(EC)/dimethyl carbonate(DMC)(1:1, v/v) as the electrolyte and metallic lithium foils as counter electrodes. The electrochemical tests were performed on a CT2001A Land battery testing systems (Hannuo Electronics Co. Ltd., China). The cells were charged and discharged at a current of 0.2C, and a temperature of 25℃ in the voltage range of 0−1.5V.

3. Results and discussion

3.1. Structure and morphology

The XRD pattern of the sample dried at 120℃ was given in Figure 1. The diffraction peaks including four major peaks (110), (101), (200) and (211) can be indexed to cassiterite phase SnO_2 (JCPDS file 41-1445), which testified that SnO_2 can be obtained by one-step electrodeposition. The reflections of samples annealed at 400, 600 and 800℃ are also corresponding well with reflections of cassiterite phase (JCPDS file 41-1445). The XRD pattern of the sample treated at 600℃ was also given in Fig.1. It is seen that the reflections of the annealed sample become sharper, indicating the annealed sample has better crystal degree.

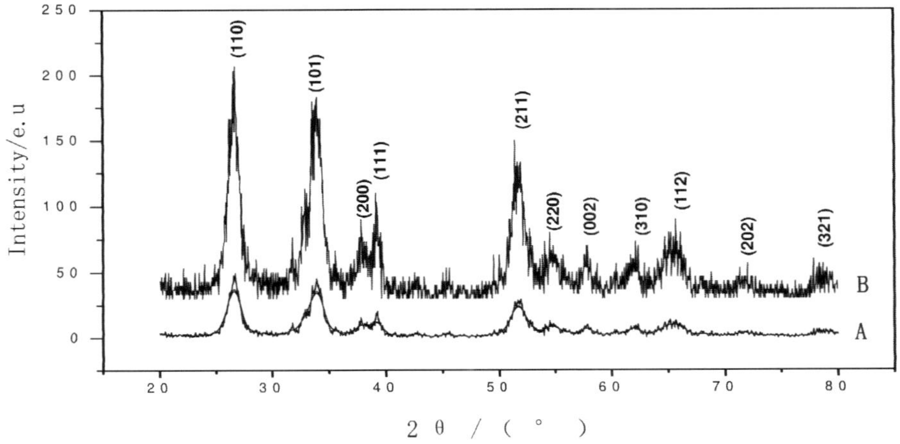

Figure 1. XRD patterns of SnO_2 samples, A: Dried at 120℃, B: Annealed at 600℃.

A typical SEM image of SnO_2 powders is given in Figure 2. It is clearly seen from the image that the powders contains sub-micron sized particles with good morphology and size. The samples consisted of uniform spherical SnO_2 crystal particles with the size in the order of 100~200nm and a few are seen to be present as agglomerates. Particles with this size and morphology can get a full contact wih electrolyte, reduce the resistance, increase the reaction area and conductivity to the insertion and deinsertion of the lithium ion, so it is suitable to be the anode material of litium ion battery.

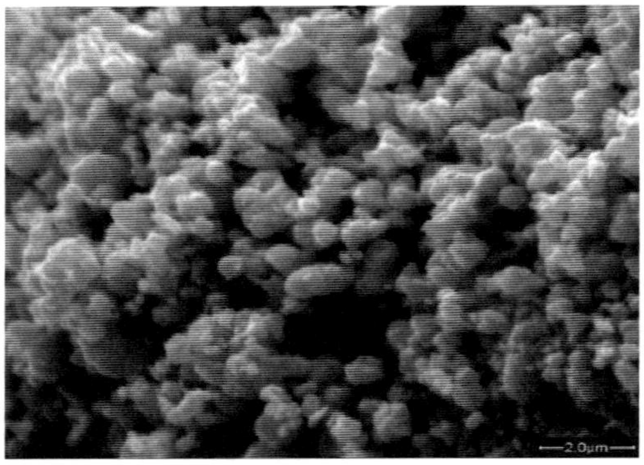

Figure 2. SEM imagine of SnO_2 (annealed at 600℃)

3.2. Electrochemical properties

The first-cycle insertion capacities for SnO_2 dried at 120℃,annealed at 400 ,600 and 800℃ are 1204.4 ,1376.8 ,1803.2, 1378.9mAh/g, respectively. The higher insertion capacity for the 600℃ sample may be due to the better refined crystallinity of the sample, which can provide easier pathways and necessary rearrangement of the constituent atoms [3]. The first-cycle deinsertion capacities for SnO_2 heat-treated at 120, 400, 600 and 800℃ are 246.4, 429.3, 723.3, 426.3mAh/g, respectively. The higher deinsertion capacity obtained with the sample heat treated at 600℃ is attributed to the lower irreversible capacity which probably is due to smaller concentrations of surface hydroxyl groups[3]. But we can see the capacity of the sample annealed at 800℃ decreased compared with that of the sample annealed at 600℃. The Brunauer-Emmett-Teller (BET) test showed that the SnO_2 treated at 600℃ has a bigger specific surface area compared with that of samples treated at 800℃. It was speculated that the insertion and deinsertion capacity will increase with the specific surface area sharply increasing.

In the tenth cycle, the deinsertion capacities of the sample dried at 120℃, annealed at 400℃, 600 ℃, 800℃ are 127.4, 206.5, 414.9 and 245.1mAh/g, respectively. The sample treated at 600℃ has a relatively high deinsertion capacity and a high capacity retention rate (57.4% at the 10th cycle). For the samples dried at 120℃, annealed at 400℃, 800℃, was not only deinsertion capacity lower than that of the former, but also the capacity retention lower than the former (51.7%, 56%, 48.7% at the 10th cycle, respectively).

Figure 3. shows the coulombic efficiency of the samples dried at 120℃ and annealed at 600℃. It is cear that the sample annealed at 600℃ has a higher efficiency in the first cycle. This is perhaps due to the perfect crystalinity of the annealed sample. As for samples annealed at 400℃ and 800 ℃, the efficiency in the first cycle was higher than that of sample dried at 120℃ but lower than the sample annealed at 600℃. After two cycles, the coulombic efficiency of all samples had reached to above 90%. It can be thought that sample nnealed at 600℃ has the best electrochemical performance. The result may be related to higher crystal degree and bigger specific surface area.

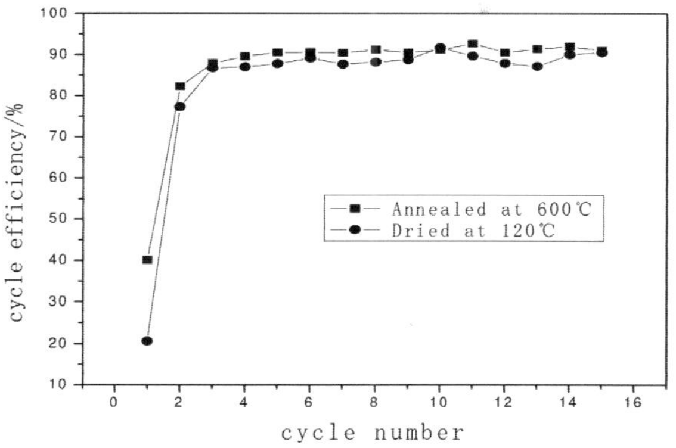

Figure 3. Coulombic efficiency of SnO_2 electrodes

4. Conclusions

Submicron-SnO_2 was prepared by one-step electrodeposition method, and the samples were characterized by X-ray diffraction (XRD), scanning electron microscopy (SEM) and N_2-adsorption-desorption experiments. The results showed that the samples have uniform spherical shape and particle size was in the order of 100~200nm. After annealed at 600℃, the electrochemical performance of samples was improved and its first discharge specific capacity could reach to 723mAh/g. It had a relatively high cycle efficiency compared with samples heat-treated at other temperatures, remained stable at around 90%. After the second cycle and after 10 cycles its capacity still preserved at 419.2mAh/g. These results showed that sample annealed at 600℃ has the best electrochemical performance. It was thought to be related to its higher crystal degree and bigger specific surface area.

Acknowledgements

The authors gratefully acknowledge the financial supports of He'nan Outstanding Youth Science Fund (No. 0612002700) and the National Natural Science Foundation of China (No. 20871107).

References

[1] Tarascon J M and Armand M 2001 *Nature* **414** 359
[2] José L and Tirado 2002 *Mater. Sci. Eng. R* **40(3)** 103
[3] Sivashanmugam A, Premkumar T, Gopukumar S 2005 *J. Appl. Electrochem.* **35** 1045

Journal of Physics: Conference Series **152** (2009) 012036 doi:10.1088/1742-6596/152/1/012036

Optical and electrical properties of ZnO nanocrystalline textured films prepared by DC reactive magnetron sputtering

W.X.Cheng, A. L.Ding*, X. S. Zheng, P. S. Qiu and X. Y. He

The State Key Laboratory of High Performance Ceramics and Superfine Microstructure, Shanghai Institute of Ceramics, Chinese Academy of Sciences, Shanghai 200050, People's Republic of China

E-mail:chwx_99@yahoo.com

Abatract. ZnO thin films were deposited by DC reactive magnetron sputtering using a metal zinc target. The crystallinity microstructure and surface morphology of ZnO thin films were investigated by X- ray diffraction (XRD) and scanning electron microscopy (SEM). The composition of ZnO thin film was characterized by XPS. The optical properties of ZnO thin film were obtained by spectroscopy ellipsometry and UV-VIS spectrometry. The optical bandgap was found to be 3.29eV of direct—transition type. The low frequency (1kHz—100kHz) dielectric properties of ZnO thin film were also discussed. The dielectric constant and dielectric loss at 10kHz are 13.4 and 0.03 respectively.

1. Introduction

ZnO thin films possess very interesting properties in the electrical and optical application fields. They have low electric resistance and high transparency in the visible wavelength range [1]. The material is composed of cheap and abundant elements, and is readily produced for large-scale coatings [2]. These films have attracted interest in many applications such as solar cells, gas and optical sensors, ultrasonic oscillators, transducers, optical waveguides and photoprotective coatings [3-6]. Various deposition techniques can be used to prepare ZnO thin films, such as reactive and ion-assisted evaporation, laser ablation, spray pyrolysis, sol - gel processing, chemical vapour deposition and sputtering [7 - 10]. In this work, ZnO films were deposited by sputtering from a metal Zn target fixed on a magnetron cathode. This deposition technique presents interesting advantages: the low substrate temperature, the good adhesion of the films on the substrates, and a high deposition rate. Moreover, this is relatively a cheap deposition method [11]. The physical properties of the films depend on the sputtering parameters, such as substrate temperature, oxygen partial pressure, the sputtering power and substrate - target distance. We have studied the structural, optical and dielectric properties of ZnO thin films.

2. Experimental procedure

ZnO thin films were prepared by DC reactive magnetron sputtering process. The metal Zn target (99.9% purity, 75mm diameter) is fixed on a magnetron-effect cathode.Pt/Ti/SiO_2/Si(100) and fused quartz glass were used as substrate. The distance between the target and the substrate was 30mm. The vacuum chamber was evacuated using a turbomolecule pump to 6×10^{-4}Pa before deposition, the sputtering gas O_2 with a purity of 99.99% were introduced to the chamber and controlled by the standard mass flow controllers. The target was cleaned by sputtering with power of 100W for 10min with pure Ar atmosphere at the same time the substrate was covered with the shield. The following sputter conditions were used in this work: DC power of 100W, ambient O_2 pressure of 0.6Pa, the substrate temperature (Ts) was 350°C. The films were prepared by sputtering on $Pt/Ti/SiO_2/Si(100)$ and quartz glass substrates, deposited thin films were annealed at 650°C for 10 min in oxygen atmosphere by rapid thermal annealing (RTA).

The microstructure of ZnO thin films were analysed by a D/Max-Rb(raguku, Japan) rotating x-ray diffractometer (XRD) with Cu Kα radiation in the angle range from 20 to 60 degree. The surface morphology and cross section properties of thin film were examined by a field emission scanning electronic microscopy (FE-SEM) (JSM-6700F).

A spectroellipsometer (VASETM, J. A. WollamCo., Inc., USA) was used to measure the ellipsometric parameter as a function of the photon energy from 0.5eV to 4.5eV. These parameters are employed to find the refractive index (n), extinction coefficient (k) and the thickness (d) of thin films, by comparison with theoretical computer-calculated spectra of the films having given values of refraction index, extinction coefficient and thickness of the thin films.

To investigate the dielectric properties of prepared ZnO thin films, the top electrode of gold (Au) with area of 0.8mm^2 was prepared on the surface of ZnO film through a shadow mask by evaporation. The capacitance and dielectric loss (tan δ) were measured using a HP 4294A LCR impedance analyzer in the frequency range from 1kHz to 100kHz.

3. Result and discussion

3.1. Crystallinity, composition and morphology of ZnO film

The crystalline quality and orientation of the ZnO thin films have been investigated by X-ray diffraction (XRD). Figure 1 (a) and (b) show the XRD patterns of ZnO thin films on fused quartz glass and $Pt/Ti/SiO_2/Si(100)$. From Fig. 1 it can be seen that only the (002)-peak is present, which indicates that the film is c-axis-oriented perpendicular to the surface. The FWHM of the (002)-peak of ZnO thin film on $Pt/Ti/SiO_2/Si(100)$ and fused quartz glass are 0.38 and 0.34 respectively.

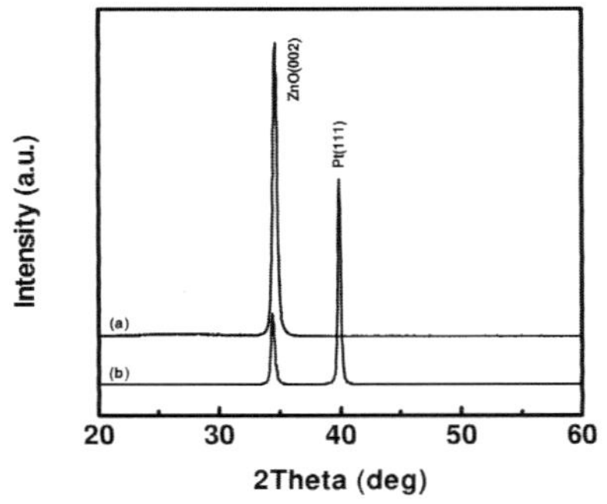

Figure 1 The XRD patterns of ZnO thin films on various substrates (a) film on fused quartz glass (b) film on $Pt/Ti/SiO_2/Si(100)$.

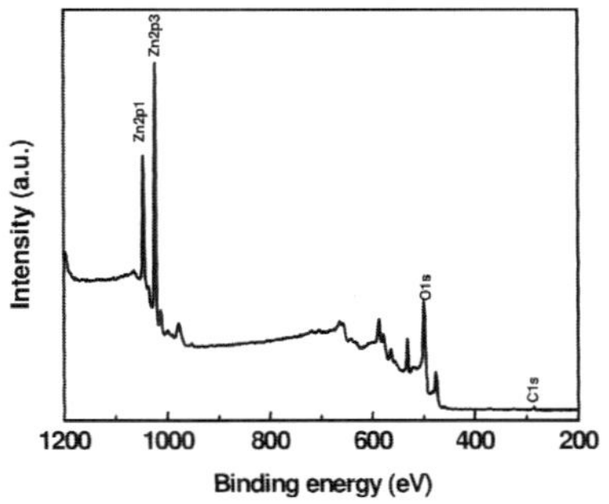

Figure 2 XPS spectra of zinc oxide film on quartz glass

That means ZnO thin film on $Pt/Ti/SiO_2/Si(100)$ has better crystal quality and larger grain size than that of ZnO film on glass. The stress in ZnO thin film on $Pt/Ti/SiO_2/Si(100)$ is derived according to the following equation:[12]

$$\sigma = -233 \times \frac{c_{film} - c_{bulk}}{c_{bulk}} \quad \text{(Gpa)} \tag{1}$$

where c_{film} is the lattice parameter of c axis for the ZnO thin films, and c_{bulk} is the lattice parameter of ZnO in bulk (or powder), namely, the unstrained lattice parameter. The numerical values of c_{film} are calculated from XRD data according to the following equation:

$$2 d_{hkl} \sin\theta = \lambda \tag{2}$$

Where d_{hkl} is lattice spacing of (hkl), λ is the wavelength of Cu $k_{\alpha 1}$ radiation (0.154056 nm) and θ is the Bragg angle (half of the peak position angle). At the same time, ZnO is a hexagonal structure, which follows the formula:

$$d_{hkl} = \frac{1}{\sqrt{4/3(h^2 + hk + k^2)/a^2 + l^2/c^2}} \tag{3}$$

where a and c are the lattice constants. The calculated c_{film} of ZnO film is 0.5204 nm. From JCPDS data (#36-1451) for ZnO c_{bulk} is 0.5206 nm. The calculated stress using the above equations for the ZnO thin film is 0.0895Gpa. It is found that ZnO thin film shows very small tensile stress along the interfaces, indicating that c-axis values of ZnO thin film are shorter in comparison with bulk ZnO.

In order to determine the composition of ZnO thin film, we carried out an XPS analysis. Figure 2 shows the wide scan spectrum, in the binding energy range of 200-1200eV, of ZnO film deposited on quartz glass substrate. From the XPS spectra we can find the binding energy of Zn 2p was at 1023.1eV in ZnO thin film, which confirmed Zn enrichment in the deposited ZnO thin film [13]. From the statistical results of XPS, we could obtain the atom ratio of O:Zn equaled 0.93:1.

Figure 3 shows SEM micrograph of ZnO film on $Pt/Ti/SiO_2/Si(100)$ substrate deposited at a substrate temperature of 350°C and annealed at 650°C for 10 min in oxygen atmosphere by RTA.

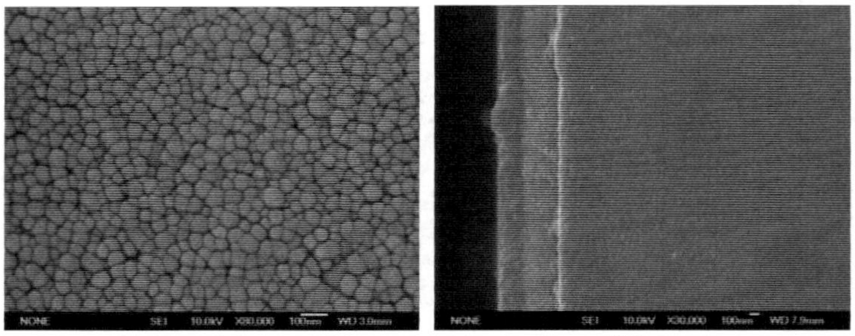

(a) (b)

Figure 3 The SEM images of ZnO thin films on $Pt/Ti/SiO_2/Si(100)$ substrate annealed at 650°C by RTA (a) surface morphology (b) cross-section.

The mean values of grain size were evaluated by a statistical analysis of this micrograph. The grain size is about 50-100nm. The growth morphology of the ZnO film was studied by SEM cross-section images of cleaved samples. The cross-sectional SEM overview micrograph of the ZnO film shows columnar growth. The columns may be regarded as slightly tilted crystallites. Each one is composed of small grains surrounded by the grain boundaries as presumed in the conductivity model proposed by wei_eurieder and Müller [14]. From the cross-sectional, it can be seen the thickness of ZnO thin film was about 480nm.

Journal of Physics: Conference Series **152** (2009) 012036 doi:10.1088/1742-6596/152/1/012036

3.2. Optical properties

Figure 4 shows the spectra of Ψ and Δ of ZnO film as a function of photon energy from 0.5eV to 4.5eV. In the lower energy range, the spectra exhibit oscillations due to the interference among multiple reflected beams within the film, which correspond to the "transparent" range of the film. The oscillating depends on the thickness of the film, in generally, the thicker the films, the higher the frequency. We analyzed these experimental data using a three-layer model (ambient/ZnO/Pt) as the Pt layers were thick enough that light could not propagate thought the Pt layers. The reflective index n and extinction coefficient k as functions of photon energy, as well as the thickness of the film, are determined by an optimization process.

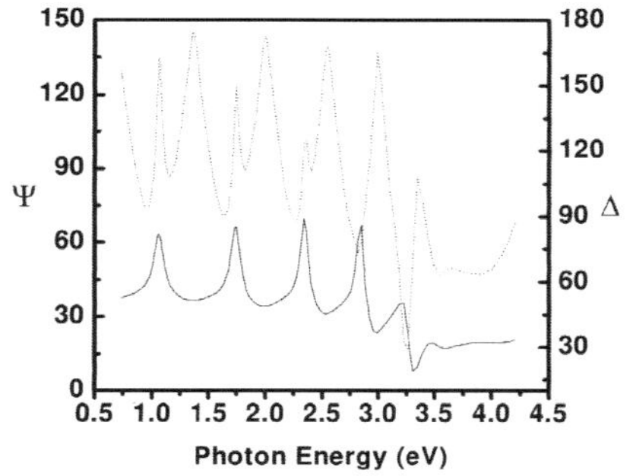

Figure 4 Spectra of the ellipometric parameter Ψ and Δ (solid lines) and (dotted lines) for ZnO films deposited on $Pt/Ti/SiO_2/Si(100)$ substrate.

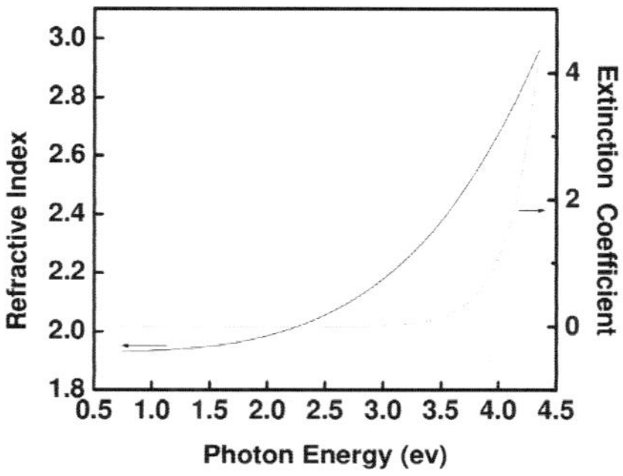

Figure 5 Refractive index n and extinction coefficient k of ZnO thin film on $Pt/Ti/SiO_2/Si(100)$ substrate.

The refractive index n and extinction coefficient k of ZnO film on $Pt/Ti/SiO_2/Si(100)$ substrate as a function of photon energy were shown in Figure 5, which is derived from model fitting the experiment spectrospic ellipsometric data. As shown in Figure 5, the refractive index of ZnO thin film increases as the photon energy increase in the range of 0.5eV to 4.5eV (normal dispersion). The extinction coefficient k increase as photon energy increases and it is very small at low energy part, where the film is transparent.

Figure 6 shows the transmission spectrum of ZnO thin film on fused quartz glass. ZnO thin film demonstrates good optical transmittance (over 90%) in the visible and near infrared spectrum.

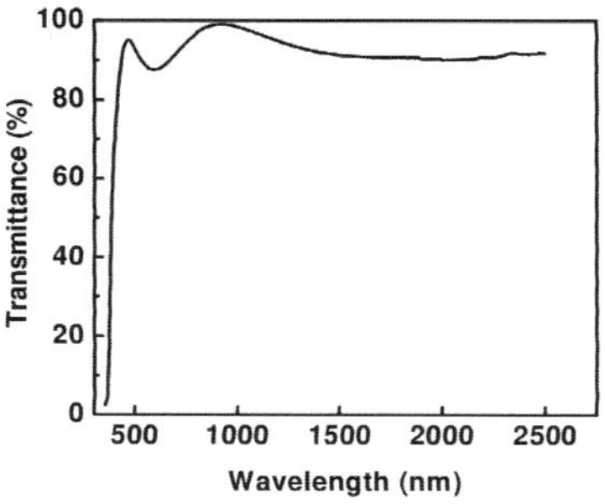

Figure 6 Optical transmittance spectra of ZnO thin film on fused quartz glass.

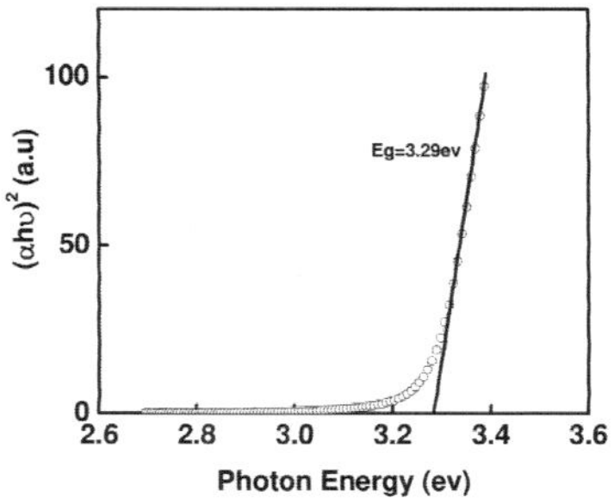

Figure 7 The $(\alpha h\nu)^2$ versus $h\nu$ curves of ZnO thin film on fused quartz glass substrate.

Considering the high absorption region, the transmittance T with absorption coefficient α follows the simple relation[15]:

$$T = A\exp(-\alpha d) \tag{4}$$

where T is the transmittance, A is nearly equal to unity at absorption edge, and d is the thickness of the films. Above the threshold of fundamental absorption, the band gaps energy (Eg) can be estimated by assuming a direct transition between valence and conduction bands. The absorption coefficient α as a function of photon energy can be expressed as [16]:

$$\alpha h\nu = C(h\nu - Eg)^{1/2} \tag{5}$$

for the allowed direct transition where C is the edge width parameter and Eg is the optical band gap. The optical band gap values are obtained by extrapolating the linear portion of the plots of $(\alpha h\nu)^2$ versus $h\nu$ to $\alpha = 0$. From Figure 7, it can be seen that the band gap of ZnO thin film is 3.29eV.

3.3. Dielectric properties

The dielectric constant and the dielectric losses are very important physical parameters for electric materials.

Figure 8 shows the dielectric constant and dielectric loss as a function of frequency from 1kHz to 100kHz. As shown in Fig. 8 the dielectric constant of ZnO thin films is about 13-14. In addition, dissipation factor measured at 10kHz was low (0.03), and the resistivity was greater than $10^{11}\Omega\cdot$cm (not shown). Therefore, we suggest that that ZnO films with high c-axis preferred orientation are suitable for the film bulk acoustic resonator (FBAR) applications.

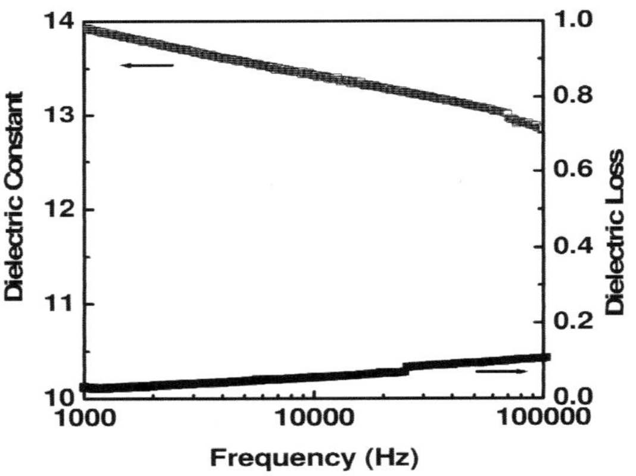

Figure 8 Dielectric constant and dielectric loss for ZnO thin film as a functio of frequency.

4. Conclusion

Highly c-axis-oriented ZnO thin films were prepared by DC reactive magnetron sputtering on Pt/Ti/SiO_2/Si(100) and fused quartz glass substrtates. The atom ratio of O:Zn equaled 0.93:1 was obtained from XPS statistical results The refractive index and extinction coefficient of ZnO thin film were obtained by spectroscopy ellipsometry. The band gap energy is found to be 3.29eV of direct-transition type. The transmission spectrum of ZnO thin film on fused quartz glass indicated that the film is highly transparent. The dielectric constant and dielectric loss at 10kHZ are 13.4 and 0.03 respectively, we suggest that that ZnO films with high c-axis preferred orientation are suitable for the film bulk acoustic resonator (FBAR) applications.

Acknowledgments
The authors would like to thank financial support by Shanghai National found (no.06IR14905).

References

[1] Vossen, G. Hass, Francombe M.H, Hoffmann R.W, 1977 Phys. Thin Films 9, 1
[2] Jeong W.J, Park G.C, Sol. 2001 Energy Mater Sol Cells. 65, 37
[3] Aly S.A, Sayed N.Z. El, Kaid M.A, 2001Vacuum. 61, 1
[4] Bachari E.M, Baud G, Amor S Ben, M. Jacquet, 1999 Thin Solid Films.348, 165.
[5] Krunks M, Mellikov E, 1995Thin Solid Films. 270, 33
[6] Olvera M.L, Maldonado A, Asomaza R, Asomoza M, 1993Thin Solid Films 229, 196
[7] Hao X, Ma J, D. Zhang, T. Yang, J. Huang, 2001 Appl. Surf. Sci. 183, 137
[8] Subramanyam T.K, Srinivasulu B, Uthanna S, 1999 Opt. Mater. 13, 239
[9] Messaoudi C, Sayah D, Abd-Lefdil M, 1995 Phys. Stat. Sol. A151, 93
[10] Schineck H, Helbig R, 1975 Thin Solid Films. 27, 101
[11] Ellmer K, 2000 J. Phys. D: Appl. Phys. 33,R17
[12] Mass J, Bhattacharya P and R Katiyar. S, 2003 Mater. Sci. Eng.B 103, 9
[13] Chen M, Wang X, Yu Y.H, Pei Z.L, Bai X.D, C. Sun, 2000 Appl. Surf. Sci. 158, 134
[14] Weibenrieder K.S, Muller J, 1997 Thin Solid Films. 300, 30
[15] Pasto A.E, Condrte R.E, 1973J. Am. Ceram. Soc. 56, 436
[16] Ramana C.V, Hussain O.M, Naidu B.S, Reddy P.J, 1997 Thin Solid Films 305, 219

The thermal decay in the IrMn-based spin valve

Guang-hong Zhou[1,2], Xian-jin Qi[1], Jie Chen[1] and Yin-gang[1]

1 School of Materials Science and Technology, Nanjing University of Aeronautics and Astronautics, Nanjing 210016, China
2 Department of Mechanical Engineering, Huaiyin Institute of Technology, Huaian 223003, China

E-mail: yingang.wang@nuaa.edu.cn

Abstract. Thermal decay of the top spin valve with a structure of // Seed Ta (5nm) / $Co_{75}Fe_{25}$ (5nm) / Cu (2.5nm) / $Co_{75}Fe_{25}$ (5nm) /$Ir_{20}Mn_{80}$ (12nm) / Cap Ta (8nm) deposited at room temperature by magnetron sputtering has been investigated by means of holding the film in its negative saturation field at various temperatures. Vibrating sample magnetometer has been used to record the magnetic hysteresis loops at room temperature. The recoil loop of the pinned ferromagnetic layer shifts towards the positive field and the exchange bias field (H_{ex}) decreases monotonously while holding the film in a negative saturation field. The decrease of H_{ex} while holding the film in a negative saturation field indicates a thermally decay process. Due to the exchange coupling at the antiferromagnetic/ferromagnetic interface, the antiferromagnetic moments reverse by thermal activation over an energy barrier distribution, which may change in some way as the temperature increases.

1. Introduction

Thermal stability is one of key factors concerned in spin valve heads for high density recording systems and related giant magnetoresistance devices. The thermal decay of exchange bias of antiferromagnetic (AFM) layer to the pinned ferromagnetic (FM) layer results from the fact that the unidirectional anisotropy of the AFM layer decreases, and thus the AFM layer gradually loses its pinning effect to the FM layer, as the temperature increases and/or holding the film at a negative saturation field of the pinned FM layer. It is important for the AFM layer in the exchange coupled structure to keep enough pinning effects at high temperature since the operating temperature of magnetoelectronic devices is sometimes higher than 100 °C [1]. Thermal decay becomes more critical for Mn alloy based spin valves because the Mn in the AFM layer can diffuse into adjacent layers easily [2]. The effect of temperature on the exchange bias field (H_{ex}) has been studied extensively [3-5]. The time dependence of the reversal of the AFM layer has also been studied by measuring the recoil loops of films spent different periods of time while the pinned FM layer maintained at its negative saturation at various temperatures [6,7].

The thermal decay is a long-term gradual degradation of exchange anisotropy under an applied magnetic field. Usually, in decay testing, an accelerated method, i.e., testing at a high temperature, is used, since testing for long time at room temperature (RT) is unpractical. The aim of this work is to find out the intrinsic stability of the spin valve by means of holding the film in a negative saturation field at various temperatures. Testing temperatures were selected properly in order to accelerate the thermal decay while not damage the film.

2. Experimental Details

The spin valve with a structure of //Seed Ta (5nm) / $Co_{75}Fe_{25}$ (5nm) / Cu (2.5nm) / $Co_{75}Fe_{25}$ (5nm) /$Ir_{20}Mn_{80}$ (12nm) / Cap Ta (8nm) was deposited by magnetron sputtering on silicon wafer substrates with a thermally oxidized surface. The deposition conditions were: base pressure $<5\times10^{-7}$ Pa, Ar sputtering pressure 7×10^{-2} Pa, deposition rate 0.03~0.12 nm/s. An electromagnetic field of 100 Oe was applied in order to induce magnetic anisotropy during FM and AFM films sputtering.

The magnetometry measurements were made using a vibrating sample magnetometer (VSM) equipped with a heater, which is used for in-suit heating in this study. The sweep-rate was kept constant at 3 Oe/s for all measurements. The magnetic field was applied parallel or antiparallel to the unidirectional easy axis (UEA), which coincides with the direction of the field applied during film growth.

Waiting time experiments have been undertaken at various temperatures. Testing temperatures are selected as RT, 50 °C, 75°C, 100°C and 125°C. Individual spin valve specimen is first held in a negative saturation field at a given temperature for various periods of time, and then the hysteresis loop is recorded at RT. The hysteresis loop was recorded from the positive saturation field of the film and the applied field was swept to a negative value, which saturates the pinned FM layer in the opposite direction to its initial state. The film was then held at negative saturation field for a given period of time t_{sat}. During this waiting time, some proportion of the AFM layer was expected to reverse as a result of the exchange coupling between the pinned FM and AFM layers and driven by thermal activation. The recoil loop was finally measured as the applied field swept to the starting point. The reversal of the pinned FM layer and therefore the loop position will be governed by the proportion of AFM layer which has reversed. The procedure was repeated for a different value of t_{sat} so that a cumulative effect of waiting time was observed.

It is evident that the hysteresis loop for an exchange biased spin valve will be complex due to the thermally activated reversal of the AFM layer. A classical seven point model of the reversal process in the pinned layer of an exchange biased system has been suggested by Goodman *et al.* [8], which describes the loop of the pinned layer by emphasizing seven important features common to a large range of systems (shown in Figure 1). These seven points will be used as an aid to discussion in our study. Also shown on the diagram are the definitions of the forward and recoil coercive fields (H_m and H_d; respectively) and the exchange field (H_{ex}).

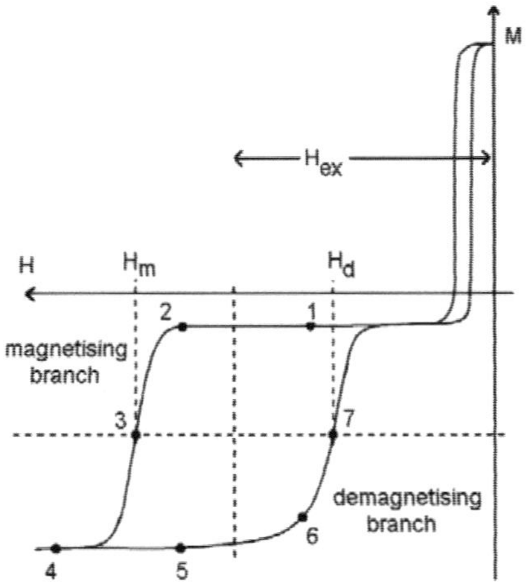

Figure 1. Schematic diagram of the hysteresis loop for a spin valve structure showing the seven-point model.

3. Results and discussion

A series of hysteresis loops for the pinned layer recorded at RT after held at its negative saturation for different periods of time, t_{sat}, are shown in Figure 2. Only the pinned layer loops have been shown for clarity. As the AFM layer is not reset in between recorded loops so the magnetizing branches of the loops are not identical. The shift in the recoil loop (H_d) is used to characterize the degree of AFM reversal, which represents the point at which 50% of the pinned FM layer has reversed and so it also represents the position on the recoil loop at which the reversal of the AFM layer ceases. Beyond this point, the AFM layer will begin to reverse in the opposite direction as the net FM moment has reversed. Furthermore, the shifting of H_d towards positive field means that some proportion of the AFM layer has reversed while the pinned layer is held at a negative saturation, i.e., waiting at point 4 the AFM layer is reordering and the shift of the loop at point 7 towards positive fields is corresponding to the degree of this reordering.

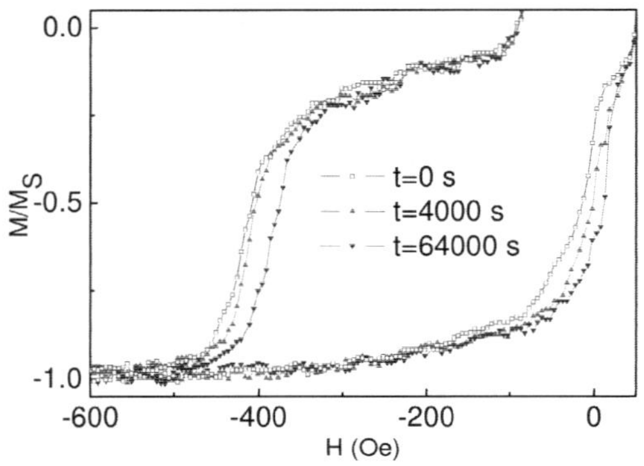

Figure 2. Hysteresis loops for the pinned layer of the spin valve, with different waiting time t_{sat} at negative saturation of the pinned layer. The lines are an aid to eyes.

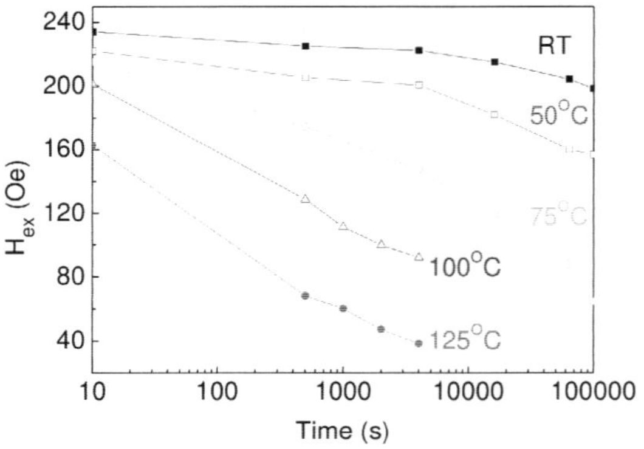

Figure 3. Exchange biasing field as a function of time for films held in a negative saturation field at various temperatures.

Figure 3 shows the exchange bias field H_{ex} as a function of time t_{sat} for samples holding in a negative saturation field at various temperatures. Firstly, one can see that the exchange field H_{ex} measured at RT decreases as the temperature, at which specimens have been experienced shortly, increases. This may arise from the interlayer interdiffusion and strain relief. This observation agrees well with ref. [9]. Secondly, the exchange bias field decreases while holding the film in a negative saturation field at a given temperature. This decay of the pinning field is believed to be a result of thermal activation occurred in the AFM layer, which has been well discussed [6,7,10,11]. Finally, one can see that increasing the temperature at which the film has been held at its negative saturation enhances the decay of exchange field.

The two energy barriers model was used to explain the thermal activation phenomena [12, 13]. In that model the energy barrier with small time constant contributes the enhanced coercivity of the pinned FM layer and that with large time constant results in the thermal decay of the exchange bias field. As a magnetic field was applied at the deposition of the spin valve, it is believed that in the deposition state, the AFM moments are along the initial pinning direction and parallel to the adjacent FM moments. Some portion of the AFM moments are easy to reverse to the opposite direction over the energy barrier with small time constant, which results in the enhanced coercivity of the pinned FM layer. Most of the AFM moments need to overcome a high energy barrier (with large time constant) to reverse to their opposite direction. As a result of that the AFM moments stay along the initial pinning direction if there is no field applied in the heating process and no change occurs in the bias field. However, when there is an applied field of 800Oe antiparallel to the initial bias field, the FM moments are at negative saturation and antiparallel to the unidirectional easy axis of the AFM layer. This state is not the lowest energy state due to the high exchange energy at the FM and AFM interface. The lowest energy state is another state, in which the AFM moments and the FM moments are parallel, i.e. both parallel to the negative magnetic field. Thus the AFM moments tend to reverse from the positive direction to the negative direction. The height of the energy barrier between these two states is determined by the anisotropy energy of the AFM layer and the exchange interaction energy between the FM layer and the AFM layer. Since the anisotropy energy of AFM layer is given by the product of the anisotropy constant K_a and the volume of the AFM grains, a smaller AFM grain has a lower energy barrier. As a result, a smaller AFM grain reverses its moment more easily than a larger AFM grain by thermal activation. The reverse process is a dynamic process and the number of reversed AFM grains increases over time, which results in the t_{sat} dependence of H_{ex}.

The different rate of the change of the H_{ex} with t_{sat} at different temperature indicates that the energy barrier distribution may be different. Coherent rotation of the AFM moments is used to be assumed, which means that all the spins in an AFM grain overcomes the highest anisotropy energy at a same time during its reversal process. However, if incoherent rotation occurs in an AFM grain, only some portion of the spins go through the highest energy site at the one time during the reversal process. Previous studies [5,14] show that a twisting pattern of spins may appear in the AFM layer in the reversal process at elevated temperature. This kind of incoherent magnetic reversal of the AFM layer needs relatively low and wide activation energy distribution than does the coherent rotation of the AFM layer. Incoherent rotation is probably dominant above 75 °C, which results in an enhanced decay of exchange bias field.

Figure 4 shows the t_{sat} dependence of the H_d. One can see that the H_d has an approximately logarithmic dependence on t_{sat} as indicated by the fitted straight lines. As t_{sat} increases, the degree of the AFM moment reverse, and thus the loop shift, increases. The logarithmic dependence observed here agrees well with previous work [15]. In reality, the thermal decay follows a logarithmic law, which results from a distribution of energy barrier over which the AFM moments reverse. One can also see that the increase of the temperature at which the film is held at negative saturation has accelerated the reversal of the AFM moments, which agrees well with the prediction that the energy barriers over which the AFM moments reverse decreases as the temperature increases.

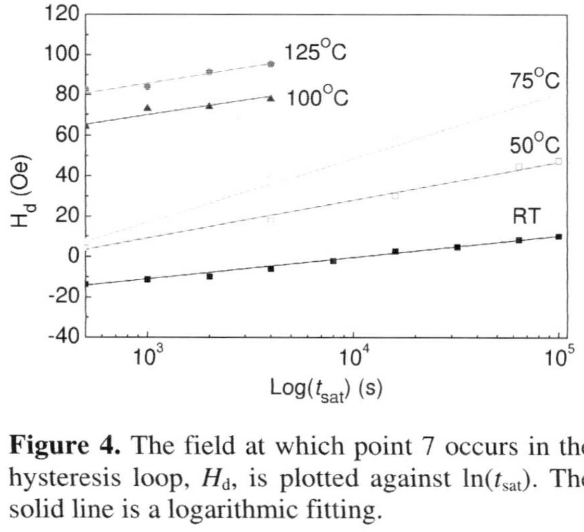

Figure 4. The field at which point 7 occurs in the hysteresis loop, H_d, is plotted against $\ln(t_{sat})$. The solid line is a logarithmic fitting.

There is another point one should note that these straight lines have different slopes, although they almost all obey the logarithmic law at temperatures studied. The slope increases with the temperature increasing up to 75°C and as the temperature increases further the slope reduces. This implies that the energy barrier may be changed in some way as the temperature increases. The origin of this phenomenon is still unclear.

4. Conclusion

It has been shown that the recoil loop of the pinned FM layer shifts towards positive fields and H_{ex} decreases monotonously while holding the film in a negative saturation field at various temperatures. The decrease of H_{ex} while holding the film in a negative saturation field indicates a thermally decay process. Due to the exchange coupling at the AFM/FM interface, the AFM moments reverse by thermal activation over an energy barrier distribution, which may change in some way as the temperature increases.

Acknowledgements

This work is supported by the National Natural Science Foundation of China under grant No.50671048. The authors wish to acknowledge Professor Xiu-feng Han in the Institute of Physics (CAS) for the support on the preparation of specimens.

Reference

[1] Prakash S, Pentek K, and Zhang Y 2001 *IEEE Trans. Magn.* **37** 1123
[2] Yoon S Y, Lee D H, Jeon D M, Kim J H , Yoon D H and Suh S J 2004 *J. Magn. Magn. Mater.* **272-276** 1879
[3] Anderson G W, Huai Y, and Pakala M 2000 *J. Appl. Phys.* **87** 5726
[4] Mao S, Mack A, Singleton E, Chen J, Xue S, Wang H, Gao Z, Li J, and Murdock E 2000 *J. Appl. Phys.* **87** 5720
[5] Nishioka K, 1999 *J. Appl. Phys.* **86** 6305
[6] Hughes T, O'Grady K, Laidler H, and Chantrell R W 2001 *J. Magn. Magn. Mater.* **235** 329
[7] Hughes T, Laidler H, and O'Grady K 2001 *J. Appl. Phys.* **89** 5585
[8] Goodman A M, Laidler H, O'Grady K, Owen N W, and Petford-Long A K 2000 *J. Appl. Phys.* **87** 6409
[9] Rickart M, Guedes A, Franco N, Barradas N P, Diaz P, MacKenzie M, Chapman J N and Freitas P P 2005 *J. Phys. D: Appl. Phys.* **38** 2151
[10] Heijden P A A van der, Maas T F M M, Jonge W J M de, Kools J C S, Roozeboom F, and Zaag P J van der 1998 *Appl. Phys. Lett.* **72** 492
[11] Heijden P A A van der, Maas T F M M, Kools J C S, Roozeboom F, Zaag P J van der, and Jonge W J M de 1998 *J. Appl. Phys.* **83** 7207
[12] Wang Y G and Petford-Long A K 2002 *J. Appl. Phys.* **92** 6699
[13] Wang Y G, Petford-Long A K, Hughes T, Laidler H, O'Grady K, and Kief M T 2002 *J. Magn. Magn. Mater.* **242-245** 1081
[14] Stamps R L 2000 *Phys. Rev. B* **61** 12174
[15] Gaunt P 1986 *J. Appl. Phys.* **59** 4129

Journal of Physics: Conference Series **152** (2009) 012038 doi:10.1088/1742-6596/152/1/012038

Ferromagnetism in sputtered Co doped CeO_2 thin film on Al_2O_3 (0001) substrate

Y Q Song[1], Q H Yang[1], H W Zhang[1*], L Peng[1], L R Shah[2]

[1] State Key Laboratory of Electronic Films and Integrated Devices, University of Electronic Science and Technology of China, Chengdu 610054, People's Republic of China

[2] Department of Physics and Astronomy, University of Delaware, Newark, DE 19716, USA

yuanqiangsong@gmail.com

Abstract. Diluted Co doped CeO_2 films with the stoichiometry of $Ce_{0.97}Co_{0.03}O_{2-\delta}$ have been prepared via long range magnetron sputtering under varied O_2 flux. Films were thoroughly characterized using X-ray diffraction (XRD), atomic force microscopy (AFM), optical transmission spectroscopy, X-ray Photoelectron Spectroscopy (XPS) and magnetic measurements. The results indicate all film are ferromagnetism (FM) at room temperature, of which the saturated magnetization moments (M_S) are changed according to the O_2 pressure adopted in deposition. This phenomenon is well consistent with the oxygen vacancy (V_O, or so called *F*-center) mediated FM mechanism which was recently proposed to interpret FM found in insulating diluted magnetic oxides (DMOs) system. For further exploration of the oxygen dependent of the M_S, oxygen non-stoichiometry was characterized by XPS. It shows that film deposited at lower O_2 pressure contains less oxygen weight or more oxygen vacancies, lead to an enhanced M_S and vice versa.

1. Introduction

Diluted magnetic semiconductors, as recently developed materials with novel functionality, are very promisingly to realize the spin injection into normal semiconductors, to facilitate new emerging spintronic devices. A great deal of investigations have been carried out on the exploration of transitional metals (TMs)-doped semiconductors or insulators such as ZnO[1, 2], TiO_2[3], SnO_2[4], HfO_2[5]. In special, the recently developed diluted magnetic oxide of Co doped CeO_2 has gained much attention. Co doped ceria either in bulk[6] or film[7-9] forms are reported with intrinsic FM at the temperature well above room temperature. The high T_C-FM obtained in these TMs-doped insulating oxides such as doped CeO_2 or HfO_2 refers to an V_O mediated FM mechanism which is comparing with the carrier induced FM mechanism usually used to explain the FM in doped semiconductors. Besides, CeO_2 itself is a promising candidate to be used in silicon-on-insulator structure for high speed devices. Successful fabrication of FM Co doped CeO_2 films bring a potential usage for this material in spintronic devices, e. g., being used in spin transistor[10, 11] or other proposed spin filter devices[12, 13] In proposed spintronic devices, Co doped CeO_2 film can act as a spin filter layer, via which much higher degree of spin polarization of injected carriers can be achieved by the tunneling of non-polarized carriers through this ferromagnetic insulator layer.

In this article Co doped CeO_2 thin films with the stoichiometry of $Ce_{0.97}Co_{0.03}O_{2-\delta}$ were deposited on $Al_2O_3(0001)$ substrate under varied O_2 flux. With the characterization of the structure and magnetic properties, results indicate that all the as deposited films are ferromagnetism (FM) at room temperature, of which the saturated magnetization moments (M_S) are changed according to the O_2 flux adopted in deposition. This phenomenon is well consistent with the *F*-center mediated FM mechanism which was recently proposed to interpret FM found in insulating DMOs system.

2. Experiment details

Co doped CeO_2 films with the stoichiometry of $Ce_{0.97}Co_{0.03}O_{2-\delta}$ were deposited on Al_2O_3 (0001) substrates by long range-magnetron sputtering method. The target was sputtered in a background vacuum of 10^{-6}Pa, on to Al_2O_3 (0001) substrates which were preheated to 723K. During the deposition process, the target-substrate distance (TSD) was kept at about 40 centimeters. This long distance, which is much larger than the usual TSD adopted in magnetron sputtering, makes the growth rate slow enough thus leading to the as deposited film in high quality. In fact, films about 1μm in thickness were obtained by sputtering for 20 hours with the power of 60W. In order to change the oxygen content in the as deposited films, different O_2 flux was injected into the chamber during films deposition. The microstructure was thoroughly characterized using XRD (Cu K_α source) and AFM. Light transmittance was also measured by optical transmission spectroscopy. The magnetic properties were measured using Vibrating Sample Magnetometer (VSM, Model BHV-525) with a resolution of 1×10^{-6} emu. XPS measurement was carried out using Al K_α source. The obtained spectra were all calibrated with C1s (E_b=284.8 eV).

3. Result and discussion

Figure 1 (a) shows the typical XRD pattern of the film deposited under 20sccm O_2. Other samples deposited with smaller O_2 flux have the similar XRD patterns. As can be seen that films deposited with varied O_2 flux are well crystallized with a preferential orientation along $CeO_2(001)$ direction, except very small amount of (111) orientation. There is no any evidence of other impurity phase rather than CeO_2 fluorite structure. The good crystallization can also be found in Figure 1 (b) of AFM characterization. It shows a typical micromorphology of the film deposited under 20sccm O_2 which demonstrates a good crystalline quality. We find an O_2 flux dependency for crystal size, i.e., grain size grows larger with the decrease of O_2 pressure, but not given here.

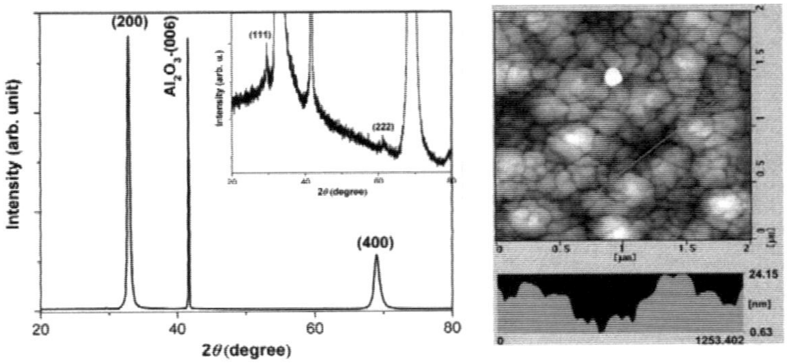

(a) (b)

Figure 1. (a) Typical XRD pattern for the as deposited $Ce_{0.97}Co_{0.03}O_{2-\delta}$ films on $Al_2O_3(0001)$ substrates under 20sccm O_2 flux. The inset shows a scaled-up part to show the low intensity of $CeO_2(111)$ orientation. (b) Typical AFM image for the $Ce_{0.97}Co_{0.03}O_{2-\delta}$ film deposited under 20sccm O_2 flux.

To further validate the phase purity of the as deposited $Ce_{0.97}Co_{0.03}O_{2-\delta}$ films, XPS characterization of Co-2p spectrum has been carried out. As shown in figure 2, the peaks for Co $2p_{3/2}$ and $2p_{1/2}$ and their shakeups are fitted using Gaussian method. The resulting Co $2p_{3/2}$ and $2p_{1/2}$ core levels for Co-O bonding are found to be at 779.98eV and 796.31 eV, respectively. This XPS features, according to Zhang[14], excludes the possibility of the Co cluster formation. Besides, two strong shake-up satellites as labeled in figure 2 emerge at high energy level of Co $2p_{3/2}$ and $2p_{1/2}$, which affords a typical evidence for bivalent cobalt ions exist in the as deposited $Ce_{0.97}Co_{0.03}O_{2-\delta}$ films.

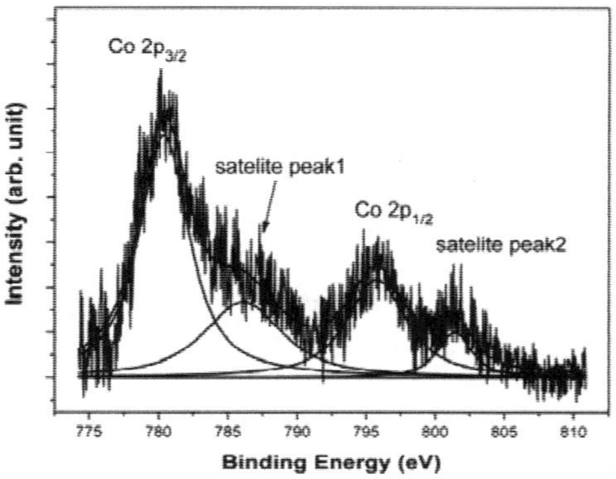

Figure 2. Co-2p XPS spectrum for $Ce_{0.97}Co_{0.03}O_{2-\delta}$ films deposited under varied O_2 flux.

The wavelength dependence of optical transmittance spectra of the films prepared under different O_2 flux are shown in figure 3. The optical transmittance is in direct ration with the adopted O_2 flux, i.e., it increases from 64% to 90% (at=600 nm) with the increase of O_2 flux from 0 to 20 sccm. The low transmittance at low oxygen pressures is due to non-stoichiometric of oxygen atom in $Ce_{0.97}Co_{0.03}O_{2-\delta}$ films. Non-stoichiometry of oxygen atom would cause many defects, which are quite prone to oxygen vacancies in CeO_2, while it still retains its fluorite crystal structure over a wide range of non-stoichiometric compositions through the inclusion of oxygen vacancies.[15-17]

Figure 4 shows magnetization versus fields (*M-H*) loops of the as deposited films with varied O_2 flux during sputtering process. It indicates that all the as deposited films are ferromagnetic with different M_S, which is essentially influenced by the value of O_2 flux. Films deposited in lower O_2 flux obtain the largest M_S which decrease consistently with the increase of O_2 flux.

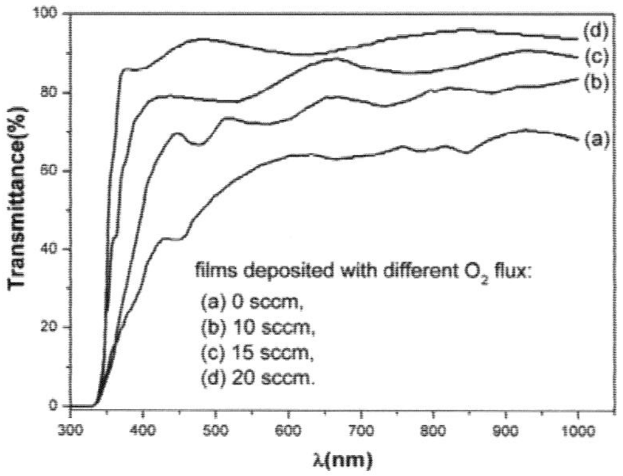

Figure 3. Optical transmittance spectra of $Ce_{0.97}Co_{0.03}O_{2-\delta}$ films deposited under various oxygen flux as indicated in the figure.

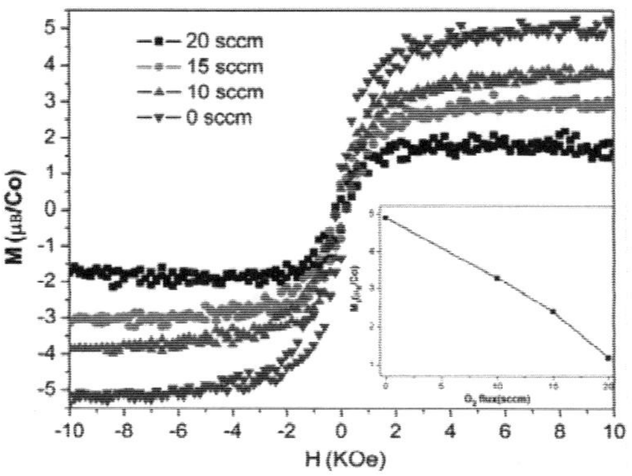

Figure 4. Room temperature M-H loops of $Ce_{0.97}Co_{0.03}O_{2-\delta}$ films deposited under various oxygen flux. The inset shows the detailed M_S dependence on the adopted O_2 flux. figure.

The dependence of M_S on adopted oxygen flux is intimate with the oxygen non-stoichiometry in as deposited $Ce_{0.97}Co_{0.03}O_{2-\delta}$ films. Here XPS characterizations of Ce 3d and O1s are employed to quantify oxygen non-stoichiometry as well as to characterize oxygen vacancies. The standard analysis techniques are applied here to the data for each of the core levels (Ce 3d, O 1s).

The change of oxygen non-stoichiometry is firstly studied by analyzing Ce 3d spectrum. As well known that with oxygen deficiency, Ce^{4+} would be reduced to Ce^{3+} in CeO_2 while still retain its fluorite structure, i.e., $2Ce^{4+} + 2e \rightarrow 2Ce^{3+} + V_O$. Thus one would obtain the density of oxygen vacancies in CeO_2 comparatively by comparing of Ce^{3+} relative weight. We firstly figure out the peak positions corresponding to Ce^{3+}, then by comparing the variation of the spectrum in these positions to qualitatively measure Ce^{3+} content. Figure 5 presents the changes in the Ce 3d XPS spectra of $Ce_{0.97}Co_{0.03}O_{2-\delta}$ films deposited under varied O_2 flux. As well known that CeO_2 has a complex Ce core photoelectron structure. Ce 3d core level spectrum contains about ten peaks, four attributed to the Ce^{3+} and six due to the Ce^{4+} valence state. The peak assignments are discussed in detail in the literature.[18, 19] But Ce 3d spectra itself is not the focus here. For the sake of simple, we only pay attentions on the main peaks which could be obviously attributed to Ce^{4+} or Ce^{3+} valence state. As shown in figure 5, the peaks corresponding to Ce^{4+} $3d_{3/2}$ and Ce^{4+} $3d_{5/2}$ are designated at binding energies of 916.9 eV and 898.3 eV, and those of Ce^{3+} $3d_{3/2}$ and Ce^{3+} $3d_{5/2}$ are indicated at 903.9 and 886.4 eV, respectively. It can be seen that the intensity of Ce^{3+} 3d peaks in films apparently increases with decreasing of the adopted O_2 flux, implying the increased amount of oxygen vacancies with the reduction of ceria.

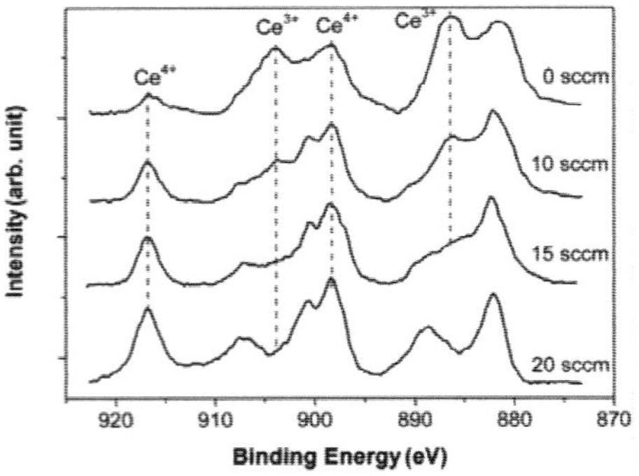

Figure 5. Ce 3d photoemission spectra of $Ce_{0.97}Co_{0.03}O_{2-\delta}$ films deposited under varied oxygen flux as indicated in the figure, in which the position corresponding to Ce^{3+} and Ce^{4+} are labeled, respectively.

The oxygen peaks are used as another source of information about the cerium oxidation state here. In $Ce_{0.97}Co_{0.03}O_{2-\delta}$, O 1s spectrum is the contributions from Ce^{4+}-O, Ce^{3+}-O, and Co^{2+}-O bonded species. We do know that the O 1s peak from Co^{2+}-O bond should be at higher binding energy than that of the O 1s peak from Ce-O bond, as Co has a higher electronegativity (1.88 on the Pauling scale) than Ce (1.12). One would expect that 1s electrons in oxygen attached to cerium atoms in the (+3) oxidation state would be more tightly bound than for cerium in the (+4) state. But the spectrum can be well fitted by two Gaussian peaks as shown in figure 6. According to the above analysis, peak A and B are from Ce^{4+}-O and Ce^{3+}-O species, respectively. The individual peak corresponding to Co^{2+}-O bond is covered up as there is very low Co concentration in the as deposited films. Examining the relative areas of peaks A and B gives an indication of the fraction of the Ce ions that are in the Ce^{3+} state and thus the relative amount of oxygen vacancies. As clearly seen from figure 6, the relative area of peak B reduces rapidly for films deposited with higher O_2 flux, indicating the reduction of the fraction of Ce^{3+} ions as well as oxygen vacancies. It should be stressed here that absorbed oxygen always come into being on the surface of some oxides such as CeO_2, leading to a confusion of the O XPS result. But it is more or less the reflection of the oxygen content variation in different samples, by the obvious dependence between peak B and O_2 flux. Angle dependent XPS could be used for more detailed investigations, and relevant work are in progress.

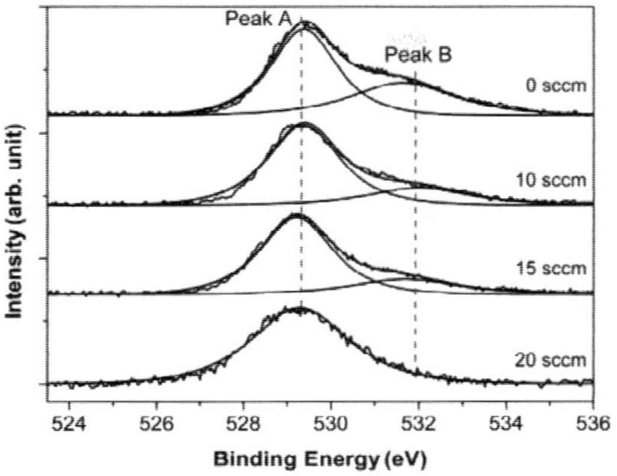

Figure 6. O 1s photoemission spectra of $Ce_{0.97}Co_{0.03}O_{2-\delta}$ films deposited under varied oxygen flux as indicated in the figure, in which peak A and B are corresponding to Ce^{4+}-O and Ce^{3+}-O contributions, respectively.

According to the XPS&VSM results, there are more Ce^{3+} ions or oxygen vacancies in films deposited with lower O_2 flux, which lead to a larger M_S and vice versa. As we know that, Ce^{3+} with the electronic structure of $[Xe]4f^1$ is paramagnetic thus has no contribution to the observed FM in $Ce_{0.97}Co_{0.03}O_{2-\delta}$, which has been proved by the electronic paramagnetic resonance measurement of reduced pure CeO_2[20]. But alternatively, the V_O dependent of M_S is just very consistent with the FCE mechanism recently proposed. According to the FCE mechanism, it is the *F*-centers which provide the coupling centers for Co^{2+} ions. In our experiments, there are more oxygen vacancies in lower O_2 flux-deposited $Ce_{0.97}Co_{0.03}O_{2-\delta}$ films. With the increase of V_O density, more Co^{2+} ions would be incorporated into the ferromagnetic coupling via the delocalized electrons trapped in V_O, leading to the enhanced FM observed in low O_2 pressure-deposited $Ce_{0.97}Co_{0.03}O_{2-\delta}$ films.

4. Conclusion

In summary, preferential oriented $Ce_{0.97}Co_{0.03}O_{2-\delta}$ films were deposited on Al_2O_3 (0001) under varied O_2 flux. All the as deposited films are FM at room temperature, and the M_S is in inverse ratio with O_2 flux. The results are well consistent with FCE mechanism. Aiming at the FM origination mechanism in diluted magnetic oxide, which is still in strong debate recently, oxygen non-stoichiometry as a function of the adopted O_2 flux has been investigated by Ce 3d and O 1s XPS spectra. It indicates an V_O-related FM, i.e., there are more oxygen vacancies in lower O_2 flux-deposited $Ce_{0.97}Co_{0.03}O_{2-\delta}$ films, which leads to the highly enhanced M_S. The result gives direct proof for the V_O or *F*-center mediated FM in $Ce_{0.97}Co_{0.03}O_{2-\delta}$ compounds, and thus very helpful in understanding the mechanism of the FM origination in many TMs-doped insulating DMOs.

Acknowledgement

Supported by the National Basic Research Program (973) under Grant No. 2007CB31407 and the International S&T Cooperation Program of China under Grant No. 2006DFA53410.

References

[1] Ueda K, Tabat H and Kawai T 2001 *Appl. Phys. Lett.* **79** 988
[2] Sharma P, Gupta A, Rao K V, Owens F J, Sharma R, Ahuja R, Osorio Guillen J M, Johansson B and Gehring G A 2003 *Nat. Mater* **2** 673
[3] Matsumoto Y, Murakami M, Shono T, Hasegawa T, Fukumura T, Kawasaki M, Ahmet P, Chikyow T, Koshihara S and Koinuma H 2001 *Science* **291** 854
[4] Ogale S B, Choudhary R J, Buban J P, Lofland S E, Shinde S R, Kale S N, Kulkarni V N, Higgins J, Lanci C, Simpson J R, Browning N D, Das Sarma S, Drew H D, Greene R L and Venkatesan T 2003 *Phys. Rev. Lett.* **91** 077205
[5] Venkatesan M, Fitzgerald C B, Coey J M D 2004 *Nature* **430** 630
[6] Wen Q Y, Zhang H W, Song Y Q, Yang Q H, Zhu H and Xiao J Q 2007 *J. Phys.: Condens.*

Matter **19** 246205
[7] Tiwari A, Bhosle V M, Ramachandran S, Sudhakar N, Narayan J, Budak S and Gupta A 2006 *Appl. Phys. Lett.* **88** 142511
[8] Fernandes V, Klein J J, Mattoso N, Mosca D H, Silveira E, Ribeiro E and Schreiner W H 2007 *Phys. Rev.* B **75** 121304(R)
[9] Vodungbo B, Zheng Y, Vidal F, Demaille D, Etgens V H and Mosca D H 2007 *Appl. Phys. Lett.* **90** 062510
[10] Johnson M 1993 *Science* **260** 320
[11] Monsma D J, Lodder J C, Popma J A and Dieny Th B 1995 *Phys. Rev. Lett.* **74** 5260
[12] Moodera J S, Hao X, Gibson G A and Meservey R 1998 *Phys. Rev. Lett.* **61** 637
[13] Worledge C and Geballe T H 2000 *J. Appl. Phys.* **88** 5277
[14] Zhang Y B, Liu Q, Sristharan T and Gan C L 2006 *Appl. Phys. Lett.* **89** 042510
[15] Inoue T, Yamamoto Y and Satoh M 1999 *Thin Solid Films* **343** 594
[16] Morshed A H, Moussa M E, Bedair S M, Leonard R, Liu S X and El-Masry N 1997 *Appl. Phys. Lett.***70** 1647
[17] Tuller H L and Nowick A 1997 *J. Phys. Chem. Solids* **38** 859
[18] Preisler E J, Marsh O J, Beach R A, McGill T C 2001 *J. Vac. Sci. Technol.* B **19** 1611
[19] Fujimori A 1983 *Phys. Rev.* B **27** 3992
[20] Vinokurov I, Zonn Z and Ioffe V 1968 *Sov. Phys. Solid State* **9** 2659

Journal of Physics: Conference Series **152** (2009) 012039 doi:10.1088/1742-6596/152/1/012039

Structural, optical and magnetic properties of Fe-doped ZnO

Wenjuan Cheng, Xueming Ma

State Key Laboratory of Precision Spectroscopy, East China Normal University, 3663 North Zhongshan Rd., Shanghai, 200062, P R China

E-mail: wjcheng@phy.ecnu.edu.cn

Abstract. $Zn_{1-x}Fe_xO$ (x =0.011, 0.029, 0.048, 0.10, 0.20) thin films have been grown on Al_2O_3 (0001) substrates by magnetron co-sputtering technique. X-ray diffraction analysis revealed that Fe atoms were successfully incorporated into ZnO host matrix, and the films exhibited wurzite structure up to x = 0.20 with reduced c-axis lattice constant compared to that of pure ZnO. There are not any secondary phases, and Fe^{2+} as well as Fe^{3+} substitutes for Zn^{2+} of ZnO host. With increasing x, the optical band gap is found to shift slightly to lower energies (66 meV for x = 0.20) than that of the pure ZnO. Magnetization measurement indicated that all samples were ferromagnetic at room temperature. An explanation for the dependence of the optical and magnetic properties of the samples on the value of x is also given.

1. Introduction

Zinc oxide, a well-known piezoelectric and electro-optic material with wide direct band gap (3.37 eV) and large exciton binding energy [1,2], is now being considered as a potential material for a wide range of applications such as optoelectronic and luminescent devices as well as chemical sensors. Extensive studies have been carried out to modify the properties of ZnO for different applications [3]. For example, a variety of elements has been proposed to intentionally manipulate its optical and electrical properties of the ZnO host. The band gap of ZnO could be increased or reduced by doping with Mg or Cd [4,5], respectively. Transparent conductive ZnO films with low resistivity as well as high visible transmittance could be realized by doping with the elements Al, Ga and In [6-8].

Since Dietl et al. [9] theoretically predicted that a high Curie temperature ferromagnetism with a large magnetization could be obtained in Mn-doped p-type ZnO, transition metal-doped ZnO has been extensively investigated as a promising diluted magnetic semiconductor (DMS) for spintronic devices. Room temperature ferromagnetism has been widely observed in many ZnO-based DMSs despite controversy over the origin of observed ferromagnetism [10-13]. Until recently, most studies have been focused on the fabrication of Co- or Mn-doped ZnO systems as well as their structural, optical and magnetic properties [14-16]. However, relatively little attention has been paid to Fe-doped ZnO film. In this research, a series of $Zn_{1-x}Fe_xO$ films were deposited on Al_2O_3 (0001) substrates using magnetron cosputtering technique. The influence of Fe concentration on the structural and magnetic properties of the samples was presented and discussed. Considering their essential importance for applications in spin-optoelectronic devices in the future, the optical properties of the films were also studied.

2. Experimental

$Zn_{1-x}FexO$ thin films were deposited on Al2O3 (0001) substrates using magnetron cosputtering technique. ZnO (99.99%) target used rf power while iron (99.99%) metal target used dc power. The composition of the film was controlled by varying the sputtering power of Fe target while ZnO target sputtering power was fixed. The distance between target and substrate was 80 mm. Ar (99.999%) and O2 (99.999%) were supplied as working and reactive gases, respectively. The sputtering chamber was evacuated to a base pressure of 7×10−5 Pa before Ar and O2 gases were introduced with the flow ratio of 2:1. During the sputtering process, total gas-flow rate, chamber pressure, and substrate temperature was kept at 40 sccm, 0.5 Pa and 400 oC, respectively. Before deposition, targets were pre-sputtered for 10 min to remove any surface contamination, while a shutter was put to prevent the substrate from being deposited. The substrate holder kept rotating during sputtering in order to obtain uniform growth. The deposited films exhibited x =0.011, 0.029, 0.048, 0.10, and 0.20, respectively.

The crystal structure of the $Zn_{1-x}Fe_xO$ films was determined by a D/max 2550V X-ray diffractometer (XRD) using a Cu Kαsource of 1.542 Å. X-ray photoelectron spectroscopy (XPS, PHI-5702XPS spectrometer using Mg Kα radiation of 1253.6 eV) was used to analyze the composition, binding energy level and chemical structure of the samples. Spectroscopic ellipsometry (SE) measurements were performed on a large spectral range ellipsometer (J.A. Woollam Co. M-2000U) to derive the refractive index and the thickness of the films using Cauchy model. Photoluminescence spectroscopy was used to measure optical emissions of the samples by using a 325 nm He–Cd laser at 25 mW. Magnetic measurements were carried out on an HH-15 vibrating sample magnetometer (VSM) with the magnetic field applied parallel to the film plane.

3. Results and discussion

Figure 1 shows the X-ray diffraction patterns of the $Zn_{1-x}Fe_xO$ films compared to that of a pure ZnO film. As we can see, besides Al_2O_3 (0006) peak relative to the substrate, only single ZnO wurtzite phase with (0002) preferential orientation is observed, suggesting that the Fe-doped ZnO thin films retain the same crystalline structure and high c-axis orientation as the un-doped ZnO films. There are no anomalous peaks related to Fe metal clusters or Fe oxides secondary phases. With increasing Fe concentration x in the films, the clear shifts of the (0002) peaks to higher angles are observed. This phenomenon indicates a reduction in the lattice constant with Fe doping, which is expected as smaller Fe ions are incorporated into the Zn sites of the ZnO lattice without changing the wurtzite structure. Also, the (0002) peak exhibits an increased line broadening as x increases, suggesting a significant lattice-defect formation by Fe doping.

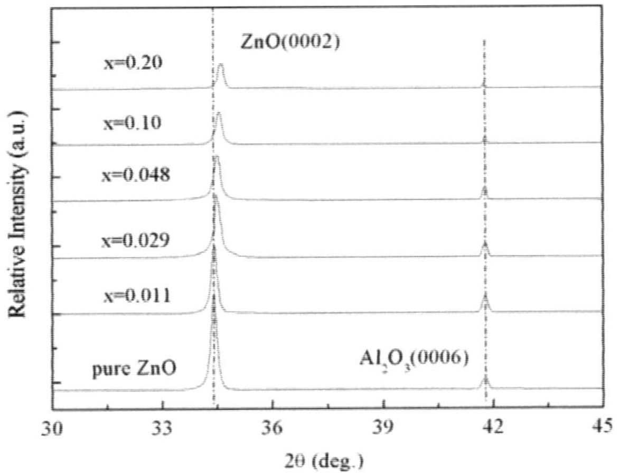

Figure 1. XRD spectra of $Zn_{1-x}Fe_xO$ films compared to that of pure ZnO film.

In order to properly substitute Zn^{2+} ionic sites with maintaining charge neutrality of the films, Fe ions need to have a valence of +2. It is noted that the ionic radius of Fe^{2+} is larger than that of Zn^{2+} by about 5%, whereas that of Fe^{3+} is smaller by about 10%[17]. The lattice parameter should not decrease with enhancing x if there is Fe^{2+} that replaces the site of Zn^{2+}. Therefore, it can be inferred that Fe^{2+} and Fe^{3+} ions coexist in the $Zn_{1-x}Fe_xO$ films, and Fe^{3+} ions are expected to distort the lattice structure for holding charge neutrality. The higher concentrations of Fe may provide more chances for Fe to substitute Zn^{2+} sites, resulting in the lattice constants further shrink with the increase of the Fe concentration x. Considering that both the valence state and the ionic radius are the major factors determining the solubility of the dopant [17], high Fe solubility (up to x=0.20) for the present $Zn_{1-x}Fe_xO$ samples is attributable to the existence of the Fe^{3+} ions with high density in the present films.

In order to verify the coexistence of the Fe^{2+} and Fe^{3+} ions in the $Zn_{1-x}Fe_xO$ films, XPS measurements were conducted. No significant changes in the chemical shift and signal were found for Zn $2p^{1/2}$ and Zn $2p^{3/2}$ (data not shown here). Fe 2*p* core evel photoemission spectrum for the x=0.10 film is shown in figure 2(a), from which Fe $2p^{1/2}$ and Fe $2p^{3/2}$ peaks located at 723.9 and 710.3 eV can be found. It is known that metallic Fe has a peak position of Fe $2p^{1/2}$ at 719.9 eV and Fe $2p^{3/2}$ at 706.5 eV, FeO at 722.3 eV and at 709.3 eV, and Fe_2O_3 at 724.9 eV and at 710.5 eV, respectively. The obtained binding energies for Fe $2p^{1/2}$ and Fe $2p^{3/2}$ peaks in our films are larger than those of Fe^{2+} and smaller than Fe^{3+}, suggesting the peak consists of more than one component. Using the Doniach-Sunjic Lorentzian line shape convoluted with the Gaussian factor, the spectra can be well fitted with two curves, as shown in figure 2(a). Thus, the XPS data gives an obvious evidence for the coexistence of the Fe^{2+} and Fe^{3+} ions in the $Zn_{1-x}Fe_xO$ films, which is in agreement with the XRD result. The Fe^{3+} ions with high density in the samples are expected to be caused by the O_2-reactive sputtering.

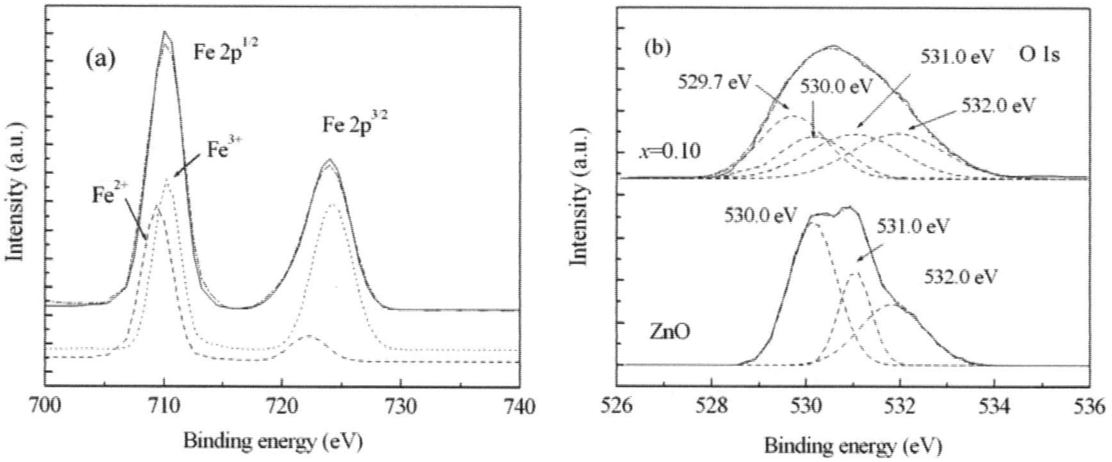

Figure 2. XPS spectra of (a) Fe 2p and (b) O 1s for pure ZnO and Fe-doped ZnO film (x=0.10).

The XPS core level spectra of O 1s peak and its deconvolution results by Gaussian fitting for pure ZnO and Fe-doped ZnO (x=0.10) film are shown in figure 2(b). For pure ZnO, the O 1s spectrum could be split into three peaks centered at 532.0, 531.0 and 530.0 eV, which were allocated to O^{-2} in ZnO structure, O^{-2} of oxygen deficiency, and the chemically absorbed oxygen sites, respectively [18]. When Fe was introduced into ZnO matrix, besides the aforementioned three types of oxygen energy levels in pure ZnO, O^{-2} state centered at 529.7 eV in Fe_2O_3 was also obtained, and meanwhile the intensity of oxygen deficiency around 531.0 eV reduced.

The effect of Fe doping on the optical properties of the samples was investigated by spectroscopic ellipsometry and photoluminescence at room temperature. For all the samples, the energy band gaps were estimated by using the following Eqs. (1) and (2) assuming a direct type transition [19].

$$(\alpha h\nu)^2 = A(E - E_g) \tag{1}$$

$$\alpha = \frac{4\pi k}{\lambda} \tag{2}$$

where A is a constant, E_g is the optical energy band gap, k represents the imaginary part of the refractive index, λ is the light wavelength, and α is the optical absorption coefficient which can be calculated from k of SE results (Eqs. (1) and (2)). As shown in figure 3, from SE results, we made a plot of $(\alpha h\nu)^2$ versus the photon energy. Extrapolation of the linear portion of $(\alpha h\nu)^2$ to zero gives the value of E_g. The dependence of E_g on Fe content x is depicted in the inset of this figure. With increasing x, the optical band gap is found to shift slightly to lower energies (66 meV for x = 0.20) than that of the pure ZnO. J. Dipuri et al. have attributed the band-gap decrease of transition-metal-doped II-VI semiconductors to the *sp-d* spin-exchange interactions between the band electrons and the localized d electrons of the transition-metal ion substituting the cation [20,21], which has been experimentally confirmed by K. Ando et al. in $Zn_{1-x}Co_xO$ [22]. Thus, in our films, the *s-d* and *p-d* exchange interactions give rise to a negative and a positive correction to the conduction- and the valence-band edges, respectively, reducing the band-gap.

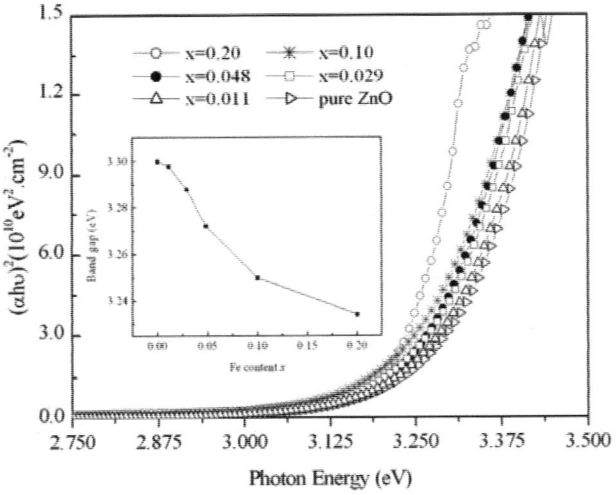

Figure 3. $(\alpha h\nu)^2$ vs photon energy plots of $Zn_{1-x}Fe_xO$ films compared to that of pure ZnO. The inset shows the variation of optical band gap with Fe concentrations.

Figure 4 shows the photoluminescence emission spectra of the pure ZnO and Fe-doped films. It can be seen all the films exhibit two kinds of emission peak at UV (~380 nm) and visible (~550 nm). The UV emission peak, whose energy is almost equal to the band gap energy of ZnO (3.37 eV), results from excitonic recombination corresponding to near-band-edge emission of ZnO [23], while the broad visible emission is attributed to the recombination of electrons trapped in singly ionized oxygen vacancies with holes and the interstitial oxygen ions (Oi^-) [24.25].

When Fe concentration in the films is enhanced, the peak intensity of the UV emission increases, whereas the visible emission intensity decreases. The result suggests that the lattice of Fe-doped ZnO films has fewer oxygen vacancies and interstitial oxygen ions compared to pure ZnO, which is consistent with the XPS data shown in figure 2(b). It also can be seen, with enhancing *x*, the UV peak position has a slightly shift from 376 nm to 383 nm in agreement with the narrowing of the optical band gap depicted in figure 3.

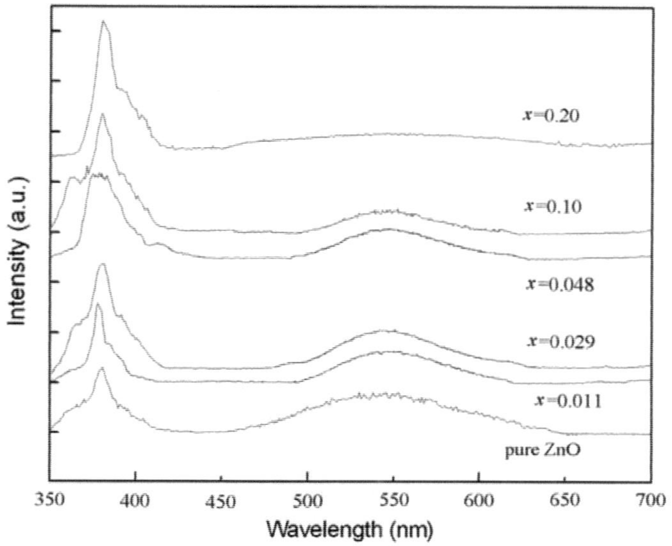

Figure 4. Room temperature PL spectra of Fe-doped ZnO films compared to that of pure ZnO.

The magnetization (*M*) as a function of magnetic field (H) for the $Zn_{1-x}Fe_xO$ films was measured at room temperature by VSM, as shown in figure 5. A magnetic field of 1500 Oe was applied parallel to the surface of the films, and the contribution from the substrate was separately measured and subtracted from the raw data. It can be seen that all these *M-H* curves corresponding to various Fe content *x* are in the shape of a distinct hysteresis loop, indicating that room-temperature ferromagnetic order has been achieved for the Fe-doped ZnO thin films. In addition, comparison of the samples with various Fe content *x* suggests that the total magnetic moment gradually increases with increasing *x* from 0.011 to 0.20. The variation of the saturation magnetization (M_S) per Fe atom with Fe concentration is shown in the inset of this figure. It can be seen that M_S per Fe atom slightly declines with increasing Fe content.

As is well known, ZnO without doping shows paramagnetic behavior, Fe is the cause of the observed ferromagnetism in the Fe-doped ZnO samples. Generally, if the Fe congregates as clusters rather than substitutes the Zn site, the lager Fe-doped concentration certainly provides the more possibilities for the formation of more Fe clusters. However in our case**,** it was difficult to form Fe cluster because the films were deposited under oxygen conditions. Furthermore, even if Fe metal segregation was present, it could not account for the dependence of the M_S per Fe atom and the c-axis lattice parameter on the Fe content observed in XRD measurement. Note that increasing *x* from 0.011 to 0.20 results in a decrease both in the value of M_S and the c-axis lattice parameter, as depicted above. In addition, if metallic iron oxides such as Fe_3O_4 or γ-Fe_2O_3 were responsible for the ferromagnetic behavior, the diminution with increasing Fe concentration would not happened. Also, the XRD patterns clearly showed that there was no any other second phase such as Fe and Fe oxides. Thus, it can be certain that the observed ferromagnetism in the $Zn_{1-x}Fe_xO$ film originates from the Fe substitution for Zn in ZnO. From the XRD and XPS results, Fe^{3+} and Fe^{2+} ions coexsist in the films by substituting on the Zn^{2+} site. In view of this, the electron double-exchange mechanism [26] has a possibility as the explanation for the magnetic behaviors mentioned above. Further research on this issue might need to be done in the future to clarify the supposition.

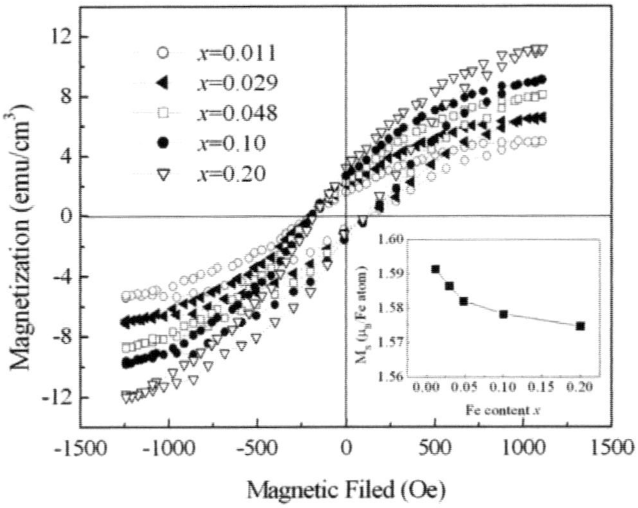

Figure 5. Magnetic field dependence of magnetization for Zn1−xFexO films at room temperature. The inset gives MS per Fe atom as a function of Fe concentration.

4. Conclusion

A series of $Zn_{1-x}Fe_xO$ thin films with the atomic fraction, x, in the range of 0.011–0.20 were deposited on Al_2O_3 (0001) substrates at 400°C by magnetron co-sputtering technique. A systematic study was done on the structural, optical and magnetic properties of the films as a function of Fe concentration. XRD and XPS results indicated that there are not any secondary phases and Fe^{2+} as well as Fe^{3+} substituted for Zn^{2+} of ZnO host. With increasing x, the optical band gap is found to shift slightly to lower energies (66 meV for $x = 0.20$) than that of the pure ZnO, and the optical properties of films are improved. Magnetization measurement indicated that all samples are ferromagnetic at room temperature. When Fe content was enhanced, the total magnetic moment increased while the saturation magnetization per Fe atom slightly declines. The observed ferromagnetism in the samples originates from the Fe substitution for Zn in ZnO, which follows the electron double-exchange mechanism.

Acknowledgment

The authors would like to acknowledge the financial support of the National Science Foundation of China under Grant no. 60606004.

References

[1] V Srikant and D R Clarke 1998 *J. Appl. Phys.* **83** 5447
[2] A Mang, K Reimann and St Rubenacke 1995 *Solid State Commun.* **94** 251
[3] V A Karpina, V I Lazorenko, C V Lashkarev, V D Dobrowolski, V D Dobrowolski, L I Kopylova, V A Baturin, S A Pustovoytov, A Ju Karpenko, S A Eremin, P M Lytvyn, V P Ovsyannikov and E A Mazurenko 2004 *Cryst. Res. Technol.* **39** 980
[4] Shan F K, Kim B I, Liu G X, Liu Z F, Sohn J Y, Lee W J, Shin B C and Yu Y S 2004 *J. APPL. PHYS.* **95** 4772

[5] Yogeeswaran G, Chenthamarakshan C R, Seshadri A, Tacconi N R de and Rajeshwar 2006 *Thin Solid Films* **515** 2464
[6] Liu W F, Du G T, Sun Y F, Bian J M, Cheng Y, Yang T P, Chang Y C and Xu Y B 2007 *Appl. Surf. Sci.* **253** 2999
[7] B Kotlyarchuk, V Savchuk and M. Oszwaldowski 2005 *Cryst. Res. Technol.* **40** 1118
[8] Ma Quan-Bao, Ye Zhi-Zhen, He Hai-Ping, Hu Shao-Hua, Wang Jing-Rui, Zhu Li-Ping, Zhang Yin-Zhu and Zhao Bing-Hui 2007 *J. Cryst. Growth.* **304** 64
[9] T Dietl, H Ohno, F Matsukura, et al., 2000 *Science* **287** 1019
[10] C N R Rao and F L Deepak 2005 *J. Mater. Chem.* **15** 573
[11] R Seshadri and Curr Opin 2005 *Solid State Mater. Sci.* **9** 1
[12] Darshan C Kundaliya, S B Ogale, S E Lofland, S Dhar, C J Metting, S R Shinde, Z Ma, B Varughese, K V Ramanujachary, L Salamanca and T Venkatesan 2004 *Nat. Mater.* **3** 709
[13] M Berciu and R N Bhatt 2004 *Phys. Rev. B* **69** 045202
[14] M Liu, A H Kitai and P Mascher 1992 *J. Lumin.* **54** 35
[15] S J Pearton, C R Abernathy, M E Overberg, G T Thaler, D P Norton, N Theodoropoulou, A F Hebard and Y D Park 2003 *J. Appl. Phys.* **93** 1
[16] K Ueda, H Tabat and T Kawai 2001 *Appl. Phys. Lett.* **79** 988
[17] Z Jin, T Fukumura, M Kawasaki, K Ando, H Saito and T Sekiguchi 2001 *Appl. Phys. Lett.* **78** 3824
[18] M Chen, X Wang, Y H Yu, Z L Pei, X D Bai, C Sun, R F Huang and L S Wen 2000 *Appl. Surf. Sci.* **158** 134
[19] K C Park, D Y Ma and K H Kim 1997 *Thin Solid Films* **305** 201
[20] J Diouri, J P Lascaray and M El Amrani 1985 *Phys. Rev.* B **31** 7995
[21] R B Bylsma, W M Becker, J Kossut, U Debska and D Yoder-Short 1986 *Phys. Rev.* B **33** 8207
[22] K Ando, H Saito, Z Jin, T Fukumra, M Kawasaki and Y Matsumoto 2001 *Appl. Phys. Lett.* **78** 2700
[23] J G Wen, J Y Lao, D Z Wang, T M Kyaw, Y L Foo and Z F Ren 2003 *Chem. Phys. Lett.* **372** 717
[24] K Vanheusden, C H Seager, W L Warren,D R Tallant and J A Voigt 1996 *Appl. Phys. Lett.* **68** 403
[25] M Liu, A H Kitai and P Mascher 1992 *J. Lumin.* **54** 35
[26] A Chattopadhyay, S Das Sarma and A J Millis 2001 *Phys. Rev. Lett.* **87** 227202

Study of Reoxidation in Heavily La-doped Barium Titanate Ceramics

Yang Wenhu, Pu Yongping, Chen Xiaolong and Wang Jinfei

School of Materials Science and Engineering, Shaanxi University of Science and Technology, Xi'an 710021, China

E-mail:dellon9960@163.com ,06yangwenhu@sust.edu.cn

Abstract. The ceramic samples of (1-x)$BaTiO_3$+½$x$$La_2O_3$+(2+$x$)mol%$TiO_2$ ($0.003 \leq x \leq 0.15$) were prepared by sintering in reducing atmosphere of H_2, subsequently reoxidized in air. The effect of heavily La dopant on the resistivity and microstructure of the reoxidized samples were investigated by means of scanning electron microscopy and electric properties testing. The results show that the critical concentration of La-doped samples increased obviously compared with the samples sintered in air. The grain size decreased with the increase of La-donor content. The positive temperature coefficient resistivity effect of the sample increased when a change reoxidizing temperature range of 600°C ~1200°C. The maximum of ρ_{max}/ρ_{min} ratio was ~700 at the reoxidizing temperature of 1100°C when x=0.8mol%.

1. Introduction

Ferroelectric $BaTiO_3$ is one of the most important ceramic materials in electronics. It has been widely used in electronic devices, such as multiplayer capacitors (MLCC), sensors with positive temperature coefficients resistivity (PTCR), piezoelectric transducers, ferroelectric thin-film memories, etc. At room temperature, $BaTiO_3$ adopts a tetragonal perovskite structure and is a ferroelectric with high permittivity. It transforms to the cubic, paraelectric state at the Curie temperature (T_c=130°C). Undoped $BaTiO_3$ is electrically insulating, but the electrical resistance can be controlled effectively by doping in either barium or titanium site with proper donor impurity ions [1, 2], for example La^{3+}, Ce^{3+} ions, etc.

Generally, the donor-doping barium titanate PTCR ceramics are prepared by a sintering reduced semiconducting barium titanate ceramics at the temperature range of 800°C ~1200°C. The grain boundary is partially oxidized under different oxygen partial pressure in cooling process, the samples show the PTCR effect [3]. The reason is that the oxygen ions escape from the lattice at high temperature in reducing atmospheres (H_2) and the donor partial substitution of A site ion, multiple ion occupation of A and/or B sites in ABO_3 compounds is expected to bring in changes in Curie temperature and other physical properties [4]. This kind of substitution can affect the lattice parameters, tetragonal distortion (c/a), polarization, ferroelectric transition and other characteristics of the samples. But the released oxygen partly enters into the sample when the samples are sintered in oxidizing atmosphere. The room temperature resistance is heavily sensitive to the degree of reoxidation, therefore, the research of reoxidation mechanism is very important to PTCR effect.

In this study, La-doped barium titanate ceramics sintered in H_2 with a range of concentration from 0.3mol% to 15.0mol% and reoxidized in air were investigated. The changes of the microstructure at a

Journal of Physics: Conference Series **152** (2009) 012040 doi:10.1088/1742-6596/152/1/012040

temperature range of 600°C ~1200°C were observed and the PTCR effect of samples was also reported.

2. Experimental

La-doped $BaTiO_3$ ceramics were prepared using conventional mixed-oxide technology according to the formulas of $BaTiO_3$+½$x$$La_2O_3$+2.0mol%$TiO_2$(0≤$x$≤15.0%).The raw materials ($BaTiO_3$, TiO_2 and La_2O_3) with TiO_2-rich (2.0mol%) in order to promote sintering in the presence of liquid phase, were mixed and milled in water for 4h inside polyethylene jars using zirconia media, dried at 80°C. After being dried, the powders were granulated using ploy-vinyl alcohol (PVA) binder and formed under uniaxial pressure of 100MPa, pressed into tablet (φ14mm×2mm). The disks were sintered in H_2 for 2h at 1350°C with a heating rate of 10K/min and a cooling rate of 5K/min. Subsequently, the sample reoxidized in air for 2h at a temperature range of 800°C ~1100°C. Fire-on silver was used as the electrode material for the measurement of electrical properties of the sintered samples. The nature surface microstructure was investigated by means of optical microscopy and scanning electron microscope. The temperature dependence of resistance was measured in a temperature-programmable furnace with a heating rate of 2°C /min from 20°C to 350°C.

3. Results and Discussion

The room temperature resistivity of La-doped $BaTiO_3$ ceramics sintered in air and H_2 as a function of dopant concentration is shown in Fig.1. From the figure, it is found that the resistivity decreased with the increase of La-doped concentration when x≤0.3mol%, subsequently increased rapidly with the increase donor content. The concentration (x=0.3mol%) was regarded as critical concentration. The resistivity drop was generally attributed to the electron compensation of the substituted cation via donor-doping.

The subsequent rise in resistivity for dopant concentration x>0.3mol% is attributed to a change in doping mechanism to ionic compensation via creation of cation vacancy. But the critical concentration increased up to 0.8mol% when the sample was sintered in H_2. It is found that the critical concentration closely depended on sintering atmosphere.

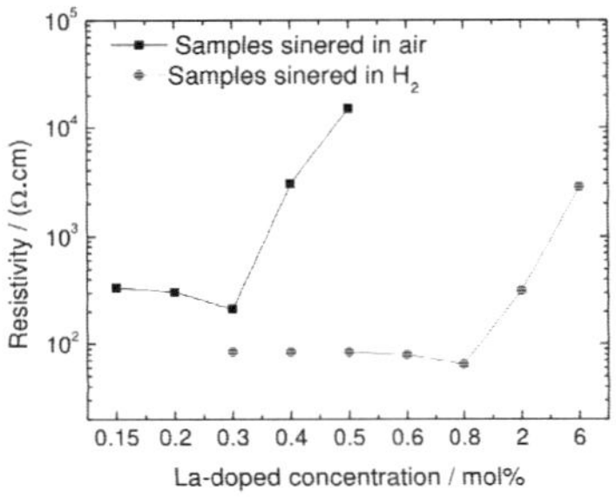

Figure 1. The room-temperature resistivity as a function of La-dopant concentration for $BaTiO_3$ ceramics sintered in air and H_2 at 1350°C.

3.1. Microstructures

Fig.2 shows SEM photographs of La-doped samples sintered at 1350°C in atmosphere of H_2. It is indicated that the grain size of the sample rapidly decreased with increasing La concentration from 0.3 mol% to 15.0 mol%.

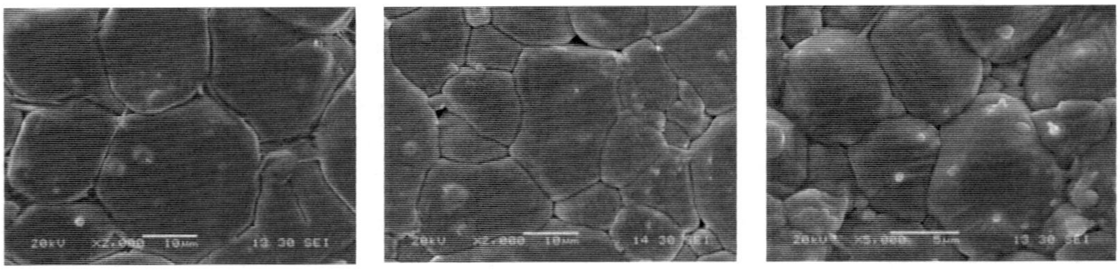

(a) x=0.3mol% (b) x=0.6mol% (c) x=0.8mol%

(d) x=0.3mol% (e) x=0.3mol% (f) x=0.3mol%

Figure 2. SEM micrograph of La-doped ceramics sintered at 1350°C in atmosphere of H_2

In Fig.2 (a) and (b), the grain size is ~30 um, however it is decreased with the La doping concentration increasing from 0.3mol% to 15mol%. There must be given two condition in the process of grain growth, one is liquid phase intervened and the other is $BaTiO_3$ grains must be dissolved [5]. In present work, the excess amount of TiO_2 present in the samples, which caused the formation of the reactive liquid phase during sintering. When the samples were sintered in 1350°C, the liquid phases $Ba_6Ti_{17}O_{40}$ was formed [6]. $BaTiO_3$ grains highly dissolved in the liquid phases, which enhanced the mass transport, promoted the small grain growth rapidly at low La-doped concentration. By comparison, the La^{3+}-donor prevented grain growth, which was related to the oxygen partial pressure of the sintering atmosphere. According to the thermodynamic model, it can be expressed by the following format (1) [7]:

$$\lg p(O_2) = \frac{A}{x_c} + B \tag{1}$$

where, $p(O_2)$ is the oxygen partial pressure, A is the stored surface energy at the phase boundary, x_c is the critical donor concentration, and B is the free energy of oxygen released during anomalous grain growth. The critical concentration is highly related to the oxygen partial pressure while A and B were constant.

3.2. Grain Boundary Reoxidation

The origin La-doped samples were black. After reoxidation at different temperature in air, the color of the samples turned to light blue gradually. It is known that the dark color of $BaTiO_3$ ceramics is related to the presence of Ti^{3+}. Oxygen uptaking occurred at outer and inner surface of the samples during the reoxidation process, oxygen diffused along with the grain boundary on the grain surface toward the center of the grains, which was related with the La-doped concentration by the format (2).

$$[V_O^{\bullet\bullet}] = [La_{Ba}^{\bullet}]^{-n} \tag{2}$$

It is shown that the concentration of oxygen vacancies $[V_O^{\bullet\bullet}]$ decreased with the increase of $[La_{Ba}^{\bullet}]$, while n was between 0.5 and 1 in a range of heating temperature 600°C ~1000°C. Thus, the reduced samples with heavily La-doped concentration were easily oxidized at low reoxidation temperature (800 ~1200°C).

In order to get the detailed knowledge about the thermodynamics and the kinetics of the several steps of the reoxidation process, Langhammer, et al [8] measured the oxygen exchange between the 1.5mol% La-doped samples and sintered atmosphere by the oxygen coulometry in $P(O_2)$=260Pa. It was found that the amount of oxygen adsorbed by the grain boundaries increased with the increase in temperature (800 ~1200°C), during the transformation from the reduced phase to oxidized phase.

3.3. PTCR effect

The resistivity dependence of reoxidizing temperature for the same samples is shown in Fig.3, there is obviously change on resistivity under the reoxidized temperature 1100°C, but it is find that the sample insulated when x=15.0 mol% at the reoxidizing temperature 600°C, as well as x=2.0mol% or 6.0mol% at 1200°C. The change of resistivity (when x=0.8mol%) and the value of ρ_{max} / ρ_{min} ratio of samples with different reoxidizing temperature are illustrated in Fig.4 and table1. It is shown that the maximum value of ρ_{max} / ρ_{min} ratio was ~700 at the reoxidizing temperature of 1100°C. Combined with the microstructure shown in Fig.2, it is concluded that the small uniform grain size formed while samples were sintering in strong reducing atmosphere, given La-doped content over the critical concentration. It is easily reoxidized in the air. The acceptor ion appeared on grain boundary surface in the grain boundary reoxidizing process, which led to the PTCR effect [9].

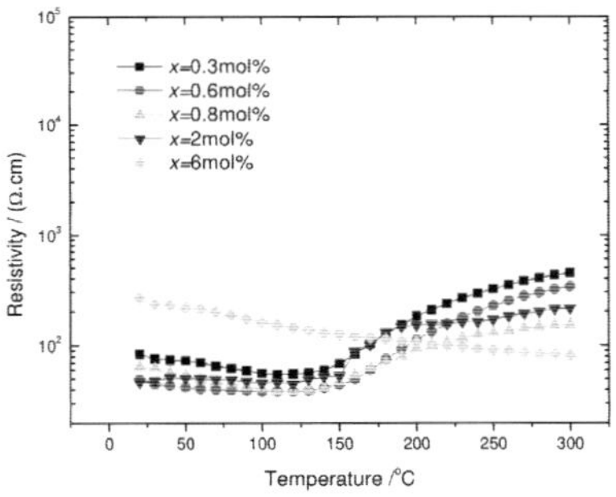

(a) 800°C

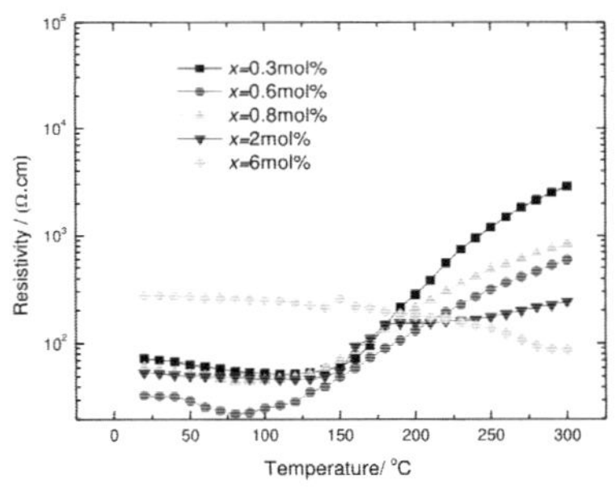

(b) 900°C

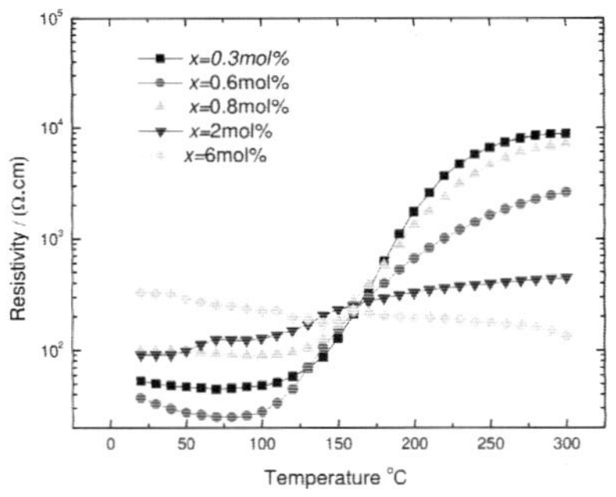

(c) 1000°C

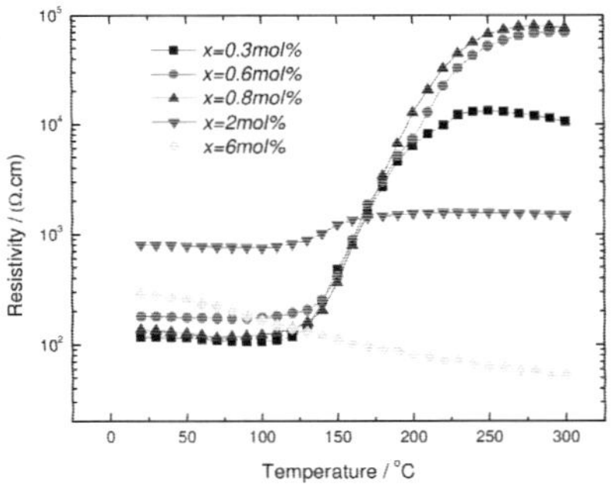

(d) 1100°C

Figure 3. Temperature dependence of resistivity for reoxidized $BaTiO_3$ samples doped with different La concentrations (sintered under reducing conditions and reoxidized in air for 2h at different temperature)

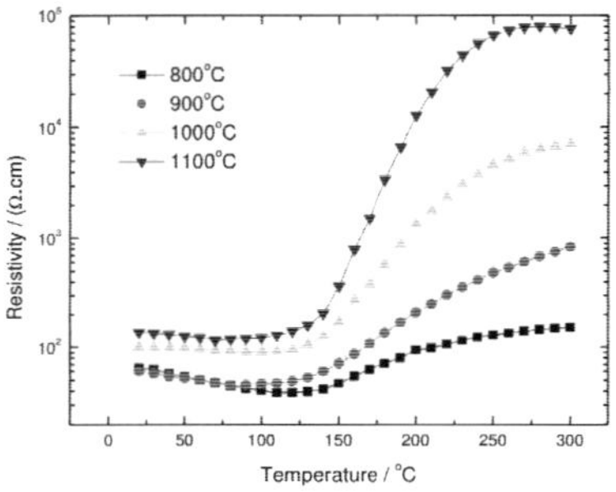

Figure 4. Temperature dependence of resistivity for reoxidized $BaTiO_3$ samples doped with La concentrations of 0.8mol% (sintered under reducing conditions and reoxidized in air for 2h at different temperature)

Table 1. Value of ρ_{max}/ρ_{min} ratio of samples with reoxidizing temperature

La-doped concentration	Reoxidizing temperature / °C			
	800	900	1000	1100
0.3mol%	8	55	198	100
0.6mol%	8	18	105	396
0.8mol%	4	18	79	662
2.0mol%	5	5	5	2
6.0mol%	3	3	3	5

4. Conclusions

The La-doped concentration and the sintering atmosphere heavily affect the microstructure and risistivity of the $BaTiO_3$ ceramics. Combined with the sample sintered in air, the critical concentration was highly improved up to 0.8mol% when ceramic samples were sintered in H_2. Meanwhile, the grain size decreased with the increase of La^{3+} donor content. It is shown that the critical concentration closely depended on sintering atmosphere. When the sample was reoxidizing in air at temperature range of 800°C ~1200°C, it is obviously found the PTCR effect exsit in some samples. This is related to the donor concentration, oxygen partial pressure and the reoxidation temperature. However, when the La-doped concentration is up to 9mol%, the sample insulated after annealed in air at 600°C. Thus, the reduced samples with heavily La-doped concentration were easily oxidized at low temperature. In the process of reoxidizing, the oxygen diffused along with the grain boundary on the grain surface toward the center of the grains or the released oxygen partly entered into the sample, which resulted in a strong PTCR effect.

References

[1] Hector Beltran, Eloisa Cordoncillo, and Purificacion Escribano. Insulating Properties of Lanthanum-Doped $BaTiO_3$ Ceramics Prepared by Low-Temperature Synthesis. J. Am. Ceram. Soc., 87 [11] 2132–2134 (2004)

[2] Finlay D. Morrison, Derek C. Sinclair, and Anthony R. West. Doping mechanisms and electrical properties of La-doped $BaTiO_3$. International Journal of Inorganic Materials, 3 1205–1210(2001)

[3] Darko Makovec and Miha Drofenik. Microstructural Changes during the Reduction/Reoxidation Process in Donor-Doped $BaTiO_3$ Ceramics. J. Am. Ceram. Soc., 83

[10] 2593–2599 (2000)

[4] Makoto Kuwabara and Hirofumi Matsuda. Shift of the Curie Point of Barium Titanate Ceramics with

[5] Sintering Temperature. J. Am. Ceram. Soc., 80 [10] 2590–2596 (1997)

[6] Finlay D. Morrison, Derek C. Sinclair, and Anthony R. West. Characterization of Lanthanum-Doped Barium Titanate Ceramics Using Impedance Spectroscopy. J. Am. Ceram. Soc., 84 [3] 531–538 (2001)

[7] Finlay D. Morrison, Derek C. Sinclair, and Anthony R. West. An Alternative Explanation for the Origin of the Resistivity Anomaly in La-Doped $BaTiO_3$. J. Am. Ceram. Soc., 84 [2] 474–476 (2001)

[8] Katsuro Hayashi, Takahisa Yamamoto, Yuichi Ikuhara, and Taketo Sakuma. Formation of Potential Barrier Related to Grain-Boundary Character in Semiconducting Barium Titanate. J. Am. Ceram. Soc., 83 [11] 2684–2688 (2000)

[9] Hans Theo Langhammera, Darko Makovec, and Yongping Pu et al. Grain boundary reoxidation of donor-doped barium titanate ceramics. Journal of the European Ceramic Society, 26[11] 2899–2907(2006)

[10] Pu Yongping, Chen Shoutian, and Langham Mer HT. Grain Boundary Reoxidation of Barium Titanate Ceramic Doped with Lanthanum. Journal of the Chinese Ceramic Society, 33[10]1237-1242 (2005)

Synthesis and characterization of magnetic nanocomposites with Fe_3O_4 core

J Chen, Y G Wang[1], Z Q Li, C Wang, J F Li and Y J Gu

College of Material Science and Technology, Nanjing University of Aeronautics and Astronautics, 29 Yudao Street, Nanjing 210016, China

E-mail: yingang.wang@nuaa.edu.cn

Abstract. To improve the infrared absorbing performance of the ferrite magnetic material which has good electromagnetic wave absorbing performance, the core-shell structured multilayer coating is introduced to achieve multiple wavelength range absorbing. Nanocomposites $Fe_3O_4/SiO_2/Al_2O_3$ and $Fe_3O_4/SiO_2/TiO_2$ have been prepared using a series of chemical methods. The results of TEM and XRD analysis show that multilayer coated structures exit in both $Fe_3O_4/SiO_2/Al_2O_3$ and $Fe_3O_4/SiO_2/TiO_2$ and that the core-double shell particles are all in nano-scale. The saturation magnetization and the coercivity of Fe_3O_4 at room temperature are 72 $A{\cdot}m^2/kg$ and 3.82 kA/m, respectively. Further coating of SiO_2 and Al_2O_3 or TiO_2 decreases the saturation magnetization and the coercivity while all nanocomposites remain ferromagnetic. There is no obvious improvement of the infrared absorbing ability of Fe_3O_4 nanoparticles in coating of SiO_2 while further coating of Al_2O_3 or TiO_2 reduces the infrared emissivity from 0.97 of Fe_3O_4 nanoparticles to about 0.83 of the core-double shell nanocomposites.

1. Introduction

Fe_3O_4 has recently been developed as an excellent electromagnetic and microwave absorbing material [1-3]. However, the infrared absorbing performance of iron oxide magnetic materials is poor [4] and there leaves an open question how to improve the infrared absorbing performance for their use, for example, in aircrafts.

Surface coating is one of the methods which have often been used for modifying or improving material performances [5]. There are several promising techniques for coating particles, most of them are chemical processes, such as the precipitation method, the sol-gel method and the electrochemistry method. However, it is difficult to control the thickness and composition of the coating layer on the particle surface using most of these techniques, which limits their application. The heterogeneous nucleation method is usually used in surface coating because of its simple procedure, low cost and, especially, easy controlling of the thickness and composition of the coating layer [6]. The proper shell thickness is favorable to keep the coated particles stable physically and magnetically. Meanwhile with this method it is convenient for surface modification of the core/shell nanoparticles.

The infrared absorbing performance of aluminium and alumina has been studied for many years [7,8] and it is believed that aluminium and alumina show excellent infrared absorbing ability at wavelength range of 2~15μm. On the other hand, TiO_2 shows excellent transparency and stability [9]. Coating of Al_2O_3 and TiO_2 is expected to be used to improve the infrared absorbing performance while not affect the microwave absorbing performance of the Fe_3O_4 magnetic material. However, Al_2O_3 and

TiO_2 layer is difficult to be coated onto Fe_3O_4 nanoparticles by chemical processes directly and a middle layer is sometimes necessary to be introduced. Ulman [10] found that, among materials employed as protection layers, SiO_2 is one of the most ideal materials for protecting Fe_3O_4 nanoparticles due to its reliable chemical stability, biocompatibility, and reactivity with various coupling agents. In this work, SiO_2 have been used as a middle layer and the $Fe_3O_4/SiO_2/Al_2O_3(TiO_2)$ nanocomposites have been synthesized using a variety of chemical methods. The structure, magnetic properties and the infrared absorbing performance have also been investigated.

2. Experimental methods

Fe_3O_4 magnetic nanoparticles were synthesized by using the titration hydrolysis method [11]. The Fe_3O_4/SiO_2 core-shell nanocomposites were then prepared by the hydrolysis process [12]. The $Fe_3O_4/SiO_2/Al_2O_3$ and the $Fe_3O_4/SiO_2/TiO_2$ core-double shell nanocomposites were finally prepared by the heterogeneous nucleation process [13] and the sol-gel process, respectively.

TEM was performed on a Tecnai-G2 20 Transmission Electron Microscope. TEM specimens were prepared by redispersing the nanocomposites in ethanol and then applied drop-wise onto carbon coated copper grids. Dried samples were submitted to Bruker D8 Advance XRD (Cu-Kα, λ=0.15405 nm) for structure analysis. For FTIR, nanoparticles and nanocomposites were mixed with KBr and pressed to the shape of tablets and the spectra were collected on a NEXUS 670 Fourier Transform Infrared Spectrometer. Magnetization curves were measured at room temperature by using a Vibrating Sample Magnetometer. Emissivity was measured at wavelength range of 1-22μm by using a Dual Wave Band Emission instrument.

3. Results and discussion

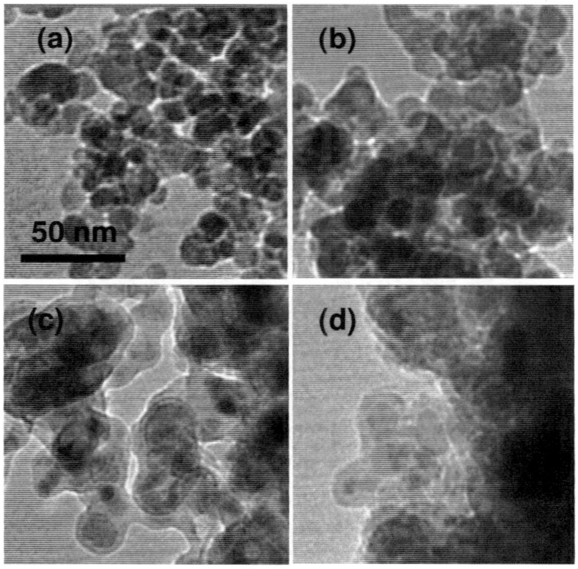

Figure 1. TEM bright field images of: (a) Fe_3O_4; (b) Fe_3O_4/SiO_2; (c) $Fe_3O_4/SiO_2/Al_2O_3$ and (d) $Fe_3O_4/SiO_2/TiO_2$.

TEM was first used to study Fe_3O_4 nanoparticles, Fe_3O_4/SiO_2 core-shell nanocomposites and $Fe_3O_4/SiO_2/Al_2O_3(TiO_2)$ core-double shell nanocomposites. Figure 1(a) shows the dispersed Fe_3O_4 particles with a diameter of 10~15 nm, which are spherical with a smooth surface. In Figure 1(b), Fe_3O_4/SiO_2 core-shell composites are also spherical with a diameter of 15~20 nm. TEM images of

Journal of Physics: Conference Series **152** (2009) 012041 doi:10.1088/1742-6596/152/1/012041

$Fe_3O_4/SiO_2/Al_2O_3(TiO_2)$ core-double shell nanocomposites are shown in Figure 1(c) and 1(d). The shape of the enlarged $Fe_3O_4/SiO_2/Al_2O_3(TiO_2)$ core-double shell nanocomposites becomes irregular. The core-double shell nanocomposites in Figure 1(c) and 1(d) have aggregated, which disagrees with the fact that the magnetic force should become weaker from Fe_3O_4 nanoparticles in Figure 1(a) to double-shell coated nanocomposites in Figure 1(c) and 1(d). It might result from the fact that the surface energy of the Al_2O_3 and TiO_2 is larger than that of Fe_3O_4 or SiO_2. The size of all particles obtained from TEM bright field images is listed in Table 1. The increased size of the nanocomposites proves the successful coating of subsequent shell layers.

Table 1. The infrared emissivity, σ_S, H_C and particles size deduced from TEM.

Material	Emissivity	$\sigma_S(A{\cdot}m^2/kg)$	$H_C(kA/m)$	Particles size (nm)
Fe_3O_4	0.97	72	3.82	10~15
Fe_3O_4/SiO_2	0.96	35	3.02	15~20
$Fe_3O_4/SiO_2/Al_2O_3$	0.82	29	2.87	20~25
$Fe_3O_4/SiO_2/TiO_2$	0.85	25	2.31	~30

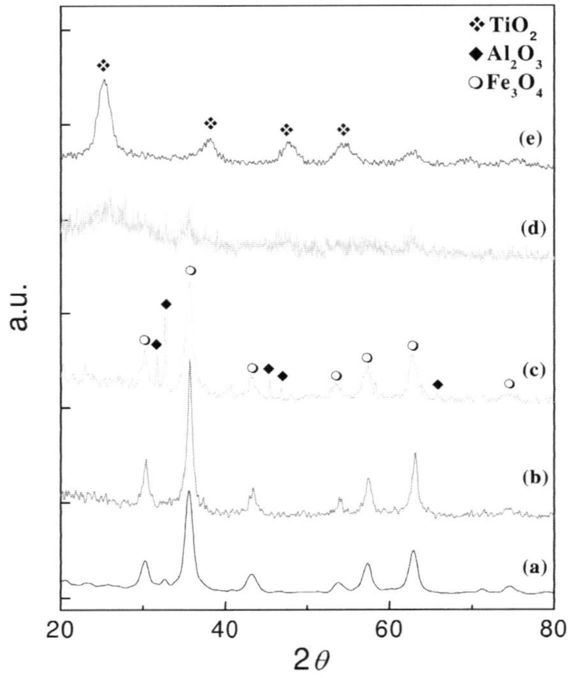

Figure 2. XRD patterns of (a) Fe_3O_4, (b) Fe_3O_4/SiO_2, (c) $Fe_3O_4/SiO_2/Al_2O_3$, (d) $Fe_3O_4/SiO_2/TiO_2$ and (e) Baked $Fe_3O_4/SiO_2/TiO_2$.

XRD was used to characterize the crystal structure of nanoparticles and nanocomposites. In Figure 2(a), the crystal structure of Fe_3O_4 nanoparticles is well indexed to the inverse cubic spinel phase. A series of characteristic peaks at 2θ =31°, 36°, 43°, 53°, 57°, 63° and 74° agree with the standard Fe_3O_4 XRD spectrum. The average particle size calculated from the full width at half-maximum of the peak at 2θ =36° using Sherrer formula is about 12 nm in good agreement with that obtained from TEM image in Figure 1(a). An additional peak is observed at 2θ = 33°, corresponding to γ-Fe_2O_3 arising from oxidation of Fe_3O_4 in air. In Figure 2(b), at 2θ =20°~25° no obvious peak of crystalline SiO_2 is found. It might result from the fact that the silica shell coating onto the Fe_3O_4 core is amorphous. In

Figure 2(c), peaks at 2θ = 32°, 33°, 45°, 47° and 66° agree with the standard γ-Al_2O_3 XRD spectrum. Fe_3O_4 characteristic peaks have been weakened because of the surface coating of shell layers. As shown in Figure 2(d), in the XRD pattern of $Fe_3O_4/SiO_2/TiO_2$ nanocomposites there appear no clear peaks corresponding to TiO_2 and the characteristic peaks of Fe_3O_4 core has also weakened as observed in $Fe_3O_4/SiO_2/Al_2O_3$ nanocomposites. In order to confirm the existence of TiO_2 shell the prepared $Fe_3O_4/SiO_2/TiO_2$ nanocomposites has been baked in vacuum at 400 °C for 15 minutes. The XRD spectrum of the baked $Fe_3O_4/SiO_2/TiO_2$ nanocomposites is shown in Figure 2(e). Peaks at 2θ =25°, 38° ,48° and 55° match the characteristic peaks of (101), (004), (200) and (211) planes of the anatase phase of TiO_2 well. Unlike that in Figure 2(c), the main phase in Figure 2(e) is no longer Fe_3O_4, but TiO_2 and the characteristic peaks of Fe_3O_4 are very weak, almost hardly to see, which indicates that the TiO_2 shell is thicker than the γ-Al_2O_3 shell and agrees well with the TEM observation. The weakened intensity of characteristic peaks of Fe_3O_4 core has also confirmed the cover-up of subsequent layers.

FTIR spectra were also used to prove the core-double shell structure, which are shown in Figure 3. In all curves in Figure 3, the bands at 1626 and 3412 cm^{-1} can be assigned to the H–O–H stretching modes and bending vibration of the free or adsorbed water, respectively. In the (a) curve, three main absorptions centered around 585, 1400 and 3140 cm^{-1} are observed, each corresponding to Fe–O, O–H bending vibrations and O–H stretching vibration, respectively. In the (b) curve, the Fe_3O_4 high-intensity bands at 1400 and 3140 cm^{-1} have disappeared, and also the Fe_3O_4 high-intensity band at 585cm^{-1} has been weakened. The broad high-intensity band at 1091 cm^{-1} is due to the asymmetric stretching bonds of Si–O–Si in SiO_2 associated with the motion of oxygen in Si–O–Si antisymmetric stretch. The band at 799 cm^{-1} is assigned to the Si–O–Si symmetric stretch, while the sharp band at 470 cm^{-1} corresponds to the Si–O–Si or O–Si–O bending modes. The band at 953 cm^{-1} is assigned to the Si–O symmetric stretch. The band at 567 cm^{-1} is an indication of the presence of Si–O–Fe. In the (c) and (d) curves, the Fe_3O_4 high-intensity band at 585cm^{-1} and the SiO_2 high-intensity band at 1091 cm^{-1} have all been weakened. In curve (c) the high-intensity band at 580 cm^{-1} is Al-O stretching vibration of Al_2O_3. In general, the coordination number of the aluminum species forming Si–O–Al bond is four or six. The six-coordinated aluminum species shift the position of Si–O stretching band (around 1091 cm^{-1}) of pure silica to lower frequency and cause the ring structure band at 799 cm^{-1} to disappear [14]. In curve (d), the band at 1070 cm^{-1} is related to the stretching vibration band of Ti-O-Si. The broad absorption bands of the stretching vibration band of Ti-O-Si and Ti-O-Ti in the region 450~755 cm^{-1} are observed. From the above information, one can see that Fe_3O_4 core has been covered by SiO_2 and $Al_2O_3(TiO_2)$ shells successfully.

Magnetization curves were recorded at room temperature and all the core nanoparticles, core-shell and core-double shell nanocomposites were proved to be ferromagnetic. The room temperature saturation magnetization (σ_S) and coercivity (H_c) of Fe_3O_4 nanoparticles, Fe_3O_4/SiO_2 and $Fe_3O_4/SiO_2/Al_2O_3(TiO_2)$ nanocomposites have been obtained from magnetization curves and data are listed in Table 1. The saturation magnetization of Fe_3O_4 nanoparticles is only about 20% lower than that of bulk Fe_3O_4 magnetite (σ_{bulk} = 92 $A \cdot m^2/kg$ [15]). The decrease of the saturation magnetization from Fe_3O_4 nanoparticles to $Fe_3O_4/SiO_2/Al_2O_3(TiO_2)$ nanocomposites is believed to result from the nonmagnetic shell layer coating. The coercivity decreases from Fe_3O_4 nanoparticles to Fe_3O_4/SiO_2 and $Fe_3O_4/SiO_2/Al_2O_3(TiO_2)$ nanocomposites. This is also believed to result from the nonmagnetic shell layer coating. Canted spin configurations may exist near the boundary between the core and the shell, resulting in a decrease of the magnetic anisotropy [16]. The presence of the amorphous shell may also affect the magnetic anisotropy. Finally there may exist exchange coupling between nearby Fe_3O_4 nanoparticles, which decreases due to SiO_2 and $Al_2O_3(TiO_2)$ shell coating.

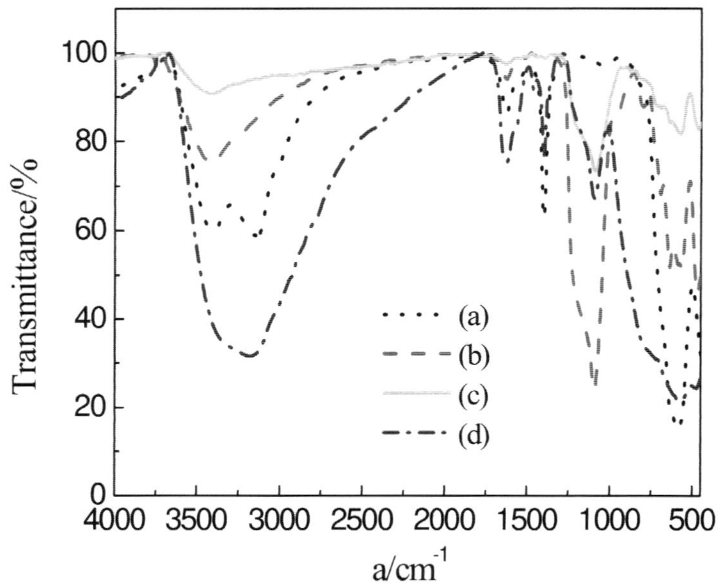

Figure 3. FTIR spectra of (a) Fe_3O_4, (b) Fe_3O_4/SiO_2, (c) $Fe_3O_4/SiO_2/Al_2O_3$ and (d) $Fe_3O_4/SiO_2/TiO_2$.

The infrared absorbing performance of Fe_3O_4 nanoparticles, Fe_3O_4/SiO_2 and $Fe_3O_4/SiO_2/Al_2O_3(TiO_2)$ nanocomposites has been investigated. The infrared emissivity absorption data are listed in Table 1. The infrared emissivity of the Fe_3O_4/SiO_2 core-shell nanocomposites is almost the same as that of Fe_3O_4 nanoparticles at wavelength range of 1~22μm while further coating of $Al_2O_3(TiO_2)$ decreases the infrared emissivity from 0.97 to about 0.83 at wavelength range of 1~22μm, which indicates that the coating of Al_2O_3 or TiO_2 has improved the infrared absorbing performance of the Fe_3O_4 nanoparticles.

4. Conclusion

In summary, the $Fe_3O_4/SiO_2/Al_2O_3(TiO_2)$ core-double shell nanocomposites were successfully synthesized using a series of chemical methods. The shape of most Fe_3O_4 nanoparticles, Fe_3O_4/SiO_2 and $Fe_3O_4/SiO_2/Al_2O_3(TiO_2)$ nanocomposites is spherical. The $Fe_3O_4/SiO_2/Al_2O_3(TiO_2)$ nanocomposites remain ferromagnetic with a saturation magnetization of 25-35 $A{\cdot}m^2/kg$ and a coercivity of 2.2-2.9 kA/m. Coating of Al_2O_3 and TiO_2 shell layers improves the infrared absorbing performance of the Fe_3O_4 nanoparticles. The results show that core-multishell structured nanocomposites have the potential for multiple wavelength range absorbing, and they are expected to become a new series of absorbing materials in the future. This work also provides a promising route to improve and modify magnetic materials with high stability, proper magnetic properties and, especially, good absorbing performance.

Acknowledgements
This work is partly supported by SRF for ROCS, SEM.

Reference

[1] Wei J, Liu J H, and Li S M 2007 *J. Magn. Magn. Mater.* **312** 414
[2] Kurlyandskaya G V, Cunanan J, Bhagat S M, Aphesteguy J C and Jacobo S E 2007 *J. Phys. Chem. Solids* **68** 1527
[3] Lima U R, Nasar M C, Nasar R S, Rezende M C, and Araujo J H 2008 *J. Magn. Magn. Mater.*

320 1666
[4] Du Y, Xing H L, and Chen S L 2007 *Paint & Coatings Industry* **37** 51
[5] Kim T K, Lee M N, Lee S H, Park Y C, Jung C K, and Boo J H 2005 *Thin Solid Films* **475** 171
[6] Ma Q 2007 *Acta. Mater.* **55** 943
[7] Supcoe R F, Greenberg M. 1982 *USP* 4289677
[8] Supcoe R F 1984 *USP* 4311623
[9] Pichot F, Pitts J R, and Gregg B A 2000 *Langmuir* **16** 5626.
[10] Ulman A 1996 *Chem. Rev.* **96** 1533
[11] Hong R Y, Zhang S Z, Han Y P, Li H Z, Ding J, and Y. Zheng 2006 *Powder Technology* **170** 1
[12] Lu Z Y, Wang G, Zhuang J Q, and Yang W S 2006 *Colloids and Surfaces A* **278** 140.
[13] Zhang J X and Gao L Q 2001 *Ceramics International* **27** 143
[14] Jiang Y Q, Ding X F, Zhao J Z, Bala H, Zhao X, Tian Y M, Yu K F, Sheng Y, Guo Y P, Wang Z C 2005 *Mater. Lett.* **59** 2893
[15] Lee J W, Isobe T, and M. Senna J 1996 *Colloid and Interface Science* **177** 490
[16] Wu M Z, Zhang Y D, Hui S, Xiao T D, Ge S H, Hines W A, Budnick J I, Yacaman M J 2002 *J. Appl. Phys.* **92** 6809

Thermal stability of low dose Ga^+ ion irradiated spin valves

Xian-jin Qi, Yin-gang Wang, Guang-hong Zhou and Zi-quan Li

College of Materials Science and Technology, Nanjing University of Aeronautics and Astronautics, Nanjing 210016, China

E-mail: qxjmsc@nuaa.edu.cn

Abstract. The thermal stability of low dose Ga^+ ion irradiated spin valves has been investigated and compared with that of the as-prepared ones. The dependences of exchange field, measured using vibrating sample magnetometer at room temperature, on magnetic field sweep rate and time spent at negative saturation of the pinned ferromagnetic layer, and training effect were explored. The training effect is observed on both the irradiated spin valves and the as-prepared ones. The magnetic field sweep rate dependence of the exchange bias field of the irradiated spin valves is nearly the same as that of the as-prepared ones. For the as-prepared structure thermal activation has been observed, which showed that holding the irradiated structure at negative saturation of the pinned ferromagnetic layer for up to 28 hours results in no change in the exchange field. The results indicate that the thermal stability of the ion irradiated spin valves is the same as or even better than the as-prepared ones.

1. Introduction

The effect of ion irradiation on magnetic properties of thin films and multilayers has recently attracted considerable interest and a wide variety of phenomena have been reported [1]. For example, in perpendicularly magnetized Co/Pt multilayers [2], ion irradiation introduces in-plane magnetization. For exchange-biased IrMn/CoFe bilayers [3], decreases in exchange field and coercivity along with a change in the magnetization reversal mechanism with increasing ion dose was reported. For NiFe/CoFe/Cu/CoFe/IrMn spin valves, on both continuous films and patterned elements, it is observed that the Ga^+ ion irradiation results in dramatic decreases in the GMR and the exchange bias field together with an increase in the film resistance [4,5]. Since the ion irradiation can modify the microstructure and thus properties of magnetic films, it can be used for maskless patterning of magnetic elements by altering magnetic properties of given regions instead of removing materials in conventional patterning such as photolithography and electron beam lithography [6,7].

In maskless patterning, the edge of the ion beam has a low dose effect on the edge of magnetic elements and will inevitably affect the thermal stability of the devices. On the other hand, spin valves are practically applicable systems in the information storage industry. In this work, we investigate the magnetization reversal of CoFe/Cu/CoFe/IrMn spin valves to gain a clear insight into how the thermal stability is affected by Ga^+ ion irradiation at a low dose.

2. Experimental procedure

The spin valve structure Seed Ta (5nm) / $Co_{75}Fe_{25}$ (5nm) / Cu (2.5nm) /$Co_{75}Fe_{25}$ (5nm) /$Ir_{20}Mn_{80}$ (12nm) / Cap Ta (8nm) was deposited by magnetron sputtering on thermally oxidized silicon wafer substrates. It is in the ''top'' configuration, where the antiferromagnetic (AFM) layer was deposited onto the FM layer. The deposition conditions were: base pressure $<5\times10^{-7}$ Pa, Ar sputtering pressure 7×10^{-2} Pa, deposition rate 0.03~0.12 nm/s. A magnetic field of 100 Oe was applied in order to induce magnetic anisotropy during FM and AFM films sputtering. Irradiation was performed in an FEI Strata 200XP FIB system using Ga^+ ions of 30 keV and ion currents of 1 nA. The stack was evenly subjected to a dose of 10^{13} ions/cm^2.

The magnetic properties of the spin valve structures were investigated using a vibrating sample magnetometer at room temperature. The field range of loop measurement is set to be -800 Oe ~ +800 Oe. In training effect and waiting time measurements, the sweep rate of magnetic field was kept constant at 3 Oe/s in order to avoid different contributions from time-dependent effects. The magnetic field was applied parallel or antiparallel to the unidirectional easy axis, which coincides with the direction of the field applied during film growth. This indicates that there is little or no spin flop coupling between the AFM and FM moments at the interface.

3. Results and discussion

Figure 1 shows two series of hysteresis loops recorded sequentially showing the training effect for both the as-prepared and the Ga^+ ion irradiated spin valves. As reported in ref. [8], the training effect on the forward branch of the loop is believed to be a result of thermal activation of AFM moment with a small time constant. Before starting the loop, all the AFM regions are aligned along or close to the positive direction. The exchange field induced by the FM layer at negative saturation results in some proportion of the AFM regions reversing and thus the exchange anisotropy decreases. As the number of magnetization cycles increases this reversed proportion approaches some equilibrium value so that the difference between one loop and the next becomes smaller.

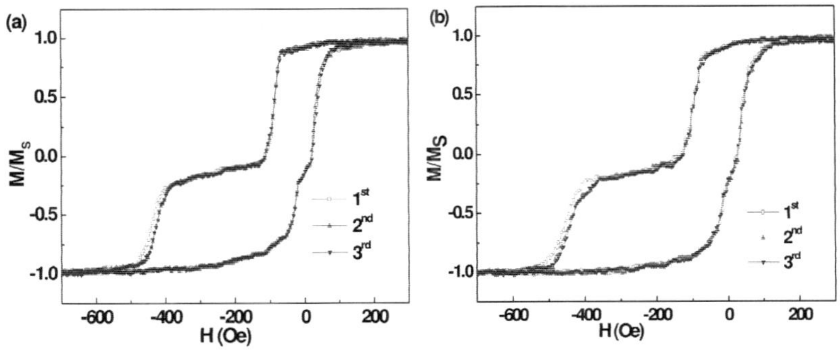

Figure 1. Series of hysteresis loops for CoFe/Cu/CoFe/IrMn top spin valves recorded sequentially. (a) as-prepared. (b) Ga^+ ions irradiated with a dose of 10^{13} ions/cm^2.

The dependences of the reverse fields H_{c1} and H_{c2} for the forward and recoil loops respectively on sweep rate of magnetic field dH/dt are shown in Figure 2. One can see similar sweep rate dependences on both the as-prepared and the Ga^+ ion irradiated spin valves. On the other hand, one can see that the coercivity increases as the sweep rate increases. A slow sweep rate permits the magnetization to decay during the measurement, driven by thermal activation. Hence, a slow sweep rate leads to a narrowing of the hysteresis loop. This dynamical behavior can be explained by a competition between the domain nucleation and domain wall propagation process [9]. At low sweep rates of magnetic field only a few inverse domains nucleate and the magnetization tends to switch by thermally activated domain wall motion. While at high sweep rates, domain wall propagation becomes relatively slow compared to the variation of the applied magnetic field, thermally activated reversal becomes less effective and therefore more domains nucleate. Since the activation energy for domain nucleation is larger than that for domain wall propagation, for both the as-prepared and ion irradiated spin valves with the dH/dt increasing the absolute values of H_{c1} increase while those of H_{c2} decrease.

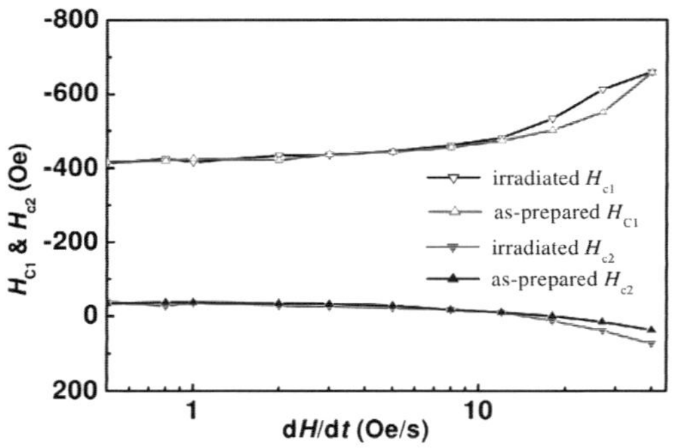

Figure 2. Sweep rate dependences of H_{c1} and H_{c2} of the as-prepared and the ion irradiated spin valves.

Figure 3 shows the exchange field of both as-prepared and ion irradiated spin valves as functions of time spent at negative saturation of the pinned ferromagnetic layer. Firstly, one can see that the exchange field of ion irradiated spin valve is slightly lower than that of the as-prepared spin valve. This can be attributed to the modification of interface between the pinned FM layer and the AFM layer caused by ion irradiation [4]. The exchange field of the as-prepared spin valve decreases as the waiting time increases while that of the ion irradiated spin valve remains nearly unchanged, which indicates the thermal stability of the ion irradiated spin valve is better than that of the as-prepared spin valve. The decrease of the exchange field of the as-prepared spin valve with the waiting time increasing is believed to result from the reversal of regions of magnetization in the AFM layer driven by thermal activation over a high energy barrier (with large time constant) [8,10,11]. With the waiting time increasing, the exchange coupling between the pinned FM and AFM layers causes more of the AFM layer magnetization to rotate antiparallel to the unidirectional easy axis, which reduces the original unidirectional anisotropy, resulting in the exchange field decrease. The stable exchange field of the Ga^+ ion irradiated spin valve is believed to result from the decrease of the degree of texture and the intermixing of atoms between various layers. Ga^+ ion irradiation modifies the interface between the pinned FM layer and the AFM layer and thus the AFM spins are tied strongly to its crystal structure. Therefore the reversal or rotation of moments in bulk regions of the AFM layer has to overcome a much higher energy barrier and the energy induced by the applied field is not high enough to overcome this barrier.

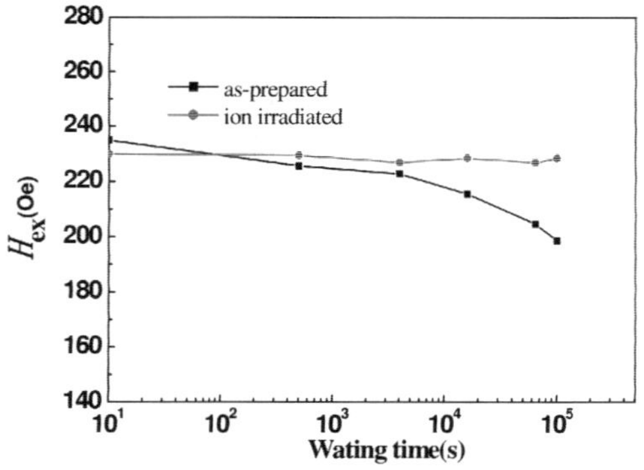

Figure 3. H_{ex} as a function of waiting time for the CoFe/Cu/CoFe/IrMn top spin valves.

4. Conclusion

In summary, the thermal stability of low dose Ga^+ ion irradiated spin valves was investigated and compared with that of the as-prepared ones. The training effect is observed on both the irradiated spin valves and the as-prepared ones. The magnetic field sweep rate dependence of the exchange bias field of the irradiated spin valves is nearly the same as that of the as-prepared ones. For the as-prepared structure thermal activation has been observed while holding the irradiated structure at negative saturation of the pinned ferromagnetic layer for up to 28 hours results in no change in the exchange field. The results indicate that the thermal stability of the ion irradiated spin valves is the same as or even better than the as-prepared ones. This work demonstrates that Ga^+ ion irradiation is a feasible technique to fabricate magnetoelectronic devices without a mask.

Acknowledgements

This work is supported by the National Natural Science Foundation of China under grant No.50671048. The authors wish to acknowledge Professor Xiu-feng Han in the Institute of Physics (CAS) for the support on the preparation of specimens.

Reference

[1] Fassbender J, Ravelosona D, and Samson Y 2004 *J. Phys. D: Appl. Phys.* **37** R179
[2] Hyndman R, Warin P, Gierak J, Ferre J, Chapman J N, Jamet J P, Mathet V, and Chapert C 2001 *J. Appl. Phys.* **90** 3843
[3] McGrouther D, Chapman J N, and Vanhelmont F W M 2004 *J. Appl. Phys.* **95** 7772
[4] Kerr E, Dijken S van, Langford R M, and Coey J M D 2005 *J. Magn. Magn. Mater.* **290** 124
[5] Guo Z B, You D, Qiu J J, Li K B, and Wu Y H 2001 *Solid State Comm.* **120** 459
[6] Fassbender J and McCord J 2008 *J. Magn. Magn. Mater.* **320** 579
[7] Klemmer T J, Ju G P, Veerdonk R J M van de, Leonhardt T D and Laughlin D E February 1 2005 *US Patent,* 6849349
[8] Wang Y G and Petford-Long A K 2002 *J. Appl. Phys.* **92** 6699
[9] Camarero J, Pennec Y, Bonfim M, Vogel J, Pizzini S, Fontaine A, Cartier M, Fettar F and Dieny B 2001 *J. Appl. Phys.* **89** 6585
[10] Hughes T, O'Grady K, Laidler H and Chantrell R W 2001 *J. Magn. Magn. Mater.* **235** 329
[11] Hughes T, Laidler H and O'Grady K 2001 *J. Appl. Phys.* **89** 5585

The properties of magnetic moment and specific heat of $La_{0.7}Ca_{0.3}MnO_3$ polycrystalline at different resistance state

Wu Z H[1], Chen L D [2*], Wang Q [2], Liu X J [2], Yu W D [2], Li X M [2]

[1] School of Urban Development and Environmental Engineering, Shanghai Second Polytechnic University, Shanghai 201209, P. R. China
2 State Key Laboratory of High Performance Ceramics and Superfine Microstructures Shanghai Institute of Ceramics, Chinese Academy of Sciences, Shanghai 200050, P.R.China

E-mail: cld@mail.sic.ac.cn

Abstract. The electric-pulse-induced resistance switching behavior was studied in bulk $La_{0.7}Ca_{0.3}MnO_3$ polycrystalline at room temperature. The relationship between the thickness and the threshold value of electric pulse on $La_{0.7}Ca_{0.3}MnO_3$ bulk was studied. It was found that the threshold value of electric pulse depends on the thickness of samples. The decrease of the magnetic moment and specific heat in the low resistance state on $La_{0.7}Ca_{0.3}MnO_3$ are opposite to that observed on $La_{0.9}Ca_{0.1}MnO_3$. These phenomena may be associated with the different ground states of $La_{0.9}Ca_{0.1}MnO_3$ and $La_{0.7}Ca_{0.3}MnO_3$. The change of magnetic configuration would change the specific heat arising from ferromagnetic spin-waves component and charge carriers.

1. Introduction

The perovskite $Re_{1-x}A_xMnO_3$-based (Re=rare earth ions, *A*=alkaline ions) manganates have attracted extensive interest for their colossal magnetoresistance (CMR) effect in the past decades [1-6]. Recently, an electric-pulse-induced switching of electrical resistance (EPIR) was discovered in $Re_{1-x}A_xMnO_3$ (Re=rare earth ions, *A*=alkaline ions). This effect provides a possibility of a new nonvolatile memory, called resistance random access memory (RRAM), which is very promising for its easy fabrication, fast switching speed, high recording density, low power consuming, and nondestructive readout. However, the mechanism of such EPIR phenomenon has still not been completely exposed. It is not clarified whether the resistance change is dominated by the bulk effect or the interface effect between the electrode and the sample [7-11]. To better understand the electric properties of these materials and to make contact with various proposed models, it is important to measure the fundamental properties of manganite perovskites with EPIR effect. Specific heat is regard as a useful method for the study of physical properties. At high temperature, specific heat data probes the nature of the magnetic, structural and electronic phase transition. At low temperature, the data yields information on lattice, electronic and magnetic excitations. We expect the analysis of the low temperature specific heat data (*C*) can provide accurate values of lattice, electronic and magnetic components in the ferromagnetic (FM) state. In this paper, the electric-pulse-induced resistance (EPIR) switching behavior and specific heat of LCMO at different resistance states was studied in bulk $La_{0.7}Ca_{0.3}MnO_3$ polycrystalline. It was found that the decrease of the magnetic moment and specific heat in the low resistance state on

$La_{0.7}Ca_{0.3}MnO_3$ is opposite to that observed on $La_{0.9}Ca_{0.1}MnO_3$. These phenomena may be associated with the different ground states of $La_{0.9}Ca_{0.1}MnO_3$ and $La_{0.7}Ca_{0.3}MnO_3$.

2. Experimental procedure

Single phase $La_{0.7}Ca_{0.3}MnO_3$ (LCMO) powder was synthesized by a chemical solution deposition (CSD) method, using $La(CH_3COO)_3 \cdot 2H_2O$ (99.9%), $Ca(CH_3COO)_2 \cdot H_2O$ (99.99%) and $Mn(CH_3COO)_2 \cdot 4H_2O$ (99.999%) as the raw materials [7]. The $La_{0.7}Ca_{0.3}MnO_3$ polycrystalline ceramics were prepared by Spark Plasma Sintering (*SPS*), with subsequent heat treatment in air at 900 °C for about 5h. Disc-shaped samples (Φ4mm × 1.6 mm) were cut out of the bulk polycrystalline. Two silver pads as electrodes 3mmin diameter were deposited on the disc-shaped sample surface. Electric pulses were provided by AVIR-3-B-QTKA Pulse Generator. Resistance was measured with a two-probe method in which a constant current (1 μA) was applied (Keithley 2410) to the junction while the voltage in the circuit was measured (Keithley 2000). The magnetization properties measurements were performed in commercial Quantum Design Physics Property Measurement System (PPMS). Specific heat measurements of $La_{0.7}Ca_{0.3}MnO_3$ polycrystalline at electric-pulse-induced different resistance states were carried out by using the thermal relaxation method from 2K to 20K. The background signal, including the exact amount of Apiezon N used to paste the sample on the platform, was recorded in a first run, and subtracted from the total heat capacity. The different resistance states used in magnetic moment and specific heat measurement were modulated at room temperature. First, a negative/positive electric pulse was applied to the sample (80V, 30ns; the resistance of the system was modulated to a stable resistance state) and then the temperature dependence of magnetic moment/specific heat was measured. After the magnetic moment/specific heat measurement, the temperature was increased to room temperature and another positive/negative electric pulse was applied.

3. Results and discussion

Effects of electric pulses on the two-terminal resistance of the LCMO systems are studied, and the typical results are shown in Fig. 1. Significant variation of the resistance is observed in the Ag/LCMO system. The structure of Ag/LCMO/Ag system is symmetrically. Then the first electric pulse leaded the resistance to increase or decrease randomly. The polarization of the electric pulse which inducing the resistance to HRS is defined as a negative voltage [7]. Reversed polarization of the electric pulse which inducing the resistance to LRS is positive voltage (seen bottom of Fig. 1). The ratio of high resistance (R_h) to low resistance (R_l) is as large as ~3. The threshold value of electric pulse (U_{th}) to change the resistance state depends on the thickness of sample. And the polycrystalline bulk sample is much thicker than thin film used in previous study, thereby, the threshold value of electric pulse (80V) employed to change the resistance state of $La_{0.7}Ca_{0.3}MnO_3$ polycrystalline bulk in our experiment is much larger than the film used (5～15V). The relationship between the thickness and U_{th} on LCMO bulk were studied. It was found that U_{th} depends on the thickness of samples (see Fig. 2). However, U_{th} also depend on original resistance value, top electrodes, number and amplitude of the voltage pulses.

The effect of grain boundary (GB) is a considerable topic. GB plays an important role on the resistance and magnetization properties of polycrystalline LCMO. However, the EPIR phenomenon has been found in $SrTiO_3$ single-crystalline by Krzysztof Szot (Nature Materials 5, 312) recently. They reported that the switching behaviour is an inherent property of the material. The phenomenon is shown to originate from local modulations of the oxygen content. And in our experiments, the results can be explained by local modulations of the oxygen content. Therefore, we think the EPIR phenomenon of polycrystalline LCMO would be related to the modulating intrinsic properties of pervskite manganites.

The *M (H)* dependence for HRS and LRS of $La_{0.7}Ca_{0.3}MnO_3$ polycrystalline at 10*K* are shown in Fig. 3. The difference of saturated magnetic moment between HRS and LRS is very small. The magnetic moment of HRS is larger than that of LRS. The saturated magnetic moment of HRS and

LRS is counted to be about 2.1μ_B/Mn in our measurement. The saturated magnetic moment of 2.1 u_B / Mn at 10 K is really a small value for the bulk LCMO. Usually, the saturated magnetic moment of bulk $La_{0.7}Ca_{0.3}MnO_3$ is reported about 3.5 u_B / Mn [12-15]. In the present experiment, the bulk sample was designed by Spark Plasma Sintering methods. The Spark Plasma Sintering (SPS) progress is a speedy sintering progress in a deoxidized atmosphere. It is often the case that there are lots of oxygen vacancies in sample synthesized by SPS. According to the reports of L. Ghivelder (PRB, 60, 12184) and Ilryong Kim (PRB, 62, 5674), oxygen plays an important role in determining the electromagnetic properties of LCMO, such as the nonuniformly distributed oxygen would induce PS with ferromagnetic-metal and antiferromagnetic-insulator coexistence. The small change of oxygen content would cause phase transition due to the double exchange interaction in manganites. It can be seen from M-H dependence of SPS synthesizing sample, the magnetic moment is not real saturation. This implies the ferromagnetic-metal and antiferromagnetic-insulator coexistence, which could be the cause of low saturated magnetic moment.

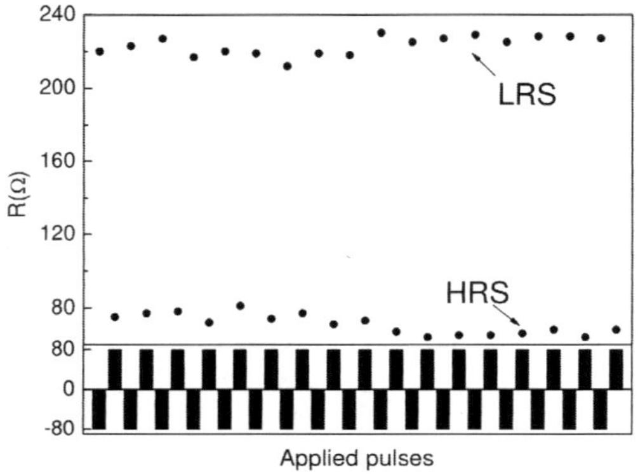

Fig 1. Resistance switching of $La_{0.7}Ca_{0.3}MnO_3$ polyctystalline bulk at room temperature as a function of electric pulse (80V and 20ns).

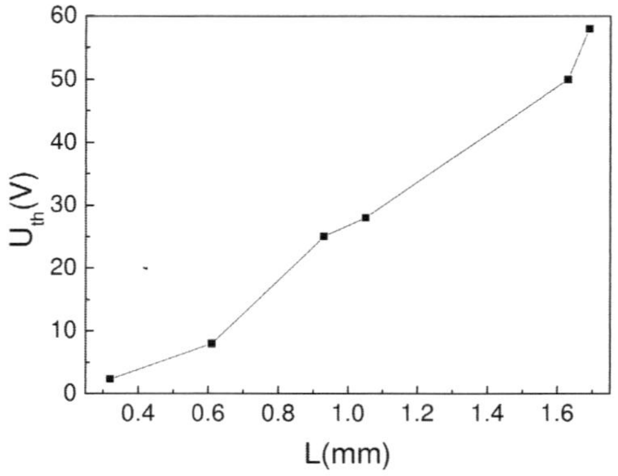

Fig 2. The relationship between the thickness and U_{th} on LCMO bulk under 1 electric pulse with 30ns length.

Ions motion driven by electric pulse current has been presented for explaining the resistance switching [16-19]. The electric pulse leads electrochemical creation or diffusion of oxygen vacancy along extended defects, such as dislocations, phase contact boundary and planar defects. During a positive pulse, the electric field drives oxygen ions moving along quasi-onedimensional extended defects under the electrodes. The vacancies defects are healed and the number of free carriers is increased. Therefore, the filamentary conducting route develops and the resistance transits to the LRS. During a negative pulse, oxygen ions escape from the extended defects and the number of carriers is decreased, and the resistance recovers to the HRS [8-11].

The decrease of the magnetic moment in the low resistance state on $La_{0.7}Ca_{0.3}MnO_3$ is opposite to that observed on $La_{0.9}Ca_{0.1}MnO_3$. These phenomena may be associated with the different ground states of $La_{0.9}Ca_{0.1}MnO_3$ and $La_{0.7}Ca_{0.3}MnO_3$. For $La_{0.7}Ca_{0.3}MnO_3$, the ground state is ferromagnetic (FM) metal, the magnetic moment expected from the spin contribution is $gS\mu_B$, where S is the spin of the ion, which is 3/2 for Mn^{4+} and 2 for Mn^{3+}, and the gyromagnetic factor is $g = 2$ in both cases. Then the increase of Mn^{4+}/Mn^{3+} would decrease the total magnetic moment of $La_{0.7}Ca_{0.3}MnO_3$. The electric pulse leads to electrochemical creation or diffusion of the oxygen vacancy. During the positive pulse, the electric field drives oxygen ions moving along quasi-one dimensional extended defects. The vacancies defects are healed and the number of oxygen ions is increased. The resistance transits to LRS. The increase of oxygen ions would increase the ratio of Mn^{4+}/Mn^{3+} and the increase of Mn^{4+}/Mn^{3+} would decrease the total magnetic moment of $La_{0.7}Ca_{0.3}MnO_3$. During a negative pulse, oxygen ions escape from the extended defects and the ratio of Mn^{4+}/Mn^{3+} would decrease. The resistance recovers to HRS and the total magnetic moment increases. Therefore, the magnetic moment of $La_{0.7}Ca_{0.3}MnO_3$ decreases in the low resistance state. On the other hand, for $La_{0.9}Ca_{0.1}MnO_3$, the ground state is magnetic multiphase coexistence insulator and the insulating character is interpreted as a disorder effect. The motion of oxygen would change the hole density of local regions. The change of hole density would still exist at low temperature and the change of hole density would lead to local AF and FM phase transition. These local ferromagnetic regions would connect ferromagnetic clusters of the nonstoichiometric samples, which are suspected to also be generated by inhomogeneous distribution of oxygen. The magnetic clusters grow to form larger clusters. The larger clusters would decrease the resistance and give rise to the value of magnetic moment. Therefore, the magnetic moment in the low resistance state of $La_{0.9}Ca_{0.1}MnO_3$ increases.

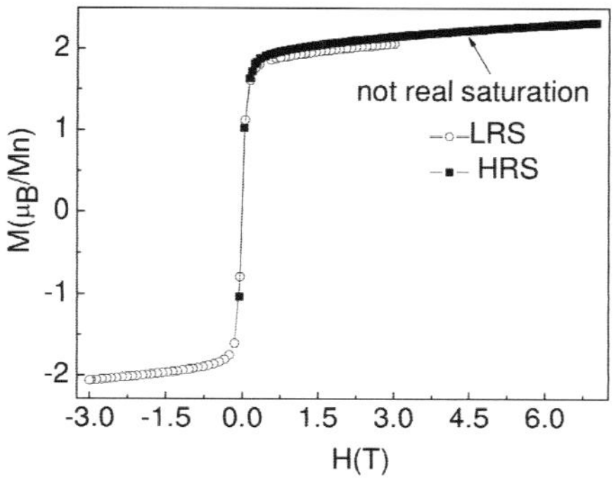

Fig 3. M-H curves for HRS and LRS of $La_{0.7}Ca_{0.3}MnO_3$ polyctystalline bulk synthesized by SPS.

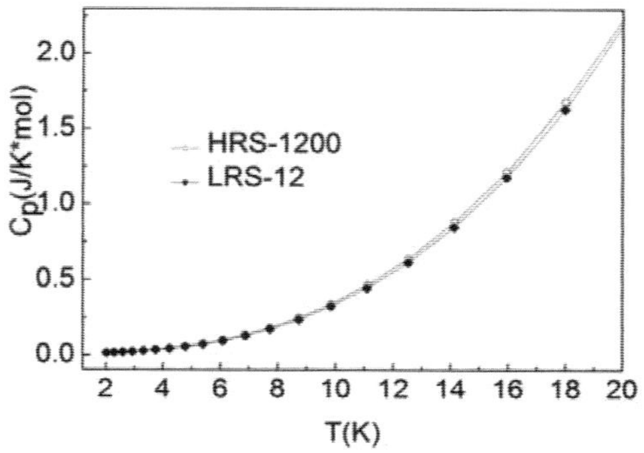

Fig 4. The low-temperature specific heat for HRS-1200, and LRS-12 from 2K to 20K. The lines are the best fits to the data as listed in Table I.

Specific heat results of LCMO at electric-pulse-induced different resistance states from 2K to 20K are shown in Fig. 4. The specific heat shows a distinct increase as the magnitude of resistance states increase. In order to evaluate the various contributions to the specific heat, results for each sample in the ferromagnetic region from 2K to 20K were fitted to the expression

$$C = \gamma T + \delta T^{3/2} + \beta T^3 \quad \text{Eq (1)}$$

where γ, δ, and β are constants. These three terms are expected to rise from, respectively, charge carriers, ferromagnetic spin waves, and the lattice. The fitting parameters obtained for all resistance stares are given in Table Ⅰ, and the fitted curves can be seen in Fig 4 as dash dot. It can be seen from the fitting data that the ferromagnetic spin-waves contribution has obvious decrease as the magnitude of resistance states decrease, whereas β almost does not change with the resistance states. And γ increases as the magnitude of resistance states decrease. These results imply that the change of specific heat may come from the change of specific heat of charge carriers and ferromagnetic spin-waves contribution with changing the magnetic configuration. To confirm the change of the specific heat is related to ferromagnetic spin-wave, the difference of specific heat between HRS-1200 and LRS-12 plotted as C/T versus $T^{0.5}$ are shown in Fig. 5. The data seem to fall on a single line within a small error for each resistance state, indicating that the above estimate of the spin-wave component is fairly accurate. From β we obtain the Debye temperature by using the standard expression

$$\theta_D = \left(\frac{12\pi^4 pR}{5\beta}\right)^{1/3} \quad \text{Eq (2)}$$

where R is the ideal gas constant and $p = 5$ is the number of atoms per formula unit. We get $\theta_D = 326$K for HRS-1200 and LRS-12, are comparable to those previously reported in manganite perovskites. There is no obvious change of θ_D as the resistance decreases. Then we would like to address the question of the spin-waves contribution to the specific heat. If the spin-wave stiffness constant is D, and there is no gap in the excitation spectrum, then we would have

$$\delta = 0.113Ra^3\left(\frac{k_B}{D}\right)^{3/2} \quad \text{Eq (3)}$$

where a is the lattice parameter of the elementary perovskite cell and k_B is the Boltzmann constant. The term 0.113 is a constant related to the cubic lattice. We get D=57meV Å^2, and 46meVÅ^2 for HRS-1200 and LRS-12, respectively.

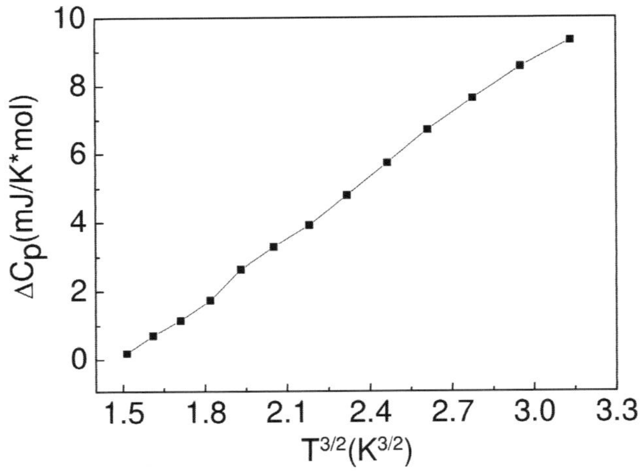

Fig 5. ΔC_P vs $T^{3/2}$. Here ΔC_P is specific heat difference as compared with LRS-12. $\Delta C_P = C_{P1200} - C_{P12}$. The lines are the fitting curves.

Table 1. Fitting results of the low-temperature specific heat of $La_{0.9}Ca_{0.1}MnO_3$ at different resistance states

	γ	β	δ
HRS-1200	3.23781	0.28134	1.26615
MRS-12	4.39612	0.27816	0.46543

The motion of oxygen induced by electric pulses would change the hole density of local regions. The change of hole density would still exist at low temperature and modulate the low temperature magnetic configuration of LCMO. At low resistance state, the regions enriched with oxygen have an enhanced hole density and the holes would establish local ferromagnetic ordering at low temperature. The magnetic clusters grow to form larger clusters where the spins become more ordered. The larger clusters should give rise to the value of D , as it is expected to increase the strength of the ferromagnetic coupling. According to Eq(3), the increase of D would decrease the spin-waves contribution to the specific heat. At high resistance state, oxygen ions escape from the extended defects, the spin-waves contribution to the specific heat increases.

4. Conclusion

In conclusion, the EPIR effects are studied in $La_{0.7}C_{a0.3}MnO_3$ polycrystalline materials. The magnetic moment and specific heat decreases when the magnitude of resistance states induced by electric-pulse decrease. The decrease of the magnetic moment and specific heat in the low resistance state on $La_{0.7}Ca_{0.3}MnO_3$ is opposite to that observed on $La_{0.9}Ca_{0.1}MnO_3$. These phenomena may be associated with the different ground states of $La_{0.9}Ca_{0.1}MnO_3$ and $La_{0.7}Ca_{0.3}MnO_3$. These results imply that the change of resistance states would be related to the electric pulse driven oxygen ion motion. The change of specific heat comes from the change of specific heat of charge carriers and ferromagnetic spin-waves contribution with changing the magnetic configuration.

Acknowledgments

This work is supported in part by the National Natural Science Foundation of China (Grant Nos. 50672116 and 50325208) and 863 Program of China (Grant No. 2006AA03Z308).

References

[1] T. Okuda, A. Asamitsu, Y. Tomioka, T. Kimura, Y. Taguchi, and Y. Tokura, Phys. Rev. Lett. 81, 3203 (1998)
[2] L. Ghivelder, I. Abrego Castillo, M. A. Gusmao, J. A. Alonso and L. F. Cohen, Phys. Rev. B 60, 12184 (1999)
[3] A. Wahl, V. Hardy, C. Martin, and Ch. Simon, Eur. Phys. J. B 26, 135 (2002)
[4] I. Kim, J. Dho, and S. Lee, Phys. Rev. B 62, 5674 (2000)
[5] T. Wu, S. B. Ogale, J. E. Garrison, B. Nagaraj, A. Biswas, Z. Chen, R. L. Greene, R. Ramesh, T. Venkatesan, and A. J. Millis, Phys. Rev. Lett. 86, 5998 (2001)
[6] S. Yunoki, J. Hu, A. L. Malvezzi, A. Moreo, N. Furukawa, and E. Dagotto, Phys. Rev. Lett. 80, 845 (1998)
[7] Wu Z H, Wang Q, Yu W D, Shang D S, Li X M, Chen L D, J. Phys D: Appl. Phys., 41, 025401, 2008. Wu Z H, Wang Q, Yu W D, Shang D S, Li X M, Chen L D, J. Phys D: Appl. Phys., 41, 115001, 2008
[8] S. Q. Liu, N. J. Wu, and A. Ignatiev, Appl. Phys. Lett. 76, 2749 (2000)
[9] X. Chen, N. J. Wu, J. Strozier, and A. Ignatiev, Appl. Phys. Lett. 89, 063507 (2006)
[10] R. Dong, Q. Wang, L. D. Chen, T. L. Chen, and X. M. Li, Appl. Phys. A: Mater. Sci. Process. 80, 13 (2005); R. Dong, Q. Wang, L. D. Chen, D. S. Shang, T. L. Chen, X. M. Li, and W. Q. Zhang, Appl. Phys. Lett. 86, 172107 (2005)
[11] L Ghivelder, I Abrego Castillo, N.McN Alford, G.J Tomka, P.C Riedi, J MacManus-Driscoll, A.K.M Akther Hossain, L.F Cohen, Journal of Magnetism and Magneti Materials 189, 274-282 (1998)
[12] Munakata K, Takubo N, Tamaru H and Miyano K 2006 Appl. Phys. Lett. 89 052105
[13] Sakai J, Yamamoto Y, Hori H and Imai S 2004 J. Appl. Phys.95 7094
[14] Park S Y, Hyun Y H, Lee Y P, Svetchnikov V L, Kim K W and Prokhorov V G 2006 Appl. Phys. Lett. 89 052502
[15] Loudon J C, Mathur N D and Midgley P A 2002 Nature 420 797
[16] K. Szot, W. Speier, G. Bihlmayer, and R. Waser, Nature Materials 5, 312 (2006)
[17] A. Baikalov, Y. Q. Wang, B. Lorenz, S. A. Sun, Y. Y. Xue, C. W. Chu, Appl. Phys. Lett. 83, 957 (2003)
[18] Y. B. Nian, J. Strozier, N. J. Wu, X. Chen, and A. Ignatiev, Phys. Rev. Lett. 98, 146403 (2007)
[19] J. J. Hamilton, E. L. Keatley, H. L. Ju, A. K. Raychaudhuri, V. N. Smolyaninova, R. L. Greene, Phys. Rev. B 54, 14926 (1996)

Giant magnetic-field-induced strain in NiMnGaSi magnetic shape memory alloy prepared by diffusion-reduction

S X Xue, S S Feng , P Y Cai, Y Zhou and W P Chen

Faculty of Physics and Electronic Engineering, Taizhou University, Taizhou 318000, China

E-mail: shuangxixue@126.com

Abstract: NiMnGaSi alloys were prepared by diffusion-reduction reacting which is a new and simple method. The effect of small mount of Si addition on the martensitic transformation temperature (T_m) and strain was investigated. It was found that the addition of Si caused a stronger magnetic exchange interaction leading to higher Curie temperature and giant magnetic field-induced strain. The typical $Ni_{43.03}Mn_{32.07}Ga_{18.48}Si_{6.43}$ alloy shows high martensitic transformation temperature (T_m) of 300K and a 0.56% giant strain measure at a magnetic field of 0.7 T.

1. Introduction

In recent years ferromagnetic shape memory alloys (FSMAS) have attracted increasing attention due to their large magnetic-field-induced strains(up to 10%) and the possibility of using these materials as actuators and sensors[1-4]. FSMAS show magnetic-field-induced strains at room temperature greater than any magnetostrictive, piezoelectric or electrostrictive material, and faster frequency response than temperature-driven shape memory alloys. Some alloy systems have developed up to now, such as Ni-Mn-Ga, Fe-Pd, Fe-Pt, Co-Ni-Ga and Co-Ni-Al alloys. Heusler alloy Ni_2MnGa is one of the most promising candidates for magnetically actuated devices due to its low hysteresis, large strains and adjustable transition temperature by composition variation.

The ferromagnetic shape memory alloy NiMnGa has been intensively investigated in recent years as a potential actuator material due to its unique giant magnetic field induced strain, based on the magnetic-field-induce rearrangement of martensitic variants. Because of the promising prospect in the design of a different kind of actuators or sensors, great effort has been devoted to giant magnetostrain in Ni-Mn-Ga Single crystals. NiMnGa alloy is quite limited in application due to a low transformation temperature and high brittles, many ways to improve combination properties of this alloy have been studied.

In general, the Ni_2MnGa alloys were prepared by a series of complicated methods such as smelting, directional solidification method[5-7]. But, it is difficult to prepared samples with big size for such as smelting, directional solidifiction method. Furthermore, in the case of high-frequency alternating applied field, it is a question for the energy loss caused by eddy current. In this paper, A new method is exploited to fabricate NI-Mn-Ga alloys. NiMnGaSi alloy powder were prepared by diffusion-reduction, then the NiMnGaSi alloy power was bonded and shaped a rod. The structure and magnetic properties of the sample were measured. the effect of little Si addition on the martensitic transformation temperature (T_m) and strain was investigated.

2. Experiment

NiMnGaSi alloys were prepared by diffusion-reduction which is a new and simple method. NiO powder, MnO_2 powder, Ga_2O_3 powder, SiO_2 powder and Ca powder were mixed in vacuum according a certain proportion and the mixture was heated in a vacuum stove. In the process of heating, the diffusion-reduction reaction was performed between NiO, MnO_2, Ga_2O_3 ,SiO_2 and the Ca powder, then Mn, Ga and Si atoms which has been reduced from the oxides diffused immediately to the reduced Ni, then the NiMnGaSi alloy power was prepared, the powder was flushed with water so that the CaO powder can be eliminated. After a series of heat treatment process, the NiMnGaSi alloy power was bonded and shaped a rod. The crystal structure of samples at room temperature was confirmed by power X-Ray diffraction using Cu-Ka radiation. The field dependence of magnetization and the temperature dependence of low-field ac magnetic susceptibility were measured by a PPMS measurement system.

3. Results and discussion

Crystal structure of $Ni_{43.03}Mn_{32.07}Ga_{18.48}Si_{6.43}$ powder was studied by X-ray diffraction(XRD) using Cu Ka radiation. Figure 1 is the X-ray diffraction pattern of NiMnGaSi alloy, showing the strong peaks of (022) ,(004) and (224) of the martensitic structure, it is suggested that the martensitic transformation temperature(T_M) of the NiMnGaSi alloy is above the room temperature. In general, NiMnGa alloy was limited in application because the martensitic transformation temperature(T_M) of the NiMnGa alloy was much lower than the room temperature, while, as shown in figure 1, the addition of Si element into NiMnGa alloy was effective for elevating martensitic transformation point. The mechanism is not clear, the possible explanation is that the smaller Si atoms diffuse into the NiMnGa crystal lattice, as a result, the crystal lattice has a small distortion, thus, the activation energy of martensitic/austenitic transformation would be increased.

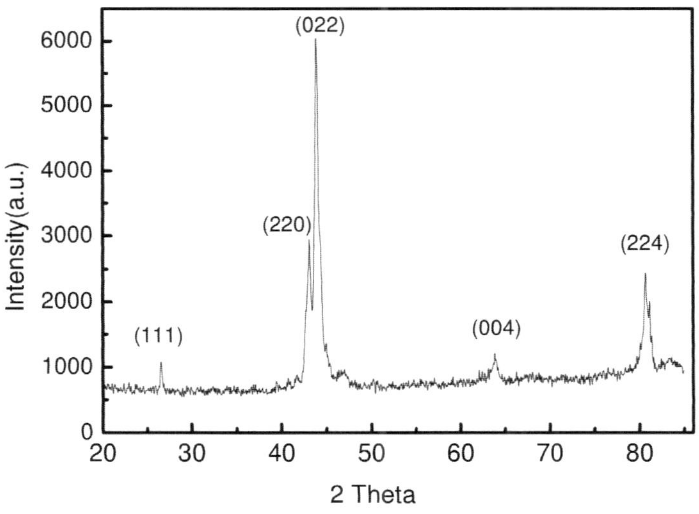

Figure 1. X-ray diffraction pattern of $Ni_{43.03}Mn_{32.07}Ga_{18.48}Si_{6.43}$ alloy

In general, the martensitic transformation temperature(T_M) and the Curie temperature (Tc) of Ni_2MnGa alloys are much low than room temperature, the low martensitic transformation temperature(T_M) and the Curie temperature (Tc) is a drawback for protential applications because the material can not be used at high temperature environment. Temperature dependence of ac susceptibility has been measured over the temperature range from 4K to 300K for $Ni_{43.03}Mn_{32.07}Ga_{18.48}Si_{6.43}$ alloy, as shown in figure 2. It is apparent that the martensitic transformation temperature(T_M) is near 300K. For some reason, the temperature dependence of ac susceptibility has not been measured at high temperature exceeds 300 K, but it is obvious that the Curie temperature (*Tc*) is much high than room temperature. This is consistent with XRD result.

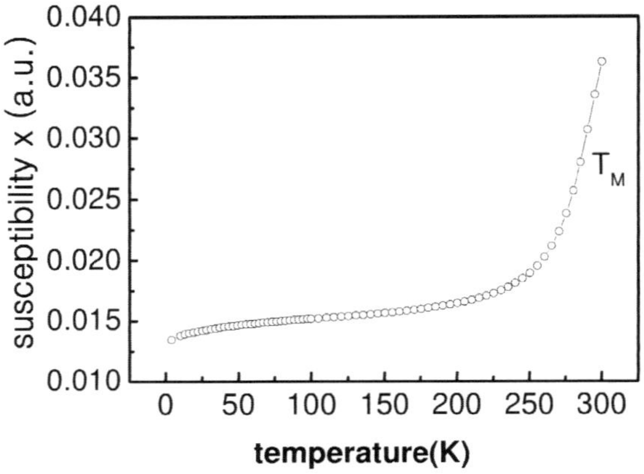

Figure 2. Temperature dependence of low-field magnetic susceptibility measured during heating

Powder samples were used for magnetization at 4K, 77K and 300K with a high applied field up to 2T by using the quantum design SQUID magnetometer. Magnetization Curves of the $Ni_{43.03}Mn_{32.07}Ga_{18.48}Si_{6.43}$ alloy powder measured at 300K was shown in figure 3. Very low hysteresis of magnetization curves and very low magnetic moments were observed. It is well known that the magnetic moments are substantially associated with Mn. The ferromagnetism mainly origiantes from the locatized moments of Mn atom, about 4.0 μB per Mn ion, and the moment of Ni is less than 0.3 μB,the contribution of Ni atoms is neglected. For the $Ni_{43.03}Mn_{32.07}Ga_{18.48}Si_{6.43}$ alloy sample, the Mn atoms are superfluous, the excess Mn atoms change the outer electronic structure of the unsubstitued Mn atoms, resulting in a greater decrease of the magnetic moment of the unsubstitued Mn atoms with the increase of Mn content[8].

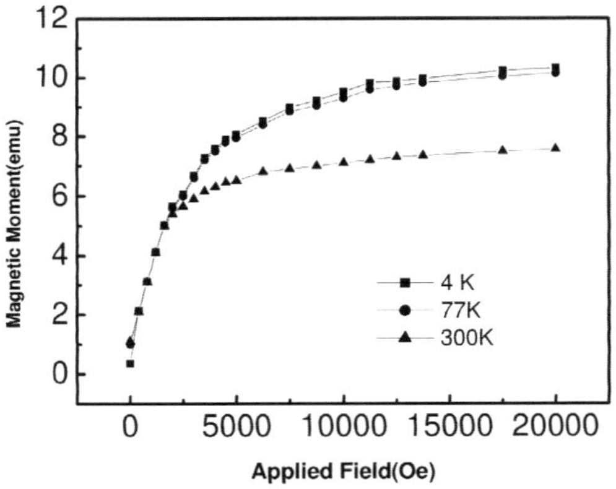

Figure 3. Magnetization Curves of the $Ni_{43.03}Mn_{32.07}Ga_{18.48}Si_{6.43}$ alloy powder measured at different temperature

Journal of Physics: Conference Series **152** (2009) 012044 doi:10.1088/1742-6596/152/1/012044

The bonded $Ni_{43.03}Mn_{32.07}Ga_{18.48}Si_{6.43}$ alloy rod was prepared for magnetic-field-induced-strain (MFIS) measurements. Fig. 4 shows the MFIS measured in the longness direction of the rod as a function of magnetic fields at room temperature. As shown in figure 4, the MFIS of $Ni_{43.03}Mn_{32.07}Ga_{18.48}Si_{6.43}$ alloy comes to 0.56% at a magnetic field of 0.7T, room temperature. More recently, Murray et. al report a 6% MFIS in $Ni_{47.2}Mn_{32.1}Ga_{20.5}$ with a compressive stress of 2 Mpa applied orthogonal to the magnetic field[9] However, Our MFIS is completely stress-free, and nearly three times as large as that of the typical magnetostriction materials Terfenol-D. In general, a large proportion of martensite variants is reoriented toward one and the same direction in the single crystal, while the grain boundaries obstruct the reorientation in the polycrystal. The improved MFIS in our sample can be attributed to the number of the grain boundaries.

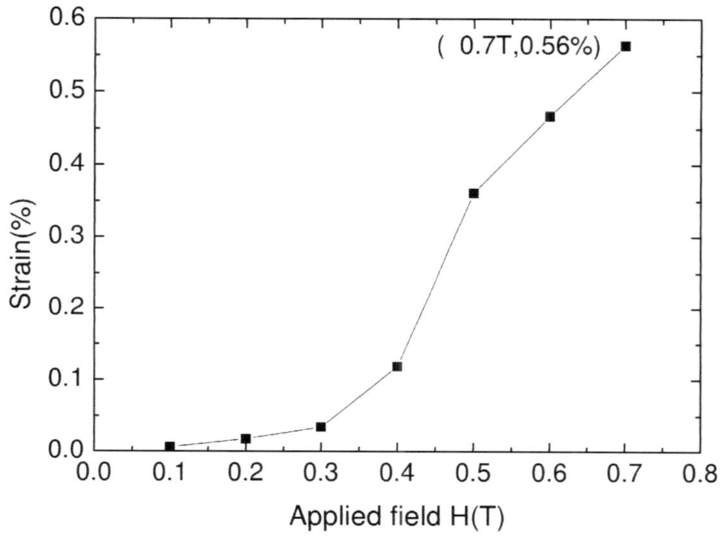

Figure 4. MFIS of the bonded $Ni_{43.03}Mn_{32.07}Ga_{18.48}Si_{6.43}$ alloy rod as a function of the magnetic field

Figure 5 shows the scanning electron microscopy photograph of the the bonded $Ni_{43.03}Mn_{32.07}Ga_{18.48}Si_{6.43}$ alloy. As shown in figure 5, the grain size was not symmetrical and many pores were observed. The $Ni_{43.03}Mn_{32.07}Ga_{18.48}Si_{6.43}$ alloy which has a high porosity due to a wash process leading to CaO grains removal and NiMnGaSi alloy dissolution by the water, exhibits coral-liked structure with many pores on the surface of the NiMnGaSi alloy grains. Inside of the bonded $Ni_{43.03}Mn_{32.07}Ga_{18.48}Si_{6.43}$ alloy rod, NiMnGaSi alloy grains were accumulated closely due to the dissymmetrical and porous structure, which results in a large MFIS of the sample. Similar results were reported by several papers[10,11].

Figure. 5 the scanning electron microscopy photograph of the the bonded NiMnGaSi alloy

4. Summary

Si additive is very effective to elevate martensitic transformation temperatures and improve the magnetic-field-induced strains of the NiMnGa alloy. The magnetic-field-induced strain up to 0.56% with a bias magnetic field of 0.7 T has been obtained in $Ni_{43.03}Mn_{32.07}Ga_{18.48}Si_{6.43}$ alloy.

Acknowledgment

This work is supported by the key project of science and technology department of Zhejiang Province (No. 2006C21085)

References

[1] Wang W H, Ren X, Wu G H 2006 *Phys. Rev. B* **73** 092101
[2] Long Y, Zhang Z Y, Wen D, Wu G H, Ye R C, Chang Y Q, Wan F R 2005 *J. Appl. Phys.* **98** 046102
[3] Jiang C B, Wang J M, Xu H B 2005 *Appl. Phys. Lett.* **86** 252508
[4] Heczko O, Ullakko K 2001 *IEEE. Trans. Mag.* **37** 2672
[5] Sozinow A, Likhachev A A, Lanska N, and Ullakko K 2002 *Appl.Phys.Lett.***80** 1746
[6] Wu G H, Wang W H, Chen J L, Ao L, Liu Z H, Zhan W S, Liang T and Xu H B 2002 *Appl.Phys.Lett.***80** 634
[7] Wang W H, Wu G H, Chen J L, Gao S X, Zhan W S, Wen G H, Zhang X X 2001 *Appl.Phys.Lett.***79** 1148
[8] Jiang C B, Muhammad Y, Deng L F, Wu W, Xu H B 2004 *Acta Materialia* **52** 2779
[9] Murray S J, Marioni M, Allen S M, and Handley R C 2000 *Appl. Phys. Lett.* **77** 886
[10] Scheerbaum N, Hinz D, Gutfleisch O, Mueller K H, Schultz L 2007 *Acta Materialia.* **55** 2707
[11] Boonyongmaneerat Y, Chmielus M, David C 2007 *Physical Review Letters.* **99** 247201

Journal of Physics: Conference Series **152** (2009) 012045 doi:10.1088/1742-6596/152/1/012045

The role of front electrode in electron emission from ferroelectric cathode

Xuan Huang, Yu-Jun Feng, Jie Cui, and Zhao-Xuan Sheng

Electronic Materials Research Laboratory, Key Laboratory of the Ministry of Education, Xi'an Jiaotong University, Xi'an 710049, People's Republic of China

E-mail address: fyj@mail.xjtu.edu.cn

Abstract. Electron emission from ferroelectric cathode (FEE) is unconventional electron emission effect. Experimental studies on ferroelectric emission (properties of PLZST antiferroelectric ceramics samples) at room temperature and in low vacuum had been carried out. Both islands electrode and stripes electrode were used in experiments. The influences of front electrode patterns on electron emission were investigated by measuring the emission current from the sample. It had been found that different patterned front electrodes yielded different emission current. The islands electrode emitted lower current than the stripes electrode.

1. Introduction

Ferroelectric cathodes had been widely studied over the last twenty years. Electron emission from ferroelectric cathode (FEE) was unconventional electron emission effect. It was well known that the application of a high- pulse of $10^5 – 10^6$ V, with $10^{-9} – 10^{-6}$ s pulse duration, between a cathode and an anode placed in vacuum, led to the generation of high current electron beams.

Although many studies about FEE had been carried out, little researches had been done to focus on the role of front electrode in electron emission. Different patterns were used by different groups, such as stripes, islands, grids and so on. Electric-field distributions over the ferroelectric cathode surface were simulated in the work of Jihoon Park et al. [1] and Rosenman et al. [2]. They found that different parameters of patterned front electrodes formed different field distributions. It was implied that different front electrodes would yield different FEE performance.

In this article, experimental results of ferroelectric emission obtained with different electrode pattern were reported. Antiferroelectric PLZST cathode placed in an electron emission configuration was used to generate electron beam. Two kinds of patterns, stripes and islands, had been investigated. We preliminarily focused on the FEE with driving pulse voltage and bias voltage of 0V~4000V. It was found that different patterns yielded different FEE currents.

2. Experiment setup

2.1. Equipment

Figure 1 showed a schematic diagram of the experimental setup. Antiferroelectric (AFE) samples were placed in the vacuum chamber with front patterned graphics electrode (GE) facing down to the flat graphite collector (GC). The front electrode was kept at ground potential, while the driving pulse was applied to the rear electrode. The distance between the sample and collector was set at 5.3 mm. The graphite collector area was 1.5 times larger than the emission area. A DC high-voltage generator was connected to the anode (graphite collector) to apply an external accelerating voltage (DC bias voltage) through the diode gap.

A cable with an impedance of 50 Ω was soldered on the electron collector and connected to the measurement circuit; the emission electrons were collected and flowed through the 50 Ω resistor. A Pearson Rogowski current monitor model 65/85 was used to measure the emission current from the surface of AFE sample. A 0.1 μF capacitor was inserted behind the anode, decoupling it from ground for the application of DC bias voltage. The decoupling capacitor acted as a short circuit for the short current pulses, thanks to the high value of its capacitance. A two-channel digitizing real-time oscilloscope (Model TDS 460, Tektronix) with a 500 MHz bandwidth was used to measure the voltage across the AFE sample and the current through the Pearson coil. To minimize electrical noise, all connector cables were electromagnetic shielded and connecting line were made as short as possible [3].

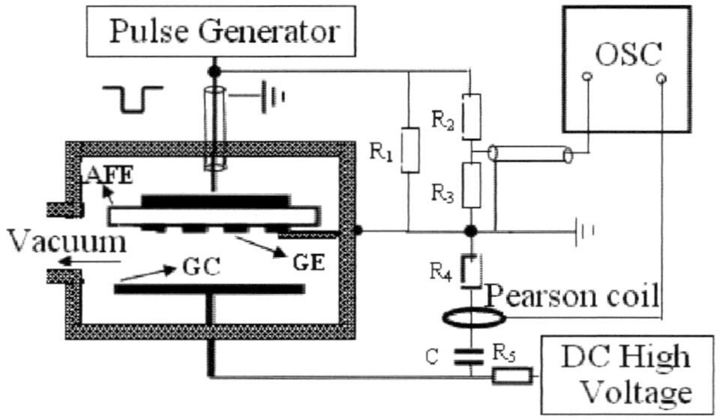

Figure 1. Schematic circuit of the experimental setup [3]

Current emission was successfully achieved with both stripes and islands patterned electrode with unipolar driving pulses applied to rear electrode. For convenience, these two samples were called stripes sample and islands sample, respectively. Only three parameters were used in this paper: driving pulse voltage, bias voltage, emission current peak. All experiments were carried out with the same positive polarity driving pulse voltage of 1200 V and the pulse duration was 1000ns. DC bias voltage ranged from 0 V to 4000 V. With every bias voltage, both stripes and islands sample were tested and the emission current was recorded.

All tests were performed in a vacuum of $5.0\sim7.0\times10^{-1}$ Pa and at room temperature of 25 ℃.

2.2. Materials

The samples used in the present study were antiferroelectric ceramic disks ($Pb_{0.94}La_{0.4}(Zr_{0.53}Sn_{0.30}Ti_{0.17})O_3$, 21.0 mm in diameter and 1.0 mm in thickness) prepared by a traditional solid-state reaction process. This composition was near the ferroelectric-antiferroelectric phase boundary.

Figure 2 showed the typical double hysteresis loop of antiferroelectrics. Hysteresis loop characteristics of PLZST were measured at room temperature by using improved Sawyer-Tower circuitry before electron-emission measurements.

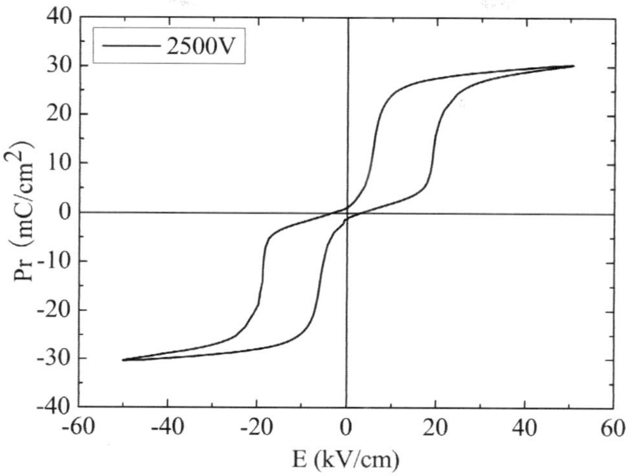

Figure 2. Hysteresis loop of PLZST measured with 1Hz sine-alternating voltage applied (2500V)

2.3. Electrode and Geometry

After ultrasonic cleaning in water, samples were patterned with a grid silver electrode on the front side and a solid electrode on the rear side. Figure 3 (a), (b), (c) demonstrated the front electrode and the rear electrode of experimental samples. Front electrode patterns consisted metal stripes or islands

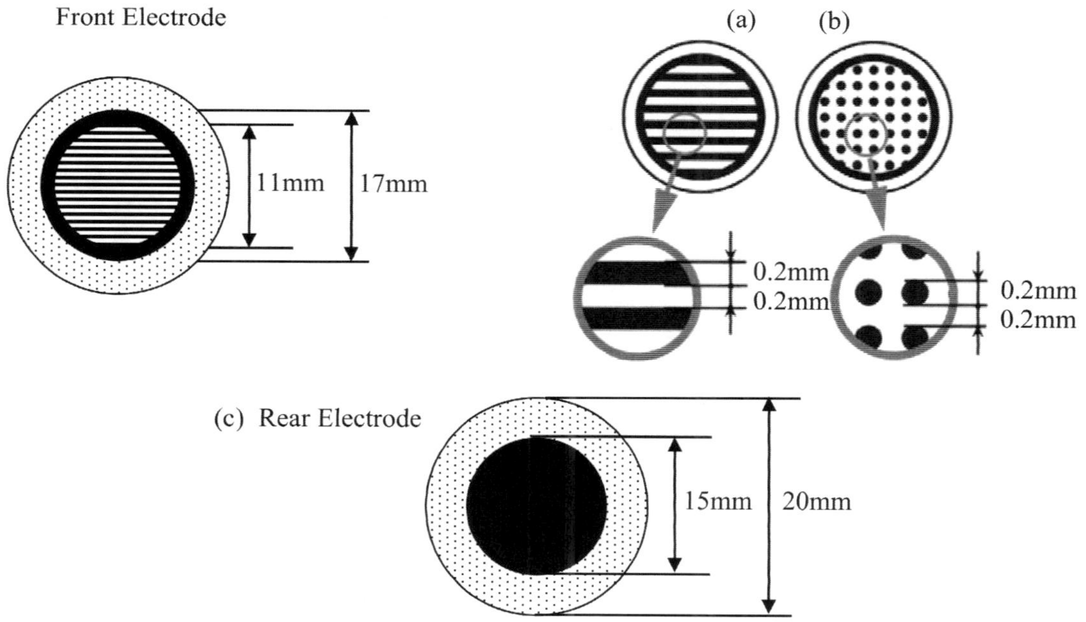

Figure 3. Sketch of the electrodes: (a) front electrode: a grating interconnected by a metal ring, (b) front electrode: a distribution of islands surrounded by a metal ring, (c) rear electrode.

surrounded by a metal ring (outer diameter 17.0 mm, inner diameter 11.0 mm) and the rear electrodes were 15.0 mm in diameter. For stripes, the wideness of silver electrode and bare ceramic spacing in the grid pattern was 0.2 mm. For islands, every metallic dot was 0.2 mm in diameter with interval of 0.2 mm. Thereby the total bare area on the surface of these two patterns was about 0.475 mm^2 and 0.764 mm^2, respectively.

The silver electrodes were obtained on the center of the disks by screen printing technique and sintered at temperature 820 ℃ for 10 minutes. And the thickness of the silver electrode was about 10 μm. Many micro protrusions might always exist on the cathode surface and they had a typical size of several microns. And the front electrode was connected to the experiment circuit through the metal ring.

3. Experiment result

3.1. Waveforms (Pulse voltage and emission current)

Figure 4 and figure 5 showed two sets of typical waveforms with low bias voltage (150V). The distinguishing feature of these two waveforms shown in figure 4 and figure 5 was the pulse voltage waveforms and emission current:

- There was a drop in the middle of pulse voltage across islands sample in figure 5 while stripes sample is not in figure 4.
- The maximal emission current of stripes sample and islands sample with a DC bias voltage of 150 V were 17.6 A and 16.1 A, respectively.

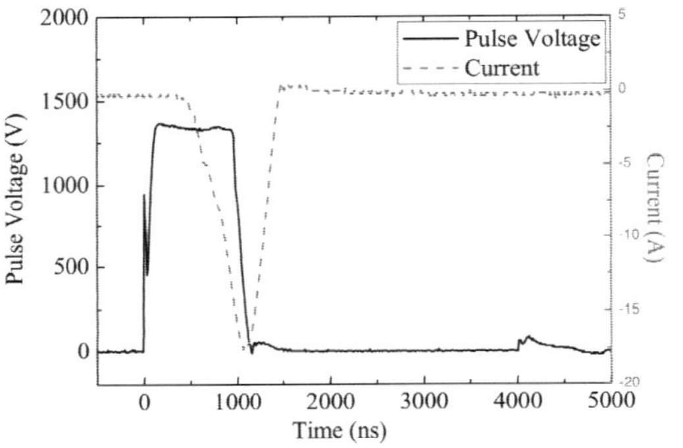

Figure 4. Waveforms of stripes sample with bias voltage of 150V

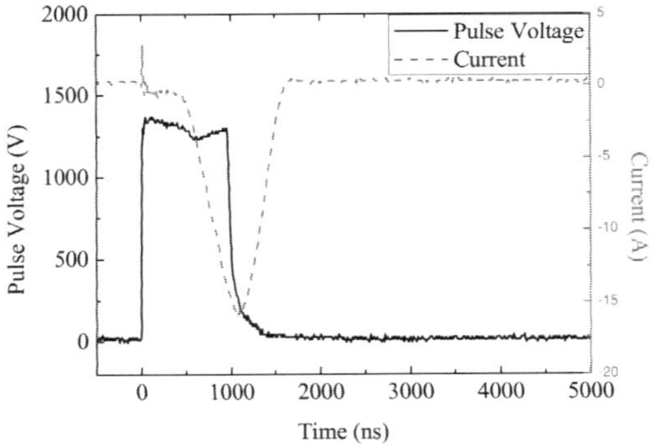

Figure 5. Waveforms of islands sample with of bias voltage of 150V

3.2. Current Peaks

Both stripes and islands sample were tested and the emission current was recorded with every bias voltage ranged from 0 V to 4000 V. Table 1 showed some of the different value of maximal currents that obtained with specific bias voltage. And figure 6 showed the complete experiment results of all maximal currents. And it could be concluded from figure 6 that the emission current was enhanced with the increasing of bias voltage. In addition, there was a linear relationship between the bias voltage and current peaks with high bias voltage.

Table 1. Different maximal currents for studied antiferroelectrics.

DC bias (V)	0	1000	2000	3000	4000
Maximal current of stripes sample(A)	0.6	257	516	785	1060
Maximal current of islands sample(A)	0.0	237	486	760	1035

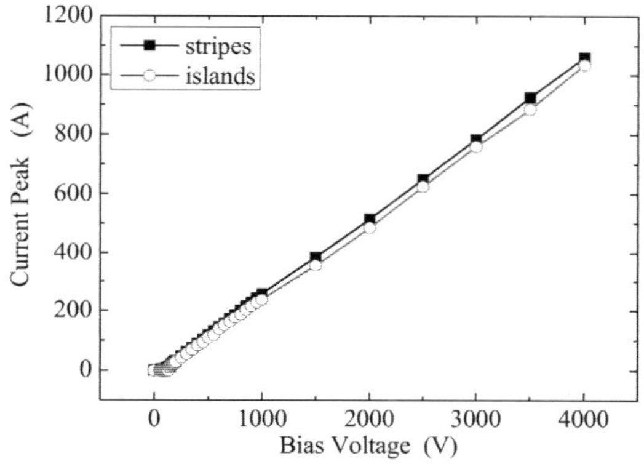

(a)

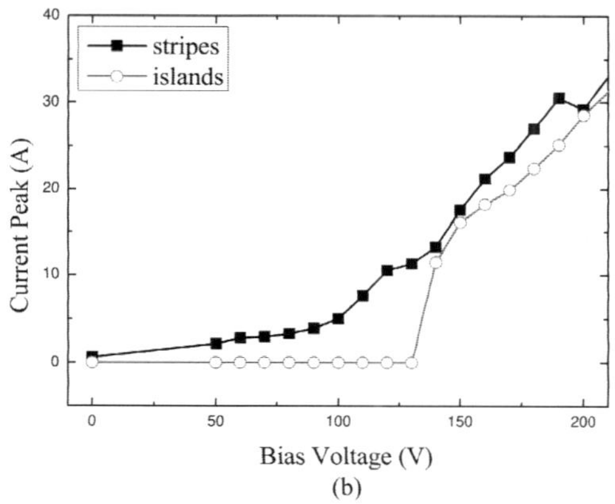

(b)

Figure 6. (a)comparison of current peak of different electrode patterns, (b) partial enlargement of (a) (0V~200V).

It could be noticed from figure 6. that there are some differences:

- The stripes sample emitted greater current than islands sample with the same bias voltage.

- The stripes samples started to emit a current of 2.4 A with the bias of only 50V. In contrast, islands samples started to emit current when a bias of 140V was applied.
- There was no significant difference in current peak between the stripes sample and the islands sample with high bias voltage.

4. Discussion

It had been shown above that there were three major differences between stripes pattern and islands pattern:

- The stripes sample and islands sample produced different maximal current of emission.
- The islands sample needed a higher bias voltage to start current emission than the stripes sample.
- There is difference between the stripes sample and islands sample in waveform of voltage.

Therefore, there were three aspects of discussion about the different of these two electrodes.

4.1. Different maximal current produced by different electrodes

Angadi et al. [4] and Shannon et al. [5] suggested that FEE was induced by polarization reversal. Electric field enhancement on the edge of metal grid resulted in weak emission to produce surface plasma. Boscolo and Cialdi [6] held that FEE was induced by the field emission in the vicinity of the metal–insulator–vacuum triple point, and emitted electrons was not high enough to ignite the surface flashover.

According to our experiment, it was obvious that there was more bare ceramic area on the islands sample. Ratio of bare area of stripes and islands was presented as in equation (1):

$$\frac{Bare\quad Area_{\text{stripes}}}{Bare\quad Area_{\text{islands}}} \approx \frac{0.475mm^2}{0.764mm^2} \approx 0.62 \tag{1}$$

If polarization reversal affected FEE, the islands sample should have emitted larger current. In fact, however, the islands sample emitted lower current.

Compared with stripes electrode, islands electrode had less effective triple points (less triple-junction area). For the stripes pattern, all stripes were interconnected by the metal ring. Therefore, both two borders of every stripes would be effective triple points. For the islands pattern, all islands could be considered as electric insulated. A brief calculation demonstrated as in equation (2) would show the difference of triple-junction area:

$$\begin{aligned}\frac{TriplePoint\quad Area_{\text{stripes}}}{TriplePoint\quad Area_{\text{islands}}} &\approx \frac{L_1 \times \Delta}{L_2 \times \Delta} \\ &\approx \frac{1}{2\pi R}\left(\frac{1}{2}\times 2\pi R + 2\times \sum l\right) \\ &\geq \frac{1}{2\pi R}\left(\frac{1}{2}\times 2\pi R + 2\times (26\times \frac{1}{2}\times 2R)\right) \\ &\approx 8.8\end{aligned} \tag{2}$$

Δ, the hypothetical width of effective triple area;
R: the radius of electrode area;
26: there are about 26 stripes in that pattern.

4.2. Threshold voltages needed to start electron emission

It could be explained by the same reason discussed in section 4.1. Application of a voltage pulse to the rear electrode leaded to the appearance of a quick polarization reversal. The strong force of repulsion caused the intense electron emission from the front electrode. Further, explosion emission took place because of the extremely high electron current density [7]. The explosion process caused the formation of a neutral cloud, which was simultaneously ionized by electrons, leading to the formation of plasma.

The plasma moved towards to the graphite anode after formed on the ceramic surface due to electron avalanching. The islands electrode formed less plasma than the stripes electrode because of less triple-junction area with the same driving voltage. So there was little plasma after plasma closure. Then very little plasma reached the anode with low bias. Therefore, the plasma could be too little to be measured by the oscilloscope.

So it is hard for islands electrode to ignite surface flashover. Higher driving voltage was needed to start electron emission. Moreover, on the same experimental condition, islands electrode produced less initial plasma near triple points. Low bias extracted less charge from the surface plasma. So less current even zero amperes were collected with low bias voltage.

4.3. Reason for drop of waveform of voltage

The differences in waveforms related to the capacitance of PLZST samples. While the driving pulse was applied to the electrode, the effective metallic electrode area of island sample was much less, than the stripes sample. Only the metallic ring was available to serve as electrode in island sample. As a result, island sample was easier to be charged by the given voltage source equipment than stripes sample.

Then the plasma extended across the sample surface with a velocity of 10 – 15 mm/μs after produced by surface flashover [8]. Therefore, plasma would cover the whole sample surface within 350 ~ 500 ns (the radius of the sample was about 21mm). The plasma might serve as a dynamic electrode to provide an additional electrode area because of the conductivity of the plasma. So the area of effective electrode expanded.

$$C = \frac{\varepsilon A}{d} = \frac{Q}{U} \tag{3}$$

The area of electrode A increased extremely fast, thereupon the capacitance C increased instantly. On the other hand, Q could not change immediately, so the voltage U across the sample had to be diminished. Then the voltage source kept on charging the sample. The voltage increased to the preset value subsequently. It was that how a drop on the waveform of voltage was produced.

5. Conclusions

The patterns of the electrodes did affect the performance of ferroelectric emission. Compared with the more common stripes electrode islands electrode was inefficient:

- Islands electrode emitted lower maximal current especially with lower bias.
- Islands electrode needed higher driving voltage to start electron emission.

Acknowledgments

The research was supported by the National Natural Science of China under grant number 10875095.

References

[1] Jihoon Park, Yong Tae Kim, Ki Hyun Yoon. [J]. *J. Appl. Phys.*, 2002, **91** (3): 1458~1463.
[2] Rosenman G, Shur D, Garb K, Cohen R, et al. [J]. *J. Appl. Phys.*, 1997, **82**: 772.
[3] Sheng Z X, Feng Y J, Cui J, et al. [J]. *Chin. Sci. Bull.*, 2008, **53**(12):.1789-1795.
[4] Angadi M, Auchiello O, Krauss A R, et al. [J]. *Appl. Phys. Lett.*, 2000, **77**: 2659.
[5] Shannon D N J, Smith P W, Dobson P J, et al. [J]. *Appl. Phys. Lett.*, 1997, **70**: 1625.
[6] Boscolo I, Ciadi B, [J]. *Appl. Phys. Lett.*, 2002, **91**: 6125.
[7] Krasik Y E, Dunaevsky A, Krokhmal A, et al. [J]. *J. Appl. Phys.*, 2001, **89**: 2379
[8] Krasik Y E, Dunaevsky A, Felsteiner J, [J]. *Appl. Phys. Lett.*, 1998, **73**: 453.

Characterizations of bismuth telluride/gallium nitride heterojunction photovoltaic detector for MWIR detection under room temperature

M Y Pang[1*], H F Lui[2], W S Li[2], K H Wong[1], and C Surya[2]

[1]Department of Applied Physics and Material Research Centre, The Hong Kong Polytechnic University, Hunghom, Kowloon, Hong Kong, China.
[2]Department of Electronic and Information Engineering, The Hong Kong Polytechnic University, Hunghom, Hong Kong, China.

E-mail: 06900904r@polyu.edu.hk

Abstract. Uncooled infrared detectors have aroused keen interest in narrow bandgap semiconductor research. We propose to use Bismuth Telluride (Bi_2Te_3) bandgap energy of about 0.15 eV for middle wavelength IR (MWIR) detection. In this work, a heterojunction structure of $Bi_2Te_3/GaN/Al_2O_3$ was fabricated to form a photovoltaic detector. The structural properties of Bi_2Te_3 thin films were characterized by high resolution x-ray diffraction spectroscopy. Crystallinity of Bi_2Te_3 thin films was found to be strongly dependent on deposition temperatures. The spectral current responsivity for this heterojunction in NIR-to-MWIR region was evaluated under back-side illumination. The devices were also characterized by studying their low-frequency noise spectra measurement. The experimental results suggest that, the noise level of the devices is correlated to the crystallinity of the Bi_2Te_3 thin films.

1. Introduction

Many physical phenomena that respond to infrared (IR) photon energy in the range of about 0.1 to 1 eV have been exploited for making IR photon detectors [1]. Mercury cadmium telluride (HgCdTe) has remained the most important semiconductor for middle to long infrared detection in the past three decades. Numerous studies attempt to replace HgCdTe by alternative materials such as mercury-based alloys (HgZnTe, HgMnTe), PbSnTe and AlGaAs [1-4]. In this respect we propose to use thin film of narrow band gap bismuth telluride for infrared detection. Bismuth telluride (Bi_2Te_3), a V-VI semiconducting alloy, is well known in thermoelectricity field for its outstandingly high Seebeck coefficient at room temperature [5]. Bi_2Te_3 and its derivatives $Bi_{2-x}Sb_xTe_3$ and $Bi_2Te_{3-y}Se_y$ are the most commonly adopted semiconductors in recent thermoelectric devices. The Bi_2Te_3 is a narrow band gap semiconductor with layered structure. The crystal structure of Bi_2Te_3-based compounds composes of atomic layers in the order of Te/Bi/Te/Te/Bi/Te/Bi/Te/Te/... oriented along the *c*-axis. The Te/Te layers are held together by weak van der Waals forces [6]. Theoretical and experimental values of band gap energy for Bi_2Te_3 are about 0.11 eV and 0.15 eV respectively [7].

In this work, a novel heterojunction photovoltaic detector structure of $Bi_2Te_3/GaN/Al_2O_3$ was fabricated for MWIR detection. High quality *n*-GaN epitaxial layers were first grown on sapphire substrates by MOCVD technique. Then Bi_2Te_3 thin films were subsequently grown on the *n*-GaN/Al_2O_3 template by pulsed laser deposition (PLD) technique. PLD is one of the most versatile physical vapour deposition techniques which is able to transfer the composition of the target materials directly on to the thin films[6, 8]. The laser plume produced under ablation ~~is~~ consists of materials very close to the stoichiometry of the target. It is feasible to use PLD to control the films quality with a variety of structures ranging from polycrystalline to textured growth and multiple stoichiometry of Bi_xTe_{1-x} thin films. At present, only few studies on PLD growth Bi_2Te_3 thin films were reported in the literature [8-10]. Our work~~s~~ demonstrates the influences of the deposition temperature on the crystallinity of Bi_2Te_3 thin films. GaN, due to its wide band gap energy of 3.4 eV at room temperature, is a good MWIR blind material. It acts as both the *n*-type semiconducting layer in the *p-n* junction and as a window for IR transmission in a back-side illuminated diode configuration. Besides, replacing the *n*-type layer by the wide band gap semiconductor will contribute to some advantages over conventional narrow band gap semiconductor homojunction. GaN has much higher thermal stability than most of the narrow band gap materials; thus, it will exhibit much lower equilibrium concentration of thermionic minority carriers under room temperature operation. As a result, lower leakage current in reverse bias branch. In addition, IR radiation is transparent to the wide band gap semiconductor; therefore, the efficiency for the photons to reach the active region (depletion layer) will also be enhanced [11].

Since the performance of a photovoltaic detector is greatly governed by its *I-V* characteristics, we have measured room temperature the *p*-Bi_2Te_3/*n*-GaN heterojunction *I-V* characteristics. The variety of *p*-Bi_2Te_3 layers were grown under control of the deposition temperature, their corresponding diode parameters were determined by employing real diode equation. Low-frequency noise is shown to-arise from the trapping and detrapping of charge carries through thermally activated process, therefore, normalized current noise power spectra were studied in order to clarify the materials quality for the devices. The characterizations of electrical and optical measurements are correlated to the structural properties of the PLD growth Bi_2Te_3 semiconductor.

2. Experimental details

Bi_2Te_3 thin films have been deposited by pulsed laser deposition (PLD) on MOCVD grown (00*n*) *n*-GaN/sapphire. The thickness of the *n*-GaN layer is about 2μm. A commercial Bi_2Te_3 (99.999% ACI ALLOYS) PLD target used. Before the deposition process, the experimental chamber was evacuated to a base pressure of ~3 x 10^{-5} Torr by diffusion pump. The *n*-GaN/sapphire substrate was slowly heated up by an ohmic heater embedded in the substrate holder. A KrF excimer laser (COMPex 205, Lambda Physik) operating at 10 Hz repetition rate was employed. The on-target pulsed laser energy density was about 5 Jcm^{-2} and each film deposition lasted for 10 min. The as-prepared films were further in-situ annealed under the same deposition ambient and temperature for 30 min. The thickness of the films was measured by imaging the cross section of the *p*-Bi_2Te_3/*n*-GaN interface with a scanning electron microscope (SEM). Typically a ~500 nm thick film is generated for a deposition time of 10 min.

The PLD growth photodiode was fabricated by using shadow mask technique. Two separate stainless steel masks containing circular holes of diameter of 2 mm and 4 mm were used to deposit ohmic contact Au on Bi_2Te_3 and Bi_2Te_3 thin films on *n*-GaN/$A1_2O_3$ respectively. The ohmic contact for *n*-GaN is obtained by depositing Ti metal with a 9 mm diameter annular aperture. The crystal structure of the heterojunction was examined by x-ray diffractometer (XRD) (Bruker D8 Discover, CuK_α radiation) configured in a θ-2θ scan mode. A JSM-6335F Field Emission Scanning Electron Microscope (SEM) operating at accelerating voltage of 3 kV was applied to study the morphology of the films. The *I-V* characteristics of the heterojunction were investigated via electrical measurements using a HP 4140B picoammeter.

Spectral responsivity of Bi_2Te_3 heterojunction detector is obtained using a-300 W halogen lamp and placed in front of an Oriel 74004 monochromator, which contains two diffraction gratings over the range from near to long infrared region. The output photocurrent was subsequently measured by an HP 617 picoammeter.

3. Experimental results and discussions

X-ray diffraction characterizations of Bi_2Te_3 thin film on GaN/sapphire substrate deposited at temperature range from 150 to 300 ℃ were conducted, in order to determine their structural parameters. The temperature dependent of X-ray diffraction pattern of the Bi rich-Bi_2Te_3/GaN/sapphire is depicted in Figure 1(a)-(d) in logarithmic scale. It is observed that all Bi_2Te_3 thin films are polycrystalline in nature[6, 10]. Crystal structure of various Bi_2Te_3 thin films are belongs to trigonal system with an equivalent hexagonal unit cell. By investigating the crystallinity (orientation and grain size) among the sample with different deposition temperature, the influence of the increase in deposition temperature to the crystallinity of the Bi_2Te_3 thin films can be pointed out. In general, there is one strong diffraction peak within each diffraction pattern in Fig. 1(a)-(d), along the plane (015). Besides, the enhancement of crystallinity of the thin films in terms of averaged grain size is also obtained from the XRD spectrum. There is only one well-defined (015) peak in Fig. 1(a). At 250℃, the crystallinity of the films is improved and the (02 10) peak is shown. However, only sample grown at 300℃ has shown peak corresponding to c-oriented (00 15) plane of Bi_2Te_3 in hexagonal phase.

From Fig. 1(d), the lattice parameters of the films were calculated using Bragg's formula. The lattice parameters deduced from the diffraction angle are a_H = 0.443nm and c_H = 2.98nm. Comparing these values to the values available in the literature[12] and ASTM data (a_H = 0.438nm and c_H = 3.048nm) [Card No. 15-0863], there is a noticeable variation in the lattice constants. The small reduction of the lattice constant c_H for the deposited thin film maybe indicates that the film is in contraction or a small amount of Te occupies Bi sites [8, 13]. Due to the large difference in vapour pressure of telluride relative to bismuth, the thin films that we obtained may demonstrate of the segregation of bismuth. A small deviation of stoichiometric composition of Bi_2Te_3 to Bi rich samples will result in an effective *p*-type doping effect on the material.

By measuring the full-width-half maximum of (015) XRD reflections in Fig. 1(a)-(d) and adopting the Scherrer formula [12], an averaged grain size of around 100 to 200 nm can be estimated for the Bi_2Te_3 films grown under 150-300℃ deposition temperatures.

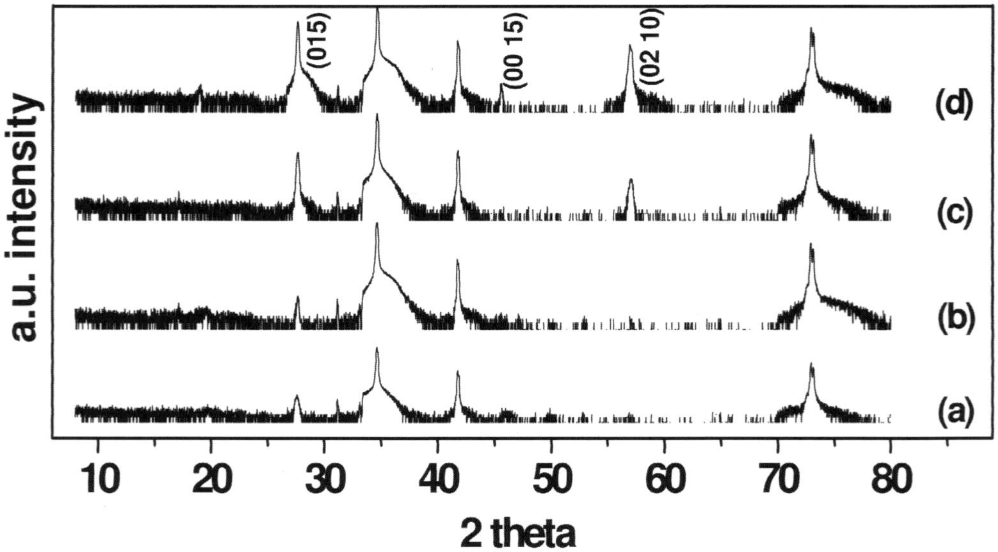

Figure 1. XRD spectra of Bi_2Te_3 films deposited at a substrate temperature of (a) 150℃, (b) 200℃, (c) 250℃ and (d) 300℃.

The temperature dependence of the surface morphology is presented in Fig. 2(a)-(d). The films deposited on GaN/Al_2O_3 at 150℃ (Fig. 2(a)) is relatively smooth with nano size crystals. When increasing the deposition temperature to 200℃, 250℃ and 300℃, the surface morphology of the films are found to be rougher than those grown at deposition temperatures and composed of high density crystals. At 200℃ the film surface is consists of 100-300 nm size grains; at 250℃ the grains size is increased to 500-1000 nm; the crystals appear on the film grown at 300℃ is quite large with few micrometer size. Moreover, it clearly exhibits the layered structure of Bi_2Te_3 with randomly oriented crystals. This is also conformed by the XRD spectra (Fig. 1(d)) where both [*m* 0 *n*] and [0 0 *n*] direction peaks are found, confirming the less preferential growth take place at such high substrate temperature.

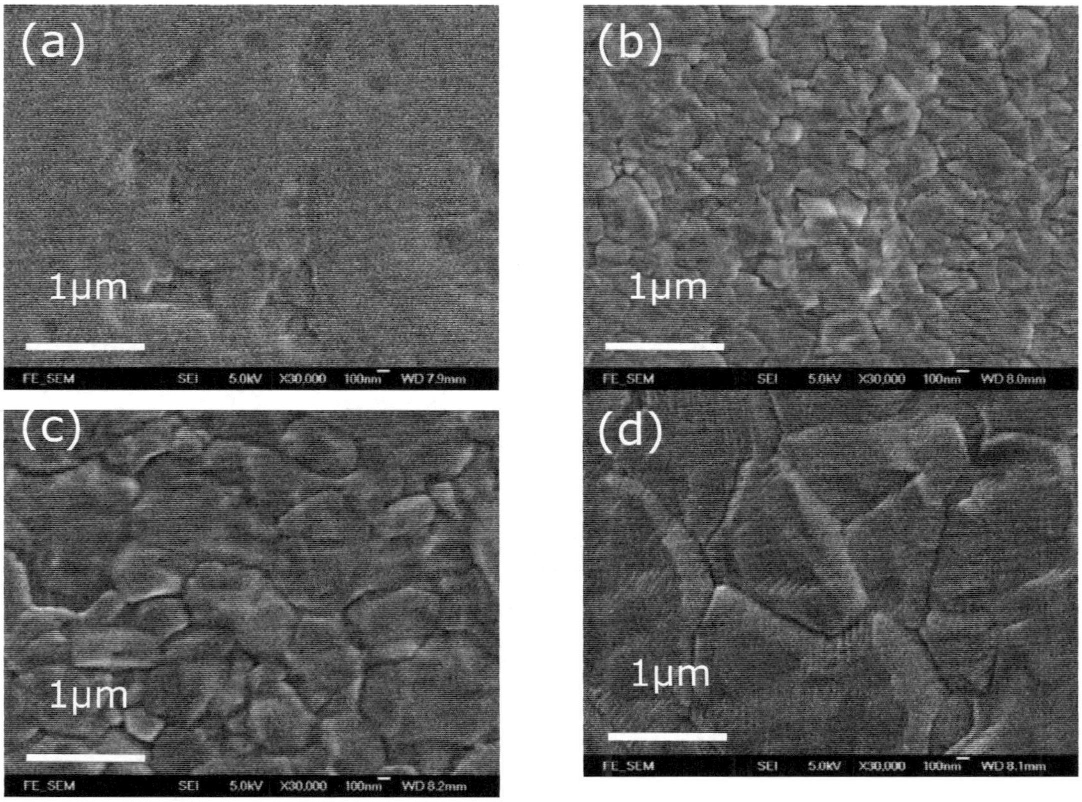

Figure 2. SEM images of Bi_2Te_3 films deposited at substrate temperature of (a) 150 ℃, (b) 200℃, (c) 250℃ and (d) 300℃.

The *I-V* characteristics of all devices were carried out in order to determine their junction parameters, the experimental results are shown in Fig. 3. Bi_2Te_3 diode were fabricated on *n*-GaN/Al_2O_3 to form a *p-n* heterojunction, which illustrates reasonably good *p-n* junction characteristics with low reverse bias current density and small series resistance in forward bias region. In Fig. 3, it is obvious that the reverse bias current increases slowly with the deposition temperature. The performance of various devices can be justify using the incremental differential resistance $R_0 = (dV/dI)_{V=0}$ [14]. The values of R_0 were obtained by differentiating the *I-V* curve at zero bias voltage. All values are summarized in table 1 for comparison. This quantitative calculation of R_0 illustrates a trend in the variation of electrical characteristics between the four devices. The incremental differential resistance varied from 2.07 to 8.34 kΩ and the R_0A product, were A is the photodiode area, from 260.1 to 1048 Ωcm^2.

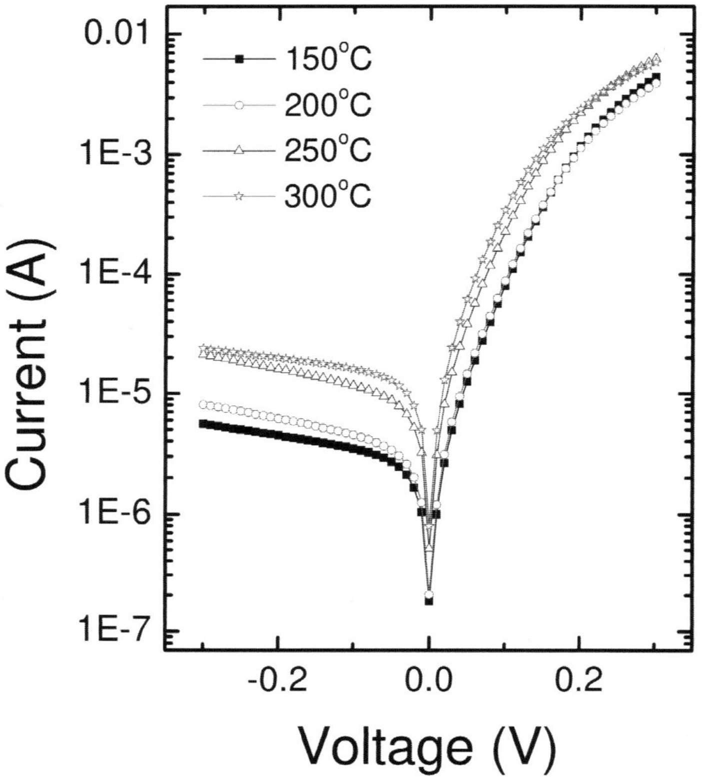

Figure 3. *I-V* characteristics of Bi_2Te_3/GaN heterojunction under different deposition conditions.

Table 1. Summary of the Bi_2Te_3/GaN heterojunction deposition conditions, incremental differential resistance R_0, R_0A product, average grain size and responsivity.

Sample ID	Substrate temperature (°C)	R_0 (kΩ)	R_0A (Ωcm^2)	Average grain size (nm)	Responsivity (A/W) @ 6 μm
S98	150	8.34	1048	94	4.34×10^{-6}
S96	200	3.21	403.4	132	4.23×10^{-6}
S94	250	2.07	260.1	161	8.46×10^{-6}
S90	300	2.14	268.9	205	10.97×10^{-6}

It has been shown that the low-frequency noise in semiconductor device is an effective tool to investigate the trapping and detrapping states of interface trap density. The low-frequency excess noise spectrum for *p*-Bi_2Te_3/*n*-GaN/Al_2O_3 photodiode fabricated under different substrate temperature is measured. The current noise power spectral density, $S_I(f)$, is given by:

$$S_I = 4\frac{I^2(\Delta R)^2}{R^2}\int_x\int_y\int_z\int_E N_T(x,y,z,E)\frac{\tau}{1+4\pi^2 f^2\tau^2}dxdydzdE\ , \quad (1)$$

where I is the dc current at constant bias voltage, R is the ac resistance of the device, ΔR is the fluctuation of the device resistance which come from the capture of one electron by an interface trap, $N_T(E)$ is the trap density in the material and τ is the activated time constant for the thermal fluctuation process. The measurement of $S_I(f)$ is useful for monitoring the change in the $N_T(E)$ inside the junction, and it is directly proportional to the trap density as well. Nearly three orders of reduction in the normalized current noise power density (Fig. 4) are obtained for *p*-Bi_2Te_3/*n*-GaN photodiode fabricated under 300℃ and 350℃. Our results indicate that the junction electronic quality is improved by choosing optimal substrate temperature, which contribute to the better crystallinity in terms of their average grain size.

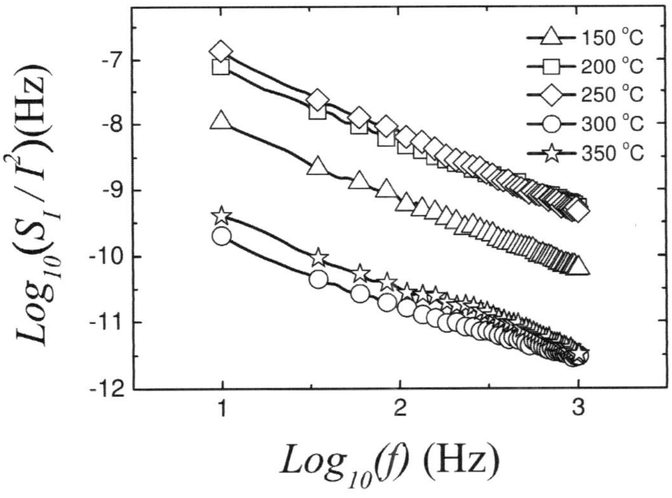

Figure 4. Room temperature normalized current noise power density for *p*-Bi_2Te_3/*n*-GaN/Al_2O_3 detectors.

The preliminary spectral response for Bi_2Te_3 photovoltaic detectors is shown in Fig. 5. Overall responsivity of these devices is improved with increasing deposition temperature. The devices are relatively more sensitive at wavelength about 6 μm. Their responsivity at this particular wavelength is summarized in table 1 for better comparison. It is important to study the correlation between the effects of the structural and electrical properties on the optical response. We found that the effect of the improved crystallinity of the *p*-Bi_2Te_3 layer may result in the improvement of responsivity in middle IR region. In contrast, higher deposition temperature may result in sacrificing the electrical properties which introduces larger reverse bias leakage current.

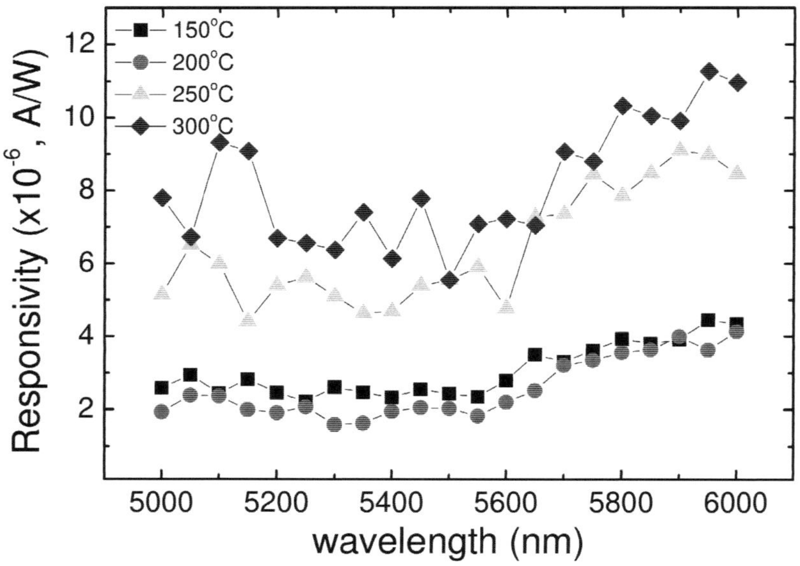

Figure 5. Responsivity curves of Bi_2Te_3/GaN photodiode fabricated under various substrate temperatures.

4. Conclusions

We performed a systematic study on the structural, electrical and optical properties of PLD-grown *p*-Bi_2Te_3/*n*-GaN/Al_2O_3 heterojunctions. The crystal quality is enhanced by increasing the deposition temperature from 150 to 300℃. By varying the deposition temperature, crystal domain size is able to have a better control. On the other hand, excessively high deposition temperature may result in the degradation of the electrical properties. The *I-V* curves reveal a larger reverse bias leakage current. In low-frequency noise measurement, the reduction in current noise power density for the device fabricated at 300 ℃ exhibit the most sensitive responsivity in middle IR region. Our results demonstrate the *p*-Bi_2Te_3/*n*-GaN/Al_2O_3 heterojunction photodetector can be utilized in middle IR detection.

Acknowledgement

This work was supported by CERG grants PolyU 5269/07E, and further support is provided by a University Research Grant G-YF18 of the Hong Kong Polytechnic University. MYP is grateful for the award of a research studentship by The Hong Kong Polytechnic University.

References

[1] Rogalski A 2003 Infrared detectors: status and trends *Progress in Quantum Electronics.* **27** 59-210
[2] Kastalsky A, Duffield T, Allen S J and Harbison J 1988 Photovoltaic detection of infrared light in a GaAs/AlGaAs superlattice *Appl. Phys. Lett.* **52** 1320-1322
[3] Boschett C, Bandeira I N, Closs H, Ueta A Y, Rappl P H O, Motisuke P and Abramof E 2001 Molecular beam epitaxial growth of PbTe and PbSnTe on Si(1 0 0) substrates for heterojunction infrared detectors *Infrared Physics & Technology* **42** 91-99
[4] Reno J, Sou I K, Wijewarnasuriya P S and Faurie J P 1985 Growth and characterization of

$Hg_{1-x}Mn_xTe$ grown by molecular beam epitaxy *Appl. Phys. Lett.* **47** 1168-1169
[5] Nolas G S 2001 *Thermoelectrics : Basic Principles and New Materials Developments.* (New York: Springer)
[6] Makala R S, Jagannadham K and Sales B C 2003 Pulsed laser deposition of Bi_2Te_3-based thermoelectric thin films *J. Appl. Phys.* **94** 3907-3918
[7] Mishra S K, Satpathy S and Jepsen O 1997 Electronic structure and thermoelectric properties of bismuth telluride and bismuth selenide *Journal of Physics: Condensed Matter* **9** 461
[8] Russo V, Bailini A, Zamboni M, Passoni M, Conti C, Casari C S, Bassi A Li and Bottani C E 2008 Raman spectroscopy of Bi-Te thin films *J. Raman Spectrosc.* **39** 131-133
[9] Bailini A, Donati F, Zamboni M, Russo V, Passoni M, Casari C S, Bassi A Li and Bottani C E 2007 Pulsed laser deposition of Bi_2Te_3 thermoelectric films *Applied Surface Science* **254** 1249-1254
[10] Dauscher A, Thomy A and Scherrer H 1996 Pulsed laser deposition of Bi_2Te_3 thin films *Thin Solid Films* **280** 61-66
[11] Krier A, Gao H H and Mao Y 1998 A room temperature photovoltaic detector for the mid-infrared (1.8-3.4 m) wavelength region *Semiconductor Science and Technology* **13** 950-956
[12] Dheepa J, Sathyamoorthy R and Subbarayan A 2005 Optical properties of thermally evaporated Bi_2Te_3 thin films *Journal of Crystal Growth* **274** 100-105
[13] Wang W, Wan C and Wang Y 2007 Composition-dependent characterization and optimal control of electrodeposited Bi_2Te_3 films for thermoelectric application *Electrochimica Acta* **52** 6502-6508
[14] Barros A S, Abramof E and Rappl P H O 2006 Electrical and optical properties of PbTe p-n junction infrared sensors *J. Appl. Phys.* **99** 024904

High and low dielectric permittivity polymer-based nanobybrid dielectric films

Zhi-Min Dang*, Hong-Tao Song,You-Qin Lin and Lan-Jie Ma

State Key laboratory of Chemical Resource Engineering, Beijing University of Chemical Technology, Beijing 100029, P.R. China

E-mail: dangzm@mail.buct.edu.cn

Abstract: It is well known that novel polymer-based functional dielectric materials are displaying a wide application in electronic devices and electrical engineering. By using for nanomaterials, people face to some new research subjects on building the polymer-based nanohybrid dielectric materials with special microstructure and subsequently leading to the outstanding dielectric properties. Here the preparation, microstructure and dielectric properties of two polymer-based nanohybrid dielectric materials are presented, which include the materials with high and/or low dielectric permittivity. By employing the in situ polymerization process, the nanohybrid dielectric materials with excellent dielectric properties are explored. It is expected to find some elemental science issues in order to bring out support on development of novel functional dielectric materials.

Keywords: Dielectric materials, Nanohybrid, Dielectric Properties, Microstructure

1. Introduction

With the development of electronic industry, some special dielectric materials with high and/or low dielectric permittivity (high k and/or low k) are attracting a remarkable attention in academic and industrial fields [1-4]. The dielectric properties of the insulative materials could be adjusted by dispersing different kinds of inorganic fillers into polymer matrices to form the polymer/filler composites [5-10]. With changing the concentration of fillers, the dielectric permittivity of the composites could be changed. When the inorganic fillers with high dielectric permittivity are dispersed into polymer, the composites with high dielectric permittivity could be acquired. While the inorganic fillers with low dielectric permittivity and special hollow channel are added into the polymer, people should acquire the composites with low dielectric permittivity. Therefore, to ensure the uniform and stable dielectric properties, the dispersion of fillers is an important factor. Surface modification with dispersants and coupling agents is often employed to improve the dispersion of fillers in polymer matrices using direct mixing technique.

In this study, by employing an in-situ polymerization processing, we fabricate two kinds of the polymer/filler nanohybrids with high and/or low dielectric permittivity, respectively. Polyimide (PI) is used as the same polymer matrix in our study because the PI is one species of the most important high performance polymers with high temperature durability, good mechanical properties and excellent insulation in wide rang of temperature. Besides, performances of the nanohybrids can be tailored by controlling the microstructure of composites and changing the volume fraction of fillers.

2. Experimental details

$BaTiO_3$ fillers with an average particle size of 100 nm were prepared by water-thermal method. Mesoporous silicate (MCM-41) particles with hollow channels (4 nm in diameter) used in this work were prepared according the reported method [11]. Other chemicals, that is, pyromellitic dianhydride (PMDA), 4, 4'-oxy dianiline (ODA) and N, N'-dimethylacetamide (DMAc) were purchased from Sinopharm Chemical Reagent Co., Ltd., Shanghai Sanaisi Reagent Co., and Beijing Yili Fine Chemical Reagent Co., respectively and were used as received. DMAc was stored over 4Å molecular sieves prior to use to minimize the adsorbed water content.

The $BaTiO_3$ or MCM-41/poly(amic acid) was prepared via in-situ polymerization. The mixture of PAA and $BaTiO_3$ or MCM-41 particles was cast onto a cleaned glass plate (using distilled water and acetone), respectively. The sticky film was kept still at room temperature for 30 min to let out the air entrapped naturally. Then PAA was converted to PI by two thermal imidization routes: one is to hold the film under 80 °C, 120 °C, 180 °C, 240 °C and 300 °C for 1h, respectively, the other is 100 °C, 200 °C and 300 °C for 1h, respectively. Then the $BaTiO_3$/PI or MCM-41/PI hybrid films with different concentrations of fillers were obtained.

The surface and cross section morphologies of films were imaged on a scanning electron microscope (SEM, JSE-6301F). The interfacial structure between $BaTiO_3$ or MCM-41 particles and PI matrix in the composites was examined using transmission electron microscope (TEM, Hitachi S-800). Thermo gravimetric analysis (TGA) and differential scanning calorimetry (DSC) analysis were also performed using a thermal analysis system (NETZSCH STA 449C).The dielectric properties of the composite samples were measured using an impedance analyzer (Agilent 4294A) both in the frequency range of 10^3 Hz to 10^6 Hz and in temperature range from –50°C to 150°C. The electrical breakdown test was carried out by breakdown voltage instrument (CS2674A). During the test, the voltage applied rose rapidly until the samples broke down with the warning current value preset as 5 mA.

3. Results and Discussion

3.1. High permittivity $BaTiO_3$/PI nanohybrid films

The present $BaTiO_3$/PI composite films face a great challenge of poor stability of $BaTiO_3$ dispersion in the matrix. Morphological images in Fig.1 show that the $BaTiO_3$ particles are uniformly dispersed in the PI matrix after imidization. Strong interfacial interaction between $BaTiO_3$ particles and PI matrix was needed to keep the stability of $BaTiO_3$ dispersion. The isoelectric point (IEP) of nanosized

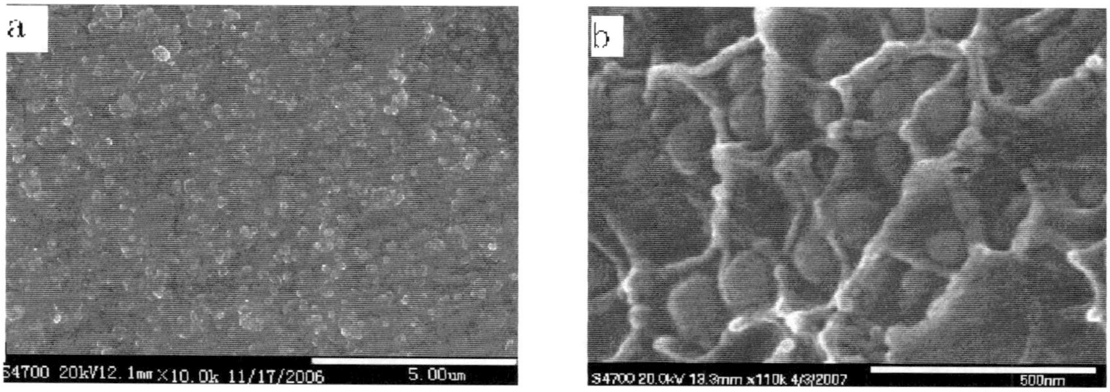

Fig 1. SEM images of the $BaTiO_3$/PI composite film prepared by in-situ polymerization with $BaTiO_3$ loading of 40 vol%. (a) fresh surface and (b) cross-sectional surface.

$BaTiO_3$ is less than 3 and the value of zeta potential is up to its maximum in the PH range of 6-11, indicating that the $BaTiO_3$ particles possess basic surface [12,13]. And many hydroxyl groups are incorporated in the perovskite lattice of hydrothermal $BaTiO_3$ particles. Besides, polyelectrolyte such as polymethyl methacrylate (PMMA) and $PAA\text{-}NH_4$ was reported to be able to be adsorbed onto the surface of $BaTiO_3$ particles in organic or inorganic solvent [14,15]. However, it was unexpected as shown in Fig.2 that a core-shell-like structure as reported in the $BaTiO_3$/PMMA system was also observed in the $BaTiO_3$/PI nanohybrid films prepared by in-situ polymerization. The special core-shell-like structure could be demonstrated by the imidization of an overlay of PAA adsorbed onto $BaTiO_3$ surface by the force of electrostatic attraction.

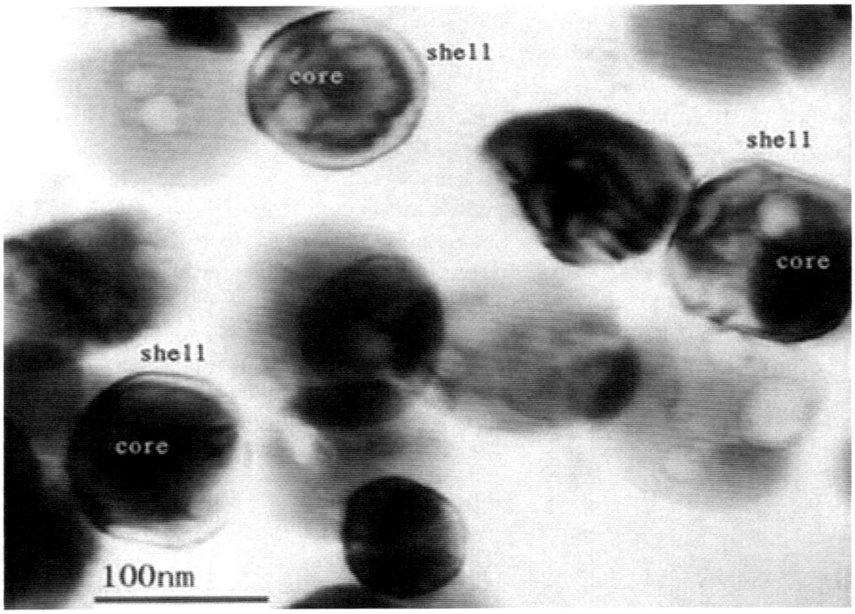

Fig 2. Cross-sectional TEM image of $BaTiO_3$/PI composite film by in-situ polymerization with $BaTiO_3$ loading of 40vol%.

Fig. 3 shows dependences of dielectric properties of the $BaTiO_3$/PI nanocomposite films on the concentration of $BaTiO_3$ and frequency at room temperature. For the sample with 40 vol% $BaTiO_3$, the dielectric permittivity was 18.35, 18.13, 17.90 and 17.77 at 10^3 Hz, 10^4 Hz, 10^5 Hz and 10^6 Hz (see Fig. 3a), respectively. Correspondingly, the loss tangent was 0.0092, 0.0081, 0.0091 and 0.0105 (see Fig. 3b). In the frequency range of 10^3~10^6 Hz, the maximum gap of the dielectric permittivity was only 0.58, which was just 3.16 % of the maximum of the dielectric permittivity mentioned above. Besides, the loss tangent of all the samples investigated was close to 0.010. Both the weak dependence of dielectric permittivity on frequency and the low loss tangent are desirable for future dielectric materials.

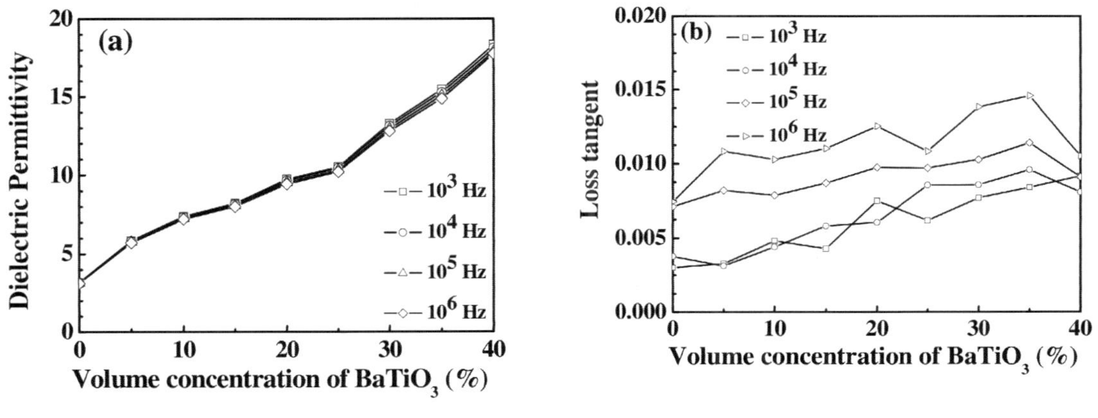

Fig 3. Dependence of (a) dielectric permittivity and (b) loss tangent of $BaTiO_3$/PI composite films by in-situ polymerization on the volume concentration of $BaTiO_3$.

Fig.4a shows temperature dependence of dielectric permittivity of the $BaTiO_3$/PI nanocomposite films in the temperature range of -50 °C ~ 150 °C. Except at 10^3 Hz, the maximum dielectric permittivity of the composite films appeared at 40 °C ~ 50°C. The result might be explained as follows. Mobility and thermal expansion of polymer composed a pair of competing factors had an effect on the dielectric permittivity [16]. When the temperature increased from -50 °C to 40 °C, the interfacial interaction by hydrogen bond between $BaTiO_3$ and PI restrained the thermal expansion of PI and the formation of interspace. At the same time, the mobility of polymer was improved. Because the PI (PMDA-ODA) is a kind of polymer with high T_g of about 352 °C, the mobility of polymer depended mainly on the change of bond length and bond angle under T_g. Interfacial interaction had little influence on the local change of bond length and bond angle under alternating current (AC) field. Therefore, an ascending trend was observed in the range of low temperature. But temperature more than 50 °C might disrupt the hydrogen bond between $BaTiO_3$ and PI, so the effect of thermal expansion was fully exhibited. The coefficient of thermal expansion (CTE) of PI (PMDA-ODA) was 50×10^{-6} °C^{-1}, while that of $BaTiO_3$ was 0.12 $\times10^{-6}$ °C^{-1} [17]. The difference of CTE between two materials is so great that it might overwhelm the effect of polymer mobility on dielectric permittivity. So a descending trend was observed in the range of high temperature. Besides, the absence of Curie peaks in the composites was clear, as the tetragonal deformation is locked in the $BaTiO_3$ nanoparticle [18]. When the temperature changed from -50 to 150°C, the dielectric permittivity of the composites exhibited weak temperature dependence. As for the sample with 40 vol% $BaTiO_3$, the maximum dielectric permittivity was 19.65, 18.75, 18.54 and 18.36 at 10^3 Hz, 10^4 Hz, 10^5 Hz and 10^6 Hz, respectively. According to the X7R EIA specification (EIA = Electronic Industries Association), the dielectric permittivity of a capacitor should vary less than ± 15% of the 25 °C value in the temperature interval from -55 to 125°C. Here, the maximum of dielectric permittivity between 20 °C and 30°C was regarded as the 25°C value of the specification. So the derived relative tolerance of the dielectric permittivity in the temperature range of -50 °C to 130 °C was 12.33%, 9.51%, 9.90% and 9.45% at 10^3 Hz, 10^4 Hz, 10^5 Hz and 10^6 Hz, respectively.

Fig.4b shows conductivity and breakdown strengths dependences of the $BaTiO_3$/PI composite films on the concentration of filler, respectively. The conductivity of the composites gradually increased with the volume concentration of $BaTiO_3$ in the composites, but the influence of frequency on the conductivity was much stronger. At 10^3 Hz, 10^4 Hz, 10^5 Hz and 10^6 Hz, the magnitude of the conductivity of the composite films was 10^{-9}, 10^{-8}, 10^{-7} and 10^{-6}, respectively. The electrical breakdown strength of pure PI had been reported to be 230 MV·m^{-1} (Semicofine LP-52, Toray, 77K) and 200~225 MV·m^{-1} (PMDA-ODA, 77K) [19-21], and the breakdown strength was nearly independent of temperature from 77 K to room temperature. By comparison, the breakdown strength of mica/PI hybrid composites with 0~20 wt% mica was 200 ~ 150 MV·m^{-1} [20], and that of silica/PI hybrid composites with 0~10wt% silica was 225 ~ 150 MV·m^{-1} [21]. The breakdown field of pure PI in our study was 283 MV·m^{-1} at room temperature, which was a little higher than reported [20, 21]. And the breakdown strengths of the PI-based nanocomposite film with high content of filler had not been reported. As for our samples, when the $BaTiO_3$ content was not more than 10 vol% (corresponding to 29.38 wt %), the average breakdown strengths ranged from 283 MV·m^{-1} to 141 MVm^{-1}. For the composites with $BaTiO_3$ not more than 25 vol% (corresponding to 55.50 wt%), the average breakdown strengths arrived at above 120 MV·m^{-1}. Even for the samples with $BaTiO_3$ content of 30 vol% to 40 vol%, the average breakdown strengths were above 67 MV·m^{-1}.

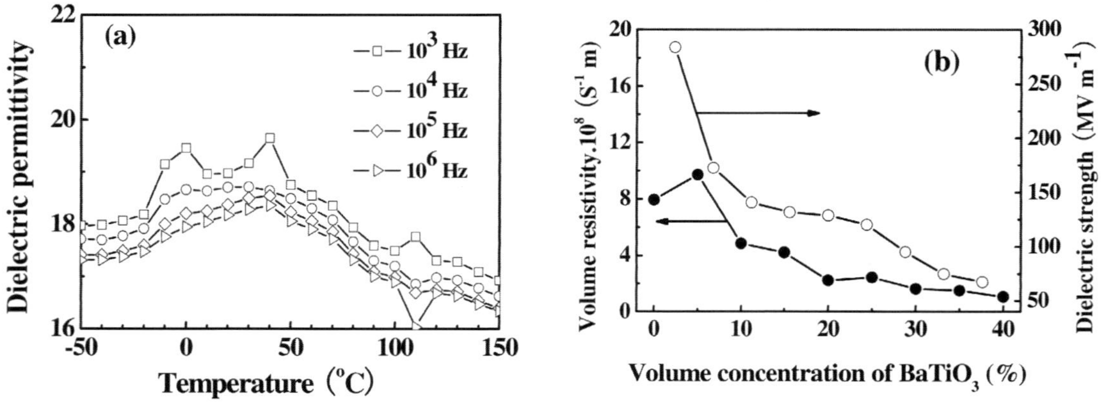

Fig 4. Temperature dependence of (a) dielectric permittivity and (b) TG and DSC curves of $BaTiO_3$/PI composite with $BaTiO_3$ loading at 40 vol%. (b) Volume resistivity (left-side) and dielectric strength (right-side) of the PI/MCM-41 nanohybrid films as the function of the concentration of $BaTiO_3$.

3.2. Low permittivity MCM-41/PI nanohybrid films

X-ray diffraction (XRD) pattern of the MCM-41 particles prepared in Fig. 5a showed four peaks in the low 2θ region, which could be indexed on a hexagonal unit cell corresponding to XRD pattern of the MCM-41 bulk sample. The hexagonal lattice parameter was 47.49 Å obtained according to the equation $a = 2d_{100}/\sqrt{3}$, in which the value of d_{100} spacing was 41.13 Å. TEM image of the MCM-41 was shown in Fig.5b. Before TEM observation, the MCM-41 was dispersed in absolute ethanol with ultrasonic vibration for 10 min, then deposited on a carbon-film-coated copper grid, and dried at room temperature. The ordered hollow channel structure of the MCM-41, which clearly showed a periodicity of about 4.0 nm in diameter, could be clearly seen from the TEM image. The results of XRD and TEM further confirmed that the prepared MCM-41 in the work possessed the ordered hexagonal array structure.

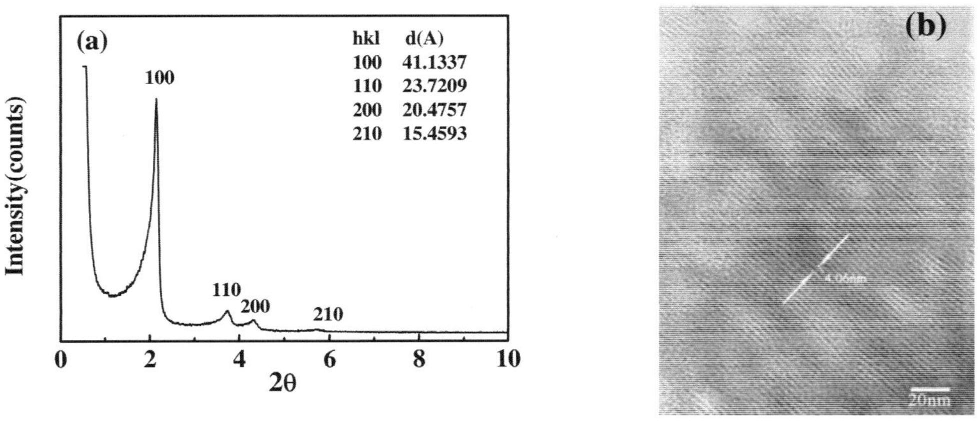

Fig 5. XRD pattern and TEM image of MCM-41.

Frequency dependence of k values of pure PI and the MCM-41/PI nanohybrid films with different concentration of MCM-41 loading at 25 °C was shown in Fig.6a. It was found that the k values of hybrid films were notably independent on frequency from 10^2 Hz to 10^6 Hz, which was very important for their practical application in electronic technology fields. The dielectric loss of pure PI and the MCM-41/PI nanohybrid films as a function of frequency ranging from 10^2 Hz to 10^6 Hz was also shown in the inset of Fig.6a.The weak increase in dielectric loss was observed when the frequency increased from 10^2 Hz to 10^6 Hz. The maximal dielectric loss was less than 0.01, which was lower than the request value (0.05). Meanwhile we observed slightly dependence of the dielectric loss on the content of MCM-41, indicating the nanohybrid polarization films were affected weakly by the incorporation MCM-41 particles. It was very interesting that the k value changed with increasing the concentration of MCM-41 as shown in Fig.6a and 6b. During the beginning stage, with increasing the

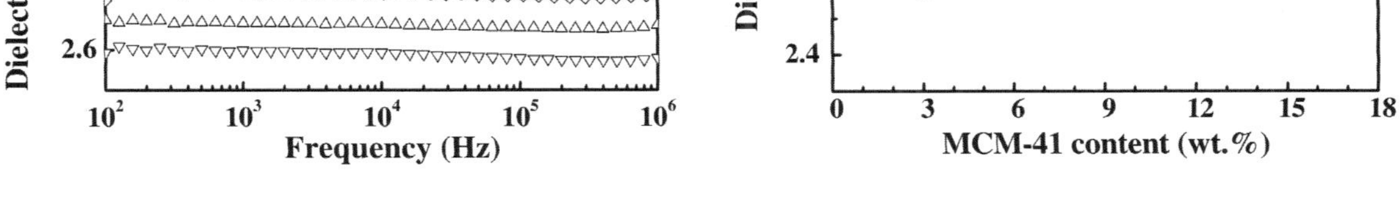

Fig 6. (a) Dielectric permittivities of pure PI and the MCM-41/PI nanohybrid films as a function of frequency from 10^2 Hz to 10^6 Hz at 25 °C, the inset was the dielectric loss versus frequency, (b) Comparison of experimental (at 10^3 Hz and 25 °C) and theoretical k values of the MCM-41/PI nanohybrid films as a function of content of MCM-41.

MCM-41 loading to 3.0 wt.%, the *k* value of the hybrid film decreased to around 2.58 at 10^3 Hz and 25 °C, which was notably lower than that of pure PI (~2.94). The decrease percent in the *k* value was 12.24% ($\frac{2.94-2.58}{2.94}\times 100\%$). However, during the subsequent stage, when the MCM-41 loading was further increased to 12.0 wt.% and 18.0 wt.%, the *k* values of the nanohybrid films increased. In this case, the *k* value was still lower than that of pure PI (see the dot line in Fig.6b).

Fig.7 showed the *k* values of pure PI and the MCM-41/PI hybrid films as a function of temperature from -50 °C to 150 °C. The results indicated that the MCM-41/PI nanohybrid films had very stable *k* values in the measurement temperature range (10^6 Hz). The temperature dependence of the dielectric loss of pure PI and the MCM-41/PI hybrid films with different concentration of MCM-41 loading was shown in the inset in Fig.7. The dielectric loss increased slightly as the MCM-41 content increased from 0 to 18.0 wt.%. Similarly, in temperature range -50 °C to 150 °C, we found that the maximal value of the dielectric loss of the MCM-41/PI hybrid film with 3.0 wt.% MCM-41 loading was only 0.011. We also observed the broad peaks of the dielectric loss at about 90 °C, which were attributed to the β relaxation due to the co-operative restricted oscillations involving both the main chain and the phenyl ring of PI matrix [22]. While the temperature was below to 0 oC, an increase in the dielectric loss was observed, which might result from the presence of the moisture and the clustering behavior of water [23]. The degradation temperature of the MCM-41/PI hybrid film with 3.0 wt.% MCM-41 loading was about 614 °C, which was little higher than that of pure PI as shown in Figure 3b. The thermal stability illustrated that the MCM-41/PI nanohybrid film with the 3.0 wt.% MCM-41 loading could sustain the high temperature (up to 425°C) as the interlay dielectric materials, which could also satisfy the IC device processing.

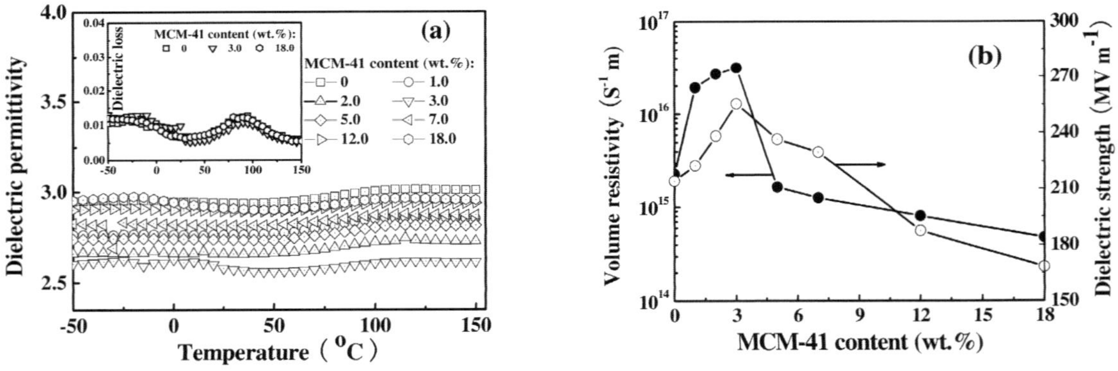

Fig 7. (a) Dielectric permittivities of pure PI and the MCM-41/PI nanohybrid films as a function of temperature from -50 °C to 150 °C at 10^6 Hz, the inset was the dielectric loss versus temperature. (b) Volume resistivity (left-side) and dielectric strength (right-side) of the MCM-41/PI nanohybrid films as the function of the concentration of MCM-41.

The other electrical properties were also very important for the use of the dielectric materials. Fig.7b showed the dielectric strength and the volume resistivity of the MCM-41/PI nanohybrid films as functions of the MCM-41 concentration. It was notedly seen that the dielectric strength and the volume resistivity of the film with the 3.0 wt.% MCM-41 loading were improved from 214.17 MV.m^{-1} and 2.29×10^{15} Ω.m to 255.45 MV.m^{-1} and 3.14×10^{16} Ω.m, respectively. However, the *k* value was only 2.58 in the composition film. The improvement of such dielectric properties could be attributed to the trapping and scattering effect of the MCM-41 with the unique hexagonal mesoporous structure.

4. Summary

High and low dielectric permittivity polymer-based nanobybrid dielectric films were realized by employing the in situ polymerization when the nanosize $BaTiO_3$ and the MCM-41 with nanosize hollow channels were incorporated into the PI polymer. Excellent dielectric properties benefited from the special microstructure of the nanohybrid films. In the $BaTiO_3$/PI films, the high dielectric permittivity was mainly from core-shell-like structure of $BaTiO_3$-PAA, which helped to obtain homogeneous dispersion of $BaTiO_3$ particles in PI matrix and reduced the voids in the nanohybrid films, and the weak voids made the high breakdown strength of the nanohybrid films. In the MCM-41/PI films, the super-low k value could be attributed to the low k value of MCM-41 itself and air gaps in the hybrid films. And for the film with the 3.0 wt.% MCM-41 loading, high volume resistivity and high breakdown strength were also observed.

Acknowledgement

This work was financially supported by NSF of China (grant No. 50677002), NSF of Beijing City (grant No. 2083031), Ministry of Education of China through Doctor Project (grant No. 20050010010), the Ministry of Sciences and Technology of China through 863-project (grant No. 2008AA03Z307), and Program for New Century Excellent Talents in University (NCET).

References

[1] J. W. Xu, C. P. Wong, *Comp. A* 2007, *38*, 1.
[2] Y. Bai, Z. Y. Cheng, V. Bhati, H. S. Xu, Q. M. Zhang, *Appl. Phys. Lett.* **2000**, *76*, 3804.
[3] T. Yagi, M. Tatemoto. J. Sako, *Poly. J.* 1980, *12*, 209.
[4] Z. M. Dang, H. Y. Wang, H. P. Xu, *Appl. Phys. Lett.* 2006, *89*, 112902.
[5] S. D. Cho, J. Y. Lee, K. W. Paik, *Mater. Sci. Eng. B* 2004, *110*, 233.
[6] Z. M. Dang, H. Y. Wang, Y. H. Zhang, J. Q. Qi, *Macro. Rap. Comm.* 2005, *26*, 1185.
[7] Z. M. Dang, Y. Shen, C. W. Nan, *Appl. Phys. Lett.* 2002, *81*, 4814.H.
[8] Treichel, G. Ruhl, P. Ansmann, R. Würl, Ch. Müller, M. Dietlmeier, *Microelectron. Eng.* **1998**, *40*, 1.
[9] T. Homma, *Mate. Sci. Eng.* **1998**, *23*, 243.
[10] B. D. Hatton, K. Landskron, W. J. Hunks, M. R. Bennett, D. Shukaris, D. D. Perovic, G. A. Ozin, *Materials Today* **2006**, *9*, 22.
[11] J. Rathouský, G. S. Ekloff, J. Had, A. Zukal, *Phys. Chem. Chem. Phys.* **1999**, *1*, 3053.
[12] M. Kosmulski, *J. Colloid & Interface Sci.* 2004, *275*, 214.
[13] D. Hennings, C. Metzmacher, B. C. Schreinemacher, *J. Am. Cera. Soc.* 2001, *84*, 179.
[14] Z. G. Shen, J. F. Chen, H. K. Zou, J. Yun, *J. Colloid & Interface Sci.* 2004, *275*,158.
[15] D. W. Kim, D. H. Lee, B. K. Kim, H. J. Je, J. G. Park, *Macro. Rap. Comm.* 2006, *27*, 1821.
[16] L. Ramajo, M. Reboredo, M. Castro, *Compo. A* 2005, *36*, 1267.
[17] S. H. Xie, B. K. Zhu, X. Z. Wei, Z. K. Xu, Y. Y. Xu, *Compo. A* 2005, *36*, 1152.
[18] C. Muralidhar, P. K. C. Pillai, *J. Mater. Sci. Lett.* 1987, *6*, 346.
[19] Y. Muramoto, M. Nagao, M. Kodaki, *Cryogenics.* 1995, *35*, 791.
[20] Y. H. Zhang, Z. M. Dang, J. H. Xin, W. A. Daoud, J. H. Ji, Y. Liu, B. F., Y. Q. Li, J. T. Wu, S. Y. Yang, L. F. Li, *Macro. Rap. Comm.* 2005, *26*, 1473.
[21] Y. H. Zhang, Y. Li, S. Y. Fu, J. H. Xin, W. A. Daoud, L. F. Li, *Poly.* 2005, *46*, 8373.
[22] O. L. Tyan, Y. C. Liu, K.-H. Wei, *Chem. Mater.* 1999, *11*, 1942.
[23] G. Xu, C. C. Gryte, A. S. Nowick， *J. Appl. Phys.* **1989**， *66*， 5290.

Simple fabrication of Si nanowire and its biological application

Y.T. Cheng[1], Y.H. Cho[1], N. Takama[1], P. Löw[2], C. Bergaud[2], and B.J. Kim[1]

[1] Institute of Industrial Science, The University of Tokyo 4-6-1 Komaba, Meguro-ku, Tokyo 153-8505, Japan
[2] LAAS-CNRS, Toulouse, France

E-mail: bjoonkim@iis.u-tokyo.ac.jp

Abstract. A novel, relatively simple and cost effective method is reported to fabricate suspended silicon nanowires. This method allows for the production of suspended silicon nanowires using anisotropic wet etching and thermal oxidation of single crystalline silicon. The dimensions of the silicon nanowires fabricated with the proposed method are evaluated. The vibration properties of the nanowires obtained by heterodyne laser doppler interferometer show the potential of being nanomechanical sensors.

1. Introduction

Nanowire has attracted much interest as essential parts for functional electronics devices because of its potential applications in nanoelectronics [1-3] and highly sensitive biosensor [4-5]. Although research focused on the fabrication of nanowire using synthesis-based bottom-up approach [3] has led to considerable achievements to date, the shortcomings have also been detailed nowadays. In contrast, top-down approaches provide many superior advantages over previously described bottom-up approaches [2, 6, 7]. For instance, top-down approaches can produce silicon nanowire precisely at designed locations in devices using nanolithography or micromachining process on a bulk silicon substrate. However, nanolithography-based top-down fabrication utilized in previous studies uses high-cost facilities, such as electron beam lithography. Therefore, because of the serial processes required for nanolithography, it is not simple to do mass production. In the case of the micromachining methods of bulk silicon substrates to fabricate nanowires, some methods have been reported: anisotropic wet etching of single crystalline silicon [8] and silicon thinning using silicon thermal oxidation [9-10]. In these methods, the silicon-on-insulator (SOI) substrates were indispensably utilized for nanowire fabrication. The top-down approaches for nanowire fabrication may limit the uniformity in the width of nanowires.

Here, we report a simple and cost-effective fabrication process to fabricate suspended silicon nanowire only using conventional MEMS processes including anisotropic wet etching and local thermal oxidation processes without nanolithography techniques. Due to the anisotropic etching of <100> single crystalline silicon substrate in KOH solution, uniform and well-defined silicon nanowires with sub-micro diameter were obtained. Because the top surface of Si nanowires is expected to remain relatively flat after finally fabricated, it is supposed to be possible to measure the resonant frequency with heterodyne laser doppler interferometer [11]. This approach also provides us a direction of applying nanowires as mechanical sensors.

Recently, silicon nanowires have been utilized as the core element of field-effect transistor and demonstrated high capabilities. G. Zheng, *et al.* developed and characterized high performance silicon nanowire field effect transistors in 2004 [15]. One promising approach for the direct label-free and electrical detection of biomolecules is to use field-effect transistors (FETs) because the conductance of a nanosized channel changes upon binding of small quantity of charged biomolecules to probe receptors grafted onto the channel surface. In addition, it should be possible to exploit the massive knowledge that exists for the chemical modification of oxide surfaces, for example, from studies of silica, planar chemical/biological sensors, to create semiconductor nanowires modified with receptors for many applications. On the other hand, nano electro mechanical systems (NEMS), particularly nanomechanical resonators vibrating at high frequency, are being actively explored for applications including, resonant sensors for ultrahigh-resolution mass sensing, force detection, quantum electromechanics, electromechanical signal generation and processing, *etc*. The development of high frequency silicon nanowire electromechanical resonators has shown a potential for applications in resonant sensing, quantum electromechanical systems, as well as biological sensing.

Therefore, although the direction of applications in this article does not focus on the electrical characterization of nanowires such as FET sensor, the measurement of mechanical property may show a potential direction of applications of utilizing this suspended silicon nanowire. In this article, the basic process of fabricating the suspended Si nanowire on silicon substrates is demonstrated. The measurement of resonant frequency of final fabricated Si nanowire is also carried out. Moreover, the fabricated Si nanowires are to be used as submicrometer local heater with additional electrodes from both sides of nanowire for joule heating. Then, high spatial resolution of surface temperature mapping in dry an even liquid environment using our fluorescent thermometry method [14] will be investigated through these suspended silicon nanowire arrays.

2. Fabrication procedure

The fabrication process is demonstrated with the capability of fabricating suspended nanowires reproducibly. The processes include photolithography, anisotropic wet etching, and local thermal oxidation of silicon and removal of silicon dioxide. Nanowires were obtained by a simple process flow where wire lines with widths of 3 μm were patterned by employing UV lithography and following anisotropic wet etching and thermal oxidation of silicon. The nanowire fabrication was performed on n-type (100)-oriented substrates being characterized by a specific resistance of 1-10 Ωcm.

Flow chart of the suspended silicon nanowire fabrication process is shown in Figure 1. The silicon substrates were prepared by depositing a 100-nm-thick silicon nitride layer (Figure 1b) by low-pressure chemical vapor deposition (LPCVD). This silicon nitride layer was used as a masking layer in the silicon anisotropic wet etching and thermal oxidation processes (Figure 1d). Then, the photolithography process is carried out to define the silicon wire patterns in micrometer scale on the silicon substrates (Karlsus aligner) (Figure 1c). The open areas on the silicon substrate obtained simultaneously in the previous process were anisotropically etched to an appropriate depth using KOH solution (40 w%, 60℃). Consequently, silicon columns aligned along the (111) direction were obtained (Figure 1e). Local oxidation of silicon (LOCOS) was then performed at 1100℃ in order to grow SiO_2, which was used as etching mask in the 2nd KOH solution for the reduced silicon wire structures, later (Figure 1f-i). The 2nd photolithography was carried out to cover and protect part of the structure (Figure 1g). After selective etching of silicon dioxide layer, additional KOH solution etching was carried out to release silicon wires from the substrate (Figure 1h-j). Finally, the rest silicon dioxide and silicon nitride was removed by BHF solution (Figure 1k).

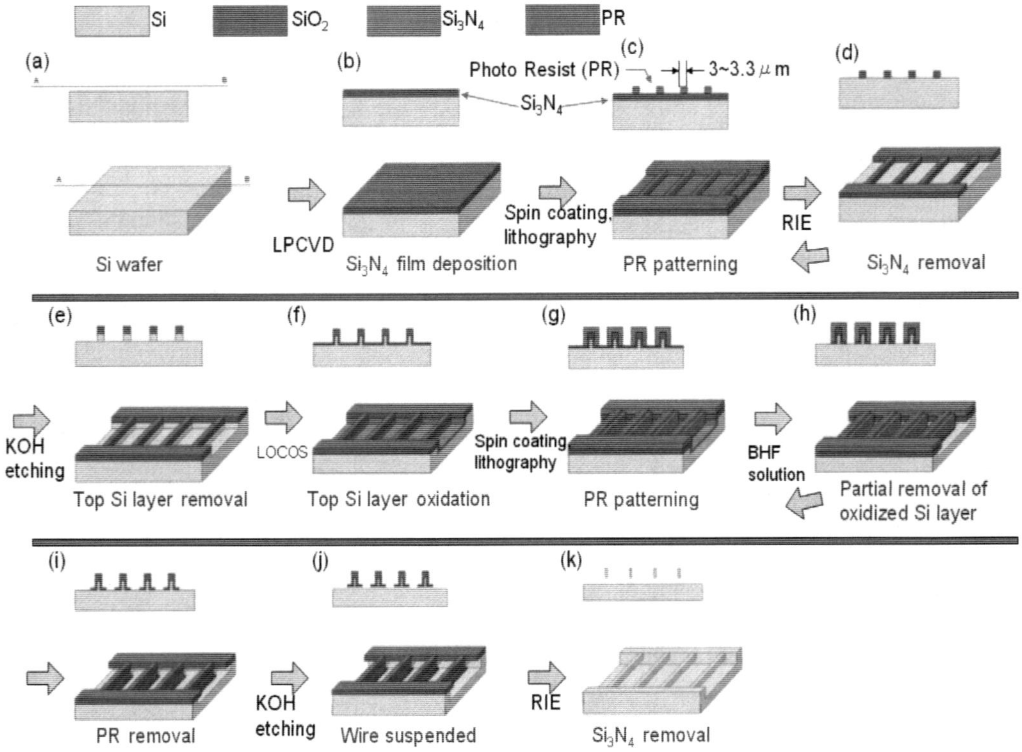

Figure 1. The schematic view of fabrication process for the suspended silicon nanowires. (a-b) preparation of the silicon substrates, (c-d) silicon columns definition by lithography and RIE, (e-f) determination of width of wires, (g) partially protecting wire structures, (h) selective etching of silicon dioxide to produce silicon dioxide mask, (i-j) determination of thickness of wires, (k) removal of rest silicon dioxide and silicon nitride.

3. Results

3.1. Dimensions of suspended Si nanowires

The dimensions of suspended silicon nanowires are decided in three processes, which are the first anisotropic KOH solution etching, thermal silicon oxidation and the second anisotropic KOH solution etching process. The field-effect scanning electron microscopy (FE-SEM) images of the suspended Si nanowires are shown in Figure 2.

In this fabrication procedure, three different sorts of length of wires were designed, which are 50, 20 and 10 μm, respectively. Due to the crystallographic characteristics and wet anisotropic etching, the final length of nanowires was shortened to be approximate 44, 10 μm and less (non-suspension). The final length (44 and 10 μm) is shorter than the designed length (50 and 20 μm). The shorten amount were around from 3 to 5 μm at one end of nanowire, as shown in Figure 3. The shortage in length of suspended silicon nanowire mainly comes from the anisotropic 2nd. KOH solution etching process and partially from the 1st. KOH etching process.

The designed width of wires on the mask is about 3 to 3.3 μm. After side-silicon etching of anisotropic KOH solution etching and thermal oxidation processes, the width of fabricated suspended silicon nanowires is expected to be 100 to 300 nm. However, due to the difficulty of keeping KOH solution etching rate stable, the width of fabricated silicon nanowires is approximately 550 nm, which becomes wider than what of was expected. Thickness of nanowire is determined by the duration of 2nd. KOH solution etching (figure 1 (i-j)). In 2nd. KOH solution etching, it was expected to etch silicon for around 5.5 μm to obtain a nanowire with 0.5 μm in thickness. However, the final thickness was around 0.35±0.2 μm. The error was supposed to be from unstable etching rate at the nano-scale reactive surface. A more stable wet etching system is supposed to be necessary in the following fabrication processes.

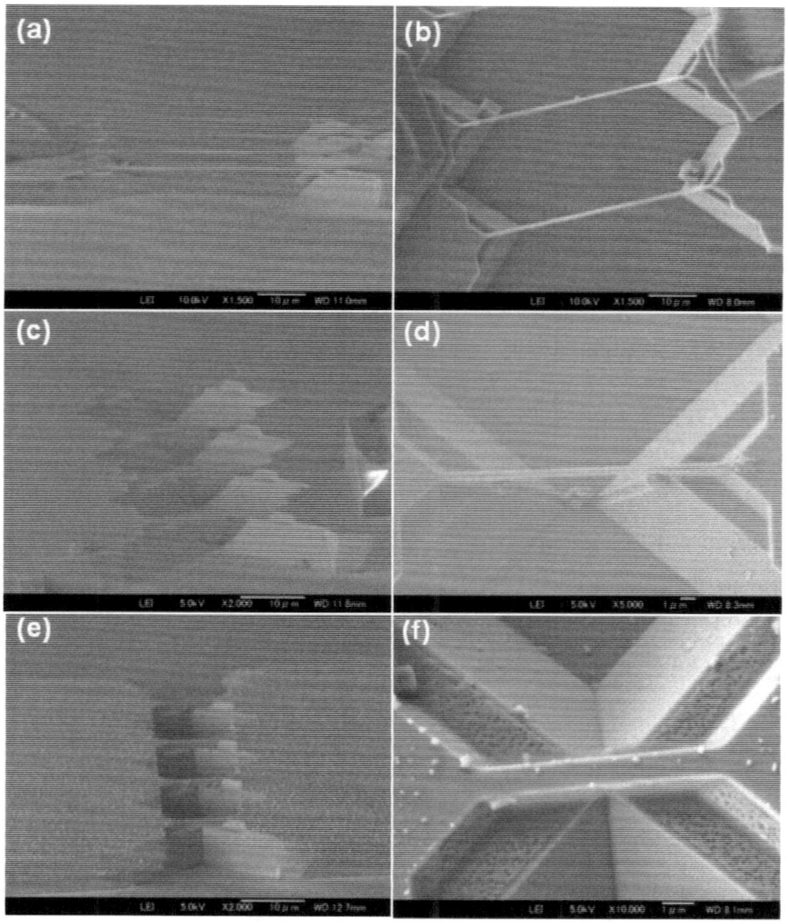

Figure 2. Field-emission scanning electron microscopy (FE-SEM) images of the fabricated suspended silicon nanowires. The designed length is (a-b) 50 μm, (c-d) 20 μm and (e-f) 10 μm. Due to the natural angle between crystallographic planes, the virtual length of suspended silicon nanowire is around 44 μm, 10 μm in the 50-μm and 20-μm cases, respectively. In the 10 μm case, the wires were not suspended because of insufficient space.

However, even using this process at this moment, the fabricated nanowire arrays have very uniform widths and thickness in the same substrates. When SOI substrate, which has already well defined thick silicon on top, is used to fabricate nanowires by our method, it can be more easily obtained to control critical dimension of nanowire arrays.

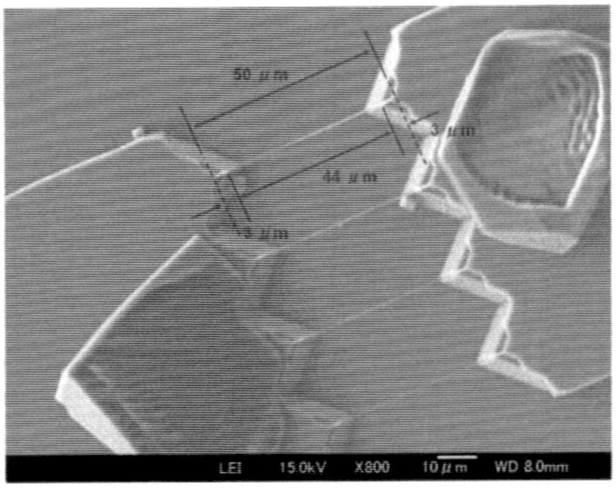

Figure 3. The length of suspended silicon nanowire is 44 μm, which is shorter than the designed length of 50 μm on mask.

3.2. Characterization of nanowires: resonant frequency

These nanowires can be used as mechanical sensors. A rectangular wire suspended across a gap is expected to have a resonance frequency f_n that scales with the dimensions of the wire and with the size of the gap according to equation (1), where the λ_1 is a constant of order unity which equals to 4.73 under the condition of two ends fixed, L is the length suspended which was measured with FE-SEM images, E is Young's modulus which equals to 110GPa, I is the moment of inertia, ρ is density which equals to 2330Kg/m^3 at 298K, and A is cross section area of nanowire. The calculated value of the 1st. mode resonant frequency of the selected wire in 44-μm (Figure 2, a) is 1.96 MHz. The calculated value of the selected 10-μm wire (Figure 2, c) is 39.1 MHz.

$$f_n = \frac{\lambda_n^{\ 2}}{2\pi L^2}\sqrt{\frac{EI}{\rho A}} \tag{1}$$

The measurement were done with a Heterodyne laser Doppler interferometer built by Prof. H. Kawakatsu' group in the University of Tokyo. [16]. The measurement with suspended Si nanowire was carried out in air condition. A HP-8752C Network Analyzer was used to control the frequency of the excitation laser diode. The nanowires were excited by a laser diode and measured with a He-Ne laser beam. The frequency of He-Ne laser light changed after being reflected by the nanowire, which was caused by the vibration of nanowire (Doppler effect). For focusing the laser beams on the nanowires, 50, 20 and 10 times of magnification lens were equipped on the sample stage. A camera and a TV set were installed and used to observe the focusing process through magnification lens. Different optical elements allowed the construction of an interference image around a base frequency of 1.08 GHz. The characteristic interference image was captured by an avalanche photodiode and the resulting electrical signal filtered and demodulated with a delay line. The resulting signal is equivalent to the speed of the cantilever. This signal was connected and transferred into the Network Analyzer that can handle with a frequency range from 300 kHz to 3 GHz. The scanning frequency range can be selected freely in this range. The data can be acquired through the General Purpose Interface Bus (GPIB) of the Network Analyzer with a personal computer running a *Labview* program.

The measurement in liquid (DI water) condition was also tried. Nevertheless, due to the limitation of measurable frequency range (max 5 MHz) of BPF (band-pass filter), the measurement in liquid condition was not available so far.

First resonant frequency of nanowires with 44 and 10 μm in length were measured in air. For the first example, the measured frequency of the 850 nm-in-width, 540 nm-in-thickness and 44 μm-in-length nanowire was 1.64 MHz. For the second example, the measured frequency of the 570 nm-in-width, 670 nm-in-thickness and 10 μm-in-length nanowire was 21.75 MHz. There is 16.3% deviation from calculated value to measured one. The calculated value of the selected 10-μm wire (Figure 2, c) is 39.1 MHz, which showed 44% deviation from calculated value to measured one, as shown in Table 1.

Table 1. Measured dimensions and measured/calculated resonant frequency of nanowires.

Measured length (μm)	Measured width (nm)	Measured thickness (nm)	Measured 1st resonant frequency (MHz)	Calculated 1st resonant frequency (MHz)	Deviation
44	850	540	1.64	1.96	16.30%
10	570	670	21.8	39.1	44.24%

Because of the low uniformity of the thickness of nanowires observed (Figure 4), the 16.3% and 39% difference was supposed to be as a result of low uniformity of thickness. The improvement of wet etching system was also thought to be one of possible approaches to have higher uniformity of wire thickness. The high performance of vibration properties shows the potential of being used as mechanical sensors [12].

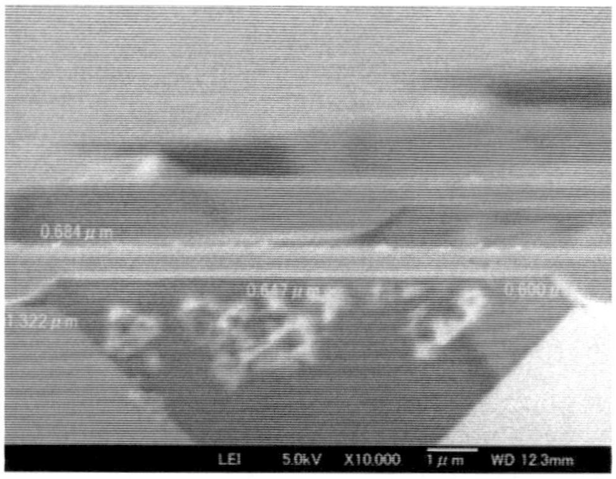

Figure 4. Side view of suspended silicon nanowire (10-μm case) clearly shows the low uniformity of thickness, especially at the two ends.

4. Further study

In various and prosperous fields of silicon nanowire applications, those applications for being used as nano-scale thermometer or sensing nano-scale temperature distribution have been supposed to have high potential in the future research. Fluorescent molecules, such as Rhodamine B or nanocrystals, have been used as monitoring interface to characterize local temperature distribution [13]. Nevertheless, nickel and silicon nanowires used in the previous research were not suspended from substrates, which resulted in obvious heat diffusion from wires to substrates. This phenomenon may cause overheating in the system and unnecessarily waste energy. Hence, our contribution will be utilized to deal with applying suspended nanowires for nanoscale heat source. By utilizing Rhodamine B as temperature monitoring interface, the localized temperature controlling will be carried out. A nano-scale temperature controlling system is also expected to be available for high efficient biosensors. Finally, this nano-scale temperature sensing and controlling system will be expected to be one of the tools to discover part of thermal-mechanical biomolecular physic or biosensors with local temperature control.

5. Summary
A novel fabrication process has been studied via anisotropic KOH solution etching and local thermal oxidation. Relatively uniform and well-defined suspended silicon nanowires were obtained by controlling the duration time of anisotropic wet etching and thermal oxidation. The 1st. mode resonant frequency of one of the 50 μm-length nanowire was measured by laser doppler interferometer way. The measured frequency of this suspended Si nanowire with 850 nm in width and 540 nm in thickness was 1.64 MHz. The measured value is in a good agreement with the calculated one. The result from the resonant frequency measurement of nanowire shows us the potential of utilizing them as nanomechanical resonators or sensors as well. In the near future, they are also supposed to be applied for nano-scale heat source and high spatial resolution mapping of temperature distribution by utilizing fluorescent molecules as monitoring interface for biomolecular research.

Acknowledgments
This research is partially supported by Sharp Corporation, Nara and KAKENHI Kiban C (*Grants-in-Aid for Scientific Research C, 19510126*) from the Japanese Society of Promotion of Science (JSPS). The authors would like to thank Prof. H. Kawakatu and Mr. K. Nakagawa in IIS, The University of Tokyo, for help with the measurement of the resonant frequency and fruitful advices. We also acknowledge the laboratories of Prof. H. Fujita for their support regarding experimental equipment and cleanroom facilities.

References
[1] X. Duan, C. Niu, V. Sahi, J. Chen, J. Parce, S. Empedocles, J. Goldman, *Nature*, **425**, 274 (2003)
[2] Y. Cui, C. M. Lieber, *Science*, **291**, 851 (2001)
[3] A. M. Morales, C. M. Lieber, *Science*, **279**, 208 (1998)
[4] Y. Cui, Q. Wei, H. Park, C. M. Lieber, *Science*, **293**, 1289 (2003)
[5] G. Zheng, F. Patolsky, Y. Cui, W. U. Wang, C. M. Lieber, *Nature Biotech.*, **23**, 1294 (2005)
[6] A. Hochbaum, R. Fan, R. He, P. Yang, *Nano Lett.*, **5**, 457 (2005)
[7] A. M. Morales, C. M. Lieber, *Science*, **279**, 208 (1998)
[8] E. Stern, J. F. Klemic, D. A. Routenberg, P. N. Wyrembak, D. B. Turner-Evans, *Nature*, **445**, 519 (2007)
[9] J. Kedzierski, J. Bokor, C. Kisielowski, *J. Vac. Sci. Technol. B*, **15**, 2825 (1997)
[10] K. Kakushima, T. Watanabe, K. Shimamoto, T. Gouda, M. Ataka, H. Mimura, Y. Isono, G. Hashiguchi, Y. Mihara, H. Fujita, *Japanese Journal of Applied Physics,* **43,** 4041 (2004)
[11] G. M. Kim, S. Kawai, M. Nagashio, H. Kawakatsu, and J. Brugger, *J. Vac. Sci. Technol. B*, **22**, 1658 (2004)
[12] N. A. Melosh, A. Boukai, F. Diana, B. Gerardot, A. Badolato, P. M. Petroff, J. R. Heath, *Science*, **300**, 112 (2003)
[13] H. F. Arata, P. Löw, K. Ishizuka, C. Bergaud, B.J. Kim, H. Noji, H. Fujita, *Sens. Actuators B*, **117**, 339 (2006)
[14] P. Löw, B.J. Kim, N. Takama, and C. Bergaud, Small, DOI 10.1002/smll.200700581, (2008)
[15] G. Zheng, W. Lu, S. Jin, C. M. Lieber, *Advanced Materials*, **16**, 21, 1890 (2004)
[16] S. Nishida, T. Sakurada, D. Kobayashi, S. Meguro and H. Kawakatsu, *Seisannkenkyu*, **58** (2) 161 (2006)

Journal of Physics: Conference Series **152** (2009) 012049 doi:10.1088/1742-6596/152/1/012049

White organic light-emitting diodes with 9, 10-bis (2-naphthyl) anthracene

Yunxia Guan and Lianbin Niu

Key Laboratory of Optical Engineering, College of physics and information technology, Chongqing Normal University, Chongqing 400047, China

E-mail: gyxybsy@126.com, niulb03@126.com

Abstract. White organic light-emitting diodes were fabricated by 9, 10-bis (2-naphthyl) anthracene (ADN) doped with Rubrene with a structure of ITO/copper phthalocyanine (CuPc) / NPB /ADN: Rubrene /Alq_3 /CsF/Mg:Ag/Ag. Multilayer organic devices using AND and Rubrene as an emitting layer produced white emissions with good chromaticity and luminous efficiency as high as 5.93 cd/A. This performance can be explained by Förster energy transfer from the blue-emitting host to the orange-emitting dopant.

1. Introduction

Organic light emitting diodes (OLEDs) have received considerable attention in recent years because of their scientific and commercial significance [1-3]. A number of strategies have been reported to make OLEDs emit white light. These include multilayer OLED structure consisting of two or more active layers [4], doping of appropriate amount of dopants in the same host [5], superposition of differently colored LEDs and smart blends [6], use of exciplex formation [7], etc. Lei et al. [8] suggested WOLED consisting of one or two emissive layers with phosphorescent dyes. However, despite their potential capabilities, no blue phosphorescent materials have yet been considered as a proper candidate for WOLED (White organic light emitting diode) due to their short lifetime and high energy gap, which causes difficulty in finding a proper host material. Efficient blue florescent materials with a long lifetime and high efficiency can be a proper candidate for WOLED using a multilayer structure [9, 10]. Several approaches to high-efficiency white OLEDs have been demonstrated, foremost among them introduces a blue emissive fluorescent dopant together with red and green emissive phosphorescent dopants in the emission layer [11, 12]. In that device, the blue fluorescent dopant harvests a majority of singlet excitons, with the remainder of lower energy triplets diffusing through the conductive host to directly excite the green and red phosphors. This architecture allows for increased power efficiency by resonant energy transfer from the conductive host into both the singlet and triplet energy levels.

In this paper, we reported our efforts at developing high efficiency white OLED with 9, 10-bis (2-naphthyl) anthracene (ADN). The white OLED structure is ITO/copper phthalocyanine (CuPc) (5 nm)/ NPB (55 nm)/ADN: 0.6% Rubrene (35 nm)/Alq_3 (55 nm)/CsF/Mg:Ag/Ag. The bright white emission is composed of an orange band from Rubrene and a blue emission from ADN. A maximum luminance of 9300 cd/ m^2 at the direct current driving voltage of 21 V, with a current efficiency of 5.93 cd/A at about 140 mA/cm^2 is achieved. The Commission Internationale de L'Eclairage (CIE) coordinates of the white emission are well within the white zone when the applied voltage is varied from 8 V to 20 V.

2. Experimental details

The molecular structure of the organic materials studied and a schematic description of the device are shown in figure 1. The device were built on ITO (sheet resistance 30Ω/□) coated glass slides, which were thoroughly cleaned. The routine cleaning procedure included ultrasonic agitation in acetone, ethanol, rinsing in de-ionized water, and isopropyl alcohol, drying with nitrogen gas and finally irradiation in oxygen plasma in a chamber. After the oxygen plasma treatment, the substrates were loaded into a vacuum chamber with a base pressure of 2.4×10^{-3} Pa. CuPc, NPB, ADN: Rubrene, Alq_3, CsF, and Mg:Ag/Ag electrode was thermally deposited without breaking the vacuum, respectively. The deposition rates were 0.01 nm/s, 0.2 nm/s, and 0.4 nm/s for CsF, the organic films and electrode layers, respectively. The deposition rate and film thickness are measured *in situ* by a quartz oscillator connected to a frequency meter near the substrates during the deposition. The thickness of CuPc, NPB, ADN: Rubrene, Alq_3, and Mg:Ag/Ag were 5 nm, 55 nm, 35 nm, 55nm, and 100nm, respectively. A computer-controlled system was used to measure the luminance–current–voltage (L–I–V) curves and EL spectra of the device. All the measurements were carried out at room temperature under ambient atmosphere.

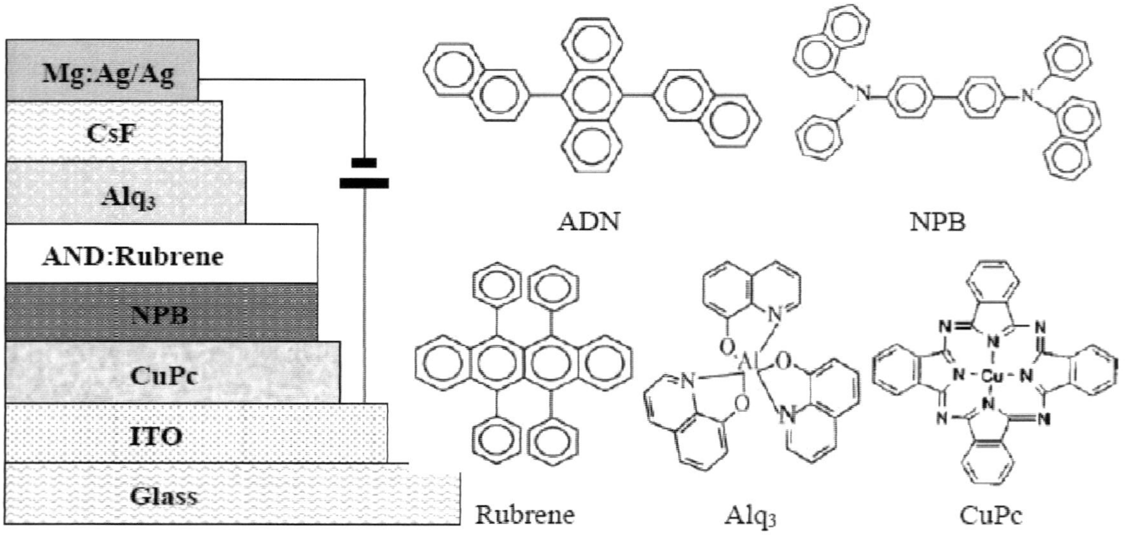

Figure 1. A schematic description of the device structure and molecular structure of the materials studied.

3. Results and discussion

In the forward biased conditions, the current increased super linearly with increasing applied voltage, and the current retains a small value in the reverse bias case in figure 2 (a). Light was emitted at above 6.0 V in the positive bias conditions for the device. Electroluminescence (EL) intensity increases with increasing injection current in the forward bias conditions in figure 2 (b). For instance, the brightness of the device was 9300 cd/m^2 when the voltage was 21 V.

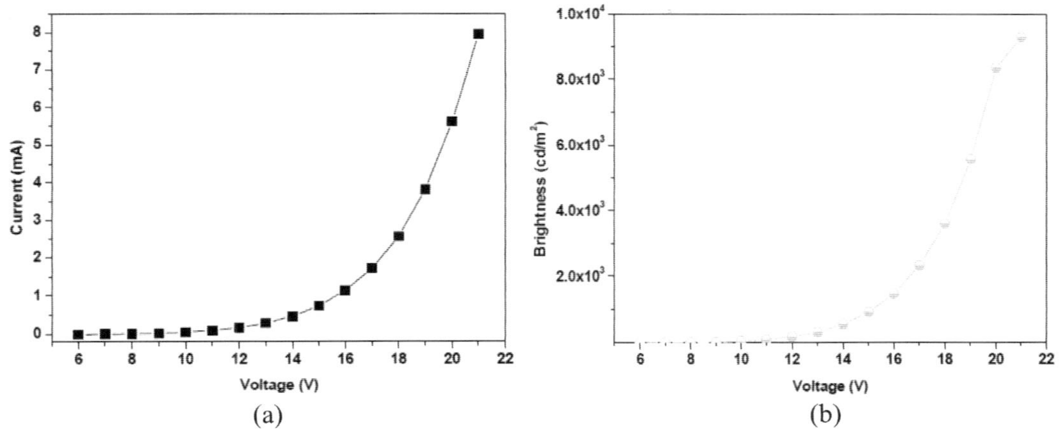

Figure 2. (a) Current density versus voltage characteristics of the device and (b) Luminance versus voltage characteristics of the device.

Figure 3 shows the normalized EL spectra of the device at different applied voltages. All the emission spectra shapes are the same and the CIE coordinates are 0.32 and 0.40. It is shown that the white light emission keep relatively stable very well with increasing bias voltage. We can observe that the EL spectrum of the white OLED including two primary emissive peaks. One is 453 nm, the other is 550 nm. Obviously, the short peak and long peak are originated from the ADN and Rubrene, respectively. It should be mentioned that the mix of the blue emission and orange emission could obtain white emission if the ration of the two-color emission is correct. So the white light emission of this device should be attributed to the mixture of the ADN emission and Rubure emission. In fact, the absorption spectrum of Rubrene overlaps to a certain extent with the emission spectrum of ADN in our preceding experiment. This means that it might be a Förster energy transfer from ADN to Rubrene when the ratio of Rubrene in ADN is well controlled. It can be concluded that the EL spectra of the devices with the white light emission due to the energy transfer from the blue-emitting host to the orange-emitting dopant.

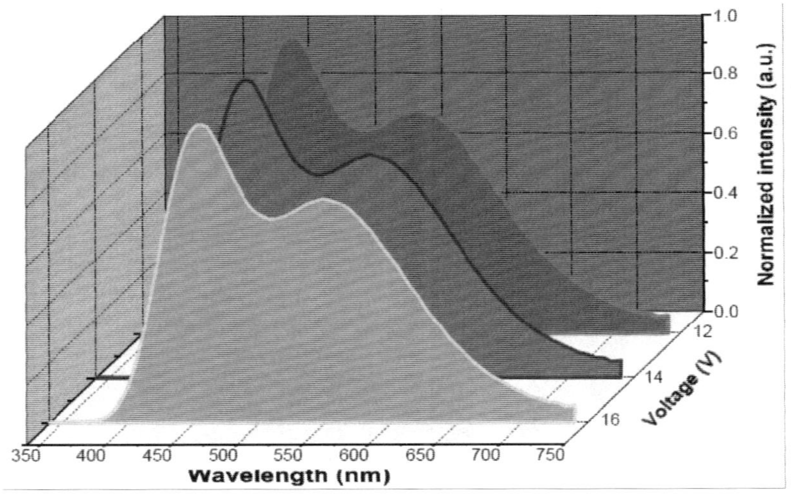

Figure 3. Normalized EL spectra of the device at different applied voltages.

Figure 4 shows the dependence of the current efficiency on the drive current density for the white OLED. The efficiency rises sharply at low current densities to a maximum of about 5.93 cd/A at about 140 mA/cm^2, then falls off slowly at higher current densities. Up to about 200 mA/cm^2, the efficiency is maintained near 4.67 cd/A.

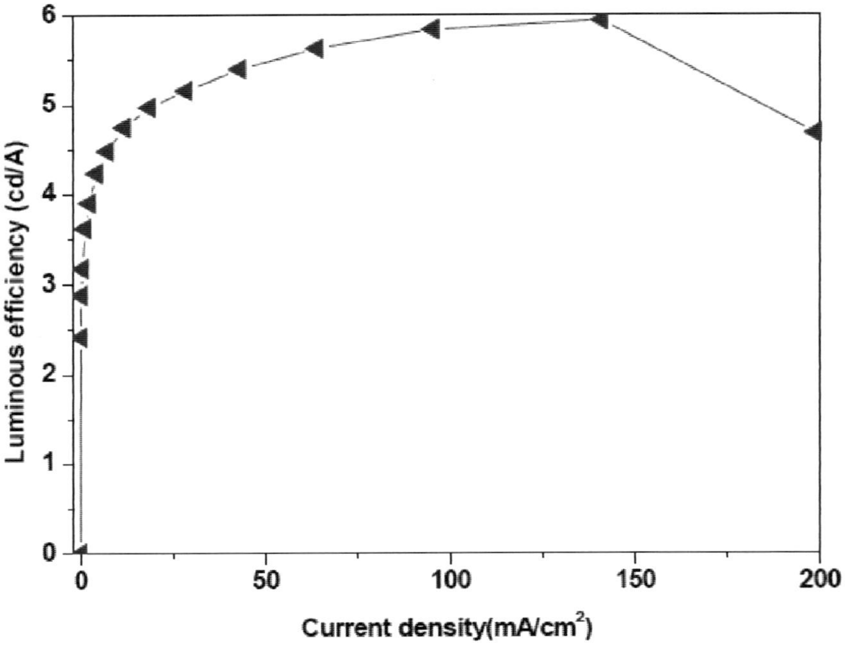

Figure 4. Efficiency-current density characteristics of the device.

4. Conclusion

In summary, we successful fabricated a high efficiency white OLED based on ADN with a simple approach. A bright white emission is achieved using a thin layer of ADN doped with Rubrene as the source of white emission. The efficiency rises to a maximum of about 5.93 cd/A when the current intensity is 140 mA/cm^2. The brightness of the device is 9300 cd/m^2 when the voltage is 21 V. The results show that the high white emission for this OLED is due to energy transfer between ADN and Rubrene.

Acknowledgements

This work is supported by the Research Programs for National Natural Science Foundation of China (No. 60806047), Research Program for Science and Technology of Chongqing Municipal Education Commission (No. KJ080816) and Natural Science Foundation of Chongqing Normal University (No. 07XLB015).

References

[1] Tang C W and VanSlyke S A 1987 *Appl. Phys. Lett.* **51** 913
[2] Vadim A, Jason B, Arnold T, Alex M A, Peter I D, Brian W D, Chihaya A, Stephen R F and Mark E T 2002 *New J. Chem.* **26** 1171
[3] DAndrade B W, Stephen R F 2004 *Adv. Mater.* **16** 1585
[4] Deshpande R S, Bulovic V and Stephen R F 1999 *Appl. Phys. Lett.* **75** 888
[5] Kido J, Shionoya H and Nagai K 1995 *Appl. Phys. Lett.* **67** 2281

[6] Thompson J, Blyth R I R, Mazzeo M, Anni M, Gigli G and Cingolani R 2001 *Appl. Phys. Lett.* **79** 560
[7] Giroa G, Cocchia M, Kalinowskib J, Marcoa P D and Fattori V 2000 *Chem. Phys. Lett.* **318** 137
[8] Gangtie L, Liduo W and Yong Q 2006 *Appl. Phys. Lett.* **88** 103508
[9] Wen S, Lee M and Chen C H 2005 *J. Display Technol.* **1** 90
[10] Liu G and Nazare H 2000 *Synth Met.* **111** 47
[11] Sun Y, Giebink N and Ma B 2005 *SPIE* (Organic Light Emitting Materials and Devices IX, San Diego) **59** 37
[12] Yiru S, Noel C G, Hiroshi K, Biwu M, Mark E T and Stephen R F 2006 *Nature* **440** 908

Investigation on photoluminescence properties of Eu3+-doped Sr2Mg1-xZnxSi2O7 (0≤x≤1) in UV-VUV region

Zhiya Zhang [1], Yuhua Wang [1, 2 *], Hailong Wang [1], Zhifeng Sun [1], Linyu Jia [1]

[1] School of Physical Science and Technology, Lanzhou University, Lanzhou, 730000, P.R. China.

[2] Key Laboratory for Magnetism and Magnetic Materials of the Ministry of Education, Lanzhou University, Lanzhou, 730000, P.R. China.

E-mail address: wyh@lzu.edu.cn

Abstract. Undoped and Eu^{3+}-doped $Sr_2Mg_{1-x}Zn_xSi_2O_7$ ($0 \leq x \leq 1$) powder crystal samples were obtained via a conventional solid-state reaction method. The XRD result and FT-IR spectra demonstrated the complete solid-solution as well as lattice distortion between $Sr_2MgSi_2O_7$ and $Sr_2ZnSi_2O_7$. The UV emission spectra showed the main emission around 616nm, which implied the non-inversion symmetry around Eu^{3+} in $Sr_2Mg_{1-x}Zn_xSi_2O_7:Eu^{3+}$ ($0 \leq x \leq 1$). The UV excitation spectra exhibited the Eu^{3+}-O^{2-} CT transitions centered at 250nm and 258nm in $Sr_2MgSi_2O_7:Eu^{3+}$ and $Sr_2ZnSi_2O_7:Eu^{3+}$, respectively. The emission of $Sr_2(Mg, Zn)Si_2O_7:Eu^{3+}$ under 147nm light excitation is very weak, which suggests that energy transfer from the host to the activator is not so efficient in $Sr_2Mg_{1-x}Zn_xSi_2O_7:Eu^{3+}$ ($0 \leq x \leq 1$) system.

1. Introduction

It is well known that Zn^{2+} is a typical transition metal ion with its $3d^{10}$ electron configuration. Despite the proximate ionic radii of Zn^{2+} (74pm, coordination number (C.N.) =4) and Mg^{2+} (71pm, C.N. =4) and the same valence state, the crystal structure usually changes in a great disparity when the substitution occurs between Zn^{2+} and Mg^{2+}. One of the typical examples is Mg_2SiO_4-Zn_2SiO_4 system: Mg_2SiO_4 exhibits an orthorhombic "forsterite" structure and Zn_2SiO_4 has a rhombohedral "willemite" structure; Mg^{2+} occupies an octahedral site in Mg_2SiO_4 and Zn^{2+} has a tetrahedral coordination in Zn_2SiO_4; the solid solution content limits of Zn^{2+} in Mg_2SiO_4 and Mg^{2+} in Zn_2SiO_4 were both observed to be as small as 5 at% [1]. As for $MgAl_2O_4$-$ZnAl_2O_4$ system, the band gap of $MgAl_2O_4$ was reported to be 7.8ev [2] while that of $ZnAl_2O_4$ was obtained to be a much smaller value of 3.8-3.9ev [3]. $Sr_2MgSi_2O_7$ and $Sr_2ZnSi_2O_7$ are very important hosts for photoluminescence (PL) materials and long afterglow phosphors [4-7]. They both belong to tetragonal akermanite-type structure with the space group $P\overline{4}21m$ (no. 113) and can form complete solid solution with each other [8]. In our recent work [9], we found an interesting result that $Sr_2ZnSi_2O_7:Mn^{2+}$ showed a strong green emission while $Sr_2MgSi_2O_7:Mn^{2+}$ had no emission under both 254nm and 147nm excitations. Since the two substances have the same type of crystal structures, the result might be correlated with their particular electronic structures. Eu^{3+} ion, possessing the $4f^6$ electronic configuration, is usually employed as a structure probe due to its particularly sensitive emission properties depending on the occupied site symmetry in the crystal structure [10-12].

In the present work, undoped and Eu^{3+}-doped $Sr_2Mg_{1-x}Zn_xSi_2O_7$ ($0 \leq x \leq 1$) were prepared and the XRD (X-ray diffraction), FT-IR (Fourier transform infrared) and PL spectroscopy methods were applied to investigate the crystal structures and PL characteristics of the as-prepared samples.

2. Experimental

Undoped and Eu-doped (5 mol%) $Sr_2Mg_{1-x}Zn_xSi_2O_7$ ($0 \leq x \leq 1$) powder polycrystalline samples were prepared by solid-state reaction method. $SrCO_3$ (99%), $Mg(NO_3)_2 \cdot 6H_2O$ (99%), ZnO (99%), H_2SiO_3 (99%), and Eu_2O_3 (99.99%) were employed as the starting materials. A stoichiometric amount of the above raw materials were weighed and mixed together in an agate mortar. Then ground down to fine powder before sintering at 1220℃ for 4h.

The XRD patterns were obtained on a Rigaku D/max-2400 X-ray diffractometer. The excitation and emission spectra were measured using an Edinburgh Instruments FLS920T. The scan speed was 30nm/m with a step of 1nm and the dwell time was 0.2s. The ultraviolet (UV) spectra were measured by steady-state spectrophotometer with Xe900 (450 W xenon arc lamp) as excitation source. The slit was 0.18nm for the excitation and 0.8nm for the emission spectra. The vacuum ultraviolet (VUV) light source of the spectrometer system was a 150W Deuterium lamp (Cathodeon Incorporated). The emission and excitation spectra were measured by the vacuum monochromator (VM504, Acton Research Corporation, ARC). The slits for the VUV excitation and emission spectra were 0.18nm and 2nm, respectively. The FT-IR spectra were performed on a Thermo Mattson Nicolet Sattellite FT-IR spectrophotometer using KBr disc technique. All measurements were carried out at room temperature.

3. Results and discussion

3.1. Sample preparation

All undoped and Eu^{3+}-doped $Sr_2Mg_{1-x}Zn_xSi_2O_7$ ($0 \leq x \leq 1$) samples appeared as single phase of tetragonal akermanite-type structure. The XRD patterns for the samples of $Sr_2Mg_{1-x}Zn_xSi_2O_7$ ($0 \leq x \leq 1$) with x=0, x=0.5 and x=1 as shown in Fig. 1 exhibits similar profile. The observed XRD peak positions almost exhibit no shift along with the x values changing. It is due to the similar ionic radii for Mg^{2+} and Zn^{2+} ions and the same type of crystal structures for the two end-members of $Sr_2MgSi_2O_7$ and $Sr_2ZnSi_2O_7$. However, the diffraction peaks for the (110) and (001) crystal planes lose their intensity gradually from x=0 to x=0.5 and almost show no intensity when x>0.5. This result seems to imply that structure adjustment takes place from $Sr_2MgSi_2O_7$ to $Sr_2ZnSi_2O_7$.

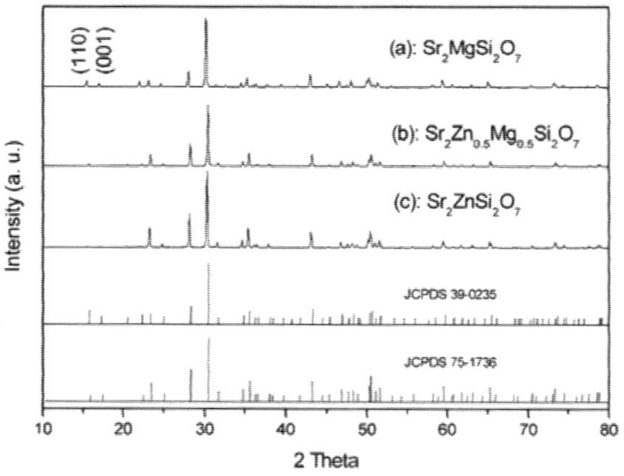

Figure 1. XRD patterns for $Sr_2Mg_{1-x}Zn_xSi_2O_7$: (a) x=0; (b) x=0.5; (c) x=1.

The FT-IR spectra in Fig. 2 further confirm the complete solid-solution between $Sr_2MgSi_2O_7$ and $Sr_2ZnSi_2O_7$. The FT-IR spectra for $Sr_2MgSi_2O_7$ and $Sr_2ZnSi_2O_7$ in Fig. 2 exhibit similar features. The FT-IR υ 1005.81, 967.18, 924.13, 839.13 cm^{-1} for $Sr_2MgSi_2O_7$ and 1006.96, 968.53, 901.16, 827.81 cm^{-1} for $Sr_2ZnSi_2O_7$ are due to the (Si-O_b-Si) and (Si-O_{nb}) strenching modes [13-16], the FT-IR υ 670.25, 617.31, 566.39 cm^{-1} for $Sr_2MgSi_2O_7$ and 666.37, 599.87 cm^{-1} for $Sr_2ZnSi_2O_7$ arise the (Si-O-Si) bending modes [13-16], and the FT-IR υ 472.15 cm^{-1} for $Sr_2MgSi_2O_7$ and 482.18, 452.15 cm^{-1} for $Sr_2ZnSi_2O_7$ are based on the (Si-O-Si) bending modes as well as the M-O (M=Mg, Zn) modes [14-15, 17]. The change in the υ values also indicates structure distortion between $Sr_2MgSi_2O_7$ and $Sr_2ZnSi_2O_7$. In $Sr_2MgSi_2O_7$ and $Sr_2ZnSi_2O_7$, Mg^{2+} and Zn^{2+} both occupy the tetrahedral sites [8]. It has been stated that Zn^{2+} is more easily accommodated in tetrahedral sites than Mg^{2+} according to Goodenough et al.'s work [18]. The same conclusion could be drawn in $Sr_2Mg_{1-x}Zn_xSi_2O_7$ ($0 \le x \le 1$) system based on the XRD result and the FT-IR spectra. As a result, $Sr_2ZnSi_2O_7$ should exhibit a more rigid crystal structure than $Sr_2MgSi_2O_7$.

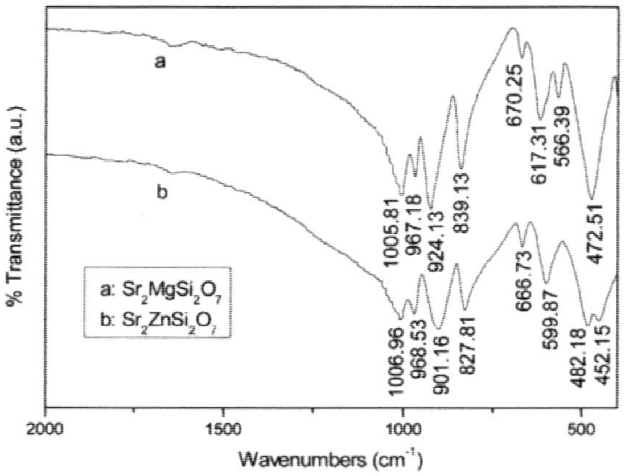

Figure 2. FT-IR spectra for $Sr_2Mg_{1-x}Zn_xSi_2O_7$: (a) x=0; (b) x=1.

3.2. PL properties

Fig. 3 shows the excitation spectra of $Sr_2MgSi_2O_7:Eu^{3+}$ and $Sr_2ZnSi_2O_7:Eu^{3+}$. The excitations around 250-260nm derive from the Eu^{3+}-O^{2-} charge transfer (CT) transitions and the peaks between 310-450nm are from the Eu^{3+} *f-f* transitions. As can be seen, the Eu^{3+}-O^{2-} CT position exhibits a red shift of up to 8nm from $Sr_2MgSi_2O_7:Eu^{3+}$ to $Sr_2ZnSi_2O_7:Eu^{3+}$.

The Eu^{3+}-O^{2-} CT shift could be explained by the different electronic configuration of the Mg^{2+} and Zn^{2+}. The O(2*p*)-Zn(3*d*) hybridization and the effect on the electronic structures have also been reported in many other systems [18-21]. In Ref. [20], role of metal *d* states were investigated and it was stated that the *p-d* repulsion and hybridization lower the band gaps and increase the valence band offset. The Eu^{3+}-O^{2-} CT position shift in the present work could also be well explained from this point of view. Eu^{3+}-O^{2-} CT transition starts from the top of valence band and ends in the Eu^{2+} ground state [22]. In $Sr_2ZnSi_2O_7:Eu^{3+}$, O(2*p*)-Zn(3*d*) repulsion and hybridization enhance the valence band offset and intensify the coupling between the Eu^{2+} ground state and the top of valence band. Once the coupling is intensified, the Eu^{3+}-O^{2-} CT energy decreases correspondingly.

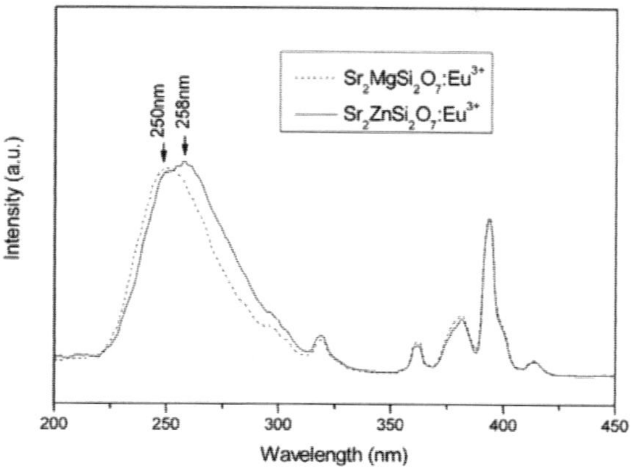

Figure 3. Excitation spectra of $Sr_2Mg_{1-x}Zn_xSi_2O_7:Eu^{3+}$: x=0 (dotted line); x=1 (solid line). (λ_{em}=616nm).

Fig. 4 (a) exhibits the typical emission spectra for $Sr_2Mg_{1-x}Zn_xSi_2O_7:0.05Eu^{3+}$ ($0 \le x \le 1$) with x=0, x=0.5 and x=1. The emissions are dominated by the electric dipole transition $^5D_0$-$^7F_2$ (616nm), which indicates that Eu^{3+} occupy the sites of non-inversion symmetry. In addition, the host emission around 390nm (as indicated in Fig. 4 (b)) can still be observed in the emission spectra. It indicates that energy transfer from the host to the activator Eu^{3+} is not efficient enough despite that there is considerable overlap between the host emission and the Eu^{3+} excitation as indicated in Fig. 4.

We calculate the chromaticity coordination and other main data of the optical properties with Zolix Color Convert program as listed in Table I. From the chromaticity coordinate change, it can be found that the red emission gets higher color purity from $Sr_2ZnSi_2O_7:Eu^{3+}$ to $Sr_2MgSi_2O_7:Eu^{3+}$. It is well known that the pure red emission derives from the hypersensitive $^5D_0$-$^7F_2$ transition of Eu^{3+}, which will obtain strong emission intensity when the crystalline lacks an inversion center [23]. So the chromaticity coordination change demonstrate that Eu^{3+} have a more symmetrical local environment in $Sr_2ZnSi_2O_7:Eu^{3+}$. The result is in a good agreement with that previously discussed in XRD and FT-IR spectra in section 3.1.

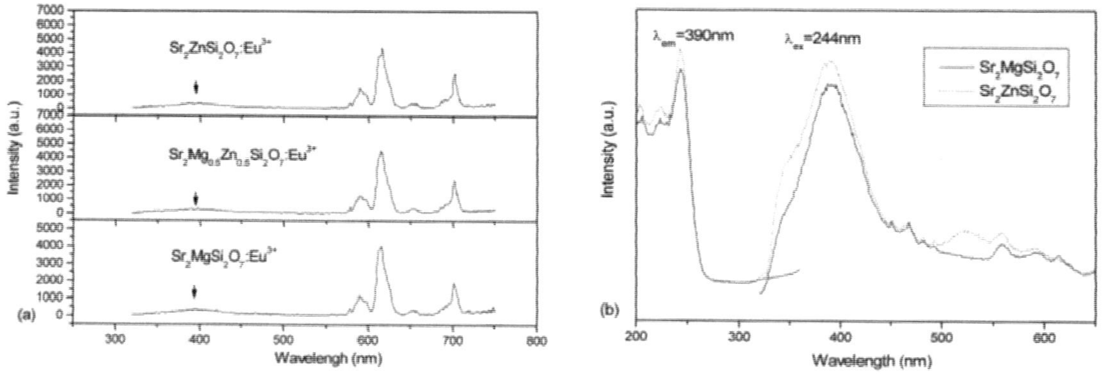

Figure 4. (a): emission spectra for $Sr_2Mg_{1-x}Zn_xSi_2O_7:Eu^{3+}$ (λ_{ex}=250-258nm); (b): excitation (λ_{em}=390nm) and emission (λ_{ex}=244nm) spectra for $Sr_2MgSi_2O_7$ and $Sr_2ZnSi_2O_7$.

Table 1. The main optical properties of $Sr_2Mg_{1-x}Zn_xSi_2O_7$:0.05Eu^{3+} (0≤x≤1).

Samples	Chromaticity coordinate	Emission peak (nm)
x=0	(0.647, 0.352)	615
x=0.25	(0.643, 0.356)	615
x=0.5	(0.640, 0.360)	616
x=0.75	(0.643, 0.366)	616
x=1	(0.634, 0.366)	616

The host absorption band around 147nm was observed in the VUV excitation spectra of $Sr_2ZnSi_2O_7$:Mn^{2+} [9], Sr_2(Mg, Ca)Si_2O_7:Eu^{2+} [24] and Sr_2(Mg, Ca)Si_2O_7:Eu^{2+}, Dy^{3+} [25]. We measured the PL spectra of Sr_2(Mg, Zn)Si_2O_7:Eu^{3+} under 147nm light excitation and found that the emission is very weak (the spectra are not shown here). The result suggests that energy transfer from the host to the activator is not so efficient in $Sr_2Mg_{1-x}Zn_xSi_2O_7$:Eu^{3+} (0≤x≤1) system under 147nm excitation and $Sr_2Mg_{1-x}Zn_xSi_2O_7$:Eu^{3+} (0≤x≤1) is not suitable to be used as VUV-applied phosphor.

4. Conclusions

$Sr_2Mg_{1-x}Zn_xSi_2O_7$ (0≤x≤1) and $Sr_2Mg_{1-x}Zn_xSi_2O_7$:Eu^{3+} (0≤x≤1) powder crystal samples were prepared. The complete solid-solution as well as the lattice distortion between $Sr_2ZnSi_2O_7$ and $Sr_2MgSi_2O_7$ was demonstrated based on the XRD and FT-IR results. The UV emission spectra showed the main emission of $^5D_0$-$^7F_2$ electric dipole transition around 616nm, which implied the non-inversion symmetry around Eu^{3+} in $Sr_2Mg_{1-x}Zn_xSi_2O_7$:Eu^{3+} (0≤x≤1). The observed host emission in the emission spectra implied the low energy transfer efficiency from the host to the activator Eu^{3+}. The UV excitation spectra presented the Eu^{3+}-O^{2-} CT position shift of up to 8nm from $Sr_2MgSi_2O_7$:Eu^{3+} to $Sr_2ZnSi_2O_7$:Eu^{3+}, which were correlated with the effect of Zn^{2+} on the electronic structure. The VUV PL properties indicated that $Sr_2Mg_{1-x}Zn_xSi_2O_7$:Eu^{3+} (0≤x≤1) was not suitable to be used as VUV-applied phosphor.

Acknowledgements
This work was supported by the Program of New Century Excellent Talents in University of China (NCET, Grant No.04-0978) and the Project of the Combination of Industry and Research by the Ministry of Education and Guangdong Province of China (Grant No.0712226100023).

References

[1] Song K.X, Chen X.M, Zheng C.W, 2008 *Ceram. Int.* **34** 917.
[2] Bortz M.L, French R.H, Jones D.J, Kasowski R.V, Ohuchi F.S, 1990 *Phys. Scr.* **41** 537.
[3] Sampath S.K, Cordaro J.F. 1998 *J. Am. Ceram. Soc.* **81** 649.
[4] Lin Y, Tang Z, Zhang Z, Wang X, J Zhang, 2001 *J. Mater. Sci. Lett.* **20** 1505.
[5] Chen Y, Liu B, Kirm M, Qi Z, Shi C, True M, Vielhauer S, Zimmerer G, 2006 *J. Lumin.* **118** 70.
[6] Lin Y, Nan C.W, Zhou X, Wu J, Wang H, Chen D, Xu S, 2003 *Mater. Chem. Phys.* **82** 806.
[7] Wang X.J, He Z.Y, Jia D, Strek W, Pazik R, Hreniak D, Yen W.M, 2005 *Microelectron. J.* **36** 546.
[8] Jiang L, Chang C.K, Mao D.L, Feng C.L, 2004 *Opt. Mater.* **27** 51.
[9] Hao Y, Wang Y.H, 2007 Mater. Res. Bull. **42** 219.
[10] Yang H.C, Li C.Y, He H, Tao Y, Xu J.H, Su Q, 2006 *J. Lumin.* **118** 61.
[11] Wang L, Wang Y, 2006 *Mater. Lett.* **60** 2645.
[12] Gou J, Wang Y, Li F, 2007 *J. Lumin.* **127** 327.
[13] Frost R.L, Bouzaid J.M, Reddy B.J, 2007 *Polyhedron* **26** 2405.
[14] Chandrappa G.T, Ghosh S, Patil K.C, 1999 *J. Mater. Syn. Process.* **7** 273.
[15] Makreski P, Jovanovski G, Kaitner B, Gajovic A, Biljan T, 2007 *Vib. Spectrosc.* **44** 162.
[16] S Abo-Naf.M, El Batal F.H, Moenis A. Azooz, 2002 *Mater. Chem. Phys.* **77** 846.
[17] Caracas R, Gonze X, 2003 *Phys. Rev. B* **68** 184102-1.

[18] Goodenough J.B, Loeb A.L, 1955 *Phys. Rev.* **98** 391.
[19] Göpel W, Pollmann J, Ivanov I, Reihl B, 1982 *Phys. Rev. B* **26** 3144.
[20] Wei S.H, Zunger A, 1988 *Phys. Rev. B* **37** 8958.
[21] Zhang H, Feng X, Kang J.Y, 2006 *Chinese J. Lumin.* **27** 750.
[22] Li L, Zhang S, 2006 *J. Phys. Chem. B* **110** 21438.
[23] Shionoya S, Yen W.M, 1999 Phosphor Handbook, CRC Press, Boca Raton, p. 190.
[24] Chena Y, Liu B, Kirm M, Qi Z, Shi C, True M, Vielhauer S, Zimmerer G, 2006 *J. Lumin.* **118** 70.
[25] Li Y. Z, Wang Y.H, Z. Zhang H, Liu W.J, 2006 *Electrochem. Solid-State Lett.* **9** J37.

The Photoluminescent Quenching Behaviors of $Na_3La_9O_3(BO_3)_8$:Tb^{3+} Phosphors

Jiachi Zhang[1], Yuhua Wang[1,2 *], Peng Xie[1]

[1]Lanzhou University, Lanzhou, 730000, P.R. China.
[2]Key Laboratory for Magnetism and Magnetic Materials of the Ministry of Education, Lanzhou University, Lanzhou 730000, P. R. China

E-mail address: wyh@lzu.edu.cn

Abstract. The quenching behaviours of $^5D_4$-$^7F_6$ emission of Tb^{3+} in $Na_3La_9O_3(BO_3)_8$:Tb^{3+} under 125-300nm excitation were systematically investigated. The results revealed that its luminescent quenching concentrations excited at particular wavelengths were dependent on corresponding excitation bands. On analyzing the luminescent quenching characteristics, all these spin-orbit 4f-5d transitions of Tb^{3+} and host absorption band in $Na_3La_9O_3(BO_3)_8$:Tb^{3+} were identified in excitation spectrum and the energy levels scheme of Tb^{3+} in $Na_3La_9O_3(BO_3)_8$:Tb^{3+} system was first established.

1. Introduction

The emission intensity of many phosphors as a function of activator concentration initially increases with activator concentration and then decreases, going through a maximum at some concentration. The decreases in emission intensity at a certain activator concentration are referred to as the phenomenon of "concentration quenching" [1,2]. And the optimal activator concentration with the highest emission intensity is considered as "quenching concentration" of phosphors. Generally, more activators in host means more emission centers and higher luminescent intensity, thus the phosphors with high quenching concentration are expected. Moreover, a complete characterization of the emission quenching behaviors in inorganic phosphors is very important not only for technological design but also for the basic understanding of the physical excitation processes [3-7].

On the other hand, the $Na_3La_9O_3(BO_3)_8$:Tb^{3+} was reported as a recently developed green phosphor with high quenching concentration and excellent photoluminescence in ultraviolet region [8-11]. Developing a model to study quenching behaviors in $Na_3La_9O_3(BO_3)_8$:Tb^{3+} was significant to understand its photoluminescence properties, excitation pathways and quenching mechanisms, even developing some novel phosphors with efficient photoluminescence.

It was our fundamental aim to clarify the quenching and excitation mechanism of $Na_3La_9O_3(BO_3)_8$:Tb^{3+} phosphor. In this work, the characteristics of quenching concentration in $Na_3La_9O_3(BO_3)_8$:Tb^{3+} under 130-290nm were investigated and the relationship between quenching concentrations and excitation bands was revealed. Based on these results, the host absorption and complete *4f-5d* transitions of Tb^{3+} in $Na_3La_9O_3(BO_3)_8$:Tb^{3+} were successfully identified and the energy levels scheme of Tb^{3+} in $Na_3La_9O_3(BO_3)_8$:Tb^{3+} was established.

2. Experimental

All the $Na_3La_9O_3(BO_3)_8:Tb^{3+}$ samples were synthesized by solid state method. The starting materials were Na_2CO_3 (A.R.), H_3BO_3 (A.R.), La_2O_3 (99.99%) and Tb_4O_7 (99.99%). The stoichiometric amount of reactants was mixed and ground homogeneously with excess of 3% H_3BO_3 to compensate the evaporation. Then, the mixed powders were sintered at 500℃ for 10h. Cooled and reground, then heated at 1100℃ for 24h twice with the intermediate grindings to obtain the phosphors. The excitation and emission spectra were measured at room temperature using an Edinburgh Instruments FLS920T combined The light source of the VUV excitation part of the spectrometer system was a 150W Deuterium lamp (Cathodeon Incorporated) and the emission and excitation spectra were measured by the vacuum monochromator (VM504, Acton Research Corporation, ARC). The VUV excitation spectra were corrected by dividing the excitation intensity of sodium salicylate at the same measurement conditions.

3. Results and discussion

3.1. Photoluminescence at 147nm and 254nm

The 147nm and 254nm was the most typical excitation sources for application in vacuum ultraviolet (VUV) and ultraviolet (UV) region, respectively. Fig.1 showed the emission spectra of typical $Na_3La_9O_3(BO_3)_8:0.1Tb^{3+}$ samples under 147nm and 254nm excitation as representative. As shown in Fig.1, the emission intensity of samples excited at 147nm was much higher than that at 254nm. The main emission peaks at about 476, 543, 570 and 621nm were due to the $^5D_4$-$^7F_6$, $^5D_4$-$^7F_5$, $^5D_4$-$^7F_4$ and $^5D_4$-$^7F_3$ transitions of Tb^{3+} respectively [12] and the $^5D_4$-$^7F_5$ at 543nm was dominant.

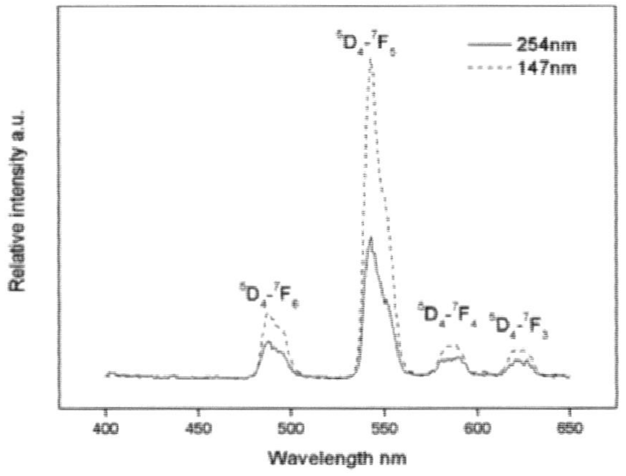

Figure 1. emission spectra of $Na_3La_9O_3(BO_3)_8:Tb^{3+}$ under 147nm and 254nm excitation, respectively.

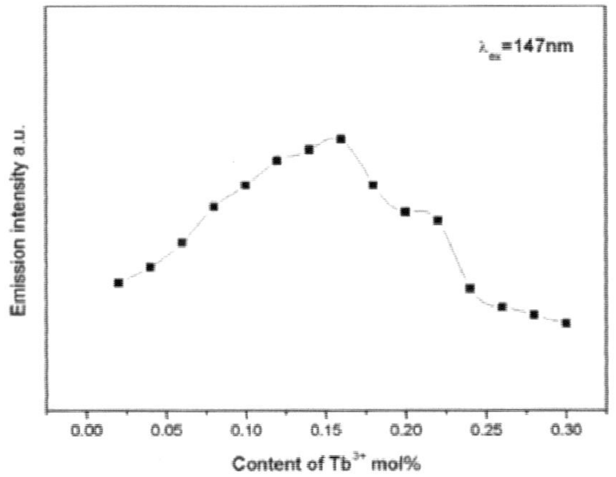

Figure 2. the quenching concentration curves of $Na_3La_9O_3(BO_3)_8:Tb^{3+}$ samples excited at 147nm.

3.2. Quenching characteristics and identification of excitation bands under 125-300nm

In order to clarify the relationship between quenching concentrations and excitation wavelengths, the quenching concentrations of $Na_3La_9O_3(BO_3)_8:Tb^{3+}$ under 125-300nm were investigated systematically. Table I exhibited the quenching concentrations of $^5D_4$-$^7F_5$ transition of Tb^{3+} in $Na_3La_9O_3(BO_3)_8:Tb^{3+}$ samples excited at 125-300nm and it was significant to find that: under 260-290nm excitation, the quenching concentration of $Na_3La_9O_3(BO_3)_8:Tb^{3+}$ was about 14% which differed from the neighbor excitation regions. As shown in excitation spectrum of $Na_3La_9O_3(BO_3)_8:Tb^{3+}$ (Fig.3), this broad excitation band at 255-293nm was also observed and it could be ascribed to the lowest spin-forbidden transition to $^9D_5$ term of Tb^{3+} based on Ref [13-15]. When excited at 220-245nm, the quenching concentrations of $Na_3La_9O_3(BO_3)_8$:Tb^{3+} varied to 10%. The high

Table 1. the quenching concentrations of $^5D_4$-$^7F_5$ transition of Tb^{3+} in $Na_3La_9O_3(BO_3)_8$:Tb^{3+} samples when excited by 125-300nm.

Excitation wavelength /nm	125	130	134	138	142	146	148
Quenching concentrations	26%	24%	16%	16%	16%	16%	16%
Excitation bands			Host absorption				

150	154	156	158	160	162	166	170	172	176
16%	16%	16%	16%	10%	10%	10%	10%	10%	10%
Host absorption				Spin-allowed transitions to ^7D term of Tb^{3+}					

180	182	184	186	188	190	196	198	200	204
10%	10%	10%	10%	10%	10%	10%	10%	10%	10%
Spin-allowed transitions to ^7D term of Tb^{3+}									

206	208	210	212	215	220	230	235	240	245
16%	16%	16%	16%	16%	10%	10%	10%	10%	10%
Spin-forbidden transitions to ^9D term of Tb^{3+}					Spin-allowed transition to $^7D_5$ term of Tb^{3+}				

250	254	260	265	270	275	280	285	290	300
24%	24%	14%	14%	14%	14%	14%	14%	14%	28%
		Spin-forbidden transition to $^9D_5$ term of Tb^{3+}							

excitation peak at around 217-246nm could also be observed in excitation spectrum in Fig.3 and it has been attributed to the lowest spin-allowed transition to $^7D_5$ term of Tb^{3+} [13-15]. In addition, these results indicated that the quenching concentration of spin-allowed transition was less than that of spin-forbidden transition. According to Ref [1,16], More energetic photons induced electronic transitions from the $4f^n$ ground state toward *d*-states belonging to the first excited configuration $4f^{n-1}5d$. These electric dipole transitions occurred between two configurations of opposite parity and were allowed at first-order. Consequently, the *4f-5d* transitions were in general sensitive to rare earth concentration resulting in less quenching concentration when they were spin allowed.

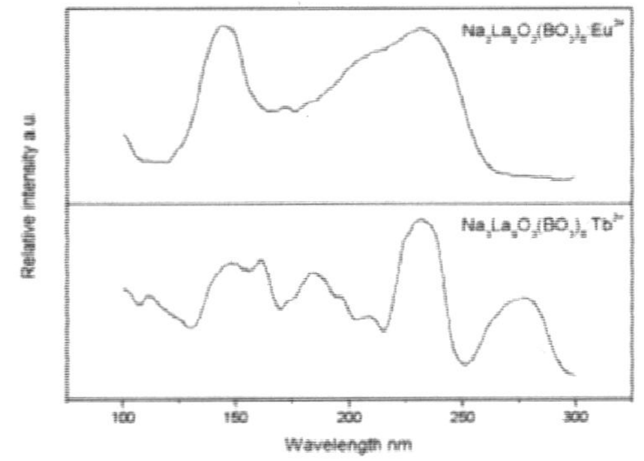

Figure 3. the comparison of excitation spectra of $Na_3La_9O_3(BO_3)_8$:Tb^{3+} and $Na_3La_9O_3(BO_3)_8$:Eu^{3+} monitored at 543nm and 611nm, respectively.

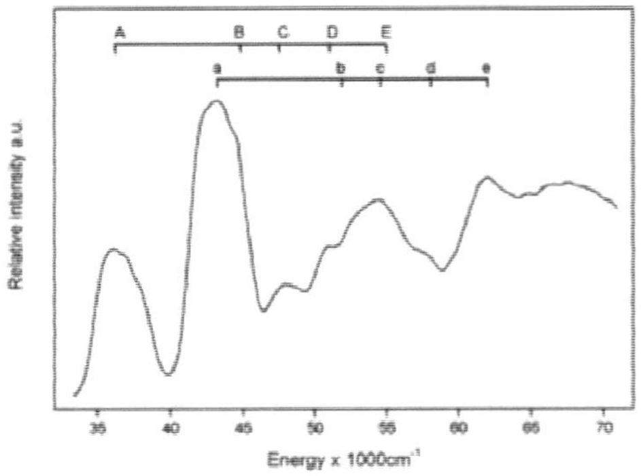

Figure 4. the identification of excitation spectrum of $Na_3La_9O_3(BO_3)_8$:Tb^{3+} sample monitored at 543nm.

As far as the excitation region in 125-215nm was concerned, it mainly consisted of host absorption band and higher spin-orbit 4*f*-5*d* transitions of Tb^{3+} which were seldom identified because of the bands overlaps. In order to understand the fundamental excitation mechanism for $Na_3La_9O_3(BO_3)_8:Tb^{3+}$, all these excitation bands should be identified. In particular, the quenching concentrations gave some useful information on these identifications. As shown in Table I , in the interval of 130-158nm, the quenching concentrations were 16% and it could be attributed to the borate host absorption band [17,18]. On the other hand, both the excitation spectra of $Na_3La_9O_3(BO_3)_8:Tb^{3+}/Eu^{3+}$ (Fig.3) showed the strong absorption band centered at 146nm and this result confirmed our above attribution of host absorption. In addition, the region of 206-215nm could be ascribed to the spin-forbidden transitions to $^9D_4$ term of Tb^{3+} (denoted as "B" in Fig.4) because of its larger quenching concentrations while the region of 160-204nm belonged to the spin-allowed $^7F_6$-^7D_J (J=4,3,2,1) transitions of Tb^{3+}.

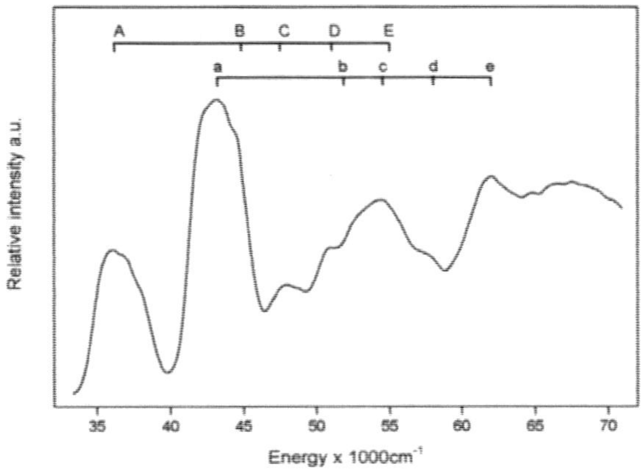

Figure 4. the identification of excitation spectrum of $Na_3La_9O_3(BO_3)_8:Tb^{3+}$ monitored at 543nm.

Based on the above results, the excitation peak centered at 161nm near host absorption in Fig.4 could be assigned to the higher spin-orbit 4*f*-5*d* transition of Tb^{3+}. In general, the energy gap between the lowest and the highest 4*f*-5*d* transitions of Tb^{3+} was about 18000 cm^{-1} [11]. Therefore, the highest spin allowed 4*f*-5*d* transition was theoretically expected to be located at around 61290 (43290+18000) cm^{-1} (163nm). In fact, a strong excitation peak at around 62112 cm^{-1} (161nm) was observed in the excitation spectrum (Fig.4) and so this peak could be attributed to the highest spin allowed 4*f*-5*d* transition ($^7F_6$-$^7D_1$, denoted as "e"). Consequently, the energy gap between the lowest and the highest 4*f*-5*d* levels of Tb^{3+} in $Na_3La_9O_3(BO_3)_8:Tb^{3+}$ was calculated to be about 18822 (62112-43290) cm^{-1} which was obviously larger than that of most Tb^{3+} doped inorganic phosphors [11].

According to the obtained energy gap of $Na_3La_9O_3(BO_3)_8:Tb^{3+}$ system, the energy of the highest spin forbidden 4*f*-5*d* transition ($^7F_6$-$^9D_1$, "E") could also be calculated to be at about 54924 (36102+18822) cm^{-1}. Considering that the energy of small excitation shoulder at 57143 cm^{-1} was higher than that of the highest spin forbidden 4*f*-5*d* transition (54923cm^{-1}, "E"), it could be attributed to be the second highest spin allowed $^7F_6$-$^7D_2$ transition (denoted as "d") of Tb^{3+}. And the broad excitation peaks at around 54645 cm^{-1} and 51813 cm^{-1} should be ascribed to the spin allowed $^7F_6$-$^7D_3$ and $^7F_6$-$^7D_4$ transition (denoted as "c" and "b") which overlaps part of the highest spin-forbidden $^7F_6$-$^9D_1$ ("E").

By comparing the same yardstick of five crystal field split levels of the d-electron in Fig.4, these weaker excitation peaks at about 45455, 47846 and 50761 cm^{-1} should belong to the spin forbidden $^7F_6$-$^9D_4$ (denoted as "B"), $^7F_6$-$^9D_3$ ("C") and $^7F_6$-$^9D_2$ ("D") transitions, respectively and the results was similar to the calculation on the 5d energy level positions of Tb^{3+} in some other inorganic compounds by P.Dorenbos [12].

Journal of Physics: Conference Series **152** (2009) 012051 doi:10.1088/1742-6596/152/1/012051

Basing on the above experimental crystal field split energies, the complete energy levels scheme of Tb^{3+} for $Na_3La_9O_3(BO_3)_8:Tb^{3+}$ system was first illustrated in Fig.5. Using this scheme, some suitable luminescence sensitizer matching the energy levels of Tb^{3+} in $Na_3La_9O_3(BO_3)_8:Tb^{3+}$ system can be selected.

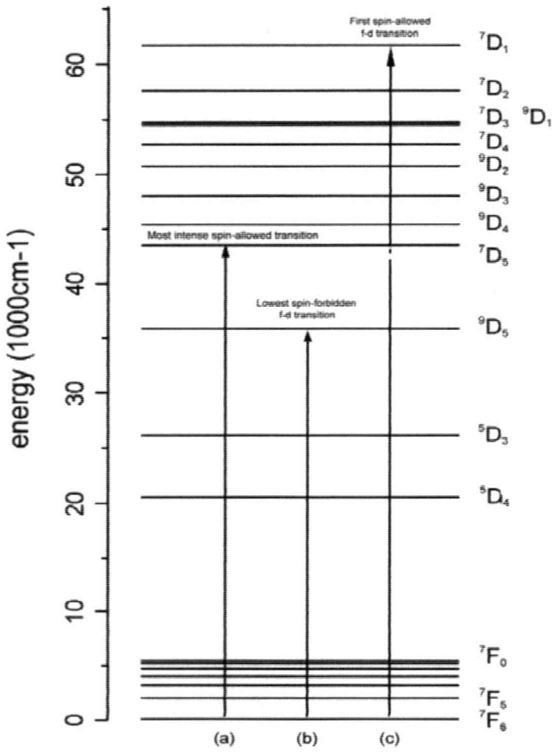

Figure 5. the energy level scheme of Tb^{3+} in $Na_3La_9O_3(BO_3)_8:Tb^{3+}$ samples. The levels are anticipated based on the results of this work.

4. Conclusions

It was revealed that the photoluminescent quenching concentrations of $Na_3La_9O_3(BO_3)_8:Tb^{3+}$ at particular excitation wavelengths were dependent on excitation bands in essence. By studying the quenching behaviors, these characteristic spin-orbit 4*f*-5*d* transitions of Tb^{3+} and host absorption band in $Na_3La_9O_3(BO_3)_8:Tb^{3+}$ were completely identified and the energy levels scheme of Tb^{3+} in $Na_3La_9O_3(BO_3)_8:Tb^{3+}$ was first established.

Acknowledgements

This study was supported by the Project of the combination of Industry and Research by the Ministry of Education and Guangdong Province (Grant No.0712226100023).

References

[1] Dexter D.L, 1953 *J.Chem.Phys.* **21** 836.
[2] Dexter D.L, and Schulman J.H. 1954 *J.Chem.Phys.* **22** 1063.
[3] John M.F and Richard C.P 1979 *Phys.Rev.B.* **19** 32.
[4] Ozawa L. 1976 *Phys.Rev.lett.* **36** 683.
[5] Broer M.M, Huber D.L and Yen W.M. 1984 *Phys.Rev.B.* **29** 2382.
[6] Klaassen D.B.M, Mulder H and Ronda C.R. 1989 *Phys.Rev.B.* **39** 42.

[7] Ozawa L and Hersh H.N, 1976 *Appl.Phys.Lett.* 2**8** 727.
[8] Gravereau P, Chaminade J.P, Pechev S, Nikolov V, Ivanova D, Pechev P. 2002 *Solid.State.Sci.* **4** 993.
[9] Zhang G, Wu Y, Li Y, Chang F, Pan S, Fu P. 2005 *J.Cryst.Growth.* **275** 1997.
[10] Zhang G C, Fu P Z, Li Y G, Wu Y C. 2005 *J.Synth.Cryst.* **34(5)** 765.
[11] Bai X Y, Zhang G C and Fu P Z. 2007 *J.Solid.State.Chem.* **180** 1792.
[12] Wang Y.H and Zhang J.C. 2007 *Electrochem.Solid-State.Lett.* **9** H26.
[13] Dorenbos P, 2000 *J.Lumin.* **91** 91.
[14] Nakazawa E, 2002 *J.Lumin.* **89** 100.
[15] Liang H.B, Tao Y, Xu J.H, He H, Wu H, Chen W.X, Wang S.B and Su Q. 2004 *J.Solid.State.Chem.* **177** 901.
[16] Lin C.K, Pang M.L, Yu M, Lin J, 2005 *J.Lumine.* 114 299.
[17] Wang Y.H, Guo X, Endo T, Murakami Y, Ushirozaw M. 2004 *J.Solid.State.Chem,* **177** 2242.
[18] Wang Y.H, Uheda K, Takizawa H, Endo T, 2001 *J.Electeochem.Soc.* **148** 6430.
[19] Blasse G and Grabmaier B.C. 1994 *Luminescent Materials*, Spinger-Verlag, Berlin Heidelberg.
[20] Dorenbos P. 2000 *J. Lumin.* **87-89** 970.

Hydrothermal synthesis and vacuum ultraviolet excited photoluminescence properties of (Y, La)VO_4:Eu^{3+} phosphors

Feng Zhang[1], Yuhua Wang[1, 2]*, Yinyan Zuo[1]

[1]Department of Materials Science, Lanzhou University, Lanzhou 730000, P. R. China
[2]Key Laboratory for Magnetism and Magnetic Materials of the Ministry of Education, Lanzhou University, Lanzhou 730000, P. R. China

E-mail address: wyh@lzu.edu.cn

Abstract. The $Y_{0.95-x}La_xVO_4$:5%Eu^{3+} ($0 \leq x \leq 0.15$) phosphors were synthesized successfully by mild hydrothermal method at 200℃ and their photoluminescence properties under ultraviolet (UV) and vacuum ultraviolet (VUV) excitation were studied. Monitoring at 619nm, a strong broad absorption was enhanced by co-doping of La^{3+} into the YVO_4:Eu^{3+} lattices at about 260nm in the UV excitation spectra. The VUV excitation spectra also showed the enhanced excitation bands at about 156nm and 200nm. Under 254nm or 147nm excitation, it was found that $Y_{0.95-x}La_xVO_4$:Eu^{3+} phosphors showed strong red emission at about 615nm and 619nm corresponding to the electric dipole $^5D_0$ -$^7F_2$ transition of Eu^{3+}, resulting in good color purity. The improvement of luminescence intensity of YVO_4:Eu^{3+} was also observed after partial substituting Y^{3+} by La^{3+} and the optimal luminescence intensity appeared with incorporation of 2.5mol% La^{3+}. The reasons of this enhancement were investigated and discussed in details.

1. Introduction

In recent research, rare earth doped phosphors have been widely studied for their numerous applications in PDPs and Hg-free lamps [1-3]. In these applications, small phosphor particles of less than 5μm in diameter are required [4]. Particularly, the nanosized phosphors have received much attention [5] for its much effect on the efficiency of luminescent devices. Hydrothermal method has been regarded as one of the most effective soft chemical routes to prepare nanomaterials at low temperature without milling or calcinations that could effectively save energy [6-8].

An excellent performance of displays and lamps requires phosphors with high quality for sufficient brightness, good colorimetric purity and long-term stability. YVO_4 has $ZrSiO_4$ structure, belonging to the space group $I4_1/amd$ [9]. It is well known that YVO_4:Eu^{3+} has been considered as a candidate of red phosphors and widely studied because of its good color purity, which is resulted from the noncentrosymmetric site of Eu^{3+} ion [10]. Moreover, nanosized YVO_4:Eu^{3+} has attracted much research, in K.Riwotzki and M.Haase's work [11], they have synthesized YVO_4:Ln (Ln=Eu, Sm, Dy) nanoparticles with size from about 10 to 30nm by a hydrothermal method. In Zhang's group [12], they have obtained YVO_4:Eu^{3+} nanopowders by sol–gel combustion method and reported that the luminescence intensity of the YVO_4:Eu^{3+} samples under 365nm excitation enhanced remarkably by the incorporation of Li^+ ions. But there are few reports about the enhanced luminescence intensity of the YVO_4:Eu^{3+} under VUV excitation.

In our previous work [13], the $YVO_4:Eu^{3+}$ nanophosphors at different pH values were synthesized by a mild hydrothermal method and their photoluminescence were evaluated under UV and VUV regions. When pH value was 7, the optimal luminescence intensity appeared. But the low efficiency of $YVO_4:Eu^{3+}$ was still a problem for the application in PDPs and Hg-free lamps. By co-doping the La^{3+}, the photoluminescence of many systems have been enhanced, such as Y_2SiO_5:Ce [14], Y_2O_3:Eu [4], and YBO_3:Eu [15].

In this study, the $Y_{0.95-x}La_xVO_4:0.05Eu^{3+}$ ($0 \leq x \leq 0.15$) phosphors were synthesized at a low temperature by hydrothermal process. The PL properties were investigated under UV and VUV excitation.

2. Experimental

The phosphor samples $Y_{0.95-x}La_xVO_4:0.05Eu^{3+}$ ($0 \leq x \leq 0.15$) were synthesized using a mild hydrothermal method. The starting materials were high purity Y_2O_3 (99.99%), La_2O_3 (99.99%), Eu_2O_3 (99.99%), and V_2O_5 (99%). Y_2O_3, La_2O_3 and Eu_2O_3 were entirely dissolved in nitric acid, and V_2O_5 was entirely dissolved in hydrochloric acid, obtaining a clear starting mixed solution by mixing the two solutions. The final pH value of the mixed solution was 7 through adding ammonia into the solution under vigorous stirring. Then, the mixture was transferred into a Teflon-lined stainless steel autoclave with a filling capacity of 40% distilled water. The hydrothermal reaction lasted 10 h at 200°C. Finally, the autoclave was cooled naturally and the product was filtered and dried in air.

The phase of synthesized phosphors were identified using a Rigaku D/max-2400 X-ray diffractometer with Ni-filtered Cu K_a (λ = 1.54178Å) radiation and the morphology and microstructure were observed by scanning electronic microscope (SEM, Hitachis-4800). The UV emission and excitation spectra were measured using an Edinburgh Instruments FLS920T. The VUV emission and excitation spectra were measured by the vacuum monochromator (VM504, Acton Research Corporation, ARC), and the VUV excitation spectra were corrected by dividing the excitation intensity of sodium salicylate. All the measurement conditions were at temperature.

3. Results and discussion

3.1. The crystal phase and SEM micrograph of $Y_{0.95-x}La_xVO_4:0.05Eu^{3+}$ ($0 \leq x \leq 0.15$) phosphors

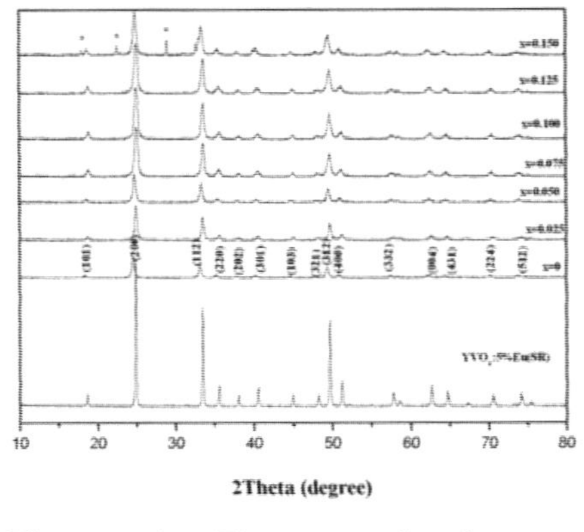

Figure 1. The crystal phase of $Y_{0.95}VO_4:0.05Eu^{3+}$ (SR) and $Y_{0.95-x}La_xVO_4:0.05Eu^{3+}$ ($0 \leq x \leq 0.15$) (HT) phosphors.

Figure 2. SEM micrograph of nanoparticles prepared by HT method.

Figure 1 showed the crystal phase of hydrothermal (HT) $Y_{0.95-x}La_xVO_4:0.05Eu^{3+}$ ($0 \leq x \leq 0.15$) phosphors and solid-state reaction (SR) $Y_{0.95}VO_4:0.05Eu^{3+}$. Except for $Y_{0.8}La_{0.15}VO_4:0.05Eu^{3+}$ with some small peaks signed by symbol "*", all the patterns of HT $Y_{0.95-x}La_xVO_4$:0.05Eu^{3+} and SR $Y_{0.95}VO_4$:0.05Eu^{3+} were consistent with a pure tetragonal phase YVO_4 (JCPDS No. 17-0341). Compared with SR $Y_{0.95}VO_4$:0.05Eu^{3+}, the broadening of the peaks were obviously observed. SEM micrograph of $Y_{0.90}V_{0.05}O_4$:5%Eu^{3+} phosphor particles prepared by HT method was given in figure 2. From the figure, it can be seen that the particles have a narrow size distribution with a slight agglomerate phenomenon, and the particles showed spherical-like shape and their main size distribution was around 40nm.

3.2. PL properties of (Y, La)VO_4:Eu^{3+} phosphors

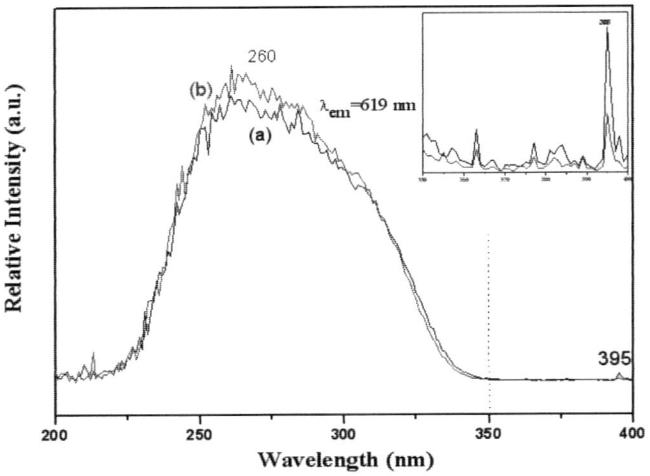

Figure 3. UV excitation spectra of (a) $Y_{0.95}VO_4$:5%Eu and (b) $Y_{0.925}La_{0.025}VO_4$:5%Eu, and the UV excitation spectra in 350-400nm regions is magnified in the insert.

Figure 3 gave the UV excitation spectra of $Y_{0.925}La_{0.025}VO_4$:5%Eu^{3+} and $Y_{0.95}VO_4$:5%Eu^{3+}. The UV excitation spectra monitored by Eu^{3+} $^5D_0$–$^7F_2$ at 619nm showed a broad band with the maxima at about 260nm and some sharp lines in the longer wavelength region. Below 350nm, the peaks ranging from 200 to 350nm were assigned to the overlap of VO_4^{3-} absorption and Eu^{3+} - O^{2-} charge transfer, the former was ascribed to a charge transfer from the oxygen ligands to the central vanadium atom inside the VO_4^{3-} group, which agreed well with Ref. [16-18]. The sharp lines in the longer wavelength region were due to the f–f transitions within $4f^6$ configuration of Eu^{3+}. Among them, the peak at 398nm with $^7F_0$–$^5L_6$ transition was most prominent, which accorded with previous report [19]. With the mol concentration of La^{3+} ranging from 0.025 to 0.15, the UV excitation spectra were similar to each other. It was observed that the peaks at about 260nm have improved with the introduction of La^{3+}. In Wang et al.'s work [15], the (Y, Ln)BO_3:Eu^{3+} (Ln=Sc^{3+}, Bi^{3+}, La^{3+}) phosphors were synthesized and the reasons of enhanced photoluminescence of YBO_3:Eu^{3+} were explained. The possible reason of the enhancement might be the different charge-to-radius ratio between Y^{3+} and La^{3+}. The charge-to-radius ratio of La^{3+} was smaller than that of Y^{3+}, when La^{3+} was substituted into the Y^{3+} site in the lattice, the tight binding of the oxygen electrons toward Y^{3+} was expected to be leased. Thus, the Eu^{3+}-O^{2-} CT band was enhanced.

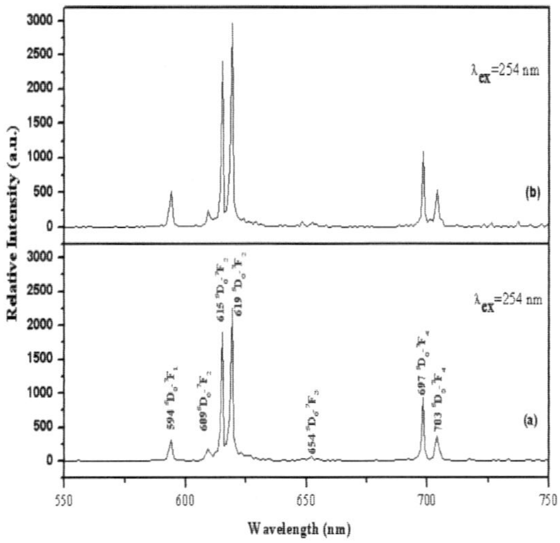

Figure 4. Emission spectra of (a) $Y_{0.95}VO_4$:5%Eu^{3+} and (b) $Y_{0.925}La_{0.025}VO_4$: 5%Eu^{3+}.

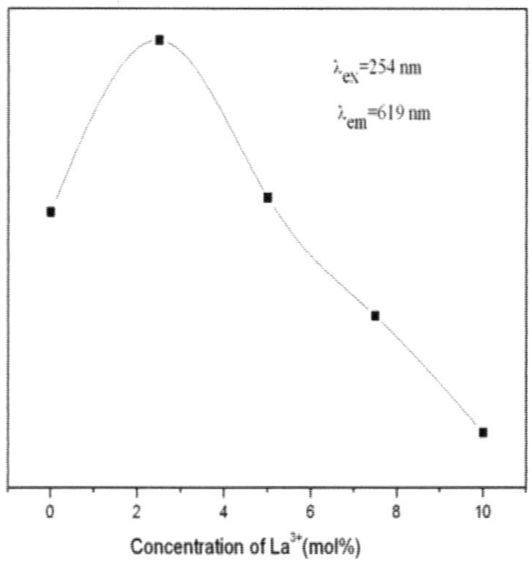

Figure 5. The emission intensity of Eu^{3+} atdifferent concentrations under 254 excitation.

Figure 4 showed the emission spectra of $Y_{0.95}VO_4$:5%Eu^{3+} and $Y_{0.925}La_{0.025}VO_4$:5%Eu^{3+} samples under 254nm excitation. The samples exhibited strong red luminescence. The emission spectra contained exclusively the characteristic emission of Eu^{3+}, namely, the $^5D_0$ -^{7} F_1 (594nm), $^5D_0$ - $^7F_2$ (609nm, 615nm, 619 nm), $^5D_0$-^{7} F_3 (654nm), and $^5D_0$-$^7F_4$ (697nm and 700nm), respectively. The emission spectra were dominated by the Eu^{3+} $^5D_0$–$^7F_2$ hypersensitive electric-dipole transition with the sharp peaks at 615nm and 619nm, which was due to the low symmetry local site of Eu^{3+} in the YVO_4 host lattices [20, 21]. Also, the dependence of the emission intensity of Eu^{3+} varied with the concentration of La^{3+}. The results were shown in figure 5, which was the curve of the $^5D_0$–$^7F_2$ emission of Eu^{3+} at 619nm upon the doping concentration in crystalline $Y_{0.95\text{-}x}La_xVO_4$:5%$Eu^{3+}$. It was found that the emission intensity of Eu^{3+} first increased with increasing the concentration of La^{3+}, reaching a maximum at x=2.5 mol%, then decreased with the doping concentrations. Thus the optimum emission intensity was obtained by substitution of 2.5mol % La^{3+} for Y^{3+} sites.

Monitoring by 619 nm emission, the VUV excitation spectra of $Y_{0.95}VO_4$:5%Eu^{3+} and $Y_{0.925}La_{0.025}VO_4$:5%Eu^{3+} were shown in figure 6. The excitation spectra were similar. Below 200 nm, the excitation spectra had a broad band with the maxima at about 156 nm. The strong broad band at about 156nm could be assigned to be the host absorption band. Also, the broad band centred at 157 and 173nm was assigned to host absorption in $LaVO_4/Eu^{3+}$ [22]. Above 200nm, the result was similar to the UV excitation spectra discussed above. In 100-200nm regions, the peaks at about 200nm improved with the incorporation of La^{3+}. This may be the absorption of La-O at about 200nm, which is consistent with Ref. [23].

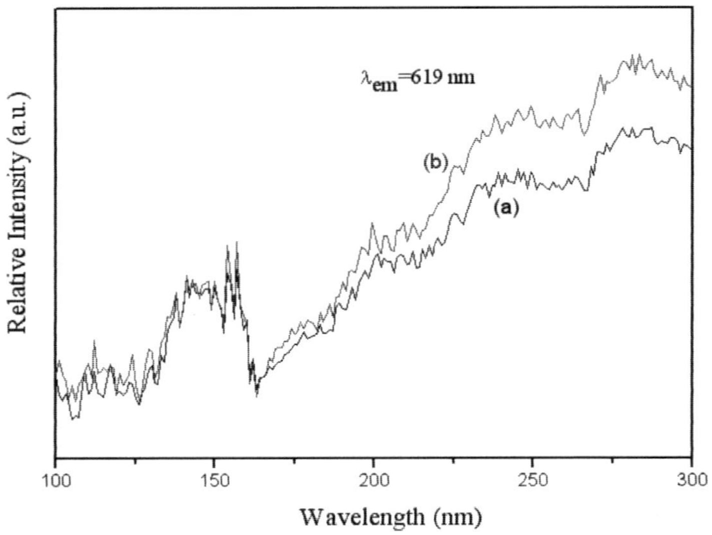

Figure 6.VUV excitation spectrum of $Y_{0.95}VO_4$:5%Eu^{3+} (a) and $Y_{0.925}La_{0.025}VO_4$:5%Eu^{3+} (b).

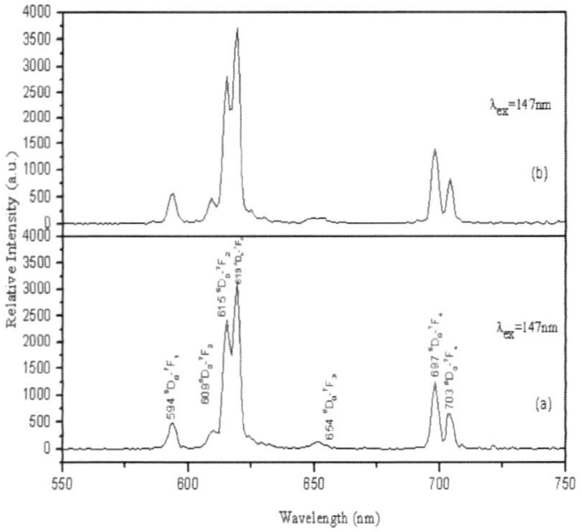

Figure 7. Emission spectra of (a) $Y_{0.95}VO_4$:5%$Eu^{3+}$and (b) $Y_{0.925}La_{0.025}VO_4$: 5%Eu^{3+}.

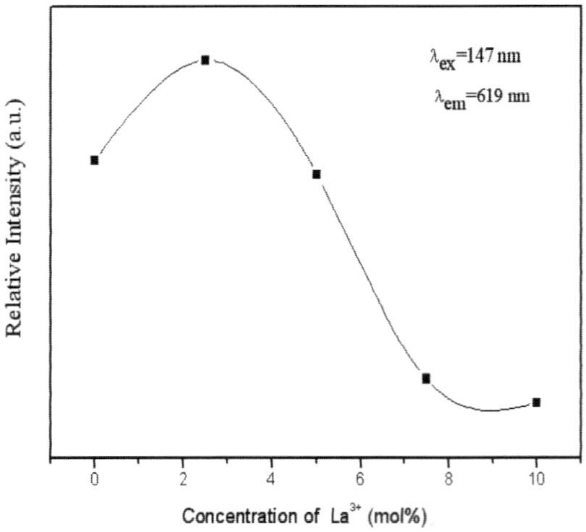

Figure 8. The emission intensity of Eu^{3+} at different concentrations under 147 excitation.

Figure 7 showed the emission spectra of $Y_{0.95}VO_4$:5%Eu^{3+} and $Y_{0.925}La_{0.025}VO_4$:5%Eu^{3+} samples under 147nm excitation, and the emission spectra consisted with the characteristic emission of Eu^{3+} discussed above. Figure 8 gave the emission intensity of Eu^{3+} at 619nm at different La^{3+} concentrations, the optimum emission intensity was also at x=2.5 mol%, and the commission international declared standards (CIE) chromaticity coordinations were calculated to be (0.671, 0.329) closing to the standard of the National Television System Committee (NTSC), which defines CIE color chromaticity coordinations as (0.67, 0.33). In addition, the enhancement may result from the contribution of La^{3+} which played an intermediate role in the energy transfer process from HA band to Eu^{3+}, the further study need to go on.

4. Conclusions

In summary, La^{3+} doped YVO_4:Eu^{3+} phosphors had been successfully synthesized at 200℃ through a hydrothermal process without further calcining treatment. SEM image observations showed that these phosphors are nanoparticles with the average size of 40nm. Their PL properties were investigated in UV/VUV region, respectively. In the excitation spectra, the broad band at about 156 nm was assigned to the host absorption band, and the peaks ranging from 200 to 350nm were assigned to the absorption of VO_4^{3-} and Eu^{3+} – O^{2-} charge transfer. The strongest emission peak of $Y_{0.95-x}La_xVO_4$:0.05Eu^{3+} was observed at about 619nm, and the optimum emission intensity was at x=2.5 mol%.

Acknowledgments

This study was supported by Key Science and Technology Project of Gansu Province (Grant no. 2GS064-A52-036-03), and the Project of the combination of Industry and Research by the Ministry of Education and Guangdong Province (Grant No.0712226100023). We are also grateful the Hitachis-4800 SEM work group for SEM micrograph.

Reference

[1] Blasse G, and Grabmaier B B 1994 *Luminescent Materials* (Berlin: Springer) p 134
[2] Okazaki C, Shiiki M, Suzuki T and Suzuki K 2000 *J. Lumin.* **87–89** 1280-1282
[3] Wu J H and Yan B 2008 *J. Alloys Compd.* **455** 485–488
[4] Imanaka N, Masui T and Chiga T 2004 *Electrochem. Solid-State Lett.* **7(6)** D7-D9
[5] Peng H S, Huang S H and Sun L D 2007 *J. Lumin.* **122-123** 847-850
[6] Ma D K, Zhang M, Xi, G C, Zhang J H andQian Y T 2006 *Inorg. Chem.* **45** 4845-4849
[7] Ma J, Wu Q S and Ding Y P 2007 *Mater. Lett.* **61** 3616-3619
[8] Haase M, Riwotzki K, Meyssamy H and Kornowski A 2000 *J. Alloys Compd.* **303–304** 191–197
[9] Bryan C Chakoumakos, Marvin M Abraham and Lynn A Boatner 1994 *J. Solid State Chem.* **109** 197-202
[10] Kang J H, Im W B, and Lee D C 2005 *Solid State Commun.* **133** 651-656
[11] Riwotzki K and Haase M 1998 *J. Phys. Chem. B* **102** 10129-10135
[12] Zhang H P, LÜ M K and Xiu Z L 2006 *Mater. Sci. Eng. B* **130** 151–157
[13] Wang Y H, Zuo Y Y and Gao H 2006 *Mater. Res. Bull.* **41** 2147–2153
[14] Wang J G, Tian S J, Li G B, Liao F H, Jing X P 2001 *J. Electrochem. Soc.* **148 (6)** H61-H66.
[15] Lingli Wang andYuhua Wang 2007 *J. Lumin.* **122–123** 921–923
[16] Yu M, Lin J and Wang S B 2005 *Appl .Phys. A* **80** 353–360
[17] Su X Q and Yan B 2006 *J. Alloys Compd.* **421** 273–278
[18] Bao A, Yang H, Tao C Y , Zhang Y and Han L L 2008 *J. Lumin.* **128** 60–66
[19] Hsu C and Powell R C 1975 *J. Lumin.* **10** 273-293
[20] Yu M, Lin J, Wang Z, Fu J, Wang S,. Zhang H J and. Han Y C 2002 *Chem. Mater.* **14** 2224-2231.
[21] Yu C C, Liu X M, Yu M,.Lin C K, Li C X, Wang H and Lin J 2007 *J. Solid State Chem.* **180** 3058–3065.
[22] Wang Y H, Zuo Y Y and Gao H 2007 *Electrochem. Solid-State Lett.* **10(6)** J71-J74
[23] You H, Wu X, Zeng X, Hong G, Kim C H, Pyun C H and Park C H 2001 Mater. Sci. Eng. B **86** 11.

Journal of Physics: Conference Series **152** (2009) 012053 doi:10.1088/1742-6596/152/1/012053

Study of transparent conducting ZnO:Al films deposited on organic substrate by reactive magnetron sputtering

J Li and Z Y Wang

Beijing Institute of Aeronautical Materials ,Beijing,100095,China

E-mail: nlijing@yahoo.com.cn

Abstract. A Zn-Al metallic target (Al 2 wt.%)has been used to prepare conductive and transparent aluminium-doped Zinc oxide(ZnO:Al) films on PI substrate by direct current reactive magnetron sputtering.The structure,crystallinity,optical properties,electrical properties and adhesion were investigated using a range of techniques,including AFM、XRD、spectrophotometry、four-point probe and adhesion tester.The optimal films were prepared with a substrate temperature of 150℃，O_2/Ar ration of 2:38 and sputtering power of 80W.The infrared emission properties of films and the feasibility for military application were also discussed in this paper. All the results to date demonstrate that magnetron sputtering is a cost-effective and easy to fabricating technique.

1. Introduction

Nowdays there is a great interest in replacing indium tin oxide (ITO) with aluminium –doping ZnO(ZAO),because of this oxide low cost,abundant raw materials,non-toxicity,relatively low deposition temperature and stability in hydrogen plasm.Extensive research work has being done on ZAO films depositied on rigid substrates.Recently preparing ZAO films on organic substrate has generated a renewed interest in it's potential use. Transparent conducting films deposited on polymer substrates have many merits compared with those deposited on glass and silicon substrate;they are lightweight,flexible, robust,of small volume and easy to carry,which are desired properties for a variety of applications such as flat panel display, heat-reflecting coating for energy saving use,low emissivity coating for stealth use etc.In this work,ZAO films were deposited on polyimides being with good heat resistance from Zn-Al metallic target.Detailed information about optical properties,electrical properties as well as structure and morphology were presented.

2. Experiment details

The substrate used in this work are polyimides foils from JinShan company with a thickness of 175μm.The main reason for choosing such polymer is that it can meet a number of different requirements including optical transparency,high temperature capability,chemical ruggedness,dimensional stability up to a maximum working temperature of 200℃..Before depositing the ZAO film,the substrates were ultrasonically cleaned in a detergent and ethanol baths,respectively.Subsequently they were rinsed with de-ionized water and dried with N_2 gas.Paralled to the target surface(φ6cm，thickness 3mm) ,the substrate with the same size was placed inside the chamber, and then evacuated until a base pressure of 1.3×10^{-3}Pa.The ZAO films were produced by direct current(D.C.) magnetron sputtering from a commercially available alloy Zn:Al(98:2 wt.%)target with 99.99% purity. Sputtering pressure was kept at 0.4Pa constantly.The flow rates of Ar and O_2 were controlled individually by a couple of mass flow controllers according to the demanded flow ratio. During the sputtering,the power was fixed as 80W and the substrate temperature was controlled at 150 ℃...The deposition time was 2h for each example.The film thickness was measured with a Sloan Dektak 3D profilometer ,while the film's sheet resistance was measured by a four-point probe system.The crystal structure of the films was determinded by X-ray diffraction(XRD) spectroscopy using a Siemens D 5000 system with CuKα radiation(λ=1.540562Å).The surface morphology was investigated by atomic force microscopy(AFM).A double-beam spectrophotometer(Varian Gary 5G) was used to measure the transmission and the reflectance spectra of the films in the wavelength range 200-3000nm.The infrared spectral emissivity was measured by a special instrument based on FTIR.

3. Rusult and discussion

3.1. Structure and morphological analysis

Compared with the sputtering power and substrate temperature,the O_2/Ar flow ration has little influence on the deposition rate.For 2h sputtering time,with different O_2/Ar flow rations,the thickness for all examples is in the range of 400 nm to 500nm.Adhesion testing showed that all films have a good adhesion on the PI substrate .

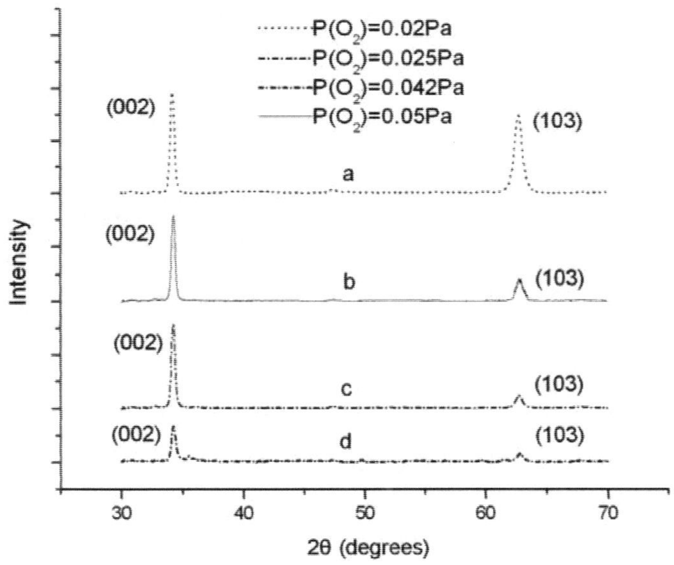

Figure 1. The XRD patterns of the ZAO films for four different O_2/Ar flow rations: (a) O_2/Ar flow ration=2:38,P(O_2)=0.02Pa; (b) O_2/Ar flow ration=2.5:37.7,P(O_2)=0.025Pa; (c) O_2/Ar flow ration=4.2:35.8,P(O_2)=0.042Pa; (d) O_2/Ar flow ration=5:35,P(O_2)=0.05Pa.

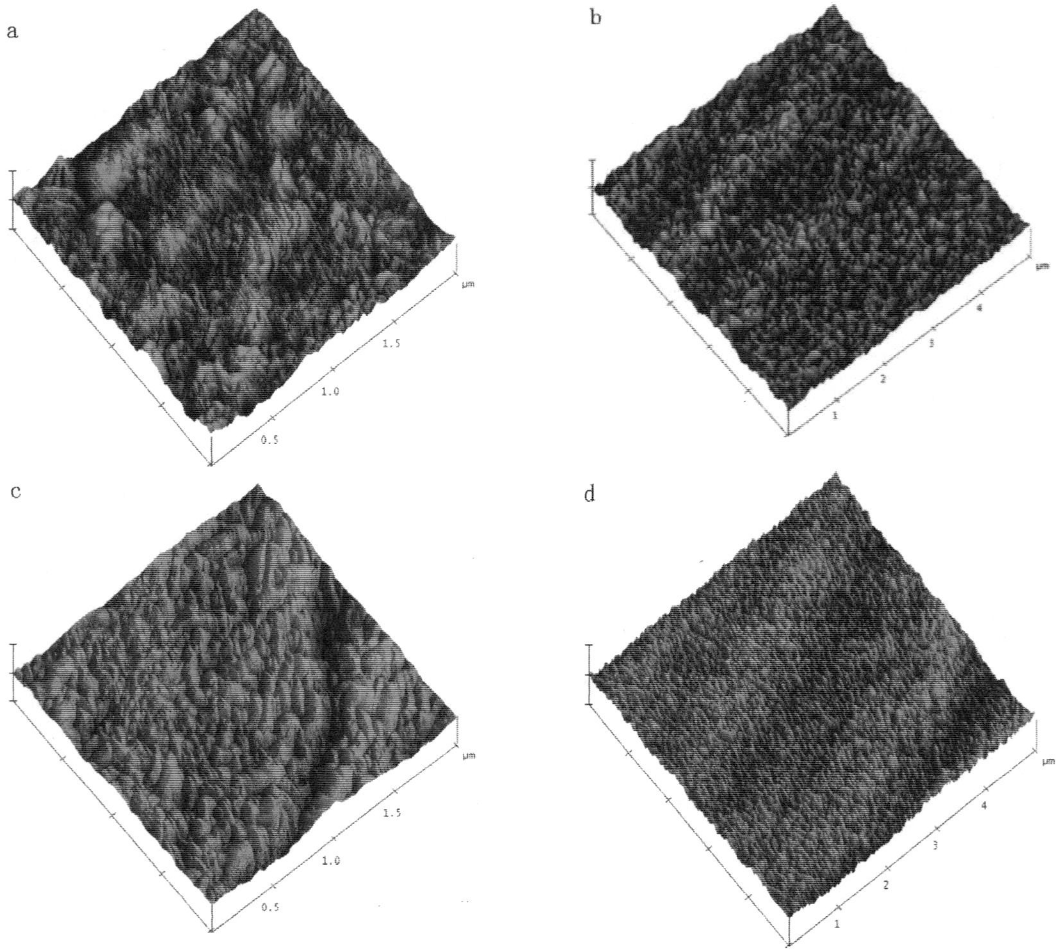

Figure 2. The AFM images of the ZAO films for four different O_2/Ar flow rations: (a) O_2/Ar flow ration=2:38,P(O_2)=0.02Pa; (b) O_2/Ar flow ration=2.5:37.7,P(O_2)=0.025Pa; (c) O_2/Ar flow ration=4.2:35.8,P(O_2)=0.042Pa; (d) O_2/Ar flow ration=5:35,P(O_2)=0.05Pa.

From figure 1,for all examples, two main peaks,corresponding to the (002)and(103)reflections are observed.The (002)diffraction peaks are observed at a 2 θ value of34.23°、34.29、34.28° 、34.26° respectively , which agree with the value reported in the literature for films grown on silicon ,glass and sapphire substrates[1,2].The c-axis length of the polycrystal tends to increase for all films,as the (002)peak angle was smaller than that of the bulk crystal(34.45°). Furthermore for all examples additionally to the (002) peak the (103)orientation is evident.The similar results are also reported by other authors[3, 4].The XRD measurements firstly reveal that these ZAO thin films are polycrystalline and retain the hexagonal wurtzite structure,having lattice parameters corresponding to that of bulk ZnO.Secondly,the XRD patterns suggest that the optimal films with improved crystallinity can be obtained at the O_2/Ar flow ration range from 2:38 to 4.2:35.8 .Particularly the (103) peak decreases evidently and the(002)peak does not change significantly,which means that at the lower oxygen content the O_2/Ar flow ration does not mainly influence the degree of crystallinity but the preferred orientation. When O_2/Ar flow ration is up to5:35,both the (002)peak and (103)peak decrease,which reveals that the ZAO film now has weak crystallinity .The full width at half-maximum(FWHM) of

(002)peak is 0.351for example a.The grain size can be estimated using the Scherrer Formula.The grain size is calculated to be of 41nm.

To identify the surface structure characteristics of films mentioned above,atomic force microscopy(AFM)perspective images are shown in figure 2.It can be easily observed that the films deposited for various O_2/Ar flow rations show a uniform grain size with a columnar structure. Comparatively example b,c and d consist of a more dense array of grains than example a,which is coincide with the XRD measurement result that film a being have a unconspicuous (002) preferred orientation .And the grain size of example d is smaller than that of the other examples.This also can be concluded from XRD resurment. Example d has a relatively weak diffraction peak

3.2. Optical properties

The optical transmittance T and reflectance R curves of ZAO films mentioned above are shown in figure.3 and figure.4 respectively. It is clear that in the given sputtering condition of this work the O_2/Ar flow ration has a very strong impact on the optical property and an insignificant influence below 1000nm. Each example has a high transmittance to visible light,with an average value over 80%,and a fundamental absorption edge lying in the near-ultraviolet part of the visible spectrum. The examples that have been determined above with relatively good crystallinity have lower IR transmittance and higher IR reflectance.

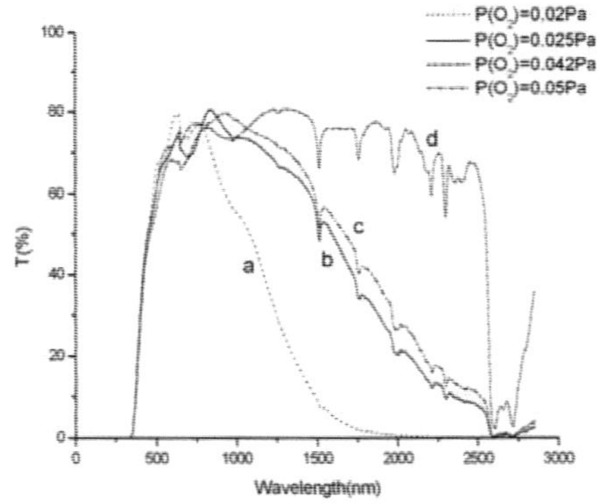

Figure 3. The transmission T of the ZAO films for four different O_2/Ar flow rations: (a)O_2/Ar flow ration=2:38,P(O_2)=0.02Pa; (b)O_2/Ar flow ration=2.5:37.7,P(O_2)=0.025Pa; (c)O_2/Ar flow ration=4.2:35.8,P(O_2)=0.042Pa; (d)O_2/Ar flow ration=5:35,P(O_2)=0.05Pa.

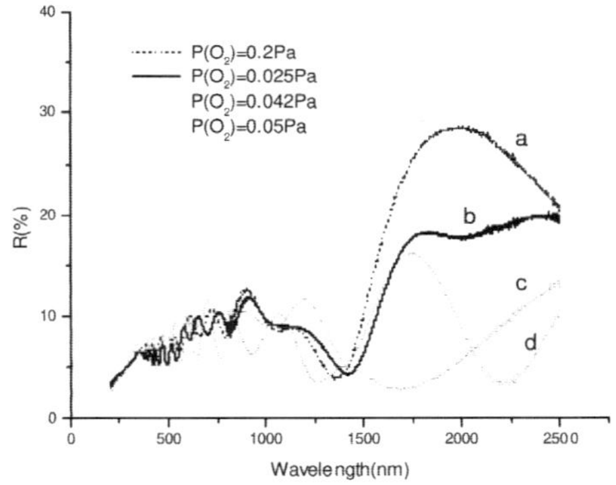

Figure 4. The reflectance R of the ZAO films for four different O_2/Ar flow rations: (a)O_2/Ar flow ration=2:38,P(O_2)=0.02Pa; (b)O_2/Ar flow ration=2.5:37.7,P(O_2)=0.025Pa; (c)O_2/Ar flow ration=4.2:35.8,P(O_2)=0.042Pa; (d)O_2/Ar flow ration=5:35,P(O_2)=0.05Pa.

3.3. Resistivity and emissivity

Figure 5 exhibits the electrical resistivity of ZAO examples.In figure 5 as the ratio O_2 increases, the conductivity of ZAO films is reduced slowly at the beginning then increased abruptly. This may result from the difference of films' crystal structure at different sputtering conditions. The electrical resistivity for all films is in the range of 10^{-3} Ω cm. And this result is also in accordance with the reflectance murerment result showed in figure 4.According to the Drude theory,there is a close relation between IR reflectance and sheet resistivity.The sheet resistivity will increase when the IR reflectance decreases.Compared figure 3with figure 5.it is observed that IR transmission increases when the ratio O_2 increases,a behaviour accompanied by an increase of the electrical resistivity. All these test results can be explained through a quantitative relation of transmission and reflection associated with free electrons[5].

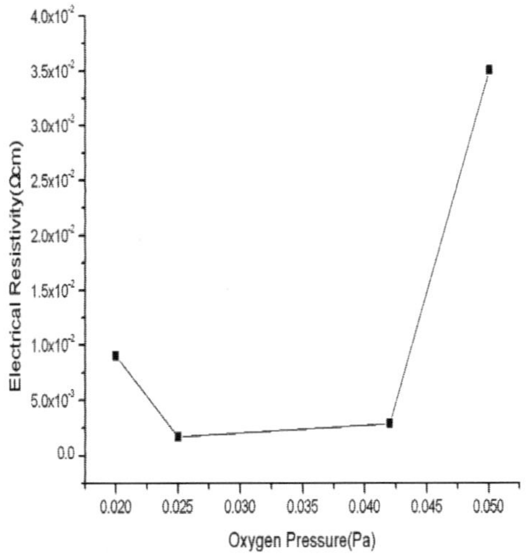

Figure 5. The electrical resistivity of the ZAO films for four different O_2/Ar flow rations: (a)ration=2:38,P(O_2)=0.02Pa; (b) O_2/Ar flow ration=2.5:37.7,P(O_2)=0.025Pa; (c) O_2/Ar flow ration=4.2:35.8,P(O_2)=0.042Pa; (d) O_2/Ar flow ration=5:35,P(O_2)=0.05Pa.

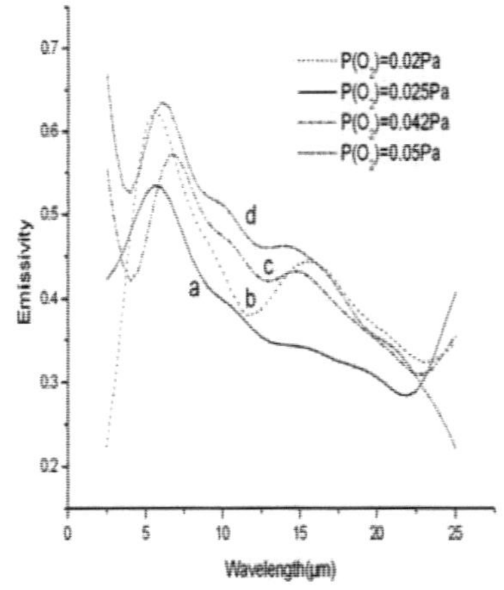

Figure 6. The IR imissivity of the ZAO films for four different O_2/Ar flow rations: (a)ration=2:38,P(O_2)=0.02Pa; (b) O_2/Ar flow ration=2.5:37.7,P(O_2)=0.025Pa; (c) O_2/Ar flow ration=4.2:35.8,P(O_2)=0.042Pa; (d) O_2/Ar flow ration=5:35,P(O_2)=0.05Pa.

Figure 6 showed the IR spectral emissivity of ZAO examples .It is well known that it is difficult to mearsure the material's spectral emissivity. In this work, a modified instrument based on FTIR was used to measure the emissivity of ZAO films.It can be seen that each example gives a lower emissivity value in the IR spectrum range. The example having lower electrical resistivity also has a lower IR emissivity. A very interesting phenomenon is that in the region of 5μm to 22μm the emissivity reduces considerably with the increase of wavelength.Detail research work should be done in the future

4. Conclusions

Transparent and conductive ZAO films were deposited on PI substrate successfully by reactive magnetron sputtering system.The effect of O_2/Ar flow ration on the structure,optical and electronic properties of ZAO films was mainly investigated.All films deposited at the given condition in this experiment showed a strong crystallographic c-axis orientation(002) perpendicular to the substrate surface although (103) diffraction peak was observed.In this case, the O_2/Ar flow ration mainly influence the orientation of crystallization. The ZAO films demonstrate transmittance of above 80% in the range of visible spectrum and a lowest electrical resistivity of $1.17\times10^{-3}\Omega$cm.The lower IR emissitivity of ZAO films obtained in this work proved that the ZAO film may be a promising material for military application.

Reference

[1] Song D Y,Per Widenborg,Winston C and Armin G A 2002 Solar *Energy Materials&Solar Cells* **1-20** 73
[2] Brehme S, Fenske F, Fuhs W,Nebauer E, Poschenrieder M, Selle B and Sieber I 1999 *Thin Solid Film* **163** 342
[3] Huang J M and Dong J H 2002 *Electroniccomponents&Materials* **7-10** 21
[4] Rodrigo F,Silva M E and Zaniquelli D 2004 *Thin Solid Films* **86-93** 449
[5] Frank G, Kauer E and Kostlin H 1981 *Thin Solid Films* **107** 77

Journal of Physics: Conference Series **152** (2009) 012054 doi:10.1088/1742-6596/152/1/012054

Colour electroluminescence with end light-emitting from ZnO nanowire/polymer film

Ying HE, Jun-an WANG[2], Wen-fei ZHANG[1], Xiao-ban CHEN[1], Zong-hao HUANG[1] and Qiu-wen GU[1]

[1] Department of Polymer Materials, School of Materials Science and Engineering, Shanghai University, 149 Yanchang Road, Shanghai 200072, China
[2] Institute of Materials, School of Materials Science and Engineering, Shanghai University, Shanghai 200072, China

E-mail: yinghe@staff.shu.edu.cn)

Abstract. The ZnO nanowires with polymer film were self-assembly grown on n-type (111) plane of the silicon substrate using polymer assisted complexing soft-template process through a simple polymer complexation and low-temperature oxidizing-sintering, which have smooth top and fine hexagonal columnar structure with average length of about 6 μm and the diameter of about 40 nm. These columnar structured ZnO nanowires had strong near-band ultraviolet emission at ~383 nm and blue electrically driven emission at ~ 400 nm with a relatively low turn-on voltage, as well as a typical diode characteristic property at room temperature. In particular, these structures, being of high aspect ratio and small tip radius of curvature, may possess a good amplified stimulated emission and lasing property. These results suggested a potential application of ZnO nanowire/polymer film as electroluminescence flat panel displays or illuminations in the future.

1. Introduction

One-dimensional (1D) zinc oxide which is an n-type wide band-gap semiconductor nanomaterial with an energy gap of 3.37 eV and high exciton binding energy of ~ 60 meV, has attracted much attention for their unique optical, electronic, and mechanical properties as well as their potential applications in nanodevices[1-4]. Nanocolumns are structures particularly well suited to lasing and directed luminescence because they can form individual Fabry-Pérot cavities and combine lateral confinement with an appreciable grain length. In particular, it is expected that they would possess a good electron field emission property due to their high aspect ratio and small tip radius of curvature. Therefore, ZnO-based 1D nanostructure could be an appropriate alternative to carbon nanotube (CNT) for field emission displays or electroluminescence flat panel displays. The field-emission properties of ZnO nanowire arrays[5], and light-emitting diode based on ZnO[6-7] have been attempted recently, and those pioneer works[8-14] demonstrated that the micro/nano-structure determines the properties of a given nanomaterial with respect to its specific application. The peculiar geometrical morphologies of ZnO ordered nanostructures may yield emission are achieved. Furthermore, measurements of electroluminescence can reflect the native band structure of the sample, contributing significantly to fundamental research.

The realization of the controlled oriented growth of ZnO nanowires always must be taken into account before the real large-scale ZnO nanowires-device fabrication[15]. Self-assembly of 1D ZnO nanostructures offers an available approach for the problem[16]. In this paper, based on the polar growth behavior of ZnO and polymer grid backbone localization model that we reported before[17,18], we prepared single crystalline ZnO nanowires with polymer film on silicon (111) substrate and the luminescence properties have been evaluated.

2. Experimental section

The ZnO nanowires were fabricated on Si (111) wafer by a simple polymer-assisted complexing nucleation method as we reported before[17]. The grown material was characterized by FE-SEM, and high-resolution transmission electron microscopy (HR-TEM). Photoluminescence (PL) measurement was performed at room temperature under the 325 nm UV fluorescent light excitation using a Hitachi F-4500 fluorescence spectrophotometer with a Xe lamp. Electroluminescence (EL) measurements were performed with diode structure in the atmosphere at room temperature. The electroluminescent device structure was ITO/(ZnO/polymer)/Si(111)/Cu.

3. Results and discussion

3.1. Morphology and crystal structure of the nanowires

The morphology of the samples grown on a Si (111) wafer is shown in figure 1a. These ZnO nanowires nearly vertically grew on the substrate, with a straight shape and a uniform diameter. The average length of the nanowires is about 6 μm and diameter of about 40 nm. The further details of morphology and atomic structure of the nanowires stripped from silicon substrate were obtained by using high resolution TEM (HRTEM). It can be seen from figure 1b that a single ZnO nanowire exhibits continuous lattice fringes, which demonstrated that the nanowire is of single crystal with wurtzite structure and its growth direction is along [0001] *c*-axis.

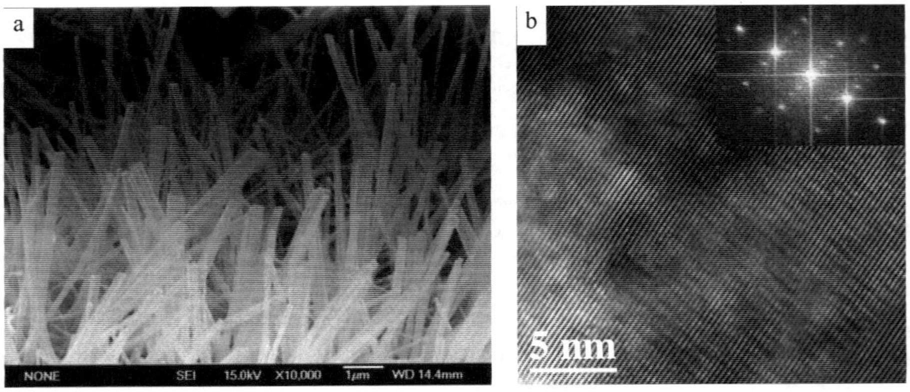

Figure 1. (a) A FE-SEM image of a well-aligned ZnO NW array; (b) HRTEM micrograph and corresponding selected-area electron diffraction pattern of a single ZnO nanowire.

Figure 2 shows an energy dispersive X-ray (EDX) spectrum of the nanowires on a silicon substrate. It reveals the presence of the Zn, O and Si. Apparently, these nanowires are composed of ZnO and rich in oxygen atoms. These results exhibit that our technique has the advantages over the other reported methods, because it solved the problem of the prevalent defect configuration which is a deficiency in O at the surface of ZnO nanowires.

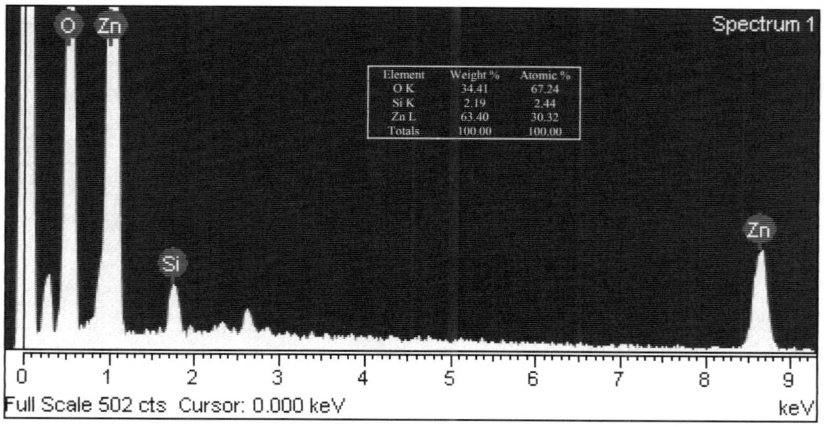

Figure 2. The EDX spectrum of the ZnO nanowires on a silicon substrate. Silicon peak comes from the substrate.

3.2. Polymer-assisted self-assembling model

The above obtained results can be explained by the polymer-assisted self-assembling preparation of ZnO nanowires via a novel process based on polymer grid backbone localization model[17]. The polymer soft-template controlled the nucleation of ZnO as and further growth into the nanowires along the [0001] direction. By using clingy polymeric membrane as self-assembling complex medium and the polar polymer (such as PVA and PAM) as suspending agent and surfactant, the surface tension of the polymer solution reduced and consequently the wettability to the substrate increased. This makes the ZnO nanowires well attached on the surface of silicon substrate. Besides, the adhesive strength between the ZnO nanowires and the substrate surface increased more efficiently by the later sintering. Therefore, the ZnO nanowires can't peel from the substrate, ensuring their photoelectricity application well and stable. This technique has solved the lattice mismatch problem of the ZnO to silicon and is more economically feasible. Further more, the ZnO nanowires are rich in oxygen atoms, which obviously resolved the problem of a deficiency in oxygen at the surface of nanowires ever reported. This makes the practical real application of the 1D ZnO nanomaterials as emission nanodevices capable break through in the future.

3.3. Photoluminescence and electroluminescence

The room-temperature photoluminescence spectrum from the ZnO nanowires excited with a wavelength of 325 nm, exhibit a sharp strong UV emission of about ~ 383 nm related to the band to band transition, and a weak blue emission of around 445 nm (shown in figure 3a). The near band-edge UV emission might be attributed to a well-known recombination of free excitons, and the blue emission might result from the oxygen vacancies and zinc interstitial defects in the ZnO nanomaterials. However, different from the PL spectrum, strong peak appears at 400 nm (corresponding to the energy of 3.11 eV) in the electroluminescence (EL) spectrum of the ZnO nanowires on silicon substrate as shown in figure 3b. With increasing alternating current, the intense blue emission was observed indeed by naked eye and the whole surface emission are highly homogeneous according to the photograph inserted in figure 3b. The luminescence mechanism is probably due to the recombination of the localized exciton emission. It could be deduced that the facets of regularly arrayed hexagonal nanowires form natural micro-cavities and these nanowire-crystallites confine the centre-of-mass motion of excitons. As a result of the quantum size effect, the oscillation strength of the excitons, which is favored to the radiation recombination of exciton at room temperature, is largely enhanced. The electronic transport for electrons and holes must occur with only a very small reduction in carrier energy over the transport path length, which corresponds to the nanowire height, that is, ~6 μm.

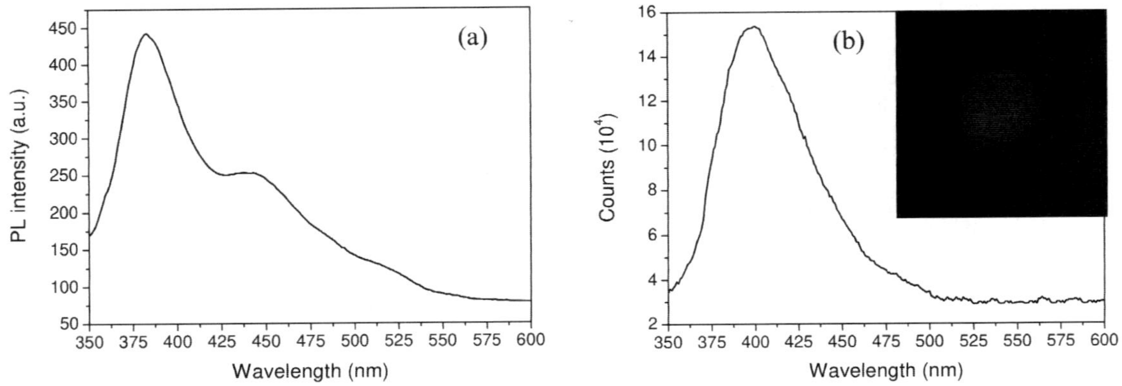

Figure 3. The PL spectrum (a) and EL spectrum (b) of the ZnO nanowires/polymer film on silicon substrate (the transparent anode image of the electroluminecence is shown in the inset picture).

Figure 4a shows the characteristics of emission current density versus applied electric voltage for the several times bias voltage sweeps. Almost identical *J–E* curves were obtained, which indicates a better stability and repetitiveness of the emission process from our samples. The current-voltage measurements and corresponding oscillograph image in figure 4b at room temperature showed a typical diode characteristic. Electroluminescence band of the ZnO nanowires device is located at the blue–violet region. The better emission homogeneity of the sample comes from a large number of ZnO nanowires as emitters on the surface of the nanowire end. The turn-on field, defined as the electric field which is required to produce a current density of 10 $\mu A/cm^2$, is estimated to be about 0.9 V/μm. Obviously, the turn-on field and threshold field are much lower than those of the aligned ZnO nanowires[5]. Low turn-on field of the sample is attributed to the sharp geometry morphology of the corners of the nanowire ends. These peculiar geometrical morphologies of ZnO nanostructures yield a large enhanced emission. It follows from these results that columnar structures of ZnO nanowires are well suited to lasing and directed luminescence due to their individual Fabry-Pérot cavities and the combined lateral confinement at an appreciable grain length. In particular, these structures, being of high aspect ratio and small tip radius of curvature, possess a good amplified stimulated emission and lasing property. As the exciton binding energy in ZnO is of 60 meV, this makes exciton stabile and provides opportunities for making highly efficient lasers operable at room-temperature.

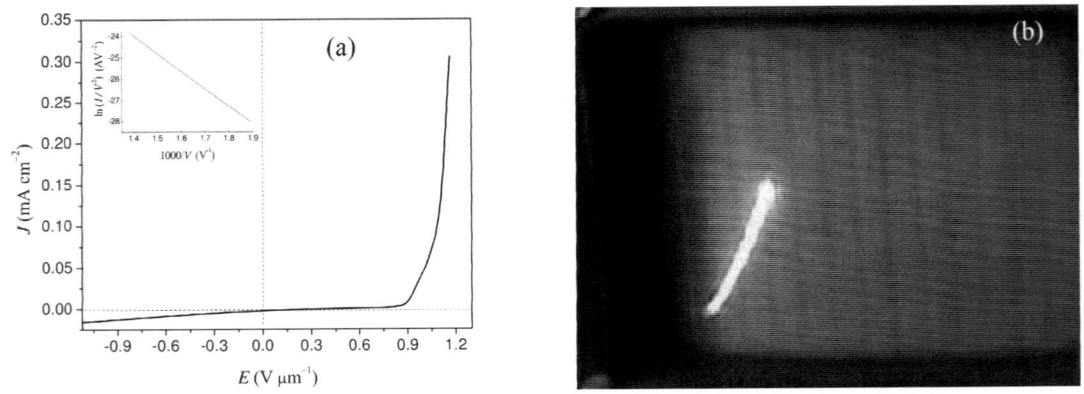

Figure 4. The current-voltage characteristics of the vertically ZnO nanowire/polymer film:(a) emission current density versus applied electric voltage and (b) corresponding temporal oscillograph image.

It is found that colour emission is achieved with low voltage, and change into yellow, violet/blue or even white emission with higher voltage. The reason may be that with increase of voltage, the blue emission is absorbed by the white's, or the blue emission stimulated the longer wavelength light such as yellow and white light. Anyhow, it is clear that the colour of the electroluminescence can be controlled by the applied voltage. These results also exhibit the potential application of ZnO nanowires as solid-state electroluminescence flat panel displays or illuminations in the future.

4. Conclusions

In conclusion, the ZnO nanowires have been fabricated vertically on silicon substrate by a polymer-assisted complexing soft-template method. The intense room-temperature blue-electroluminescence of the ITO/(ZnO/polymer)/Si(111)/Cu LED was observed indeed by naked eye. In the hybrid system, the polymer is used not only as binder in the device of LED, but also a dispersion medium in the luminescence layer, which stabilize the ZnO nanowires and passivate the surface of ZnO nanocrystals and therefore prevent them from aggregating and quenching of luminescence. By controlling the content and dimension of the ZnO nanowires in the polymer matrix, the emission strength and the emission wavelength could be adjusted. These results also indicated that the ZnO nanowires might be a promising candidate as electroluminecence flat panel displays or illuminations in the future.

Acknowledgments

The authors would like to acknowledge support from the Key Scientific Research Projects of Shanghai Municipal Commission of Education, Grant No. 07ZZ13, China.

References

[1] Hu J, Ouyang M, Yang P and Lieber C M 1999 *Nature* **399** 48
[2] Bagnall D M, Chen Y F, Zhu Z, Yao T, Koyama S, Shen M Y and Goto T 1997 *Appl. Phys. Lett.* **70** 2230
[3] Huang M H, Mao S, Feick H, Yan H, Wu Y, Kind H, Weber E, Russo R and Yang P 2001 *Science* **292** 1897
[4] Duan X, Huang Y, Agarwal R and Lieber C M 2003 *Nature* **421** 241
[5] Ham H, Shen G, Cho J H, Lee T J, Seo S H and Lee C J 2005 *Chemical Physics Letters* **404** 69
[6] Tsukazaki A, Ohtomo A, Onuma T, Ohtani M, Makino T, Sumiya M, Ohtani K, Chichibu S F, Fuke S and Segawa Y 2005 *Nature Materials* **4** 42
[7] Ramamoorthy K, Arivanandhan M, Sankaranarayanan K and Sanjeeviraja C 2004 *Materials Chemistry and Physics* **85** 257
[8] Xing Y J, Xi Z H, Xue Z Q, Zhang X D, Song J H, Wang R M, Xu J, Song Y, Zhang S L and Yu D P 2003 *Appl. Phys. Lett.* **83** 1689
[9] Wang X H, Zhao D X, Liu Y C, Zhang J Y, Lu Y M and Fan X W 2004 *J. Cryst. Growth* **263** 316
[10] Zheng M J, Zhang L D, Li G H and Shen W Z 2002 *Chemical Physics Letters* **363** 123
[11] Wang Z and Li H L 2002 *Appl. Phys. A* **74** 201
[12] Kovtyukhova N I, Martin B R, Mbindyo J K N, Mallouk T E and Cabassi T S 2002 *Materials Science and Engineering C* **19** 255
[13] Ni Y H, Wei X W, Hong J M and Ye Y 2005 *Materials Science and Engineering B: Solid-State Materials for Advanced Technology* **121** 42
[14] Guo L, Ji Y, Xu H, Simon P and Wu Z 2002 *J. Am. Chem. Soc.* **124** 14864
[15] Xia Y, Yang P, Sun Y, Wu Y, Mayers B, Gates B, Yin Y, Kim F and Yan H 2003 *Adv. Mater.* **15** 353
[16] Glotzer S C 2004 *Science* **306** 419
[17] He Y, Sang W, Wang J A, Wu R and Min J 2005 *Materials Chemistry and Physics* **94** 29
[18] He Y, Sang W, Wang J A, Wu R and Min J 2005 *J Nanopart Res.* **7** 307

Journal of Physics: Conference Series **152** (2009) 012055 doi:10.1088/1742-6596/152/1/012055

Combustion synthesis of TiO_2-Al-C/Al_2O_3 mixture in the presence of oxygen

R Rahbari G[1], L H Saw[1], M Hamdi[1] and R Yahya[2]

[1] Department of Engineering Design and Manufacture,
[2] Department of Chemistry,
University of Malaya, Malaysia.

E-mail: rahbarg@gmail.com

Abstract. Quasi-adiabatic combustion synthesis of optimized $3TiO_2$+4Al+(3+x)C mixture in the presence of oxygen was carried out to produce TiC-Al_2O_3 composite. The influence of Al_2O_3 diluent in combusted ($3TiO_2$+4Al+3.5C+yAl_2O_3) reactant mixture was investigated. Thermodynamics studies of the system shown that self-sustaining (SS) mode of combustion wave propagation can occur for large amounts of diluents (y = 2.5). Diluent reduced combustion temperature and rate of reaction kinetics. Oxide (Ti_2O, Ti_3O_5) and intermetallic ($AlTi_2$) phases were detected when extra Al_2O_3 was added to the mixture. Semi-quantitative XRD analysis was employed to study the phase formation of products for various amounts of diluent content in the mixture. The boundary between SS and non-SS modes of wave propagation was determined. Adding Al_2O_3 diluent slightly increased density that resulted in better thermal conductivity.

1. Introduction

Combustion synthesis (CS) uses chemical reaction energy to produce various high temperature materials. The mechanism of structure and phase formation of this rapid, high temperature synthesis method requires appropriate control to produce the desired product.

Adding inert diluents to initial mixture absorb the heat and decrease the violence of combustion reaction. A diluent can be an inert component to the reactant mixture and/or one of the specific products of the combusted mixture. Adding one of the products, such as Al_2O_3, to the reactant ($3TiO_2$+4Al+3C) mixture is commonly used to control combustion process and grains grow of the product [1-3]. The diluent reduces the adiabatic temperature and decreases kinetic rate of reaction to achieve desired microstructure and properties.

TiC-Al_2O_3 composite is a promising material in industry as an abrasive tools, wear resistant coating, magnetic head recording substance as well as heatable valves and nozzles due to its attractive high temperature mechanical, chemical and electromagnetic properties. Conventionally, TiC-Al_2O_3 is produced by hot pressing of (TiC+Al_2O_3) powders consuming extensive amount of heat energy. On the contrary, combustion synthesis provides an alternative route to synthesize TiC-Al_2O_3 composite in a simple, energy saving and economic way.

Wide range of investigation has been performed on combinations of two or more components in TiO_2-Al-C system to produce different types of materials such as metal-ceramic and ceramic-ceramic composites [4-12]. In addition, considerable efforts have been made to investigate the reactions of heterogeneous TiO_2-Al-C and TiO_2-Al-C-Al_2O_3 mixtures in self-sustaining reaction mode [4, 7, 9, 10, 11]. However, setup of an adiabatic system for combustion synthesis might raise technical difficulties and result in higher cost, so studying the synthesis under quasi-adiabatic conditions is a practical need.

Furthermore, lack of detail studies on quasi-adiabatic combustion of reactant mixtures in the presence of oxygen necessitates more efforts in this direction. In previous work [13], it was discussed in details that additional powders of Al and/or C is needed for production of desired TiC-Al_2O_3 composite from TiO_2-Al-C mixture.

This research is continued the work and focused on the combustion of TiO_2-Al-C mixture under quasi-adiabatic conditions with the addition of inert diluent-Al_2O_3 in the presence of oxygen. Different impacts of various diluent-Al_2O_3 fractions in the mixture on macrokinetics, and thermo-mechanical behavior of the product were studied in the project. Semi-quantitative XRD analysis revealed the phase formation and mechanism of the process. Al_2O_3-diluent absorbed heat of the exothermic reaction, so that decreased the combustion temperature. Moreover, extra Al_2O_3 reduced the contact points between reactants, decreased the reaction rate and lead the formation of intermetallic phase.

However, thermodynamic studies revealed that self-sustaining propagation of combustion wave was occurred even by adding large amounts of Al_2O_3-diluent into the reactant mixture. Experimental results showed that extra diluent decreased the porosity and consequently increased thermal conductivity of the product, slightly.

2. Experimental

Titanium (IV) Oxide (rutile; <5 μm; 99.9% purity; Sigma Aldrich), Al (<10 μm; 99.9% purity; Lab Chemical), Carbon black (<0.02 μm; 99.9% purity, Juhalim Co., Malaysia) and Aluminum Oxide (<10 μm; 99.9% purity; Sigma Aldrich) were mixed and combusted to produce TiC-Al_2O_3 composite. At first, the mixture of TiO_2-Al-C powders was prepared according to equation (1) to achieve the desired product under quasi-adiabatic conditions in the presence of oxygen [13]. For this purpose, various contents of carbon, C (x = 0, 0.5), were examined to optimize the reaction and achieve the final TiC-Al_2O_3 product. Then, the influence of inert diluent, Al_2O_3, on the combustion of the optimized mixture was studied. In this stage, mixtures were prepared according to equation (2). All the reactants were mixed in a planetary ball mill for 6 hours. Cylindrical pellets with 20 mm in diameter and about 5 mm thickness were produced under a constant compact load of 100 MPa using uniaxial hydraulic press.

$$3TiO_2 + 4Al + (3 + x)C \qquad x = 0, 0.5 \tag{1}$$

$$3TiO_2 + 4Al + 3.5C + yAl_2O_3 \qquad y = 0, 0.5, 1.0, 1.5, 2.0, 2.5 \tag{2}$$

The combustion of TiO_2-Al-C green mixture was carried out in a tube furnace in argon and oxygen gas environment. The pellet heated up to 1100 °C and the holding time was 90 minutes. Then, the furnace was turned off and the sample was kept inside the furnace until the temperature dropped to the room temperature. On this way, possible thermo-mechanical stresses were minimized.

The combusted TiO_2-Al-C-Al_2O_3 compact powders were characterized by using scanning electron microscope (SEM), and x-ray diffraction (XRD) with CuKα radiation. Semi-quantitative composition XRD analysis was performed to study the phase formation of the products in details. Density of the products was measured by using Archimedes method. Finally, the thermal conductivity of the products was investigated using Koh and Fortini model [14].

3. Results and Discussion

It is known that the combustion synthesis of (3TiO2+4Al+3C) mixture includes two stages. In the first stage aluminothermic reduction of TiO2 occurs according the following equation:

$$3TiO_2 + 4Al \rightarrow 3Ti + 2Al_2O_3 + Q\uparrow \tag{3}$$

Exothermic reaction between Ti and C occurs in the next stage of the synthesis as follows:

$$3Ti + 3C \rightarrow 3TiC + Q\uparrow \tag{4}$$

The proposed mechanism in (3) and (4), occurs in adiabatic closed system. But in this investigation the synthesis carried out under quasi-adiabatic conditions in the presence of oxygen. Here, mechanism of the process is more complicated. Therefore, the optimum amount of C was found experimentally to produce the desired product [13].

The XRD result of combusted ($3TiO_2+4Al+3C$) and optimum reactant ($3TiO_2+4Al+3.5C$) mixtures are shown in figure 1. The figure shows that the combustion of ($3TiO_2+4Al+3C$) mixture in quasi-adiabatic conditions of the open system did not produce the desired product. However, it was observed that the optimum value of $x = 0.5$ would produced the desired TiC-Al_2O_3 composite. This result is in agreement with the previous work [13]. Therefore, in equation (2) the coefficient of C is considered to be equal to the optimum value, $x = 3.5$.

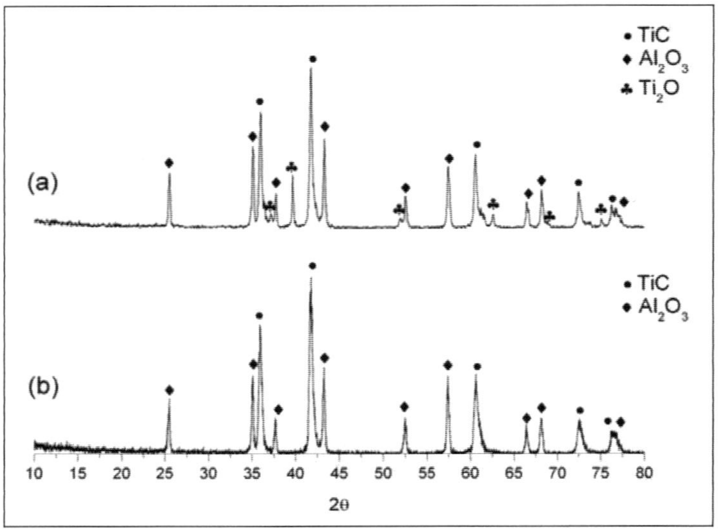

Figure 1. XRD pattern of product for (a) ($3TiO_2+4Al+3C$), (b) ($3TiO_2+4Al+3.5C$) mixtures.

3.1. The influence of diluent content on the adiabatic temperature

The calculated theoretical adiabatic temperature ($T_{ad.}$) from the thermo-chemical parameters of the reaction can be used to predict the combustion reaction mode for a given system [15-18]. The thermo-chemical parameters of the reactants and products are given in table A1 in appendix A. The calculated adiabatic temperature ($T_{ad.}$) for various amounts of Al_2O_3 diluent content in the mixture is shown in figure 2. It can be seen that for all the samples (y = 0-2.5), $T_{ad} \geq 1800K$. The horizontal line corresponds to the melting point of Al_2O_3 (T = 2303 K). Experimental investigations of combustion temperature for various Al_2O_3 diluent added into TiO_2-Al-C system were performed elsewhere [10]. The dependence of combustion temperature on diluent content showed the similar characteristics as shown in figure 2.

3.2. The influence of diluent content on the combustion wave propagation

Propagation mode of combustion wave is an important parameter of the synthesis. Heat generation due to chemical reaction as well as heat loss due to system condition and environment are defining parameters. If heat loss is not considerable comparing with heat generation (adiabatic conditions), self-sustaining mode (SS) can occur. If heat loss is considerable comparing with heat generation unsteady propagation can occur.

Investigation showed that thermal properties of reactants in heterogeneous powder mixtures are the limiting factor [20]. Bowen C R et al. [9] studied the finite-difference model of combustion in TiO_2-Al-C system based on heat generation and heat loss competition. The temperature profile throughout the sample shows that the significant temperature gradient (300 K to 2500 K) is in a small zone between reactants and products. However, the temperature in the reactant zone is almost constant due to low thermal conductivity of reactants. High temperature product which is very near to the maximum combustion temperature, results in negligible temperature gradient in product zone. It is shown that self-sustaining (SS-mode) and non-SS mode occurs depending on the ratios in (5). In other words, the point at which the combustion reaction becomes extinguished i.e. SS-mode converts to non-SS regime, can be determined by examining the equation (5). Density and thermal conductivity are not in the equation as an approximation. SHS mode will occur when the following condition is satisfied [9]:

$$\frac{(T_{ad} - T_{ig})}{(T_{ig} - T_0)} \geq \frac{\bar{C}_P(\text{reactants})}{\bar{C}_P(\text{products})} \tag{5}$$

Where, $\bar{C}_P$ is average specific heat capacity in that temperature range.

Munir Z A [21] proposed another model based on activation energy of the combustion reaction to study influence of the diluent content and initial temperature on wave propagation. To estimate the SS/ non-SS behavior of the TiO_2-Al-C system with Al_2O_3 diluent, at first, the heat capacity of the reactant and products were calculated for various amounts of diluent. Then, the ratio of C_p(reactant)/C_p (diluent) was calculated to find the right hand side of the equation (5). Ignition temperature was assumed to be constant $(T_{ig} \approx 1144K)$ based on thermal analysis results. Furthermore, the adiabatic temperature (T_{ad}) was calculated for initial temperature of T_0 = 298 K. Finally, the corresponding diluents content was determined from figure 2. The results are shown in figure 3. The figure shows that for initial temperature of T_0 = 298 K, the minimum amount of Al_2O_3 diluent that converts SS to non-SS mode is about 39.6 wt% ($y \approx 2.46$). It needs more diluents content to convert the SS to non-SS mode for initial temperatures greater than 298 K. Therefore, the SS-mode was occurred in all samples, even though the experiment was carried out under quasi-adiabatic condition in the presence of oxygen. The optical image of combustion product is shown in figure 4. All the samples retained its initial shape without any cracks or fractures after the combustion.

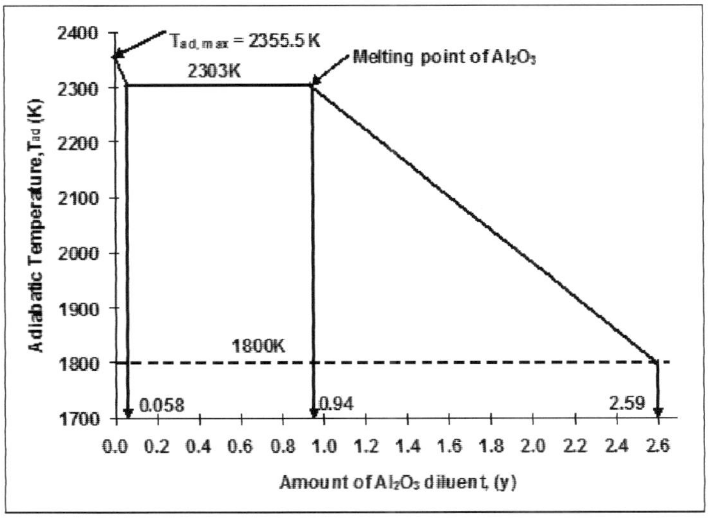

Figure 2. Calculated adiabatic temperature for various diluent - Al_2O_3 content in ($3TiO_2$+4Al+3.5C+yAl_2O_3) mixture.

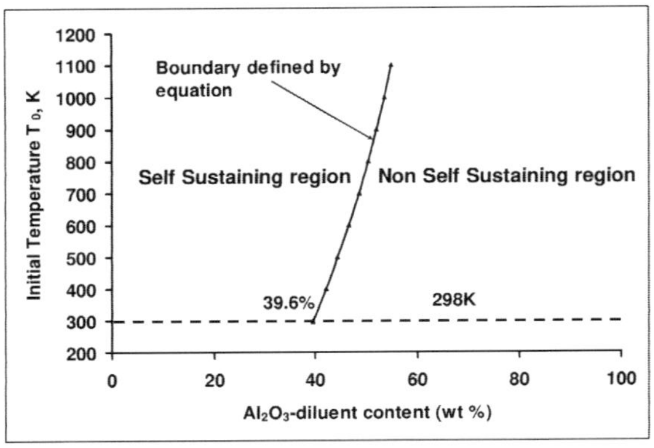

Figure 3. Combustion wave propagation dependence on initial temperature (T_0) and Al_2O_3-diluent content (wt %).

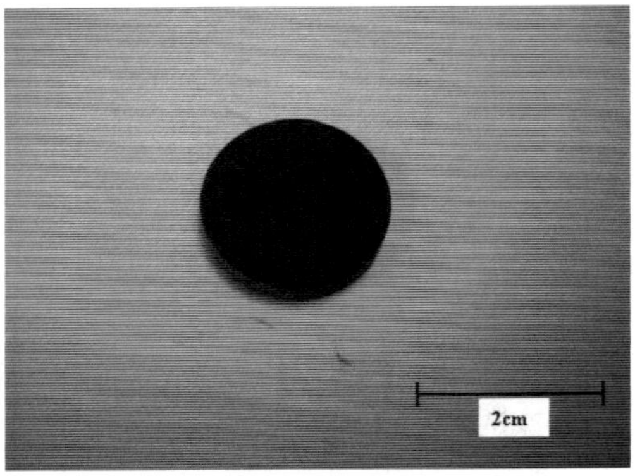

Figure 4. Typical image ofthe combustion product.

3.3. Reactant mixtures

XRD pattern of the green compact mixture without diluent is presented in figure 5. The XRD composition analysis shows the presence of three components TiO_2, Al, and C in the green mixture. SEM micrographs of green compact mixture without diluent (x = 3.5, y = 0) and with diluent (y = 2) are shown in figures 6a and 6b, respectively. In figure 6a the biggest particles probably belong to Al powder, while the smaller bright particles represent TiO_2 powder. The particles of carbon black are very small and cannot be seen easily. In figure 6b two types of big particles represent Al and Al_2O_3 powders. Smaller particles belong to TiO_2.

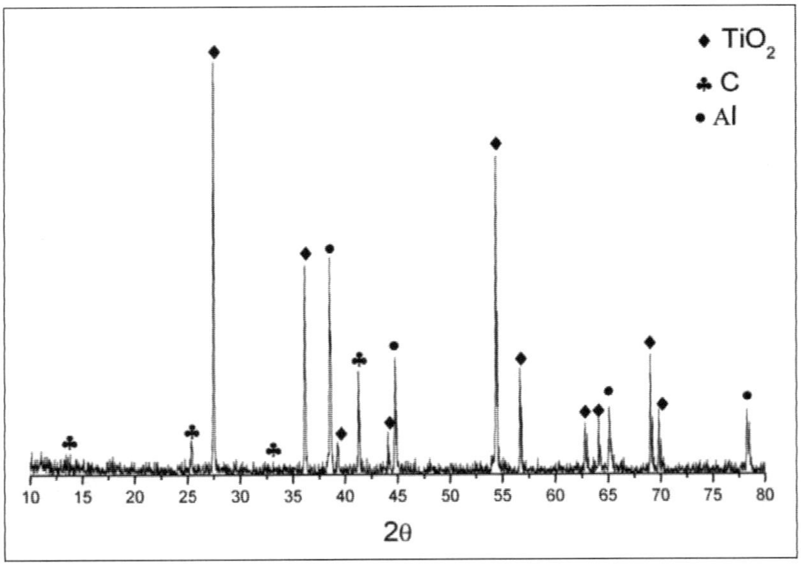

Figure 5. X-Ray Diffraction pattern of green specimen without diluent.

3.4. Product composition of combusted reactant mixture without diluent

The combustion of the reactant powder for x = 0 as in (1) did not produce the desired TiC-Al_2O_3 composition. It is due to the quasi-adiabatic condition and the presence of oxygen in the system. Therefore, the additional Ti_2O phase was detected in the product along with TiC-Al_2O_3 (figure 1a). Adding extra carbon to the mixture (x = 0.5) directed the mechanism of the combustion to produce expected TiC-Al_2O_3 composite (figure 1b). The reaction mechanism can be explained by considering the presence of oxygen. Al particles can be easily oxidized due to the presence of adsorbed oxygen or available oxygen in the environment. This resulted in deficiency of Al in aluminothermic reduction of TiO_2. Thus, carbothermal reduction of TiO_2 completed the reduction stage at higher temperature proposed by Woo Y C et al. studies [22]. The carbothermal reduction process is as in equation (6) [22]:

$$TiO_2(s) + 3C(s) \rightarrow TiC(s) + 2CO(g) \quad (6)$$

Hence, the addition of extra carbon (x = 0.5) yielded the desired composition of the TiC-Al_2O_3 product. Microstructure of the desired TiC-Al_2O_3 composite is shown in figure 7. It features various grain size distributions of dark and bright grains with pores.

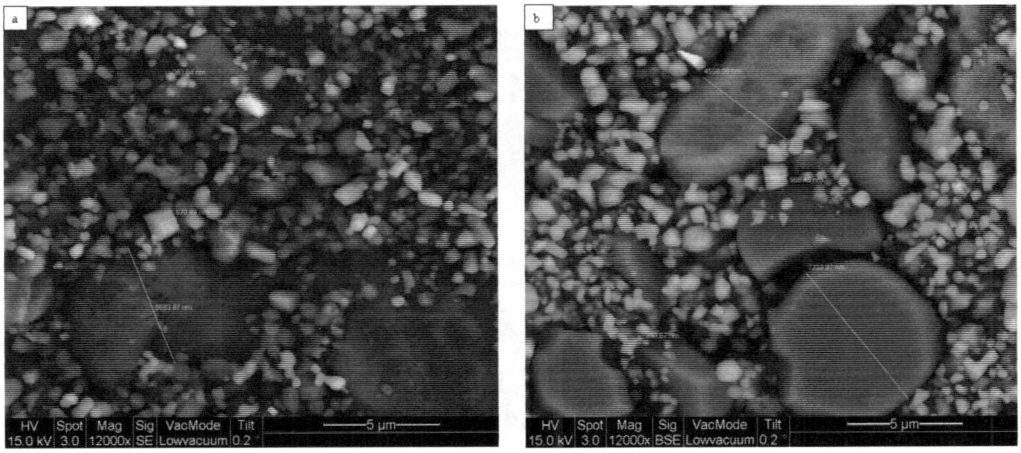

Figure 6. (a) SEM micrograph of ($3TiO_2$+4Al+3.5C) mixture. (b) SEM micrograph of (3TiO2+4Al+3.5C+2Al2O3) mixture.

3.5. Semi-quantitative phase analysis of combusted reactant mixtures with diluent

XRD patterns of combusted ($3TiO_2$+4Al+3.5C+yAl_2O_3) powder mixtures were collected from diffracted CuKα radiation. Semi-quantitative composition analysis of the products was performed using "Reference Internal Standard Method" proposed by Chung [23]. The method includes 5-10% of error.

The semi-quantitative phase analysis of products for various values of diluent (y = 0, 0.5, 1.0, 1.5, 2.0 and 2.5) is shown in table 1. The results showed that after adding extra carbon (x = 0.5, y = 0), the combustion product contained only two phases: TiC (45 wt %) and Al_2O_3 (55 wt %). Adding Al_2O_3-diluent up to about (y = 1.0) into the initial mixture increased the Al_2O_3 fraction in the same amount that consequently decreased TiC phase fraction in the product. This could be predicted theoretically as proposed by Moore J J et al. [11]. Moreover, it can be explained based on the calculated adiabatic temperature given in figure 2. According to thermodynamic estimation, when the diluent content $y \leq 0.058$, combustion temperature could reach above melting point of Al_2O_3 (figure 2). At that temperature, Al_2O_3 is fully melted.

In the region of about (y < 1) the horizontal line is corresponding to the melting of Al_2O_3 in the system. It means that combustion temperature was stabilized at melting point of Al_2O_3 (T = 2303 K). This was not limited to newly formed Al_2O_3 but included Al_2O_3 from diluent. Additional diluent (1 < y < 2) decreased the adiabatic temperature (figure 2) that resulted in detection of a new phase Ti_2O in the product which is illustrated in the table 1.

The lowering of adiabatic temperature results in decreasing the propagation velocity. Diluent slows the reaction kinetics and decreases interfacial contact area between Al and TiO_2 which results in the formation of intermetallic compounds [24]. The reaction heat was absorbed by extra diluents and consequently another oxide phase (Ti_3O_5) was formed in the product. Finally, adding more diluent powders to the reactant mixture directed the mechanism of the combustion to produce intermetallic compound ($AlTi_2$). From the previous reports, the intermetallic compounds were formed by diffusion of Ti into molten Al and were existed as intermediate phase during the reaction [24-25]. Obviously, these variations of composition influence on the microstructure and properties of the final product.

Figure 7. Micrograph of TiC-Al_2O_3 composite product produced by adding extra carbon ($x = 0.5$).

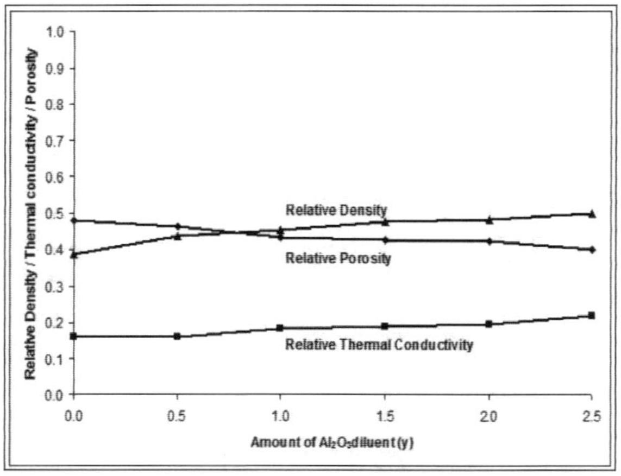

Figure 8. Relative density, thermal conductivity and porosity of combusted mixture at room temperature for various diluent content (y).

Table 1. Semi-quantitative phase analysis of combusted ($3TiO_2+4Al+3.5C+yAl_2O_3$) mixture.

Amount of Al_2O_3 diluent,	Weight fraction of product				
(y)	Al_2O_3	TiC	Ti_2O	Ti_3O_5	$AlTi_2$
0.0	0.55	0.45	0.00	0.00	0.00
0.5	0.59	0.41	0.00	0.00	0.00
1.0	0.62	0.37	0.01	0.00	0.00
1.5	0.67	0.27	0.06	0.00	0.00
2.0	0.68	0.29	0.03	0.00	0.00
2.5	0.69	0.20	0.00	0.09	0.02

3.6. Thermo- Mechanical properties

Adding Al_2O_3 diluent into the reactant mixture directed the mechanism of the combustion synthesis to produce small amount of oxide and intermetallic phases along with the main TiC-Al_2O_3 product. Furthermore, additional diluent powder not only changed the density of the mixture, but also influenced on the porosity of the compact pellet. Thermal conductivity of reactant and product depends on the porosity. Heat transfer in the system occurs in three mechanisms of conduction, convection and radiation. Any changes in the porosity vary the contribution part of each mechanism of heat transfer inside the sample. It is obvious that under quasi-adiabatic conditions of the experiment and the presence of oxygen porosity has more significant influence in the process.

A semi-empirical model for evaluation the thermal conductivity was proposed by Koh and Fortini [14]. In this model the effective thermal conductivity is given as a function of porosity and thermal conductivity of dense material as follows:

$$\lambda_{eff} = \lambda_0 \frac{1-P}{1+11P^2} \tag{7}$$

Where,
λ_{eff} - Effective thermal conductivity, $Wm^{-1}K^{-1}$
λ_0 - Thermal conductivity of dense material, 20 $Wm^{-1}K^{-1}$
P - Porosity,

Semi-empirical results of thermal conductivity, porosity and measured density of the products for various amounts of diluent (y) at room temperature are shown in figure 8. The trend of changes in relative porosity, density, and thermal conductivity demonstrates tight correlations. Overall density value obtained is less than theoretical density ($4.436 gcm^{-3}$). The figure illustrated a slight change in thermal conductivity.

4. Conclusion

Combustion synthesis of $3TiO_2$+4Al+(3+x)C mixture in the presence of oxygen was resulted the optimum value of x = 0.5 to produce desired TiC-Al_2O_3 product. Addition of Al_2O_3-diluent into the mixture directed the mechanism of the process to form oxide (Ti_2O, Ti_3O_5) and intermetallic ($AlTi_2$) phases. Thermodynamics studies of the system shown that self-sustaining mode of combustion wave propagation can occur for large amounts of diluents (y = 2.5). Presence of diluent in the initial mixture reduced combustion temperature and reaction rate. Adding Al_2O_3 diluent slightly increased density that resulted in better thermal conductivity.

5. Acknowledgements

This work was funded via Vot-F internal research fund from University of Malaya. Through research grant No. F0159/2007B.

6. Appendix A

Table A1. Thermo-chemical data for reactants and products of (TiO_2-Al-C-Al_2O_3) system [19].

Element	Temp. Range (K)	ΔH^0_{298} kcal/mol	ΔH_m kcal/mol	ΔH_{tr} kcal/mol	C_P (cal/Kmole)
TiC	298 - 3290	-44.0	17.0	0	$11.939+0.234x10^{-3}T-3.531x10^5T^{-2}+0.451x10^{-6}T^2$
Liquid	3290 – 3500	.			15.0
Al_2O_3 alpha	298 – 1273	-400.0	26.0		$27.43+3.06x10^{-3}T-8.47x10^5T^{-2}$
gamma	1273 – 2303			5.24	$25.48+4.25x10^{-3}T-6.82x10^5T^{-2}$
Liquid	2303 – 3000				34.623
TiO_2	298- 2143	-225.8	16.0	-	$15.023+2.715x10^{-3}T-2.380x10^5T^{-2}$
Al	298- 932	0		-	$4.94+2.96x10^{-3}T$
Liquid	932- 1650		2.6		7.6
C	298- 1100	0	-	-	$0.026+9.307x10^{-3}T-0.354x10^5T^{-2}-4.155x10^{-6}T^2$
	1100- 4073				$5.841+0.104x10^{-3}T-7.559x10^5T^{-2}$
Ti alpha	298- 1155	0	4.45	0.99	$5.296+2.458x10^{-3}T$
beta	1155- 1933				$4.739+1.894x10^{-3}T$
liquid	1933- 3575				8.5

References

[1] Moore J J and Feng H J 1995 *Prog. Mater. Sci.* **39** 243-273.
[2] Sytschev A E 1991 Gasless Combustion: Regularities and Mechanisms, in Physical-Chemical and Technological Base of Self-propagating High Temperature Synthesis (Moscow).
[3] Gladoun G 1994 *Inter. J. SHS.* **3(1)** 51.
[4] Lee J H, Ko S K and Won C W 2001 *Mater. Res. Bull.* **36** 1157-1167.
[5] Lee J H, Ko S K and Won C W 2001 *Mater. Res. Bull.* **36** 989-996.
[6] Kitiwan M and Atong D 2006 *Inter. J. JSME.* A **49** 85-90.
[7] Choi Y and Rhee S W 1995 *J. Am. Ceram. Soc.* **78** 986-992.
[8] Ashitani T, Tomoshige R, Ueno T and Sakai K 2003 *J. Ceram. Soc..Jpn.* **111** 1294, 372.
[9] Bowen C R and Derby B 1995 *Acta Metall. Mater.* **43** 3903-3913.
[10] Yang B, Zhang E, Zheng S, An G and Ma M 1999 *J. Mater. Eng.* **4(4)** 43-46.
[11] Feng H J and Moore J J 1995 *Metall. Mater. Trans. B.* **26B** 265-273.
[12] Cutler R A, Hurford A C 1988 *Mater. Sci. Eng. A.* **105/106** 183-192.
[13] Saw L H, Rahbari R G, Hamdi M and Rosiyah Y 2007 *3rd Int. Grad. Cong. on Mathematics and Physical Sciences* (Kuala Lumpur) P4CB-3.
[14] Koh J C Y and Fortini A 1973 *Inter. J. Heat Mass Transfer.* **16** 2013-2022.
[15] Novikov N P, Borovinskaya I P, Merzhanov A G 1975 Thermodynamic analysis of Self-Propagating high-Temperature Synthesis reaction. *Combustion in Chemical Engineering and Metallurgy* ed Merzhanov A G (Chernogolovka, Moscow) 174-188.
[16] Merzhanov A G 1975 Technological combustion problems. *Comb. Proc. Chem. Tech. Metall.* **1**. 5-28. (Russian).
[17] Holt J and Munir Z A 1986 *J. Mater. Sci.* **21** 251-259.
[18] Martinez Pacheco M 2007 Self–sustained high–temperature reactions: Initiation, propagation and synthesis: Doctorial dissertation (Netherlands) 71.
[19] Barin I and Knacke O 1973 *Thermochemical properties of inorganic substances* (New York: Springer-Verlag Berlin Heidelberg).
[20] Mukasyan A S, Rogachev A S, Mercedes M and Varma A 2004 *Chem. Eng. Sci.* **59** 5099-5105.
[21] Munir Z A 1992 *Metall. Trans. A.* **23** 7-13.
[22] Woo Y C, Kang H J and Kim D J 2007 *J. Eur. Ceram. Soc.* **27** 719-722.
[23] Chung F H 1974b *J. Appl. Cryst.* **7** 526–531.
[24] Horvitz D, Gotman I, Gutmanas E Y and Claussen N 2002 *J. Eur. Ceram. Soc.* **22** 947-954.
[25] Yu P, Mei Z and Tjong S C 2005 *Mater. Chem. Phys.* **93** 109-116.

Journal of Physics: Conference Series **152** (2009) 012056 doi:10.1088/1742-6596/152/1/012056

Effect of Li-Halides on the Morphology of Cuprates Ceramics and Their Properties under Neutron Irradiation

V Sandu, G. Aldica[1], S. Popa[1], and E. Sandu[2]

[1] Department of Low Temperature Physics and Superconductivity, National Institute of Materials Physics, Magurele, Ilfov, 077125 Romania;
[2] National Institute of Nuclear Physics and Engineering "Horia Hulubei", Magurele, Ilfov, 077125 Romania

E-mail: vsandu@infim.ro

Abstract. Superconducting $YBa_2Cu_3O_{7-\delta}$ ceramics were prepared with small addition (2-10 mol%) of lithium halides (LiF and LiCl). In all cases, the sample structure became less compact but the grains got better connected. Therefore, there is an optimal content of Li halide, around 4 mol%, which provides the best transport properties and the highest critical temperature. When submitted to neutron irradiation, the normal state resistivity shows an overall increase while the critical temperature slightly increases ($\delta T_c \leq 1$ K) in the more homogeneous samples and decreases in the stronger Li-doped samples ($\delta T_c \leq -2.8$ K).

1. Introduction

The exciting advantage of cuprate superconductors of carrying supercurrent at high temperatures has a fundamental drawback arising from the nature of vortex pinning in these materials. The small coherence length ξ of these materials, of order of few nanometres, restricts the efficient pinning centres only to defects of extension comparable to ξ, i.e., to small structural defects like vacancies or interstitials. Naturally, in cuprate superconductors, the most frequent defects are the oxygen vacancies that provide a pinning relief where the vortex lines meander in order to minimize their total free energy. Some special defects, with one extended size (twin boundaries, point defect loops, particle tracks, etc) but keep the others of order ξ, provide very strong pinning force as the vortex is aligned along the longer size. As expected, the density of this kind of defects is extremely low; hence, a functional amount of such strong pinning centres requires special processes of fabrication. Therefore, the first challenge for these materials is the creation of artificial pinning centers with increased pinning strength.

A way to create strong pinning centres is particle irradiation but only the use neutrons is effective in bulk samples because the charged particles are promptly stopped in the matter. The main drawback of neutron is the low effective cross section, of order unity, of the atomic species in cuprates. It is possible to overcome this shortcoming by the use of Li-based additives to cuprates because the isotope ^6Li has an effective cross section $\sigma = 936$ barn, a natural abundance of 17%, and is not poisonous for superconductivity [1]. Appropriate vectors (additives) for Li were found to be Li-halides [2] because both Li and the halogen have beneficial effects on superconductivity when used in reasonable amount. However, the morphology of the ceramic sample stays sensitive to the amount of the halide which in turn controls the defect accumulation during neutron irradiation. In this paper, we present the effects of lithium halides on the structural and superconducting properties of virgin and neutron irradiated $YBa_2Cu_3O_{7-\delta}$ (Y-123) ceramic samples.

2. Experimental

$YBa_2Cu_3O_{7-\delta}(LiX)_x$ ceramic samples X = F and Cl, x = 0.02; 0.04; 0.08, and 0.10, were prepared by solid state reaction from high purity reagents: Y_2O_3, CuO, $BaCO_3$, and LiX. The preparation is already trivial and involves calcinations and sintering processes in flowing oxygen for 24 h at temperatures in the range of 940-960°C.

The samples were further submitted to complex investigations regarding the magnetic and transport properties. Transport investigation was made using the four points method whereas the *ac*-susceptibility was measured with a homemade susceptometer. The morphology of the samples was investigated with a Hitachi S-2600N scanning electron microscope.
The irradiation of the samples was performed at the INR Pitesti 14MW TRIGA reactor at a fluence of 5×10^{17} neutrons/cm^2. The samples were sealed in quartz ampoules and suspended in the reactor core. After irradiation, the samples were stored for one month to reduce the residual activity at the level accepted for public (1 mSv/year).

3. Result and discussion

X-ray diffraction investigation revealed only single phase Y-123 for all concentrations of Li-halide, in agreement with other reports [3]. The investigations by scanning electron microscopy revealed a continuous diminish of the sample compactness as the amount of Li-halide is inserted. The most homogeneous and compact structure is displayed by the samples with x = 0.02 (see Fig. 1a for LiF and Fig. 1b for LiCl) that have polyhedral grains but their size is clearly larger in the sample with LiCl (Fig. 1b), possible due to the lower melting temperature of LiCl that provides local flux for growth. The pores have size of about one micron and are situated both between and within grains.

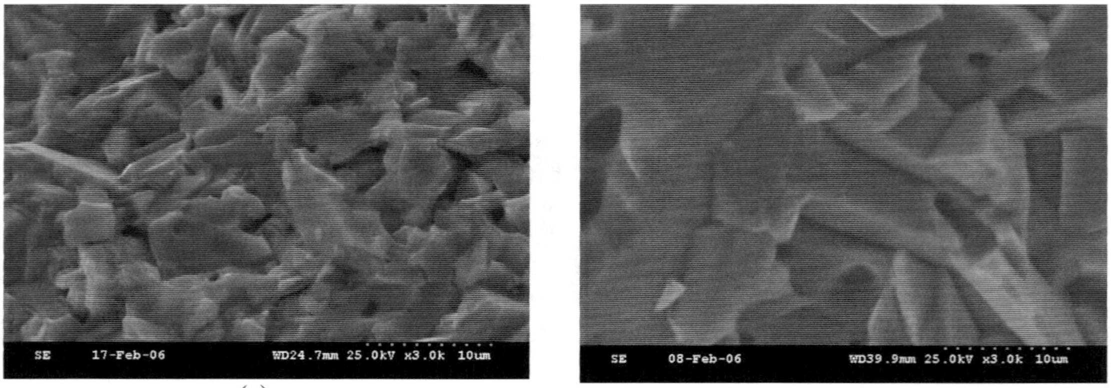

(a) (b)

Figure 1. SEM micrograph of $YBa_2Cu_3O_{7-\delta}(LiF)_{0.02}$ (a) and $YBa_2Cu_3O_{7-\delta}(LiCl)_{0.02}$ (b) (×3000).

The increase of the halide contents decreases the degree of homogeneity with predominantly smaller grains (about 2.5 μm). Grains of size of 10-25 μm are also frequent. The inhomogeneity is emphasized by an increased density of voids of order of 10 microns. However, the intergrain contacts are perceptibly improved. Some grains appear to be almost soldered in clusters (see Fig. 2 for x = 0.04). As the halide content increases, the samples get almost porous, but display different characteristics. The sample with LiF has large voids (10-30 μm) and often large grains (up to 25 μm) but also molten areas (Fig. 3a for x = 0.08). Contrarily, the grains of the sample $YBa_2Cu_3O_{7-\delta}(LiCl)_{0.08}$ are tiny (about 5 μm), of nonuniform shape and size, like at the incipient stage of growth (Fig. 3b).

The morphology of the samples has, in turn, influence on the transport properties as reflected in the resistivity ρ vs. temperature T measurements. At low temperatures, both types of doping induce similar behaviour, i.e., an improvement of the transition characteristics (T_{c0} and transition width) for x = 0.04 followed by a depression at higher halide content.

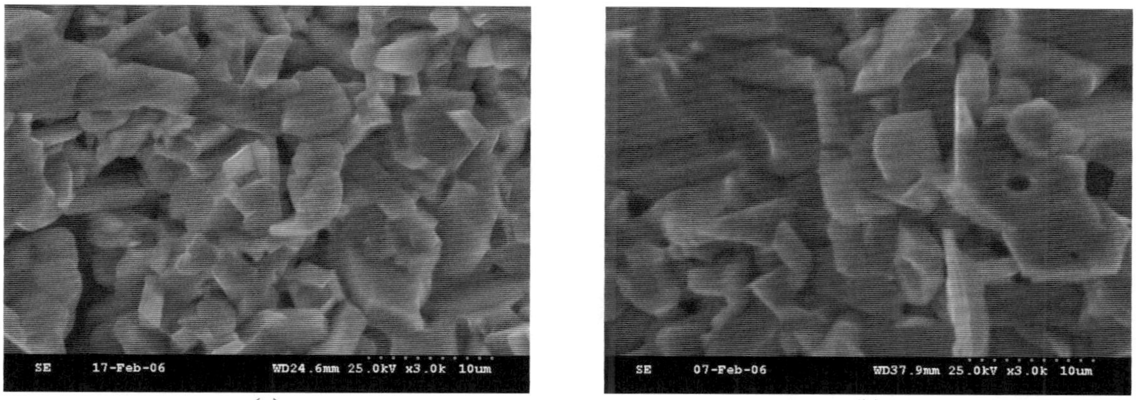

(a) (b)

Figure 2. SEM micrograph of $YBa_2Cu_3O_{7-\delta}(LiF)_{0.04}$ (a) and $YBa_2Cu_3O_{7-\delta}(LiCl)_{0.04}$ (b) (×3000).

(a) (b)

Figure 3. SEM micrograph of $YBa_2Cu_3O_{7-\delta}(LiF)_{0.08}$ (a) and $YBa_2Cu_3O_{7-\delta}(LiCl)_{0.08}$ (b) (×3000).

At high temperatures $T > 140$ K, the evolution is different. The doping with LiF decreases the normal state resistivity (with the minimum for x = 0.04), whereas for LiCl the doping increases continuously the resistivity. A reason would be the better intergrain contacts which visible for the $YBa_2Cu_3O_{7-\delta}(LiF)_{0.04}$ sample (Fig. 2a). An enhanced critical temperature was also previously reported [4]. It is considered that F substitutes O atoms mainly at O(4) sites [5, 6] increasing the ionic contribution within the Cu(1)-O(4)-Cu(2) bond, hence, the decrease of the latter and the increase of the free carrier density in the CuO_2 planes. The occupancy of O(2) and O(3) sites is also possible [7]. For chlorine, the increase of the critical temperature suggests a similar mechanisms though extended Hückel tight binding calculations indicate the O(2) sites as preferred within the CuO_2 plane [8]. Additionally, the larger size of chlorine not only limits the possibility of occupation of different oxygen sites but also induces mechanical strains and dislocations promoting granular behaviour (see Fig. 3b) [9]. On the other hand, the substitution of Li for Cu reduces the hole doping. Therefore, there is an optimal value for x that we find to occur at $x \approx 0.04$, for which superconducting properties display optimal values.

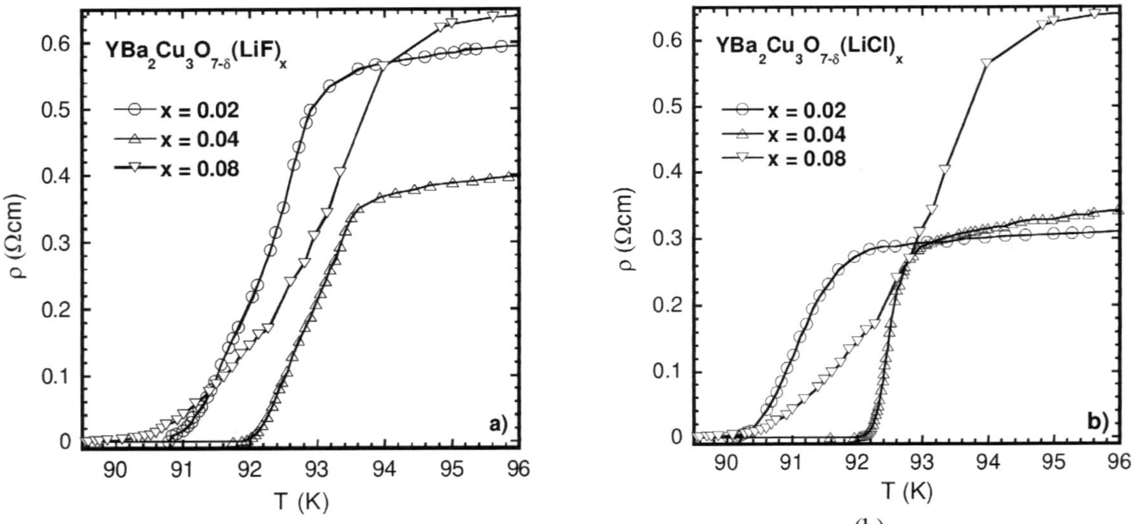

(a) (b)

Figure 4. Temperature dependence of the resistivity of the virgin samples of $YBa_2Cu_3O_{7-\delta}$ doped with Li via LiF.

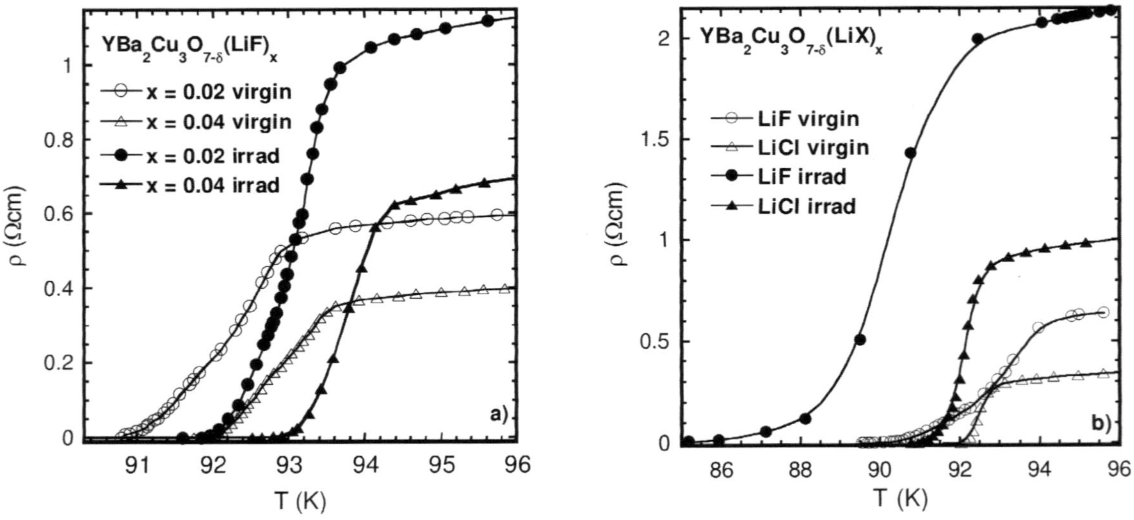

(a) (b)

Figure 5. Temperature dependence of the resistivity before (empty symbols) and after neutron irradiation at 5×10^{17} cm^{-2} (filled symbols). a) $YBa_2Cu_3O_{7-\delta}$ $(LiF)_x$, $x = 0.02$ and 0.04; b) $YBa_2Cu_3O_{7-\delta}$ $(LiF)_{0.08}$ and $YBa_2Cu_3O_{7-\delta}$ $(LiCl)_{0.04}$

After irradiation, the samples with a higher structural homogeneity show a 0.9 K increase of the post-irradiation critical temperature (Fig. 5a) accompanied by a slight decrease of the transition width.

The residual resistivity doubles up, as well as the slope dρ/dT. The improvement of the superconducting transition after irradiation in the more homogenous samples is not in contradiction with the general increase of the resistivity in normal state. In a ceramic sample, both the grains and the intergrain area contribute to the resistivity. Usually, the resistivity of the grain is lower than the intergrain space. The two contributions behave differently under neutron irradiation. The border area that has the largest resistance plays the role of a sink for the irradiation defects, hence, it always increases. The transition, however, is controlled by the grain. Therefore, the improvement of T_{c0} and of the transition width is a grain effect. The sample with LiF for $x \geq 0.08$ and the sample with LiCl for $x \geq 0.04$ have a different dependence on irradiation. The normal state resistivity is multiplied by two after irradiation whereas the critical temperature is noticeably depressed (ΔT_c = -2.8 K for $YBa_2Cu_3O_{7-\delta}$ $(LiF)_{0.08}$ and 1.2 K for $YBa_2Cu_3O_{7-\delta}$ $(LiCl)_{0.04}$) (Fig. 5b). The degradation of the transition temperature suggests that even the mechanisms of superconductivity are disturbed by the energy deposited by neutron irradiation.

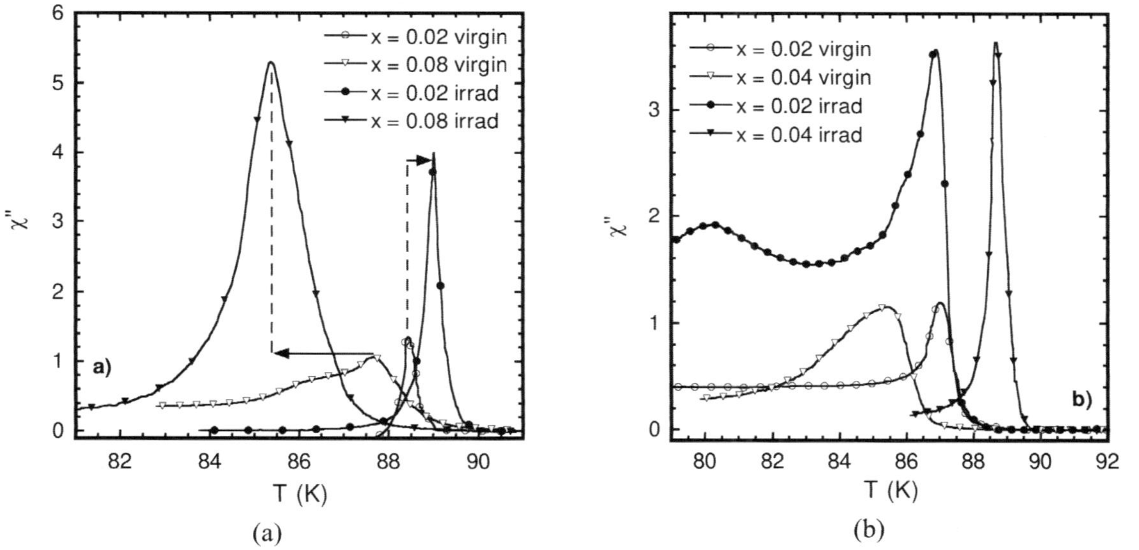

Figure 6. Temperature dependence of the imaginary part of *ac*-susceptibility before (empty symbols) and after neutron irradiation at 5×10^{17} cm^{-2} (filled symbols). a) $YBa_2Cu_3O_{7-\delta}$ $(LiF)_x$, x = 0.02 (circles) and 0.08 (triangles); b) $YBa_2Cu_3O_{7-\delta}$ $(LiCl)_x$, x = 0.02 (circles) and 0.04 (triangles)

This response to neutron irradiation is the result of the interplay between defect generation, their diffusion, capture, release and recombination in the complex landscape of the ceramic sample. All the above processes have a rate that dependent on the defect size and density. It was demonstrated that, for a low density of network dislocations, the uniform distribution of defects may become unstable relative to the pattern formation [10]. The pattern formation is enhanced in the anisotropic systems most likely in the basal planes where the mobility of the interstitial is larger. They consist in walls made of clusters of dislocations separating dislocation free areas. When the wavelength of the pattern is of the order of the grain size, the grain remain almost defect free expelling the walls in the vicinity of the grain border. The effect is important in the samples with good and large grains, i.e., at low LiX content, where it is expected that defect self-organization occurs within grains. The shape and size of the grains impose the position of the walls where the defects build up, most likely at the grain border. The presence of the walls increases both the residual and normal state resistivity (Fig. 5). Collecting the defects, these walls leave intragranular defect-free areas where superconductivity gets improved. In addition, the defects within wall strongly pin the flux lines. These assumptions are confirmed by the evolution of the imaginary part of the *ac*-susceptibility. The intragranular peak narrows and shifts to higher temperatures (Figs. 6).

For higher content of LiX, the grains are smaller, more disordered, with a high density of defects. These samples do not satisfy the requirements for self-organization of the irradiation induced defects. Therefore, the latter get evenly distributed within sample and disturb the basic interactions responsible for superconductivity. The critical temperature is depressed (Figs. 6), but the defect accumulation allows an increase of the intragranular pinning.

In summary, we have found that the addition of small amounts of LiF or LiCl to the $YBa_2Cu_3O_{7-\delta}$, controls the sample morphology (shape, size, and size distribution of the grains, size and density of the voids, etc) which, in turn, influences the transport and magnetic properties of the sample after neutron irradiation. When the amount of pre-irradiative defects is small enough the irradiation produces an enhancement of the intragrain superconductivity due to a self-organization of the defects. At high density of defects (larger LiX amount), the density of pre-irradiative defects is high and the defects density is uniformly distributed within grain, the superconducting properties are depressed, but the pinning properties remain sensitively enhanced.

Acknowledgment
The research was supported by the Romanian NARS under the Projects CEEX 45 and CEEX 73 at NIMP.

References

[1] Sandu V, Jaklovszky J, Aldica G, Cimpoiasu E, and M. C. Bunescu M C, 1998 *Balkan. Phys. Lett.* **6** 238 and therein references.
[2] Sandu V, Jaklovszky J, Cimpoiasu, Bunescu M C 1998 *J. Supercond.* **11** 653.
[3] Goren S D, Frenkel Ben-Yakar L, Shames A, Pandyopadhayay B, Korn C, Shaked H, Massiot H, Perrin C, Gallier J and Privalov A 1999 *Physica* C **313** 127.
[4] Deng H Q, Gao X H , Wang X G, Zhao L H 2000 *Trans. Nonferrous Met. Soc. China* **10** 176.
[5] Sacuto A, Julien C, Shchukin V A, Perrin C, and Mokhtari M, 1995 *Phys. Rev.* B **52** 7619.
[6] Cho S K, Kim D, Choi J S and Kim K H, 2002 *Solid State Commun.* **100** 365.
[7] Fedorenko V V, Galakhov V R, Elokhina L V, Finkelstein L D, Naish V E, Kurmaev E Z, Butorin S M, Nordgren E J, Tyagi A K, Rao U R K, Iyer R M, 1994 *Physica* C **221** 71
[8] Mokhtari M, Perrin C, Sergent M, Furet E, Halet J-F, Saillard J-Y, Ressouche E, and Burlet P, 1995 *Solid State Commun.* **93** 487
[9] Peña O, Perrin C, and Sergent M, 1993 *Supercond. Sci. Technol.* 6 183
[10] Walgraef D and Ghoniem N M 1989 *Phys. Rev.* **B 39** 8867

Journal of Physics: Conference Series **152** (2009) 012057 doi:10.1088/1742-6596/152/1/012057

Interplay between spin-singlet and spin-triplet order parameters in a model of an anisotropic superconductor with cuprate planes

Mateusz Krzyzosiak[1,2], Ryszard Gonczarek[2], Adam Gonczarek[3] and Lucjan Jacak[2]

[1]Department of Physics, Beijing Normal University, Beijing, China
[2]Institute of Physics, Wroclaw University of Technology, Wroclaw, Poland
[3]Faculty of Fundamental Problems of Technology, Wroclaw University of Technology, Wroclaw, Poland

E-mail: mateusz.krzyzosiak@pwr.wroc.pl

Abstract. Model of a two-dimensional anisotropic superconductor with a relatively wide, partially filled conduction band is considered. Attractive nearest neighbor interaction with the amplitude V_1, and on-site interaction (with the amplitude V_0) taken either as repulsive or attractive are included in the model, along with the tight-binding dispersion relation implying singularities in the density of states. The analytic study is based on the conformal transformation method, which can be regarded as an extension of the Van Hove Scenario, plausible for anisotropic systems such as high-T_c cuprates. Employing the Fourier harmonics as a complete and orthogonal basis for irreducible representations of the cuprate plane symmetry group (C_{4v}), the symmetry of the superconducting order parameter is identified for various values of the model parameters. A wide range of parameters $\eta = 2t_1/t_0$ (the ratio of the second and the first nearest neighbor hopping integrals) and n (the carrier concentration) is studied. The obtained diagrams allow one to identify and mark the areas of stability for the superconducting spin-singlet *s*- and *d*-wave and the spin-triplet *p*-wave order parameters. In particular the problem of coexistence of the s-, *p*- and *d*-wave order parameters is addressed and solved for selected values of the ratio V_0/V_1 . A possible island of stability of the *d*-wave order parameter in the *s*-wave order parameter environment for a relatively strong on-site interaction is revealed. The triple points, around which the *s*-, *d*-, and *p*-wave order parameters coexist, are identified on the diagrams. It is shown that results of the calculations performed for the two-dimensional tight-binding band model are dissimilar with some obtained within the standard BCS-type approximation. The influence of the on-site interaction on the stability of the *s*-wave order parameter is explained in detail. The developed and widely illustrated formalism can be easily applied to some more composed models.

1. Introduction

In spite of the fact that it has been already over twenty years since the discovery of high-T_c superconductors, they still attract attention due to their unusual properties at relatively high temperatures. In particular, recent solid state implementations of quantum information processing take advantage of these properties allowing one to define the superconducting two-level system (qubit) either on charge or on phase degrees of freedom [1,2]. In theoretical description of high-T_c superconducting copper-oxides, one usually treats the carriers within the CuO_2 planes as a quasi-two-dimensional system of fermions [3-5]. This is due to the characteristic, strongly anisotropic and layered structure of these materials with the symmetry of the CuO_2 planes corresponding to the point group C_{4v} [6-8].

Assuming that, in general, a superconducting system can be described by an effective Hamiltonian with a given dispersion relation $\xi_{\mathbf{k}}$ and anisotropic pairing interaction $V(\mathbf{k},\mathbf{k}')$, one can derive the two fundamental equations for high T_c superconductors: the gap equation

$$\Delta_{\mathbf{k}} = \frac{1}{N}\sum_{\mathbf{k}'} V(\mathbf{k},\mathbf{k}')\frac{\Delta_{\mathbf{k}'}}{E_{\mathbf{k}'}}\tanh\frac{E_{\mathbf{k}'}}{2T}, \tag{1}$$

where $E_{\mathbf{k}} = \sqrt{(\xi_{\mathbf{k}}-\mu)^2+\Delta_{\mathbf{k}}^2}$, and the equation

$$2n = \frac{1}{N}\sum_{\mathbf{k}}\left(1-\frac{\xi_{\mathbf{k}}-\mu}{E_{\mathbf{k}}}\tanh\frac{E_{\mathbf{k}}}{2T}\right), \tag{2}$$

which allows one to determine the total chemical potential $\mu = \mu_0 + \mu(T)$ for a given conduction band filling $0 < n < 1$ fixed for the normal metallic phase at $T = 0$. Here μ_0, defined for the metallic phase at $T = 0$, fixes the shift of the Fermi level due to doping, whereas $\mu(T)$ is the temperature correction.

The experimental data and studies of the electron-phonon interaction in strongly correlated superconducting cuprates within the *t-J* model supply a direct evidence of a phonon contribution to the pairing interaction [9], similarly to the mixed phononic and electronic model, where the phononic pairing occurs between holes strongly dressed by antiferromagnetic fluctuations [4]. However, we have to emphasize that there are also some other approaches where the pairing mechanism can be of magnetic origin, as in the magnetic interaction model which is mathematically analogous to the conventional electron-phonon problem with the generalized magnetic susceptibility playing the role of the phonon propagator [3]. Moreover, according to the idea of the tight binding description one can assume that the overlap of the orbitals in different unit cells is small compared to the diagonal overlap values. Thus the matrix element of the pairing interaction $V(\mathbf{k},\mathbf{k}')$ may contain the on-site, the nearest neighbors and the further neighbors terms, which give rise to diverse wave-symmetry channels [10].

This paper is aimed to point to a possibility of realization of a spin-singlet (i.e. $S = 0, M = 0$) *d*-wave symmetry state or a spin-triplet (i.e. $S = 1, M = 0$) *p*-wave symmetry state. The results presented here, allow us to define stable regions of the order parameter for given values of model parameters, and identify some potentially global conditions of realization and coexistence of distinct superconducting states with a fixed symmetry.

2. Model

Let us consider a system with a dispersion relation of the tight-binding band model

$$\xi_{\mathbf{k}} = -2t_0(\cos k_x + \cos k_y + \eta\cos k_x \cos k_y) \tag{3}$$

where $\eta = 2t_1/t_0 < 1$ and t_0, t_1 denote the nearest-neighbor and the next-nearest-neighbor hopping integrals, respectively.

The boson-mediated pairing interaction corresponding to on-site and nearest-neighbors pairing allows one to obtain the attractive interaction in the reciprocal space as [11,12]

$$V(\mathbf{k},\mathbf{k}') = -V_0 - V_1\left[\cos(k_x - k'_x) + \cos(k_y - k'_y)\right] \qquad (4)$$

where $V_0, V_1 > 0$ are the amplitudes of the isotropic and the anisotropic channel, respectively. The interaction (4) can be expressed as $V(\mathbf{k},\mathbf{k}') = V^s(\mathbf{k},\mathbf{k}') + V^a(\mathbf{k},\mathbf{k}')$, where $V^s(\mathbf{k},\mathbf{k}')$ and $V^a(\mathbf{k},\mathbf{k}')$ are the spin-singlet and the spin-triplet pairing potentials, respectively. The pairing interaction is assumed to be strongly anisotropic in both spin-symmetric and spin-antisymmetric channels. The maximal boson energy ω_c is identified with the cut-off parameter imposed on the one particle energy: $-\omega_c \le \xi_{\mathbf{k}} \le \omega_c$. The assumed form (4) of the interaction is sufficiently general to allow us to derive stable superconducting states from the spin-singlet and the spin-triplet ones, taking into account the *s*-wave and the *d*-wave symmetry states, as recognized in high-T_c superconductors [6-8,13]. Below we focus on the anisotropic (forward scattering) interaction, which has a very rich structure and is responsible for *d*-wave symmetry pairing. Within the applied formalism we are also able to explain the role of the on-site interaction in creation of stable superconducting states [5,11-12]. In the following section we will briefly overview an analytical approach, which can be employed to extract the essential features of the discussed model.

3. Conformal transformation method

Recently, we have developed a formalism based on the conformal transformation of the momentum space, which can be treated as an extension of the standard Van Hove Scenario (VHS) [14]. In the standard VHS the summation over momenta $\mathbf{k}$ (cf. equations (1) and (2)) is replaced by the integration over the particle energy ξ and large values of the transition temperature are attributed to singularities or narrow maxima in the density of states (DOS) function $\nu(\xi) = 4\pi^{-2}\int df \left|\nabla_{\mathbf{k}}\xi_{\mathbf{k}}\right|^{-1}$ (df is an element of the Fermi surface), corresponding to flat regions of the dispersion relation, where $\nabla_{\mathbf{k}}\xi_{\mathbf{k}} = 0$. The proposed approach [14] overcomes a weakness of the standard VHS which is the reduction of the number of degrees of freedom in a two-dimensional space only to one, namely the particle energy ξ. The main idea [14] of the conformal transformation method is to construct a new orthonormal 2D space (ξ,ϕ) in which ξ stands for one of the coordinate axes. Hence, in this method the actual number of degrees of freedom is employed. The conformal transformation $(k_x,k_y) \mapsto (\xi,\phi)$ introduces local changes of the DOS in the (ξ,ϕ) space described by the Jacobian $J(\xi,\phi) = \left|\frac{\partial(k_x,k_y)}{\partial(\xi,\phi)}\right|$.

Moreover, the coordinate ϕ can be identified with the angular variable $0 \le \varphi < 2\pi$ of the standard polar coordinate system. If any additional transformations are performed, the effective Jacobian is a product of Jacobians of successive transformations. Therefore, replacing the summation over the momentum in equations (1) and (2) by the integration over the particle energy, one has to include also the integration over the angle φ with the Jacobian $J(\xi,\varphi)$. Only in the case when the integrand does not depend on the angle φ (e.g. for s-wave pairing) the summation can replaced by the integration solely over the particle energy with the DOS defined as $\nu(\xi) = \int_0^{2\pi} d\varphi (2\pi)^{-1} K(\xi,\varphi)$, where $K(\xi,\varphi) = 2\pi^{-2}J(\xi,\varphi)$ is the kernel of the DOS. Hence the standard VHS cannot be successfully applied to other than the pure *s*-wave pairing [14]. It is worth to emphasize at this point that the conformal transformation method allows one to study analytically many thermodynamic characteristics such as e.g. the critical temperature, specific heat or isotope exponent in a consistent and systematic manner [14]. Similar formalism has been also used very recently to study Fermi liquid instabilities in two-dimensional lattice models [15].

4. Interplay between spin-singlet and spin-triplet superconductivity

Performing the conformal transformation for the discussed model, we can re-express the gap equation (1) and the equation for the chemical potential (2) in terms of the variables ξ and φ. The conformal transformation method offers all the advantages of the polar coordinate system (ξ,φ). For the system under discussion, characterized by the dispersion relation (3), the kernel of the DOS can be found explicitly [6-18, 13]. Moreover, the basis functions of irreducible representations of the group C_{4v}, which is the symmetry group of the problem, can be chosen as the Fourier harmonics in the variable φ [7]. Hence, the method allows one to expand the pairing potential (4), re-expressed in the new variables (ξ,φ), in the double Fourier series. Both components (spin-symmetric and spin-antisymmetric) of the pairing potential (4) can be rewritten in terms of specifically chosen basis functions of irreducible representations of the group C_{4v} [7]. If a single Fourier component dominates the others, the pairing potential for the corresponding state of pure pairing symmetry can be represented in the form given in table 1. Consequently, the order parameter has to be taken as $\Delta(\xi,\varphi;l,\eta)=\Delta(T)v_l(\xi;\eta,n)D_l(\varphi)$, where its angular structure $D_l(\varphi)$ is given by: 1 (for $l=0$), $\sqrt{2}\cos(\varphi+\beta_1)$ (for $l=1$ with $\beta_1=0,\pm\pi/4,\pi/2$) or $\sqrt{2}\cos 2\varphi$ (for $l=2$).

Table 1. Pairing potential for pure *s*-, *p*-, and *d*-wave pairing states. The coefficients $v_l(\xi;\eta,n)=\chi_l(\xi;\eta)/\bar{\chi}_l(\eta,n)$, where $\bar{\chi}_l(\eta,n)=(2\omega_c)^{-1}\int_{-\omega_c+\mu_0}^{\omega_c+\mu_0}d\xi\chi_l(\xi;\eta)$ is the averaged coefficient of the expansion of the pairing potential into the Fourier series [6, 7].

l	*pairing potential*
0	$-V_0-U_0(\eta,n)v_0(\xi;\eta,n)v_0(\xi';\eta,n)$
1	$-2U_1(\eta,n)v_1(\xi;\eta,n)v_1(\xi';\eta,n)[\cos\varphi\cos\varphi'+\sin\varphi\sin\varphi']$
2	$-2U_2(\eta,n)v_2(\xi;\eta,n)v_2(\xi';\eta,n)\cos 2\varphi\cos 2\varphi'$

Replacing the summation over the momentum by the integration in the (ξ,φ) space according to the conformal transformation method, one can rewrite the two fundamental equations (1) and (2) in the limit $T=T_c$ as [13]

$$1=\left[V_0\delta_{0,l}+U_l(\eta,n)\right]\int_0^{2\pi}\frac{d\varphi}{2\pi}\int_{-\omega_c}^{\omega_c}d\xi K(\xi+\mu_0,\varphi;\eta)\frac{D_l^2(\varphi)}{\xi-\mu(T_c)}\tanh\frac{\xi-\mu(T_c)}{2T_c} \tag{5}$$

and

$$2n=\frac{1}{N}\int_0^{2\pi}\frac{d\varphi}{2\pi}\int_{-\omega_c}^{\omega_c}d\xi K(\xi,\varphi;\eta)\left[1-\tanh\frac{\xi-\mu_0-\mu(T_c)}{2T_c}\right], \tag{6}$$

where the functions $U_l(\eta,n)=V_1\bar{\chi}_l^2(\eta,n)$ correspond to the coupling constant in the BCS-type factorizable pairing interaction [6-7] and it has been assumed that $v_l(\xi+\mu_0,n)\cong 1$. On the other hand, employing equation (2) in the opposite limit ($T=0$), we can find the relation among n, μ_0 and η. It then turns out that the following symmetry relation holds $n(-\eta,\mu_0)=1-n(\eta,-\mu_0)$ [7,8,13].

In order to identify the domains of stability for *s*-, *p*-, and *d*-wave superconducting states on the (η,n) plane for chosen values of the ratio $\nu=V_0/V_1$, we compare the values of the transition temperature $T_c(l,n,\eta)$, found as a self-consistent solution of equations (5) and (6) [7-8,13]. According to the experimental data, e.g. [16,17] (cf. also references in [14]), for cuprate materials $|\eta| < 0.9$, $t_0 = 0.24$ eV or 0.35 eV and ω_c takes values from 26 meV up to 65 meV. The numerical results presented in this section have been obtained for $\omega_c/2t_0 = 0.1$, and $\omega/\omega_c = 11$.

4.1. Critical temperature. BCS-type approximation

In order to provide a reference to our results it is instructive to calculate first the transition temperature within the standard BCS approach. Replacing the kernel of the DOS $K(\xi,\varphi;\eta)$ by the standard DOS $\nu(\xi;\eta)$ and putting $\nu_l(\xi;\eta) \equiv 1$ one can reduce the gap equation (5) to the BCS-type form with the actual dimensionless pairing coefficients of the form $\nu_0(\eta)U_l(\eta)/2$, where the mean value of the DOS $\nu_0(\eta) = (2\omega_c)^{-1}\int_{-\omega_c}^{\omega_c} d\xi\nu(\xi;\eta)$. The critical temperature can be then found in a standard way as

$$T_{c0}(l,\eta,n) = \frac{2e^{\gamma}}{\pi}\omega_c \exp\left(-\frac{2}{\nu_0(\eta)\left[V_0\delta_{l,0} + U_l(\eta,n)\right]}\right), \tag{7}$$

where γ is the Euler constant, or $T_{c0}(l,\eta,n) = 0$ if $V_0\delta_{l,0} + U_l(\eta,n) < 0$. The results obtained within the BCS-type approximation reveal a strong domination of the *s*-wave state in the presence of an attractive on-site interaction. Results of numerical calculations presented in figure 1 clearly illustrate

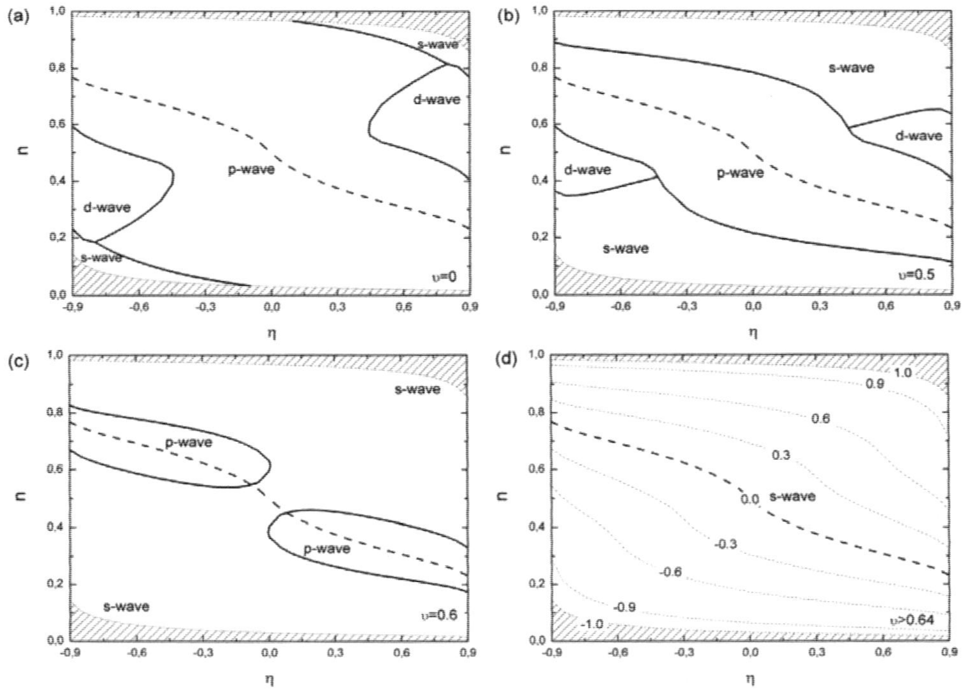

Figure 1. The diagram of stable spin-singlet and spin-triplet superconducting states obtained within the BCS-type approximation: (a), $\nu = 0$ (b), $\nu = 0.5$ (c), $\nu = 0.6$, (d) $\nu \geq 0.64$. The half-filled band concentration ($\mu_0/2t_0 = 0$) is denoted by a dashed line. In figure (d) the equi-concentration lines are additionally displayed as thin dashed lines labelled with the values of the ratio $\mu_0/2t_0$ [8,13].

how the stability areas of the *s*-wave order parameter gradually dominate the diagram on the (η, n) plane with increasing attractive on-site interaction. For $\nu \geq 0.64$, i.e. in the case of a strong attractive on-site interaction, only the *s*-wave order parameter survives (cf. figure 1d). On the other hand, in the case of a repulsive on-site interaction $\nu < 0$, our calculations [8,13] point out to the fact that the order parameter of the *s*-wave symmetry cannot be stable at all for $\nu < -0.61$. For moderate values of the parameter ν various symmetry states (*s*-, *p*-, and *d*-wave) can be observed (cf. figures 1a-c). In particular there exist some characteristic points on the (η, n) around which the order parameters of the various symmetry can be realized. These points can be referred to as the 'triple points'. Moreover, the symmetry relations for the carrier concentration mentioned at the beginning of the present section imply that for l = *0, 1, 2* we have $T_{c0}[l, -\eta, n(-\eta, \mu_0)] = T_{c0}[l, \eta, n(\eta, -\mu_0)]$ as reflected in the shape of the diagrams in figure 1 [8,13].

4.2. Critical temperature. Conformal transformation method

In order to derive the actual values of the transition temperature $T_c(l, n, \eta)$ within the tight-binding approach discussed in the present paper, we self consistently solve equations (5) and (6) using recently developed methods [7,8,13,14]. Note that also in this case the relation $T_c[l, -\eta, n(-\eta, \mu_0)] = T_c[l, \eta, n(\eta, -\mu_0)]$ holds for l = *0, 1, 2*. The resulting diagrams are presented in figures 2 and 3. In these figures the supreme value of the superconducting transition temperature within the *d*-wave stability area is marked with a dotted line, which always goes through the origin of the coordinate system. The characteristic feature of the 'triple' points is also present in these diagrams (figures 2 and 3a-3c). The location of the triple points on the diagrams in figure 2 are presented in table 2.

Table 2. The location of triple points on the diagrams in figure 2 [8].

ν	0	0.5
(η, n)	(-0.75, 0.26)	(-0.24, 0.28)
(η, n)	(0.75, 0.84)	(0.24, 0.72)

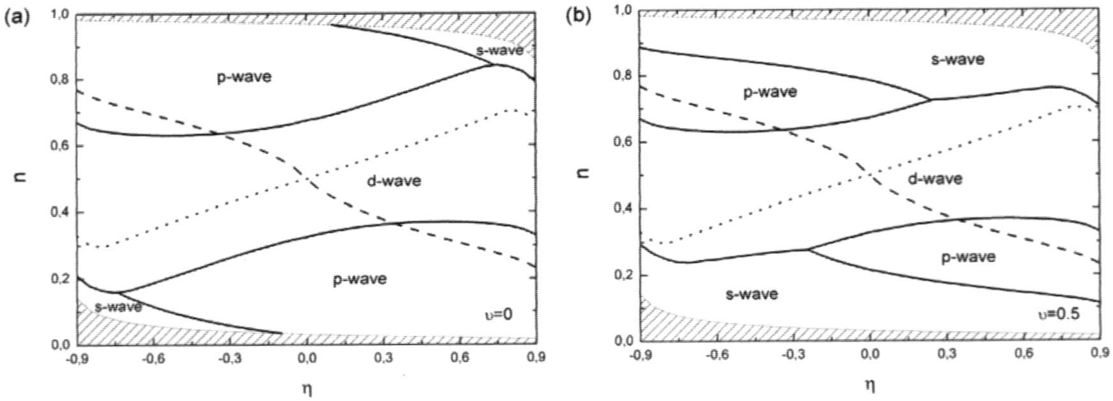

Figure 2. The diagram of stable spin-singlet and spin-triplet superconducting states obtained within the tight-binding approach: (a) $\nu = 0$, (b) $\nu = 0.5$. The half-filled band concentration ($\mu_0 / 2t_0 = 0$) is denoted by a dashed line [8,13].

From the diagrams in figure 2, and especially 3, we can conclude that the stability areas for the *s*-wave state expand, when the positive ratio ν of the pairing amplitudes increases up to the value 0.64.

Then the p-wave symmetry is eliminated from the diagram (cf. figures 3 a, b). Further increase in the ratio ν results in shrinking of the d-wave stability domain, which eventually forms an island occupying the central section of the diagram (figure 3c). Finally, when the ratio is greater than 0.9 the island vanishes and only the s-wave state is stable in the system (figure 3d). Again, in the case of a repulsive on-site interaction ($\nu < 0$), our calculations [8, 13] prove that the order parameter with the s-wave symmetry cannot be stable at all for $\nu < -0.61$.

5. Conclusions

In the present paper, employing the conformal transformation method [14], we have studied the interplay between the spin-singlet and spin-triplet order parameters, identifying the domains of stability for s-, d- and p-wave superconducting states on the diagrams in the variables (η, n).

The location of stability regions (figure 1) implies that within the BCS-type approximation the s-wave state is preferred by the system for low concentration n if $\eta < -0.4$ and for high concentration if $\eta > 0.4$. The d-wave symmetry is expected to be realized for sufficiently large $|\eta|$ and moderate n. The other parts of the diagrams are occupied by the p-wave states. On the other hand the stability domains found in the tight-binding approach are distributed according to the following scenario (figures 2, 3): for low concentration n if $\eta < -0.4$ and for high concentration if $\eta > 0.4$ the spin-singlet s-wave state is the preferred one, as in the BCS-type approximation; the spin-triplet p-wave symmetry dominates in two separated areas. However now, the d-wave state is realized for moderate values of n and η. It is also worth to emphasize that the evolution of the diagrams with the increasing ratio ν of the pairing amplitudes in these two cases is different. In the BCS-like approximation the p-wave state coexist with the s-wave in the final stage of the evolution (large values of ν), whereas in the tight-binding approach it is the d-wave symmetry which forms an 'island' in the s-wave environment.

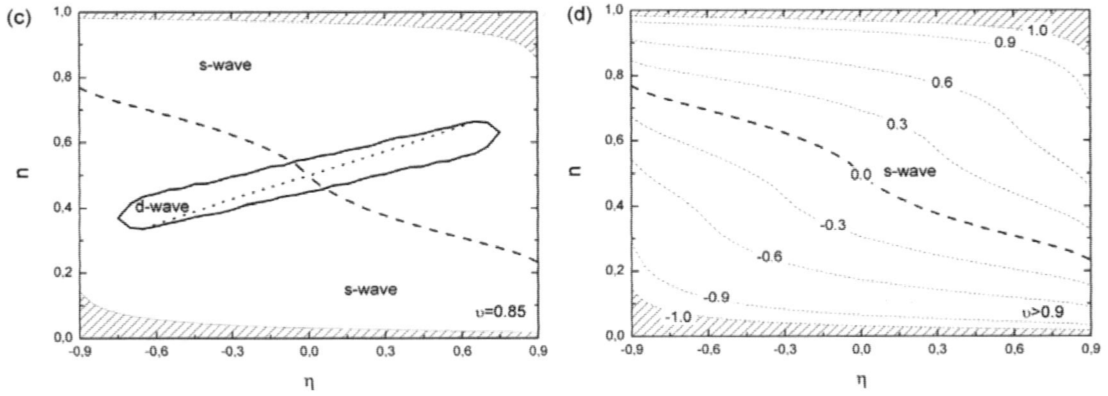

Figure 3. Evolution of a d-wave stability 'island' on the diagrams of spin-singlet and spin-triplet superconducting states obtained within the tight-binding approach: (a) $\nu = 0.6$, (b) $\nu = 0.7$, (c) $\nu = 0.85$, (d) $\nu \geq 0.9$. The half-filled band concentration ($\mu_0/2t_0 = 0$) is denoted by a dashed line. In figure (d) the equi-concentration lines are additionally displayed as thin dashed lines labelled with the values of the ratio $\mu_0/2t_0$ [8, 13].

Regarding the characteristic feature of triple points and coexistence of states with various symmetry of the order parameter, we can state that since all the diagrams have been evaluated in the limit $\Delta(T) \to 0$, the pure-symmetry states exist only in areas, which are separated from each other by phase transition lines (cf. Figures 2, 3a-3c). However, one can expect that in the superconducting phase, if the temperature T is very close to $T_c(l, n, \eta)$, the phase transition lines broaden proportionally to $|T - T_c(l,n,\eta)|$. Then it is possible to find regions along the phase transition lines, where the orders parameters of various symmetry could coexist.

We do not insist that the results obtained within the tight-binding model are quantitatively complete. Our aim was rather to show that applying the presented method to some other more complex models of high-T_c superconductors, one should obtain qualitatively comparable results. In particular, the triple point effect and coexistence of various symmetries of the order parameter should be possible to observe. Certainly, one can not expect that the model parameters (n,η,ν) can be covered in their full range discussed in this paper by a single superconducting system. There are however many possibilities in the experimental search for the discussed effects. The most promising candidates seem to be doped superconducting compounds such as spinel- and perovskite-type structures of the type $A_{2-y}B_yCuO_{4-x}$ and $A_yB_{3-}yCu_3O_{6+x}$, where *A* is a trivalent rare-earth and *B* is a divalent alkali earth ion or organic superconductors with a controlled bandwidth and band-filling (cf. [13] and references therein).

Acknowledgements

One of the authors (MK) is grateful to Professor Shiping Feng for his kind hospitality at the Department of Physics, Beijing Normal University. MK was supported by the China Scholarship Council (CSC No. 2007616032). A research scholarship from Wroclaw University of Technology is also gratefully acknowledged.

References

[1] Nakamura Y, Pashkin Yu A and Tsai J S 1999 *Nature* **398** 786
[2] Chiorescu I, Bertet P, Semba K, Nakamura Y, Harmans C J P M and Mooij J E 2004 *Nature* **431** 159
[3] Monthoux P and Lonzarich G G 2002 *Phys. Rev. B* **66** 224504
[4] Nazarenko A and Dagotto E 1996 *Phys. Rev. B* **53** R2987
[5] Xing D Y, Liu M, Wang Y-G and Dong J 1999 *Phys. Rev. B* **60** 9775
[6] Gonczarek R and Krzyzosiak M 2005 *Physica C* **426** 278
[7] Gonczarek R, Jacak L, Krzyzosiak M and Gonczarek A 2006 *Eur. Phys. J. B* **49** 171
[8] Gonczarek R, Krzyzosiak M, Jacak L and Gonczarek A 2007 *phys. stat. sol. (b)* **244** 3559
[9] Ishihara S and Nagaosa N 1987 *Phys. Rev. B* **35** 3359
[10] Ghosh H 1999 *J. Phys.: Condens. Matter* **11** L371
[11] Shimahara H and Hata S 2000 *Phys. Rev. B* **62** 14541.
[12] Gonzalez J 2000 *Phys. Rev. B* **63** 024502
[13] Gonczarek R, Krzyzosiak M and Gonczarek A 2008 *Eur. Phys. J. B* **61** 299
[14] Gonczarek R, Krzyzosiak M and Mulak M 2004 *J. Phys. A* 37, 4899
[15] Lamas C A, Cabra D C and Grandi N 2008 *Preprint* cond-mat/0804.4422
[16] Tsuei C C, Newns D M, Chi C C and Pattnaik P C 1990 *Phys. Rev. Lett.* **65** 2724; 1991 *Phys. Rev. Lett.* **67** 409; 1992 *Phys. Rev. Lett.* **68** 1091
[17] Dagotto E 1994 *Rev. Mod. Phys.* **66** 763

Singular Fermi liquid as a model of unconventional superconductivity: thermodynamic and magnetic properties

Mateusz Krzyzosiak[1,2], Ryszard Gonczarek[2] and Adam Gonczarek[3]

[1]Department of Physics, Beijing Normal University, Beijing, China
[2]Institute of Physics, Wroclaw University of Technology, Wroclaw, Poland
[3]Faculty of Fundamental Problems of Technology, Wroclaw University of Technology, Wroclaw, Poland

E-mail: mateusz.krzyzosiak@pwr.wroc.pl

Abstract. Superconductivity in the system of strongly correlated to electrons treated as a singular Fermi liquid (SFL) is studied. In the system the scattering amplitude for particles with the same momenta is divergent. Thermodynamics of the system is discussed within the Bogolubov method. The gap equation and the equation for the chemical potential are solved. It is found that the energy gap and the chemical potential are doubly-valued in the subcritical region, however only the upper branches are thermodynamically stable. The obtained results prove that the SFL system displays some features of the superconducting state also in the supercritical temperature range. It is proposed to relate the observed phenomena to the emergence of the pseudogap. The phase transition to the superconducting state is found to be of the first order with a pronounced discontinuity in the entropy. The supercritical temperature range is considered in detail. Some of the magnetic field induced effects are also discussed: the critical magnetic field and the the magnetic field penetration depth are found. Possibilities to generalize the presented results are also pointed out.

1. Introduction

Since the discovery of high-T_c superconductors (HTSC) over twenty years ago much effort has been put into explanation of their phase diagram. Sophisticated experimental methods – in particular in the area of HTSC spectroscopy [1] – revealed a fine structure of the phase diagram, especially in the case of systems near to the optimal doping. In this area of the phase diagram many anomalies have been found [2] pointing to the fact that the normal phase of HTSC should not be treated as a conventional Fermi liquid, but rather as a more complex kind of a quantum liquid (cf. [3] and references therein). The richness of the phase diagram has pushed research on the theoretical front to go beyond the limits set by the theory of Bardeen, Cooper and Schrieffer (BCS). A number of non-Fermi or singular Fermi liquid models have been proposed to deal with the anomalies observed experimentally in nearly optimally doped structures [3]. These include models with additional non-standard terms introduced into standard Hamiltonians [4-6] in order to grasp the essential properties of the HTSC. In the present paper we follow one of these paths and consider a model of a singular Fermi liquid with divergent scattering amplitude for particles with the same momenta, widely discussed in the literature [4-6].

2. Model and method

Our starting point is the effective SFL Hamiltonian, established in Refs. [4-8], with the dispersion relation $\varepsilon_{\mathbf{k}}$ (obtained e.g. from the 2D tight-binding band model), a separable pairing potential $-\eta\zeta_{\mathbf{k}}\zeta_{\mathbf{k}'}$ and the interaction $R_{\mathbf{k}}n_{\mathbf{k}\uparrow}n_{\mathbf{k}\downarrow}$ responsible for scattering of particles with the same momenta $\mathbf{k}$ and opposite spins. The scattering interaction can be separated into regular and singular parts and it is assumed that the effects of particles interaction due to the regular part, which renormalize the one-particle dispersion relation by means of the mass operator, have been included by a particular choice of the dispersion relation [9].

For an arbitrary value of the scattering amplitude $R_{\mathbf{k}}$, Hamiltonian (1) acts in a 16-dimensional Fock space spanned upon the basis

$$\psi_{\sigma\sigma'}(\mathbf{k}) = \left| \begin{matrix} \mathbf{k},\sigma \\ -\mathbf{k},\sigma' \end{matrix} \right\rangle,$$

with $\sigma,\sigma' = 0,\uparrow,\downarrow$ or $\uparrow\downarrow$ denoting the spin projection. However, when the scattering amplitude becomes divergent $(R_{\mathbf{k}} \to \infty)$, the contribution of terms corresponding to the doubly occupied states $\psi_{\sigma\uparrow\downarrow}(\mathbf{k})$ and $\psi_{\uparrow\downarrow\sigma'}(\mathbf{k})$ to the Hamiltonian is infinitely large. Hence, these terms become energetically forbidden and the Fock space effectively reduces to a 9-dimensional one with the basis $\psi_{\sigma\sigma'}(\mathbf{k})$, where $\sigma,\sigma' = 0,\uparrow,\downarrow$ [7,8,12]. Consequently the SFL Hamiltonian acting in the reduced, 9-dimensional space reads

$$H_{\mathrm{SFL}} = \sum_{\mathbf{k}} \xi_{\mathbf{k}}(n_{\mathbf{k}\uparrow} + n_{\mathbf{k}\downarrow}) - \eta \sum_{\mathbf{k},\mathbf{k}'} \zeta_{\mathbf{k}}\zeta_{\mathbf{k}'} a^{+}_{\mathbf{k}\uparrow} a^{+}_{-\mathbf{k}\downarrow} a_{-\mathbf{k}'\downarrow} a_{\mathbf{k}'\uparrow}, \tag{1}$$

where $\xi_{\mathbf{k}} = \varepsilon_{\mathbf{k}} - \mu$ and μ denotes the chemical potential.

The problem of superconductivity in the SFL characterized by the Hamiltonian (1) can be studied in a systematic manner within the Bogolubov method [10]. The main idea is to introduce the approximating Hamiltonian

$$H^{0}_{\mathrm{SFL}} = \sum_{\mathbf{k}}{}' \left[\xi_{\mathbf{k}}(n_{\mathbf{k}\uparrow} + n_{\mathbf{k}\downarrow} + n_{-\mathbf{k}\uparrow} + n_{-\mathbf{k}\downarrow}) - \Delta_{\mathbf{k}}\left(a^{+}_{\mathbf{k}\uparrow}a^{+}_{-\mathbf{k}\downarrow} + a^{+}_{-\mathbf{k}\uparrow}a^{+}_{\mathbf{k}\downarrow} + \mathrm{H.c}\right) + \frac{\Delta^2_{\mathbf{k}}}{\eta\zeta^2_{\mathbf{k}}} \right] \tag{2}$$

where the order parameter $\Delta_{\mathbf{k}}$ is taken in a self-consistent manner and the (primed) sum over the momentum runs over a half of the reciprocal space only. In the next step the partition function and the corresponding thermodynamic potential are found for the approximating Hamiltonian (2) [9,11,12]. The Bogolubov method ensures that in the thermodynamic limit, the thermodynamic potential of the approximating Hamiltonian (2) converges to the thermodynamic potential of the original SFL Hamiltonian (1). The advantage of the method is that the calculations can be done in a quite general form and the results can be easily extended, e.g. including the effects of external fields or composed dispersion relations [7,11,13].

Although it is possible to consider a very general case [7,8], in order to present some crucial features of the model, the results presented and discussed in this paper have been obtained for the isotropic s-wave paired SFL with $\varepsilon_{\mathbf{k}} = \xi = v_{\mathrm{F}}(k - k_F)$, $\eta\zeta_{\mathbf{k}}\zeta_{\mathbf{k}'} = \lambda$, and $\Delta_{\mathbf{k}} = \Delta(T)$. The pairing interaction has been assumed to cover the whole nearly half-filled conduction band of the width 2ω [7,8,11-14].

Journal of Physics: Conference Series **152** (2009) 012058 doi:10.1088/1742-6596/152/1/012058

3. Thermodynamics of the superconducting phase transition

The thermodynamics of the system is fully contained in the thermodynamic potential Ω, which is given by the Bogolubov method [7,8,10]. The thermodynamic potential is related to the free energy of the system by the equation $F = \Omega + \mu N$. At non-zero temperature the stable equilibrium state of the system is determined by the usual equilibrium conditions, which require the free energy to be minimal. In the superconducting phase, the free energy of the SFL is a function of the order parameter and the chemical potential. Consequently, the equilibrium conditions imposed on the free energy yield the gap equation and the equation for the chemical potential, whose solutions provide the temperature dependence of these two quantities. Once the temperature dependence is known, one can find the temperature dependence of the free energy and other thermodynamic characteristics, such as the entropy $S(T) = -dF(T)/dT$ and the specific heat $C(T) = -T\,d^2F(T)/dT^2$. The same arguments apply to the normal phase, the only difference being the absence of the superconducting order parameter ($\Delta_{\mathbf{k}} \equiv 0$).

The two fundamentals equation have been solved numerically for the s-wave paired SFL [7,8,12] and the results are presented in Figure 1a. The striking feature of these solutions is the appearance of two branches of solutions above a certain temperature, which we will denote as T_0. This is in apparent contrast with the usual BCS shape of the order parameter amplitude $\Delta(T)$, cf. Figure 1b. Therefore, the temperature range $T_0 \le T \le T^*$, where T^* denotes the highest temperature at which a solution of these two equation exists, has to be discussed in detail.

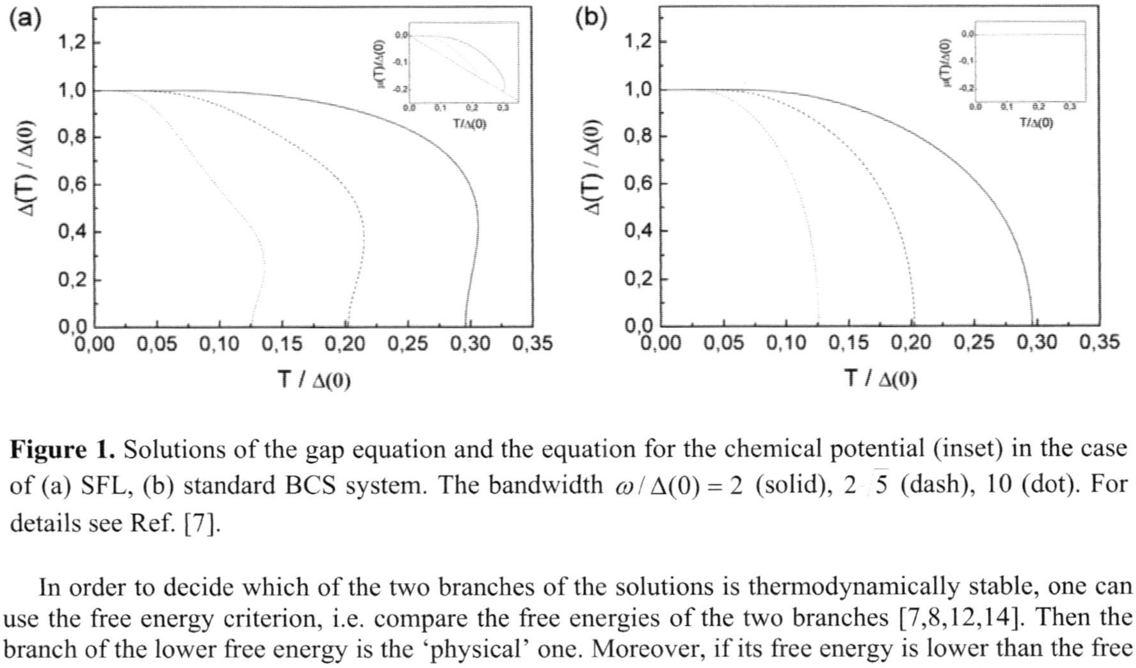

Figure 1. Solutions of the gap equation and the equation for the chemical potential (inset) in the case of (a) SFL, (b) standard BCS system. The bandwidth $\omega / \Delta(0) = 2$ (solid), 2.5 (dash), 10 (dot). For details see Ref. [7].

In order to decide which of the two branches of the solutions is thermodynamically stable, one can use the free energy criterion, i.e. compare the free energies of the two branches [7,8,12,14]. Then the branch of the lower free energy is the 'physical' one. Moreover, if its free energy is lower than the free energy of the normal phase, the SFL is in the superconducting phase. The temperature T_c at which the free energies of both (the superconducting and the normal) phases are equal, can be then identified with the temperature of the superconducting phase transition.

After a detailed analysis of the free energy (cf. figure 2 as an example), it turns out that the free energy of the superconducting phase corresponding to the lower branch of the energy gap (an the chemical potential) is greater than the free energy of the normal phase in the whole temperature range $T_0 < T \le T^*$. Therefore in this temperature range this branch is not physically significant. On the other hand, the free energy corresponding to the upper branch of the solutions is smaller than the free energy of the normal phase for $0 \le T < T_c$. For $T_c < T \le T^*$ the free energy of the normal phase is smaller and the system stops to be superconducting. The temperature T_c, at which the free energies of both phases are equal, is therefore identified with the temperature of the phase transition between the normal and the superconducting phase [7,8,12-14].

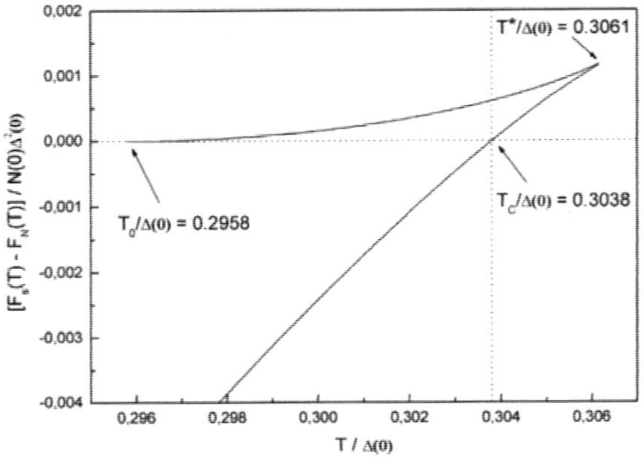

Figure 2. The difference between the free energy of the superconducting and the free energy of the normal phase for the SFL in the critical region (the case $\omega/\Delta(0)=2$ has been chosen as an example). The three characteristic temperatures are marked. The vertical dotted line is plotted to indicate the temperature of the phase transition.

These observations can be summarized as follows [7,13]. Although in the supercritical temperature range $T_c < T \le T^*$ there is a non-zero solution of the equation for the chemical potential and the gap equation (which is a unique feature of the superconducting state) the system stays normal, because the free energy criterion chooses the normal phase as the stable one. The existence of a non-zero solution of the energy gap (chemical potential) equation in spite of the fact that the system as a whole is in the normal phase, points to a possibility of Cooper pairs formation even above the phase transition temperature T_c. This fact can be related to the emergence of a pseudogap within the following scenario. At the temperature T^* phase incoherent Cooper pairs can be formed and the pseudogap opens. As a result of cooling the system further down to T_c – when the normal phase becomes (in terms of the free energy) disadvantageous for the system – the preformed Cooper pairs gain the phase coherence and the systems undergoes a phase transition to the superconducting state. The values of the three characteristic temperatures $T_0 < T_c < T^*$ involved in this scenario are quoted in table 1.

Table 1. The values of the three characteristic temperaturesin the pseudogap scenario (see also Ref. [7]).

$\frac{\omega}{\Delta(0)}$	$\frac{T_0}{\Delta(0)}$	$\frac{T_c}{\Delta(0)}$	$\frac{T^*}{\Delta(0)}$
2	0.2958	0.3038	0.3061
$2\sqrt{5}$	0.2022	0.2124	0.2151
10	0.1259	0.1339	0.1359

The pseudogap-like behaviour is not the only unusual signature of the superconducting phase in the SFL system. As already mentioned, having known the free energy of both phases, we are able to find other thermodynamic quantities characterizing the phase transition (cf. table 2). In particular, calculation of the entropy reveals its pronounced discontinuity at the phase transition, what could be already inferred from the form of the free energy presented in figure 2. Hence, the superconducting phase transition observed in the SFL system is of the first order [7,8,12].

Table 2. Some thermodynamic characteristics at the phase transition temperature [7].

$\frac{\omega}{\Delta(0)}$	$\frac{T_c}{\Delta(0)}$	$\frac{\Delta(T_c)}{\Delta(0)}$	$\frac{\mu_S(T_c)}{\Delta(0)}$	$\frac{T_c\Delta S(T_c)}{N(0)\Delta^2(0)}$
2	0.3038	0.5320	0.1691	0.1719
$2\sqrt{5}$	0.2124	0.4704	0.1301	0.1380
10	0.1339	0.3324	0.0888	0.0689

4. Magnetic field induced effects

In the present section we wil briefly discuss some properties of the SFL placed in an external magnetic field. We will limit our discussion to two characteristics, namely the critical magnetic field and the magnetic field penetration depth. These two quantities, which are important in practical applications, represent the two opposite limits: of strong and weak magnetic fields, respectively. For a more detailed review of the effects induced by the magnetic field, including the spin susceptibility and the critical current, the reader is referred to our papers [7,11].

Having know the free energy values of both phases for $0 \le T \le T_c$ in the absence of any external fields [7,8] we are able to find the critical magnetic field H_c as [7,11,13]

$$H_c(T) = \sqrt{\frac{F_N(T) - F_S(T)}{8\pi}} . \tag{3}$$

The temperature dependence of the critical magnetic field is presented in Figure 3.

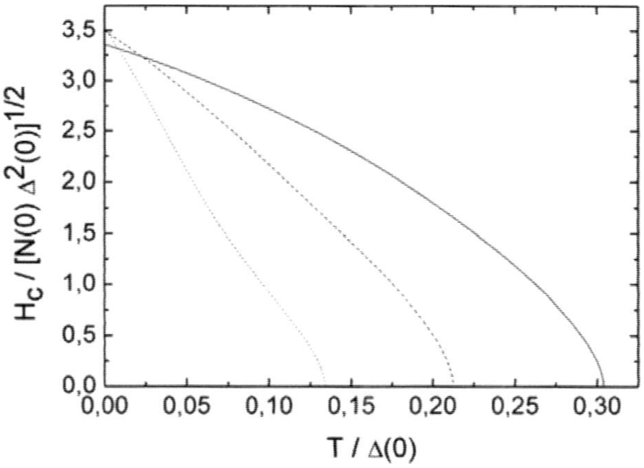

Figure 3. Critical magnetic field for the *s*-wave paired SFL with $\omega/\Delta(0) = 2$ (solid), $2\sqrt{5}$ (dashed), 10 (dotted) [7,11,13].

Rewriting equation (3) as $F_N(T) - F_S(T) = (8\pi)^{-1} H_c^2(T)$ and differentiating both sides with respect to the temperature, one obtains the formula

$$S_S(T) - S_N(T) = \frac{1}{4\pi} H_c(T) \frac{dH_c(T)}{dT} . \tag{4}$$

Note that the rhs of equation (4) is negative for all $0 \le T < T_c$. Hence, the superconducting phase is the phase of the lower entropy. Moreover, the heat of the magnetic-field-driven phase transition to the normal phase is non-zero in the discussed temperature range. Hence the phase transition is of the first order, similar to the situation for $T = T_c$ and $H = 0$ discussed in the previous section.

The analysis of the penetration depth, which is a quantitative measure of the Meissner effect, is more complicated. In order to proceed with it, one introduces the gauge-invariant superflow $\mathbf{v} = \nabla\varphi - e\mathbf{A}/m$, where $\mathbf{A}$ is the vector potential, φ is the phase of Cooper pairs wave-function and m is either the bare or the effective mass. The superflow has a physical meaning of the Cooper pair velocity [15] and its inclusion into the Hamiltonian is performed by the transformation $\mathbf{k} \mapsto \mathbf{k} + m\mathbf{v}$, where $k >> mv$. Note that the total momentum of the Cooper pair is now $2m\mathbf{v}$, unlike in the absence of the magnetic field when it is zero. Therefore the basis of the 9-dimensional Fock space should be chosen as

$$\psi_{\sigma\sigma'}(\mathbf{k}) = \left| \begin{matrix} \mathbf{k} + m\mathbf{v}, \sigma \\ -\mathbf{k} + m\mathbf{v}, \sigma' \end{matrix} \right\rangle$$

where $\sigma, \sigma' = 0, \uparrow, \downarrow$. Moreover, the chemical potential $\mu \mapsto \mu - mv^2/2$, i.e. it is constant up to terms $O(v^2)$.

However, in order to find the complete expression for the thermodynamic potential one should also take into account the energy of the magnetic field itself. The variation of the thermodynamic potential due to the vector potential turns to be $\delta\Omega = \mathbf{j} \cdot \delta\mathbf{A}$, where $\mathbf{j}$ is the current of normal particles in the SFL system given (up to linear tems in the superflow) by the formula [7,11,15]

$$\mathbf{j} = \frac{e\beta}{2m} \sum_{\mathbf{k}} \mathbf{k}(\mathbf{kv}) \frac{4}{4 + 3e^{-\beta\xi_{\mathbf{k}}} + 2\cosh\beta E_{\mathbf{k}}},$$

where $E_{\mathbf{k}} = \sqrt{\xi_{\mathbf{k}}^2 + 2\Delta_{\mathbf{k}}^2}$ and $\beta = (k_B T)^{-1}$. The superconducting (Cooper pairs) current, the superflow and the current of normal particles obey the relation $\mathbf{j}_S = eN\mathbf{v} - \mathbf{j}$, which can be rewritten as $\mathbf{j}_S = eN\mathbf{v}(1 - N_N(T)/N)$. Here N_N is the temperature-dependent number of normal (unpaired) particles, related to the magnetic field penetration depth $\lambda(T)$ by the formula $[\lambda(0)/\lambda(T)]^2 = 1 - N_N(T)/N$ [16]. Note that $N_N(0) = N$, as at $T = 0$ all particles are paired. For the *s*-wave paired SFL the temperature dependence of the magnetic field penetration depth is presented in figure 4 [7,11,14].

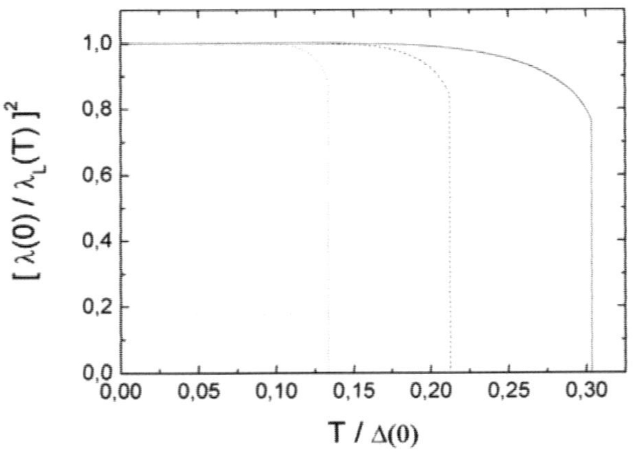

Figure 4. Magnetic field penetration depth for the *s*-wave paired SFL with $\omega/\Delta(0) = 2$ (solid), $2\sqrt{5}$ (dashed), 10 (dotted) [7,11,14].

5. Conclusions

The detailed study of the critical temperature region presented in the paper has revealed a number of features which point to the fact that the superconducting phase of the SFL is not a standard one. In the system a kind of pseudogap behavior can be observed with a possibility of pre-formation of phase-incoherent Cooper pairs above the temperature T_c of the superconducting phase transition. Moreover, unlike in a standard BCS system, the phase transition has been found to be of the first order with a pronounced discontinuity in the entropy.

The brief discussion of the effects induced by a weak external magnetic field showed that in the SFL system at the temperature infinitesimally smaller than T_c, the fraction of normal particles is small (and so is the penetration depth). This is in contrast with the typical BCS behavior, where the fraction of normal particles is infinitesimally close to one for a system infinitesimally below T_c.

Acknowledgements

One of the authors (MK) is grateful to Professor Shiping Feng for his kind hospitality at the Department of Physics, Beijing Normal University. MK was supported by the China Scholarship Council (CSC No. 2007616032). A research scholarship from Wroclaw University of Technology is also gratefully acknowledged.

References

[1] Damascelli A, Hussain Z and Shen Z-X 2003 *Rev. Mod. Phys.* **75** 473
[2] Misochko O V, Georgiev N, Dekorsy T and Helm M 2002 *Phys. Rev. Lett.* **89** 067002
[3] Varma C M, Nussinov Z and Van Saarloos W 2002 *Phys. Rep.* **361** 267
[4] Czerwonko J. 1994 *Physica C* **235–240** 2337
[5] Spałek J. and Wójcik W 1988 *Phys. Rev. B* **37** 1532
[6] Maćkowiak J.and Tarasewicz P. 2000 *Physica C* **331** 25
[7] Gonczarek R. and Krzyzosiak M 2008 *Model of superconductivity in a singular Fermi liquid*, in: *Progress in Superconductivity Research* (edited by O.A. Chang), chapter 6, pp. 163-198
[8] Gonczarek R. and Krzyzosiak M 2003 *Int. J. Mod. Phys. B* **17** 5683
[9] Micnas R., Pedersen M H, Schafroth S, Schneider T, Rodríguez–Núñez J J and Beck H 1995 *Phys. Rev. B* **52** 16223; 1996 *Phys. Rev. B* **54**3662
[10] Bogolubov N N Jr 1972 *Method for Studying Model Hamiltonians*, Vol. 43, Pergamon Press
[11] Krzyzosiak M and Gonczarek R 2004 *Int. J. Mod. Phys. B* **18** 4143
[12] Gonczarek R and Krzyzosiak M 2004 *phys. stat. sol. (b)* **241** 352
[13] Gonczarek R and Krzyzosiak M 2006 *Physica C* **445-448** 158
[14] Krzyzosiak M and Gonczarek R 2005 *Physica C* **426-431** 273
[15] Svidzinskii A V 1982 *Spatially Inhomogeneous Problems in the Theory of Superconductivity* [in Russian], Nauka, Moscow
[16] Fetter A L, Walecka J D 2003 *Quantum Theory of Many-Particle Systems*, Dover Publications, New York

Journal of Physics: Conference Series **152** (2009) 012059 doi:10.1088/1742-6596/152/1/012059

Non Invasive, Multiscale 3D X-Ray Characterization of Porous Functional Composites and Membranes, with Resolution from MM to Sub 50 NM

S H Lau[1], Wilson K S Chiu[2], Fernando Garzon[3], Hauyee Chang[1], Andrei Tkachuk[1], Michael Feser[1], Wenbing Yun[1]

[1]Xradia Inc, Concord, CA, United States

[2]Dept. of Mechanical Engineering, University of Connecticut, Storrs, CT, USA.

[3] Los Alamos National Lab., Los Alamos, NM, USA

Email: shlau@xradia.com

Abstract. We describe a novel x-ray computer tomography (CT) system for high contrast, non invasive 3D imaging of internal structures of functional ceramics, composites and polymeric membranes. System is capable of multi-length scale imaging from mm to sub 50 nm spatial resolutions and requires little or no sample preparation or staining. Relatively large samples with thickness from several mm to several microns may be imaged at high resolution. Examples using functional composites and membranes from SOFC (solid oxide fuel cell) and PEM (proton exchange membranes) fuel cell to derive direct information such as porosity and catalytic membrane degradation will be discussed. The key to the novel laboratory CT technology lies in utilizing proprietary x-ray optics with Fresnel Zone plates, and innovative high resolution, high contrast detectors.

1. Background

A variety of porous functional ceramics and membranes materials are actively being developed for energy conversion, microelectronics and biomedical applications. These includes fuel cell membranes and ceramic composites which can convert chemical energy directly into electrical energy without combustion, membranes that are responsive to their environment, breathable fabric that can neutralize biological and chemical agents. While conventional imaging tools such as optical microscopy, electron microscopy and AFM are adequate to visualize surface structures of these materials, it has been difficult to accurately characterize their internal 3D arrays, porosity and functionalities. To do that, destructive sample preparation through physical or chemical cross section must be performed. This approach can be tedious and introduces artifacts. Optical and confocal microscopy suffers from diffraction limits with spatial resolution no better than 200 nm. While electron microscopy can achieve spatial resolution in the nm scale, sample preparation can be very elaborate, including the need to be compatible with high vacuum and be electrically conductive. Moreover, conventional imaging modalities will not easily characterize functional and structural changes of materials and sensors in 3D at the multiscale level.

We describe a novel multiscale lab-based x-ray CT (computed tomography) system for rapid non invasive 3D characterization of internal structures and porosity of advanced functional materials- from polymers, ceramics and composites to microelectronic sensors, with mm to sub 50 nm spatial resolution. Using x-ray lens and proprietary optics, the system is capable of imaging samples with dimensions from several cm to microns and at relatively large working distance, opening up the possibility to set up experimental fixtures for in situ characterization of degradation phenomena of materials under certain operating conditions. With phase contrast optics, low Z composite materials and polymers can be imaged with excellent contrast.

In this paper we will describe the use of the novel CT for non invasive 3D structural characterization using examples from two representative functional materials used in Fuel Cell technologies. The first illustration is porosity characterization of a ceramic composite from Solid Oxide Fuel cell (SOFC), a promising technology for power plants, while the next example shows the power of high contrast imaging to characterize polymer membrane degradation in Proton Exchange Membrane (PEM) Fuel cells. PEM is actively being research for automotive and portable power sources. Both materials are difficult to characterize in 3D with conventional imaging modalities such as optical, SEM (scanning electron microscopy) or TEM (transmission electron microscopy).

2. Method

By combining the novel MicroXCT™ [1] and nanoXCT™ [2,3] from Xradia Inc (Concord, CA), a laboratory based multiscale CT solution may be configured. For example, relatively large samples (such as 15 cm x 15 cm x few mm thick polymeric or composite panels) may be imaged at variable resolution, including the ability to zoom in on selected region of interest for tomography with high voxel resolution of 1 to 3 microns. At the other end of the spectrum, nanoscale imaging with resolution to 50 nm can be achieved with smaller sample sizes.

2.1. MicroCT system configuration

The schematic of the MicroXCT™ system configuration is depicted below (Figure 1).

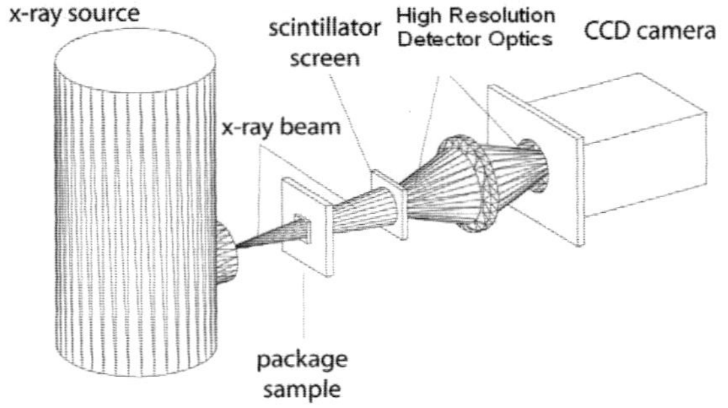

Figure 1. Schematic of the MicroXCT
The MicroXCT™ uses a commercial microfocus x-ray source. The sample can be translated in x-y-z directions in a high resolution precision rotation stage. High resolution and high contrast capability is achieved with proprietary *PhaseEnhanced*™ x-ray imaging and detector optics- with effective detector pixel size < 1 micron.

Unlike other commercial 2D or 3D x-ray systems in the market, which uses geometric projection thereby limiting resolution to the spot size of its source, the novel microCT configuration does not require the use of small spot size x-ray source nor the need for the sample to be placed very close to the source to achieve high resolution imaging. High resolution and high contrast images of a number of polymeric and composite materials with relatively large sample dimensions and with thickness of several mm can be imaged to 1 to 3 micron resolution, at large working distance, several mm from the source.

2.2. nanoCT system configuration

X-ray imaging using conventional x-ray point projection with a submicron spot sized, transmission x-ray laboratory source is commercially available. However, the smallest spot size that can be generated by a commercial x-ray source is currently around 0.25 micron, hence spatial resolution is theoretically no better than 0.25 micron in 2D. However submicron 3D resolution has only been demonstrated for certain low Z materials provided sample diameter is < 1 mm and there is sufficient image contrast within the sample material. Coupled with the poor flux and spot drift when such x-ray tube is operating at this spot size, nanoscale x-ray imaging using this approach is not practical.

Compared to conventional point projection technique, the novel lens based nanoCT provides far superior spatial resolution (currently at 50 nm) and signal to noise, with 2 to 4x greater x-ray photons reaching the detector for imaging. This configuration is similar to the high resolution x-ray microscopes at synchrotron light sources which utilizes Fresnel zone-plates as objective lenses [4]. Fresnel zone plates (Figure 2) are diffractive optical elements which focus x-rays by means of diffraction.

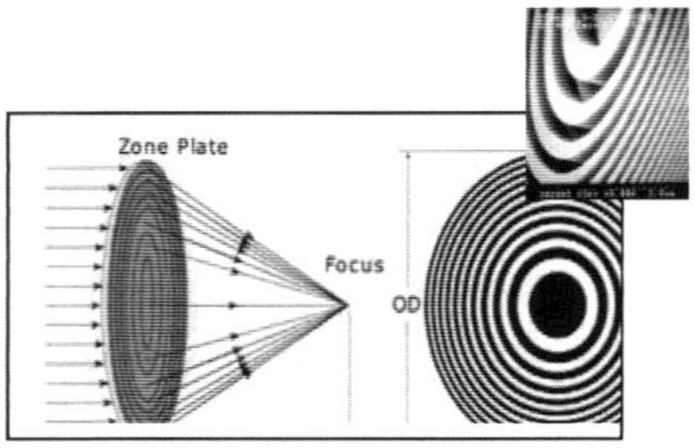

Figure 2. Fresnel Zone plates as focusing optics for x-rays. The Rayleigh resolution δ of a x-ray microscope is proportional to the outermost zone width Δr_n of the zone plate: $\delta = 1.22\Delta r_n$

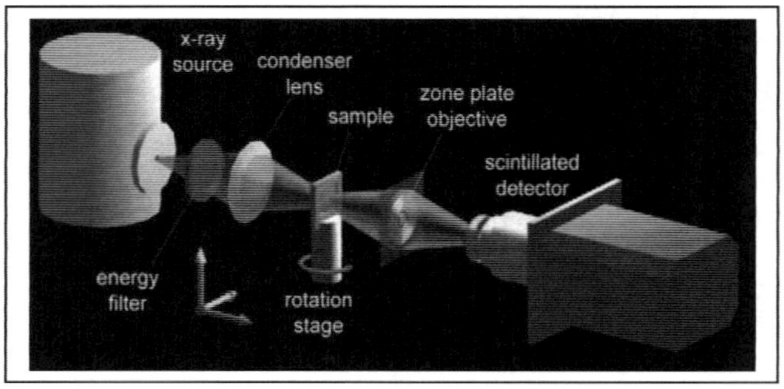

Figure 3. Schematic of sub-50 nm nanoXCT 3D X-ray microscope with specialized highly efficient reflective condenser, Fresnel Zone plate optics and high resolution imaging detector.

The Rayleigh resolution δ of a x-ray microscope is a function of the outermost zone width Δr_n of the objective zone plate:

$$\delta = 1.22\Delta r_n \quad (1)$$

The resolution of the zone-plate based x-ray microscope is independent of x-ray source spot size and is ultimately limited by the outermost zone width of the zone plate (finer zones give higher resolution).

To compensate for the lack of flux from laboratory sources, for nanoscale imaging the novel nanoCT uses a very high-efficiency reflective capillary x-ray condenser lens matched to a numerical aperture (NA) of the Fresnel zone plate objective [Figure 4] [3]. This design is similar in concept to that of a transmission light microscope where glass refractive lenses are substituted with x-ray optics. Operating in the hard x-ray regime at 5.4 or 8.0 keV, the laboratory system can, for example, penetrate through as much as 100 microns of silicon (or equivalent), making it possible to obtain 3D images of the interiors of dense samples.

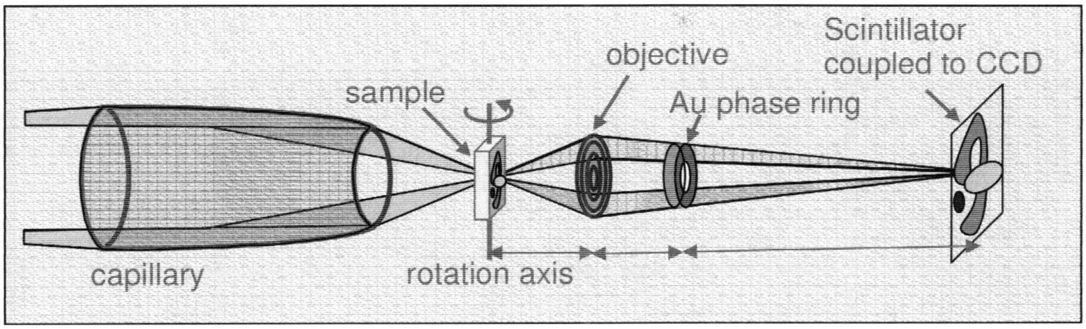

Figure 4. Imaging geometry of nanoCT with X-ray condenser lens (capillary optics), objective lens (Fresnel zone plate), Phase ring and detector

Contrast in transmission x-ray imaging is mainly based on absorption differences between different materials within the sample. For most composites, polymer or biomaterials, the x-ray attenuation length for these low Z materials can be very long, resulting in little x-ray absorption and therefore poor imaging contrast. To overcome this problem, the novel x-ray microscope can dramatically increase imaging contrast by operating in the Zernike phase contrast mode by means of a phase ring in its optics

3. Results and Discussion

Examples of non invasive multiscale 3D characterization are illustrated with two representative functional materials from Fuel Cell technologies. The first example is porosity characterization of a ceramic composite from Solid Oxide Fuel cell (SOFC), a promising technology for power plants, while another example shows degradation of the polymeric catalytic membrane from Proton Exchange Membrane (PEM) Fuel cells under different test conditions. PEM is actively being research for automotive and portable power sources. Both materials are difficult to characterize in 3D with conventional imaging modalities such as optical, SEM or TEM.

3.1. Characterization of Functional Ceramics:

3.1.1. Porosity Studies in Ceramic composites of Solid Oxide Fuel Cell (SOFC)

A critical step in making SOFC more commercially viable for large scale generation of electricity in power plants is to improve the resistance of the electrode and electrolyte components to aging. The electrochemical reactions occur in a narrow zone along the *three-phase boundary* (TPB), where the electrode material, the solid electrolyte, and the gas are in contact. It is crucial to be able to non-destructively image these nano-porous structures in 3D to study the size, distribution, and connectivity of the pores and channels to model gas transport and ionic charge transport electrochemistry. As an example of the application of the novel nanoCT system, we show CT data of a fragment of a hollow tubular solid oxide fuel-cell (SOFC), with a solid YSZ electrolyte shell and a much thicker porous Ni-YSZ anode layer on the inside [Fig 5]. Raw projection data is shown in Figure 5 as it appears on the computer screen of the microscope. Data collection is fully automated with 5 min/projection exposure time in order to obtain data with a good signal to noise ratio. Since the electron density of the solid and porous material are very similar, the solid YSZ electrolyte appears darker on the left image because of the greater projected thickness the x-rays had to penetrate at this particular imaging angle.

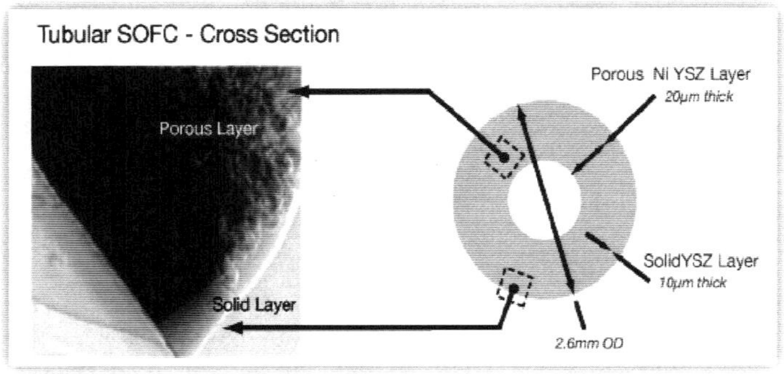

Figure 5. CT imaging of porous YSZ layer of SOFC in 3D at 50 nm resolution. Sample is imaged with 8 kV excitation using Cu anode lab source

A 2D radiograph of the sample incident shows evidence of pore structures within the sample (Figure 6a). As 2D data comprise several overlapping features, it is impossible to derive any quantitative porosity data. A major advantage of tomography is the ability to combine and mathematically reconstruct several such 2D projected images into a 3D virtual data set, where surface and buried information can be analyzed in different orthogonal planes through CT slices. Four CT slices in the depth direction from the rendered volume are shown in Figure 6(b-e). In particular, Figure 6(b) shows a CT section in anode layer, Figure 6(c) and (d) show the transition between the anode and electrolyte, and Figure 6(e) shows the electrolyte layer. Figure 6(f) shows a perspective of the rendered volume.

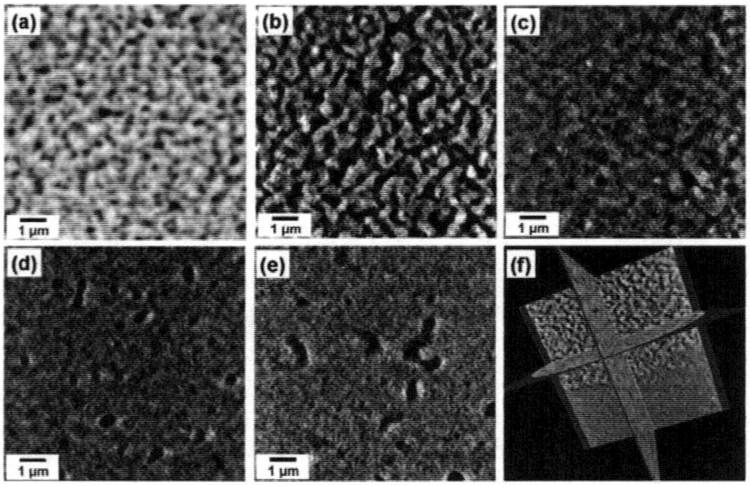

Figure 6 (a) A two-dimensional radiograph of a SOFC sample imaged through the cathode, electrolyte, and anode layers. (b) CT slice through the anode layer, (c-d) transition between anode and electrolyte, and (e) electrolyte. Each CT slice has a thickness of 30 nm. (f) A perspective of the rendered volume.

Furthermore, a 3D tomography data set can lend itself to further image analysis to derive important parameters such as porosity, open, connected and close pores, and the degree of tortuosity, which are critical for modeling electrochemistry, gas transport, electronic and ionic charge transfer [5].

While SEM can provide surface details in very high resolution to a few nm, it is not a practical imaging modality to derive porosity data in 3D especially for hard materials such as ceramics, since sample preparation to provide cross sections at the various depths and imaging planes are too tedious, time consuming and prone to artifacts. To calibrate for porosity of materials, the Mercury Intrusion

Porosity (MIP) is often used as the standard tool. While this technique is ubiquitous in the laboratory for such measurement, it does not give actual physical structure, the size of the pores at various depths, nor does it provide information on the degree of connected and close pores in the sample, which are readily available with the novel CT measurements. Validation of the novel CT data with that of the MIP for the above sample is shown in Figure 7.

3.1.2. Pore size distribution: Validation with Hg Intrusion Porosimeter

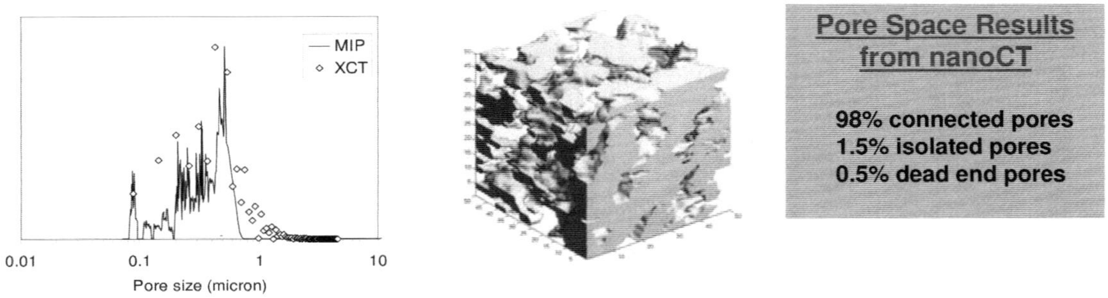

Figure 7. Pore size distribution profile comparing Mercury Intrusion Porosimeter (MIP) with novel nanoCT. However, CT results provide actual physical 3D and 2D structures & additional information such as the extent of pore connectivity [5]

The ability to non invasively characterize porosity and pore connectivity in 3D could provide critical validated data for simulation models used currently in the design of next generation SOFC micro and nano structures.

3.2. Characterization of Functional Polymeric Membranes

3.2.1. Degradation study in Proton Exchange Membrane (PEM) Fuel cell

The next example is the characterization of buried structures within functional polymeric membranes, which are also difficult to image with conventional imaging modalities such as visible or electron microscopy. Some advantages of using the novel CT for this class of functional materials includes: its non-invasive technique, it requires little or no sample preparation, does not require sample to be conductive and does not disturb the microstructure compared to TEM or SEM. Moreover, the novel CT instruments operate under ambient conditions rather than high vacuum, thus producing less morphological changes due to severe dehydration of the ionomer. X-rays generate much less sample beam damage for polymers compared to charged particles used in electron imaging techniques.

Proton exchange membrane (PEM) fuel cells offer great promise for direct conversion of hydrogen energy to electricity and are being currently developed for light transportation applications. A major requirement for this technology to be commercially viable is to improve the durability of these cells to at least 5000 operating hours with minimal degradation. Such durability benchmarks must be met under the conditions of transient power cycle operation, many start/stops, and over a range of vehicle operating temperatures (-40° to +40° C). Unfortunately the varying operating conditions have been shown to greatly exacerbate fuel cell degradation. Membrane degradation caused by Pt electrocatalyst sintering and consequent loss in active catalyst surface area and the corrosion of the supporting carbon are some of the key problems affect durability of the PEM fuel cells.

To better understand changes in the internal structure of fuel cell membrane electrode assemblies (MEAs) the novel CT is used to non invasively characterize the changes in morphologies across lengthscales, from micro to nanoscale and at high contrast [6].

3.2.2. Multilengthscale Characterization of PEM membrane degradation

3 MEAs (membrane electrode assembly) sample with dimension of 22 mm x 22mm were used for this study. They comprised 0.2mg 20 wt% Pt supported on Vulcan XC-72 carbon (ETEK) anodes (GDL:E-Tek ELAT Version 2.0 w/double-sided MPLs) and cathodes (GDL:E-Tek ELAT Version 2.22 w/single-sided MPL) and Nafion 1135.

The 3 membranes are subjected to drive cycles with start stops and without stops until failure with one control (unused membrane). Results shown in Figures 8 and 9.

3.2.3. Results: microCT slices at 0.7 micron pixel resolution

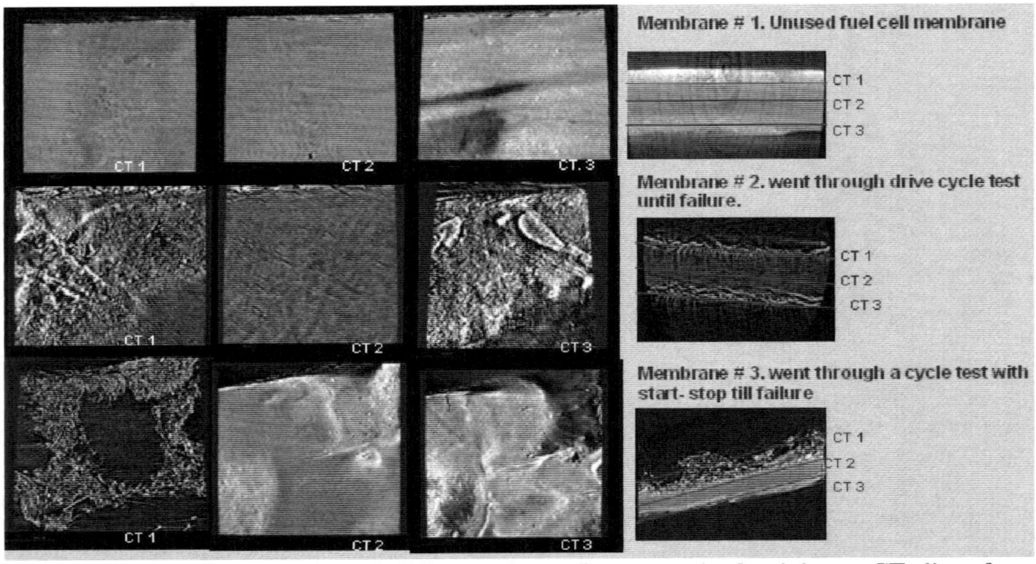

Figure 8: MicroCT slices from 2 different planes. Images at the far right are CT slices from side and the first 3 images from the left are CT slices from top of the flat membrane

Membrane #1. Unused; sample electrode density is uniform; the interface is smooth

Membrane #2. Failed membrane: went through a drive cycle without stops until failure. Electrode interfaces rough with membrane thinning, significant C corrosion at cathode with Pt compaction

Membrane #3. Undergone drive cycle with start stops until failure: Cathode Electrode interface is rough and anode interface is smoother and more uniform density; C corrosion at cathode with Pt compaction

3.2.4. *Results: Multiscale CT slices at 1 micron and 50 nm resolution*

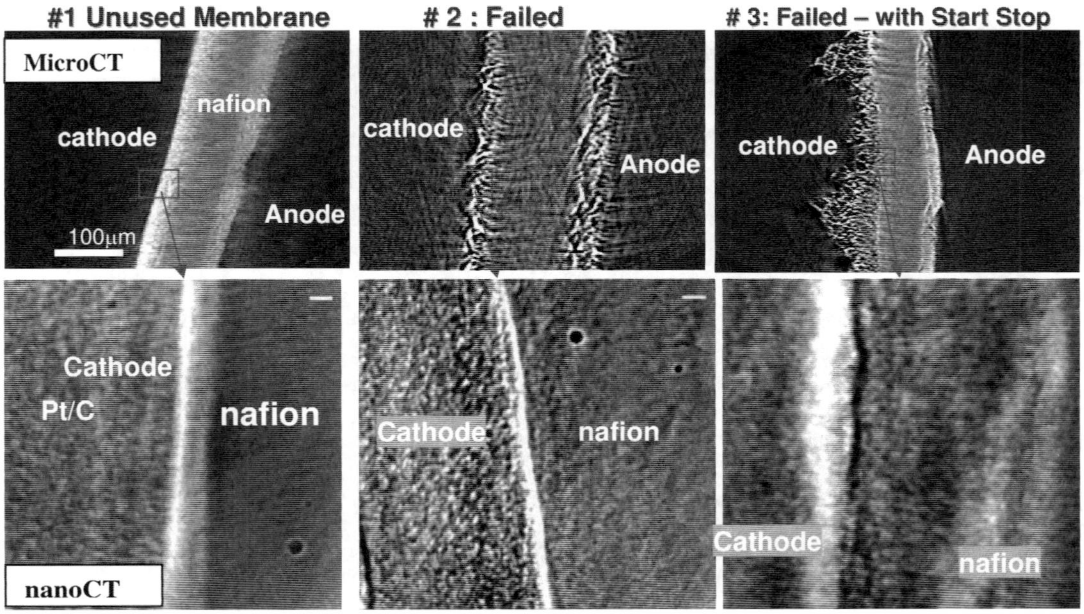

Figure 9*:* Comparison of CT slices from MicroCT imaged at 0.7 micron pixel resolution (~ 1 micron spatial resolution),and the zoomed region for higher resolution imaging with nanoCT imaged at 25 nm pixel resolution (50 nm spatial resolution).

From both micro and nanoCT imaging, it is evident that transient operation of hydrogen fuel cells produced large redistributions of cathode catalyst and compaction of the cathode layer. The compaction of the cathode layer and thinning of membrane appears to be due to carbon loss, Pt migration and agglomeration, which are more readily apparent with the nanoCT images at 50 nm spatial resolution (Membrane # 2 and 3). Nafion membrane texture also assumes a preferred orientation in the start-stop experiment (Membrane # 3). Results were in agreement with earlier studies using TEM, XRF and XRD [7], with a TEM micrograph of Pt agglomeration shown below [Figure 10].

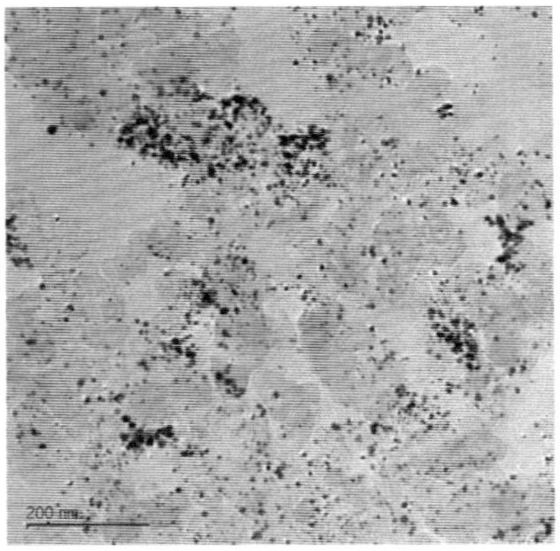

Figure 10. TEM image of MEA showing Pt agglomeration after drive cycle testing

The set of preliminary results of such non invasive CT imaging of relatively large samples at high resolution shows a lot of promise, as the study may be further extended to in situ characterization of intact PEM fuel cells for observing changes in the internal morphology of the cell under operating condition; or under controlled environmental conditions, such as freezing, heat and different humidity conditions which are encountered under normal vehicle operating environment.

4. Conclusions

Using representative examples of both hard and soft composite material, such as ceramic composites and polymeric catalyst membranes from Fuel cells as illustrations, we describe the potential applications and advantages of the novel multiscale CT system for rapid non invasive structural and failure characterization of functional materials in 3D. As little or no sample preparation is required, it is envisioned that such a lab based CT will find widespread adoption within the research community for a variety of advanced materials, biomaterials and sensors in the years to come
ome text.

References

[1] Scott, D, et al.,: "High Resolution 3D Tomography for Advanced Package Failure Analysis", *ISTFA Proceedings (2003)*

[2] S H Lau, et al., : "Virtual Non Invasive 3D Imaging of Biomaterials and Soft Tissue with a Novel High Contrast CT, with Resolution from mm to sub 30 nm", *Symposium on Advanced Functional Biomaterials, Proceedings of ICMAT, Singapore 2007*

[3] A. Tkachuk, F. Duewer, H. Cui, M. Feser, S. Wang, W. Yun," X-ray tomography in Zernike phase contrast mode @ 8 keV with 50 nm resolution, using Cu rotating x-ray source", *Z. Kristallogr.* ***222*** *(2007) 650-655*

[4] Attwood, D, : "Nanotomography comes of age", *Nature* 442, 642-643 (2006)

[5] J. R. Izzo, Jr., A. S. Joshi, A. A. Peracchio, K. N. Grew, W. K. S. Chiu, A. T. Tkachuk, S. H. Wang and W. Yun, "Nondestructive Reconstruction and Analysis of Solid Oxide Fuel Cell Anodes using X-ray Computed Tomography at Sub-50 nm Resolution," *J. Electrochem. Soc* **155** (2008) B504-B508.

[6] F. H. Garzon, S.H. Lau, J. R. Davey, and R. L. Borup: "Micro And Nano X-Ray Tomography Of PEM Fuel Cell Membranes After Transient Operation", *ECS Transactions, Proton Exchange Membrane Fuel Cells 7-Part 2, Vol* ***11****, No 1, pp 1139-50 (2007)*

[7] Borup, R.; Davey, J.; Garzon, F.; Wood, D.; Welch, P.; More, K., *ECS Transactions* **3**, 879 (2006)

Journal of Physics: Conference Series **152** (2009) 012060 doi:10.1088/1742-6596/152/1/012060

Investigation of the effect of different dopants on the trap states of ZnO-based and SnO_2-based varistors

Jiwei FAN[1], Zhenguo ZHANG[1], Huimin TIAN[1], Huijun ZHAO[1] and Robert Freer[2]

[1]School of Materials Science and Chemical Engineering, Zhongyuan University of Technology, 41 Zhongyuan Road Middle, Zhengzhou, 450007, China

[2]School of Materials Science, The University of Manchester, Manchester, M1 7HS, UK

E-mail: jwf@zzti.edu.cn

Abstract. The electrical properties of ZnO varistors and novel SnO_2 varistors have been attributed to their grain boundary barriers. In this report, the techniques of capacitance-voltage (C-V) measurement and deep level transient spectroscopy (DLTS) were employed to investigate the effect of different dopants on the electronic states of ZnO-based and SnO_2-based varistors. Two trap levels were determined in typical ZnO varistors: trap L_1 at E_c-0.15±0.01eV and trap L_2 at E_c-0.25±0.01eV. The different dopants affected the parameters of those traps significantly. Two deeper trap levels, trap S_1 at E_c-0.30±0.03eV and trap S_2 at E_c-0.69±0.03eV, were identified in both SnO_2-CoO-Nb_2O_5 and SnO_2-CoO-Nb_2O_5-Cr_2O_3 varistors. The variations in the donor density and trap density appear to be associated with the addition of trivalent Cr^{3+}. The features of these trap levels and the defect theory related to the SnO_2 varistors need to be further studied.

1. Introduction

Varistor (variable resistor) ceramics have become technologically important because of their highly non-ohmic I-V characteristics and excellent energy handle capabilities. Therefore they are widely used as surge absorbers and stabilizers in electric power systems and electronic circuits. The most commercial varistors are based on the ZnO ceramics system; the development of ZnO varistors has been one of great successes of functional ceramics [1-3].

The typical composition of ZnO varistors is based on ZnO with small amount of basic additives, such as Bi_2O_3, Sb_2O_3, Co_2O_3, MnO_2 and Cr_2O_3. It has been found that the electrical properties of ZnO varistors depend strongly upon the type and amount of additives [1-5], especially monovalent and trivalent metal oxides. Amongst the available additives, it is known that Al_2O_3 can improve the non-linearity of ZnO varistors significantly [4, 5], but it also tends to lead to a degradation of device stability[5]. In contrast, Ag_2O additions may decrease the non-linearity, but improve the stability [4]. However, an optimum combination of Al and Ag doping can improve both the non-linearity and the stability of ZnO varistors [5].

Recently, there has been the demonstration that highly non-ohmic I-V characteristics can also be achieved with SnO_2 based ceramics [1, 6, 7]. SnO_2 is an n-type semiconductor. Due to its low

Journal of Physics: Conference Series **152** (2009) 012060 doi:10.1088/1742-6596/152/1/012060

densification behaviour during sintering, this oxide is primarily used as porous gas and humidity sensors. Pianaro et al [6] reported that doping by CoO could drastically improve the sinterability of SnO_2. In addition, they observed that SnO_2 doped with 1.0mol% CoO and 0.05mol% Nb_2O_5 showed varistor behaviour with a non-linearity coefficient of 8. Further added 0.05mol% Cr_2O_3 to above composition, the non-linearity coefficient could be improved to about 40; this non-linearity level is similar to that of commercial ZnO-based varistors. Bueno et al [7-9] proposed that the nature of varistor behaviour in SnO_2-based varistors are due to the Schottky-like potential barriers at SnO_2 grain boundaries, as is that frequently reported for the conventional ZnO-based varistor systems [1-3], although the microstructure of SnO_2-based varistors are quite different from that of ZnO-based varistors.

Since the electrical properties of both ZnO varistors and SnO_2 varistors have been attributed to their grain boundary barriers (Schottky barriers) [1-7], the effect of additives must be related to the changes they cause in the grain boundary barriers. Three decades ago, Lang [10] demonstrated that deep level transient spectroscopy (DLTS) is an effective technique to investigate the trap states present in the p-n junctions or Schottky barriers in semiconductors. DLTS measurement can be used to identify the traps in barrier depletion regions and provide the information of energy levels of the traps as well as the trap densities. Most previous DLTS studies ZnO varistors have focused on the assignment of the traps present in the depletion region and sought to identify the relationship that existed between the traps, processing methods and electrical properties [11, 12]. However, the effects of additives on trap energy levels and trap densities are still not fully understood, especially for novel SnO_2 varistors. Therefore, the objective of this article is to the present the results of DLTS obtained from typical ZnO varistors doped with aluminium (from $Al(NO_3)_3{\bullet}9H_2O$) and/or silver (from Ag_2O), and the trap states obtained from SnO2-based varistors. Obviously, such DLTS information can aid our understanding of the effects of additives on the electrical properties of ZnO varistors and SnO_2 varistors.

2. Experimental Procedure

The ZnO varistors samples were based on a typical commercial formulation containing Bi_2O_3, Sb_2O_3, Co_2O_3, MnO_2 and Cr_2O_3 (but without Al or Ag) [4,5], it was designated as ZNR. The other samples were: (i) Al-doped ZNR (powder batch containing 500ppm $Al(NO_3)_3{\bullet}9H_2O$, and designated as AL); (ii) Ag-doped ZNR (powder batch containing 1000ppm Ag_2O, and designated as AG sample); (iii) combined Al- and Ag-doped ZNR (powder batch containing 500ppm $Al(NO_3)_3{\bullet}9H_2O$ and 1000ppm Ag_2O, and designated as AA). The SnO_2 varistor samples were based on the compositions:(i) 98.95mol%SnO_2+1.0mol%CoO+0.05mol%Nb_2O_5 (it was designated as SCN) and (ii) 98.90mol%SnO_2+1.0mol%CoO+ 0.05mol%Nb_2O_5+0.05mol% Cr_2O_3 (it was designated as SCNCr). All starting oxides were analytic grade (from Aldrich). The batches were mixed in deionised water for 20 hours by ball milling. After drying, the powders were pressed into disks and sintered in air at 1180°C (ZnO varistor samples) and 1350°C (SnO_2 varistor samples) for 1 hour, respectively.

The average grain size was determined by a linear intercept method using SEM micrographs. For the measurement of electrical properties, silver paste was applied on the both sides of the sintered disks to form electrodes by firing at 650°C. The non-linearity coefficients α was determined from I-V characteristics of samples between 0.1mA/cm^2 and 1mA/cm^2, according to Equation (1).

$$\alpha = \log (J_2/J_1) / \log (E_2/E_1) \tag{1}$$

where E_1 and E_2 represent the electrical fields across the sample, when the current densities were J_1 and J_2, respectively. The DC degradation rate efficient K_T was determined according to Equation (2):

$$I_L = I_{L0} + K_T t^{1/2} \tag{2}$$

where I_L is the leakage current of the sample, at time t, and I_{L0} is the leakage current when t = 0.

Capacitance-voltage (C-V) characteristics were determined at room temperature using an LCR meter (Hewlett Packard 4275A) at 10kHz together with an adjustable DC power supply (Brandenburg 475R). By assuming that the donor density (N_d) is uniform in the bulk of the ZnO or SnO_2 grains, then N_d can be determined from C-V data according to Equation (3) [13]:

$$(1/C_b - 1/2C_{b0})^2 = 2\,(\Phi+V_g)/(q\varepsilon_r\varepsilon_o N_d) \qquad (3)$$

where C_b is the capacitance per unit area of a grain boundary; q is the electronic charge; ε_o and ε_r are the vacuum permittivity and relative permittivity (ZnO: ε_r = 8.5 and SnO_2: ε_r =14), respectively; Φ is the barrier height; V_g is the applied voltage per grain boundary, and C_{bo} is the value of C_b when V=0.

The DLTS measurement was conducted by using the DLTS system at the Centre for Electronic Materials, Devices and Nanostructures, the University of Manchester. The specimens used for DLTS investigations were ground down to about 400μm thick (ZnO varistors samples) and about 100μm thick (SnO_2 varistors samples, due to they have smaller grain sizes), respectively. The specimens with electrodes were mounted in a rig capable of operating at temperature between 77 and 400 K. The capacitance of each specimen was measured using a capacitance meter operating at 1MHz. The required bias pulses (reverse bias: -35V; filling voltage: 10V; width of filling pulse: 7ms) were supplied by a dedicated DLTS unit via the capacitance meter.

The time constant of the capacitance transient of the specimen (which is the reciprocal of trap electron emission rate e_n) was evaluated by the DLTS system by comparing the time constant with one predetermined by the system sampling pulses applied at times t_1 and t_2. These pulses define a so called 'rate window' with a time constant given by $\tau = (t_1\text{-}t_2) / \ln (t_1/t_2)$. Whilst scanning across a temperature range the capacitance change ΔC was measured; at a particular temperature the time constant of the capacitance transient and the system rate window coincide. Under such circumstances the capacitance change ΔC is a maximum, and the reciprocal for the rate window time constant is equal to the trap emission rate e_n at that particular temperature. By employing a variety rate window the activation energy of the trap ΔE can be obtained by Equation (4) [10]:

$$e_n / T^2 \approx A \exp(-\Delta E / kT) \qquad (4)$$

where A is a constant, k is Boltzmann's constant and T is absolute temperature. The trap density, N_t, was determined via Equation (5) [10]:

$$N_t \approx 2N_d\Delta C_0 / C \qquad (5)$$

where N_d is the donor density, C_0 is the transient capacitance change at time t = 0 due to a saturating injection pulse, and C is the specimen capacitance.

3. Results and Discussion

3.1. ZnO Varistor Samples

Table 1 summarizes the nonlinear coefficients and degradation rate coefficients for the samples. There are clear differences and trends in α and K_T values. DLTS spectra were collected for all specimens as a function of temperature, as shown in Figure 1. The data suggests the existence of two electron traps (L_1 and L_2) at temperature of approximately 125 and 185K, respectively, in ZNR, AL and AA samples (Fig.1a). In contrast, there only appears to be one obvious electron trap (L_1) in the AG samples (Fig.1b). However, there is the suggestion of a very shallow L_2 peak for the AG sample at ~185K (the same as for ZNR, AL and AA samples) but its density must be at least two or three orders of magnitude less than that of the L_2 traps in the other samples. These results suggest that trap levels L_1 and L_2 are unrelated and are not due to the same native defect.

Table1. Non-linearity coefficient α and degradation ratecoefficient K_T for the ZnO varistor samples.

Samples	α	K_T (mA h $^{-1/2}$)
ZNR	38±2	1.4 x 10^{-2}
AL	60±2	2.7 x 10^{-2}
AG	22±1	~0
AA	58±2	~0

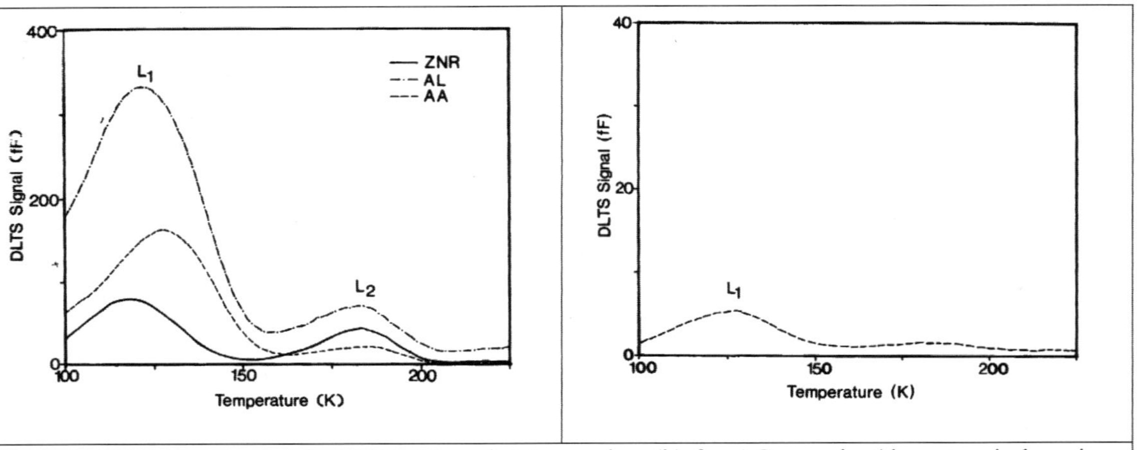

Figure 1. DLTS spectra (a) for ZNR, AL and AA samples; (b) for AG sample. (the rate window time constant, τ = 5ms).

From Arrhenius plots for the traps detected in each samples, in terms of e_n / T^2 as a function of reciprocal temperature, it was concluded tha t trap L_1 is located at E_c-0.15±0.01eV and trap L_2 at E_c-0.25±0.01eV. The details of trap levels, trap densities and donor densities for the samples are given in Table 2. It is noted that the densities of the two trap levels (L_1 and L_2) in the AL samples are much greater than in the other types of samples. In contrast, the L_1 and L_2 trap densities in the Ag-doped samples (AG) are the smallest amongst the samples studied (L_2 trap density is too small to be measured).

Table 2. Trap Parameters and Donor Densities of ZnO Varistor Samples

Sample	Density of electron traps, N_t (cm^{-3})		Donor density
	L_1 E_c-0.15±0.01eV	L_2 E_c-0.25±0.01eV	N_d (cm^{-3})
ZNR	1.5x10^{15}	9.0x10^{14}	2.2x10^{17}
AL	1.6x10^{16}	3.0x10^{15}	1.3x10^{18}
AG	3.0x10^{13}	—	2.5x10^{16}
AA	5.0x10^{15}	6.0x10^{14}	5.8x10^{17}

On the basis of the combined data, it can be seen that trivalent Al-doing increased the densities of both the L_1 and L_2 traps; such samples (AL) have the best nonlinearity characteristics, but are less stable. In contrast, doping with monovalent Ag decreased the densities of the two traps significantly; Ag-doped samples (AG) have inferior nonlinearity, but better stability. An optimum combination of Al- and Ag-doping gives a high L_1 trap density and a low L_2 trap density; such samples (AA) exhibit good performance in terms of both nonlinearity and stability. The L_1 trap level can be associated with the nonlinearity of ZnO varistors. The L_2 trap level appears to be closely associated with the stability of ZnO varistors. Since the migration of zinc interstitials should be responsible to the instability of ZnO varistors [14], this trap density may reflect the concentration of zinc interstitials. Recently, Cheng et al [15] reported that the activation energy level for electronic relaxation of zinc interstitial is 0.26eV, which is consistent with our finding.

3.2. SnO_2 Varistor Samples

SnO_2 based varistors are a new type of varistors. They are still in research stage and have not been commercialized [16]. Therefore to fully understand the electrical and electronic properties of SnO_2 varistors is quite important for their development. Table 3 shows the relative sintered densities, average grain sizes and the non-linearity coefficients α of two type SnO_2 varistor samples (SCN and SCNCr). It can be seen that those results are in good agreement to the findings by Bueno et al [7, 8] and Pianaro et al [17], except a lower α value obtained on SCNCr samples which may be caused by the different cooling rates.

Table 3. Relative sintered density ρ_r, average grain size d and non-linearity coefficient α for the SnO_2 Varistor samples.

Sample	Relative Density ρ_r (%)	Average Grain Size d (μm)	Non-linearity Coefficient
SCN	98.42±0.19	8.6±2.8	10
SCNCr	98.30±0.28	7.9±2.6	25

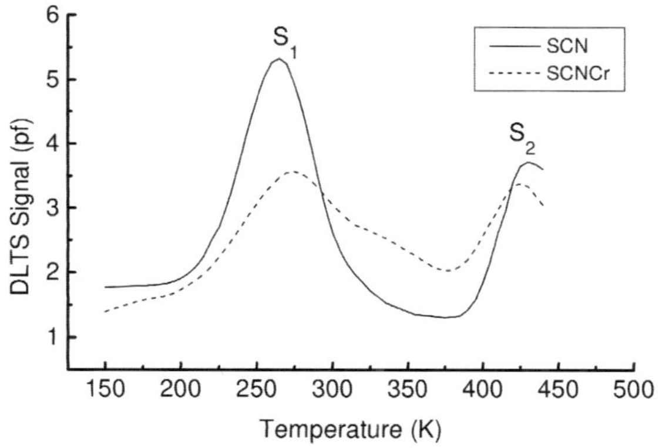

Figure 2. DLTS spectra for SnO_2 based SCN and SCNCr samples. (the rate window time constant, τ = 5ms).

Figure 2 shows the typical DLTS spectra for both of SCN and SCNCr samples. The data suggest the existence of two electron traps (S_1 and S_2) at temperature of approximately 270K and 428K, respectively. If comparing the magnitude of DLTS signals, the magnitude of S_1 peak for SCN sample is higher than that of SCNCr sample. These differences must relate to the doping of chromium. From Arrhenius plots for the traps detected in each samples, it can be considered that trap S_1 is located at E_c-0.30±0.01eV, and trap S_2 is located at about E_c-0.69±0.03eV. Both trap levels are deeper than those found on ZnO Varistor Samples. The details of donor densities, trap levels and trap densities of SCN and SCNCr samples are shown in Table 4. Since two deep level traps (S_1 and S_2) are determined on both SCN and SCNCr samples, these two traps could be related to intrinsic defects of SnO_2 and extrinsic defects. However, it can be confirmed that both trap levels are not attributed to the chromium doping, since Cr_2O_3 doping is only change the donor density and trap densities. Cr^{3+} doping plays an acceptor role; therefore the lower donor density obtained on SCNCr sample is reasonable.

Table 4. Trap Parameters and Donor Densities of SnO_2 Varistor Samples

Sample	Density of electron traps, N_t (cm^{-3})		Donor density
	S_1 E_c-0.30±0.01eV	S_2 E_c-0.69±0.03eV	N_d (cm^{-3})
SCN	$6.8x10^{16}$	$4.8x10^{16}$	$2.7x10^{18}$
SCNCr	$3.6x10^{16}$	$3.4x10^{16}$	$1.5x10^{18}$

The possible intrinsic defects of SnO_2 are oxygen vacancies $V_O^{\bullet}$, $V_O^{\bullet\bullet}$, tin vacancies V_{Sn}'', V_{Sn}'''', and tin interstitials $Sn_i^{\bullet\bullet}$ $Sn_i^{\bullet\bullet\bullet\bullet}$. The presence of dopants CoO, Nb_2O_5 and Cr_2O_3 lead to the formation of extrinsic defects. Bueno et al [18] reported that cobalt atoms could affect the trapping state at grain boundary and modify the potential barrier. The segregated or precipitated cobalt atoms can become richer in oxygen through oxidation of Co atom (e.g. CoO to Co_2O_3). Considering that the ionic radii of Co^{2+} (0.072nm), Co^{3+} (0.063nm), Nb^{5+} (0.069nm) and Cr^{3+} (0.063nm) are close to the radius of Sn^{4+} (0.071nm), the possible extrinsic defects could be Co_{Sn}'', Co_{Sn}', $Nb_{Sn}^{\bullet}$ and Cr_{Sn}' [18].

In practice both natural and synthetic SnO_2 crystals are oxygen deficient (n-type semiconductors), leading to the formation of a shallow donor level (approximately Ec-0.1eV). This shallow donor level is ascribed to the first ionization energy of oxygen vacancy $V_O^{\bullet}$ [20]. Chang [21] studied SnO_2, reported the activation energy of $Sn_i^{\bullet\bullet}$ is about 0.06eV. Therefore both $V_O^{\bullet}$ and $Sn_i^{\bullet\bullet}$ energy level are shallow, they could not be the species of two deep levels reported at this article. Since Cr_2O_3 doping is only change the donor density and trap densities, Cr_{Sn}' is also not the species of these two deep levels.

By analyzing the admittance-frequency characteristics, Bueno et al [18] reported the presence of deep trap states at about E_c-0.42±0.03eV on SnO_2-CoO-Nb_2O_5-Cr_2O_3 (SCNCr) based varistor ceramics doped with La_2O_3, Pr_2O_3 and CeO_2, respectively. From same research group, Orlandi et al [22] studied the admittance and dielectrics spectroscopes of SnO_2-MnO-Nb_2O_5 (SMN) based varistor ceramics. They obtained the activation energies of higher frequency loss peaks were about 0.49eV, while the activation energies of the low frequency dispersions were about 0.67eV. They believed that the values of high frequency process are, as similar as the values (~0.42eV) which Bueno et al [18] obtained on SCNCr based varistor ceramics, attributed to a deep trapped level in the potential barrier. They believe that the deep level (~0.42-0.49eV) determined by admittance spectroscopy is associated to the second ionization energy of oxygen vacancies $V_O^{\bullet\bullet}$ [18,22]. In their proposed model for potential barrier, Bueno et al[17] indicate that Co_{Sn}' and $Nb_{Sn}^{\bullet}$ belongs to the shallow level species, and V_{Sn}'', V_{Sn}'''' and Co_{Sn}'' should be at the grain boundary interface.

From above discussion, it could be argued that the two deep trap levels determined by DLTS on SCN and SCNCr varistor ceramics may be associated with those defects, such as the second ionization energy of oxygen vacancies $V_O^{\bullet\bullet}$, and ionized tin interstitial $Sn_i^{\bullet\bullet\bullet\bullet}$. However, the features of these trap levels and the defect theory related to the SnO_2 varistors need to be further studied.

4. Conclusions

(1) DLTS is a very useful technique to investigate the trap states presented in ZnO and SnO_2 based varistors.

(2) Two deep level electron traps have been exist in ZnO varistors of a typical commercial composition (ZNR) and specimens doped with Al and /or Ag. The traps L_1 and L_2 are located at 0.15±0.01eV and E_c-0.25±0.01eV below the conduction band, respectively.

(3) The L_1 trap level can be associated with the non-linearity of ZnO varistors; the higher the L_1 trap density the higher the non-linearity exponent. The L_1 trap may be caused by extrinsic donor defects, including monovalent and trivalent dopants.

(4) The L_2 trap level appears to be closely associated with the stability of ZnO varistors; the higher the L_2 density the lower the stability of the samples. The L_2 level is due to zinc interstitials. The trap density may reflect the concentration of zinc interstitials.
(5) Two deeper traps were detected in typical SnO_2 varistors (SCN and SCNCr). The traps S1 and S2 are located at 0.30±0.01eV and E_c-0.69±0.03eV below the conduction band, respectively.
(6) It can be confirmed that both trap levels determined in SCN and SCNCr are not attributed to the chromium doping, since Cr_2O_3 doping is only change the donor density and trap densities.
(7) The two deep trap levels exist determined by DLTS may be associated with ionized oxygen vacancies $V_O^{\bullet\bullet}$ and tin interstitial $Sn_i^{\bullet\bullet\bullet\bullet}$. The features of these trap levels and the defect theory related to the SnO_2 varistors need to be further studied.

Acknowledgments
The Authors would like to thank Professor A. Peaker and Dr. I. D. Hawkins of the Centre for Electronic Materials, Devices and Nanostructures, the University of Manchester for access to the DLTS measurement facility and helpful advice.

References
[1] Clarke D R 1999 *J. Am. Ceram. Soc.* **82** 485
[2] Gupta T K 1990 *J. Am. Ceram. Soc.* **73** 1817
[3] Levinson L M and Philipp H R 1986 *Am. Ceram. Soc. Bull.* **65** 639
[4] Fan J and Freer R 1993 *J. Mater. Sci.* **28** 1391
[5] Fan J and Freer R 1993 *Br. Ceram. Trans.* **92** 221
[6] Pianaro S A, Bueno P R, Longo E and Varela J A 1995 *J. Mater. Sci. Lett.* **14** 692
[7] Bueno P R, Pianaro S A, Pereira E C, Bulhoes L O S, Longo E and Varela J A 1998 *J. Appl Phys.* **84** 3700
[8] Bueno P R, Cassia-Santos M R, Leite E R, Longo E, Bisquert J, Garcia-Belmonte G and Fabregat-Santiago F 2000 *J. Appl. Phys.* **88** 6545
[9] Bueno P R, Leite E R, Oliveria M M, Orlandi M O and Longo E 2001 *Appl. Phys. Lett.* **79** 48
[10] Lang D V 1974 *J. Appl. Phys.* **45** 3023
[11] Arakawa T, Kurachi H, Tsujita M and Shiokawa J 1985 *J. Mater. Sci. Lett.* **4** 1442
[12] Rohatgi A, Pang S K, Gupta T K, and Straub W D 1988 *J. Appl. Phys.* **63** 5375
[13] Mukae K, Tsuda K and Nagasawa I 1979 *J. Appl. Phys.* **50** 4475
[14] Gupta T K and Carlson W, 1985 *J.Mater.Sci.* **20** 3487
[15] Cheng P, Li S, Zhang L and Li J 2008 *Appl. Phys. Letter.* **93** 012902
[16] Fan J, Huang H and Xia L 2007 *Funcational Materials,* **30** sp.issue, 557 (in Chinese)
[17] Pianaro S A, Bueno P R, Longo E and Varela J A 1999 *Ceram.Inter.* **25** 1
[18] Bueno P R, Oliveria M M, Bacelar-Junior W K, Leite E R, Longo E, Garcia-Belmonte G and Bisquert J 2002 *J. Appl. Phys.* **91** 6007
[19] Antunes A C, Antunes S R M, Pianaro S A, Rocha M R, Longo E, and Varela J A 1998 *J. Mater. Sci. Letter.* **17** 577
[20] Moulson A J and Herbert J M 1995 *Electroceramics* (London: Chapman & Hall) p123.
[21] Chang S C 1980 *J. Vac. Sci. Technol.* **17** 366.
[22] Orlandi M O, Bomio M R D, Longo E and Bueno P R 2004 *J. Appl. Phys.* **96** 3811

Journal of Physics: Conference Series **152** (2009) 012061 doi:10.1088/1742-6596/152/1/012061

Preparation and dielectric properties of Nb_2O_5-BaO-Na_2O-SiO_2 glass-ceramic for energy storage capacitors

Changhui MAO[1,2], Xudong SUN[1], Jun DU[2] and Qun TANG [2]

[1] North-east University, Shenyan, 110004, China

[2] General Research Institute for Non-ferrous Metals, Beijing,100088,China

Email: mao@grinm.com

Abstract. Full density Nb_2O_5-BaO-Na_2O-SiO_2 Glass-ceramics, which could be used as the dielectric energy-storage materials to fabricate high energy density devices, were prepared by means of rapid quenching and succeeding annealing under different temperature. DTA and X-ray diffraction analysis showed that $NaBa_2Nb_5O_{15}$ with tungsten bronze structure and $NaNbO_3$ with perovskite structure were formed as the dielectric phases from the glass matrix at above 750°C. Dielectric constant of the crystallized glass-ceramics were measured at the frequency from 100 Hz to 100 kHz under the testing temperature from -40 to 150°C. The results indicated that theoretical energy storage density of the material crystallized at 800°C could reach up to $1.87x10^6 J/m^3$, which would be suitable to be used as the innoxious dielectric media for high-energy storage capacitors.

1. Introduction

Controlled crystallization full density glass-ceramic composites, which have high dielectric constant and breakdown strength simultaneously, are recently attracting more attention as the dielectric materials in fabricating high energy density devices[1][2][3]. Until now glass-ceramics with Pb content, such as PbO-Na_2O-Nb_2O_5-SiO_2, are still the most common-used materials due to their high Curie point and dielectric constant[2]. However, because of the high temperature during the melting of glass-ceramic mixture, mostly above 1400°C, the volatilization of Pb would be harmful to human health. So how to replace Pb with some other elements is the main challenge to solve the above problems. Moreover, the instability of the temperature dependence of dielectric constant is another disadvantage for the further application of this glass-ceramic material system[3]. In this study, Ba was used to replace Pb in a niobate based glass-ceramic material based on the high dielectric constant and easy-controlled crystallization of barium niobate phase. Full density Nb_2O_5-BaO-Na_2O-SiO_2 glass-ceramics were prepared by means of rapid quenching and succeeding annealing under different temperature. Then the influence of crystallization temperature, testing temperature and frequency on dielectric properties were mainly investigated. Since energy storage density of the materials is proportional to dielectric constant (ε) and square of breakdown strength (E_b), another purpose of this study is to balance ε and E_b through controlled crystallization of the glass-ceramic, which would be suitable for the preparation of high energy storage density capacitors.

2. Experimental

$35.8Nb_2O_5$-$16.2Na_2CO_3$-$10.4BaO$-$37.6SiO_2$ in weight percentage was selected as the nominal composition in this study. Well-mixed powder mixtures were melted in a platinum crucible for 2 h at 1450°C. Then the melting was poured to a rapid melt-quenching device containing two cold rotating stainless rollers to form 0.5~1 mm thick glass sheets. The glass sheets were immediately annealed at 400°C for 2 h to eliminate internal stress before furnace-cooling to room temperature.

About 0.5g as-quenched glass sheet was used for differential thermal analysis (DTA, WCT-2C), which was heated from room temperature to 1000°C at a heating rate of 5°C/sec. According to the crystallization temperature determined by DTA, controlled crystallization was carried out in air at 600°C, 700°C, 750°C, 800°C and 900°C for 2h, respectively. Phase structure of the samples was identified by powder X-ray diffraction (XRD, MSAL XD2 Diffract Meter) patterns with Cu Kα radiation over the range of 2θ from 20°~80°.

Dielectric properties of the glass-ceramics were tested by 4284A precision LCR meter (Agilent) in the temperature range of –40~150°C and frequency range of 10K~100KHz. Gold electrodes were sputter-deposited on both sides of the glass-ceramic sheets.

Breakdown strength testing was performed by a self-assembly clamp which was fully immersed in glycol to avoid electric breakdown of the air under high voltage, which was provided by TREK 30 high voltage amplifier with 0.1 sec triangle wave pulses.

3. Results and discussion

3.1. Crystallization

Controlled crystallization of dielectric phases under different heat-treatment temperature is an important prerequisite in the preparation of glass-ceramics for high energy storage density capacitors. Since controlled crystallization temperature could be guided by DTA test, figure1 shows the DTA curve of $35.8Nb_2O_5$-$16.2Na_2CO_3$-$10.4BaO$-$37.6SiO_2$ glass-ceramic in the range from room temperature to 1000°C at a heating rate of 5°C/sec. As temperature is increased to about 720°C, an endothermic dip appears in the DTA curve, which is attributed to the softening of the glass. With the increase of temperature, two exothermic peaks, which is related to the crystallization of dielectric phases, are observed with maximum intensities located at 790 and 970°C, respectively. Therefore, five temperatures, which are 600°C, 700°C, 750, 800°C and 900°C, are chosen as the isothermal treatment temperatures to investigate the crystallization behavior of the glass-ceramic materials.

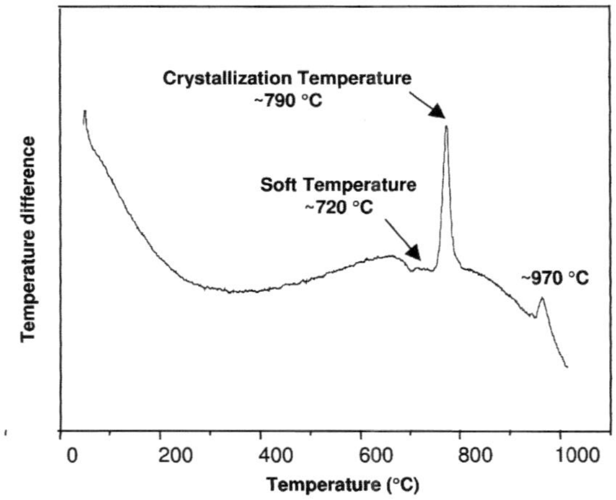

Figure 1. DTA profile of as-quenched Nb_2O_5-Na_2CO_3-BaO-SiO_2 glass

XRD patterns for Nb_2O_5-Na_2CO_3-BaO-SiO_2 glass-ceramic materials crystallized at 600, 700, 750, 800 and 900°C are shown in figure2. Typical amorphous feature with a broad peak at about 30° appears in the samples crystallized under 700°C. As the temperature is increased to 750°C, a little trace of $NaBa_2Nb_5O_{15}$ with tungsten bronze structure is observed. When the temperature is further increased to above 800°C, large amount of $NaNbO_3$ with perovskite structure and $NaBa_2Nb_5O_{15}$ precipitate from the glass matrix. It seems that $NaNbO_3$ and $NaBa_2Nb_5O_{15}$ phases almost appear simultaneously with the increase of crystallization temperature. For the crystallization process of the glass-ceramic is so complex that the exact composition of the dielectric phases would be very difficult to be evaluated. Figure3 shows the phase diagram of $NaNbO_3$ and $NaBa_2Nb_5O_{15}$ [5]. As the amorphous structure of the as-quenched glass could also be likely regarded as the liquid phase, figure3 was used to analyze the crystallization process and composition of the precipitated dielectric phases. Since a little $NaBa_2Nb_5O_{15}$ precipitates firstly and then followed by the co-precipitate of $NaNbO_3$ and $NaBa_2Nb_5O_{15}$ during crystallization as indicated in figure2, the final ratio of the two dielectric phases would locate at the position which is a little higher than the eutectoid point (~78mol% $NaNbO_3$) shown in figure3. However, further research is still necessary to confirm this analysis.

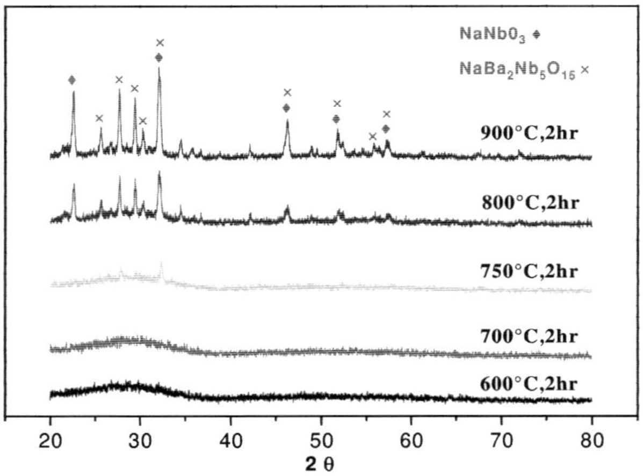

Figure 2. XRD patterns of Nb_2O_5-Na_2CO_3-BaO-SiO_2 glass-ceramic

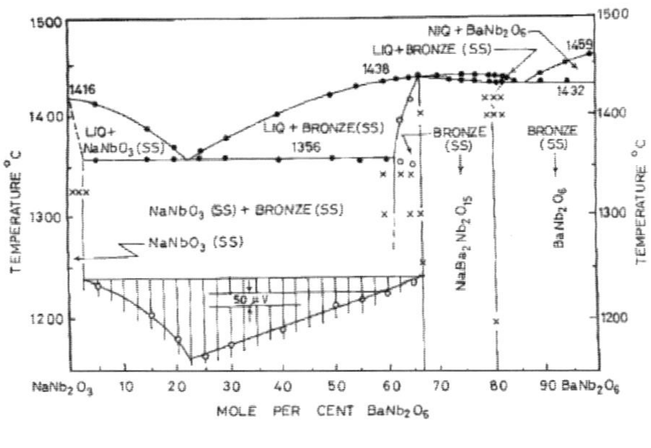

Figure 3. Phase diagram of $NaNb_2O_3$ and $BaNb_2O_6$

3.2. Dielectric properties

Figure 4(a) shows the dielectric constant of the annealed glass-ceramic samples tested at room temperature from 100Hz~100KHz. The initial dielectric constant of the as-quenched glass is about 30, which is relatively higher compared with ordinary glass probably due to the high content of polarizable ions[4]. With the increase of crystallization temperature, the dielectric constant increases with the formation of dielectric phases. Firstly, the dielectric constant is improved significantly from 600°C to 750°C. This is mainly due to the formation of high-ε_r $NaBa_2Nb_5O_{15}$ phase ($\varepsilon_r \approx 240$) at this temperature range[5]. Upon the crystallization temperature is further increased to 800°C and 900°C respectively, the gradual increase of dielectric constant comes from two facts: One is the result of the co-crystallization of $NaBa_2Nb_5O_{15}$ and $NaNbO_3$($\varepsilon_r \approx 500\sim600$), and the second is the increase of the content of dielectric phases. Moreover, from figure 4(a), it is also indicated that the dielectric constant of each crystallization sample keeps steady from 100Hz to 100KHz. The difference of the dielectric constant value is less than 10% at this frequency range.

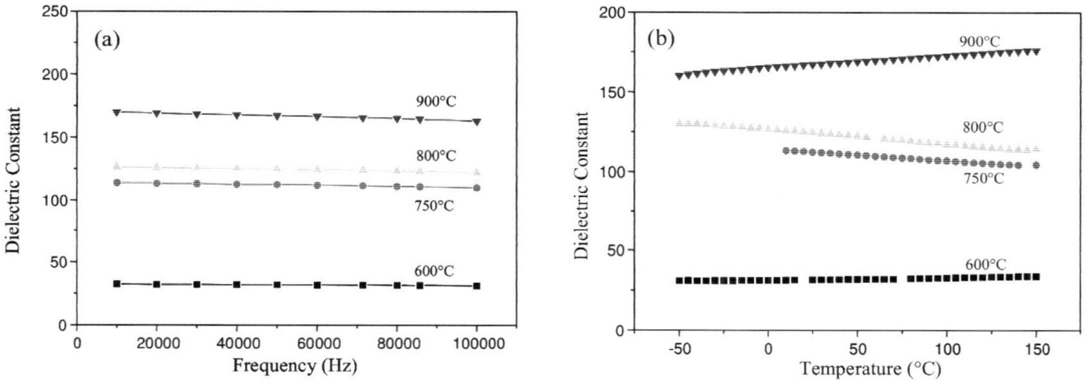

Figure 4. (a)Frequency and annealing temperature dependence of dielectric constant.(b) Dielectric constant as a function of testing temperature for the samples at different crystallization temperature (f = 50kHz)

Figure 4(b) shows the dielectric constant as a function of testing temperature for the samples at different crystallization temperature. The dielectric constant of the initial glass (600°C) almost keeps constantly at this testing temperature range because there is not any dielectric phase occurred in the material. With the crystallization temperature is increased to 750°C and 800°C respectively, dielectric constant of the glass-ceramic reduces slightly with the increase of testing temperature. This trend may come from the large ratio of $NaBa_2Nb_5O_{15}$ phase at the beginning of crystallization process with Curie temperature of ~560°C[5], which is too much higher than the testing temperature to affect the temperature dependence of the dielectric constant. However, as the crystallization temperature reaches to 900°C, dielectric constant increases with testing temperature, which may due to the increasing content of $NaNbO_3$ phase with a lower Curie temperature of 365°C[4][7].

3.3. Breakdown strength and energy storage density

Five Nb_2O_5-Na_2CO_3-BaO-SiO_2 glass-ceramic samples crystallized at 800°C were chosen to be finely polished to flakes with thickness of about 0.3~0.4mm. Then, the flakes were immersed in glycol to test the breakdown strength. Table 1 shows the result of the tests. The difference of individual value of the breakdown strength appears to be about 30%. This result suggests that the breakdown strength of the glass-ceramic is a structure-sensitive parameter ascribed to the distribution of dielectric phases and glass matrix. Moreover, It seems that the breakdown strength is reduced with the increase of sample thickness.

Table 1. Breakdown strength of Nb_2O_5-Na_2CO_3-BaO-SiO_2 glass-ceramic material (800°C)

Sample thickness(mm)	0.310	0.325	0.335	0.355	0.370
Breakdown strength(KV/mm)	63.9	59.7	54.3	57.6	47.2

From Table 1 the average breakdown strength of 56.5 kV/mm could be obtained. In this study, the following formula was used to evaluate the theoretical energy storage density of glass-ceramic.

$W=\varepsilon_0\varepsilon_r E_b^2/2$

where W is energy storage density (J/m^3), ε_0 is the vacuum permittivity (8.85×10^{-12} F/m), ε_r is the relative permittivity. Then the energy storage density of the sample crystallized at 800°C could reach to $1.87\text{x}10^6$ J/m^3, which is suitable to be used as the innoxious dielectric media for high-energy storage capacitors.

4. Conclusion

In this study, fully densified Nb_2O_5-BaO-Na_2O-SiO_2 glass-ceramics were prepared by rapid quenching and controlled crystallization process. The influence of crystallization temperature, testing temperature and frequency on dielectric properties were investigated. Theoretical energy storage density was evaluated. The results show that:

(1) $NaNbO_3$ with perovskite structure and $NaBa_2Nb_5O_{15}$ with tungsten bronze structure precipitate from the glass matrix during controlled crystallization process.

(2) Dielectric constant increases with crystallization temperature, and the difference of the dielectric constant is less than 10% at the frequency range from 100 Hz to 100 kHz.

(3) Dielectric constant of the materials crystallized at 750°C and 800°C has negative linear relationship at the temperature range from -40°C to 150°C, which is contrary to the material crystallized at 900°C.

(4) Theoretical energy storage density of the material crystallized at 800°C could reach to $1.87\text{x}10^6$J/m^3, which is suitable to be used as the innoxious dielectric media for high-energy storage capacitors.

References

[1] Fletcher N H, Hilton A D and Ricketts B W Optimization of energy storage density in ceramic capacitors. 1996 J. Phys. D: Appl. Phys. 29(1996) 253-258.

[2] Gorzkowski E P and Pan M J 2007 Glass-ceramics of barium strontium titanate for high energy density capacitors, 2007 J. Electroceram. 18 (3–4) 269–276.

[3] Slenes K M, Winsor P, Scholtz T. Pulse power capability of high energy density capacitors based on a new dielectric material. 2001 IEEE Trans., 37(1) 324–327.

[4] Jun Du, Beth Jones and Michael Lanagan. Preparation and characterization of dielectric glass-ceramics in Na_2O–PbO–Nb_2O_5–SiO_2 system. 2005 Materials Letters 59 2821 – 2826

[5] Sambasiva Rao K and Ki Hyun Yoon, Review of electrooptic and ferroelectric properties of barium sodium niobate single crystals, 2003 J. of Materials Science 38 391-400

[6] Pisarski M, 1980 Acta Phys. Pol., A 57 693

[7] Cross L E and Nicholson B T, 1955 Philos. Mag. 46 453.

Journal of Physics: Conference Series **152** (2009) 012062 doi:10.1088/1742-6596/152/1/012062

Influence of CuO or $MnCO_3$ Additive on the Dielectric Properties of $Ca_{0.65}Ti_{0.65}La_{0.35}Al_{0.35}O_3$ Ceramics

Jiansen ZHENG[1], Chong ZHANG[1], Zhaoxian XIONG[1], Fen XIAO[2], Guoshan YANG[2], Hao XUE[1]

1.Department of Materials Science and Engineering, College of Materials, Xiamen University, Xiamen 361005, China;

2.Department of Physics, Xiamen University, Xiamen 361005, China

E-mail: zxxiong@xmu.edu.cn

Abstract. The influence of CuO or $MnCO_3$ additive on the sinterability, microstructure and microwave dielectric properties of $Ca_{0.65}Ti_{0.65}La_{0.35}Al_{0.35}O_3$, i.e., CTLA, ceramics was investigated in this paper. The results indicated that the density and dielectric properties of CTLA ceramics can be improved by both low-level doping of CuO (up to 1 wt%) and $MnCO_3$ (up to 0.75 wt%). The sintering temperature of CTLA ceramics with addition of CuO/$MnCO_3$ can be effectively reduced from 1450°C to 1320°C due to the liquid phase effect. No secondary phase was observed for the addition at the amount of 0.5-1 wt% CuO and 0.25-0.75 wt% $MnCO_3$. With the addition of CuO/$MnCO_3$, the dielectric constant (ε_r) and the temperature coefficient of resonant frequency (τ_f) of the samples were not significantly affected, while the unloaded quality factors Q were effectively promoted. Excellent microwave dielectric properties were achieved for the ceramic $Ca_{0.65}Ti_{0.65}La_{0.35}Al_{0.35}O_3$ + 0.75wt% CuO/ $MnCO_3$ sintered at 1320°C for 2 h, with values of $Q\times f$ = 32,800 GHz, ε_r = 43 and τ_f = +10 ppm/°C.

1. Introduction

With the advance of wireless technologies, it is of key importance to develop microwave dielectric ceramics (MWDC).The requirements for microwave ceramics are a high dielectric constant (ε_r), a low dielectric loss (high Q value) and a temperature-stable resonant frequency ($\tau_f \approx 0$ ppm/°C). By increasing ε_r we can reduce the size of the ceramic component and therefore reduce the overall size of the telecommunications device, and the signal quality of filter is improved by increasing the Q value of the resonators [1].

$CaTiO_3$ is an orthorhombic distorted perovskite [2] at room temperature with ε_r=170, $Q\times f$ =3,500 GHz and $\tau_f \approx$+800 ppm/°C [3]. In contrast $LaAlO_3$ is a rhombohedral perovskite at room temperature with ε_r=23.4, $Q\times f$ =68,000GHz and τ_f=-44ppm/°C [4]. Ceramic material based on the solid solution of $CaTiO_3$-$LaAlO_3$ is a promising candidate for microwave frequencies because of its high dielectric constant ($\tau_f \approx 44$) with high quality factor ($Q\times f \approx$32,000GHz) and small or zero τ_f.

Moon J H et al. [5] first studied the dielectric and sintering properties of (1-x)$CaTiO_3$-x$LaAlO_3$ ceramics. A high sintering temperature of 1600°C was required to obtain dense ceramics. Additions of Bi_2O_3-$3Al_2O_3$ or Bi_2O_3-NiO lowered the sintering temperature to 1450°C. However, the additives reduced the $Q\times f$ value by at least 15%. The microwave dielectric properties of $Q\times f$ 32,500 GHz, $\varepsilon_r \approx$

41 and $\tau_f \approx +5$ ppm/°C were reported for the undoped $0.65CaTiO_3$-$0.35LaAlO_3$. Jancar B et al. [6] studied the dielectric and sintering properties of (1-x)$CaTiO_3$-x$ReMO_3$（Re=La, Nd, Sm；M=Al, Ga）ceramics. They reported dielectric properties of $Q\times f$ =30,000 GHz, ε_r=44 and τ_f=-2ppm/°C for $0.65CaTiO_3$-$0.35LaAlO_3$. However, the high calcinations temperature in the range of from 1350 to 1450°C for 40h and high sintering temperature of 1450°C for 10h were reported.

In this paper, the effects of CuO and $MnCO_3$ additive on the sinterability, microstructure and microwave dielectric properties of $Ca_{0.65}Ti_{0.65}La_{0.35}Al_{0.35}O_3$ ceramics are investigated.

2. Experimental procedure

Samples were prepared by conventional solid-state ceramic route using high purity (>99%) of $CaCO_3$, TiO_2, La_2O_3 and Al_2O_3 as the raw materials. They were weighed according to composition of $Ca_{0.65}Ti_{0.65}La_{0.35}Al_{0.35}O_3$ and ball-milled with deionized water in a planetary mill for 2h with zirconia balls. The mixed powders were dried and calcined at 1200°C for 2 h in air. The resultant powders added with 0.5-1 wt% CuO (>99%) and 0.25-0.75 wt% $MnCO_3$ (>99%), and then ball milled again in deionized water with zirconia balls for 2h. After drying, the powders were blended with 5% PVA solution, and then granulated, screened through a sieve with 40-mesh, and pressed uni-axially at 200MPa into disks with 13mm in diameter and 2-7mm in thickness with a steel die. The green pellets were then sintered at 1280°C to 1340°C for 2h in air at a rate of 5°C/min, and furnace-cooled to ambient temperature.

Bulk densities of the ceramics were determined by the Archimedes method. Crystalline structures were identified by powder X-ray diffraction (Panalytical X'pert) in the 2θ range from 10° to 90°. The ceramics were observed with scanning electron microscopy (SEM) (Philips XL30). Microwave dielectric properties were measured using an Agilent E8362B vector network analyzer in the range of 1-20GHz. Dielectric constant (ε_r) was measured by the Hakki-Coleman dielectric resonator method [7]. The value of τ_f was obtained in the range from 25°C to 80°C, and was calculated with following expression.

$$\tau_f = \frac{f_2 - f_1}{f_1(T_2 - T_1)}$$

Where f_1 and f_2 were the resonant frequencies at temperatures of T_1 and T_2, respectively.

3. Results and discussion

3.1. Phase identification

3.1.1 *Effect of the amount of CuO on CTLA ceramics.* X-ray diffraction patterns of CuO-doped CTLA ceramics sintered at 1320°C for 2h are shown in Figure 1. It is observed that CTLA belongs to the tetragonal-type crystal structure. No obvious peak of secondary phase is found for the samples with additive CuO in weight percent of 0.5-1.0 due to the fact that detection of a minor phase by XRD is extremely difficult. Since the ionic radius of Ti^{4+} (0.068nm) and Al^{3+} (0.072nm) is similar to that of Cu^{2+} (0.073nm) [8], they might be substituted by Cu^{2+} ion in the ceramics. The total charge of two ions of Cu^{2+} is moreover equal to that of one ion of Ti^{4+}, and no charge distortion is induced. Hence, the additive CuO might be substituted into the main structure of CTLA ceramics.

3.2. *Effect of the amount of $MnCO_3$ on CTLA ceramics.* X-ray diffraction spectra for CTLA ceramics prepared with different levels of $MnCO_3$ sintered at 1320°C for 2h is displayed in Figure 2. The single phase with tetragonal structure is observed for CTLA ceramics doped with 0.25-0.75wt% $MnCO_3$. The results suggest that Mn^{2+} maybe incorporate into the lattices of $Ca_{0.61}Ti_{0.61}La_{0.39}Al_{0.39}O_3$ structure, since the ion radius of Mn^{2+} (0.080 nm) is similar to that of Ti^{4+} (0.068 nm) and Al^{3+} (0.072 nm) [8].

Therefore, $MnCO_3$ can be used as a useful additive to lower the sintering temperature of CTLA ceramics without changing the crystalline structure of major phase.

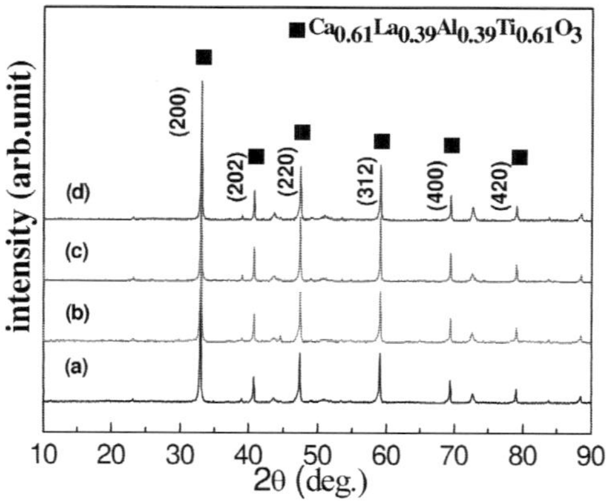

Figure 1. The XRD patterns of CTLA with CuO additives sintered at 1320℃ for 2h (a)0.0wt% (b)0.50wt% (c)0.75wt% (d)1.00wt%

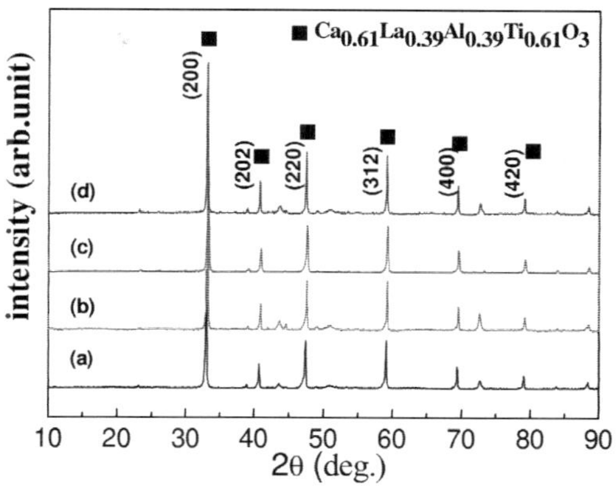

Figure 2. The XRD patterns of CTLA with $MnCO_3$ additions sintered at 1320℃ for 2h (a)0.0wt% (b)0.25wt% (c)0.50wt% (d)0.75wt%

3.2. Densification

3.2.1. *Effect of the amount of CuO on CTLA ceramics.* Bulk density of CTLA ceramics with CuO as a function of sintered temperature is illustrated in Figure 3. The results demonstrate that the relative density of the samples steadily increase with increasing sintering temperature. The density is increased with 0.5wt% CuO addition, then decreased with more CuO addition above 1300°C.The densification might caused by the liquid phase effect of CuO. At higher temperature, the densitiy decreased might due to the low density of liquid phase and the vaporization of CuO. Maximum relative density is found to be 93.93% with 0.5wt% CuO addition at 1320°C.

3.2.2. *Effect of the amount of $MnCO_3$ on CTLA ceramics.* Bulk density of CTLA ceramics with $MnCO_3$ as a function of sintered temperature is given in Figure 4. It can be observed that the relative density of the samples steadily increase with increasing sintering temperature. The density is increased with 0.25-0.5wt% $MnCO_3$ additive, then decreased with more addition of $MnCO_3$ above 1300°C. Maximum relative density is found to be 93.98% with the addition of 0.25wt% $MnCO_3$ at 1320°C.

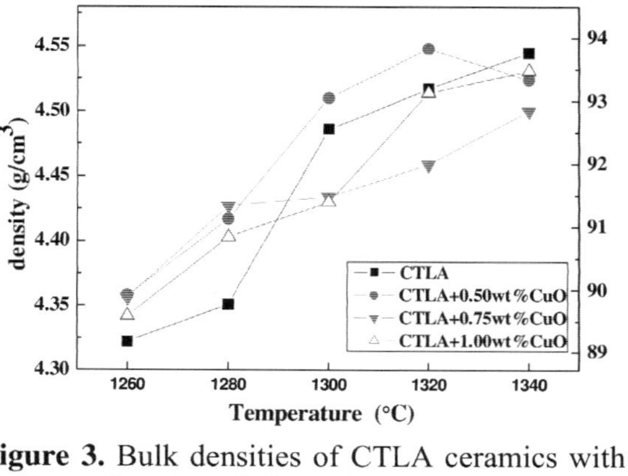

Figure 3. Bulk densities of CTLA ceramics with CuO as a function of sintered temperature

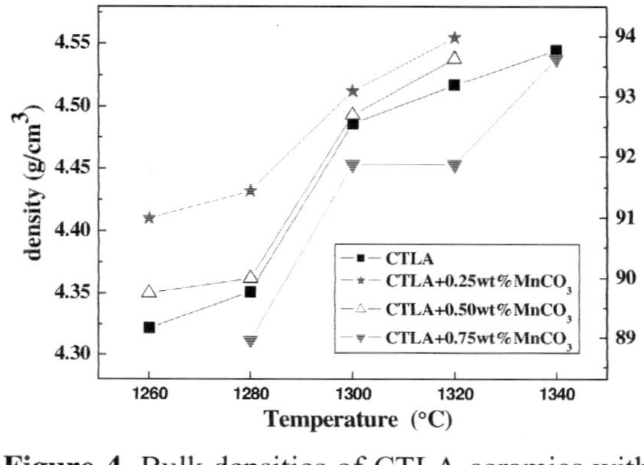

Figure 4. Bulk densities of CTLA ceramics with $MnCO_3$ as a function of sintered temperature

3.3. Microstructure

3.3.1. Effect of the amount of CuO on CTLA ceramics. Observation of microstructure for as-sintered CTLA ceramics reveals difference in the samples with different levels of CuO, as displayed in Figure 5(a)-(f). It can be seen that the grain size is about 1-2μm and there are few pores in the samples. For undoped and CuO-doped specimens sintered at 1320°C, there are lots of quadrate grains and the ratio of length to width is about 1:1. Moreover, for additions of 0.50 wt% and 1.00wt% CuO the liquid phase effect is observed in the microstructure, but not for that with 0.75wt% CuO additive. The liquid phase effect may be caused by eutectic of CuO-Cu_2O-TiO_2 (Cu_3TiO_4) [9] at 1070°C. Especially in the case of 1.00wt% CuO additive, the vaporization of CuO is presented as evidence for the liquid phase.

3.3.2. Effect of the amount of $MnCO_3$ on CTLA ceramics. SEM micrographs of $MnCO_3$-doped CTLA ceramics sintered at 1320°C are present in Figure 5(g)-(i). It can be seen that the grain size is about 1-2μm and there are few pores in the samples. Moreover, a little of glass phase is found in the triple region of grains of $MnCO_3$-doped specimens. Though the size of grains is almost the same between the undoped and $MnCO_3$-doped specimens, the distribution of glass phase of the samples doped with 0.75 wt% $MnCO_3$ was more extensive than other samples. From the SEM images, it is observed that the $MnCO_3$ additive can lower the sintering temperature of CTLA ceramics, but dose not enhance the grain growth.

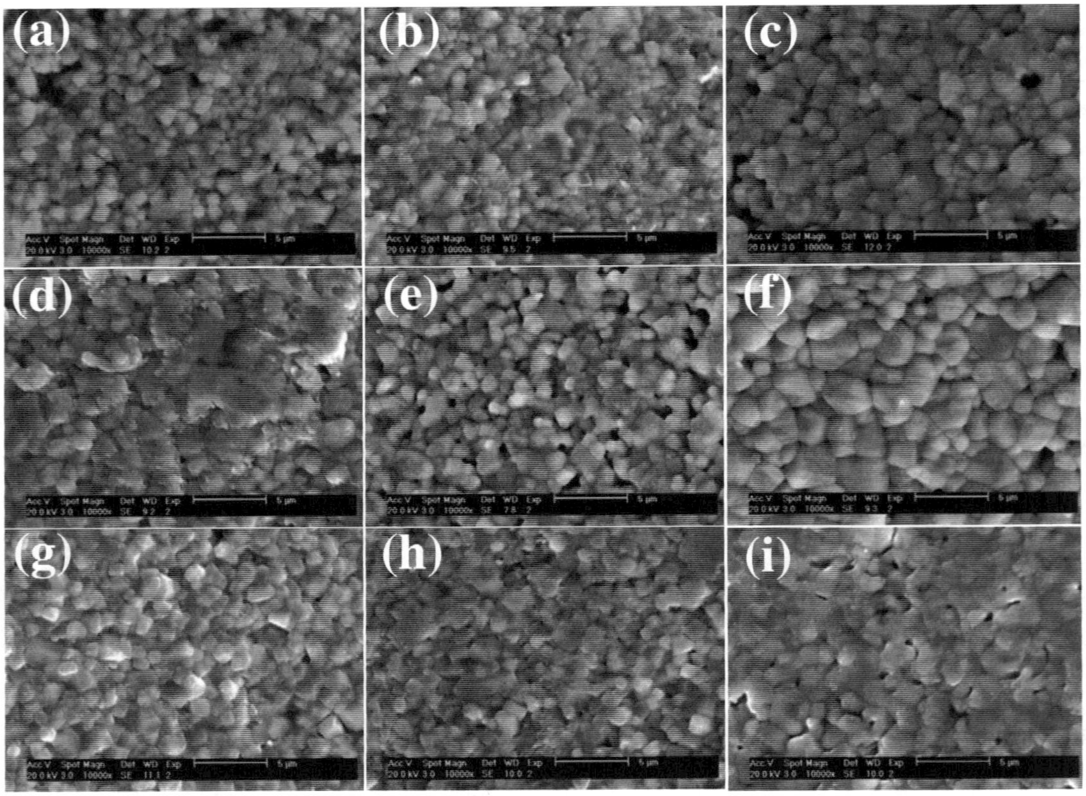

Figure 5. SEM micrographs of CTLA ceramics with CuO additive sintered at 1320°C for 2h (a)0.00wt% CuO (b)0.50wt% CuO (c)0.75wt% CuO (d)1.00wt% CuO; with 0.75%wt CuO additive sintered at 1300°C for 2h (e) and sintered at 1340°C for 2h (f); with $MnCO_3$ addition sintered at 1320°C for 2h (g) 0.50wt% $MnCO_3$ (h) 0.75wt% $MnCO_3$ (i) 1.00wt% $MnCO_3$

Journal of Physics: Conference Series **152** (2009) 012062 doi:10.1088/1742-6596/152/1/012062

3.4. Dielectric properties

3.4.1. *Effect of the amount of CuO on CTLA ceramics.* The dielectric constants of CTLA ceramics with different amounts of CuO additive as functions of their sintering temperatures are displayed in Figure 6. The relationships between ε_r values and sintering temperatures reveal the same trend as those between densities and sintering temperatures, because higher density means lower porosity (ε_r =1). The decrease of ε_r values could be explained as being due to lower densities as well as to the liquid phase effect. The ε_r values of the well-sintered CTLA ceramics range from 43 to 45.

The $Q\times f$ values of CTLA ceramics with various amounts of CuO additive as functions of their sintering temperatures, are shown in Figure 7. With the increasing of sintering temperature, the $Q\times f$ values of CTLA ceramics doped with different amounts of CuO increase firstly, after reach the maximal value at a certain temperature, then decrease. The $Q\times f$ values of CTLA ceramics increase with the increasing amount of CuO additive, after reach the maximum value at the 0.75wt% CuO addition, and then decrease slightly with more addition of CuO. It is expressed that the addition of 0.75wt% CuO would be the most suitable amount for CTLA ceramics.

It has been reported [10] that the microwave dielectric loss is mainly caused not only by the lattice vibrational modes, but also by the pores and the secondary phases. Relative density plays an important role in controlling dielectric loss, as has been shown for other microwave dielectric materials. However, the $Q\times f$ value has been reported [11] to be independent of the density and the porosity for a theoretical density greater than 90%. Because the secondary phase and charge unbalance are not possible, the grain morphology is suggested to dominate the $Q\times f$ values of CTLA ceramics with CuO additive.

The grain size of CTLA ceramics with respect to the sintering aid content is critical for understanding the dielectric loss. Grain size is similar for different amounts of CuO additive in this study. However, the $Q\times f$ values vary dramatically with CuO additive. The $Q\times f$ value with 0.75wt% CuO addition is much higher than that with either 0.5wt% or 1wt% CuO addition at 1320°C. From Figure 5(b)-(d) it is observed that the grain morphology for 0.75wt% CuO addition is more uniform than that for the 0.5wt% or 1wt% addition. This would reveal reductions in lattice imperfection as well as dielectric loss for the ceramics [10].

The τ_f values of CTLA ceramics with different amounts of CuO addition as functions of their sintering temperatures are present in Figure 8. No any significant changing is observed in the τ_f value with fixed addition of CuO at different sintering temperature. The τ_f value is known as to be related with the composition and the secondary phase of the material. Because CTLA ceramics are temperature stable, and addition of CuO does not cause a secondary phase, the τ_f values do not change much in the experiment. With the addition of 0.75wt% CuO, the CTLA ceramics sintered at 1320°C get excellent microwave dielectric properties of an ε_r value of 43.51, a $Q\times f$ value of 32,705 GHz (at 6 GHz), and a τ_f value of +9.48 ppm/°C.

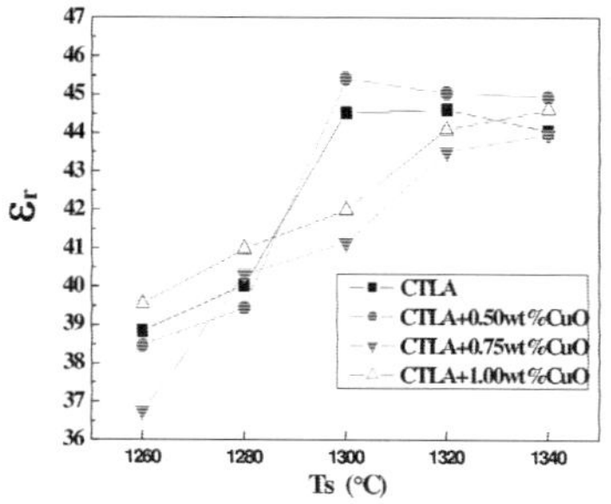

Figure 6. Dielectric constants of CTLA ceramics with CuO additives sintering at different temperature

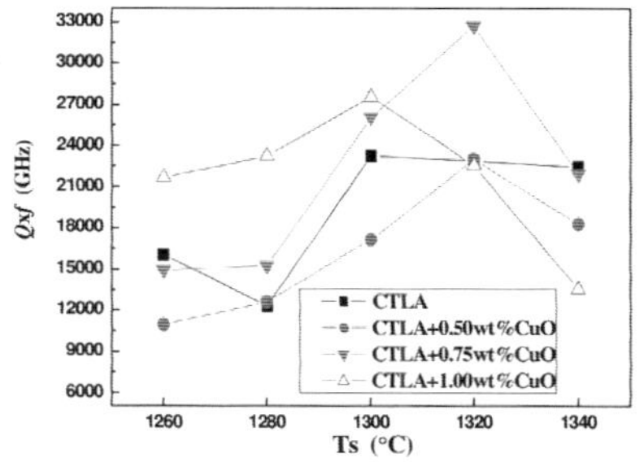

Figure 7. Quality factors of CTLA ceramics with CuO additives sintering at different temperature

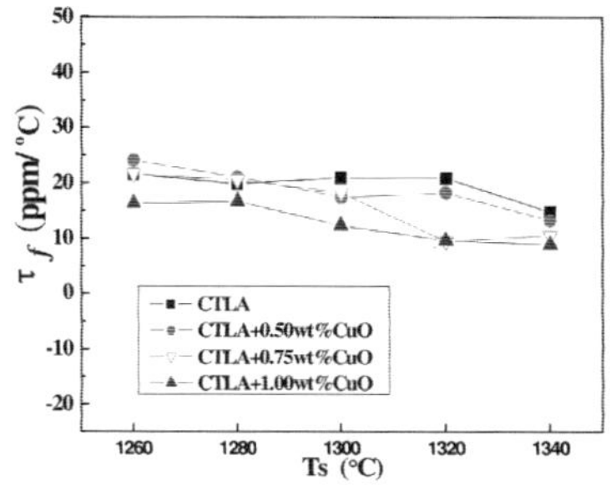

Figure 8. Temperature coefficients of resonant frequency of CTLA ceramics with CuO additives sintering at different temperature

3.4.2. *Effect of the amount of $MnCO_3$ on CTLA ceramics.* The dielectric constants of CTLA ceramics with different levels of $MnCO_3$ additive of sintering temperature are demonstrated in Figure 9.The relationship between ε_r values and sintering temperature is the same as that between relative densities and sintering temperature. Those phenomena are attributed to the increase of relative densities which depended on the sintering temperature. Maximal ε_r values of CTLA ceramics doped and undoped are almost the same, which is about 44.5. Therefore, it suggests that ε_r values of CTLA ceramics have not significant correlation with the amount of $MnCO_3$ additive. The phenomenon is similar to the CTLA ceramics with different amounts of CuO. The $Q\times f$ values of CTLA ceramics doped with different amounts of $MnCO_3$ as a function of sintering temperature are revealed in Figure 10. With the increasing of sintering temperature, the $Q\times f$ values of CTLA ceramics doped with different amounts of $MnCO_3$ increase firstly, after reach the maximal value at a certain temperature, then decrease. Maximal $Q\times f$ value of CTLA ceramics undoped is 23,241GHz, however, maximal $Q\times f$ values of CTLA ceramics doped with $MnCO_3$ of 0.75wt% is 32,896GHz. The $Q\times f$ values of CTLA ceramics increase with the increasing amount of $MnCO_3$ addition, after reach the maximum value at the 0.75wt% $MnCO_3$ addition. It is expressed that the addition of 0.75wt% $MnCO_3$ would be the most suitable amount for CTLA ceramics. It might be due to variable valence of Mn ion in $MnCO_3$ which decrease effectively the leak loss of conduction current.

The τ_f values of CTLA ceramics with different amounts of $MnCO_3$ addition as functions of sintering temperature are displayed in Figure 11. Not any significant change is observed in the τ_f values with fixed $MnCO_3$ addition at different sintering temperature. Because CTLA ceramics are temperature stable, and $MnCO_3$ additions do not cause a secondary phase, the τ_f values do not change much in the experiment. With 0.75wt% $MnCO_3$ addition, the CTLA ceramics sintered at 1320°C obtain excellent microwave dielectric properties of an ε_r value of 43.05, a $Q\times f$ value of 32,896GHz (at 6GHz), and a τ_f value of +10.59ppm/°C.

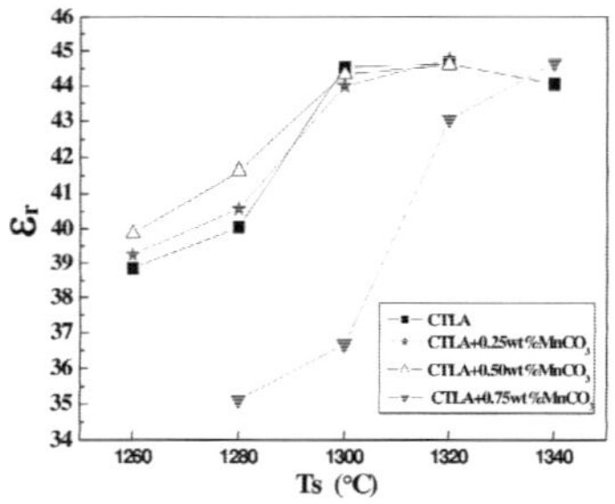

Figure 9. Dielectric constants of CTLA ceramics with $MnCO_3$ as a function of sintered temperature

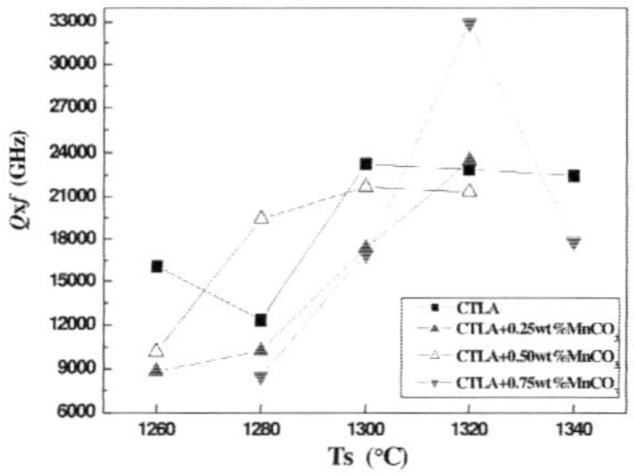

Figure 10. Quality factors of CTLA ceramics with $MnCO_3$ as a function of sintered temperature

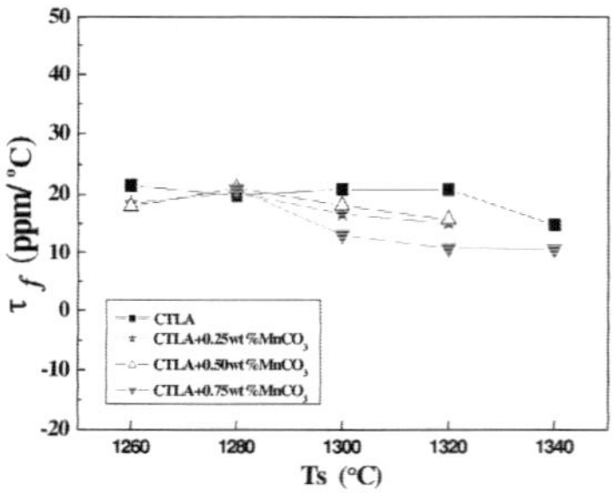

Figure 11. Temperature coefficients of resonant frequency of CTLA ceramics with $MnCO_3$ as a function of sintered temperature

4. Conclusion

The effects of CuO and $MnCO_3$ additives on the microwave dielectric properties and the microstructures of CTLA ceramics were investigated. CTLA ceramics with proper additions of CuO/$MnCO_3$ could be well sintered to approach 94% theoretical density at 1320°C due to the liquid phase effect. The reduction of the dielectric loss of CTLA ceramics with CuO additive might be dominated by the uniform grain morphology, while the reduction of the dielectric loss of CTLA ceramics with $MnCO_3$ additive might be attributed to variable valence of Mn ion in $MnCO_3$ which decreased effectively the leak loss of conduction current.

CTLA ceramics with the addition of 0.75wt% CuO/$MnCO_3$ sintered at 1320°C for 2h resulted in a ε_r of 43, a $Q\times f$ value of 32,800 GHz, and a τ_f value of +10 ppm/°C. Therefore, the sintering

Journal of Physics: Conference Series **152** (2009) 012062 doi:10.1088/1742-6596/152/1/012062

temperature of CTLA ceramics can be lowered effectively with $CuO/MnCO_3$ additive without decreasing the microwave dielectric properties.

References

[1]. Wakino K,Nishikawa T, Ishikawa Y, *et al.Dielectric resonator materials and their application for mobile communication systems*.Br. Ceram. Trans. J.,1990,89(2):39-43

[2]. More references Jancar B, Suvorov D, Valant M, *et al.Characterization of* $CaTiO_3$-$NdAlO_3$ *dielectric ceramics*.J. Eur. Ceram. Soc.,2003,23:1391-1400

[3]. Suvorov D,Valant M, Jancar B, *et al.*$CaTiO_3$*-based ceramics:microstructural development and dielectric properties*.Acta Chim.Slov.,2001,48 :87-99

[4]. Cho S Y, Kim I T and Hong K S.*Microwave dielectric properties and applications of rare-earth aluminates*.J.Mater. Res.,1999,14(1):114-119

[5]. Moon J H,Jang H M, Park H S, *et al.Sintering behavior and microwave dielectric properties of* $(Ca,La)(Ti,Al)O_3$ *ceramics*.Jpn. J. Appl. Phys.,Part 1,1998,38(12A):6821-6826

[6]. Jancar B, Suvorov D and Valant M.*Microwave dielectric properties and microstructural characteristics of aliovalently doped perovskite ceramics based on* $CaTiO_3$.Key Eng. Mater.,2002,206-213:1289-1292

[7]. Hakki B W and Coleman P D.*A dielectric resonator method of measuring inductive capacitance in millimeter range*.IEEE Trans. Microwave Theory Tech.,1960,8:402-410

[8]. Shannon R D and Prewitt C T.*Effective ionic radii in oxides and fluorides*.Acta. Cryst.,1969, B25:925- 946

[9]. Gadalla A M M, Ford W F and White J.*Equilibrium relationships in* CuO-Cu_2O-TiO_2.Trans. Br. Ceram,Soc.,1963,62(1):57

[10]. Silverman B D. *Microwave absorption in cubic strontium titanate*.Phys. Rev.,1962,125:1921-1930

[11]. Kim W S, Hong T H, Kim E S, *et al.Microwave dielectric properties and far infrared reflectivity spectra of the* $(Zr_{0.8}Sn_{0.2})TiO_4$ *ceramics with additives*.Jpn. J. Appl Phys., Part 1,1998,37(9B):5367-5371

Influence of Mo Addition on Dielectric Properties of AlN Ceramic Matrix Composites

Yan ZHANG, Zhimin YANG, Huina MA and **Jun DU**

Institute of Advanced Electronic Material, General Research Institute for Non-ferrous Metals, Beijing 100088, P. R. China

E-mail:zhy126111@163.com

Abstract. AlN-Mo composite ceramics were prepared by spark plasma sintering (SPS) with CaF_2 as sintering aids. Effect of Mo addition on the thermal conductivity and dielectric properties of the composite ceramics had been studied. The results show that the room temperature thermal conductivity increases with increasing the content of Mo, and the value begins to decrease slightly when the Mo concentration exceeds 20 vol. %. Analyses indicate that the key factors to dielectric properties are the metal phase concentration and the microstructure of Mo particles. 1 vol. % Ni has been added into the composite ceramics to change the distribution of the Mo phase. The elongated shape particles which link with each other have a tendency to acquire rounded forms which are thermodynamically more stable. Consequently, the dielectric constant and loss of the composite ceramics could be adjusted and the material becomes an electrical conductor in the case of Mo volume fraction of more than 23%. Furthermore, the dielectric properties could be improved to a large extent by transforming the microstructure of the metal particles when the concentration of Mo is fixed.

1. Introduction

Microwave lossy materials are widely used in vacuum electronic devices to suppress unwanted electromagnetic modes, prevent self-oscillation, enhance the bandwidth of resonant cavities, and provide matched electromagnetic terminations [1]. Electromagnetically lossy ceramic materials based on BeO matrix have traditionally been the best choice of materials due to their high thermal conductivity. However, its field of application with BeO-containing materials is restricted because of health issues. Aluminum Nitride ceramics attract growing interests for applications which need high thermal conductivity, good electrical properties, low density value and a low thermal expansion coefficient. According to the microstructure and composition of the sintered ceramics, it presents a good thermal stability at temperatures higher than 2673 K, whereas thermal conductivity of polycrystalline AlN varies between 80 and 200 $Wm^{-1}K^{-1}$[2-4]. New AlN-based materials are developed as the lossy materials instead of traditional material BeO. The way of improving microwave absorbing property of the AlN-based composites could be the addition of a metallic phase with higher electronic and thermal properties [5-6]. Molybdenum was chosen due to its sufficiently high melting point (2883 K), high thermal conductivity (138 $Wm^{-1}K^{-1}$), good electrical properties and close to AlN thermal expansion coefficient (~$7x10^{-6}$ K^{-1} between 300 and 1200 K, against ~$6.5x10^{-6}$ $K^{-1}$for pure AlN in the same range of temperature) [7-9].

In the present study, for the first time, composites of AlN and Mo were sintered by spark plasma sintering to evaluate the density of the composites. The objective of this work was to study the influence of molybdenum addition on the dielectric properties of AlN-Mo composites in order to investigate the thermal conductivity, the percolation threshold, and to explore the factors which were vital to dielectric properties.

2. Experimental procedure

Commercially available AlN, Mo and CaF_2 were used as the starting powders. The average particle sizes of AlN and Mo were 0.1～1 μm and 1～10 μm, respectively. The combined powders were prepared by wet ball milling in alcohol for 24 h with zirconium dioxide balls. After drying at 80℃, the mixtures were blended by dry milling for a period of 12 h. Each of the resulted powders was put into a graphite die with a pressure of 20 MP in vacuum for 5 min to produce a ceramic disk. A thin BN film was deposited on the inner wall of the mould in order to avoid any interaction between the powder and graphite. The sintering temperature employed for SPS varied from 1500℃ to 1700℃.

The crystalline phases were characterized by X-ray diffraction (XD-2) with Cu Kα radiation. The microstructure was analyzed by BRIGHT A100 digital microscope. DRP-4W thermal conductivity equipment and Agilent 4284A were employed to test the thermal conductivity and dielectric properties, respectively.

3. Results and discussion

3.1. Thermal Conductivity

The values of room temperature thermal conductivity of AlN-Mo composites, as a function of Mo concentration, are presented in figure 1. It is obvious that the thermal conductivity variation has been due mainly to the variation in volume fraction of metal phase. It could be seen that the value for pure AlN reaches 89 $Wm^{-1}K^{-1}$. The relatively low value is mainly attributed to two causes: the presence of oxygen impurities within the AlN lattice and on the AlN grain surface [10-11], besides, the low compact structure because of the nature of AlN. Its value decreases when a little portion Mo had been added in and increases to about 105 $Wm^{-1}K^{-1}$ with increasing the content of Mo to 20 vol. %. However, the value begins to decrease slightly to about 101 $Wm^{-1}K^{-1}$ for a composite containing 23 vol. % Mo. Because of the ductile nature of AlN, the addition of Mo is supposed to improve the sintering performance as well as the thermal conductivity. Presence of Mo played an important role to develop a perfect contact with the matrix and to reduce the thermal contact resistance at the interface. Due to the addition of CaF_2, densification was further increased. Therefore, the values of room temperature thermal conductivity increase rapidly. However, high proportion of Mo reduces the value because of its lower thermal conductivity compared with AlN.

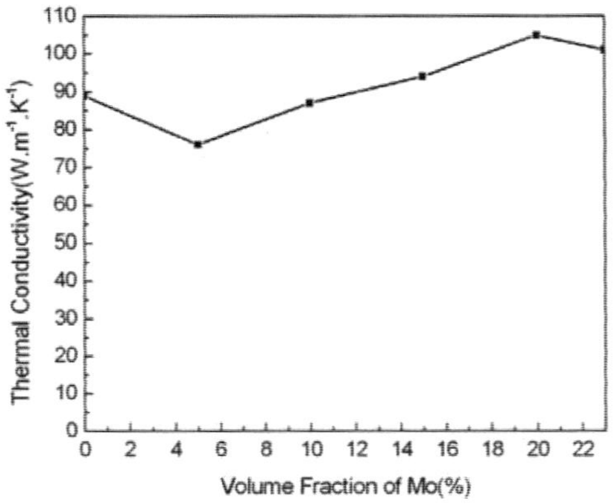

Figure 1. Room temperature thermal conductivity of composites as a function of Mo volume fraction

3.2. Microstructure Characterization

Earlier studies prove that when two surfaces having different work functions are in contact, a charge transfer takes place in such a manner that the surface having higher value of work function is enriched in electrons. This transfer of charge at the interface between metallic particle and the ceramic matrix is responsible for the formation of a space charge layer at the particle surface [12-13]. The dielectric properties of the system depend upon the electronic concentration in the space charge layer, the thickness of the space charge layer and the distance between the two particles. Analyses indicate that the key factors to dielectric properties are the metal phase concentration and the microstructure of Mo particles. According to the results, the efforts to increase the Mo concentration and to control the microstructure are essential.

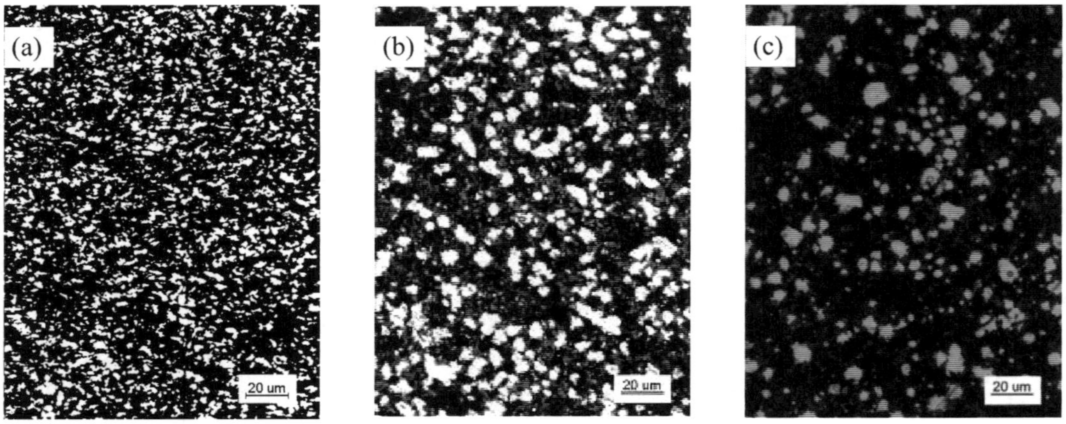

Figure 2. The microstructure of AlN-Mo composite ceramics with different ratio of Ni (a)Without Ni (b) 1% Ni by volume (c) 2% Ni by volume

Figure 2(a) shows the typical microstructure of the sintered AlN-Mo sample containing 20 vol. % Mo (AM20). The black areas are AlN, and the white phases with clear contrast are of Mo. The structure is relatively homogeneous, showing metal particles quite uniformly distributing in the AlN matrix. The metal particles which contact with each other present slim strip shape. The sample becomes a conductor without any dielectric property. Figure 2 (b) and (c) illustrates the structures of AM20 doped with 1 vol. % Ni and 2 vol. % Ni, respectively. Three samples were prepared under same processing with a pressure of 20 MPa and the temperature of 1500℃. Compared to figure 2(a), the angular slim structure of initial Mo powder particles is not preserved in the sintered structure. The particles have a tendency to acquire rounded forms which are thermodynamically more stable and the particle size is increased wherever a contact is established. With the increasing content of Ni, the size of metal particles becomes larger. Presence of Ni makes for forming liquid phase which improves sintering process. It is obvious that a little liquid phase leads to rounded metal particles to keep from contacting with each other during the sintering processes. This phenomenon makes for increasing the Mo concentration and controlling the microstructure of metal phase which are key factors for dielectric properties.

Figure 3 shows the X-ray diffraction analysis of the AM20 doped with 1 vol. % Ni after sintering at 1500℃ for 5 min. It can be seen that no trace of Ni phase was found in the sintered specimens. Mo_2C are newly formed phases, and are resulted from the reaction of Mo with graphite although 1.5mm cemented layer has been removed to void carbide for each side. It can be concluded that Mo solid solution with Ni has been formed during sintering process owing to high temperature and the magnetic properties won't be influenced.

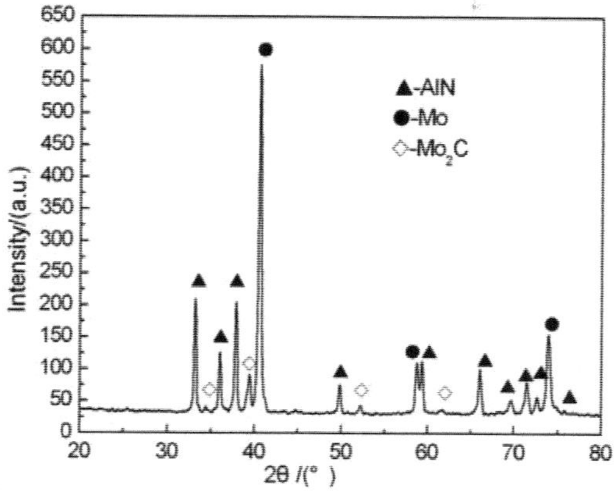

Figure 3. XRD analysis of AM20 doped with 1% Ni

3.3. Dielectric Properties

Figure 4(a) and (b) give the variation of the dielectric constant and loss as a function of Mo volume fraction at 1MHz, respectively. The results show that the values of dielectric constant and loss increase with increasing the volume fraction of Mo in the composite to 23%. The dielectric constant εr reaches 73.3 and loss tgδ reaches 0.26 when the Mo content is 23%. When the distance between metal phase particles in the solid material is decreased due to an increase in the metal phase concentration, the free electron population in the region between the two particles is increased. This insulator to conductor transformation can be attributed to a percolation phenomenon due to the presence of metal particles in the ceramic matrix. The metal particles are in contact with each other for a volume fraction bigger than 23%. Critical concentration corresponding to AlN-Mo composites can be viewed as 23% by volume. It can be seen in figure 4, both the dielectric constant and loss increase rapidly before the volume fraction of Mo reaching 20%, the values don't show an remarkably improvement when molybdenum concentration in the AlN matrix is increased from 20% to 23%.

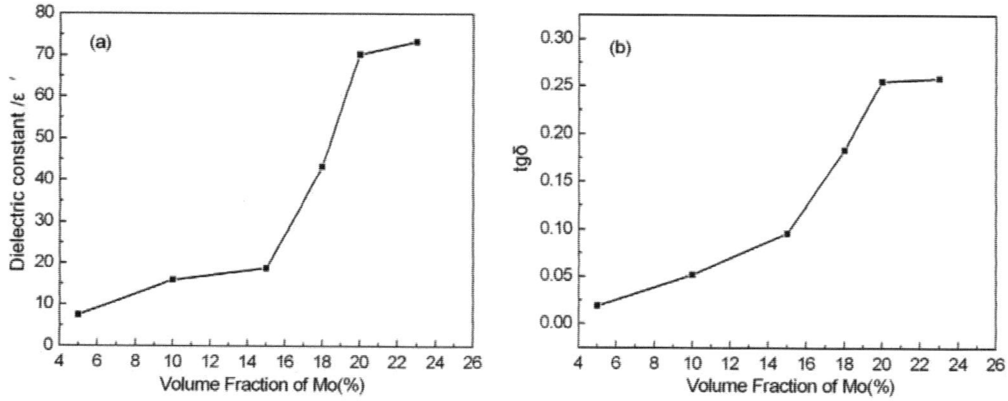

Figure 4. Dielectric constant and loss of composites as a function of Mo volume fraction

Figure 5 show the variation of the dielectric constant and loss as a function of frequency under different sintering temperatures for AM20. The results show that dielectric constant and loss can be changed evidently by different sintering performance when the Mo concentration is fixed. The phenomenon can be the result of the changed microstructure of metal phase.

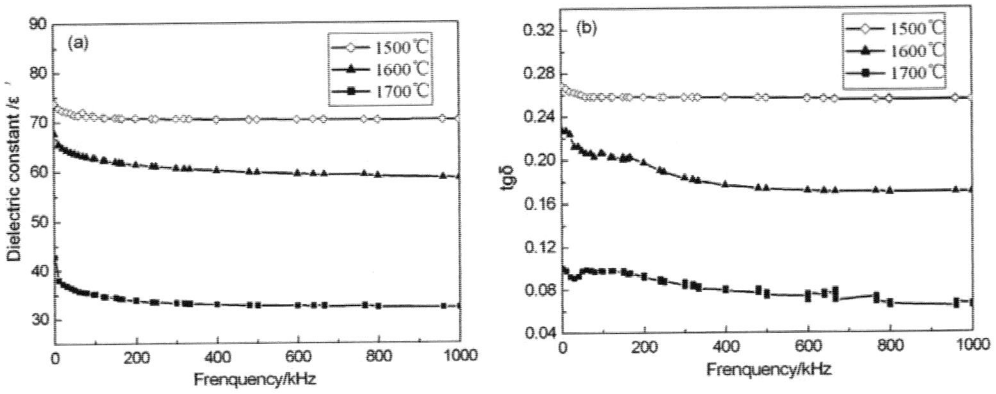

Figure 5. The dielectric properties of AlN-Mo composite ceramics under different sintering temperature

4. Conclusions

The effects of Mo addition on the thermal conductivity and dielectric properties of the composite ceramics have been studied. The value of room temperature thermal conductivity increases with the increasing content of Mo, and the value begins to decrease slightly when the Mo concentration exceeds 20 vol. %. Increasing the Mo concentration and controlling the microstructure of metal phase are key factors to increase dielectric properties, adding 1vol. % Ni into the composite ceramics plays an important role to control the properties. The dielectric properties also could be adjusted by transforming the microstructure of the metal particles when the concentration of Mo is fixed.

References

[1] Biljana Mikijelj, David K. Abe and Ron Hutcheon 2003 *J. Euro. Ceram. Soc.* **23** 2705-2709
[2] G. A. Slack, R. A. Tanzilli, R. O. Pohl and J. W. Vandersande 1987 *J. Phys. Chem. Solids.* **48** 641-647
[3] A.V. Virkar, T. B. Jackson and R. A. Cutler 1989 *J. Am. Ceram. Soc.* **72** 2031-2042
[4] D. G. Böcker, R. W. Hamminger, J. Heinrich, J. Humber and A. Roosen 1987 *Adv. Mater.* **131** 169-178
[5] CAI Delu and LIU Chengjun 1985, *Beijing Sci pub press* 54-65
[6] XU Gengfu 2003 *J. Mater. Res.* **18** 66-76
[7] Amir Azam Khan and Jean Claud Labbe 1997 *Materials Science and Engineering*, **A 230** 33-38
[8] Amir Azam Khan and Jean Claud Labbe 1997 *Journal of the European Ceramic Society*, **17** 1885-1890
[9] Amir Azam Khan and Jean Claud Labbe 1996 *Journal of the European Ceramic Society*, **16** 739-744
[10] K. Watari, K. Ishizaki and T. Fujikawa 1992 *J. Mater. Sci.*, **27** 2627-2630
[11] K. Watari, M. Kawamoto and K Ishizaki 1991 *J. Mater. Sci.*, **26** 4727-4732
[12] Kieffer, J. and Wagner, J. B. 1988 *J. Electrochem. Soc.*, **135** 198-205
[13] Dudney, N. J. 1985 *Journal Am. Ceram. Soc.*, **68** 538-545

Journal of Physics: Conference Series **152** (2009) 012064 doi:10.1088/1742-6596/152/1/012064

Influences of Ti^{4+} and Mg^{2+} substitutions on the properties of lithium ferrites

Hua SU, Huaiwu ZHANG, Xiaoli TANG, Baoyuan LIU

State Key Laboratory of Electronic Thin Films and Integrated Devices, University of Electronic Science and Technology of China, Chengdu 610054, China

Email: uestcsh@163.com

Abstract. The Ti^{4+} and Mg^{2+} co-substituted lithium ferrites with different compositions of $Zn_{0.1}Li_{0.45}Mn_{0.1}Fe_{2.35-2x}$ $(TiMg)_xO_4$ (x=0.0-0.5) were prepared by the ceramic standard processing. The magnetic properties and microstructure of the samples were investigated. A single phase spinel structure was confirmed by XRD in substituting range. Sintering densities continuously decreased with the increase at x value, which was attributed to the fact that the heavier Fe^{3+} ions were replaced by the relatively lighter Ti^{4+} and Mg^{2+} ions. However, relative density of the samples had no obvious relationship with the substituting value. Saturation magnetization continuously decreased with x value, which was attributed to the decrease of resultant magnetic moment between A and B sub-lattice. Remanence decreased monotonously with x value due to the decrease of saturation magnetization and magnetocrystalline anisotropy constant. But the effect of Ti^{4+} and Mg^{2+} substitutions on the Br/Bs ratio values was not obvious. Coercive force was mainly determined by the microstructure and magnetocrystalline anisotropy constant of the ferrites. In this research, with the increase of Ti^{4+} and Mg^{2+} substitutions, the advantageous influence by the decrease of magnetocrystalline anisotropy constant was more significant than the disadvantageous influence caused by the increase of closed pores. As a result, coercive force of the ferrites also decreased monotonously with the increase at x value.

1. Introduction

Li-based spinel ferrites have been widely used in high frequency magnetic applications, such as circulators, phase shifters and isolators, due to their high Curie temperature, high saturation and hysteresis loop properties offer performance advantages over other spinel structure [1-5]. The properties desired for specific applications have been provided by controlling the preparation conditions or by adopting appropriate substitutions. Till now, the conventional way of producing these ferrite materials is by the solid state reaction of oxide/carbonate and calcined at high temperature. Furthermore, some different soft-chemic methods, such as coprecipitation, sol-gel synthesis, alkoxide hydrolysis, hydrothermal reaction and citrate route have also been reported to prepared lithium ferrites [6-8]. As for the influences of substitutions, probably the substitutions of Zn and Mn are best known to improve the magnetization and electrical resistivity respectively [9]. In addition, some sintering aids [10-11], such as cordierite glass, which can lower sintering temperature, have also been introduced with beneficial influences on magnetic properties. To meet requirements of different microwave magnetic devices, which possibly work at different frequency band, properly adjusting the saturation magnetization of the ferrites is necessary. At the same time, other magnetic properties, such as Br/Bs (remanence/Saturation flux density) ratio, Hc (coercive force), can't be depressed. In this paper, we substituted the Fe^{3+} ion with the Ti^{4+} and Mg^{2+} ions synchronously to adjust saturation magnetization of the lithium ferrites, and prepared $Zn_{0.1}Li_{0.45}Mn_{0.1}Fe_{2.35-2x}(TiMg)_xO_4$ (x=0.0, 0.1, 0.2, 0.3, 0.4 and 0.5) ferrites by the conventional ceramic processing. To lower sintering temperature and prevent lithium element volatilizing, 0.5wt% Bi_2O_3 additive was added in all samples. And the influences of Ti^{4+} and Mg^{2+} co-substitution on the magnetic properties and microstructure of the lithium ferrites were studied in detail.

2. Experimental details

The lithium ferrite samples were prepared by the solid state reaction method. Analytical grade Fe_2O_3, ZnO, Li_2CO_3, $MnCO_3$, TiO_2 and MgO were weighed following the compositions of $Zn_{0.1}Li_{0.45}Mn_{0.1}Fe_{2.35-2x}(TiMg)_xO_4$ (x=0.0, 0.1, 0.2, 0.3, 0.4 and 0.5). The oxides were wet-mixed in planetary mills firstly. After drying, the mixtures were calcinated at 800℃ in air. Then the calcinated powders were mixed with 0.5wt% Bi_2O_3 and wet-milled in planetary mills again. Following the addition of polyvinyl alcohols (PVA), the milled powders were granulated and pressed into toroidal shaped samples, which were finally sintered at varying temperatures from 900℃ to 1000℃ in 50℃ intervals, and left to cool inside the electric furnace.

Phase purity were checked by X-ray diffraction (XRD), using Cu K_α radiation. Densities of the sintered specimens were calculated by mass/volume ratio. Relatively density values were obtained by comparing sintering density and X-ray density of the samples. Microstructures were obtained using scanning electron microscopy (SEM). Saturation magnetization ($4\pi M_s$), remanence (Br) and coercive force (Hc) of the samples were tested by a B-H analyzer (IWATSU, SY8232).

3. Results and discussion

The diffraction patterns of the 900℃ sintered samples with x=0.0, 0.3, 0.5 are presented in Figure 1. A single phase spinel structure is observed for all samples, no second phase is detected. So it can be concluded that the Ti^{4+} and Mg^{2+} substitutions fully enter into the spinel lattice in the substituting range.

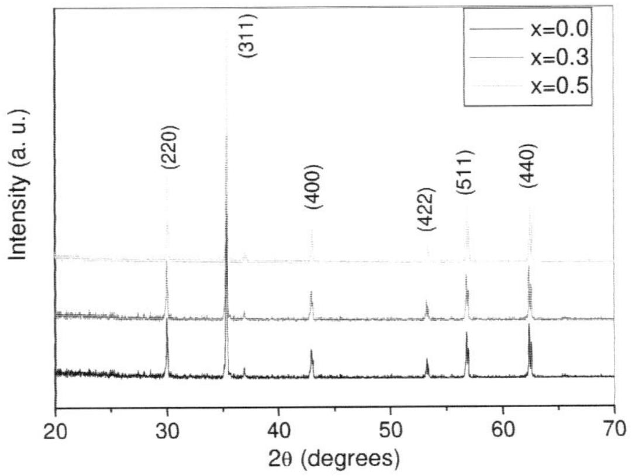

Figure 1. X-ray diffraction patterns of the samples with different substitutions

Figure 2 shows the sintering density and relative density values as a function of the x value for the samples sintered at 900℃, 950℃ and 1000℃. It is seen that both sintering densities and relative densities gradually increase with increasing sintering temperature for all samples. Furthermore, with increasing content of Ti^{4+} and Mg^{2+} substitutions, sintering densities decrease monotonously. This fact

Journal of Physics: Conference Series **152** (2009) 012064 doi:10.1088/1742-6596/152/1/012064

is attributed to that the heavier Fe^{3+} ions are replaced by the relatively lighter Ti^{4+} and Mg^{2+} ions in the present case. However, the relative densities have no obvious relationship with the substituting values, which means that the Ti^{4+} and Mg^{2+} substitutions don't markedly influence the sintering characteristic of the samples.

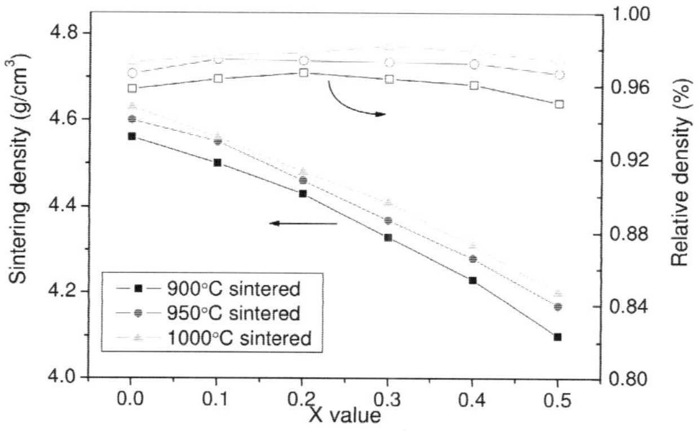

Figure 2. Sintering Densities and relative densities of the samples. The curves with open circles are the relative densities. The curves with solid circles are the sintering densities

Relationship between saturation magnetization and x value for the samples sintered at different temperatures is shown in Figure 3. It clearly indicates that the saturation magnetization values present an approximatively linear decrease with the increase of x value. On the basis of site preference energy, it is known that Ti^{4+} and Mg^{2+} ions preferably occupy the octahedral site by replacing Fe^{3+} ions. In view of this, the cation distribution of the ferrites can be written as

$$(Zn_{0.1}^{2+}Fe_{0.9}^{3+})_A[Li_{0.45}^{1+}Mn_{0.1}^{3+}Ti_x^{4+}Mg_x^{2+}Fe_{1.45-2x}^{3+}]_B O_4^{2-}$$

Both Ti^{4+} and Mg^{2+} are non-magnetic ions. With increasing x value, the resultant magnetic moment between A and B sub lattice gradually decreases. As a result, saturation magnetization values of the lithium ferrites decrease. However, the saturation magnetization values show no significant dependence with the sintering temperature, especially for the samples sintered at 950℃ and 1000℃. This fact is attributed to that with increasing sintering temperature; the advantage influence of increased density on saturation magnetization may be counteracted by the disadvantage influence of lithium volatilization on saturation magnetization.

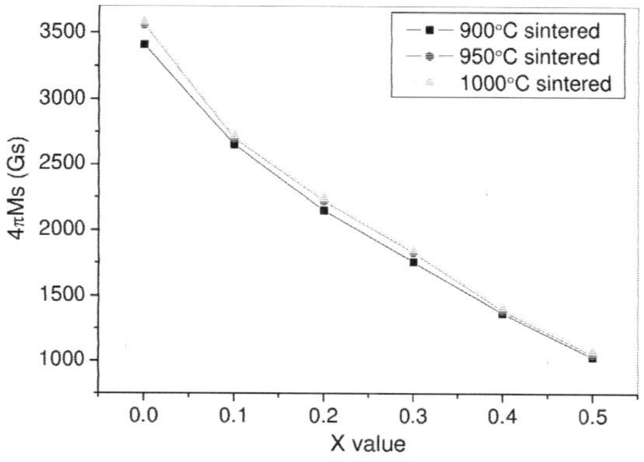

Figure 3. Saturation magnetization of the samples with different content of substitutions and sintering temperatures

Influences of the Ti^{4+} and Mg^{2+} substitutions on Br and Br/Bs ratio properties of the samples are shown in Figure 4. It can be seen that remanence values also continuously decrease with increasing x value, despite different sintering temperature. First reason leaded to this fact is due to the decrease of saturation magnetization; the other reason is attributed to that non-magnetic Ti^{4+} and Mg^{2+} substitutions decrease the magnetocrystalline anisotropy constant of the lithium ferrites. Furthermore, remanence first increases and then decreases with the increase of sintering temperature. The highest remanence values are obtained when sintered at 950℃, regardless different x value. This fact may be attributed to that the volatilization of lithium element enhances when sintered at higher temperature and leads to the increase of closed pores in grains. And the 950℃ sintered samples also obtain the highest Br/Bs ratio. But the effect of Ti^{4+} and Mg^{2+} substitutions on the Br/Bs ratio is values not obvious.

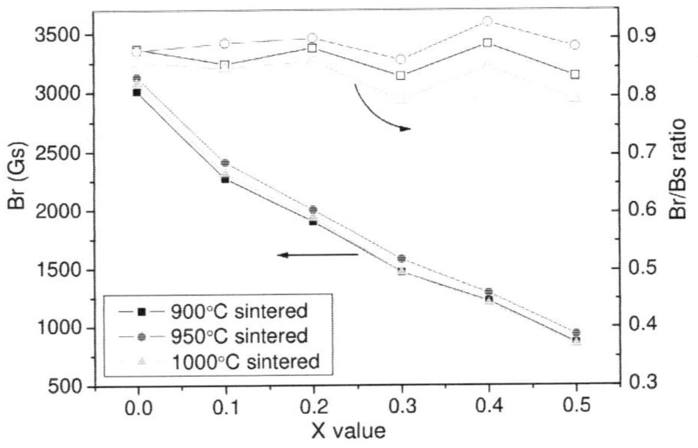

Figure 4. Br and Br/Bs ratio values of the samples. The curves with open circles are the Br/Bs ratio values. The curves with solid circles are Br values

The coercive force values of the samples sintered at different temperatures versus Ti^{4+} and Mg^{2+} substitutions are plotted in Figure 5. With increasing sintering temperature, coercive force decreases monotonously, in spite of different x value. This fact is attributed to that coercive force tends to be inversely proportional to the density and grain size. With increasing sintering temperature, both sintering density and grain size increase, so the coercive force decreases. To further determine the influence of Ti^{4+} and Mg^{2+} substitutions on the coercive force characteristic of the lithium ferrites, SEM photos of several samples sintered at 900℃ were examined and presented in Figure 6.

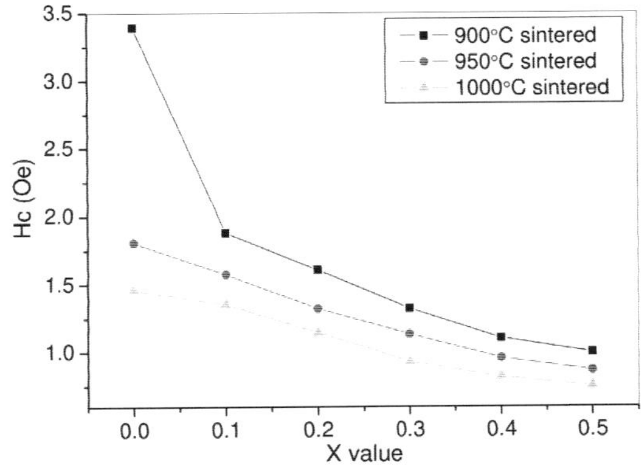

Figure 5. Coercive force of the samples with different content of substitutions and sintering temperatures

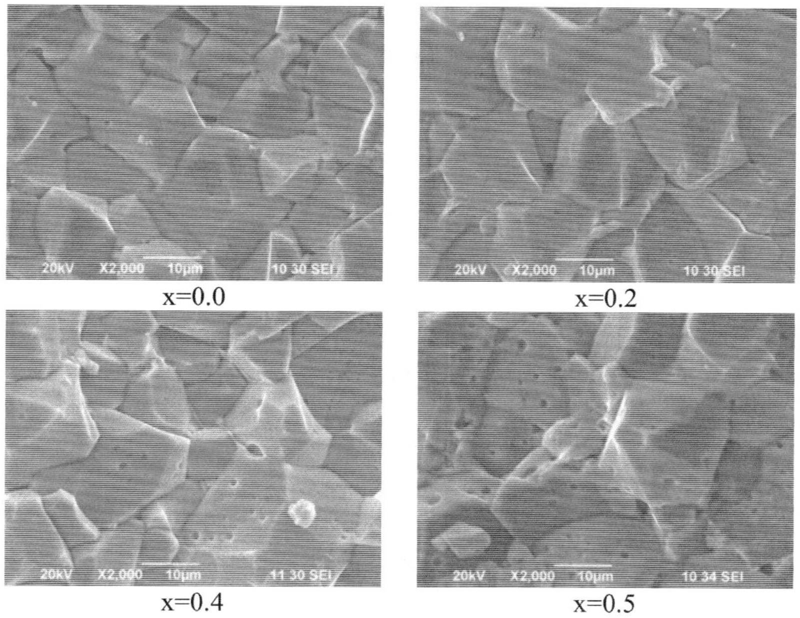

Figure 6. SEM photos of the samples with different substitutions

It is found that average grain size has no obvious connection with the Ti^{4+} and Mg^{2+} content. However, when the substitutions increase too much, closed pores in big grains increase obviously. As expected, coercive force is inversely proportional to grain size and directly proportional to magnetocrystalline anisotropy constant [10]. With increasing Ti^{4+} and Mg^{2+} substitutions, average grain size has no prominent change, but magnetocrystalline anisotropy constant gradually decreases, which is in favor of decreasing coercive force. It is also known that closed pores lead to increase of coercive force. From Figure 5, we can conclude that the advantageous influence leaded by the decrease of magnetocrystalline anisotropy constant is more predominant than the disadvantageous influence leaded by the increase of closed pores. As a result, coercive force also gradually decreases with increasing x value.

To sum up, Ti^{4+} and Mg^{2+} co-substitution is an effective method to adjust the saturation magnetization of the lithium ferrites, which also doesn't obviously influence the sintering and Br/Bs ratio characteristics of the ferrites. At the same time, some magnetic properties, such as coercive force, can be improved.

4. Conclusions

In this experiment, sintering density, microstructure and magnetic properties of the Ti^{4+} and Mg^{2+} co-substituted lithium ferrites were investigated, and the following conclusions were drawn:

1. Sintering density values of the lithium ferrites gradually decreased with increasing Ti^{4+} and Mg^{2+} substitutions, which was attributed to that the heavier Fe^{3+} ions were replaced by the relatively lighter Ti^{4+} and Mg^{2+} ions in the present case. However, the relative densities of the ferrites had no obvious relationship with the substituting value.
2. Saturation magnetization decreased with increasing x value, which was attributed to the decrease of resultant magnetic moment between A and B sub lattice.
3. Remanence decreased monotonously with x value due to the decrease of saturation magnetization and magnetocrystalline anisotropy constant. But the influence of Ti^{4+} and Mg^{2+} substitutions on the Br/Bs ratio values was not obvious.
4. Ccoercive force decreased monotonously with increasing x value due to the predominant influence of the decrease of magnetocrystalline anisotropy constant.

Acknowledgements

This work is supported by the National Natural Science Foundation of China ((Nos. 50702011, 60425102 and 60721001).

References

[1] Yen-Pei Fu, Cheng-Hsiung Lin, Chung-Wen Liu, Yeong-Der Yao, J. Alloy. Comp. 395 (2005) 247.
[2] Nutan Gupta, Mukesh C. Dimri, Subhash C. Kashyap, D. C. Dube, Ceram. Inter. 31 (2005) 171.
[3] H. M. Widatallah, C. Johnson, F. J. Berry, A. M. Gismelseed, E. Jartych, J. F. Marco, F. S. Gard, M. Pekala, J. Phys. Chem. Solid, 67 (2006) 1817.
[4] H. M. Widatallah, C. Johnson, F. J. Berry, E. Jartych, A. M. Gismelseed, M. Pekala, J. Grabski, Mater. Lett, 59 (2005) 1105.
[5] S. A. Mazen, H. A. Dawoud, Mater. Chem.. Phys, 82 (2003) 557.
[6] N. Ramachandran, A. B. Biswas, J. Solid. State. Chem, 30 (1979) 61.
[7] T. Shirane, R. Kanno, Y. Kawamoto, Y. Takeda, M. Takano, T. Kamiyama, F. Izumi, Solid. State. Ionics, 79 (1995) 227.
[8] M, Tabuchi, K. Ado, H. Kobayashi, I. Matsubara, H. Kageyama, M. Wakits, J. Solid. State. Chem, 141,(1998) 554.
[9] J. Nicolas, Microwave Ferrites, North-Holland Publishing, New York, 1980, Chap. 4.
[10] G. M. Argentina, P. D. Baba, IEEE. Trans. Microwave. Theor. Tech, 20 (1974) 652.
[11] Yong S. Cho, Jung W. Lee, Vasantha R. W. Amarakoon, Mater. Lett, 59 (2005) 710.

Journal of Physics: Conference Series **152** (2009) 012065 doi:10.1088/1742-6596/152/1/012065

Dielectric and Ferroelectric Properties of (1-x)$Na_{0.5}Bi_{0.5}TiO_3$-x$SrTiO_3$ Lead-free Piezoceramics System

Ting WANG, Huiling DU, Xiang SHI

College of Materials Science and Engineering, Xi'an University of Science and Technology, 58YanTa Middle Road , Xi'an 710054, China

E-mail: hldu@xust.edu.cn

Abstract. Lead-free piezoelectric ceramics (1-x)$Na_{0.5}Bi_{0.5}TiO_3$-x$SrTiO_3$ (NBST100x), where x=0.02-0.90, were prepared by the fine ceramic sintering technique in air. It was found that all compositions showed single perovskite phase and the crystal structure changed from rhombohedral to cubic at 0.20<x<0.50. The experiment results showed that with the increasing of ST content, the piezoelectric parameters $d_{33,}$ electromechanical coefficient k_P and coercive field E_c both decreased. When x=0.25, T_c of 210°C, E_c of 1.26 kV/mm, ε_m of 4984.84 and d_{33} of 60 pC/N were obtained, respectively. In addition, the samples with lower ST content experienced two phase transitions, the diffuse phase transition had typical relaxor-ferroelectric characteristic. While, with the increasing of ST content (≥40%), the samples had only one phase transition peak, the temperature dependence curve of dielectric constant moved towards lower temperature, which could be mostly ascribed to the function of paraelectric ST, and explained by the micro-macro domain transition theory.

1. Introduction

Lead-based piezoelectric ceramics have dominated the commercial market for the past several years, owing to their outstanding piezoelectric, dielectric, ferroelectric and electro-mechanical properties. However, these ceramics contain more than 70 wt% lead oxide. The high lead content can cause damage to the human beings and environment during preparation, use and reclaim process. With the increasing demand for environmental protection and sustainable development, therefore, it is ultimate way to develop environmental friendly lead-free piezoelectric ceramics.

Lead-free piezoelectric materials have already got primary development since 1960s and formed several mainly studying systems. In recent years, growing attention has been given to the new ABO_3 perovskite structure-sodium bismuth titanate ($Na_{0.5}Bi_{0.5}TiO_3$, NBT) and NBT based solid solutions. As NBT shows strong ferroelectricity with large remnant polarization P_r = 38 μC/cm^2 and high Curie temperature T_c = 320℃, it has been considered to be a promising candidate for lead-free piezoelectric materials. However, this ceramic has drawbacks such as high conductivity and large coercive field, causing problems in poling process. As a result of those, a lot of work have been done by substitution of $BaTiO_3$, $SrTiO_3$, $K_{0.5}Bi_{0.5}TiO_3$, $NaNbO_3$, etc[1-5], to modify and improve NBT ceramics piezoelectric properties. Among these NBT-based solid solutions, (1−x)($Na_{0.5}Bi_{0.5}$)TiO_3–x$SrTiO_3$ system has attracted much attention due to its strong dispersion of the permittivity with a relaxor-like behavior. $SrTiO_3$ is a typical perovskite compound possessing cubic structure at room temperature. It is not until $SrTiO_3$ is cooled to 110 K when its cubic structure distorts and becomes tetragonal. Thus, $SrTiO_3$

exhibits paraelectric behavior at room temperature, although recent studies suggest that the material is indeed an incipient ferroelectric whose ferroelectricity is suppressed by quantum fluctuations[6]. In this paper, we present the results of a comprehensive investigation of the dielectric, piezoelectric and ferroelectric properties of $(1-x)(Na_{0.5}Bi_{0.5})TiO_3$-$xSrTiO_3$ materials within almost the whole composition range of the solid solutions.

2. Experimental

Reagent grade oxide or carbonate powders of Bi_2O_3, Na_2CO_3, $SrCO_3$, TiO_2 were used as starting raw materials. A conventional powder synthesis method was adopted to synthesize $(1-x)(Na_{0.5}Bi_{0.5})TiO_3$-$xSrTiO_3$ (NBST100x, x=0.02-0.90). Raw materials were mixed using planetary milling with agate media and alcohol for 4h. The mixed powders were calcined at 800~850℃ for 2h. After second ball-milled for 4h, the powders dried and granulated with PVA as a binder. The granulated powders were pressed into disc with diameter 12 mm and thickness 1-1.2 mm. After binder burnout, these pressed disks were sintered at 1100-1200℃ for 2h soaking period in air. Silver paste was fired on both sides of the disks at 800℃ as electrodes. The specimens for measurement of piezoelectric properties were poled in silicon oil at 80℃ under 3-6kV/mm for 15 min.

X-ray diffraction (XRD) patterns were taken on a D/MAX -2400 X-ray diffractometer with Cu Kα radiation (λ = 0.15406 nm). The microstructure evolution was observed using a scanning electron microscopy (JSM 6460LV, Japan). Dielectric measurements were carried out from room temperature to 600℃ using Agilent 4284A Precision LCR meter at 1-100Hz. Piezoelectric constant d_{33} of the samples were measured by means of quasistatic d_{33} meter (ZJ-3A) based on Berlincourt method at 110 Hz. The remanent polarization P_r and coercive field E_c were determined from P–E hysteresis loops measured by aixACCT TF-2000 ferroelectric analyzer.

3. Results and Discussion

3.1. Crystal structure

Figure 1(a) shows the X-ray diffraction patterns of the NBST100x ceramics. A single phase perovskite structure is observed for all compositions, with no trace of TiO_2-nonstoichiometry induced second phase. As increasing the ST content, the peaks shift toward lower angle slightly, showing the lattice parameters changed. The d-spacing was found a little increased because the host Na^+ and $Bi^{3+}$are replaced by doped Sr^{2+}.Figure 1(b) compares the XRD patterns (2θ= 38°–50°) of five typical NBST100x compositions.It is evident that the crystal structure of NBST4 exhibits a high degree of rhombohedral with the splitting of the(210) and (012) characteristic peaks at the 2θ about 40°. With increasing ST concentration, the symmetry gradually changed from rhombohedral to cubic. When x=0.50, only the (111) peak and (200) peak are observed at 2θ about 40°, 47° respectively, suggesting a cubic structure. Mixtures of the (200), (020), (210) and (012) peaks are observed in the diffraction patterns, suggesting that rhombohedral and cubic phase coexist in the ceramics when 0.20<x<0.50.

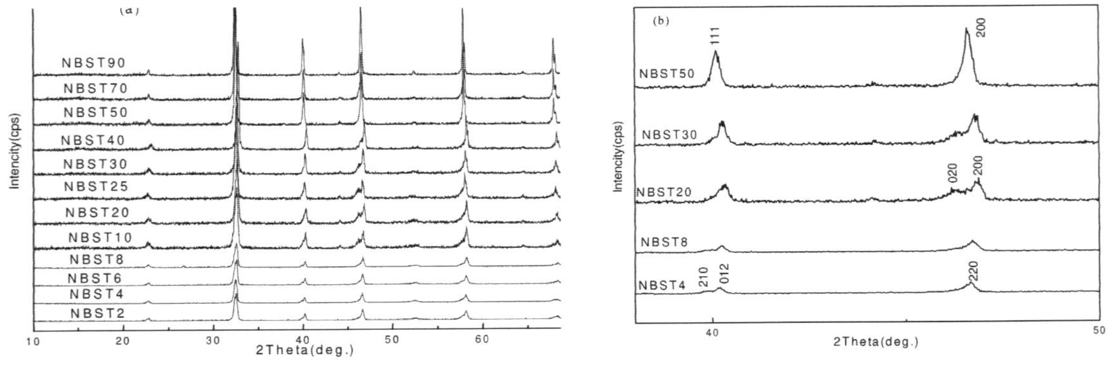

Figure 1. XRD patterns of the NBST100X ceramic.

3.2. Microstructure

Figure 2 shows the SEM images of natural surface for NBST100x ceramics. The sintering temperature increases with the addition of ST resulting from the melting point of ST higher than NBT. All sample surface grains present regular geometry form, the average grain size is about 2 μm, structure is compact and almost without pore and glass phase. It is distinct that the grain size becomes smaller when x>0.40. The reason may be that Sr^{2+} abound in crystal boundary, which prevents the ion from migrating and restrains the growing of grains. Furthermore, none of the doped samples exhibit deliquescence against water, indicating that the ST addition can reduce the formation of unstable secondary phases enhancing the stability.

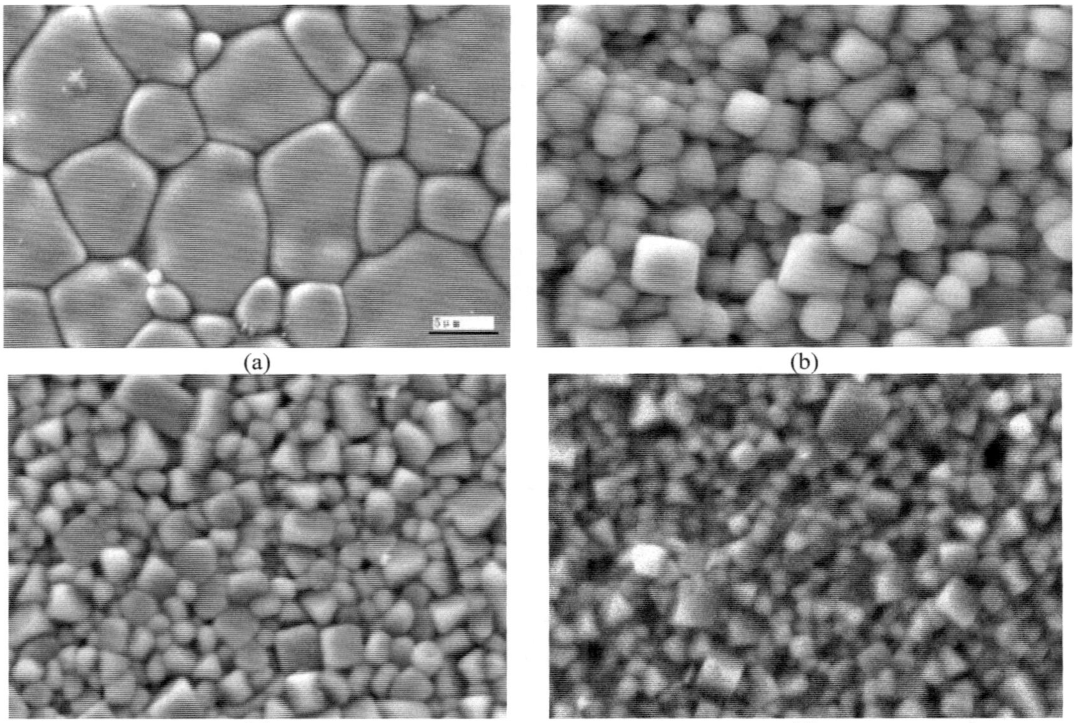

(a) (b) (c) (d)

Figure 2. SEM micrographs of natural surface for NBST100x ceramic: (a)x=0.04; (b) x=0.08; (c) x= 0.25; (d) x=0.50.

3.3. Dielectric properties

Figure 3 shows the temperature dependences of ε and $tan\delta$ for NBST100x ceramics with x = 0.10, 0.25, 0.40 and 0.70. Phase transitions are all observed obviously and the peaks become broader as the increase of ST content. It's apparently that the transition temperatures decrease with increasing ST additives. When x<0.40, the curves exhibit two dielectric anomalies at T_d and T_m. T_d is the depolarization temperature which corresponds to the transition from ferroelectric state to anti-ferroelectric state, while T_m is the Curie temperature at which ε reaches maximum value and corresponds to a transition from anti-ferroelectric state to paraelectric state. The dielectric behavior of specimens are changed by doping ST, ε exhibits a strong frequency dependence at the temperature of T_m, namely the T_m move towards higher temperature as frequency increases while corresponding the maximum value of ε decreases, suggesting that the ceramics are relaxor ferroelectrics. In the NBST100x ceramics, Na^+, Bi^{3+}, Sr^{2+} are randomly distributed in the 12-fold coordination sites, so the observed relaxor behavior is reasonably attributed to the disordering of A-site cations[7]. Besides, $SrTiO_3$ is perovskite structure paraelectric with Ti^{4+} occupies centre location, for the ion state is different that produces oxygen vacancy, the motion of domain wall is restrain with introducing of ST content, so when x up to 0.40, only one dielectric peak can be seen.

It also can be seen in the curve of loss tangent versus temperature, a loss peak appears during the temperature of 100～200℃. With the temperature increasing, the loss falls to the minimum value and sharply increases above $T_{d.}$ This phenomenon may connect with the existence of polar micro-region[8]. After the materials occur a transition from ferroelectric state to anti-ferroelectric state, the anti-ferroelectric phase changes from macro-domain to micro-domain until polar micro-region. The dielectric loss which induced by ferroelectric domain will decreases and shows the minimum value in the curve of loss tangent versus temperature. When the temperature is kept on rising, the leakage current becomes dominant in the materials, which results in the sharply increase of the dielectric loss.

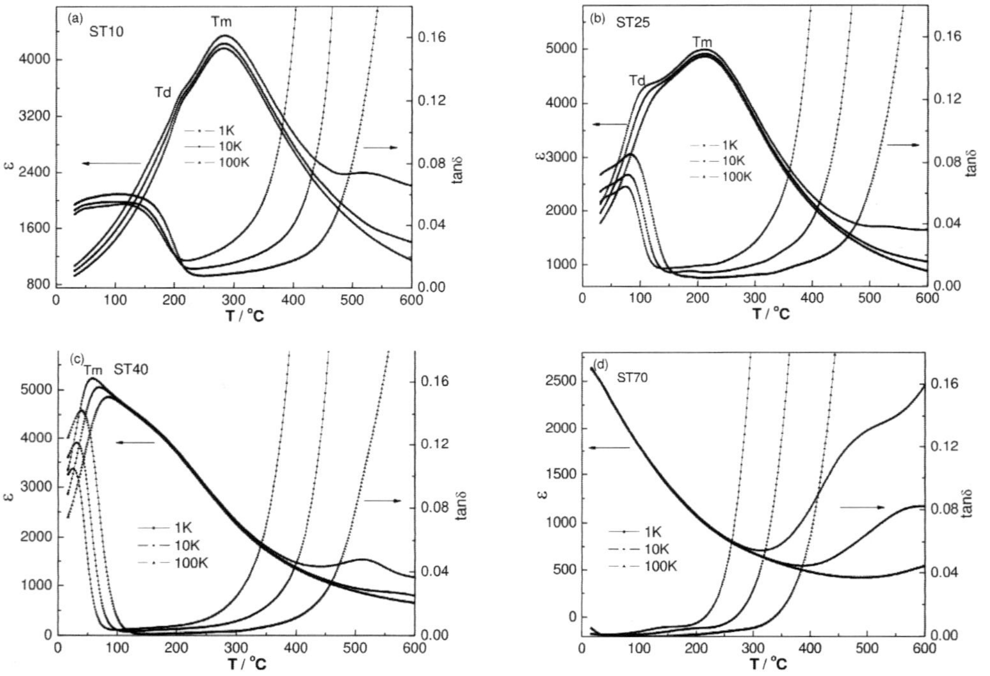

Figure 3. Temperature dependence of dielectric constant and loss of NBST100x ceramic: (a) x=0.10; (b) x=0.25; (c) x= 0.40; (d) x=0.7.

For relaxor ferroelectrics, the diffuseness in the phase transition can be described by the equation$1/\varepsilon - 1/\varepsilon_m = C^{-\alpha}(T - T_m)^{\alpha}$, where ε_m is the maximum value of dielectric constant at T_m, α is the degree of diffuseness and C is the Curie-like coefficient. α can have a value ranging from 1 for anormal ferroelectric to 2 for an ideal relaxor ferroelectric[9–12]. Based on the temperature plots of ε at 1 kH,the graphs of $\log(1/\varepsilon - 1/\varepsilon_m)$ versus $\log(T - T_m)$for the NBST100x ceramics were plotted, giving the results shown in Figure 4. All the samples exhibit a linear relationship. By least-squared fitting the experimental data to the equation, α was determined, giving the results shown in Table 1. All the fittings give a large regression coefficient, indicating that the experimental data are in very good agreement.All the ceramics have a large α value (>1.7), confirming that they are relaxor ferroelectrics.It is also seen that the diffuseness increases in general with increasing x.

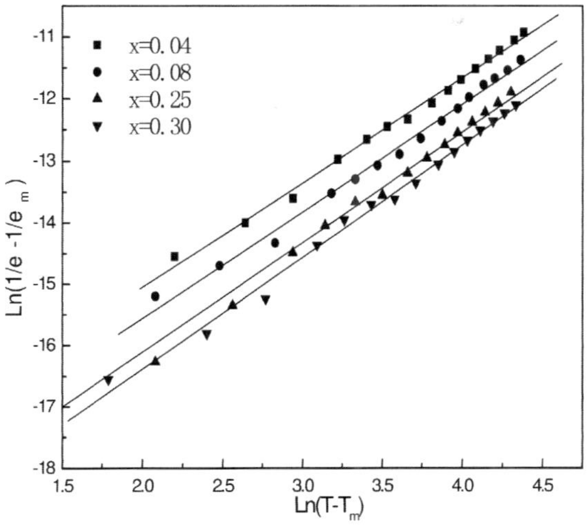

Figure 4. Plot of ln (1/ε − 1/εm) versus ln (T − Tm) at 1 kHz for the NBST100x ceramics at x = 0.04, 0.08,0.25 and 0.30.

Table 1. Least-squared fitting results for the NBST100X ceramics.

x	T_m(℃)	ε_m	α	R
0.04	305	4276.58	1.70	0.9961
0.08	286	5394.69	1.72	0.9975
0.25	212	4988.12	1.76	0.9954
0.30	158	5155.12	1.81	0.9963

3.4. Piezoelectric and Ferroelectric properties

Dependence of d_{33} and k_p of the NBST100x ceramics with the substitution level of ST is shown in Figure 5. k_p increases gradually with increasing x and then decreases, giving a maximum value of 0.375 at x=0.10. Addition of small amounts of ST yields to larger electromechanical coupling factors than those of pure NBT samples. The promotion may be attributed to increased density, lowering the leakage current and enhancing the poling process. However, when the ST content exceeds 0.40, the d_{33} and k_p both decrease sharply since the ST belongs to symmetry structure and has no piezoelectricity. On the other hand, the d_{33} decreases from NBST2 96 pC/N to NBST25 60 pC/N, which remains higher than the pure NBT ceramic (58 pC/N), demonstrating that forming a multicomponent system is an efficient approach for improving the piezoelectric properties of the ceramics.

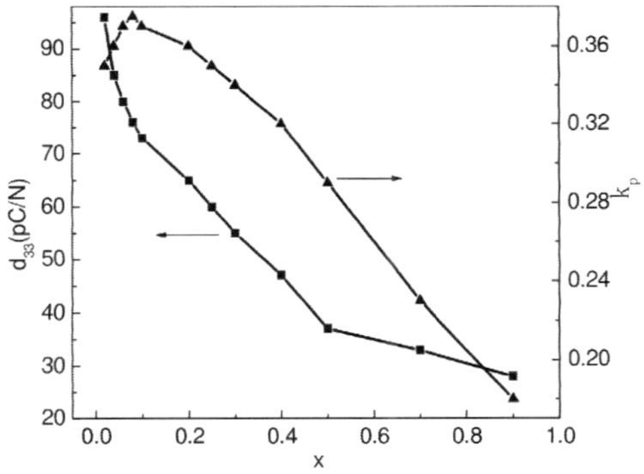

Figure 5. Compositional dependence of d_{33} and k_p for theNBST100x ceramics.

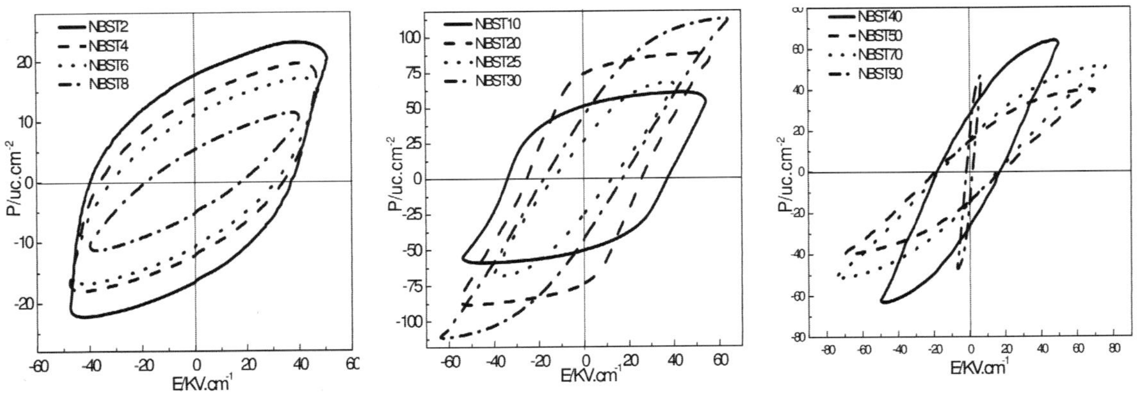

Figure 6. Hysteresis loop of NBST100x ceramics

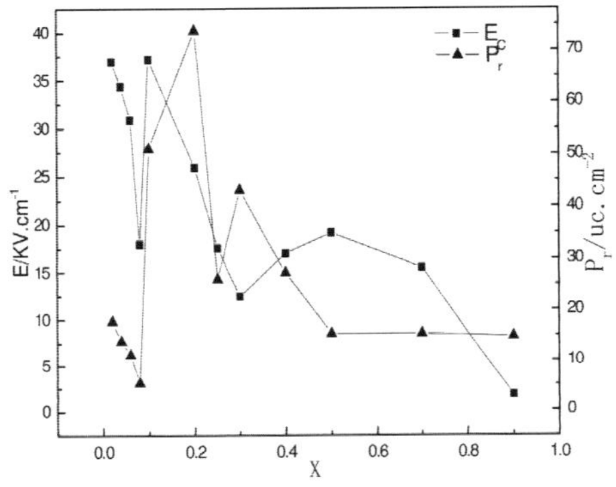

Figure 7. Variations of P_r and E_c with content for the NBST100x ceramics

The *P–E* hysteresis loops the NBST100x ceramics are shown in Figure 6, while the variations of P_r and E_c at different contents are plotted in Figure7. All the ceramics exhibit a typical ferroelectric*P–E* loop. However, because of the high resistivity, the *P–E* loop of the NBST2-NBST8 is not well saturated. Seen from the Figure7, the optimum properties of samples is at x=0.20, the P_r and E_c are 73.81 and 25.93, respectively. The P_r and E_c gradually decrease with increasing of x as a whole. When the x>0.40, the curve becomes deformed from loop to linear and different from the typical ferroelectriccharacteristics. The radius of Sr^{2+} is bigger than the radius of Na^{+} and Bi^{3+}, the ion enter into the A place, which has weakened the coupling between TiO_6 octahedron and promoted their revolved, as a result, the ferroelectric properties of samples degenerated.

4. Conclusion

(1-x)$Na_{0.5}Bi_{0.5}TiO_3$-x$SrTiO_3$ lead-free piezoelectric ceramics have been prepared by the fine ceramic sintering technique in air, and their structural, piezoelectric, dielectric and ferroelectric properties have been studied. The ceramics can be well sintered at 1100–1200℃. Our results reveal that $SrTiO_3$ diffuse into the $Na_{0.5}Bi_{0.5}TiO_3$ lattices to form a solid solution with a single-phase perovskite structure, it changed from rhombohedral to cubic at 0.20<x<0.50. The experiment results showed that with the increasing of ST content, the piezoelectric parameters d_{33}, electromechanical coefficient k_P and coercive field E_c both decreased, the temperature dependence curve of dielectric constant moved towards low temperature, diffuse phase transition had typical relaxor-ferroelectric characteristic and suggested that this should result from the cation disordering in the 12-fold coordination sites. While, with the increasing of ST content(≥40%), the samples had only one phase transition and shifted to lower temperature ,which could be mostly ascribed to the function of paraelectric ST, and explained by the micro-macro domain transition theory.

Acknowledgement
This work is financially supported by the National Natural Science Foundation of China under Grant No. 50702044 and Xi'an Applied Materials Innovation Fund under Grant No. XA-AM-200715.

References

[1] A. Herabut and A. Safari 1997 *J Am Ceram Soc* **80(11)** 2954-2958
[2] T. Takenaka and H. Nagata 1998 *Japan J Appl Phys* **37** 5311-5314
[3] H. Nagata and T. Takenaka 2001 *Eur Ceram Soc* **21** 1299-1302
[4] Hui-dong Li , Chu-de Feng and Wen-long Yao 2004 *Mater Lett* **58(7-8)** 1194-1198
[5] Chune Peng, Jing-Feng Li and Wen Gong 2005 *Mater Lett* **59** 1576-1580
[6] Y. X. Wang 2005 *Solid State Commun* **135** 290
[7] G. A.Smolenskii, V.A. Isupov, Agranovskaya， A.1.and Kxainik 1961 *Sov Phys Solid State* **2** 2651-2654
[8] VIEHLAND D, TANG S J and CROSS L E 1990 *J Appl Phys* **68(30)** 2916-2921
[9] Li Y M, Chen W, Xu Q, Zhou J, Gu X and Fang S 2005 *Mater Chem Phys* **94** 328–32
[10] Karthik C, Ravishankar N and Varma K B R 2006 *Appl Phys Lett* **89** 1760–2
[11] Li Z F, Wang C L, Zhong W L, Li J C and Zhao M L 2003 *J Appl Phys* **94** 2548–52
[12] Guo Y, Kakimoto K and Ohsato H 2004 *Solid State Commun* **129** 279–84

The effect of Lead-glass Additives on Densification and Dielectric Properties of $Ba_{0.4}Sr_{0.6}TiO_3$ Ceramics

Q.M. ZHANG, J.D. CUI, G.X. DONG and J. Du

Advanced Electronic Materials Institute, General Research Institute for Nonferrous Metals, Beijing 100088, china

E-mail: dujun@grinm.com

Abstract. Effects of lead-glass additives on the densification, microstructure and dielectric properties of $Ba_{0.4}Sr_{0.6}TiO_3$ (BST) were investigated. Three simple glass systems including $PbO\text{-}SiO_2$, $PbO\text{-}SiO_2\text{-}B_2O_3$, and $PbO\text{-}ZnO\text{-}B_2O_3$ were employed. 4 wt% lead-containing glass was added to BST ceramics, which accelerated densification with limited grain growth. The sintering temperatures of lead glass added BST ceramics were lowered to 1000 °C~1050 °C with high density above 96%, which were all higher than that of the pure BST sintered at 1350 °C. The dielectric constant of BST ceramics with $PbO\text{-}SiO_2$ glass additives was higher than that of the pure BST ceramics and dissipation factor of all BST samples with lead glass additives was lower than 1×10^{-3} at 1 kHz

1. Introduction

Barium strontium titanate ($Ba_xSr_{1-x}TiO_3$) is the solid solution of $BaTiO_3$ and $SrTiO_3$ with excellent dielectric properties, which has been used in electronic, electro-optical, optical, acoustic, and microwave devices [1-4]. Most of the above mentioned applications exploit the properties of the ferroelectric phase. Comparatively little attention has been paid to the dielectric characteristics of the paraelectric state, especially for $x\leq0.4$, which has relatively high dielectric constant, lower dielectric loss, and higher breakdown strength [5-7]. Dielectric loss is often important for dielectrics applications, which is usually lower in the paraelectric state than in the ferroelectric state owing to the disappearance of hysteresis. It is desirable that the Curie temperature of the ferroelectric materials be lower than the operating temperature in order to minimize the loss factor.

In addition, Barium strontium titanate has a high sintering temperature of about 1350 °C, which causes excess grain growth and increases manufacturing costs, especially for multilayer ceramic capacitors (MLCC) [8, 9]. Coarse grains are also bad for dielectric properties, especially for the breakdown strength [10-12]. Therefore, much effort has been directed towards the investigation of the decrease of the sintering temperature by controlling the sintering conditions and adding sintering aids. One effective method of decreasing the sintering temperature is to add glasses that have low melting points to form liquid phase at the sintering temperature [13-16]. The liquid phase promotes sintering process at a relatively low temperature.

The main purpose of this paper is to investigate the effects of addition of three lead-containing glasses on densification and dielectric properties of $Ba_{0.4}Sr_{0.6}TiO_3$ ceramics. Three lead-glass systems used in the present study were 60 mol% PbO-40 mol% SiO_2 (PS), 40 mol% PbO-20 mol% SiO_2-40 mol% B_2O_3 (PSB), and 36 mol% PbO-18 mol% ZnO-46 mol% B_2O_3 (PZB), respectively. 4wt% glass was added to BST ceramics.

2. Experimental procedure

$Ba_{0.4}Sr_{0.6}TiO_3$ powders used in this study were prepared by a conventional solid state reaction. Reagent-grade $BaCO_3$ (≥99%), $SrCO_3$ (≥99%) and TiO_2 (≥99%) powders were used as starting materials. The stoichiometric amounts of powder mixtures were ball-milled in ethanol medium using zirconia balls for 24 h. The slurry was then dried, ground well and calcined at 1150 °C in air for 4 h.

Calculated amounts of reagent-grade chemicals for the various glass compositions were weighed, mixed, and melted in quartz crucible in an air atmosphere at 1100 °C for 2 h. The molten glass was quenched in water and then crushed in an agate bowl. Each crushed glass powder was sieved to pass through the 400 mesh screen.

4 wt% glass powders and 96 wt% BST powders were mixed thoroughly as mentioned earlier, dried and then pressed into pellets. Pure BST was sintered at 1350 °C for 3 h, and glass-added pellets were sintered at different temperatures ranging from 900 °C to 1150 °C for 3 h.

Density measurements were made using the Archimedes technique. The crystal structure and phase purity of the samples were analyzed by X-ray diffraction. Microstructures of the ceramics were studied on as-fired surfaces using SEM. Dielectric measurements were performed on plane parallel disks approximately 1 mm thick, which were coated uniformly with a silver paste on both faces, and then fired at 600 °C for 20 min. Room temperature dielectric properties such as dielectric constant and dissipation factor were measured using Hp 4284 LCR meter in the frequency range of 1 kHz to 1 MHz.

3. Results and discussion

Figure 1 shows the change of relative density of the BST ceramics with different lead-containing glass additives and sintering temperatures. It is seen that the melted glass phase accelerated the densification of sintered bodies and lowered the sintering temperature. Addition of 4 wt% PS and PSB glass decreased the sintering temperature to 1050 °C with the maximum relative density of 97.1% and 96.4%, respectively. Addition of 4 wt% PZB glass decreased the sintering temperature to 1000 °C, and the maximum relative density was 96.5%. The relative density of all samples at their optimum sintering temperature was higher than that of the pure BST ceramics sintered at 1350 °C. Throughout the rest of this study, all the glass added samples were sintered at their optimum temperature. From this study, liquid-phase sintering is shown to decrease the sintering temperature by 300 °C ~350 °C with 4 wt% lead-containing glass additives, which can in turn lower the processing costs of BST ceramics and allow for the use of less expensive electrode such as Ni.

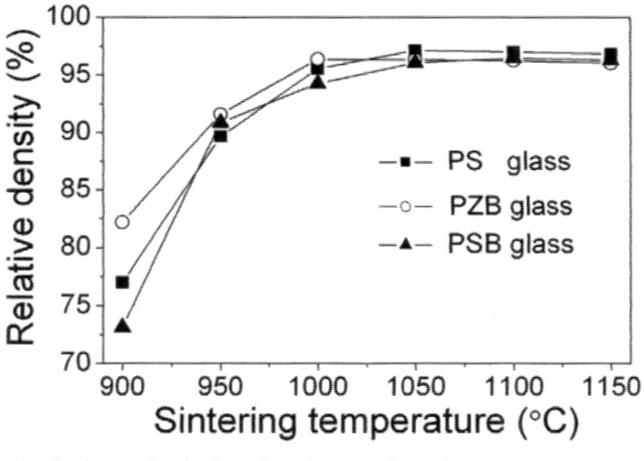

Figure 1. Variation of relative density as sintering temperature increases.

Figure 2 shows the XRD patterns of pure BST and BST ceramics with 4 wt% PS, PSB and PZB glass additives. There were no impurity peaks observed in figure 2(a), and all peaks were identified to be due to perovskite BST phase with cubic structure. In figure 2(b)-(d), in addition to the main phase of BST, very small amounts of silicate phases could be founded, which showed reaction between BST and glasses occurred slightly during the sintering process. The lattice constant was calculated by XRD data. The lattice constants of pure BST and BST with 4 wt% PS, PSB and PZB glass additives were 3.9423, 3.9395, 3.9415 and 3.9407, respectively. With the addition of glasses, the lattice constants were slightly decreased. This indicated that the Pb^{2+} ion was substituted into the BST lattice with the chemical reaction between BST and glasses. The microstructure of pure BST and lead-glass added BST is shown in figure 3. Pure BST had a wide grain size distribution, with the average size of about 20 μm. Figure 3(b)-(d) shows the microstructure of 4 wt% PS, PSB and PZB glass added BST ceramics, respectively. The grain size obviously decreased with lead glass addition, of which the average size was only several microns, and the grain size distribution was narrower, which was mainly attributed to lower sintering temperature.

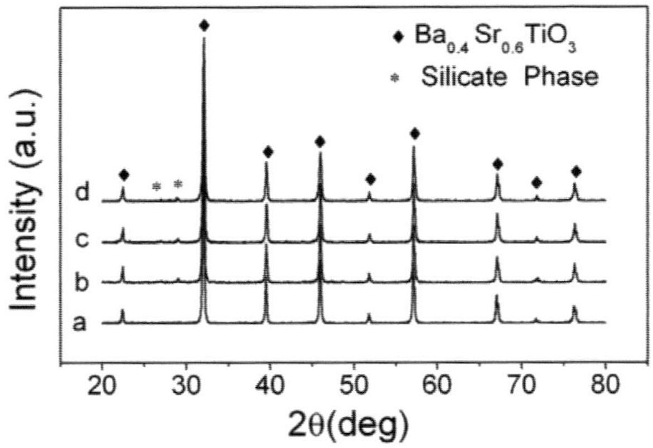

Figure 2. The XRD patterns of (a) pure BST, (b) BST + 4 wt% PS glass, (c) BST+4 wt% PSB glass, (d) BST+4 wt% of PZB glass.

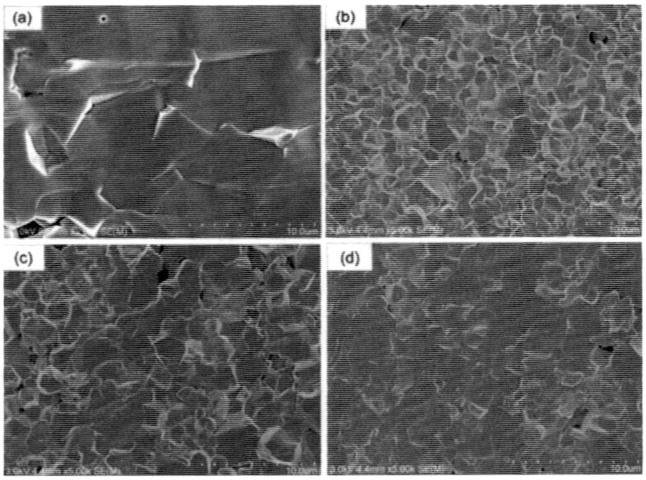

Figure 3. The scanning electron micrographs of (a) pure BST sintered at 1350 °C, (b) 4 wt% PS glass added BST sintered at 1050 °C, (c) 4 wt% PSB glass added BST sintered at 1050 °C, (d) 4 wt% PZB glass added BST sintered at 1000 °C.

Journal of Physics: Conference Series **152** (2009) 012066 doi:10.1088/1742-6596/152/1/012066

Change of room temperature dielectric constant and dissipation factor as a function of measurement frequency for the BST ceramics with lead glass additives at their optimum sintering temperatures is shown in figure 4. The dielectric constant of pure BST sintered at 1350 °C was about 1087. Dielectric constant of the PS glass added BST sintered at 1050 °C was about 1143, which was higher than that of the pure BST. This increase was attributed to lower porosity and a refinement of grain size [17]. The dielectric constant of the PSB glass-added BST was 1065, which was slightly lower than that of the pure BST. The PZB glass-added BST had the lowest dielectric constant of about 965. The dissipation factor of BST samples with addition of PS and PZB glasses was lower that of pure BST, which was about 1×10^{-3} at 1 kHz.

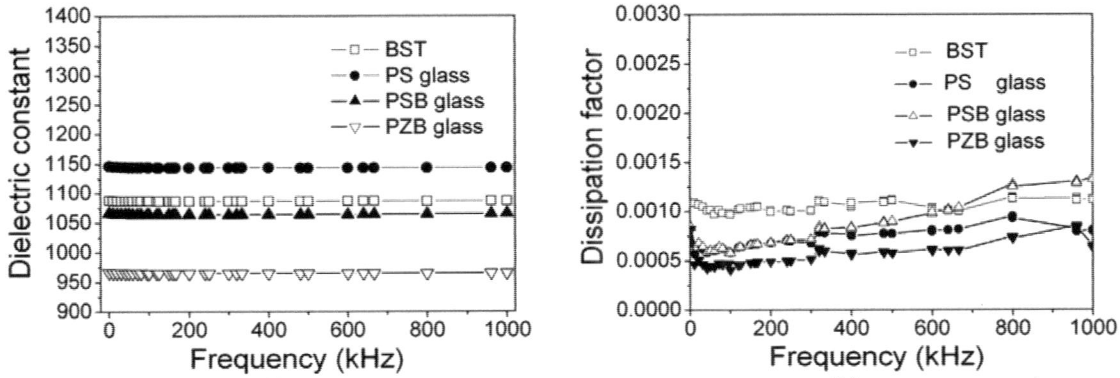

(a) (b)

Figure 4. Variation of (a) dielectric constant and (b) dissipation factor as a function of measurement frequency.

Temperature dependence of dielectric properties of the specimens is denoted in figure 5. The Curie temperature of pure BST was found to be about -68 °C. The addition of three lead glasses led to a slight shift in Curie temperature towards higher temperature, which indicated Pb^{2+} ion diffused into BST lattice and formed the solid solution. Pb^{2+} ion present in the glasses was known to form substitutional solid solution with perovskite compounds, replacing either the *A* or *B* site cations in the lattice, which resulted in a shift in Curie temperature [18]. The dielectric loss also showed a peak at the transition temperature, which indicated a well-defined phase transition. The dielectric loss decreased with increase in the temperature beyond the loss peak and reached a minimum at about room temperature. Further increase in temperature increased the dielectric loss again.

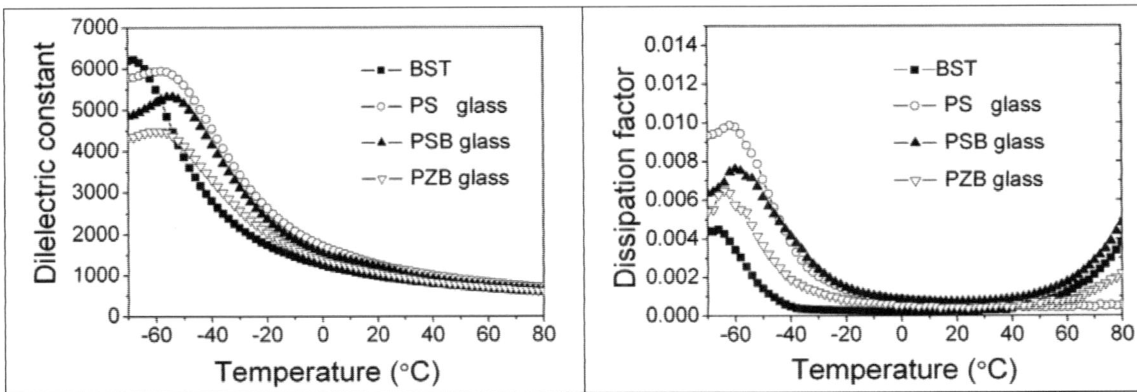

(a) (b)

Figure 5. Variation of (a) dielectric constant and (b) dissipation factor as a function of measurement temperature.

4. Conclusions

Three lead glasses were added to BST ceramics in the amount of 4wt%. The relative density of all samples at their optimum sintering temperature was higher than that of the pure BST ceramics sintered at 1350 °C. Dielectric constant of PS glass added BST ceramics sintered at 1050 °C were higher than that of the pure BST ceramics, and the dielectric loss of all samples with glass addition was lower than 1×10^{-3} at 1 kHz. It was observed that a small amount of glasses effectively lowered the preparation temperature and improved the dielectric properties of BST ceramics, especially for PS glass.

References

[1] Thakur O P, Chandra Prakash, and Agrawal D K 2002 *Materials Science and Engineering* B **96** 221-5
[2] J. W. Liou and B. S. Chiou 1997 *J. Am. Ceram. Soc.* **80** 3093-9
[3] Ali N J and Milne S J 1993 *J. Am. Ceram. Soc.* **76** 2321-6
[4] Chen Y C and Huang C L 2006 *Materials Letters* **60** 451-4
[5] Liou J W and Chiou B S 1997 *Materials Chemistry and Physics* **51** 59-63
[6] Pohanka R C, Freiman S W, and Bender B A 1978 *J. Am. Ceram. Soc.* **61** 72-5
[7] Tomoko Kito and John U T 1995 *J. Mater. Res.* **10** 1411-7
[8] Jeon H P, Lee S K, Kimb S W, and Choi D K 2005 *Materials Chemistry and Physics* **94** 185–9
[9] Prakash D, Sharma B P, Rama Mohan T R, and Gopalan P, 2000 *Journal of Solid State Chemistry* **155** 86-95
[10] Yang Y, Zhang S C, and Dogan F 2003 Pulsed Power Conference *Digest of Technical Papers* **1** 719–22
[11] Beauchamp E K 1971 *J. Am. Ceram. Soc.* **54** 484–7
[12] Tunkasiri T and Rujijanagul G 1996 *Journal of Materials Science Letters* **15** 1767 -9
[13] Priya Rani B R and Sebastian M T 2008 *J. Mater. Sci: Mater Electron* **19** 39–44
[14] Armstrong T R, Young K A and Buchanan R C 1990 *J. Am. Ceram. Soc.* **73** 700-6
[15] Zhang B,Yao X, and Zhang L Y, 2004 *Ceramics International* **30** 1767 –71
[16] Zhai J w, Yao X, Cheng X G and Zhang L Y 2002 *Journal of Materials Science* **37** 3739 – 45
[17] Arlt G, Hennings D, and Dewith G, 1985 *J. Appl. Phys.* **58** 1619
[18] Hirata A and Yamaguchi T 1997 *J. Am. Ceram. Soc.***80** 79-84

Journal of Physics: Conference Series **152** (2009) 012067 doi:10.1088/1742-6596/152/1/012067

Second Phases of (La,Nb)-Codoped TiO_2 Varistor Ceramic

Jikang YAN, Guoyou GAN, Jinghong DU, Jialin SUN

Department of materials, Faculty of Materials and Metallurgical Engineering, Kunming University of Science and Technology, Kunming, Yunnan, 650093, China.

E-mail:scyjk@163.com

Abstract. Crystal structure, chemical composition and phase transition of second phases in (La,Nb)-Codoped TiO_2 varistor ceramics was investigated. SEM, XRD, XPS and EDS were carried out the analytic studies on the microstructure, chemical composition and crystal structure of TiO_2 ceramics. The results showed that the second phases existed in (La,Nb) doped TiO_2 varistors ceramics. With sintering temperature increasing, second phases will transit from $LaNbO_4$ to $LaNbTiO_6$. Ternary phase diagram of La_2O_3-Nb_2O_5-TiO_2 was plotted based on binary phase diagram of La_2O_3-Nb_2O_5, Nb_2O_5-TiO_2 and La_2O_3-TiO_2. According to the ternary phase diagram, the relative content of second phases in TiO_2 ceramics sintered at different temperature was estimated. It is suggested that increasing of both second phases and titanium vacancies V_{Ti}'''' will result in a higher varistor voltage V1mA and nonlinear coefficient α.

1. Introduction

Titanium Dioxides (TiO_2) ceramic is a double functional material with varistor and capacity properties [1]. For more wide applications, TiO_2 ceramics with better properties were developed by doping to improve microstructure, barrier structure and electrical properties. Some dopants, including Ca, Sr, Y, Ba, Ta, Mn, Ce, Co, Nb, V, La and etc, were added into TiO_2 ceramics. For instance, the Nb doped TiO_2 ceramic exhibits lower varistor voltage and nonlinear coefficient [2]. On the contrary, La doped TiO_2 ceramic possesses higher varistor voltage and nonlinear coefficient [3]. In order to get a kind of ceramic with low varistor voltage and high nonlinear coefficient, a feasible method is to codope La and Nb into TiO_2.

Usually, there exist second phases besides main crystal phase rutile titanium dioxide in doped-TiO_2 ceramics. The variety of secondary phases may be dependent on ionic diameter and content of dopants, sintering temperature. The doped ion with ionic diameter close to diameter of Ti^{4+}, such as Nb^{5+}, has a higher solid solubility in TiO_2 lattice. For the doped ion with ionic diameter far greater than Ti^{4+}, such as Y^{3+} and La^{3+}, its solid solubility limit is very low. When doping ion concentration is greater than its solid solubility limit, secondary phases is formed[4-5]. Secondary phases will grow up with increasing of sintering temperature.

Studies about secondary phases were principally focused on formation mechanism, detection method, characteristics of secondary phases and their influences on structure and properties of matrix ceramics. The formation of secondary phases generally originated from segregation of dopants or impurities. Pentavalent cation Nb^{5+} close in ionic size to Ti^{4+} will segregate at Ti site in grain boundary of TiO_2 ceramics using electrostatic potential as driving force [6-7]. Trivalent cation Y^{3+} far larger in ionic size than Ti^{4+} will segregate at interstitial site in grain boundary using elastic strain energy as driving force[5]. In the current works, most of Nb^{5+} ions were dissolved into TiO_2 lattices and the remaining Nb^{5+} cation segregated at grain boundary of TiO_2 ceramics. However, almost complete cation La^{3+} segregated at grain boundaries. When concentrations of Nb^{5+} and La^{3+} in grain boundary reached a certain values chemical reaction $Nb_2O_5+La_2O_3=2LaNbO_4$ occurred and then second phase $LaNbO_4$ was formed. With sintering temperature increasing, reaction products was observed between $LaNbO_4$ and rutile TiO_2 resulting in another second phase $LaNbTiO_6$[8]. Chih-Chung T Yang studied formation mechanism of secondary phases at the interface between Sr-doped $LaMnO_3$ and Y-stabilized ZrO_2[9]. The result showed that second phases $SrZrO_3$ and $La_2Zr_2O_7$ indeed occurred at the interface between LSM and YSM co-fired at 1400°C while no interfacial reaction were found at the interface between LSM and YSZ by the modified process. Moreover, the long-term stability results showed that no reaction products can be detected either made by co-fired or modified process when annealed at 1000°C for 120h. The segregation and diffusion of La, Sr, Mn into YSZ had a significant effect on microstructure leading to the formation of a denser region.

Characterization techniques of second phases mainly included SEM, XRD, EDAX, TEM, EDS, Raman and etc. E. Brzozowskia researched the secondary phase of $BaTiO_3$ ceramic doped with Nb_2O_5 by means of EDAX and SEM [10]. Jianying Li utilized SEM, XRD and EDAX to study the secondary phase of (Nb,Ce,Si,Ca)-doped TiO_2 ceramic[11]. R. Parra made use of SEM, TEM, EDS and ED to investigate the secondary phase existed in SnO_2 varistor ceramic [12]. Hong Peng He inspected the secondary phase of Sr doped $LaInO_3$ through XRD and Raman [13]. Zhi-Cheng Li used series SAED to get 3-D spatial points of secondary phase, and determine its spatial group and crystal lattice [14].

Secondary phase influences ceramic on microstructure, chemical composition, and mechanics and electrical performance. The second phase segregates at the grain surface during sintering process and makes an insulating layer which result in a higher V_{1mA} [11]. N.S.Hari studied PTRC ceramic $BaTiO_3$ whose secondary phase $BaAl_6TiO_{12}$ formed segregation layer and changed the charge distribution around $BaTiO_3$ crystal grains. In this case, V''_{Ba} become main electron traps and induced an acceptor energy level which influences n-type BaTiO3 ceramic PTRC on positive temperature coefficient [15].

For current study, SEM, EDS, XRD, XPS and phase diagram are carried out to investigate crystal structure, chemical composition and the transition of the secondary phase and its effects on varistor properties of（La, Nb）codoped TiO2 ceramic sintered at different temperature.

2. Experiments

TiO_2 varistor samples with a molar composition $0.98TiO_2 + 0.008Nb_2O_5 + 0.007La_2O_3$ were prepared by the classical electronic ceramic procedure and single sintering technique. The oxides raw materials used in this study were analytical-grade TiO_2 (99.5%), Nb_2O_5 (99.5%), SiO_2 (99.5%) and La_2O_3 (99.5%). The chemicals were mixed in proportion and wet-milled in agate jar with agate balls for 12h in deionized water. The milled powders were dried and granulated with a PVA binder. The granulated powder was pressed into discs 13 mm in diameter and 1.0 mm thickness under the pressure of 140MPa. After burning out PVA at 650℃, the discs were placed in Al_2O_3 crucibles and sintered in air at 1300℃ ~1400℃ for 1h, and then cooled to room temperature automatically. Sliver paste was coated on two sides of TiO_2 samples to form electrodes by firing at 600℃ for 30 min. XRD analysis of the sample phase was observed by Bruker D8 XRD. The microstructure characterization and chemical compositions were measured by Philips XL30-ESEM with EDS attachment. XPS spectra were measured by Phi5500 XPS meter. The varistor properties of TiO_2 ceramics were measured by CJ1001 varistor test meter.

2. Results and discussions

2.1. Microstructure and composition

Figure 1 gives the SEM image and EDS profile of the sample mentioned above, which corresponds to sintering temperature 1300℃, 1350℃ and 1400℃. In Figure . 1 (a), (b) and (c), the middle picture is backscattered electron image, and the left picture is the chemical composition of secondary phase obtained via energy spectrum instrument corresponding to high light area in SEM image, and then the right picture is the chemical composition of TiO_2 grain.

From Figure 1, with the rising of sintering temperature, TiO_2 grains grow up and the TiO_2 sample becomes denser. There exist some highlight areas in microstructure photos of TiO_2 ceramic. The highlight areas may be caused by the segregation of La and Nb which have larger atomic number.

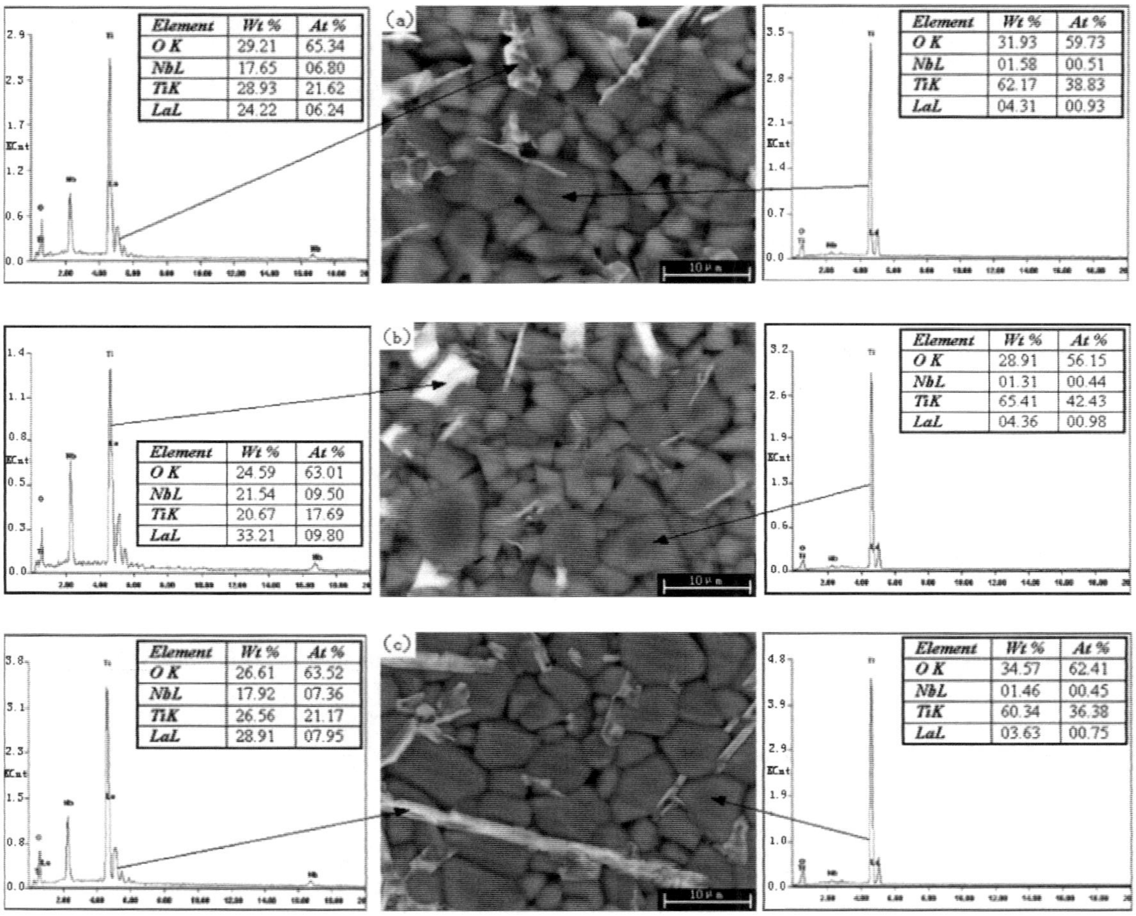

Figure 1. SEM images and EDS of TiO_2 varistor ceramic samples at different sintering temperature (a) 1300℃, (b) 1350℃ and (c) 1400℃.

To determine the element and content of highlight area, point analysis method of EDS is used to measure chemical composition of highlight area and TiO_2 grain, measured results shown in Figure. 1. According to EDS, the compositions of TiO_2 grain are the same basically at different sintering temperature. However, the compositions are very different in corresponding highlight area. Obviously, there exist serious segregation of La and Nb. La and Nb mole percent are, respectively, 6.8 and 6.24 at the temperature 1300℃, 9.5 and 9.8 at 1350℃, 7.36 and 7.95 at 1400℃. It can be noted that the molar ratio of La and Nb in highlight area is almost one.

2.2. Crystal structure

To confirm whether secondary phase is formed in TiO_2 ceramic because of La and Nb segregation, X-ray diffraction is carried out to observe crystal structure of TiO_2 ceramic, shown as Figure .2. From the Figure .2, peaks corresponding to TiO_2 were detected; other crystalline phase besides rutile (TiO_2) was also observed, such as $LaNbO_4$ and $LaNbTiO_6$. X-ray diffraction analysis indicated that there exist secondary phase in TiO_2 varistor ceramic besides main crystal phase rutile TiO_2. $LaNbO_4$ is only secondary phase of TiO_2 ceramics sintered at 1300℃ while for the sample sintered at 1350℃, secondary phases contains $LaNbO_4$ and $LaNbTiO_6$. $LaNbTiO_6$ is only secondary phase of TiO_2 ceramics sintered at 1400℃. The molar ratio of La /Nb in the chemical composition of secondary phase $LaNbO_4$ and $LaNbTiO_6$ is one. These accords with the EDS analysis in section 3.1. XRD, SEM and EDS synthetically demonstrate that the secondary phase $LaNbO_4$ and $LaNbTiO_6$ corresponding to highlight area in SEM images is formed through the segregation of La and Nb. With the rising of sintering temperature, the secondary phase will change from $LaNbO_4$ to $LaNbTiO_6$.

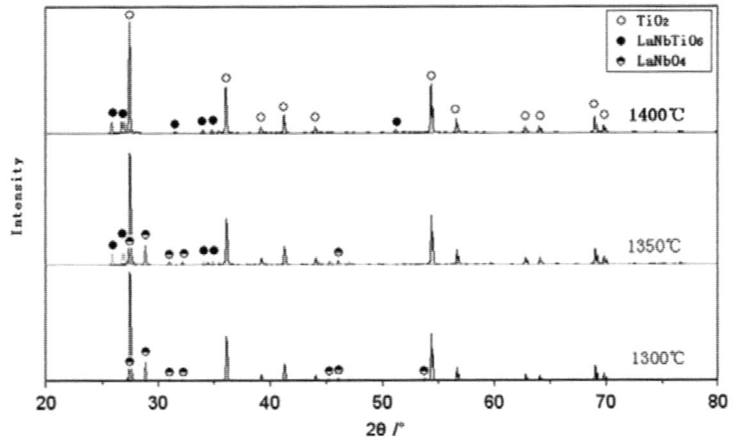

Figure 2. X-ray diffraction spectrum of samples at different sintering temperature

In current works, TiO_2 raw materials is anatase TiO_2, which will transform into rutile TiO_2 at about 900℃. Rutile TiO_2 is tetragonal system with lattice parameter a=b= 4.594 Å, c= 2.959 Å, α= β=γ=90. Secondary phase $LaNbO_4$ is monoclinic system, a=5.5735Å, b=11.5418 Å, c=5.2159 Å, α=90,β= 94.070, γ=90. Secondary phase $LaNbTiO_6$ is Orthorhombic system, a= 10.934 Å, b= 7.572 Å, c= 5.446 Å, α=β= γ=90. Figure .3 shows crystal structures of rutile TiO_2, $LaNbO_4$ and $LaNbTiO_6$. At high temperature, there maybe exists a chemical reaction as equation (1). So rutile TiO_2 and $LaNbO_4$ maybe produce $LaNbTiO_6$. Figure . 4 presents the binary phase diagram of $LaNbO_4$-TiO_2, which also indicates formation mechanism of secondary phases $LaNbTiO_6$ according to equation (1).

$$TiO_2 + LaNbO_4 \rightarrow LaNbTiO_6 \quad (1)$$

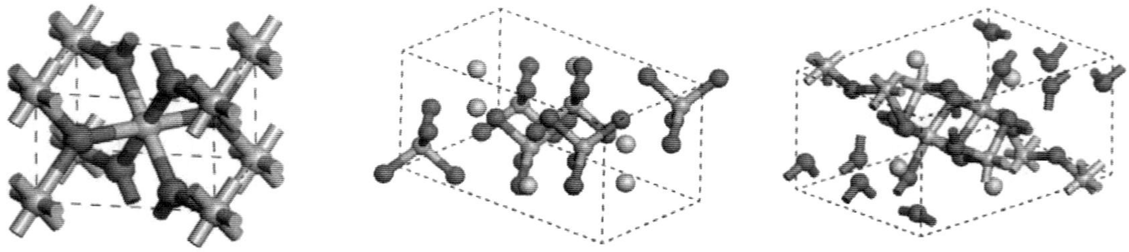

(a) (b) (c)

Figure 3. Crystal structures of (a) Rutile TiO_2; (b) $LaNbO_4$ and (c) $LaNbTiO_6$

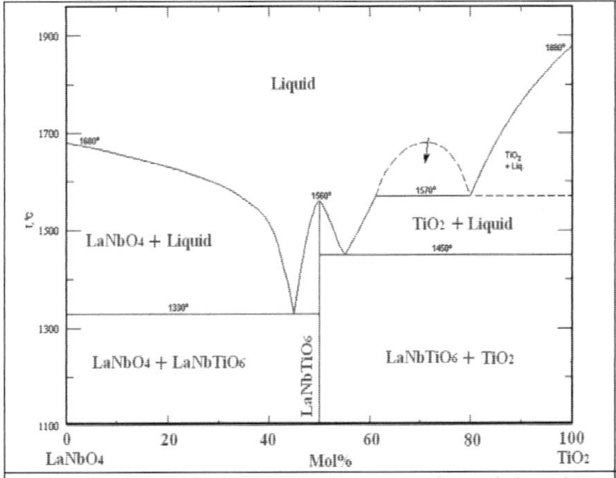

Figure 4. The binary phase diagram of $LaNbO_4$-TiO_2

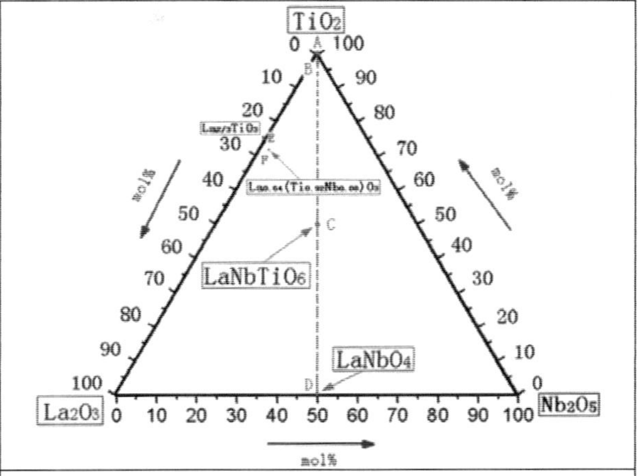

Figure 5. La_2O_3—Nb_2O_5—TiO_2 ternary phase diagram

2.3. Varistor properties

In order to investigate the change mechanism and the relative content of secondary phase $LaNbO_4$ and $LaNbTiO_6$, the ternary phase diagram of La_2O_3-TiO_2-Nb_2O_5 is drawn based on the binary phase diagram of La_2O_3-TiO_2, La_2O_3-Nb_2O_5 and Nb_2O_5-TiO_2 [16-18], seeing Figure . 5. It is reported that the ternary oxide of La_2O_3-TiO_2-Nb_2O_5 includes $LaNbTiO_6$ (C point in Figure .5) and $La_{0.64}$ $(Ti_{0.92}Nb_{0.08})$ O_3 (F point in Figure .5). The latter is produced by the binary oxide $La_{2/3}TiO_3$ (E point in Figure . 5) doped with Nb. Its molar composition is $0.25La_2O_3$-$0.72TiO_2$-$0.03Nb_2O_5$ [19]. The molar ratio of La /Nb in $La_{0.64}$ $(Ti_{0.92}Nb_{0.08})$ O_3 is eight, which have a big difference from EDS analysis in Figure .1, so that the current experiment samples don't contain ternary oxide $La_{0.64}$ $(Ti_{0.92}Nb_{0.08})$ O_3 but $LaNbO_4$ or $LaNbTiO_6$. XRD analysis in Figure .2 also confirms this.

The composition of sample is presented at B point in Figure .5, which is $0.98molTiO_2 + 0.007molLa_2O_3 + 0.008molNb_2O_5$. The main crystal phase in TiO_2 ceramic is rutile TiO_2, shown as A point in Figure .5. The secondary phase is $LaNbO_4$ and $LaNbTiO_6$, corresponding to D and C points, respectively, in Figure .5. Thus, the contents evaluation of different solid phases in (La, Nb) doped TiO_2 varistor ceramic can be worked out via straight line ABCD.

According to Figure .2, there only exist TiO_2 and $LaNbO_4$ at 1300℃. Via ABD line and the basic principle of ternary phase diagram, the molar content of TiO_2 is evaluated as 97.22%, and secondary phase $LaNbO_4$ is 2.78%. At 1400℃, there only exist TiO_2 and $LaNbTiO_6$. From line ABC, the content of TiO_2 is calculated as 94.48% and $LaNbTiO_6$ is 5.52%. At 1350℃, there coexist ternary phase TiO_2, $LaNbO_4$ and $LaNbTiO_6$. Their relative contents are difficult to evaluate. $LaNbO_4$ is a ferroelastic material with domain structure and also is a insulator with larger band energy of 4.8ev[20-21]. $LaNbTiO_6$ may be one of $ReTiNbO_6$ microwave dielectric ceramics with low dielectric constant and high electrical resistivity[22]. As the rising of sintering temperature, the TiO_2 grains become larger and the content of secondary phase increase. Larger grains make α and V_{1mA} reduce while insulating second phases make α and V_{1mA} rise.

XPS profile of O1s in TiO_2 varistor ceramics consist of two peaks as showed in Figure .6. The peak at 529.7ev is assigned to lattice oxygen in TiO_2. The peak at 531.8ev is corresponding to adsorption oxygen. Quantity of adsorption oxygen increases with concentration of vacancy oxygen increasing. Yang Seng-lu[23] reported that mass transport during sintering process mainly depend on oxygen vacancies and that chief point defect of rutile TiO_2 is vacancy oxygen under atmosphere and high temperature. With sintering temperature increasing, XPS spectra peak of absorbed oxygen was enhanced. This indicated that concentration of oxygen vacancies increased. Titanium vacancies and oxygen vacancies were intrinsic defect of rutile TiO_2[24]. So increment of oxygen vacancies implied that increase of Titanium vacancies. On the other hand, V_{Ti}'''' is main electron potential trap and it will form acceptor energy level, which will leads to the increase of varistor voltage and nonlinear coefficient [11, 15]. The varistor properties presented in Figure .7 is the result of combined action of secondary phase, point defect and microstructure.

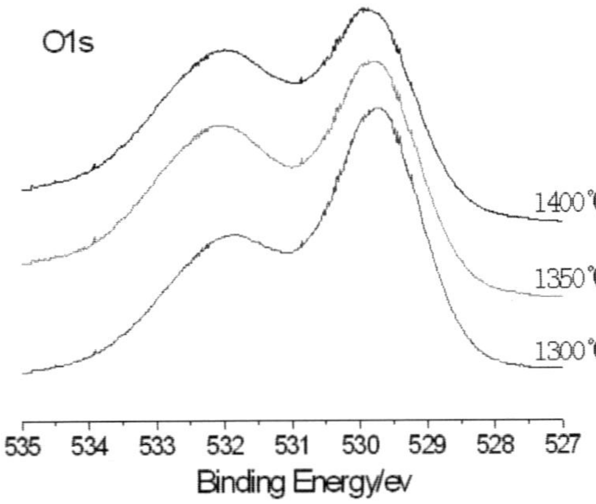

Figure 6. XPS profile of O1s in TiO_2 varistor ceramics

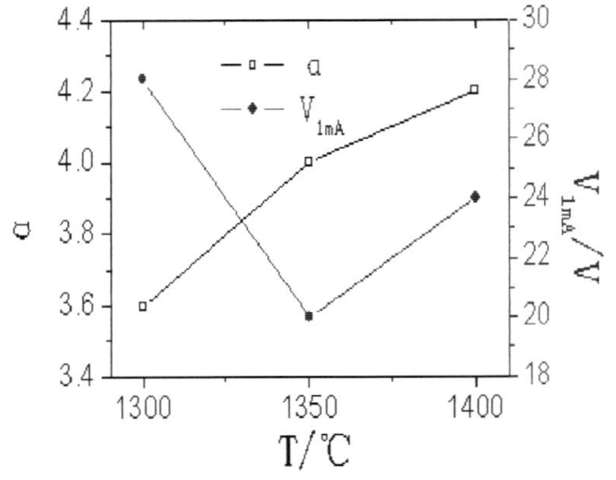

Figure 7. Influence of sintering temperature on V_{1mA} and α

3. Conclusions

(1) SEM analysis denoted that La and Nb will segregation in (La, Nb) codoped TiO_2 sample. EDS analysis showed that the chemical composition and content ratio in TiO_2 grain were kept the same. In the segregation area of La and Nb, the contents of La and Nb are very different from TiO_2 grain. Their molar ration La/Nb of the segregation area is almost one.

(2) XRD research indicated that secondary phase $LaNbO_4$ and $LaNbTiO_6$will be formed in (La, Nb) codoped TiO_2 ceramic at different sintering temperature. With the rising of temperature, the secondary phase can change from $LaNbO_4$ to $LaNbTiO_6$. the secondary phase can make V_{1mA} and α to be increased.

(3) XPS spectra showed that with sintering temperature increasing, the concentration of Ti vacancy increased. As a main electron potential well, V_{Ti}'''' can form acceptor energy level and lead to the increase of varistor voltage and nonlinear coefficient.

4. Acknowledgement

The present work has been supported by the plan of Yunnan Province Science and Technology Commission under grant No.2002GG09, and the Science Research Foundation of Kunming University of Science and Technology under Grant No.2007-15 and the Analysis & Testing Foundation of Kunming Universitty of Science and Technology under Grant No.2008-002.

References

[1] Yan M and Rhodes W 1982 *Appl. Phys. Lett.* **6** 536
[2] Yan Jikang, Gan Guoyou and Du Jinghong 2003 *J. Kunming University of Science and Technology.* **2** 30
[3] Luo Jianjun, Fang xiangyi and Wu Mingtang 1997 *Journal of Xi'an Jiaotong University.* **9** 62
[4] Xiangyi Fang and J.T. Oh 2007 *Materials Science and Engineering B.* **136** 15
[5] Qinglei Wang, Guoda Lian and Elizabeth C. Dickey 2004 *Acta Materialia.***52** 809
[6] Jeri Ann S. Ikeda and Yet-Ming 1993 *J.Am.Ceram.Soc.***10** 2437
[7] Ying Pang, Noel T.Nuhfer and Paul Wynblatt 2005 *Microsc Microanal Suppl* **2** 11
[8] Chen Hai-fang, Gan Guo-you, Yan Ji-kang and Zhang Xiao-wen 2006 *Semiconductor*

Technology. **2** 102

[9] Chih-Chung T Yang, Wen-Cheng J Wei, Andreas Roosen and H P Buchkremer 2002 *Ceramic Engineering and Science Proceedings* 23 3 ABI/INFORM Trade & Industry. p 733-740.

[10] E. Brzozowski, M.S. Castro, C.R. Foschini and B. Stojanovic 2002 *Ceramics International* **28** 773

[11] Jianying Li, Shaohua Luo, Weihua Yao and Zhongtai Zhang 2003 *Materials Letters* **57** 3748

[12] R. Parra, M. S. Castro and J. A. Varela 2005 *J. Euro. Ceram. Soc.* **25** 401

[13] Hong Peng He, Xue Jie Huang and Li Quan Chen 2001 *J. Phys. Chem. Solids.* **62** 701

[14] Zhi-Cheng Li, Hong Zhang, Xiaodong Zou and Bill Bergman 2005 *Materials Science and Engineering B.***116** 34

[15] N.S.Hari and T.R.N. KUTTY 1998 *J. Mater. Sci.* **33** 3275

[16] J.B. Braue and MacChesney, H. A 1962 *J. Am. Ceram. Soc.* **9** 416

[17] R.S.Boughanour and Roth, L.W 1955 *J, Res .Natl. Bur. Stand. (U.S.)* **4** 209

[18] E.P. Bodina Savchenko and N.A.Celer E.K 1965 *Pro. All-Union Conf, 2nd, Leningard, USSR;*, Chem. High-Temp Mater.p.108-113.

[19] MizukiMORI, Masatomo YASHIMA and Rowshown ALI. High-resolution synchrotron X-ray powder diffraction study of the orthorhombic-tetragonal phase transition in $La_{0.64}Ti_{0.92}Nb_{0.08}O_3$. http://pfwww.kek.jp/acr2001pdf/PartB/PF01UR130.pdf.

[20] P.M. Kelly, and C.J. Wauchope 1996 *Materials Letters.* **27** 7

[21] Y.J. Hsiao, T.H. Fang,, Y.S. Chang, Y.H. Chang, C.H. Liu, L.W. Ji and W.Y. Jywe 2007 *J. Luminescence* **126** 866

[22] Tang Rongmei Gan Guoyou Yan Jikang and Dong Weixia 2008 *J. Ceram.* **1** 58

[23] YANG Seng-lu and wu Jenn-ming 1995 *J. Mater. Res.* **2** 345

[24] Feng Shao-Xin, Li Bao-Hui, Jin Qing-Hua, Guo Zhen-Ya and Ding Da-Tong 2000 *ACTA PHYSICA SINICA.* **7** 1307

Journal of Physics: Conference Series **152** (2009) 012068 doi:10.1088/1742-6596/152/1/012068

Fabrication and Characteristics of Relaxor Ferroelectric PZN-PZT (53/47) Thin Films by a MOD Process

Xiyun HE, Xia ZENG, Xinsen ZHENG, Pinsun QIU, Wenxiu CHENG, Aili DING

Shanghai Institute of Ceramics, Chinese Academy of Sciences, 1295 Dingxi Rd., Shanghai 200050, China

xiyunhe@hotmail.com

Abstract. Relaxor ferroelectric (1-x)Pb(Zn1/3Nb2/3)O3 - xPb(Zr0.53Ti0.47)O3 (abbreviated as PZN-PZT (53/47)) (x=0.4-1.0) thin films were prepared by a metallic-organic decomposition (MOD) process on Pt/Ti/SiO2/Si substrates. The influence of PZT (53/47) content on the film phase component has been investigated and analyzed. The ferroelectric and piezoelectric properties of the PZN-PZT(53/47) thin films have been examined and discussed. The lattice spacing d111 of the thin film was found to have a maximum value when the PZT (53/47) content x is about 0.7~0.8; which was attributed to the film lattice distortion. Compared with the pure PZT (53/47) thin film, the 0.2PZN-0.8PZT (53/47) thin film was found to have a much lower coercive electric field (Ec) while the spontaneous and remanent polarization was not decreased obviously. Also a more intensive piezoresponse in micro-area of the 0.2PZN-0.8PZT (53/47)) thin film has been observed by a modified AFM.

1. Introduction

The solid solution of Pb(Zn1/3Nb2/3)O3 (PZN) and Pb(Zr,Ti)O3 (PZT) (abbreviated as PZN-PZT) is a relaxor ferroelectric material which has a rhombohedral/tereagonal morphotropic phase boundary (MPB) near ≥70%(mol.) PbZr0.53Ti0.47O3 (PZT (53/47)). It exhibits a famous piezoelectric property at the MPB, which is superior to those of the most widely used PZT. Therefore, perovskite PZN-PZT films have the potential to replace PZT films for applications in microelectronic devices such as micro-actuators and transducers [1-6].

Several works have been reported about the fabrication and characteristics of PZN-PZT films from MOD, sol-gel and MOCVD processes [3, 7-9], etc. However, PZN-PZT thin films with high quality are seldom to be present and the piezoelectric property of the film is seldom to be reported. In this study, the relaxor ferroelectric (1-x)PZN-xPZT (53/47)) (x=0.4-1.0) thin films were prepared from a metallic-organic decomposition (MOD) process on Pt/Ti/SiO2/Si substrates. The ferroelectric and piezoelectric properties of the films were examined and analyzed. The 0.2PZN-0.8PZT (53/47) thin film is found to have a lower coercive electric field (Ec) with a high remanent polarization, and an intensive piezoresponse was also observed. The 0.2PZN-0.8PZT (53/47) thin film is promising for applications in the microelectronic devices.

2. Materials and experimental

Lead acetate [Pb(CH3COO)2·3H2O, 99.5%], zinc acetate [Zn(CH3COO)3·2H2O, >99.0%], niobium ethoxide [Nb(CH3CH2O)5, 99.99%], zirconium n-propoxid [Zr(O(CH2)2CH3)4, 70%] and tetrabutyl titanate [Ti(OC4H9)4, >98%] were the starting materials. 2-methoxyethanol was selected as solvent. The PZN-PZT (53/47) precursor solutions were prepared by mixing the 2-methoxyethanol based lead acetate, zinc acetate, zirconium n-propoxid, tetrabutyl titanate and niobium ethoxide solutions according to the stoichiometry. A small amount of acetylacetone [C5H8O2, >98.5%] was employed as the solution stabilizer. 10 mol.% excess lead acetate was added. The concentration of solution was adjusted to 0.4 M. The PZN-PZT (53/47) thin films were deposited on the Pt/Ti/SiO2/Si substrates by spinning coating technique. The wet film was pre-heated at 450oC for 3 min on a hot plate. This step was repeated several times to obtain the desired film thickness (~250 nm). Finally the PZN-PZT (53/47) thin films were annealed at 750oC for 20 min in an oxygen atmosphere in an optic thermal annealing furnace to obtain highly crystalline PZN-PZT (53/47) thin films. In order to investigate the electric properties of the films, the electrodes of gold (Au) were evaporated on the film surface by a vacuum evaporation method.

The phase structure of the film was evaluated by the X-ray diffraction (XRD). Ferroelectric polarization versus electric field (P-E) hysteresis loop of the film was examined using a RT-66A ferroelectric tester. The microstructure and the micro-area piezoresponse of the film were investigated by a modified Atom Force Microscopy (AFM).

3. Results and discussion

The XRD patterns of the (1-x)PZN-xPZT (53/47) (x=0.4, 0.5, 0.6, 0.7, 0.8, 0.9, 1.0) thin films grown on the Pt/Ti/SiO2/Si substrates are shown in Fig. 1. When the x is larger than 0.8, the (1-x) PZN-xPZT (53/47) thin films exhibit almost pure perovskite structure with (111) preferred orientation which is

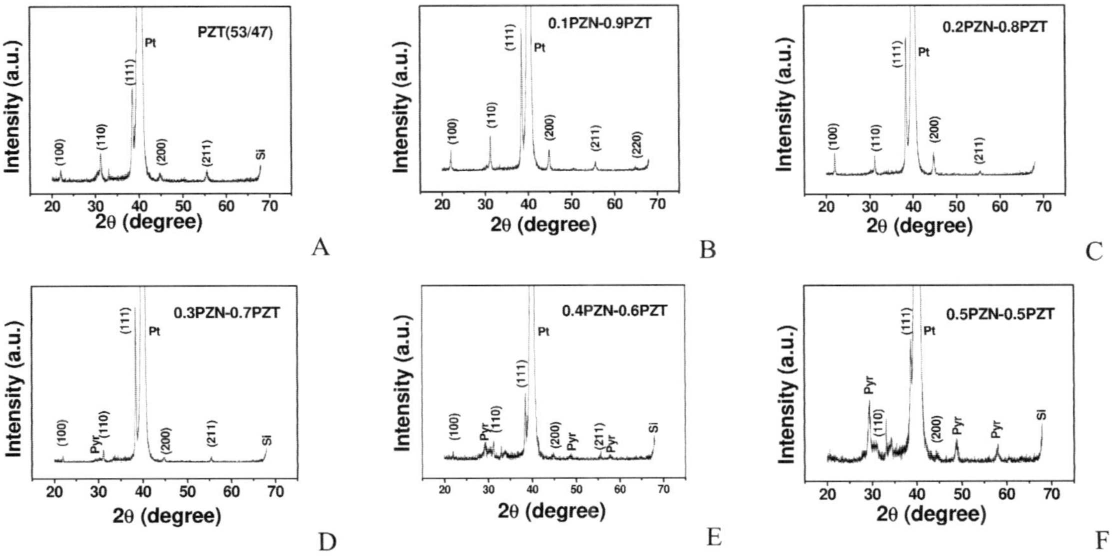

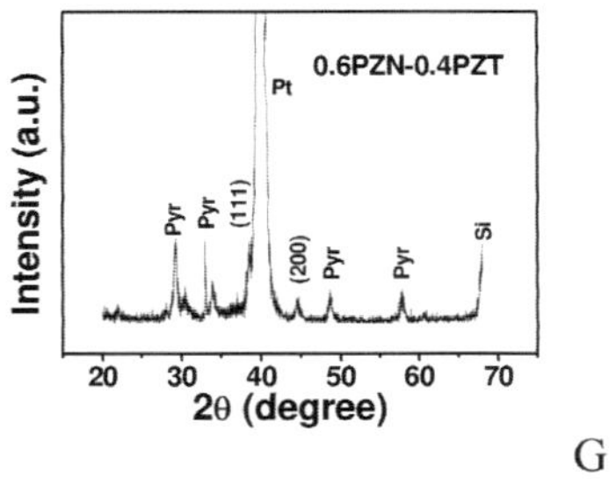

Figure 1. XRD patterns of (1-x)PZN-xPZT(53/47) thin films on Pt/Ti/SiO2/Si substrates.(a) x=1.0; (b) x=0.9; (c) x=0.8; (d) x=0.7; (e) x=0.6; (f) x=0.5; (g) x=0.4.

probably initiated by the surface Pt (111) orientation. At x=0.6 and x=0.7, a small amount of pyrochlore phase appears in the films. When PZT (53/47) content is decreased to x=0.5 and x=0.4, the films present a large amount of pyrochlore phase. It has been reported that the pure perovskite structure can be obtained in PZN-PZT (53/47) bulk ceramics when PZT (53/47) content is ≥30% (mol.) [10]. This indicates that a higher PZT (53/47) content is required to stabilize the perovskite phase in the PZN-PZT (53/47) thin films than that of the bulk ceramics. This result agreed with those reported by Y. Yang and K. Kim [11]. The great internal stress in the film produced from substrate restraint and lattice mismatch is thought to be the main causes to decrease the perovskite phase stability.

From the XRD patterns of the PZN-PZT (53/47) thin films, it also can be found that the 2-theta angle 2θ corresponding to the preferred (111) peak shows a small shifting as the PZT (53/47) content decreasing from 1.0 to 0.6. The lattice spacing d(111) of the film was estimated using the 2θ(111) values according to the equation $d(111)=\lambda/2\sin\theta(111)$, where $\lambda=1.54060$Å, denote the x-ray wavelength. Fig. 2 shows the variation of lattice spacing d(111) of the film as the PZN content (1-x) increasing. The d(111) value increases first as the PZN content increases. When the PZN content is about 0.2~0.3 (PZT (53/47) content is about 0.7~0.8), it reaches to the maximum, and then it decreases quickly. This result implies that the lattice of the PZN-PZT (53/47) thin film has been distorted seriously when the PZN component is added. The lattice constants of the films have been changed greatly then result in the variation of lattice spacing d(111). So it can be deduced that the PZN-PZT (53/47) thin film will have a great variation in their ferroelectric and piezoelectric properties.

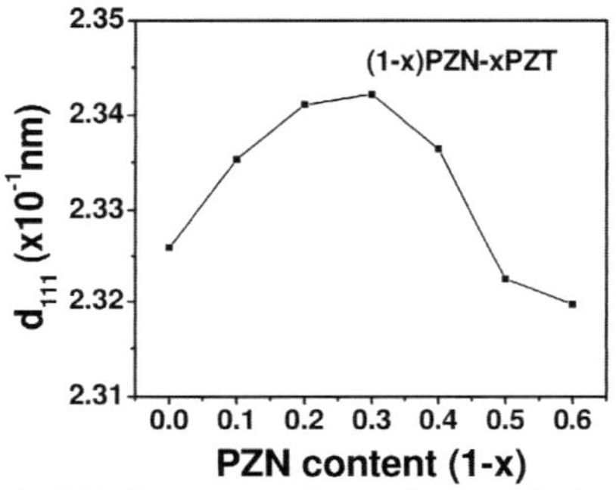

Figure 2. PZN content dependence of the lattice spacing d111 of PZN-PZT(53/47) thin films.

The ferroelectric hysteresis loops of (1-x)PZN-xPZT(53/47) thin films are shown in Fig. 3. The PZT(53/47) content x is 0.6,0.7,0.8,0.9,1.0 respectively, The thin films all present perfect hysteresis loops with high remanent polarization value (≥42 μC/cm2) except the film with x=0.6. It has been found that the ferroelectric property of the PZN-PZT(53/47) thin film deteriorated rapidly when the PZT(53/47) content is ≤ 60% (mol). The degradation is thought to be caused by the great amount of pyrochlore phase in the films.

The spontaneous polarization (Ps), remanent polarization (Pr) and coercive electric field (Ec) of (1-x)PZN-xPZT(53/47) thin films (x=0.6,0.7,0.8,0.9,1.0) are summarized in Fig.4. As the PZN content (1-x) increasing, the Ps, Pr and Ec plots of the film all show a wave-shape variation. Here it deserves to notice that when the PZT(53/47) content is 0.8, the Ec value of the 0.2PZN-0.8PZT(53/47) thin film is only 116Kv/cm. It is almost the half of that of the pure PZT(53/47) thin film 210v/cm. And in the same time, the Pr value of the 0.2PZN-0.8PZT(53/47) thin film does not decrease obviously compared with the PZT(53/47) thin film. This result can be explained by the increasing of lattice spacing d111 in the 0.2PZN-0.8PZT(53/47) thin film. A longer lattice spacing d111 is always thought to make the lattice be looser and the lattice polarization be easier to reverse. On the other hand, although the d111 value of 0.3PZN-0.7PZT(53/47) thin film is a little higher than that of 0.2PZN-0.8PZT(53/47) thin film, it does not present the optimum ferroelectric property because of the existing of a small amount of pyrochlore phase in it as discussed in the XRD analysis.

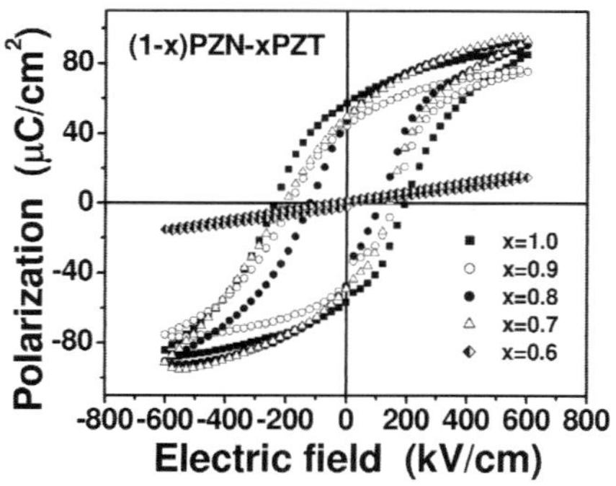

Figure 3. The P-E loops of PZN-PZT(53/47) thin films.

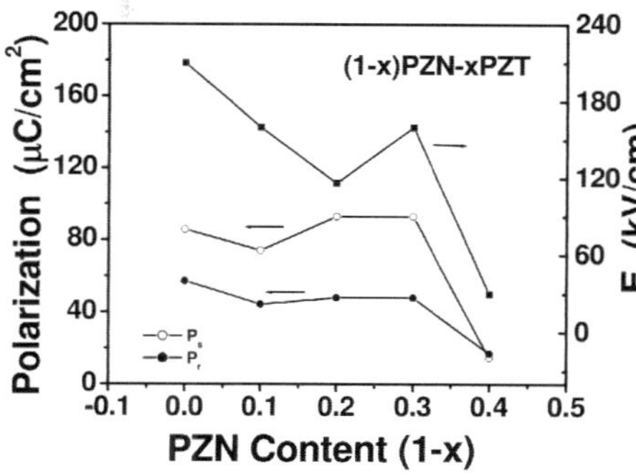

Figure 4. PZN content dependence of Ps, Pr and Ec of PZN-PZT(53/47) thin films.

The piezoresponse in the micro-area of PZT(53/47) and 0.2PZN-0.8PZT(53/47) thin film has been studied by a modified AFM. Fig. 5 shows the corresponding microscope images and piezoresponse images of PZT(53/47) thin film (a) (b) and 0.2PZN-0.8PZT(53/47) thin film (c) (d), respectively. The dark and bright bands in the piezoresponse images (b) (d) are corresponding to the regions with opposite polarization directions [12]. The 0.2PZN-0.8PZT(53/47) thin film presents a high contrast image with a more intensive piezoresponse signal, see Fig.5 (d). This implies that the 0.2PZN-0.8PZT(53/47) thin film shows a larger piezoresponse in micro-area than that in PZT(53/47) thin film. It is also attributed to the increasing of lattice spacing d111 in the 0.2PZN-0.8PZT(53/47) thin film.

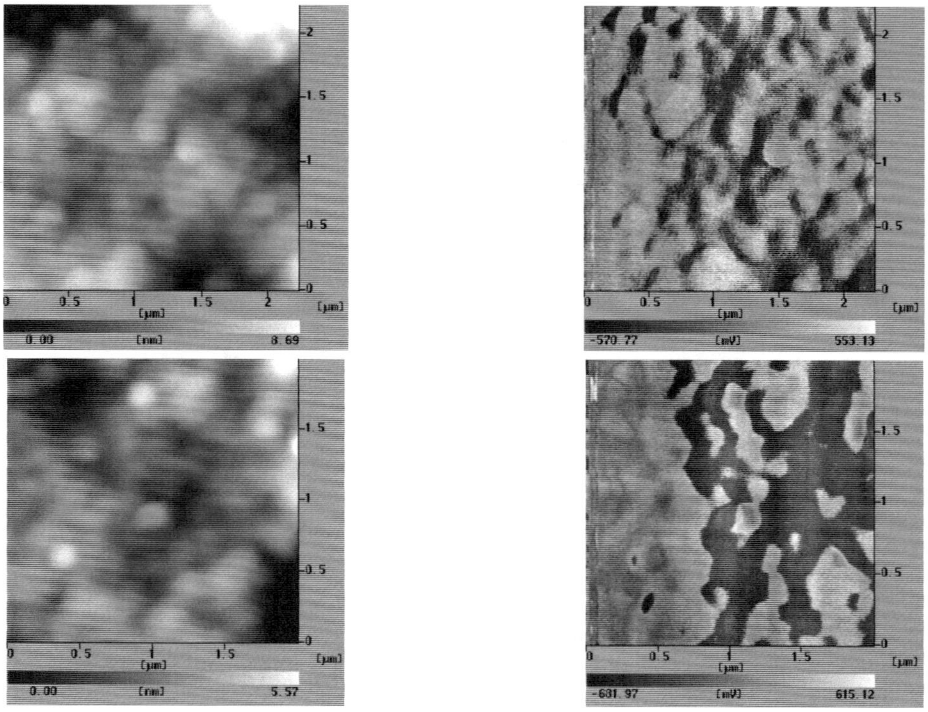

Figure 5. Microscope images and piezoresponse images of PZT(53/47) thin film (a) (b) and 0.2PZN-0.8PZT(53/47) thin film (c) (d).

4. Conclusion

(1-x)PZN-xPZT(53/47) thin films with different PZT(53/47) contents have been prepared by a MOD process on Pt/Ti/SiO2/Si substrates. Pure perovskite PZN-PZT(53/47) thin films with (111) preferred orientation have been obtained when PZT(53/47) content x is ≥ 80% (mol). The lattice spacing d111 of the film was found to have a maximum value near the x=0.7~0.8. When the PZT(53/47) content is ≥ 70% (mol), the PZN-PZT(53/47) thin films can present perfect hysteresis loops with high Pr value (≥42 μC/cm2). Especially the 0.2PZN-0.8PZT(53/47) thin film presented a lower Ec (116Kv/cm) while the important parameter Pr did not decrease obviously. It is almost the half of that of the pure PZT(53/47) thin film. And a more intensive piezoresponse has been also observed in the 0.2PZN-0.8PZT(53/47) thin film. These results can be explained by the increasing of lattice spacing d111 of the film. This optimized 0.2PZN-0.8PZT(53/47) thin film would be a good application potential in the microelectronic device.

Acknowledgements

The authors thank Natural Science foundation of Shanghai (06ZR14095) for financial support.

References

[1] J. E. Spanier, M. Levy, I. P. Herman, , et al. 2001 *Appl. Phys. Lett.* **79[10]** 1510-12
[2] J. K. Lee, J. Y. Yi, K. S. Hong 2004 *J. Appl. Phys.* **96[12]** 7471-75
[3] F. Qin, X. Yao, X. Zeng, Q. Rao, et al. 2005 *J. Crystal Growth* **275** e2487-91
[4] S.-E. Park, T. R. Shrout 1997 *J. Appl, Phys* **82[4]** 1804-11
[5] R. Bertram, G. Reck and R. Uecker 2003 *J. Crysal Growth* **253** 212-20
[6] J. J. Choi, J. H. Lee, B. D. Hahn, et al. 2008 *Materials Research Bulletin* **43** 483-90
[7] Y. Kamisuki, T. Taniguchi, T. Takcnaka 1995 *Microelectronic Engineering* **29** 169-72
[8] Y. Hu 1996 *J. Mater. Sic.* **31** 4255-59
[9] S. H. Yu, K. Yao, F. E. Tay 2004 *Chem. Mater.* **16** 346-350
[10] Y. Zhang 2004 Thesis of Chinese Academy of Science: *Modification of PLZT transparent ceramics* 85-87 (Chinese)
[11] Y. Yang, K. Kim, H. Choi 2001 *Thin Solid Film* **396** 97-102
[12] H. R. Zeng, G. R. Li, et al. 2003 *Appl. Phys.* A **76** 401-04

Synthesis of coiled carbon nanofibers and Y-shaped carbon fibers

Qian ZHANG and Zuolin CUI

Key Laboratory of Nanomaterials, Qingdao University of Science & Technology, Qingdao, 266042, China

E-mail: Cuizl@public.qd.sd.cn, zhangqian@qust.edu.cn

Abstract. In this paper, the synthesis of carbon fibers (CFs) with different morphology, that is, coiled carbon nanofibers with symmetrical structure and Y-shaped carbon fibers was presented by using copper tartrate as catalyst precursor at low temperature, 279 °C. Hydrogen and commercial acetylene or commercial acetylene only was introduced into the reaction tube, respectively, during the synthesis reaction, the introduction of hydrogen will be the key factor to the formation of Y-shaped carbon fibers. According to the analysis, the growth model for carbon fibers has been established. X-ray diffraction, field emission scanning electron microscopy, transmission electron microscopy were used to carry out the product characterization.

1. Introduction

Similar to carbon nanotubes [1-4], carbon fibers with different morphologies have been extensively investigated by researchers [5-12], which have a wide potential application in thermal, mechanical, electric areas, due to their superior properties. At the same time, the growth mechanism of carbon fibers and carbon tubes has particularly attracted the attention of many scientists, and lots of growth models for carbon fibers have been established. These one-dimension carbon nanostructures are usually prepared by chemical vapor deposition (CVD) method, in which hydrocarbon (C_2H_2, C_2H_4, CH_4) and small metal particles (Fe, Co, Ni, Cu or their alloys) were used as carbon sources and catalyst, respectively.

In this paper, coiled carbon nanofibers and Y-shaped carbon fibers with high purity were successfully prepared by CVD using copper tartrate as catalyst precursor at 279 °C and the effect of hydrogen on the morphology of carbon fibers were also examined.

2. Experimental

In this paper, copper tartrate was used as catalyst precursor for the preparation of carbon fibers. The preparation method of copper tartrate is as follows: potassium sodium tartrate aqueous solution (0.0586 mol/l) was slowly added to 0.0586 mol/l copper dichloride aqueous solution under vigorous stirring, copper tartrate (light blue precipitate) was formed quickly. At last, copper tartrate was filtered and washed with distilled water, anhydrous ethanol, respectively, and then dried for 1 h at 100 °C.

In the present work, a horizontal quartz tube (9 cm in diameter and 90 cm in length) was used as the reaction tube. The preparation method of carbon fibers is as follows:

2.1. Coiled carbon nanofibers

A ceramic plate containing copper tartrate was placed in the reaction tube. After the tube was heated to 279 ºC in a vacuum (5Pa) and the catalyst precursor was pushed into reaction area (holding at 279 ºC for 30 min in order that all the precursor particles decomposed to form metallic copper particles), acetylene was introduced into the reactor for a selected reaction time. The pressure within the reaction tube was atmospheric. After the reaction, the quartz tube was cooled in a vacuum (5 Pa) to room temperature.

2.2. Y-shaped carbon fibers

A ceramic plate containing copper catalyst was placed in the reaction area of the reaction tube in a hydrogen atmosphere. After the tube was heated to 279 ºC, hydrogen was immediately replaced by acetylene and the reaction started subsequently. After the reaction, the quartz tube was cooled in a vacuum (5 Pa) to room temperature.

FE-SEM micrographs were obtained with a cold field emission scanning electron microscope (FE-SEM, JEOL, JSM-6700F); transmission electron microscopy (TEM) was carried out on a JEOL JEM-2000EX; X-ray powder diffraction (XRD) patterns were recorded on a Philips X'Pert MPD diffractometer with Cu Kα radiation (λ = 1.5418 Å).

3. Results and discussion

The FE-SEM image (Fig.1a) shows Y-shaped carbon fibers with high purity, in which the bright parts (indicated by arrows) in the junction of Y-shaped carbon fibers represent copper catalyst particles. The TEM image of Y-shaped carbon fibers was shown in Fig.1b, in which the inset is selected area electron diffraction (SAED) pattern of the copper particle in the junc- tion of carbon fibers, which indicates that

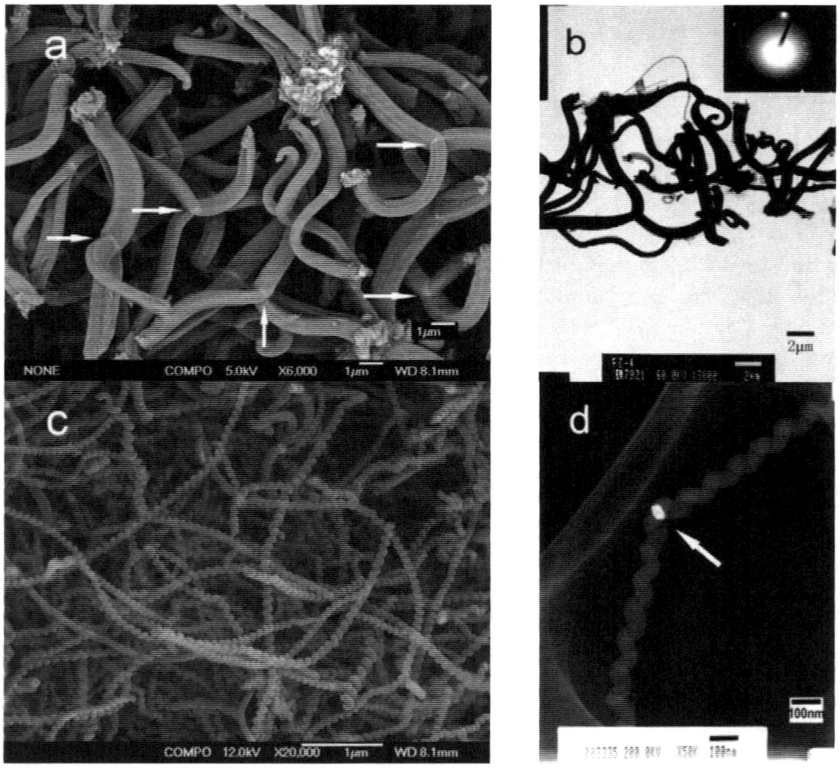

Figure 1. FE-SEM and TEM images of coiled carbon nanofibers and Y-shaped carbon fibers

these copper particles are polycrystalline structure. On the other hand, the FE-SEM image of coiled carbon nanofibers was shown in Fig.1c and copper catalyst particles included in carbon nanofiber are clearly observed in Fig.1d (indicated by arrow). Through TEM analysis, we know that these two pieces of carbon nanofibers, grown from a single copper nanocrystal, have identical fiber diameter, coil diameter, coil number, coil pitch, and different helical sense.

It is well known that the diameter of carbon fibers is decided by the size of copper catalyst particles. The size of copper catalyst particle is in accord with the diameter of carbon fibers, whether Y-shaped carbon fibers (0.8~1 μm) or coiled carbon nanofibers (~100 nm).

In this study, we proposed that the growth model of carbon fibers is decided by the existing structure of copper catalyst particles, that is, monocrystalline or multicrystalline. In the synthesis reaction of Y-shaped carbon fibers, many small copper particles, decomposed from catalyst precursor with reaction temperature increasing, slowly aggregated together to form larger copper particles through Van der Waals force in a hydrogen atmosphere. Then each of these larger copper particles as a whole possesses three carbon fibers growth faces after a period of surface reconstruction [13-15], and can synthesize catalytically three pieces of carbon fibers simultaneously, and then Y-shaped carbon fibers have been obtained. In a word, it is the reconstruction of crystal face of bigger copper catalyst particles due to the effect of hydrogen that lead to the formation of Y-shaped carbon fibers.

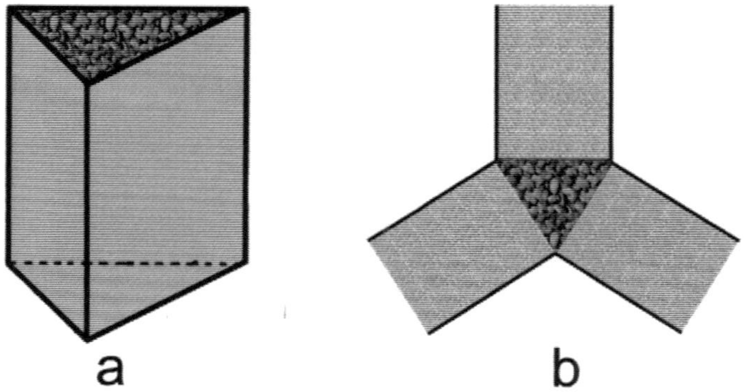

Figure 2. The growth model for Y-shaped carbon fibers

The growth model for Y-shaped carbon fibers was shown in Fig.2. Fig.2a represents a bigger copper catalyst particle in Y-shaped carbon fibers, which consist of several monocrystalline copper particles. These grey faces in Fig.2a represent carbon fibers growth face of catalyst particle. The model of Y-shaped carbon fibers was displayed in Fig.2b, in which there are three pieces of carbon fibers grown from the same copper catalyst particle, and then Y-shaped carbon fibers have been obtained.

In the synthesis reaction of coiled carbon fibers, the catalyst precursor was located in the reaction area when the reaction temperature arrived. Therefore, lots of small copper catalyst particles were formed quickly in an acetylene atmosphere and acted as catalyst to synthesis coiled carbon nanofibers. Due to the anisotropy of copper monocrystalline for carbon fibers growth, there is carbon fiber growth rate difference between different crystal faces, and the considerable dispersion of growth rate over the surface of catalyst particle induced the formation of coiled carbon nanofibers [16]. The SAED pattern gives evidence of the presence of well-developed monocrystalline structure in small copper particle located in coiled carbon fibers.

Through XRD analysis (see Fig.3), we know that both of Y-shaped carbon fibers and coiled carbon nanofibers have the same components. The sharp peaks and the disperse peak correspond to the metal copper and amorphous carbon. In addition, copper oxide (78-0428) was also found due to their weak signals. No copper carbide solution peaks were detected in these two kinds of carbon fibers.

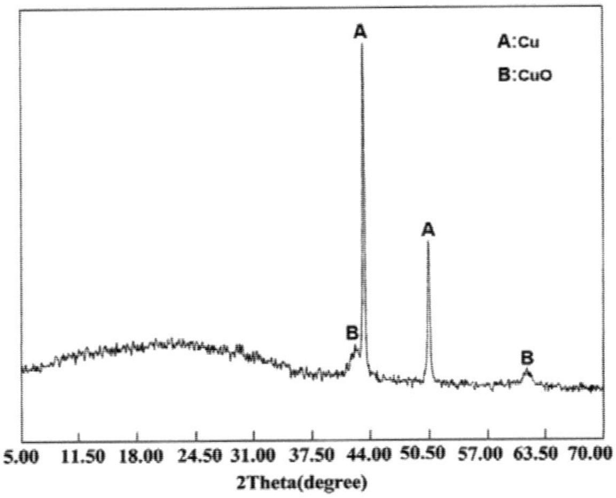

Figure 3. XRD patterns of carbon fibers

4. Conclusions

Coiled carbon nanofibers and Y-shaped carbon fibers with high purity have been successfully obtained, respectively. The formation of Y-shaped carbon fibers is related to the introduction of hydrogen before the start of the reaction. It is the participation of hydrogen that leads to the formation of the growth faces of catalyst particles for carbon fiber and Y-shaped carbon fibers subsequently. The copper catalyst particles included in Y-shaped carbon fibers exist as polycrystalline structure. The anisotropy of monocrystalline copper particles included in carbon nanofibers lead to the formation of coiled morphology, that is, the growth rates of nanofibers on different crystal faces are different.

Acknowledgement

This research was supported by the Natural Scientific Foundation of China (No. 50872059), the Natural Science Foundation of Shandong Province (Z2005F04).

References

[1] Iijima S 1991 Nature **354** 56
[2] Wei D, Liu Y, Cao L, Fu L, Li X, Wang Y, Yu G, and Zhu D B 2006 Nano Lett. **6** 186
[3] Gan B et al. 2001 Chem. Phys. Lett. **333** 23
[4] Wang Z Y, Zhao Z B, Qiu J S 2006 Carbon **44** 1321
[5] Baker R T K 1989 Carbon **27** 315
[6] Gothard N, Daraio C, Gaillard J, Zidan R, Jin S, and Rao A M 2004 Nano Lett. **4** 213
[7] Sharon M and Pradhan D 2005 J. Nanosci. Nanotech. **5** 1718
[8] Pradhan D, Sharon M, Kumar M, Ando Y 2003 J. Nanosci. Nanotech. **3** 215
[9] Xi G C, Zhang M, Ma D K, Zhu Y C, Zhang H B, Qian Y T 2006 Carbon **44** 734
[10] Cuong P H, Ricardo V, Benoit L, Alain C, Julien A, Thierry D, Marc J L 2006 J. Catal. **240** 194
[11] Tao X Y, Zhang X B, Zhang L, Cheng J P, Liu F, Luo J H, Luo Z Q, Geise H J 2006 Carbon **44** 1425
[12] Qin Y, Zhang Q, Cui Z L 2004 J. Catal. **223** 389
[13] Barteau M A, Ko E J, Madix R J 1981 Surf. Sci. **102** 99
[14] Van Hove M A, Koestner R J, Stair P C, Bibérian J P, Kesmodel L L, BartoŠ I, Somorjai G A 1981 Surf. Sci. **103** 189
[15] Heinz K, Hammer L 1997 Phys. stat. sol. (a) **159** 225
[16] Amelinckx S, Zhang X B, Bernaerts D, Zhang X F, Ivanov V, Nagy J B 1994 Science **265** 635

Journal of Physics: Conference Series **152** (2009) 012070 doi:10.1088/1742-6596/152/1/012070

Preparation of Nanocrystalline CeO_2 by nanocasting with mesoporous silica

H. ZHANG, S.C. WANG, D. XUE, Q. CHEN and Z.C. LI

School of Materials Science and Engineering, Central South University, 410083 Changsha, China

Email: hzhang@mail.csu.edu.cn

Abstract. Nanocasting is one of the methods that can be used to prepare nanostructured materials. The mesoporous silica with $Ia\bar{3}d$ structure was used as template for the nanocrystalline CeO_2 synthesis. The as synthesized CeO_2 powder was characterized with XRD and TEM. The crystal size was determined to be about 8.3 nm, a little smaller than that of the pore size of the mesoporous silica used.

1. Introduction

Nanostructured materials, a new branch of materials research, are attracting a great deal of attention because of their potential applications in the areas such as electronics, optics, catalysis, ceramics, magnetic data storage and nanocomposites. The unique properties and the improved performances of nanomaterials can be determined by their sizes, surface structures and inter-particle interactions [1]. The development of synthetic strategies to obtain nanocrystals has progressed impressively in the last decade. Nanocasting is one of the methods that could be used to obtain nanostructured materials. In this case, the nanocrystals are formed in the confined space of a host material. The host can be soft, for instance micelles formed by the self assembly of diblock copolymers [2] or hard, for instance anodic alumina membranes [3], single-wall carbon nanotubes [4], crystalline microporous or mesoporous solids. Among the variety of potential hard hosts, mesoporous silicas, have several advantages: their uniform mesopore dimensions that can be adjusted; several channel structures (ranging from layers to 3D) are available; after crystallization of the nanostructures, the silica template can be removed by acidic or basic chemical treatments.

Basically, the nanocasting route comprises three steps: Infiltration of the porosity of the template with a solution containing the precursors of the synthesized material; Heat treatment under a controlled atmosphere of the impregnated template to convert the infiltrated precursor into the synthesized material and the third to remove the template framework, by dissolution or by oxidation at high temperatures. Due to the fact that the synthesis takes place in a confined nanospace, the sintering of the particles is restricted and the preparation of high surface area materials (nanostructures or nanoparticles) is achieved [5].

Crystalline CeO_2 materials are often desired for potential applications, e.g. with respect to their catalytic [6], gas sensing [7, 8], or ion conducting properties [9]. In this paper, nanocrystalline CeO_2 was synthesized by nanocasting with mesoporous silica as template.

2. Experimental

The mesoporous silica used as template in this experiment was synthesized by Ting Yu, the Ph.D student from DY Zhao's group at Fudan University, China. The synthesis route of the mesoporous silica is referred to the publication [10], just the sodium dodecyl sulfate (SDS) was replaced by FC80. Impregnation experiment was performed by dissolving 1.5 g of mesoporous silica in 8 g ethanol, then 0.5 g $Ce(NO_3)_3{\bullet}6H_2O$ was added to this solution and stirred for 2 h. The resultant mixture was dried at 40°C and calcined at 300°C for 3 h. 8 g ethanol was added to the dried powder again and stirred, then 0.3 g $Ce(NO_3)_3{\bullet}6H_2O$ was added to the solution. The resultant mixture was dried at 40 °C and calcined at 550°C for 5 h. The obtained CeO_2–silica composite powder from above procedure was washed in 20 ml of 2 mol/L NaOH solution in order to dissolve the mesoporous silica wall.

The CeO_2 powder was performed the XRD analysis with D/Max 2550 diffract meter. The sample for TEM observation is prepared with the powder dissolving in ethanol and then dried on a holy carbon film grid. The TEM work is done in the Tecnai G^2 20 ST TEM with 200 kV accelerating voltage.

3. Results and discussions

The TEM observation of mesoporous silica is shown in Figure.1. The unit cell of the mesoporous silica was determined to be about 21 nm, with $Ia\bar{3}d$ structure and the pore size is about 9 nm. The image in Fig. 1 is obtained along the [111] direction of the mesoporous silica structure. It can be clearly seen that the mesoporous silica used in this experiment is well ordered. The CeO_2 powder was synthesized by taking this mesoporous silica as template with the route described in the experimental section.

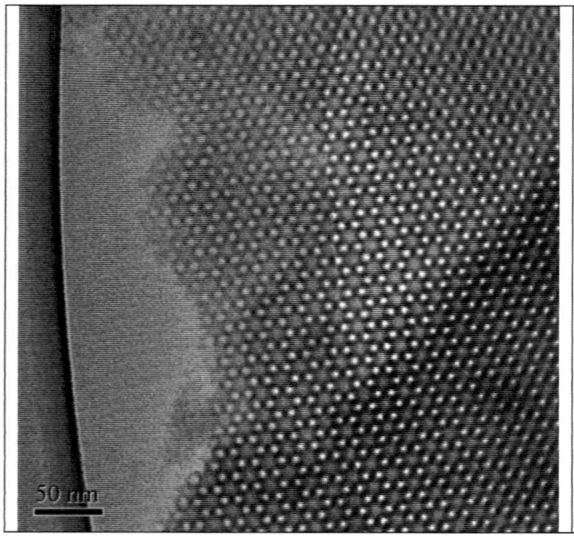

Figure 1. The image of $Ia\bar{3}d$ mesoporous silica taken along [111] direction

The XRD pattern of the obtained CeO_2 powder is shown in Figure 2. From the peaks presented in the figure, one can certain that the CeO_2 powder synthesized with nanocasting method has Fluorite structure with the space group Fm-3m and the lattice parameter a = 5.4104 Å. From the broaden peaks in the figure, we can measure that the half width for the (111) peak is about 0.948°. In Debye-Scherrer equation d is the average grain size, K is set to be 0.89 for normal, λ is the wavelength of X ray (1.54184 Å), β is the half width of the diffraction peak and θ is the Bragg diffraction angle.

$$d = \frac{K \times \lambda}{\beta \times \cos\theta}$$

From this equation, the calculated grain size for the synthesized CeO_2 powder is about 8.3 nm, which is a little smaller than the pore size of mesoporous silica.

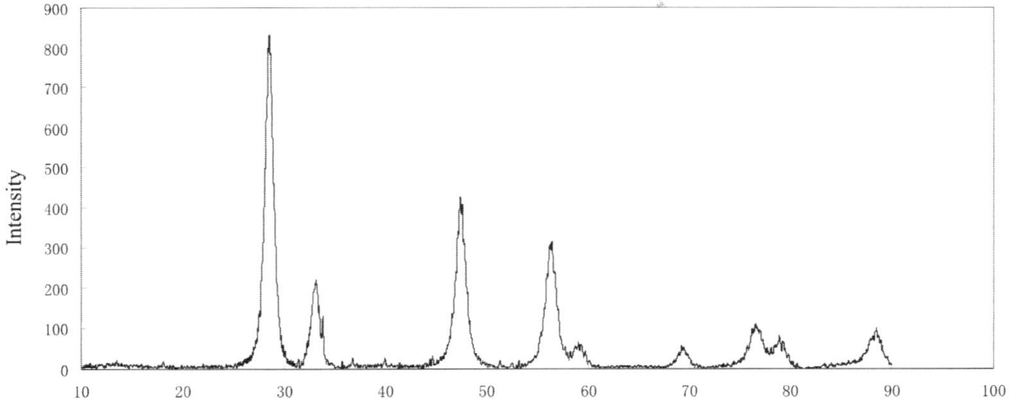

Figure 2. The XRD pattern of the synthesized CeO_2 powder

The TEM image of the synthesized CeO_2 powder is shown in Figure 3. From the image, we can see that the powder size is about 9 nm, similar to that calculated by the Debye-Scherrer equation from the XRD result. The particles are mostly with unique size and quite well separated. It could be concluded that nanocasting is a very powerful and useful way for synthesis of nano particles. Also the nano particles from nanocasting could be separated and with very narrow particle sizes distribution.

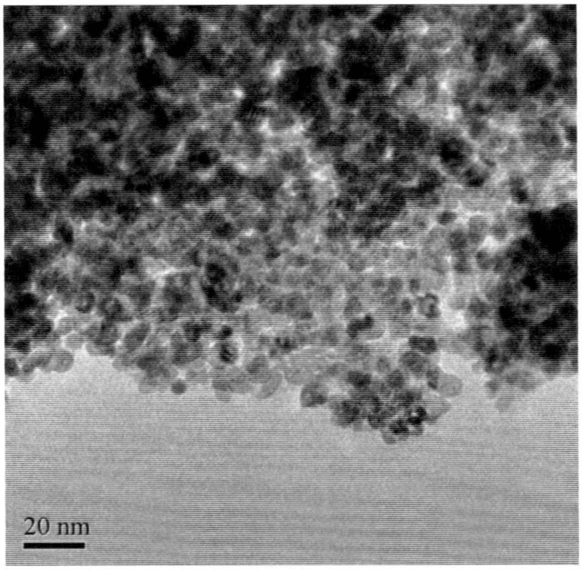

Figure 3. The TEM image of synthesized CeO_2 powder

4. Conclusions

By controlling the synthesis condition, nanocrystalline CeO_2 powder with unique particle size can be obtained. By XRD and TEM analysis, the particle size of CeO_2 powder obtained from the nanocasting by using the ordered mesoporous silica as template is a little smaller than that pore size of the mesoporous silica. It was testified that the synthesis of CeO_2 powder is confined to the nanospace of mesoporous silica. So nanocasting is very powerful and useful method that can be used to prepare nanostructured materials, especially nanomaterials, which is difficult obtained by other synthetic method.

Acknowledgement
Many thanks to Dr. Ting Yu and Prof. DY Zhao at Fudan University for providing the mesoporous silica used in this experiment.

References
[1] Characterization of Nanophase Materials, Edited by Zhong Lin Wang, Wiley-VCH Verlag GmbH, 2000.
[2] Thurn-Albrecht T, Scotter J, Kastle G A, Emley N, Shibauchi T, Krusin-Elbaum L, Guarini K, Black T, Touminen M S, Russel T P, 2000 *Science* **290** 2126.
[3] Zhang XY, Zhang L D, Chen W, Meng G W, Zheng M J, Xao L X, 2001 *Chem. Mater.* **13** 2511.
[4] Pham-Huu C, Keller N, Estournes C, Ehret G, Ledoux M J, 2002 *Chem. Commun.* 1882.
[5] Valdés-Solís T, Fuertes A B, 2006 *Materials Research Bulletin* **41** 2187.
[6] Kašpar J, Fornasiero P, Graziani M, 1999 *Catal. Today* **50** 285.
[7] Beie H J, 1991 *Sens. Actuators* B **4** 393.
[8] Jasinski P, Suzuki Z, Anderson H U, 2003 *Sens. Actuators*, B **95** 73.
[9] Tschöpe A, 2005 *J. Electroceram.* **14** 5.
[10] Zhang H, Yu T, Oleynikov P, Zhao D Y, Hovmöller S and Zou X D, CRISP and eMap: the software for determining 3D pore structures of ordered mesoporous materials by electron crystallography, 5th International Mesostructured Material Symposium, August 5-7, 2006, Shanghai, China.

Effects of Ni catalyzer on growth velocity and morphology of vapor growth SiC nano-fibers

Xianfeng XU[1, 2], Peng XIAO [1], Xiang XIONG [1] and Boyun HUANG [1]

[1] State Key Lab. for Powder Metallurgy, Central South University, Changsha, 410083, China

[2]College of Mechanical and Electrical Engineering, East China Jiaotong University, Nanchang, 330013, China

E-mail：xu-xianfeng@163.com

Abstract. Using Ni granules as a catalyzer, composite felts reinforced by both SiC nano-fibers (SiC-NFs) and carbon fibers were prepared at 1273K. SiC-NFs were deposited on the surface of the carbon fibers in-situ by chemical vapor growth. The microstructure and morphology of the fibers after being electroplated and deposited were characterized by scanning electron microscopy and transmission electron microscopy. The phase of the fibers after being deposited was characterized by X-ray diffractometry. The results show that the SiC-NFs produced by chemical vapor growth are single crystals of β-SiC. It was found that Ni catalyzer had important effects on the growth of SiC-NFs. Smaller nano-granules were more active as catalyzers, resulting SiC-NFs appeared more spindle-like and had a more homogeneous dispersion. The weight change of the samples before and after deposition shows that a longer electroplating times results in SiC-NFs having a faster growth velocity.

1. Introduction

Since the discovery of carbon nanotubes (CNTs) by Iijima in 1991 [1], researchers have been very actively trying to gain controls on length, diameter, etc. [2-3]. CNTs and carbon nano-fibers (CNFs) are interesting candidates for realizing novel composite materials due to their almost one-dimensional shape coupled with the exceptional mechanical, thermal and electrical properties [4-5]. The transition elements are promising candidates for use as catalyzers for preparing CNTs and CNFs [6-10]. The method of preparing CNTs and CNFs is also used for preparing SiC nano-fibers (SiC-NFs) [11]. SiC-NFs or SiC nanoparticles have some special properties such as high strength, high modulus, high corrosion and oxidation resistance and so on. So, experiments on SiC-NFs deposited on the surface of carbon fibers in-situ was perfectly developed in carbon fiber (CF) reinforced carbon composites (C/C composites) and CF reinforced carbon composites (C/SiC composites) [12]. The configuration and distribution of the catalyzer plays a key role in the process of vapor growth. During our research, we applied electroplating Ni granules on the surface of the CFs to afford a random distribution of the Ni catalyzer. Then, SiC-NFs were grown on the surface of the carbon fibers in-situ by chemical vapor growth. By the method of chemical vapor growth, it is possible to achieve a large scale synthesis of these composite felts reinforced by both SiC-NFs and CFs. The resulting composite felt material is the subject of intensive research, and appears to be one of the most promising candidates for the future preparation of a reinforced framework for C/C and C/SiC composites. The effect of different electroplating times on the morphology and growth velocity of the SiC-NFs is discussed in this work.

2. Materials and methods

2.1. Raw materials

Carbon fibers were produced by Toyor Co.; Japan (T700, 12K). Needled integer felts (0.4 g/cm^3) were produced by the Tianniao Company, Yixing city, Jiangsu Province. Ni_2SO_4 was produced by Qinxiang Chemical Co.; Ltd in Shanghai. The electroplating equipment produced by Antaixin Electronic Co.; Ltd in Shenzhen comprises a direct current power supply with a symmetrical output, the type of which is PMR155.

2.2. Electroplating Ni on carbon felts to form catalyst nano granules

The felt was cut into spindle samples of 6×6×50mm in the same direction, soaked in acetone for 24 hours, washed in deionized water repeatedly, and then dried. Such samples were placed into an electroplating solution (Ni_2SO_4, 10 wt %), and then electroplated for 2.5, 5, 7.5 and 10min. The electronic current intensity was 100mA. Then, the samples were soaked in deionized water, washed, dried and weighed. The microstructure and surface morphology of the carbon fibers after being electroplated with Ni were observed by scanning electron microscopy (SEM, TSM-6360LA, and Japan).

2.3. Growth of SiC-NFs by chemical vapor growth

These samples were first electroplated with Ni for different periods of time, and then SiC was deposited on the electroplated samples in a chemical vapor growth furnace at 1273K for 2, 4, 6, 8 and 10h respectively. The resulting samples were cooled and then removed from the furnace and weighed. Methyltrichorosilane (MTS, 95%) was used as the source of carborundum, Argon (99.99%) was used as the carrier gas, and hydrogen (99.99%) was employed in the chemical vapor growth process in a dilute form. The pressure inside the furnace was 200~400pa. After growth of the SiC-NFs, SEM was used to characterize the morphology of the carbon fiber integer felt. Transmission electron microscopy (TEM, JEOL 2010F, Japan) was used to characterize the diameter, morphology and type of crystal formed. Composite felts first plated with Ni for 10 min and then coated with SiC by vapor growth for 10h were used as samples for powder X-ray diffractometry (XRD, PANAlytical Company in Holland) to ascertain the phase of the composite felts.

3. Results and discussion

Figure 1 shows the SEM morphology of the CFs after electroplating with Ni for different times and Figure 2 shows the CFs after coating with SiC-NFs by chemical vapor growth. As shown in figure 1 (a) and (b), the electroplated Ni features are smaller than those in figure 1 (c). The nano-fiber shown in figure 2 (a) and (b) is more spindle-like than that in figure 2 (c) and also disperses much more homogeneously.

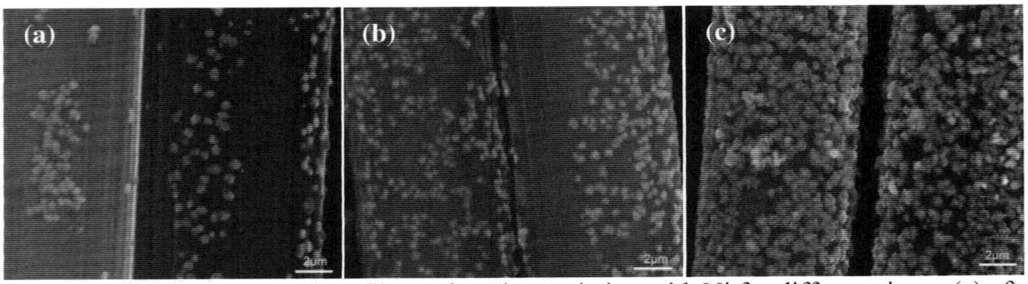

Figure 1. SEM images of carbon fibers after electroplating with Ni for different times. (a) after electroplating Ni for 5min, (b) after electroplating with Ni for 7.5min and (c) after electroplating with Ni for 10min.

Journal of Physics: Conference Series **152** (2009) 012071 doi:10.1088/1742-6596/152/1/012071

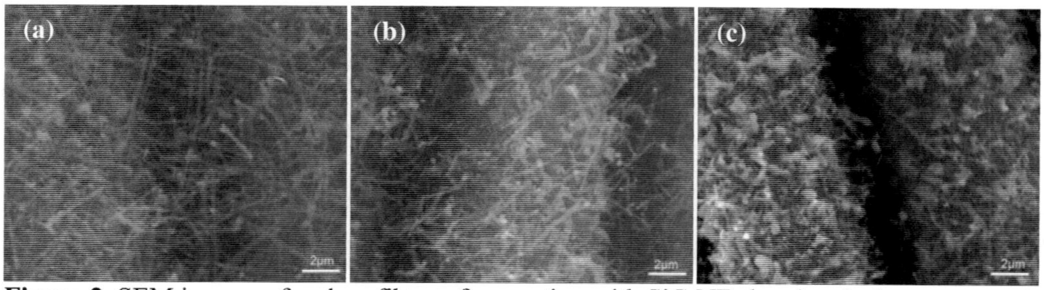

Figure 2. SEM images of carbon fibers after coating with SiC-NFs by chemical vapor growth for 2h. (a) after chemical vapor growth for 2h with plating Ni for 5 min, (b) with Ni for 7.5min and (c) with Ni for 10min.

Figure 3 shows the XRD patterns of carbon felts after coating with SiC by chemical vapor growth for 6h with plating Ni for 10min. In the diffraction graph, the diffraction peaks of C, β-SiC and Ni were observed. The weight content of β-SiC was found to have reached 10 percent as determined by phase analysis of XRD.

Figure 4 shows the TEM morphology of the SiC-NFs and diffraction spot of a selected section on the sample surface. As shown in figure 4(a) and (b), the SiC-NFs are unbent, have a single crystal diffraction spot and Ni catalyst nanogranules were found at the point of the arrow in Figure 4 (c).

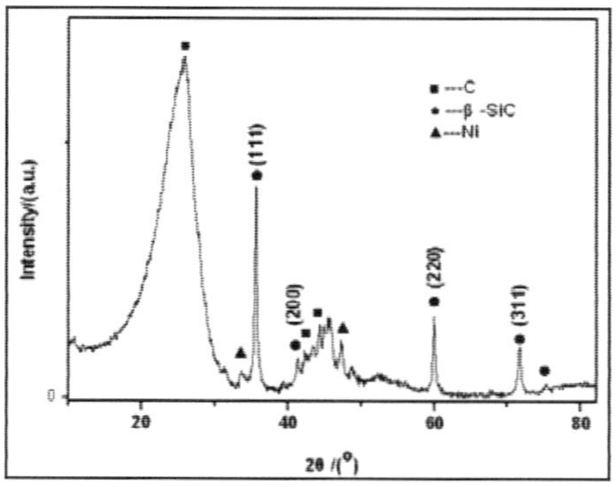

Figure 3. XRD patterns of carbon felts produced after electroplating with Ni for 10min and then coating with SiC for 6

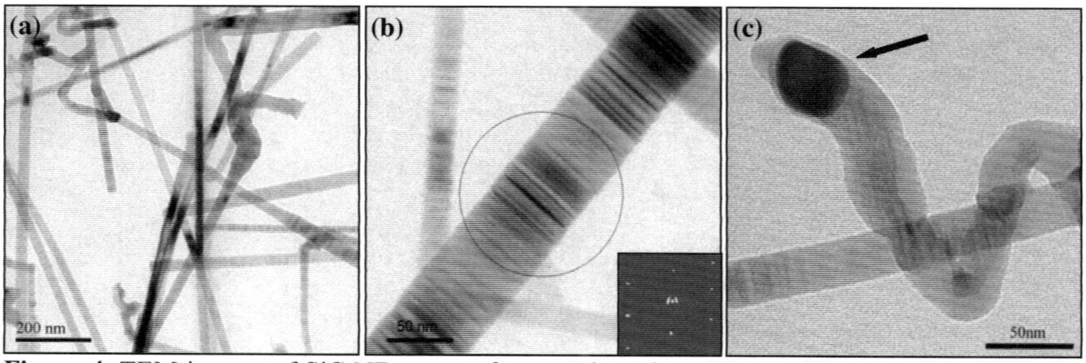

Figure 4. TEM images of SiC-NFs grown from a Ni catalyst. (a) overall aspect of the SiC-NFs; (b) diffraction spot of a selected section on the surface of the SiC-NFs; (c) tip of an individual SiC-NF with a Ni particle.

3.1. Preparation of SiC-NF/carbon fiber composite felts

As shown in figure 1 (a) and (b), the Ni-nanogranules that are produced have diameters of several nanometers. But the Ni granules in figure 1 (c) are bigger than those of figure 1 (a) and (b), and have diameters in the range of hundreds of nanometers, and the whole surface of the CFs is almost covered with Ni granules. In this respect, by keeping the current density at 100mA, the size, morphology and the amount of plating Ni on the surface of the CFs would be controlled by adjusting the plating time. As shown in figure 2, after plating the surface of the CFs with Ni, the nanofibers were deposited on the surface of the CFs in-situ.

In figure 2 (a) and (b), the carbon fibers are bridged by chemical vapor growth SiC-NFs, making each of them not be a single fiber as reinforcement phase in composite materials. In addition, the traditional carbon felts, whether planar, two-dimensional or three-dimensional, invariably cause anisotropy after compositing. As described above, the effect of bridging and growing random nanofiber composite felts will improve this anisotropy. Moreover, it is also expected to improve the properties of the fiber-reinforced composites such as strength, modulus, corrosion and oxidation resistance, electric and heat conduction performance, etc.

3.2. Phase analysis of SiC-NF/CF composites

As shown in figure 3, the observed phases of the composite felts after chemical vapor growth include C, β-SiC and Ni. The C diffraction peak is broad; this corresponds with the fact that the carbon felts contain polycrystalline graphite. The β-SiC diffraction peak is sharp and is characteristic of the different 2θ angle accorded with the JCPDS CARD (No.29-1129), where the crystal content is close. From above, the β-SiC nanofibers are grown on the CF surface in-situ in the carbon felts after first plating with Ni and then coating with SiC using chemical vapor growth.

Figure 4(b) shows that the as-produced SiC-NFs are single crystals and have a single crystal diffraction spot.

The β-SiC fibers behave as an intensifier for the composite material resulting in properties of high specific strength, high specific modulus, high-temperature stability and high corrosion and resistance durability [13]. Moreover, the β-SiC nanofibers can improve the compatibility between the carbon fibers and the matrix material (pitch, resin or metal) [12]. Thereby, the β-SiC nanofibers improve the properties of the resulting composite material prepared with C/C or C/SiC composites [14, 15].

3.3. Effect of plating time on growth velocity and morphology of SiC-NFs

The weight gain of as-received samples and the different-timed electroplating Ni samples are converted into weight increase percent. Figure 5 was obtained by taking the electroplating time coordinate as the X axis and the weight increase percent as the Y axis. In figure 5, comparing with the non-electroplated Ni samples, electroplating with Ni can improve the deposition velocity of SiC. The longer of electroplating times are, the more SiC-NFs produced by chemical vapor growth are. But if the electroplating time surpasses 5 min, the corresponding weight increase is negligible. Moreover, by keeping the same electroplating time, and prolonging the vapor growth time, more SiC-NFs are observed on the CF surface and the weight increases accordingly. Therefore, carbon felts electroplated with Ni and coated with SiC by chemical vapor growth, the density velocity in chemical vapor infiltration (CVI) C/C and C/SiC composite materials is found to increase.

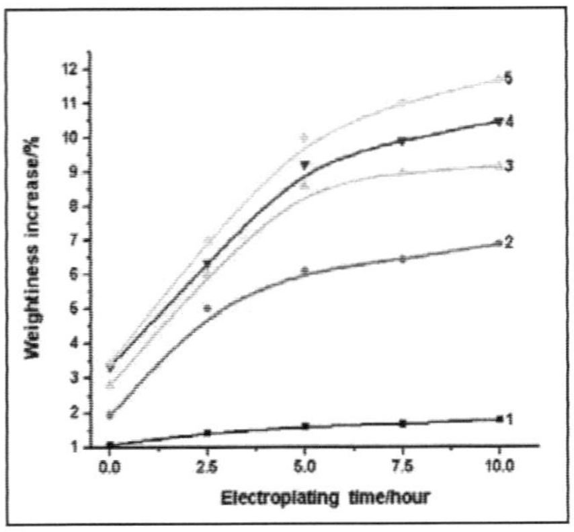

Figure 5. The relation between electroplating time and weight increase percent: 1-CVD for 2h; 2- CVD for 4h; 3- CVD for 6h; 4- CVD for 8h; 5-CVD for 10h.

4. Conclusions

The SiC-NF/CF composite felts are prepared by the methods of chemical vapor growth SiC-NFs in-situ. The results show the SiC-NFs are single crystals of β-SiC. The catalyzer of Ni granules have a great effect on the growth of SiC-NFs. Smaller and well-distributed Ni granules result in more spindle-like SiC-NFs with a better distribution. Longer electroplating times result in more Ni nanogranules growing on the surface of the CFs, and a faster growth velocity of SiC-NFs.

Acknowledgements
This project was supported by the State Key Development Program for Basic Research of China (Grant No. 2006CB600904).

References

[1] Iilima S 1991 Helical micro-tubules of graphitic carbon *Nature* **354** 56-8
[2] Ren Z F, Huang Z P, Xu J W, Wang J H, Bush P, Siegal M P 1998 Synthesis of large arrays of well-aligned carbon nanotubes on glass *Sci.* **282** 1105-7
[3] Huang S, Woodson M, Smally R, Liu J 2004 Growth mechanism of oriented long single walled carbon nanotubes using "fast-heating" chemical vapor deposition process *Nano lett.* **4** 1025-8
[4] Lau K T and Hui D 2002 The revolutionary creation of new advanced materials-carbon nanotube composites *Composites Part B* **33** 263-77
[5] Thostenson E T, Ren Z F and Chou T W 2001 Advances in the science and technology of carbon nanotubes and their composites: a review *Compos. Sci. Technol.* **61** 1809-912
[6] Wunderlich, W 2007 Growth model for plasma-CVD growth of carbon nano-tubes on Ni-sheets *Diamond and Related Mat.* **16(2)** 369-78
[7] Xiong G Y, Wang D Z, and Ren Z F 2006 Aligned millimeter-long carbon nanotube arrays grown on single crystal magnesia *Carbon* **44** 969-73
[8] De Riccardis M F, Carbone D, Makris T D, Giorgi R, Lisi N and Salernitano E 2006 Anchorage of carbon nanotubes grown on carbon fibers *Carbon* **44** 671-4
[9] Hu M H, Yoichi M, Masaru O, Shigeo M and Tatsuya O 2004 Morphology and chemical state of Co-Mo catalysts for growth of single-walled carbon nanotubes vertically aligned on quartz substrates *J. Catal.* **225(1)** 230-9
[10] Ci L J, Wei J Q, Wei B Q, Liang J, Xu C L and Wu D H 2001 Carbon nanofibers and single-walled carbon nanotubes prepared by the floating catalyst method *Carbon* **39** 329-35
[11] Xu X F, Xiao P, Xu L, Xiong X and Huang B Y 2008 Preparation of Carbon Felts Reinforced by both SiC nano-fibers and carbon fibers *J. Funct. Mater.* **39(4)** 692-4
[12] Kristoffer K, Zmago S and Tomaz K 2007 Preparation and properties of C/C-SiC nano-composites *J. Eur. Ceram. Soc.* **27(2-3)** 1211-6
[13] Takuya AOKI, Toshio OGASAWARA and Takashi ISHIKAWA 2004 Mechanical behavior and oxidation-resistance of an orthogonal 3D β-SiC fiber/carbon matrix composite *Ceramic engineering and Science* **25(4)** 147-52
[14] Jayaseelan, DD, Lee, W E, Amutharani, D, Zhang SW, Yoshida K and Kita, H 2007 In situ formation of silicon carbide nanofibers on cordierite substrates *J. Am. Ceram. Soc.* **90(5)** 1603-6
[15] Vladimir P 2004 Nanostructured superconductors: from granular through wire towards high-Tc nanotubular 2D composites *International Journal of Nanotechnology* **1(1-2)** 170-92

Journal of Physics: Conference Series **152** (2009) 012072 doi:10.1088/1742-6596/152/1/012072

Synthesis of SiC nanowires from gaseous SiO and pyrolyzed bamboo slices

Cui-yan LI, Jianfeng Huang, Liyun CAO, Jing LU

School of Materials Science and Engineering, Shanxi University of Science & Technology, Xi'an 710021, P.R. China

E-mail: licuiyan@sust.edu.cn

Abstract. Bamboo slices have been pyrolyzed by heat treatment in an inert atmosphere at 1000°C. The reaction of gaseous SiO and pyrolyzed bamboo slices at 1300°C resulted in the formation of SiC nanowires. Microstructure characterization of the SiC nanowires was carried out by X-ray diffraction (XRD), transmission electron microscopy (TEM) and scanning electron microscopy (SEM). XRD analysis showed that the reaction product is crystalline β-SiC. The diameter and the length of the SiC nanowires is ranging from 20~70nm to 10~20μm. Most of the SiC nanowires possess straight and smooth morphology. SiC nanowires have a preferential growth axis parallel to the [111] direction.

1. Introduction

SiC nanowires are very promising because of their large elasticity and strength compare to SiC whiskers or bulk SiC. Such materials can be widely used to reinforce ceramic, metal, and polymer materials [1-4]. In addition, SiC nanowires have great potential for use in electron emitting filaments because of their unique optical and field emitting properties [5, 6]. Therefore, considerable efforts have been devoted to the synthesis of SiC nanowires.

Carbothermal reduction is a useful method for the synthesis of SiC nanowires. Kinds of reductant have been developed for the synthesis of SiC nanowires, such as carbon nanoparticles, carbon nanotubes (CNTs) and amorphous activated carbon (AAC) [6-8]. However, these reductants have some disadvantages which may limit the synthesis and application of SiC nanowires. It may cause quantity of SiC particles mixed in the products by fabrication of SiC nanowires from carbon nano-particles or amorphous activated carbon. Employing the CNTs as templates in the formation of SiC nanowires can obtain high purity nanowires, but the cost of CNTs restrain its application.

To overcome these disadvantages, SiC nanowires were synthesized via carbothermal reduction of SiO with the pyrolyzed bamboo slices in our experiment. There are many advantages to apply pyrolyzed bamboo slices for the synthesis of SiC nanowires. Firstly, the pyrolyzed bamboo presents the character of amorphous activated carbon. The large specific surface area of the pyrolyzed bamboo is benefit to the growth of SiC nanowires [9]. Secondly, the pyrolyed bamboo contents the elements Fe, Si and K, which might be catalyst in the synthesis of SiC nanowires [10]. Thirdly, the slices structure of the pyrolyzed bamboo can avoid the formation of the SiC particle in the products, which was generated from the particle AAC. In addition, the growth cycle of bamboo is short, which is considered as renewable forest resources. The cost of raw materials for the synthesis of SiC nanowires is low.

In this work, the SiC nanowires was prepared by the carbothermal reduction of SiO and pyrolyzed bamboo. The morphology of SiC nanowires was characterized by SEM. The mircrostructure of the nanowires was analyzed by XRD and TEM. The growth mechanism of SiC nanowires was discussed.

2. Experimental

2.1. Preparation of pyrolyzed bamboo slices

Bamboo slices were in dimensions of 80mm×10 mm×3 mm. The specimens were rinsed with water and then dried at 70°C for 24 h. Subsequently, the specimens were pyrolyzed at 1000°C for 10h in flowing argon atmosphere with slow heating rate of 1°C/min up to 400°C and a higher rate of 5°C/min up to 1000°C.

2.2. Carbothermal reduction

Carbothermal reduction of the gaseous SiO was performed in a vacuum sintering furnace. A graphite crucible, which is 60 mm in diameter and 100 mm in height, was put in the furnace. The graphite crucible was fed with the mixture powder of Si and SiO_2 and the pyrolyzed bamboo slices. The slices were inserted in the mixture powder of Si and SiO_2. Then the furnace was heated to 1300°C and maintained at this temperature for 1h under flowing argon atmosphere. Finally, the furnace was naturally cooled to room temperature.

2.3. Characterization of Samples

The phases of the samples were identified by X-ray diffractometer (XRD, Cu-Kα, X'Pert HighScore). The microstructure of the samples was characterized by means of scanning electron microscopy (SEM, JSM-6064) and high resolution transmission electron microscopy (TEM, JEM-2010).

3. Results and Discussion

3.1. Pyrolyzed bamboo slices

After pyrolysis, there is significant volume shrinkage for the bamboo slices. The weight loss was found to be about 70~80 wt% for the samples. The bio-structure of bamboo was still retained, as shown in Figure 1.

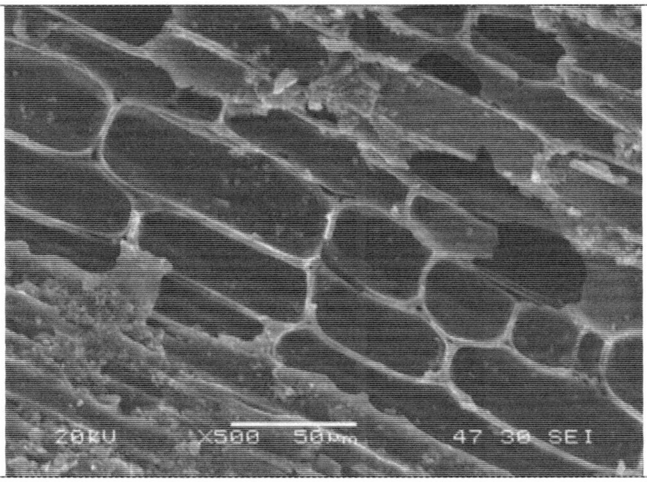

Figure 1. SEM images showing lateral view of the bamboo bio-template

3.2. SiC nanowires

Figure 2 (a) presents the XRD diffraction result of the pyrolyzed bamboo slices. The two broad peaks located at 22° and 44° corresponded to (0 0 2) and (1 0 0) peaks of the carbon. It is indicated that the bamboo slices had absolutely converted to charcoal. Figure 2 (b) presents the XRD diffraction result of the pyrolyzed bamboo after carbothermal reduction reaction. The XRD pattern containing peaks at 35.67°, 60.11°, and 71.87° indicated that these β-SiC generated after carbothermal reduction. Apart from these three peaks, one extra shoulder peak located at 33.73° was a characteristic peak indicating typical stacking faults on the (1 1 1) planes in β-SiC.

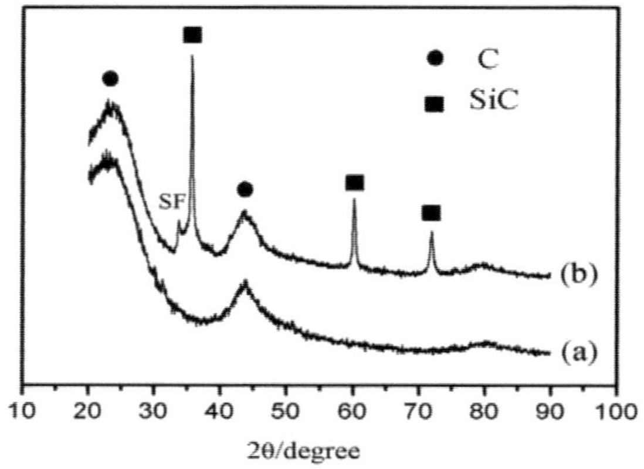

Figure 2. XRD patterns of (a) the pyrolyzed bamboo and (b) the pyrolyzed bamboo after carbothermal reduction reaction.

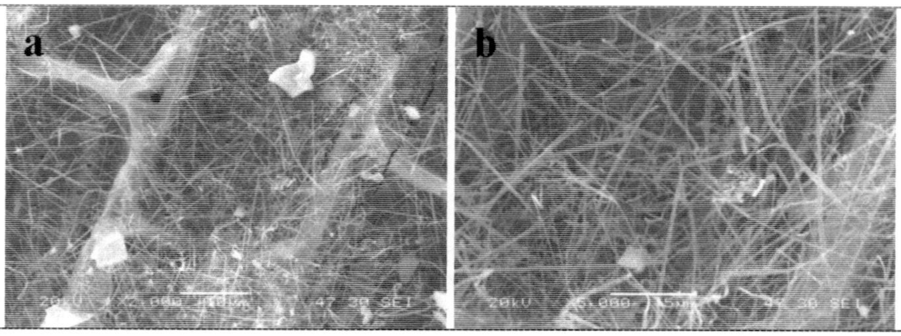

Figure 3. SEM image of the pyrolyzed bamboo slices after carbothermal reduction reaction.

The SEM images of the pyrolyzed bamboo slices after carbothermal reduction reaction are shown in figure 3. It can be seen that a large quantity of SiC nanowires has been obtained on the surface of the pyrolyzed bamboo slices. The SiC nanowires randomly distribute on the surface of the pyrolyzed bamboo slices, as shown in figure 3 (a). Most of the SiC nanowires are straight and smooth, as shown in figure 3 (b). Some of the SiC nanowires were curved and twisted. The diameter and the length of the typically SiC nanowires is ranging from 20~70nm and 10~20μm.

The TEM images, as shown in figure 4, were the morphology of SiC nanowire on surface of the pyrolyzed bamboo slices. Most of the SiC nanowires are straight and smooth, as shown in figure 4 (a). Some of these nanowires were found to have different morphologies, such as curved and twisted nanowires as shown in figure 4 (b). The high density of striations can be observed in the nanowires, as shown in figure 4 (c). The SAED pattern (inset in figure 4 (c)) exhibited a set of diffraction spots along the (111) zone axis. It is indicated that the striations normal to the axes in figure 4 (c) are stacking faults normal to the [111] direction.

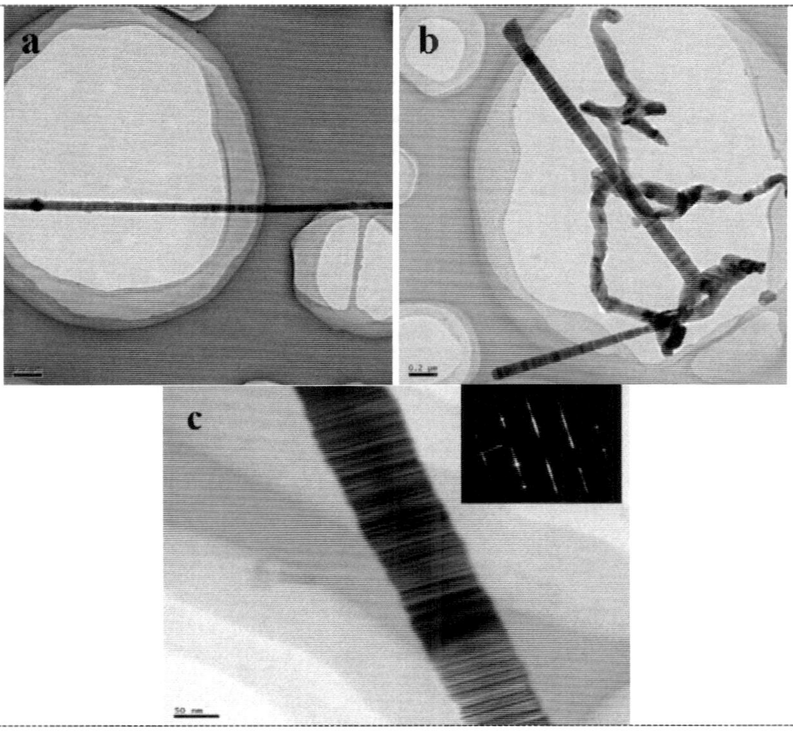

Figure 4. TEM image of the SiC nanowires.

3.3. Growth Mechanism of SiC Nanowires

In our experiment, a mixture powder of Si/SiO_2 was used as the SiO(g) source. SiO(g) was produced from a mixture of Si and SiO_2. Generally, the following reactions are considered in the the SiO-C system [3, 9].

- $Si_{(s)} + SiO2_{(s)} \rightarrow SiO_{(g)}$ (1)
- $SiO_{(g)} + C_{(s)} \rightarrow SiC_{(s)} + CO_{(g)}$ (2)
- $SiO_{(g)} + CO_{(g)} \rightarrow SiC_{(s)} + CO_{2(g)}$ (3)
- $CO_{2(g)} + C_{(s)} \rightarrow CO_{(g)}$ (4)

First, gaseous SiO react with carbon nanoparticle on the surface of the pyrolyzed bamboo slices, the products are SiC nanograins and gaseous CO; and then, reaction (3) is a gaseous phase reaction between SiO and CO, which can occur under a supersaturated condition of CO vapor. Reactions (3) and (4) have a positive feedback characteristic to each other and lead to a supersaturation of CO vapor ultimately. The SiC nanowires can grow under the supersaturation condition of CO. The SiC nanowires start their growth from SiC nanoparticles, which is serves as a SiC nucleus. The growth direction is along the (111) plane for SiC. Some curved nanowires were found as shown in Fig.4 (b). The growth of curved nanowires is due to the growth direction changes in the growth condition [9].

4. Conclusions

SiC nanowires with 20~70nm diameter and 10~20μm length have been synthesized by a reaction between SiO and pyrolyzed bamboo at 1300°C. Most of the SiC nanowires possess straight and smooth morphology. SiC nanowires has a preferential growth axis parallel to the [111] direction.

Acknowledgment

This work has been supported by the Ph. D. Programs Foundation of Ministry of Education of China under Grant No. 20070708001 and the Natural Science Foundation of Shanxi University of Science & Technology under Grant No.ZX07-15.

Reference

[1] Meng G W, Zhang L D, Mo C M, Zhang S Y, Qin Y, Feng S P and Li H J 1998 *J. Mater. Res.* **13** 2533
[2] Shin Y, Wang C, Samuels W D and Exarhos G J 2007 *Mater. Lett.* **61** 2814
[3] Kholmanov I N, Kharlamov A, Barborini E, Lenardi C, Bassi A L, Bottani C E, Ducati C, Maffi S, Kirillova N V and Milani P 2002 *J. Nanosci. Nanotech.* **2** 1
[4] Li Y B, Xie S S, Zou X P, Tang D S, Liu Z Q, Zhou W Y and Wang G 2001 *J. Crys. Grow.* **223** 125
[5] Gao Y, Bando Y, Sato T and Kitami Y 2001 *Jpn. J. Appl. Phys.* **40** 1065
[6] Wu R B, Yang G Y, Pan Y and Chen J J 2007 *J. Mater. Sci.* **42** 3800
[7] Meng G W, Zhang L D, Qin Y, Mo C M and Phillipp F 1999 *NanoStructured Mater.* **12** 1003
[8] Gundiah G, Madhav G V, Govindaraj A, Seikh M M and Rao C N R 2002 J. *Mater. Chem.* **12** 1606
[9] Gao YH, Bando Y, Kurashma K and Sato T 2002 *J. Mater. Sci.* **37** 2023
[10] Cheung T, and Ng D 2007 *J. Am. Ceram. Soc.* **90** 559

Journal of Physics: Conference Series **152** (2009) 012073 doi:10.1088/1742-6596/152/1/012073

V and Ga Co-doping Effect on Optical Absorption Properties of TiO_2 Thin Films

Q.R. DENG[1], Y. GAO[1], X.H. XIA[1], R.S. CHEN[1], L. WAN,[1] G. SHAO[2]

[1] The Faculty of Physics and Electronic Technology, Hubei University, Wuhan,. China

[2] Centre for Materials Research and Innovation, University of Bolton, Bolton BL3 5AB, U. K.

E-mail: gaoyun@hubu.edu.cn

Abstract. In this work, V and Ga co-doped TiO_2 thin films were obtained by pulsed laser deposition (PLD) using sintered ceramic targets, followed by short term annealing at 1000℃. The structures of the thin films were characterized by X-ray diffraction (XRD) and the optical absorption properties of the films were investigated using UV–VIS spectrophotometer. A narrow energy band gap of about 1.5 eV was obtained in a typical co-doped film with 1 at.% V and 0.5 at.% Ga as dopants. The analysis on the chemical states of Ti and Ga elements by X-ray photoelectron spectroscopy (XPS) revealed that the Ga doping plays an important role in the band gap narrowing.

1. Introduction

Among various oxide semiconductor materials, titanium dioxide (TiO_2) is most widely studied because of its promising application in photocatalysis [1], solar cells [2], and so on. However, as it has a energy band gap of 3.0 eV (rutile TiO_2) and 3.2 eV (anatase TiO_2), it is activated only under UV light irradiation which accounts for only a small fraction (~5%) of the solar irradiance compared to the visible region (~45%) [3]. In recent years, many studies focused on the process of shifting the optical response of active TiO_2 from the UV to the visible spectral range. An efficient approach is doping TiO_2 with metal [4], [5] and nonmetal elements [6], [7]. Anpo et al. modified TiO_2 catalysts by bombarding it with high-energy metal ions, such as V, Cr, Mn, Fe, etc. They found that among them V ion implantation had the highest effectiveness in the red shift of TiO_2 absorption edge [8]. Wu and Chen narrowed the band gap of TiO_2 to 2.8 eV with 8% V-doping [9]. Likewise, simultaneous incorporation of V and Pd in TiO_2 by magnetron sputtering results in the decrease of the band gap of TiO_2 to 2.09 eV [10].

Our previous theoretical calculation showed that the bandgap of TiO_2 could be reduced gradually by increasing the Gallium doping level. However, gallium has very low solubility in the crystal lattice of TiO_2, which leads to considerable difficulty to increase the Ga doping amount. In order to overcome this problem, we deliberately co-doped TiO_2 thin films with both V and Ga elements, expecting to enhance redshift in the band gap of TiO_2.

2. Experimental details

The TiO_2 thin films were prepared by PLD-IV (KrF laser wave length: 248 nm) using ceramic targets. The targets were made by pressing TiO_2 powder with various V and Ga ratios added as V_2O_5 and Ga_2O_3 powder into disks and then baked at 1200℃ for 2 h.

Deposition processes were performed on quartz substrates at room temperature for 60 min. The substrate was placed at a distance of 5 cm from the target. The laser output power was 200 mJ per pulse at a frequency of 5 Hz. The base pressure of the vacuum chamber was kept at lower than 3×10^{-4} Pa in all deposition process. After deposition, the samples were thermal treated at 1000 °C in air for 10 min.

X-ray diffraction spectra of the samples were measured by Bruker D8 Advance X-ray Diffractometer (Cu Kα, $2\theta = 20\sim60°$). Transmission spectra of the TiO_2 films were obtained by a Lambda UV-VIS spectrometer (UV-2500, Shimadzu, Japan) in the range of 326–1500 nm to study the optical absorption properties. X-ray photoelectron spectroscopy (XPS) measurements were carried out using a Kratos XSAM800 XPS system with Cu Kα source and a charge neutralizer. All the binding energies were referred to the C 1s peak at 284.6 eV of the surface adventitious carbon.

3. Results and Discussion

3.1. XRD analysis

Figure1 shows the XRD spectra of the films without and with V, Ga doping, the peaks were identified by comparing them with JCPDS cards. Only one characteristic XRD peak at $2\theta = 27.3°$ corresponding to (110) plane of rutile phase TiO_2 was detected. With the addition of V 1% and Ga 0.5%, the intensity of the (110) diffraction peak became sharper than that of the pure TiO_2 film, suggesting that the film with V and Ga co-doping experienced better crystallization. Other co-doped samples were also rutile TiO_2 and no sign of metal or oxides of Vanadium or Gallium was detected.

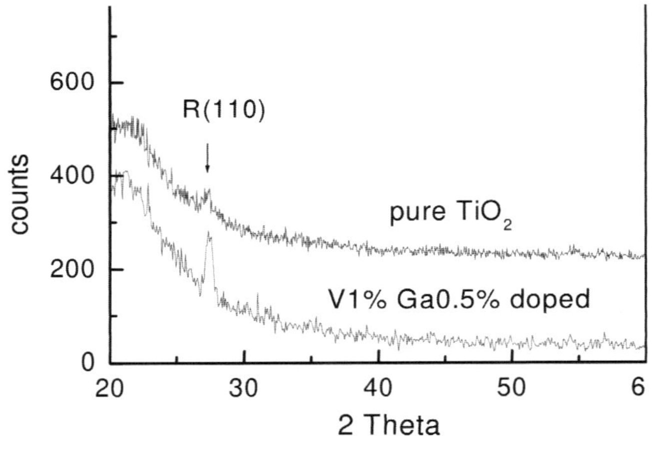

Figure 1. XRD spectra of the pure and V1%-Ga0.5% co-doped TiO_2 films

3.2. Absorption properties

Transmission spectra of the TiO_2 films were measured by UV-VIS spectrophotometer to study the optical absorption properties. The absorption coefficient α of the films was calculated as follows:

$$\alpha(\lambda)d = \ln[1/T(\lambda)]$$

where d is the thickness of the thin film, and T(λ) is the transmittance directly measured by UV-VIS spectrometer.

Figure 2 presents the optical absorption spectra of the pure and doped TiO_2 films. The linear dependence of α^2 on the photon energy over an adequate energy range indicates that all these films have a direct band property, which is consistent with other reports about rutile TiO_2 [11]. The band gap of the co-doped rutile TiO_2 film can be estimated by extrapolating the linear part of the α^2 versus photon energy curve to zero area as shown in figure 2. The band gap of pure rutile TiO_2 and the films with V 1% or Ga 0.5% solely doped is approximately 3.5 eV, while the V 1% and Ga 0.5% co-doped film has a forbidden band energy as small as 1.5eV as shown in figure 2a. The results reveal that V and Ga co-doping rather than sole V or Ga doping can alter the band gap of TiO_2 significantly. Figure 2b compares the α^2-$h\nu$ curves of the co-doped V1%-Ga0.5% film with other diverse V and Ga contents (V 0.5% -Ga 0.5%, V 0.5% -Ga 1%, and V 1% - Ga1%). The band gap of the co-doped films with other V and Ga contents appears at around 3.5 eV. These results indicate that the content and proportion of V and Ga play a dominant role for narrowing the forbidden band gap of TiO_2.

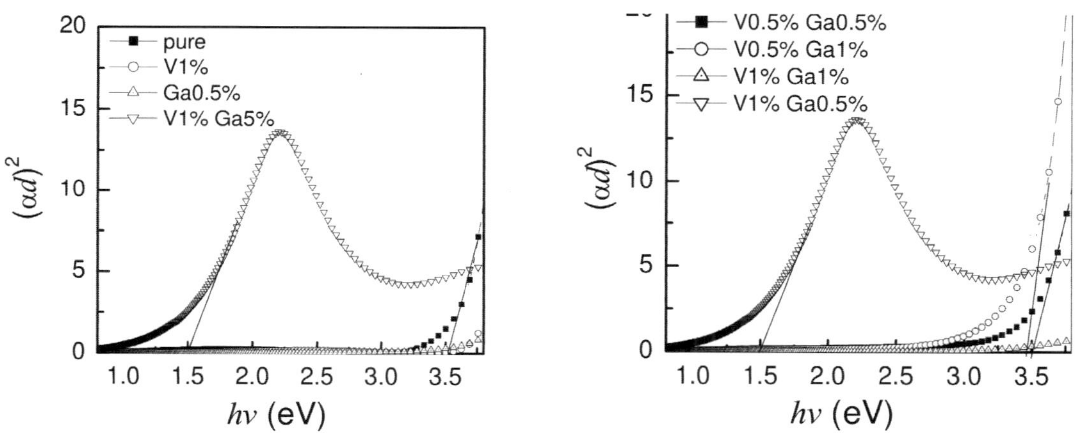

(a) (b)

Figure 2. Dependence of α^2 on photon energy for pure, V, Ga co-doped and solely doped TiO_2 thin films

3.3. XPS analysis

X-ray photoelectron spectroscopy was used to analyze the states of Ti and Ga on the surface of the as-obtained films. The peaks were determined by Gaussian fitting after a Shirley background subtraction. There are two sets of Ti core levers observed in the pure TiO_2 film as shown in figure 3a. One set locates at 458.9 and 464.6 eV, related to the Ti $2p_{3/2}$ and Ti $2p_{1/2}$ binding energies of Ti^{4+} in TiO_2 [12]. The other set of core binding energies at 457.0 and 462.6 eV are assigned to Ti^{3+} as reported in previous studies [13], showing that there are O vacancy in the film due to the oxygen deficient environment during deposition. The simultaneous Ti^{4+} and Ti^{3+} chemical states were observed in the V1%-Ga0.5% co-doped film in figure 3b. Figure 3c shows the corresponding Ga $2p_{3/2}$ XPS spectrum. Two Ga $2p_{3/2}$ peaks located at 1116.5 and 1118.8 eV corresponding to Ga^{3+} and Ga^{4+}, respectively. There are about 80% Ga ions taking Ga^{4+} state, which means that most of the Ga ions were doped at Ti^{4+} sites rather than separated out to form Ga_2O_3. This is consistent with XRD result that no sign of Ga metal or oxide phase was observed due to the little content.

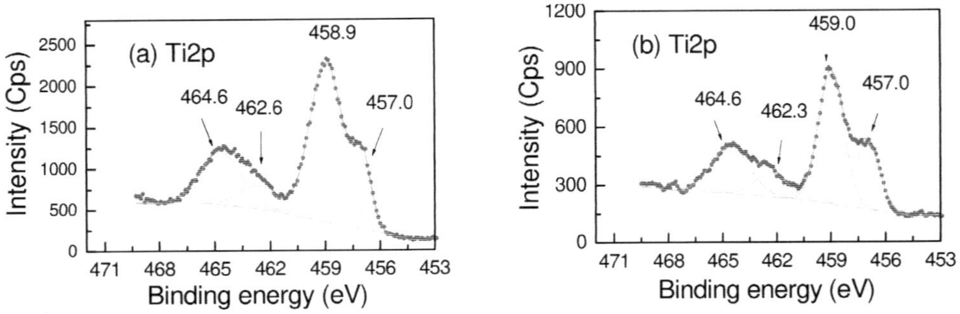

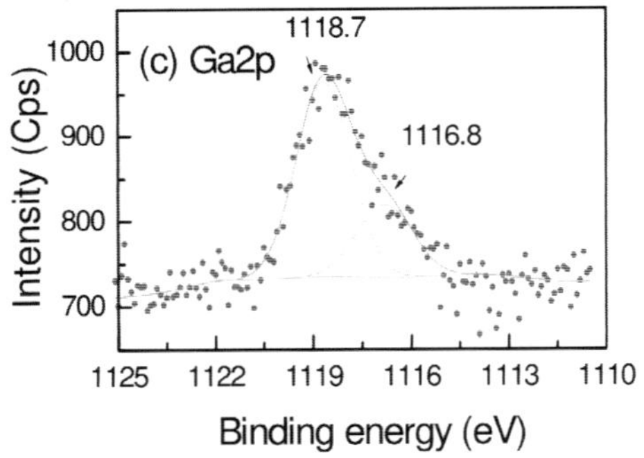

Figure 3. XPS spectra of (a) Ti2p in pure TiO_2 film, (b) Ti2p in V1%-Ga0.5% co-doped film, (c) Ga $2p_{3/2}$ in V1%-Ga0.5% co-doped film.

According to the previous report, V doping has effect on the TiO_2 band narrowing, but 1% V can redshift the band edge for only 0.2 eV [9]. Combined with the above discussion, it is reasonable to attribute the large band narrowing effect to Ga ions doped at Ti^{4+} sites instead of oxygen vacancy observed in both pure and doped films or films doped solely by V.

4. Conclusion

Rutile TiO_2 thin films with and without V and Ga doping were obtained by the PLD method. The addition of V and Ga improved the crystallization of the TiO_2 film. V and Ga co-doping rather than V or Ga sole doping can induce significant reduction of the band gap of TiO_2, from 3.5eV to 1.5eV. XPS revealed Ti^{4+} and Ti^{3+} chemical states in both the pure and V-Ga co-doped films, showing that there are oxygen vacancies in the films due to the oxygen deficient environment during deposition. The dominance of the Ga^{4+} states indicates that Ga ions are doped at Ti^{4+} sites rather than separated out to form Ga_2O_3, which accounts for the band narrowing effect.

Acknowledgments

This work is partially supported by the by the NSFC (Ref. No:50702019) and Educational Commission of Hubei Province of China (Ref. No: D20071006).

References

[1] Yamaki T, Sumita T, Yamamoto S and Miyashita A 2002 *J. Cryst.Growth* **237–9** 574–9
[2] Tripathi M, Pandey K and Kumar S D 2007 *Solar Energy Materials and Solar cell* **91** 1663-8
[3] Prabakara K, Takahashia T, Nezukaa T, Takahashia K, Nakashimab T, Kubotac Y and Fujishima A 2008 *Renewable Energy* **33** 277-81
[4] Lu A, Li Y, Lv M, Wang C, Yang L, Liu J, Wang Y, Wong K and Wong P K 2007 *Solar Energy Materials and Solar Cell* **91** 1849-55
[5] Yang X, Cao C, Hohn K, Erickson L, Maghirang R, Hamal D and Klabund K 2007 *Journal of Catalysis* **252** 296-302
[6] Somekawa S, Kusumoto Y, Ikeda M, Ahmmad B and Horie Y 2008 *Catal. Commun.* **9** 437-40
[7] Umebayashi T, Yamaki T, Ito H and Asai K. 2002 *Appl. Phys. Lett.* **81** 454

[8] Anpo M, Dohshi S, Kitano M and Hu Y 2005 *Annu. Rev. Mater. Res.* **35** 1
[9] Domaradzki J 2006 *Thin Solid Films* **497** 243
[10] Zhang H and Ji T 2008 *Acta. Phys.Chim. Sin.* **24** 607-11
[11] Mane R S, Joo O S, Min S K, Lokhande C D, Han S H, 2006 *Appl. Sur. Sci.* **253** 581–5
[12] Wang J, Ni X 2008 *J. Appl. Poly. Sci.* **108** 3552-8
[13] Pouilleau J, Devilliers D, Groult H and Marcus P 1997 *J. Mater. Sci.* **32** 5645

Journal of Physics: Conference Series **152** (2009) 012074 doi:10.1088/1742-6596/152/1/012074

CIS(CIGS) thin films prepared for solar cells by one-step electrodeposition in alcohol solution

Fei LONG[1,2], Weiming WANG[1], Jingjing DU[1], Zhengguang ZOU[2]

[1] State Key Laboratory of Advanced Technology for Materials Synthesis and Progressing, Wuhan University of Technology, Hubei, Wuhan 430070

[2] Key Laboratory of Nonferrous Materials and New Processing Technology of Ministry of Education, Guilin University of Technology, Guangxi, Guilin 541004

E-mail:long.drf@gmail.com

Abstract: The precursors of CIS and CIGS thin films were prepared by one-step electrodeposition in alcohol solution and then annealed in Ar atmosphere at 550℃ for 30min. The influences of deposition potentials and salt concentrations on the morphology and composition of the thin films were studied in details. The films were characterized by X-ray diffraction(XRD), scanning electron microscopy(SEM), and energy-dispersive X-ray spectroscopy(EDS). The results show that annealed films have pure CIS or CIGS phases with good crystallization and the morphology of the films are very uniform and dense. The high quality precursor films were obtained using optimized deposition parameters. Moreover the electrodeposition mechanism of CIS thin films was briefly discussed.

1. Introduction

Thin layers of $CuInSe_2$ (CIS) and $CuIn_{1-x}Ga_xSe_2$ (CIGS) are considered as the most promising material for low cost and high-efficiency solar cells, because of their high absorptivity ($>10^5 cm^{-1}$) and stability against photo-degradation. Solar cells based on $CuInSe_2$ (CIS) and Cu(In,Ga)Se_2 (CIGS) have reached conversion efficiencies as high as 18.8% and 19.5% [1]. Several methods for preparation CIS(CIGS) absorber films have been reported, such as co-evaporation [2], sputtering and selenization [3], and electrodeposition [4]. Until now, co-evaporation is the most successful technique for the preparation of CIS(CIGS), which achieved the highest efficiency CIGS-based solar cells, but it seems difficult to scale up nowaday. In all these methods, Electrodeposition technique is potentially suitable to obtaining good-quality, large-area CIGS precursor films. CIS and CIGS films have been reported by several groups with the so-called one-step electrodeposited route [5], and the NREL group has also demonstrated an efficiency of 15.4% used a hybrid approach consisting of the "adjustment" of the electrodeposited precursor by evaporating 300nm Ga and 700nm In [6]. However, in aqueous deposition baths, In and Ga deposition efficiencies are limited by H^+ reduction, which causes composition inhomogeneity and the pinholes in the film and hence limits the cell efficiencies. For this reason, the present study examined the feasibility of using the alcohol as solvent for the deposition of Cu-In-Se and Cu-In-Ga-Se precursor films.

Journal of Physics: Conference Series **152** (2009) 012074 doi:10.1088/1742-6596/152/1/012074

2. Experimental details

Electrodeposition of CIS and CIGS precursors was carried out potentiostatically in a three electrode cell in which the substrate was Mo foil, the counter was Pt gauze, and reference electrode was saturated calomel electrode(SCE). A Zahner IM6ex Electrochemical workstation was used for the preparation. The electrodeposition was performed from a bath containing 0.005M $CuCl_2$, 0.05M $InCl_3$(0.03M $GaCl_3$), 0.025M H_2SeO_3, and 0.1M LiCl dissolved in alcohol. In the experiments, the applied potential was -0.6V for CIS deposition and -1.6V for the CIGS deposition. The pH of the chemical bath was adjusted between 1.9 and 2.2 by adding drops of concentration HCl. The obtained thin films were rinsed with bi-distilled water and dried under a nitrogen flux, then thermal annealed at 550℃ for 30min in Ar atmosphere.

The X-ray diffraction(XRD) patterns of the precursors and annealed thin films were characterized by glancing angle X-ray diffraction (GAXRD) using X'Pert PRO X-Ray MPD employing CuK_α radiation(λ=1.54 Å). The surface morphology was investigated using a JEOL JSM-6380LV SEM. The composition of the film was analyzed with an OXFORD-IE350 energy dispersive spectrometer (EDS) attached to SEM.

3. Electrodeposition studies in the CIS(CIGS) system

The electrodeposition of multinary semiconductors containing more than two constituent elements was expected to be more complex owing to a large difference in the deposition potentials of the elements. During CIS and CIGS film formation, the general reaction occurring at the working electrode surfaces was as follows:

$$Cu^{2+} + In^{3+} + 2H_2SeO_3 + 13e^- + 8H^+ \rightarrow CuInSe_2 + 6H_2O \quad (1)$$

$$Cu^{2+} + (1-X)\ In^{3+} + XGa^{3+} + 2H_2SeO_3 + 13\ e^- + 8H \rightarrow Cu\ (In,Ga)\ Se_2 + 6H_2O \quad (2)$$

This is an ideal situation but in practice it is a combination of electrochemical and chemical reactions that occur at certain pH values. The electrochemical aspects are becoming even more complex, with the possibility of forming either the elements in their elemental form, or as binary compounds(Cu_xSe, In_2Se_3, Ga_2Se_3) in addition to the desired ternary CIS phase and CIGS. It can be seen in XRD patterns of precursor (Fig.1), several binary compounds have been founded. These binary compounds formed compliance Kroger's mechanism [7] .

For long time the formation of the quaternary phase with gallium by one step electrodeposition was a difficult problem, because the Ga^{3+} was far more difficult deposition from others. In our study, this serious bottleneck seems to have been overcomed by electrodeposition in alcohol solution, the alcohol solution provide a wide electrochemical window, so CIGS precursor films could be deposition at -1.6V. alcohol also act a surface modifiers which affect markedly the quality of the films or electrochemical kinetics.

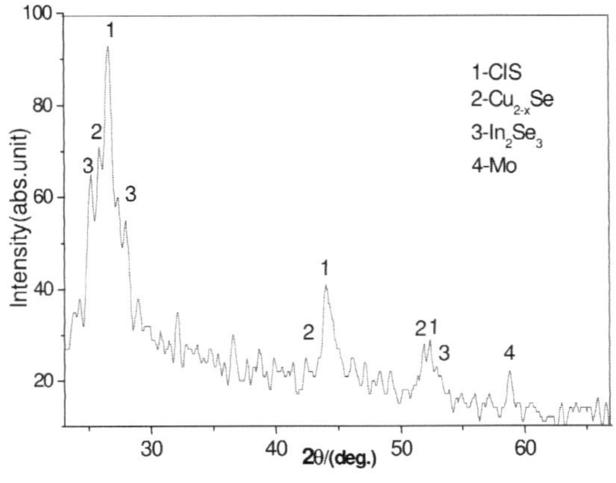

Figure 1. XRD patterns of as-deposited CIS thin film

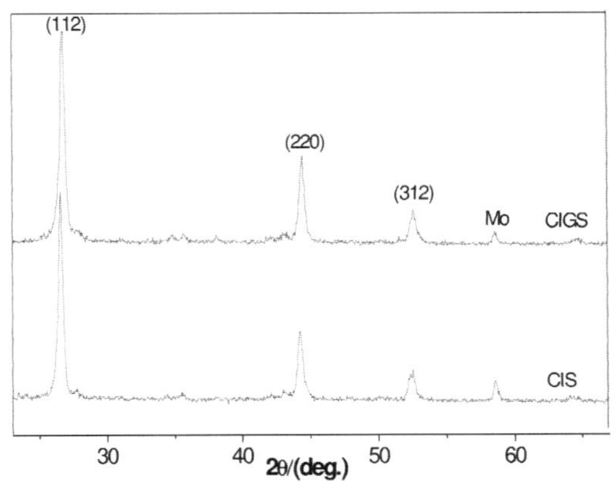

Figure 2. XRD patterns of CIS and CIGS films after annealing

4. Results and discussion

Fig.2 shows the XRD patterns of CIS and CIGS films after annealing at 550℃ for 30min in Ar atmosphere, and they were indexed with the tetragonal chalcopyrite structure, No detectable secondary phase could be seen. The strongest diffraction peak around 2θ = 26.74° corresponds to diffraction from the (112) planes while the other peaks at 2θ = 44.35°, and 52.74° are due to diffraction from the (220) plus (204), and the (312) planes, respectively. The small positive shift in 2θ value of (112) peak for the CIGS sample indicated that gallium takes partly the place of indium in the tetragonal CIS phase and then results in the tetragonal CIGS phase.

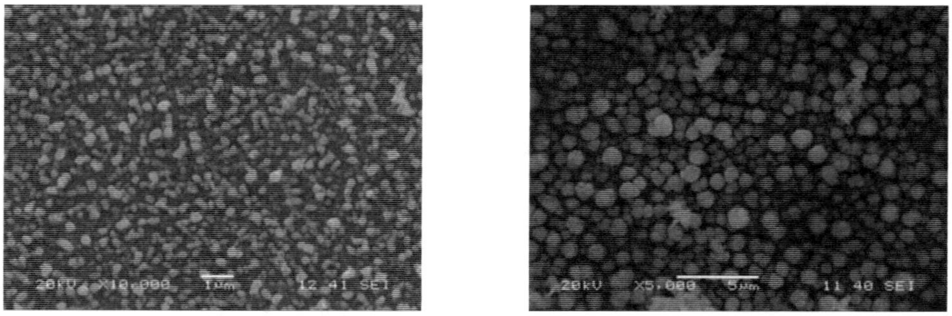

Fig.3. SEM images of CIS and CIGS film before annealing (a.CIS, b.CIGS)

Fig.4. SEM images of CIS and CIGS film after annealing (a.CIS, b.CIGS)

The as-deposited CIS film consists of grains with the order of 500nm, The grain size is generally improved with a second processing step, where the material is annealed at higher temperature. And the thin film was dense and compact and shows a smooth surface morphology(Fig.4). The SEM images of CIGS thin films (Fig.4) showed large grains (~2 μm). From these images we can infer that the surface roughness appear to decrease as grain size get larger with higher temperature. From the above analysis it is concluded that the annealing process is important to fabricate CIS(CIGS) films.

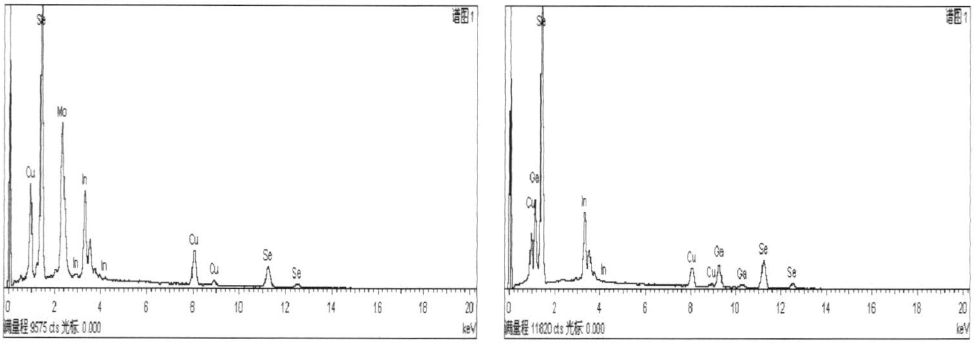

Fig.5. EDS patterns of CIS and CIGS films after annealing

Fig.5 shows EDS patterns of CIS(CIGS) thin films after annealing. The EDS results show that the films composition were only slightly Cu-poor with respect to the anticipated composition of CIS(CIGS). The [Ga] to ([Ga]+[In]) ratio for the CIGS film is 0.25, This ratio span the range necessary to achieve CIGS band gaps which are optimized for the terrestrial solar spectrum.

5. Conclusions

One-step electrodeposition of CIS(CIGS) thin films in alcohol solution was studied in this article, satisfactory control of film composition was achieved by the choice of deposition potential and simple combinatorial variation in bath concentrations. An annealing process in Ar atmosphere was developed for CIS(CIGS) growth. The results indicate CIS(CIGS) thin films little deviate from the ideal stoichiometric one and single chalcopyrite structure. The SEM pictures show CIS(CIGS) thin films have uniform grain size and compact. Further work is in progress to achieve the incorporation of higher levels of gallium and better definition of the electrochemical and annealing process so that the process can be developed to produce absorber layers for efficient cells.

References

[1] Contreras, Miguel A. Romero, Manuel J. Noufi, R. 2005 *Thin Solid Films*. 480 p.2.
[2] Moharram A H, Hafiz M M, Salem A. 2001 *Applied Surface Science*. 172 p.61.
[3] Muler J, Nowoczin J, Schmitt H. 2006 *Thin Solid Films*. 496 p.364.
[4] Calixto M E, Sebastian P J, Bhattacharya R N, Noufi R. 1999 *Solar Energy Materials and Solar Cells*. 59 p.75.
[5] Bhattacharya R N. Solution growth and electrodeposited CuInSe2 thin films. 1983 *J Electrochem Soc.* 130 p.2040.
[6] R.N. Bhattacharya, W. Batchelor, K. Ramanathan, M.A. Contreras and T. Moriarty, 2000 *Solar Energy Materials and Solar Cells*. **63** p. 367.
[7] M.P.R. Panicker, M. Knaster and F.A.Kroger. 1978 *J.Electrochem.soc.*125 p.566.

Microstructures and dielectric properties of $BaZr_{0.2}Ti_{0.8}O_3$ ceramics

Chunlin FU[1,2], Fusheng PAN[1], Wei CAI[1,2], Xiaoling DENG[2] and Xiaowei LIU[2]

[1] School of Materials Science and Engineering, Chongqing University, Chongqing 400030, China

[2] School of Metallurgical and Materials Engineering, Chongqing University of Science and Technology, Chongqing 400050, China

E-mail address: chlfu@126.com

Abstract. The crystalline microstructure, surface morphology and dielectric properties of $BaZr_{0.2}Ti_{0.8}O_3$ (BZT) ceramics were investigated. The *c*-axis lattice parameter (c=0.4107nm) of the BZT ceramics is larger than the *a*-axis one (a=0.4101nm). It was found that the crystal structure of the ceramics is to be tetragonal phase, and close to cubic phase. The temperature dependence of the relative permittivity showed that $BaZr_{0.2}Ti_{0.8}O_3$ ceramics is to be relaxor ferroelectrics with the diffuseness constant equal to 1.93. As the applied maximum voltage increased from 500V to 1000V and 1500V, the remanent polarization increased from 1.69μC/cm^2 to 3.10μC/cm^2 and 3.60μC/cm^2, furthermore the coercive field increased from 0.88kV/cm to 1.19kV/cm and 1.35kV/cm.

1. Introduction

Barium titanate ($BaTiO_3$) ceramics have been used extensively as capacitor dielectrics for decades. Pure $BaTiO_3$ undergoes a paraelectric to ferroelectric phase transition at 130℃, which is accompanied by a sharp peak in the permittivity curve. Isovalent dopants, such as strontium and zirconium, are often employed to shift the Curie point for particular applications. The $Ba_xSr_{1-x}TiO_3$ (BST) has high dielectric constant, low leakage current and low dielectric dispersion against frequency [1-3]. However, the leakage current often abruptly turns up in orders of magnitude as the applied field exceeds a critical value of several hundreds kV/cm, and breakdown occurs at a field around 2MV/cm [2,3]. Recently, $BaZr_xTi_{1-x}O_3$ (BZT) has been chosen as an alternative to BST in the fabrication of ceramic capacitors. The Zr^{4+} ion (0.087nm) is chemically more stable than the Ti^{4+} (0.068nm) and has a larger ionic size to expand the lattice. Therefore, the conduction by electron hopping between Ti^{4+} and Ti^{3+} would be depressed by the substitution of Ti with Zr [4].

In the past years, some preliminary results on the microstructure and properties of BZT ceramics were reported [5-10]. However, it is necessary to improve the dielectric properties of BZT ceramics. In this paper, the crystalline microstructure, surface morphology and dielectric properties of $BaZr_{0.2}Ti_{0.8}O_3$ ceramics were investigated.

2. Experiments

2.1. Ceramics preparation

The starting chemicals were 99.5% $BaCO_3$, $ZrCO_3$, and TiO_2 powders purified from merchandises. A series of samples of $BaZr_{0.2}Ti_{0.8}O_3$ ceramics were prepared by the conventional mixed-oxide method. The raw material was weighed out in stoichiometric proportions, ball-milled in water, dried and then calcined at 1200℃ for 2h. The obtained powders were pressed at 20MPa into disks with a diameter of 10.0mm and thickness of 1.5mm prior to sintering at 1350℃ for 4h.

2.2. Ceramics Characterization

After sintering, x-ray diffraction (XRD) with Cu Kα radiation (λ=0.1541nm) was performed to examine the phase constitution of the specimens at room temperature. Scanning electron microscopy (SEM) was used to investigate the morphology of $BaZr_{0.2}Ti_{0.8}O_3$ ceramics.

In order to measure the dielectric properties, silver paste was painted on the polished samples as the electrodes and fired at 830℃ for 15min. The dielectric properties of the ceramics were determined with a HIOKI 3532-50 LCR at 1V/mm during heating the samples from -55℃ to 200℃ with 0.5℃ /min. The dielectric constant was calculated from the capacitance using the following equation:

$$\varepsilon = Cd/\varepsilon_0 A \tag{1}$$

where C is the capacitance (F), ε_0 the free space dielectric constant value (8.85×10^{-12} F/m), A the capacitor area (m^2) and d the thickness (m) of the ceramics. The polarization-electric voltage (P-V) hysteresis characteristics were performed out using a Radiant Precision LC material analyzer.

3. Results and discussion

3.1. Crystalline structure

The x-ray diffraction pattern of $BaZr_{0.2}Ti_{0.8}O_3$ ceramics is shown in Fig. 1. It is evident that the ceramics is of perovskite crystal structure. No other phases are observed in the spectra. The lattice constants of BZT ceramics were calculated by the following equation:

$$d_{\mathrm{hkl}} = [(\frac{\mathrm{h}}{a})^2 + (\frac{\mathrm{k}}{a})^2 + (\frac{\mathrm{l}}{c})^2]^{-\frac{1}{2}} \tag{2}$$

where d_{hkl} is interplanar distance, h, k and l are indices of crystallographic. The c-axis lattice parameter (c=0.4107nm) of the BZT ceramics is more than the a-axis one (a=0.4101nm). It is calculated that the ratio of c/a is 1.001. It is suggested that the crystal structure of the ceramics belongs to tetragonal phase and is close to cubic phase.

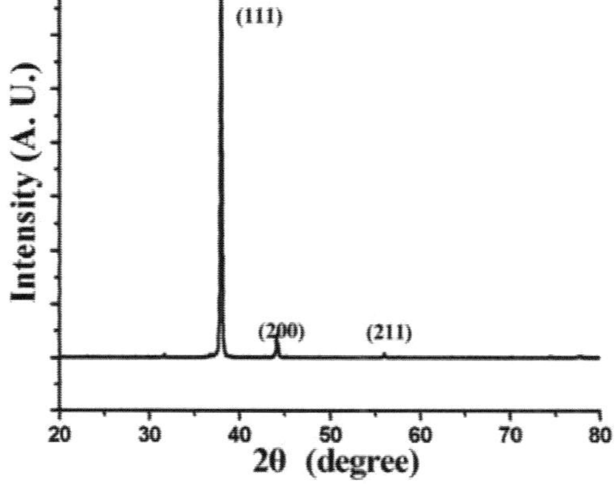

Figure 1. XRD pattern of $BaZr_{0.2}Ti_{0.8}O_3$ ceramics.

3.2. Surface morphology

Fig. 2 shows the typical SEM micrographs of $BaZr_{0.2}Ti_{0.8}O_3$ ceramics. It is seen that the ceramics were very dense and crack-free. The average grain size of the ceramics is ~45μm.

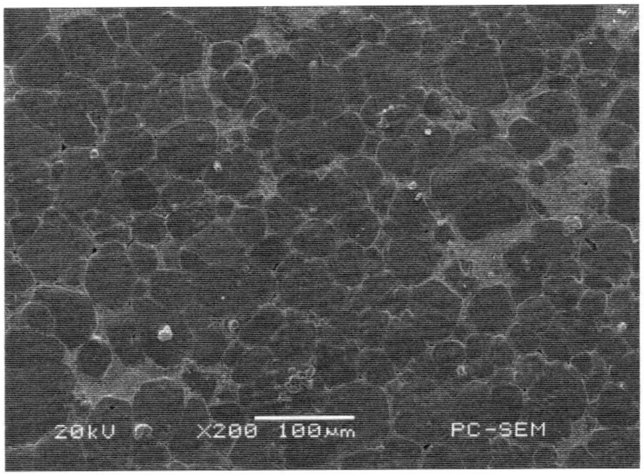

Figure 2. Scanning electron micrograph of $BaZr_{0.2}Ti_{0.8}O_3$ ceramics.

3.3. Temperature dependence

The temperature dependence of the relative permittivity of $BaZr_{0.2}Ti_{0.8}O_3$ ceramics is shown in Fig. 3. It was found that there is a maximum value of dielectric constant at 50℃. Comparing with BST ceramics [11], the dielectric constant of BZT ceramics exhibited broad peaks. The broadness indicates the diffuse phase transition from ferroelectric to paraelectric phase. That is to say, there is diffuse transition behavior in $BaZr_{0.2}Ti_{0.8}O_3$ ceramics.

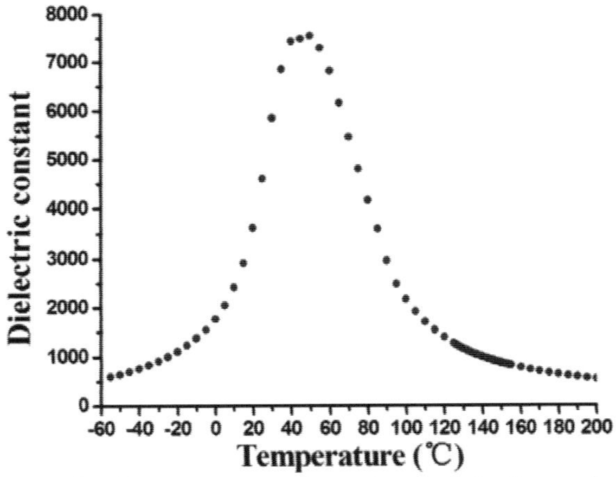

Figure 3. Temperature dependence of the relative permittivity of $BaZr_{0.2}Ti_{0.8}O_3$ ceramics at 1kHz.

For a normal ferroelectric, above T_C the dielectric behavior follows Curie-Weiss law:

$$1/\varepsilon = (T - T_0)/C \tag{3}$$

where T_0 is the Curie-Weiss temperature and C the Curie-Weiss constant. The change of inverse dielectric constant with respect to temperature is shown in Fig. 4. It can be seen that the dielectric behavior doesn't completely follow Curie-Weiss law at temperatures above the critical point (T_C).

According to Curie-Weiss law, the Curie-Weiss temperature can be obtained from linear extrapolation of inverse dielectric constants in the high-temperature region.

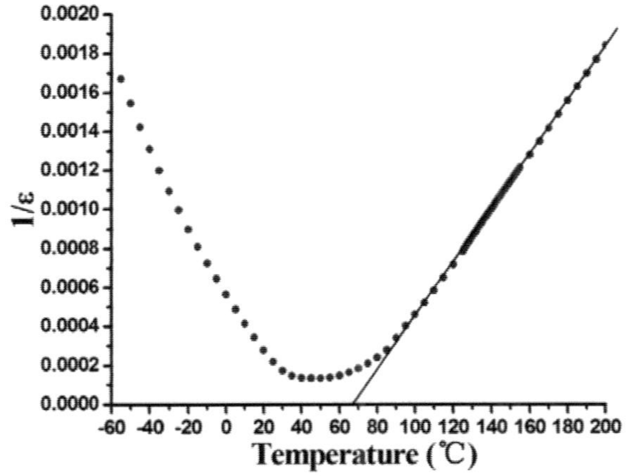

Figure 4. Temperature dependence of the inverse permittivity for $BaZr_{0.2}Ti_{0.8}O_3$ ceramics at 1kHz.

A modified Curie-Weiss law was proposed to describe the diffuseness of the ferroelectric phase transition as [5,6]:

$$1/\varepsilon - 1/\varepsilon_m = (T - T_m)^{\gamma}/C_1 \quad (4)$$

where ε_m and T_m represent dielectric constant maximum and the corresponding temperature, moreover γ and C_1 are assumed to be constant. The parameter γ gives information on the character of the phase transition: for γ=1, a normal Curie-Weiss law is followed, whereas γ=2 describe a complete diffuse phase transition.

The T_m and $1/\varepsilon_m$ of $BaZr_{0.2}Ti_{0.8}O_3$ ceramics are 50℃ and 1.325×10^{-4}. The plot of $\ln(1/\varepsilon-1/\varepsilon_m)$ as a function of $\ln(T-T_m)$ is shown in Fig. 5. According to Eq. (4), the slope gives the diffuseness constant (γ). From Fig. 5 a near linear relationship is observed and the γ value equals to 1.93.

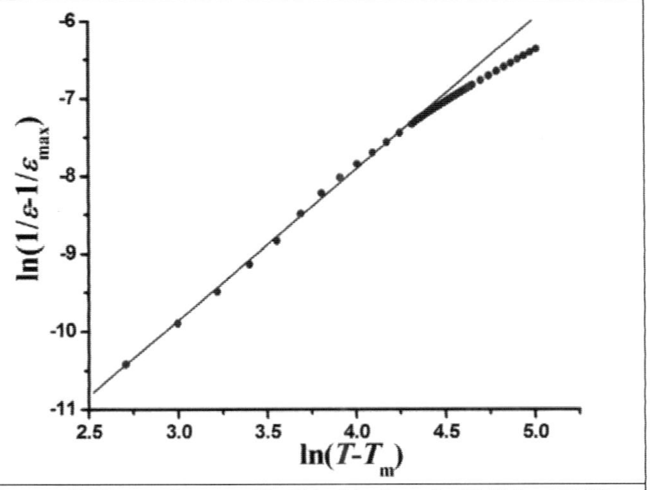

Figure 5. Dependence of $\ln(1/\varepsilon-1/\varepsilon_m)$ on $\ln(T-T_m)$ at 1kHz.

3.4. Hysteresis loop

Fig. 6 shows the polarization hysteresis curves of the same ceramics at a frequency of 1 kHz. It is found that well-behaved hysteresis loops can be observed. This suggests that at room temperature the crystal structures are in the ferroelectric phase, which is consistent with the XRD patterns and the temperature dependence of the relative permittivity. Moreover, it is found that the remanent polarization ($2P_r$) and the coercive field ($2E_C$) of $BaZr_{0.2}Ti_{0.8}O_3$ ceramics both increased as the applied maximum voltage increased. As the applied maximum voltage increased from 500V to 1000V and 1500V, the remanent polarization increased from 1.69μC/cm^2 to 3.10μC/cm^2 and 3.60μC/cm^2, furthermore the coercive field increased from 0.88kV/cm to 1.19kV/cm and 1.35kV/cm.

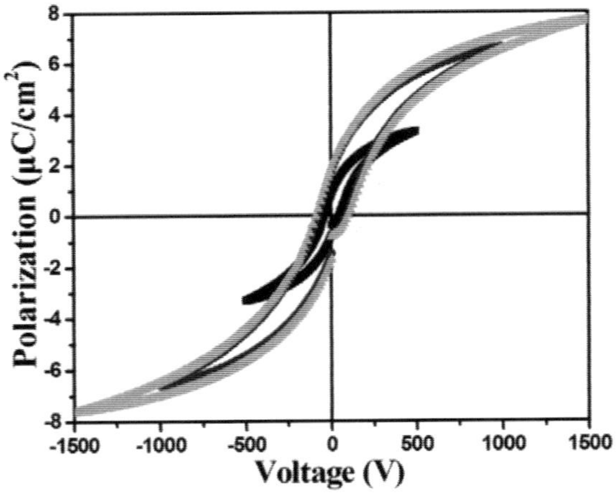

Figure 6. Hysteresis loops for $BaZr_{0.2}Ti_{0.8}O_3$ ceramics at 1kHz.

4. Conclusions

The microstructures and dielectric properties of $BaZr_{0.2}Ti_{0.8}O_3$ ceramics were investigated. The *c*-axis lattice parameter (*c*=0.4107nm) of the BZT ceramics is larger than the *a*-axis one (*a*=0.4101nm). It was found that the crystal structure of the ceramics belongs to tetragonal phase and is close to cubic phase. There was diffuse transition behavior in $BaZr_{0.2}Ti_{0.8}O_3$ ceramics and the diffuseness constant (γ) equals to 1.93. It was found that both remanent polarization ($2P_r$) and the coercive field ($2E_C$) of $BaZr_{0.2}Ti_{0.8}O_3$ ceramics both increased as the applied maximum voltage increased.

Acknowledgements

This work is supported by the National Science Foundation for Post-doctoral Scientists of China (Grant No. 20070410774), the Natural Science Foundation of Chongqing, China (Grant No. CSTC2007BB4212) and the Research Foundation of Education Bureau of Chongqing, China (Grant No. KJ071401 and No. KJ081411).

References

[1] Ioachim A, Toacsan M I, Banciu M G, Nedelcu L. VasiliuF, Alexandru HV, BerbecaruC and Stoica G 2007 Barium strontium titanate-based perovskite materials for microwave applications *Prog. Solid State Chem.* **35** 513-20

[2] Fu C, Pan F, Chen H, Wei C and Yang C 2007 Effect of annealing on leakage current characteristics of Pt/$Ba_{0.6}Sr_{0.4}TiO_3$/Pt thin-film capacitors *J. Mater. Sci.: Mater. Electron,* **18** 453-6

[3] Lee S Y and Chiou B S 2007 Improving dielectric loss and mechanical stress of barium strontium titanate thin films by adding chromium layer *Jpn. J. Appl. Phys.* **46** 6550-3

[4] Choi W S, Jang B S, Lim D G, Yi J and Hong B 2002 Characterization of $Ba(Zr_{0.2}Ti_{0.8})O_3$ thin

films deposited by RF-magnetron sputtering *J. Cryst. Growth* **237-239** 438-42

[5] Tang X G, Chen K H and Chan H L W 2004 Diffuse phase transition and dielectric tunability of $Ba(Zr_yTi_{1-y})O_3$ relaxor ferroelectric ceramics *Acta Mater.* **52** 5177-83

[6] Tang X G, Wang J, Wang X X and Chan H L W 2004 Effects of grain size on the dielectric properties and tunabilities of sol-gel derived $Ba(Zr_{0.2}Ti_{0.8})O_3$ ceramics *Solid State Commun.* **131** 163-8

[7] Chou X, Zhai J and Yao X 2007 Relaxor behavior and dielectric properties of $MgTiO_3$-doped $BaZr_{0.35}Ti_{0.65}O_3$ composite ceramics for tunable applications *J. Am. Ceram. Soc,* **90** 2799-801

[8] Liang R H, Dong X L, Chen Y, Cao F and Wang Y 2007 Dielectric properties and tunability of $Ba(Zr_xTi_{1-x})O_3$ ceramics under high DC electric field *Ceram. Int.* **33** 957-61

[9] Shan D, Qu Y F and Song J J 2007 Dielectric properties and substitution preference of yttrium doped barium zirconium titanate ceramics *Solid State Commun.* **141** 65-8

[10] Cao W, Xiong J and Sun J 2007 Dielectric behavior of Nb-doped $Ba(Zr_xTi_{1-x})O_3$ *Mater. Chem. Phys.* **106** 338-42

[11] Fu C, Yang C, Chen H, Wang Y and Hu L 2005 Microstructure and dielectric properties of $Ba_xSr_{1-x}TiO_3$ ceramics *Mater. Sci. Eng. B* **119** 185-8

Symposia D, E and F from MRS International Materials Research Conference IOP Publishing
Journal of Physics: Conference Series **152** (2009) 012076 doi:10.1088/1742-6596/152/1/012076

Preparation and electrical characterization of carbon nanotube /ZrO_2 composite ceramics

Yan SHAN[1], Lian GAO[2], Xuegang YU[1], Xia LI[1]and Kezheng CHEN[1]

[1] College of Materials Science and Engineering, Qingdao University of Science and Technology, 53 Zhengzhou Road, Qingdao 266042, China

[2] State Key Laboratory of High Performance Ceramics and Superfine Microstructure, Shanghai Institute of Ceramics, Chinese Academy of Sciences, 1295 Dingxi Road, Shanghai 200050, China

E-mail:shanyan@qust.edu.cn

Abstract. Carbon nanotubes (CNTs) in a kind of ideal reinforcements for structural materials, as well as the promising components to prepare multifunctional composites. In this paper, multi-walled carbon nanotube /ZrO_2 functional composite ceramics were investigated and successfully fabricated via the spark plasma sintering technique. The structures of nanocomposite ceramics were characterized by XRD, SEM and EDS, and their mechanical and electrical properties were also studied. It was found that there is an increasing of 18% in ceramics toughness compared with pure ZrO_2 ceramics prepared by the same method. Interestingly, the results of the AC complex impedance spectra at 25~325 °C showed that the electrical resistance of the ceramics increased with the temperature under 150 °C, and will decrease dramatically when it is higher than 150 °C.

1. Introduction

Due to their high mechanical strength (strength and flexibility), excellent electrical and thermal conductivities, carbon nanotubes (CNTs) are not only ideal reinforcements for structural materials, but also promising components to prepare multifunctional composites. In recent years, much attention has been paid to the fabrication of novel CNT composites since the large scale production of multi-walled carbon nanotubes (MWNTs) has been realized. By far, various materials, including CNT-polymer composites [1-3], CNT-metal composites [4-6], and CNT-ceramic composites [7-10] have been successfully prepared. Comparing with polymer and metal matrix composites, CNT-ceramics composites are not extensively investigated, and most of them are SiO_2 [11-12], Al_2O_3 [8, 13-15], SiC [7, 16] ceramics matrix with CNT as a strengthening phase. However, because CNTs are usually not dispersed in the matrix and there is insufficient bonding across the nanotubes/host interface, the improvement in mechanical properties of these composites is not very satisfying. It is still a big challenge to fabricate ceramic matrix-CNT composites with high mechanical properties and other properties.

Zirconia is an important ceramic materials and has a broad range of applications, including catalysts, oxygen sensors, and fuel cells etc [17]. Zirconia and its composites have been studied extensively during the past years, but few covers the CNT/ZrO_2 ceramic composites. Sun et al [18] prepared 3 mol% yttria-doped zirconia matrix composites using multi-wall and single-wall carbon nanotubes as reinforcements, respectively. But their results showed that no prominent improvement on the mechanical properties, which results from the weak bond between 3Y-TZP matrix and CNTs. Here, a new method was introduced to prepare MWNT/ZrO_2 ceramic composites, and their mechanical and electrical properties were investigated.

2. Experimental procedure

2.1. Modifying MWNTs with ZrO_2 nanoparticles

MWNTs were kindly provided by Shenzhen NanoPort Company. The method of MWNTs coated with ZrO_2 nanoparticles was reported previously by our research [19]. The typical experimental procedure employed here was described as follows. First, the MWNTs were treated by concentrated nitric acid at 140 °C for 6 h, rinsed with distilled water and dried at 100 °C for 12 h. Second, 70 mg of acid treated MWNTs were added into vigorously stirred 30 mL of $ZrOCl_2 \cdot 8H_2O$ (0.03 M) aqueous solution. After being ultrasonicated for 0.5 h, the black solution was hydrothermal treated at 150 °C for 12 h. when it was washed and dried, the obtained products were denoted as A.

2.2. Preparation of MWNT/ZrO_2 composite powders

Some of the as-synthesized sample A was added into the mixed solution formed by $ZrOCl_2 \cdot 8H_2O$, 3 mol% $Y(NO_3)_3$, and superfluous NH_4HCO_3. After stirring for 10 mins, the solution was first heated by microwave for 3 min and then dropped into ethanol to form a precipitation. The precipitation was calcined at 600 °C for 2 hours in a nitrogen atmosphere. A series of MWNTs/ZrO_2 composite powders with different content of sample A (0.5 wt%, 1 wt %, 3 wt %, 5 wt %) were prepared.

2.3. Fabrication of MWNT/ZrO_2 composite ceramics

The as prepared MWNT/ZrO_2 composite powders were sintered by spark plasma sintering (Dr. Sinter 2080 SPS apparatus) at 1150 °C for 3-4 min. A pressure of 15 MPa was applied at the beginning of the sintering process and it increased to 35 MPa after the temperature of 800 °C. After sintering, the obtained ceramics were denoted as ZrO_2-x wt% A (x is the content of A in the composite ceramics). For comparison, those acid treated MWNTs that without modification with ZrO_2 were used to prepare MWNTs/ZrO_2 composites followed the same procedure.

2.4. Characterization and properties measurement

The products were characterized by using X-ray powder diffraction (XRD, D/max 2550V) with Cu Ka radiation (1.5406 nm). The density of composite ceramics was measured by the Archimedes' method. Hardness and fracture toughness were measure using a Vickers indenter (Akashi-A). Scanning electron micrographs were obtained using a Field Emission SEM (JSM-6700F, 15 kV). The impedance spectra (40 Hz to 6000 kHz) were measured at HP 4294 Impedance Analyzer.

3. Results and discussions

Figure 1 shows the X-ray diffraction patterns of pure ZrO_2 (a) and MWNT/ZrO_2 composites (b, c, d) with different contents of sample A. The most intense peaks of pure ZrO_2 (Figure 1(a)) correspond to the (101), (112), and (211) reflection of tetragonal ZrO_2 (t-ZrO_2, JCPDS card No. 79-1770), and a minor of monoclinic ZrO_2 (m-ZrO_2) are also detected. While in the curves of MWNT/ZrO_2 composites (b, c, d), they show that both t-ZrO_2 and m-ZrO_2 coexist in the composites and the t-ZrO_2 is dominant, but with the increase of sample A content, the quantity of m-ZrO_2 increases. It also found that it is very hard to elicit the characteristic peaks of the MWNTs, which mainly results from the low content of MWNTs in the composites.

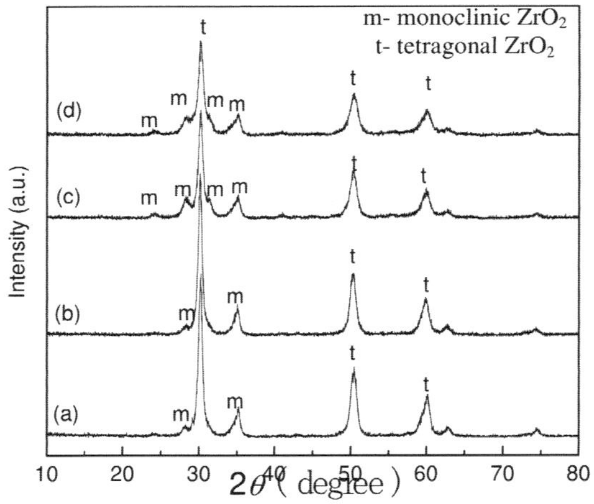

Figure 1. XRD patterns of pure ZrO_2 (a) and MWNT/ZrO_2 composite powders: (b) ZrO_2/1wt% A, (c) ZrO_2/3wt% A, and (d) ZrO_2/5wt% A.

The morphology of the MWNT/ZrO_2 composite powders was examined by TEM, as shown in Figure 2. The typical TEM micrograph in Figure 2(a) clearly shows that the surface of MWNTs was coated by zirconia particles and it is hardly to see the shell of MWNTs. EDS measurement was performed to investigate the composition of these nanoparticles. The EDS spectrum (Figure 2(b)) indicates that these particles anchored on the nanotubes consist of C, O, Zr and Y. The carbon peak come from the carbon nanotubes, and the peaks of Zr, Y and O are attributed to yttria-doped zirconia. So MWNTs coated with 3Y-ZrO_2 nanoparticles were successfully prepared by the method. The phases of ZrO_2 in the composites were mainly t-ZrO_2, and a minor of m-ZrO_2, but the more content of sample A in the composites, the more m-ZrO_2 were.

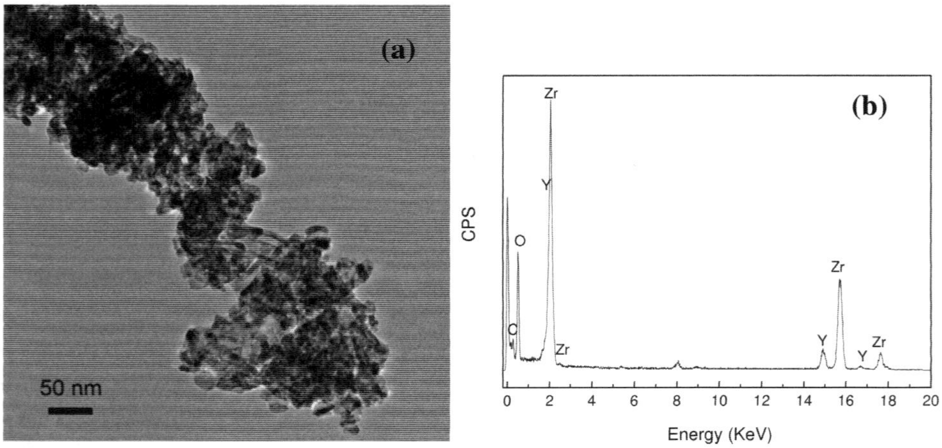

Figure 2. TEM image (a), and EDS spectrum (b) of MWNT/ZrO_2 composite powders.

Table 1 shows the mechanical properties of MWNT/ZrO_2 composites. It can be seen that the hardness increased from 11.36 to 11.67 GPa, and the fracture toughness increased from 4.56 to 5.56 MPa·$m^{1/2}$ when the addition of A is 1 wt%. All the other samples' mechanical properties decreased with the addition of A. When the weight content A is increased to 5%, the hardness and fracture toughness decreased by 50%. The decreases may result from the different phase of ZrO_2 in pure zirconia and the composites. XRD patterns of zirconia and MWNT/ZrO_2 composite ceramics (Figure 3) reveals that ZrO_2 mainly exists in tetragonal phase in zirconia ceramics, while in the composites, m-ZrO_2 is dominant. It is estimated by the following relationship [20]:

$$V_m = \frac{PX_m}{1+(P-1)X_m} \tag{1}$$

where V_m is the volume fraction of m-ZrO_2, and the volume fraction of t-ZrO_2 is given by V_t=1-V_m. P= 1.340, X_m is the integrated intensity ratio defined by:

$$X_m = \frac{I_m(\bar{1}11)+I_m(111)}{I_m(\bar{1}11)+I_m(111)+I_t(101)} \tag{2}$$

where I_x *(l,m,n)* represent the relative intensity diffraction from *(l,m,n)* plane of x phase in XRD spectrum, the crystalline contents of t-ZrO_2 in Figure 3(a), (b) and (c) were 60.7 vol%, 22.23 vol% and 21.76 vol%, respectively. Tetragonal ZrO_2 can improve the mechanical properties due to the transformation of t-ZrO_2 to m-ZrO_2. The content of t-ZrO_2 in composites is far below that of pure zirconia ceramics, so their hardness and toughness decreased greatly. Comparing with the products in which MWNTs were not modified with ZrO_2, the modified composites have better properties, which proved that after being coated with a layer inorganic species, the wettablility of CNTs surface with the matrix has been improved.

Table 1. the mechanical properties of ZrO_2 and MWNT/ZrO_2 composites

Specimens	Relative density (%)	HV(GPa)	K_{IC} (MPa·m$^{1/2}$)
ZrO_2	96	11.36	4.56
ZrO_2/1wt%A	96	11.67	5.56
ZrO_2/1.5wt%A	95.3	9.233	3.64
ZrO_2/3wt%A	95.4	-	-
ZrO_2/5wt%A	94.5	5.47	2.28
ZrO_2/0.5wt% acid treated MWNTs without modifications	94.3	4.17	2.44

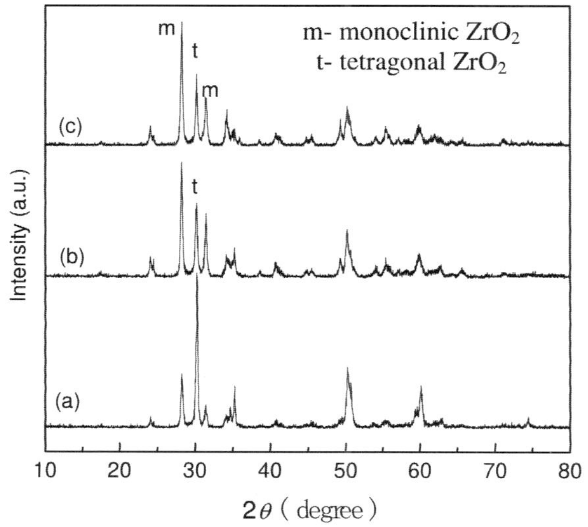

Figure 3. XRD patterns of 3Y-ZrO_2 (a) and MWNT/3Y-ZrO_2 composite ceramics: (b) 3Y-ZrO_2/0.5wt% A, and (c) 3Y-ZrO_2/5wt% A.

Figure 4 is the SEM images of the fracture surface of 3Y-ZrO_2/0.5wt% A (a), and 3Y-ZrO_2/5wt% A (b). It is scarcely to see MWNTs in ZrO_2/0.5 wt% A due to the low content of MWNTs. With more addition of sample A, most MWNTs are located at the zirconia grain boundary. All the samples' fracture surface exhibited intergranular fracture.

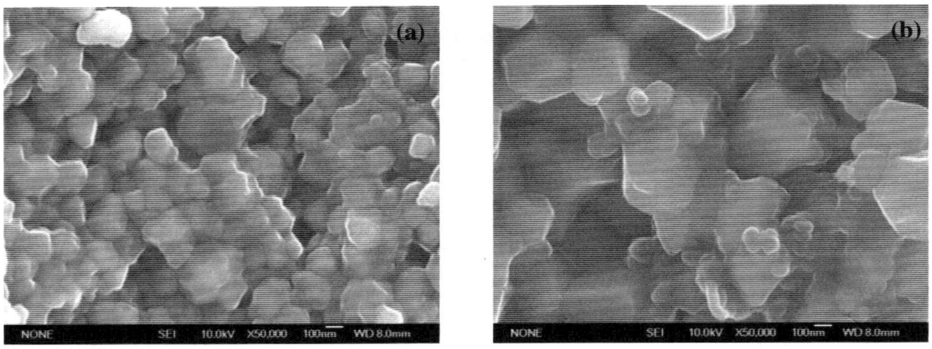

Figure 4. SEM images of the fracture surface: (a) ZrO_2/0.5wt% A, and (b) ZrO_2/5wt% A.

The AC complex impedance spectra (40 Hz to 6000 kHz) of MWNT/ZrO_2 ceramics were investigated. The electrical resistivity of ZrO_2 ceramics at room temperature decreases after addition ZrO_2 modified MWNTs, but it shows no regularity with increasing of the modified MWNTs. Figure 5 (a) is a typical ac complex impedance specta of ZrO_2/1.5 wt% A at different temperature. A single semicircle is seen in the complex impedance, which means the grain boundary resistance dominates the overall impedance. The dependence of electrical resistivity on temperature (as shown in Figure 5 (b)) shows that the electrical resistance of the composite ceramics increases with the temperature under 150 °C, and decrease dramatically when it is higher than 150 °C. The further conductive mechanism is under way.

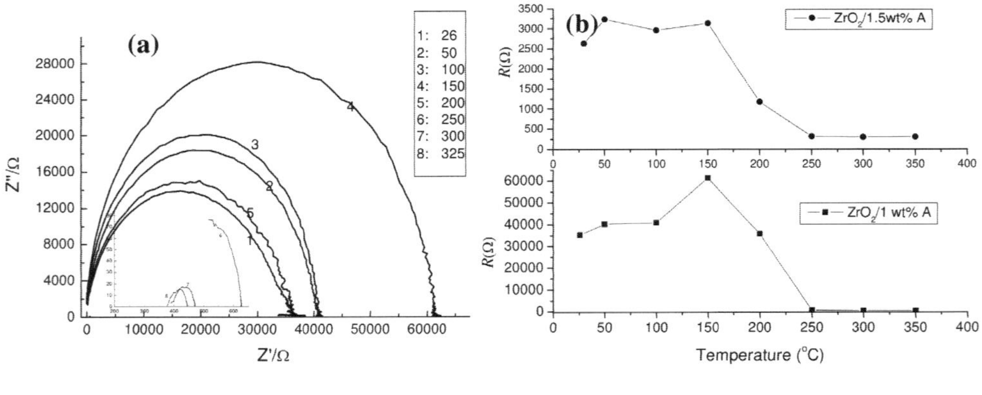

Figure 5. (a) the AC complex impedance spectra of ZrO_2/1.5wt% A at different temperature, (b)the dependence of electrical resistivity in composites on temperature.

4. Conclusions

MWNT/ZrO_2 composite ceramics were successfully fabricated via the spark plasma sintering at 1150 °C. It was found that, no obvious incensement in mechanical properties was achieved after addition of ZrO_2 modified MWNTs, which is because more m-ZrO_2 was formed after sintering comparing with pure ZrO_2 ceramics. The results of the AC complex impedance spectra showed that the electrical resistance of the composite ceramics increased with the temperature under 150 °C, and will decrease dramatically when above 150 °C.

Acknowledgment

This work was supported by the National Natural Science Found of China (No.50672039, No.50772051), and the Research Fund for the Doctoral Program of Qingdao University of Science Technology.

References

[1] Qian D, Dickey E C 2000 *Appl. Phys. Lett.* **76** 2868
[2] Zengin H, Zhou W, Jin J, Czerw R, Smith Jr D W, Echegoyen L, Carroll D L, Foulger S H and Ballato J 2002 *Adv. Mater.* **14**, 1480
[3] Xu M, Zhang T, Gu B, Wu J, Chen Q 2006 *Macromolecules* **39** 3540
[4] Kuzumaki T, Miyazawa K, Ichinose H and Ito K 1998 *J. Mater. Res.* **13** 2445
[5] Barrera E V, Sims J, Callahan D L 1995 *J. Mater. Res.* **10** 366
[6] Xu C L, Wei B Q, Ma R Z, Liang J, Ma X K and Wu D H 1999 *Carbon* **37** 855
[7] Ma R Z, Wu J, Wei B Q, Liang J, Wu D H 1998 *J.Mater. Sci.* **33** 5243
[8] Zhan G D, Kuntz J D, Wan J and Mukherjee A K 2003 *Nat. Mater.* **2** 38
[9] Flahaut E, Peigney A, Laurent Ch, Marlière Ch, Chastel F, and Rousset A 2000 *Acta Mater.* **48** 3803
[10] Huang Q, Gao L, Liu Y Q, Sun J 2005 *J. Mater. Chem.* **15** 1995
[11] Gan L H and Kuo C K 2001 *J. Mater. Chem.* **11** 1722
[12] Wang J, Kou H, Liu X, Pan Y, Guo J 2007 *Ceram.s Inter.* **33** 719
[13] Balani K, Zhang T, Karakoti A, Li W Z, Seal S, Agarwal A 2008 *Acta Materialia* **56** 571
[14] Peigney A, Laurent C, Dumortier O and Rousset A 1998 *J. Eur. Ceram. Soc.* **18** 1995
[15] Sun J, Gao L, Jin X 2005 *Ceram. Inter.* **31** 893
[16] Morisada Y, Miyamoto Y, Takaura Y, Hirota K, Tamari N 2007 *Inter. J. Refra.y Metals and Hard Mater.* **25** 322
[17] Mueller R, Jossen R and Pratsinis S E 2004 *J. Am. Ceram. Soc.* **87** 197
[18] Sun J, Gao L, Iwasa M, Nakayama T, Niihara K 2005 *Ceram. Inter.* **31** 893
[19] Shan Y and Gao L 2005 Nanotechnology **16** 625
[20] Toraya H, Yoshimura M and Somiya S 1984 *J. Am. Ceram. Soc.* **6** C-119

Preparation and Characterization of Lead Zirconate Titanate Ceramic Fibers with Alkoxide-Based Sol–Gel Route

Manfang MAI[1], Cheng LIN[1], Zhaoxian XIONG[1, 2], Hao XUE[1] and Lifu CHEN[1]

[1] Department of Materials Science and Engineering, College of Materials, Xiamen University, Xiamen 361005, P.R. China

E-mail: zxxiong@xmu.edu.cn

Abstract. Lead zirconate titanate (PZT) fibers with diameters from 10μm to 40μm were prepared via a sol-gel route. Several kinds of chemicals were used, including lead acetate trihydrate, zirconium n-butoxide and titanium n-butoxide, in addition to butanol as a solvent. As a water source for hydrolysis reaction, $Pb(CH_3COO)_2 \cdot 3H_2O$ was directly used without further adding of water or catalyst conventionally. Acetylacetonate and acetate were added as stabilization agents to obtain stable sols for fiber drawing. The gel-to-ceramic conversion was characterized with thermal analysis, infrared spectroscopy and X-ray diffraction. A pure perovskite phase was obtained after heat treatment at 650°C for 1h. By using scanning electron microscope (SEM), it was observed that a lower rate of heat treatment resulted in a denser microstructure of the fibers. The relative permittivity and the P-E hysteresis loop of the crystalline PZT fibers were also measured and discussed in the paper.

1. Introduction

Lead zirconate titanate (PZT) materials have been widely used in electric-mechanical applications such as transducers and motors for their excellent ferroelectric and piezoelectric properties. However, PZT bulk material has high density, low hydrostatic piezoelectric coefficients of charge and voltage, along with high stiffness, which limit its performance in certain applications including hydrophones and ultrasonic devices. Therefore, it has been focused on composite fabrication of PZT-based piezoelectric materials for sensing and actuating. PZT fiber is a good candidate for 1-3 piezoelectric composite, which has been prepared by extrusion[1], impregnation[2], viscous suspension-spinning process (VSSP)[3] and sol-gel[4-7]. Meanwhile, sol-gel technique offers the advantages of homogenous chemistry in multicomponent system and easy adjustment of composition. Moreover, it produces fibers with less crack in fine-scale diameter less than 100μm. Methoxythanol (MOE) is the most commonly used solvent, which was introduced in 1985 by Budd et al.[8]. Methoxythanol route provides high-quality precursor, but toxic and unfriendly for use. Recently, carboxylic acid route[9] and diol route[10] have been reported for the preparation of PZT based systems. However, no report has been brought out on directly drawing PZT fibers using butanol as solvent by sol-gel method.

In this study, several kinds of raw chemicals were used, especially butanol as solvent, acetylacetonate and acetate ligands as stabilizers to control the processing of hydrolysis and condensation. Emphasis was paid on the preparation process of the fibers.

2. Experimental Procedure

2.1. Preparation of gel fibers

Starting reagents were used to prepare the precursor sol for fiber drawing with lead acetate trihydrate ($Pb(CH_3COO)_2 \cdot 3H_2O$, 99.5% purity), titanium n-butoxide ($Ti(O{\cdot}n\text{-}C_4H_9)$, 97% in butanol), zirconium n-butoxide ($Zr(O{\cdot}n\text{-}C_4H_9)$, 80% in butanol), and n-butanol (C_4H_9OH, AR reagent grade) as the solvent. A flow-chart is present in figure 1, showing the processing for PZT precursor sol. The atomic ratio in the starting solution was Pb:Zr:Ti = 1:0.52:0.48, no excess lead ion was introduced. Acetylacetonate was firstly added to the Zr-Ti mixtures as a chelating agent to reduce alkoxide hydrolysis with molar ratio between 1:1 and 2:1 for acetylacetonate to (Ti+Zr) alkoxide. Acetic acid was then added to the mixture of lead acetate and butanol to increase the solubility of lead acetate. As hydration resource, $Pb(OAc)_2 \cdot 3H_2O$ was directly used without any distillation or addition of water. The precursor solutions remained stable for a year.

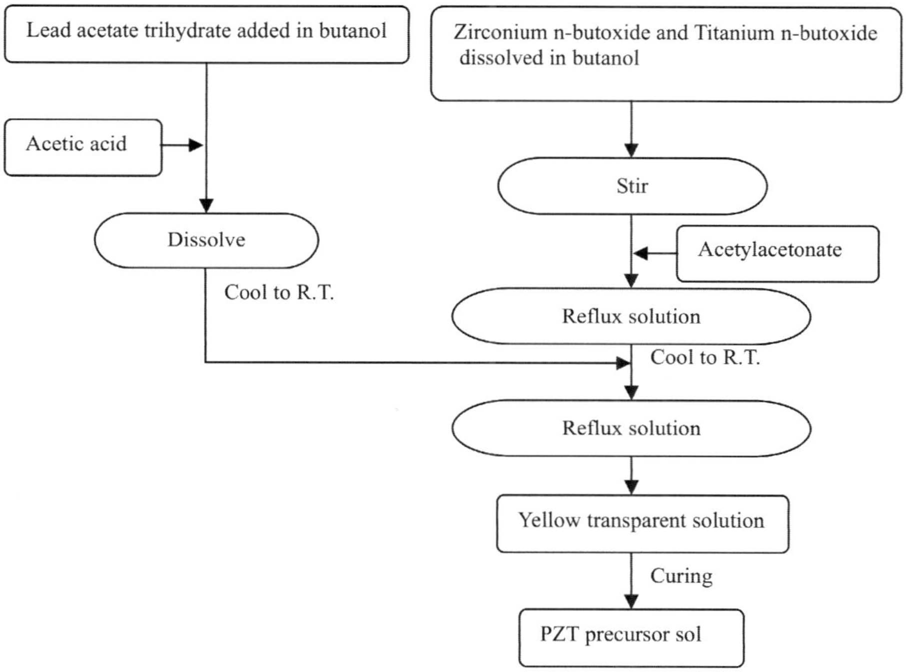

Figure 1. Flow-chart for PZT precursor sol

It was conducted that the solvent was evaporated and the solution was concentrated at various curing temperatures from 25°C to 80°C, the time to gelation was significantly reduced from 60 hours at 25°C to 4 hours at 80°C. Therefore, all of the precursor solutions were efficiently cured at 80°C in this paper.

When the solution achieved the desired viscosity, it was poured into a spinnerette with a 100μm pinhole at the bottom. Fiber extrusion was carried out at ambient temperature. The gel fibers were collected on a take-up wheel, and then they were dried for 2 days at ambient temperature.

2.2. Heat treatment of gel fibers

Different samples of the fibers were sintered between 850°C and 1050°C for 1h at 50°C increments, each at a heating rate of 5°C/min. By using of crucible with PbO, lead atmosphere was conducted during the pyrolysis and sintering to compensate the lead volatilization.

2.3. Characterization of fibers

Thermal gravimetric analysis (TGA) and differential temperature analysis (DTA) were employed to monitor the organic burn-out temperature of the PZT fibers and the chemical reactions at a heating rate of 5°C/min in air. The transitions of the molecular and phase structure of the gel fibers were investigated by FTIR (AVATAR360, Nicolet Ltd.) in the wave number ranging from 4000cm^{-1} to 400cm^{-1} using KBr disk method. Crystallographic evolution was identified with X-ray diffraction (XRD) using CuK$_\alpha$ radiation (X'pert PRO, Panalytical Ltd.). The morphology of the fibers was observed by a scanning electron microscope (SEM, Philips Ltd.).

For the electrical characterization of PZT fibers, bundles of twenty fibers cut for a length of 3mm were used to amplify the signals, as the output signal of a single fiber was quite low. Electrodes were applied by air-dried silver paste painted on both ends of the fibers. The cross sections of the fibers were observed by SEM. The dielectric constant and loss were measured using an Agilent 6500B precision impedance analyzer at 10kHz. The P-E hysteresis loop of the fibers was obtained using a ferroelectric precision tester (RT6000HVS, Radiant Ltd.) in a silicon oil bath at room temperature.

3. Results and Discussions

3.1. Characteristics of gel fibers

Continuous gel fibers were successfully produced, as shown in figure 2. As-drawn gel fibers were yellowish transparent and homogeneous. The diameter of the gel fibers was affected by the precursor viscosity and the drawing speed[4]. Gel fibers with 40-100μm diameter were easily obtained by carefully controlling these two factors. The gel fibers were very flexible just as drawn, and became brittle after aging for several days.

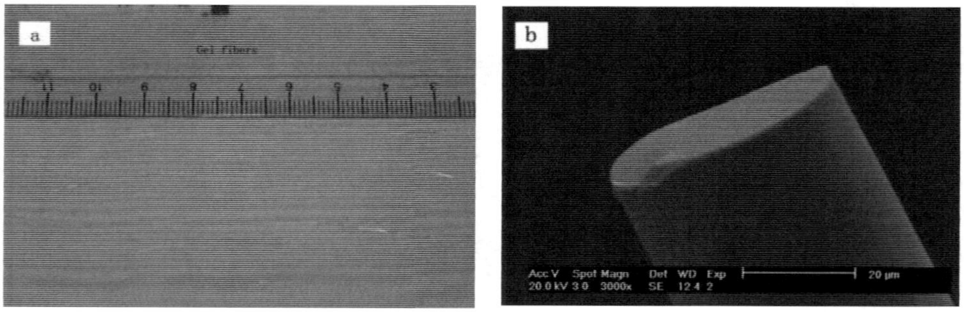

Figure 2. (a) Digital photo of gel fibers as-drawn, (b) SEM image of a gel fiber aged

3.2. Thermal Behavior of PZT Gels

TGA/DTA curves of the PZT gel powders were displayed in figure 3. Three main stages in the decomposition reactions were shown in the TGA curve. The first stage of weight loss below 180°C resulted from the removal of molecular water and butanol with boiling point at 117°C. The second drastic one of weight loss between 180°C and 370°C corresponded to the decomposition of acetate groups. After that, there was a slight weight loss due to the residual organic groups and carbon compositions. The gel powders consisted of about 70% solid components.

Exothermic and endothermic reactions which related to solvent removal and crystallization of PZT were shown in DTA curve. At temperature 100°C, there was an endothermic peak corresponding to the removal of solvent. The broad exothermic event at temperature 360°C was responsible for the acetate burn-out and the removal of residual carbon and decomposed organics, as well as the primary crystallization. The exothermic peak in 450-550°C region was due to phase transformation, which was also observed in the XRD patterns.

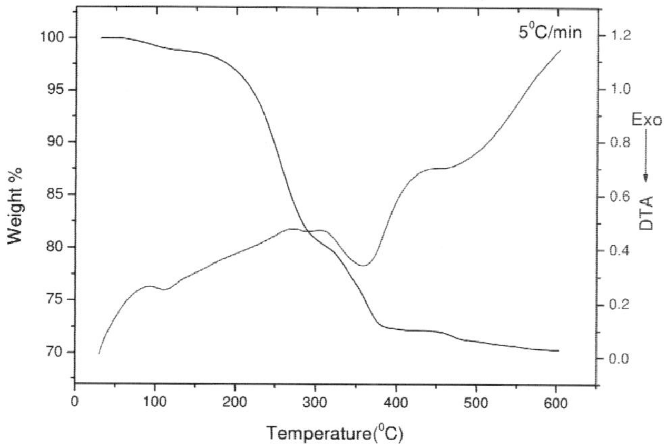

Figure 3. TGA/DTA curves for PZT gel powders

3.3. Molecular and phase evolution of PZT Gels

The FTIR transmittant spectra of the PZT gel powders heat-treated at a rate of 5°C/min to various temperatures was shown in figure 4. The spectrum of PZT gels had characteristic doublets at 1553 and 1413cm^{-1}, which were assigned to the asymmetric and symmetric COO vibration, respectively; the 1337cm^{-1} band, was attributed to the deformation vibration of CH_3 groups; the peaks at 1050 and 1018cm^{-1} were corresponded to the stretching of the C-O bond of butanol. 3400cm^{-1} band was ascribed to O-H groups and the shoulder at 2900cm^{-1} was related to the stretching vibration of C-H bonds[11]. These bands decreased in intensity as the heat treatment temperature increased and disappeared at 600°C. The band at 460-600cm^{-1} was corresponded to Ti-O and Zr-O stretching vibration[12], which shifted to higher wave-numbers and became sharp as the heat treatment temperature increased.

Sol-gel powders were heat-treated in the temperature range of 400°C-700°C for 1h to observe the development of the phase structure of PZT, which was displayed in figure 5. Diffraction peaks attributed to metallic lead and pyrochlore phase were observed at 400°C. When the heat treatment temperature was up to 650°C, pyrochlore phase disappeared and only the perovskite phase was obtained.

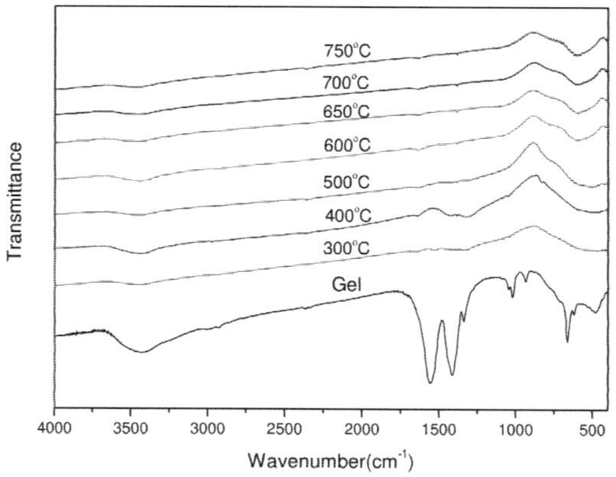

Figure 4. FTIR spectra of PZT gels heat-treated to various temperatures for 1h

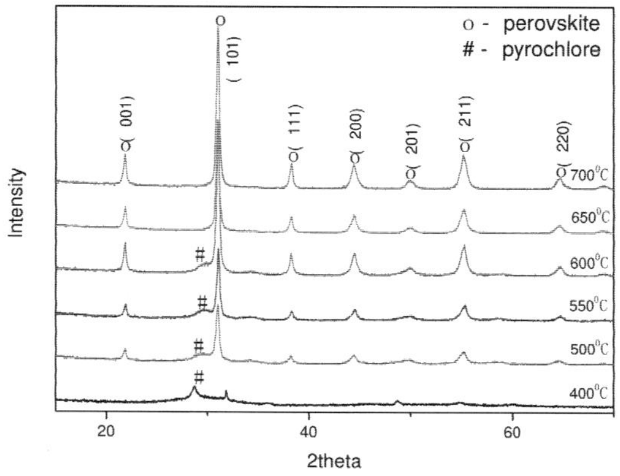

Figure 5. XRD pattern of PZT gel powders heat-treated at different temperatures for 1h

3.4. Morphology of PZT fibers

The cross section of the fiber and longitudinal surface were shown in figure 6, with heat treatment temperature ranges from 650°C to 1050°C. It can be seen that the surface of the fibers was very smooth; the interior of the fiber was dense, free from large pores or other defects. The grain size of the fibers sintered at 900°C and 950°C was 0.2-0.5μm. When the sintering temperature was higher than 1000°C, grain size was up to 1μm, which was also in accordance with Meyer's report [7].

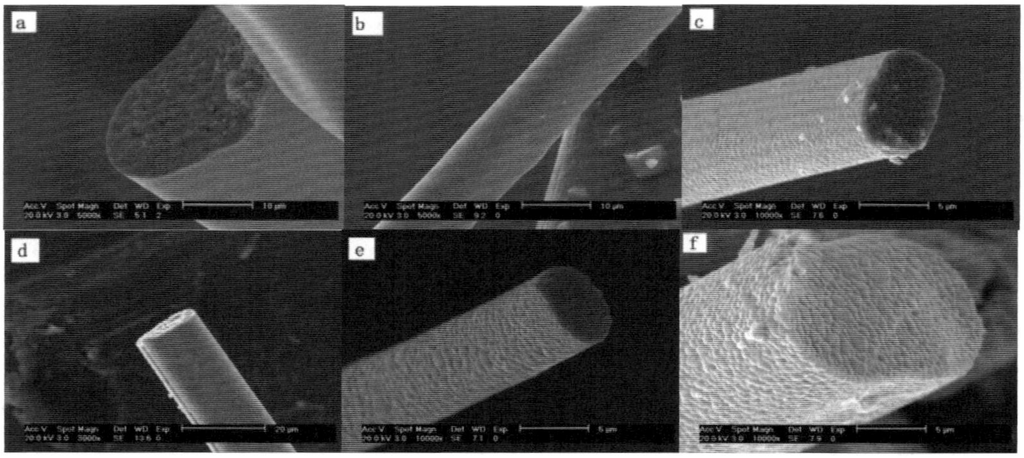

Figure 6. SEM micrographs of ceramic PZT fibers heat-treated at (a) 650°C for 1h; (b) 850°C for 1h;(c) 900°C for 1h; (d) 950°C for 1h;(e) 1000°C for 1h; (f) 1050°C for 1h

3.5. Dielectric and ferroelectric properties measurement

Ferroelectric hysteresis loop of the fibers heat-treated at 1000°C for 1h were measured as shown in figure 7. A typical ferroelectric hysteresis loop was observed using 20Hz triangular waves. The hysteresis loop was square with the remnant polarization (P_r) of 25μC/cm^2 and coercive field (E_c) of 9kV/cm. However, well saturated hysteresis loop could not be obtained even when large electric field was applied. This may due to: (1) organic content remained in the fibers by insufficient burnout that may retard densification of the fiber resulting in the large leakage current density; (2) non switching linear response ascribed to space charge polarization which is originated from pores in the fibers[13]. The permittivity and dielectric loss angle of the PZT fibers heat-treated at 1000°C for 1h measured at 10kHz were 581 and 0.01, respectively. The value of dielectric constant here is lower than those of R. Steinhausen[14] and Yong-Il Park[13], but similar to that of Meyer[5]. Such a low permittivity was attributed to the presence of pores left inside the fibers during the removal of organics in the heat treatment process.

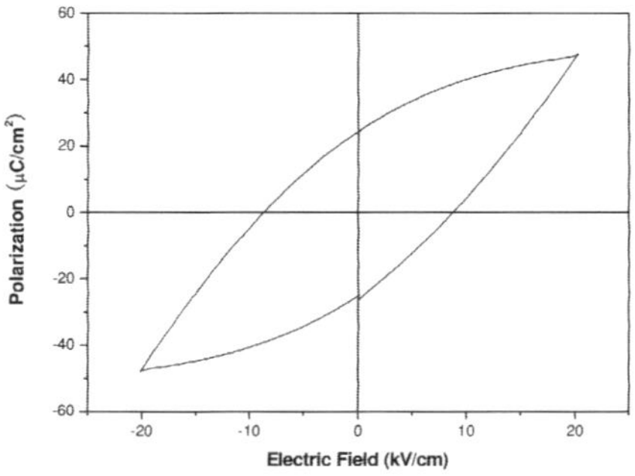

Figure 7. Hysteresis loop of PZT fibers heat-treated at 1000°C for 1h

Journal of Physics: Conference Series **152** (2009) 012077 doi:10.1088/1742-6596/152/1/012077

4. Conclusions

The PZT ceramic fibers were successfully prepared by the sol-gel technique from the solutions of lead acetate trihydrate, zirconium n-butoxide and titanium n-butoxide, using butanol as solvent. As a water source for hydrolysis reaction, $Pb(OAc)_2 \cdot 3H_2O$ was directly used. Acetylacetonate and acetate were added as stabilization agents to obtain stable sols. The suitable viscosity of the sols was achieved for fiber drawing.

Single perovskite phase was obtained after the fibers were heat-treated at 650°C for 1h. The diameter of the PZT ceramic fibers was from 10μm to 40μm. Lower heating rates were employed to avoid cracking with dense and smooth surface. The grain size of the fibers increased with the heat treatment temperature, which was up to 1μm when heat-treated at 1000°C.

It was confirmed that the PZT fibers heat treated at 1000°C had ferroelectric property. The fibers showed square-shaped hysteresis loop with the remnant polarization (P_r) of 25μC/cm^2 and coercive field (E_c) of 9kV/cm. The permittivity and dielectric loss angle of the PZT fibers heat-treated at 1000°C were 581 and 0.01 at 10kHz, respectively.

References

[1] Yoon C B, Jun S H, Lee S M and Kim H E 2006 *J. Am. Ceram. Soc.* **89** 1333–36
[2] Walker D J, Safari A, Card R J and O'Toole M P 1990 *J. Am. Ceram. Soc.* **73** 3503-06
[3] Cass R B 1991 *Am. Ceram. Soc. Bull.* **70** 424-29
[4] Hu Y 2000 *J. Sol-Gel Sci. Tech.* **18** 235–47
[5] Zhang M, Wanga X D, Wanga F M, Miranda Salvadob I M, Vilarinhob P M and Li W C 2005 *Ceram. Inter.* **31** 281–86
[6] Binkle O and Nass R 1998 *J. Sol-Gel Sci. Tech.* **13** 1023– 26
[7] Meyer R J, Shrout T and Yoshikawa S 1998 *J. Am. Ceram. Soc.* **81** 861–68
[8] Budd K D, Bey S K and Payne D A 1985 *Br. Ceram. Proc.* **36** 107-21
[9] Yi G and Sayer M 1996 *J. Sol-Gel Sci. Tech* **6** 65
[10] Calzada M L, Sirera R, Carmona F and Jimenez B 1995 *J. Am. Ceram. Soc.* **78** 1802
[11] Kitaoka K, Kozuka H and Yoko T 1998 *J. Am. Ceram. Soc.* **81** 1189–96
[12] Tarte P 1965 *The Determination of Cation Co-ordination in Glasses by Infra-Red Spectroscopy* ed J A Prims (Amsterdam: The Netherlands) pp 549–565
[13] Park Y I, Nagai M, Miyayama M and Kudo T J 2001 *Mat. Sci.* **36** 1995–2000
[14] Steinhausen R, Hauke T, Beige H, Watzka W and Lange U 2001 *J. Eu. Ceram. Soc.* **21** 1459-62
[15] Etin A, Shter G E and Grader G S 2007 *J. Am. Ceram. Soc.* **90** 77-83
[16] Li K 2002 Piezoelectric Ceramic Fibre/Polymer 1-3 Composites for Transducer Applications PhD thesis (Hong Kong: The Hong Kong Polytechnic University)

The structure and oxygen permeation performance of $Ba_{0.5}Sr_{0.5}Ga_{1-x}Fe_xO_{3-\delta}$ perovskite-type oxides

J. HU, H. CHEN and R.R. ZHANG

Department of Material and Chemical Engineering, Henan Institute of Engineering, Zhengzhou 450007, PR China

E-mail: huj@hntc.edu.cn

Abstract. Solid-state reaction was used to prepare mixed-conducting $Ba_{0.5}Sr_{0.5}Ga_{1-x}Fe_xO_{3-\delta}$ (BSGF) perovskite-type membranes. They were characterized by SEM, XRD and DSC and their oxygen permeation performance was examined. The experiments showed that the doping content of the Ga can play an important role in the structure of BSGF. It was found that the perovskite structure was sustained when Ga doping was lower than 0.4 or a little higher, but when Ga doping reached 0.6, the samples will not be perovskite structure. The oxygen permeation experiment showed that the temperature can affect the oxygen permeation of the BSGF membrane remarkably. At high temperature (about 940°C), its oxygen permeation property was higher and its oxygen permeation property will decrease with the temperature. The measured oxygen desorption activity energy of the $Ba_{0.5}Sr_{0.5}Ga_{0.2}Fe_{0.8}O_{3-\delta}$ measured is about 306.314 $kJ\cdot mol^{-1}$.

1. Introduction

Partial oxidation of methane (POM) to syngas is a potential alternative to steam reforming processes. Since mixed-conducting oxygen permeation membranes can separate oxygen from air at an elevated temperature, using oxygen permeation membranes as reactor can combine the separation of air with POM into a single step, thus simplifying the process and reducing the cost by about 20~30%[1-3]. Over past years, extensive efforts have been focused on developing new mixed-conducting oxygen permeation membranes and improving their performance in POM processes. Many mixed conductors with the perovskite-type have shown notably high oxygen permeability[4, 5]. Unfortunately, most of them are unstable under methane conversion conditions. This is an inevitable concomitant of the expansion of the perovskite unit cell volume with reduction of the transition metal ions to their larger lower-valent forms as the oxygen partial pressure is lower[6, 7]. The membrane will eventually crack due to the volume change. The perovskite-type $Ba_{0.5}Sr_{0.5}Co_{0.8}Fe_{0.2}O_{3-\delta}$ (BSCFO) is one of the mixed conductive materials with high permeability[8-10]. However the stability of this material is poor in reduced atmosphere. Hendriksen et al.[11] reported that a strong contraction was observed when exposing $SrFeCo_{0.5}O_{3-\delta}$ even to mildly reducing conditions. It is obvious that the cobalt-containing perovskite oxides suffer from stability problems in a syngas environment, which might cause membrane failure during long-term operation. This forms a major incentive in several recent studies for either lessening the relative amount of cobalt in the perovskite phase to co-dope the material with less reducible ions, such as Zr^{4+}, Ga^{3+} and Al^{3+}, or to add a toughening material such as zirconium or gallium[12-14].

Journal of Physics: Conference Series **152** (2009) 012078 doi:10.1088/1742-6596/152/1/012078

It was reported that the oxygen permeation membranes with gallium-containing perovskite exhibit excellent stability in the POM process[15, 16]. The average metal–oxygen bonding energy of Ga ion in perovskite structure is -143 kJ·mol^{-1}[17], smaller than that of Co ion. The Ga ions only take the +3 valence state in perovskite structure. The lower metal–oxygen bonding energy is beneficial to promote the conductivity of oxygen ions. The stable valence state could avoid the expansion of the perovskite unit cell volume. It is conjectured that replacing Co by Ga may improve the stability of BSCFO membrane in the POM process. Therefore, the main objective of the present work is to develop new membrane materials $Ba_{0.5}Sr_{0.5}Ga_{1-x}Fe_xO_{3-\delta}$ and investigate their structure and oxygen permeation performance.

2. Experimental

The $Ba_{0.5}Sr_{0.5}Ga_{1-x}Fe_xO_{3-\delta}$ (BSGFO, 0<x<0.8) samples were prepared by the common solid-state reaction method. Briefly, $BaCO_3$, $SrCO_3$, Ga_2O_3 and Fe_2O_3 were mixed and ground in an agate mortar. Then the mixture was pressed into disk, which was then sintered at 900~1000°C for 10~15 h. The disk was then ground into fine powder, and pressed into disk again at a diameter of 20 mm by a thickness of 1.0~1.4 mm. The disk was sintered at 1000~1200°C for 5 h with a heating rate of 2 °C·min^{-1} once again and then annealed at 500°C for 5 h. The phases of membranes were determined by X-ray diffraction (XRD, Model D/MAX-3B, RIGAKU, Japan) with Cu-Kα radiation. The morphology of membranes was examined by scanning electron microscope (SEM, Model JSM-5610LV, JEOL, Japan).

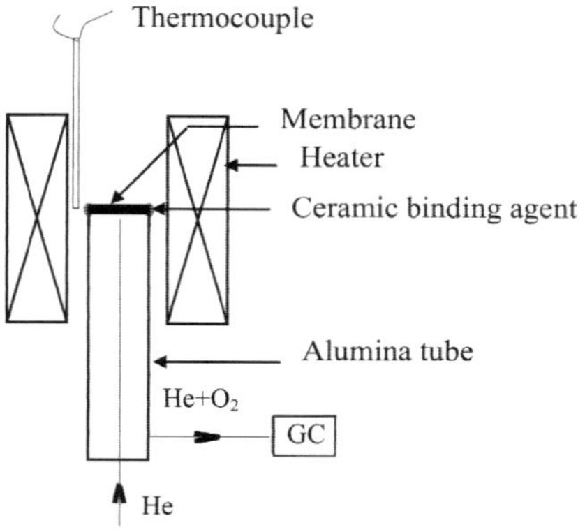

Figure 1. Schematic diagram of ceramic membrane reactor

The oxygen permeation experiments were performed in a vertical high-temperature gas permeation system as shown in figure 1. BSGFO disks (efficient surface area 2.0 cm^2) were sealed onto the top of alumina tubes by using a homemade ceramic binder[18]. High purity helium was introduced into alumina tube at atmospheric pressure. The effluent gas was analyzed by an on-line gas chromatograph (Agilent 1790, TCD detector). Helium was used as carrier gas and a molecular sieve 5A column was used to separate oxygen and nitrogen. The densities of the sintered membranes were determined by the Archimedes method using ethanol.

3. Results and Discussion

3.1. The influence of Ga doping on the phase structure

The influence of Ga doping on the phase structure of $Ba_{0.5}Sr_{0.5}Ga_{1-x}Fe_xO_{3-\delta}$ was investigated. Figure 2 shows the X-ray diffraction patterns of $Ba_{0.5}Sr_{0.5}Ga_{1-x}Fe_xO_{3-\delta}$ powders with varying Ga-contents obtained from calcining the precursors at high temperatures under stagnant air atmosphere. Pure phase perovskite was formed at x=0.2, besides the basic perovskite, a few additional peaks differed were also observed at x=0.4, and obviously different diffraction patterns from that of perovskite were found at x≥0.6. If the Ga and Fe ions take the valence state +3 and +4 respectively, the effective ionic radii of Ga and Fe are 0.62 and 0.585 Å, respectively. Furthermore, the calculation of the tolerance factor showed that the tolerance factor decreased from 0.946 to 0.936 with increasing Ga-content (See Table 1). Besides the ionic radii requirement, another condition to be fulfilled is electroneutrality[19], i.e., the sum of charges of A and B equals the total charge of oxygen anions. It is evident that with the increasing of Ga content, an anion-deficient perovskite structure tends to failure eventually due to the increasing of oxygen ions vacancy.

Table 1. The tolerance factor of materials

Ga content x	r_A(Å)	r_B(Å)	r_O(Å)	t
0.2	1.265	0.592	1.40	0.946
0.4	1.265	0.599	1.40	0.943
0.6	1.265	0.606	1.40	0.939
0.8	1.265	0.613	1.40	0.936

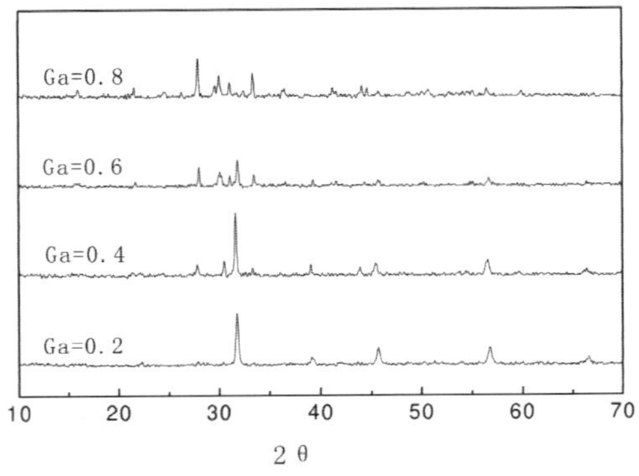

Figure 2. X-ray diffraction patterns of BSGFO

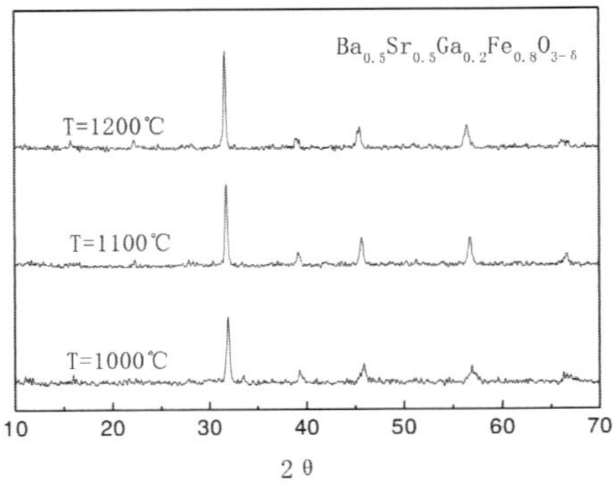

Figure 3. XRD patterns of BSGFO sintered at different temperatures

3.2. The influence of sintering temperature on the phase structure

Figure 3. shows the XRD patterns of $Ba_{0.5}Sr_{0.5}Ga_{0.2}Fe_{0.8}O_{3-\delta}$ membranes sintered at different temperatures. Almost a pure phase perovskite was observed for temperature above 1100°C. The intensity of XRD diffraction peaks at 2θ of about 32°, 46° and 56° increases directly with the sintering temperature, which reflects the content of the corresponding phase. However, some other peaks could also be observed when temperature is 1000°C. Therefore, it can be concluded that the increases of sintering temperature are beneficial to the formation of perovskite phase. In this case proper sintering temperature is 1100°C.

3.3. The influence of sintering condition on the micrographs of membrane

A dense body is desirable for an oxygen permeation membrane. The morphology of membrane sintering with a heating rate of 2 °C·min^{-1} was showed in figure 5(a). We can see that there are some pores on the surface of membrane. The formation of pores may be the volatilization of organic reagent (PVA) added as molding of membrane and the secondary recrystallization. Figure 4 shows the DSC/TG data of the composite precursor powders and BSGFO powders. Endothermic peaks present at about 220°C showing the decomposition of PVA and 796~1070°C showing the formation of mixed oxides. In order to volatilize organic reagent completely before sintering and form mixed oxides at mild condition, the modified sintering condition should be heating to 220°C with a heating rate of 2 °C·min^{-1} and hold for 2 h first, then heating to 800°C and hold for 2 h and finally sintering at 1100°C for 5 h. The morphology of membrane with new condition was shown in figure 5(b), from which it can be seen that for the fresh membrane, ceramic grains with clear grain boundaries are visible and there are few pores on the surface of membrane. The relative density of membrane reaches to 5.538 g·cm^{-3}, close to that of theoretical value.

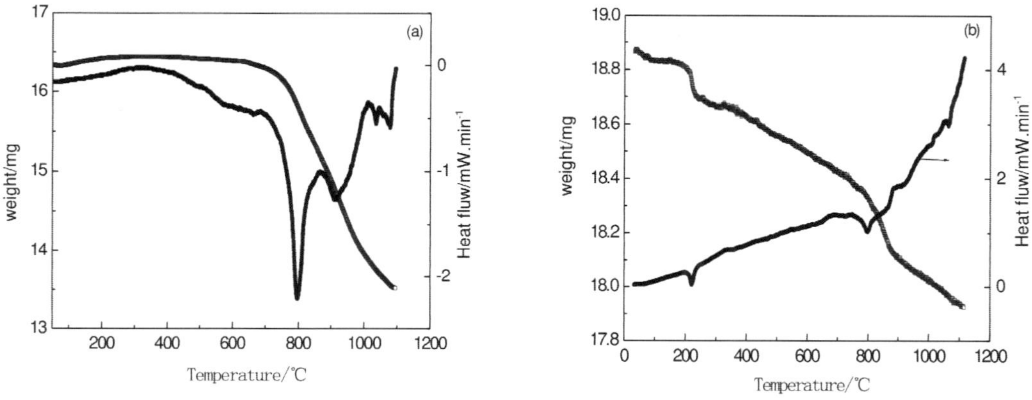

(a) (b)

Figure 4. DSC curves of the composite precursor powders (a) and BSGFO powders (b)

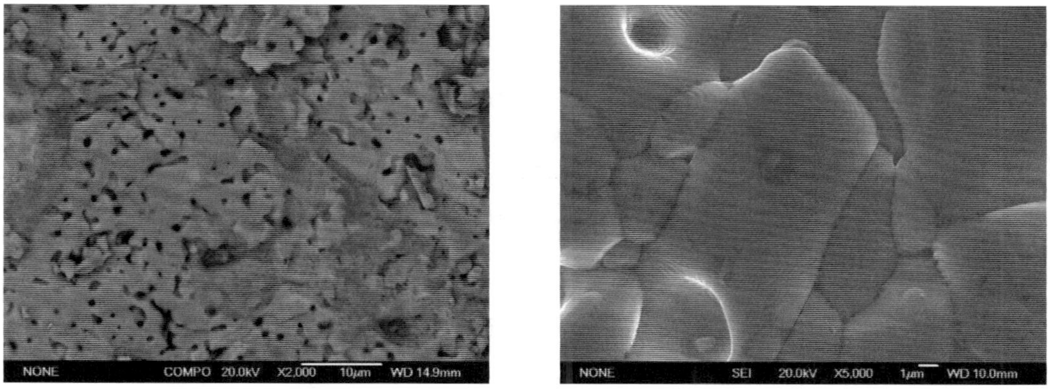

(a) (b)

Figure 5. SEM pictures of BSGFO membrane with different sintering conditions

3.4. The oxygen permeation performance of $Ba_{0.5}Sr_{0.5}Ga_{0.2}Fe_{0.8}O_{3-\delta}$

The value of oxygen permeation flux is a key parameter for an oxygen permeable membrane to obtain real applications. Figure 3 presents the dependencies of oxygen permeation fluxes of $Ba_{0.5}Sr_{0.5}Ga_{0.2}Fe_{0.8}O_{3-\delta}$ membranes on temperature under atmosphere. It can be seen that the oxygen permeation flux is about 0.04 ml·min^{-1}·cm^{-2} only at 800°C. With the increase of temperature the oxygen permeation flux reaches to about 0.11 ml·min^{-1}·cm^{-2} at 940°C. The activation energy of BSGFO estimated by Freemen-Carroll method using the TG data[20] was 306.314 kJ·mol^{-1} for 800~900°C. That explained the reason of the oxygen permeation at lower temperature.

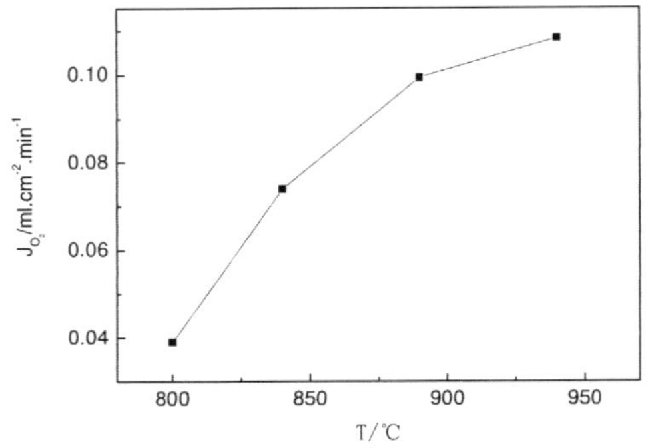

Figure 6. Temperature dependence of oxygen permeation fluxes

4. Conclusions

The doping content of the Ga can affect on the structures of $Ba_{0.5}Sr_{0.5}Ga_{1-x}Fe_xO_{3-\delta}$ significantly. It was found that the perovskite-type ceramic membrane was sustained only when Ga content $x \leq 0.4$. Indicating a stable perovskite-type structure depended on not only the geometric structure but also the oxygen ions deficiency. The temperature can affect the oxygen permeation of the BSGF membrane remarkably. At a high temperature of 940°C, its oxygen permeation reaches to about 0.11 $ml \cdot min^{-1} \cdot cm^{-2}$. However, at a lower temperature of 800°C, it can also exhibit some performance of oxygen permeation. The oxygen desorption activity energy of the $Ba_{0.5}Sr_{0.5}Ga_{0.2}Fe_{0.8}O_{3-\delta}$ calculated using TG data is about 306.314 $kJ \cdot mol^{-1}$.

Acknowledgement

This work was supported by the Natural Science Foundation of Henan Province and the Doctor Fund of Henan Institute of Engineering.

References

[1] Tan L, Gu X H, Xu N P, *et al* 2003 *J. Membrane Sci.* **212** 157–165
[2] Bouwmeester H J M 2003 *Catal. Today* **82** 141–150
[3] Balachandran U, Dusek J T, Mieville R L *et al* 1995 *Appl. Catal. A-Gen* **3** 19–29
[4] Diethelm S and Van herle J 2004 *J. Eur. Ceram. Soc.* **24** 1319–23
[5] Hu J, Hao H S, Chen C P, Yang D L and Hu X 2006 *J. Membrane Sci.* **280** 809–14
[6] Mitchell B J, Rogan R C, Richardson J W *et al* 2002 *Solid State Ionics* **146** 313–21
[7] Wang H H, Cong Y and Yang W S 2002 *J. Membrane Sci.* **209** 143–52
[8] Lu H, Tong J H, Cong Y and Yang W S 2005 *Catal. Today* **104** 154–59
[9] Wang H H, Tablet C, Feldhoff A *et al* 2005 *J. Membrane Sci.* **262** 20–26
[10] Wang H H, Wang R, Liang D T and Yang W S 2004 *J. Membrane Sci.* **243** 405–15
[11] Hendriksen P V , Larsen P H and Mogensens M 2000 *Catal. Today* **56** 283–95
[12] Wu Z T, Jin W Q and Xu N P 2006 *J. Membrane Sci.* 279 320
[13] Tong J H, Yang W S, Cai R, Zhu B C and Lin L W 2002 *Catal. Lett.* **78** 129–37
[14] Kima J M, Hwanga G J, Leea S H, Parka C S, Kima J W and Kimb Y H 2005 *J. Membrane Sci.* **250** 11–16
[15] Schwartz M, White H J and Sammells FA 1999 *International Pat* : WO 99/ 21649, 05–06
[16] Kharton V V, Sobyanin V A, Belyaev V D *et al* 2004 *Catal. Commun.* **5** 311–16
[17] Shao Z P, Xiong G X and Yang W S 2001 *J. Inorg. Mater.* **16** 23–31
[18] Hu J, Xing T L, Jia Q C, Yang H Z, Sun H W and Hu X 2007 *Catal. Commun.* **8** 1301–04
[19] Pena M A and Fierro J L 2001 *Chem. Rev.* **101** 1981–2017
[20] Zhu Z L, Yang D L, Guo Y Q, Hu X *et al* 2002 *Physica C* **383** 169–74

Journal of Physics: Conference Series **152** (2009) 012079 doi:10.1088/1742-6596/152/1/012079

Fabrication of Transparent Yttrium Aluminum Garnet Ceramic

Xia LI

College Materials Science and Engineering, Qingdao University of Science & Technology, Qingdao, 266042, China

lix@qust.edu.cn

Abstract. In this paper, the synthesis of Nd-doped yttrium aluminum garnet (Nd:YAG) spherical nano-crystallites was investigated by using the solvothermal method, and the optimum processing conditions for processing the transparent ceramic preparation was determined. Powder consisting of nanosized particles obtained by the solvothermal method displays significantly less crystallite agglomeration, indicating a high degree of sinterability. The phase structure and the morphology of the YAG crystallites were depended on the reaction conditions, the optimum temperature is 300 centigrade for 1h, at which the pure phase of spherically shaped YAG nanoparticles can be obtained. Microstructure evolution at different sintering stages demonstrated that fully transparent YAG ceramic can be fabricated by vacuum sintering at 1750 centigrade for 5h by using the as-synthesized powders. The ceramic has a relative density of about 99.98% of the theoretical value, and an average grain size of about 3-5μm. The transmittance of the ceramic is 55% in the visible range, and 70% in the infrared range.

1. Introduction

Yttrium aluminum garnet ($Y_3Al_5O_{12}$) is an oxide material with a number of attractive properties. Neodymium-doped YAG is an important solid-state laser material [1]. When doped with Tb, YAG and substituted garnets derived there from are useful phosphors in the blue green region of the visible spectrum, particularly for high brightness cathode ray tubes, where extremely rugged phosphors are required. YAG is stable at high temperatures and has the highest creep resistance of any oxide material known [2,3]. Transparent polycrystalline YAG is therefore currently of interest for use in the preparation of solid laser matrices [4-7]. Compared to the growth technique for Nd:YAG single crystals, the fabrication of polycrystalline transparent Nd:YAG ceramic does not require any special technique, and large size, high concentration (>1%) neodymium-doped samples can be fabricated. Moreover, multi-layer active elements and multi-functional ceramics can easily be fabricated together without having to employ the single-crystal growth method. The electrical conductivity of YAG is also reported to be lower than any other polycrystalline oxide [8]. For electronic and optical applications, control of the crystalline microstructure of the YAG powder is important.

Reproducible fine control over microstructure is seldom achievable using traditional ceramic approaches. Unfortunately, the high temperatures required to overcome solid-state diffusion and long reaction times do not allow for control of microstructure, grain size, and grain size distribution [9,10]. Some low-temperature synthesis methods of fabrication of nano-sized YAG crystallites were studied to overcome the difficulties encountered in traditional solid-state reactions, including different wet-chemical processes, i.e. the sol-gel process [11,12] homogeneous precipitation [13,14] and the combustion method [15,16] have been reported. In the case of homogeneous precipitation and the alkoxide method, direct crystallization of YAG below 1000℃, and even at 800℃ was reported, but the systematic evolution of the microstructure, sinterability and grain growth phenomena have not yet been studied. Although YAG nanoparticles synthesized by chemical methods have some advantages, i.e., high purity, homogenous composition and fine grain size, additional heat treatment is required at higher temperatures (>1000℃) to obtain well-crystallized products. Hard aggregation of the resulting powder is easily caused during post heat treatment processing. Spray pyrolysis is known to be one of the feasible processes for the preparation of phosphor particles of improved morphology [17,18].

In this work, we report on the preparation of spherical YAG nano-particles by the solvothermal method. Non-agglomeration and higher sinterability powder was obtained by processing at 300℃ for 1h. Fully transparent YAG ceramic can be fabricated by sintering at 1750℃ for 5h under vacuum using the as- prepared powders.

2. Experimental procedure

Y_2O_3 (99.99% pure, supplied by Shanghai Yuelong Non-Ferrous Metals Limited), Nd_2O_3 (99.99% pure, supplied by Shanghai Yuelong Non-Ferrous Metals Limited) and $Al(NO_3)\cdot 9H_2O$ (A.R. grade, supplied by Tianjin Fine Chemicals) were used as starting materials. The ammonium hydrogen carbonate was used as a precipitant. The stock solution of mother salts was made by dissolving aluminum nitrate in distilled water; yttrium oxide was dissolved in 1mol L^{-1} HNO_3 by stirring the mixture for 2 h at 80-90℃in a beaker covered with a watch-glass. The precipitate was produced by adding 100mL of the salt mixture (the stoichiometric molarity ratio of the $Y^{3+}$and Al^{3+} ions was maintained at 3:5) at a speed of 2mL min^{-1} into a 200mL ammonium hydrogen carbonate solution contained in a beaker under mild agitation. Precursor precipitates were separately subjected to solvo-thermal conditions in an autoclave and were heated to the desired temperature at a rate of 2℃min^{-1}, then held there for 1 h and 2h. During the reaction period, the pressure gradually increased to 10MPa. After being cooled to room temperature, the resulting products were again separately washed with distilled water and dried under an infrared lamp.

Tetraethyl orthosilicate (TEOS, 0.5% mass) was added to calcined YAG powder as a sintering aid. The mixture was milled in ethanol for 6h with a high purity zirconia ball. The milled slurry was dried and calcined at 800℃ in oxygen to remove residual organic materials. The granulated, spherical powder was isostatically pressed at 150MPa into a disk of 20mm in diameter. The green compacted material was sintered for 5h at 1750℃ under vacuum. The pressure in the furnace was maintained below 1×10^{-2} Pa during the holding period.

Phase identification was performed on an X-ray diffractometer (D/max-III, Rigaku, Tokyo, Japan) using CuKα radiation (λ=1.54178A). Observation of particle morphology and size was performed on a transmission electron microscope (TEM) (Model H-8010, Hitachi, Tokyo, Japan). The microstructure of the YAG powder was analyzed by high-resolution transmission electron microscopy (HRTEM), which was performed with a Philips Tecnai 20v-Twin transmission electron microscope using an accelerating voltage of 200KV. Bulk densities after polishing with diamond paste were determined by Archimedes' method. The sintered specimens were thermally etched at 1500 ℃ for 2h. The microstructure of the polished surface of the specimens was observed using a scanning electron microscope (FEI QUANTA200). Specimens for transmittance measurement were cut into 1mm thick slices, and the transmittance of the transparent YAG ceramic was measured over the wavelength region from 400 to 1000nm using a spectrometer (UV757CRT). A Nd:YAG single crystal was used as an optical reference.

3. Results and Discussion

Fig.1a shows the X-ray diffraction pattern of YAG sintered at 300℃ . The diffraction pattern was identical to that of the standard sample (JCPDS Card 33-40), indicating that single phase YAG had been successfully prepared at a temperature as low as 300℃. It is noted that no intermediate phases co-exist at such low temperatures, implying that the derived precursors were directly transformed into the crystalline YAG phase.

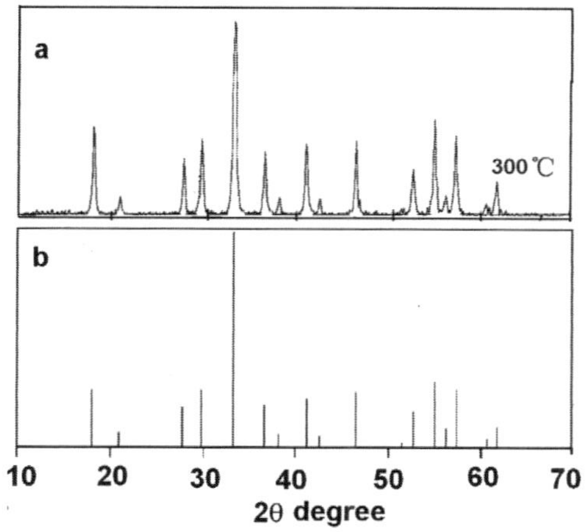

Figure 1. XRD pattern of the synthesized powder.

Figure 2 a,b show TEM photographs of samples prepared at 300℃for 1h and 2h, respectively. Sample a is mainly composed of well-dispersed fine grains with a relatively narrow grain size distribution. The mean size of the particles measured from the TEM photograph is about 40nm. These crystallites are discrete, fairly uniform, and have a spherical shape.

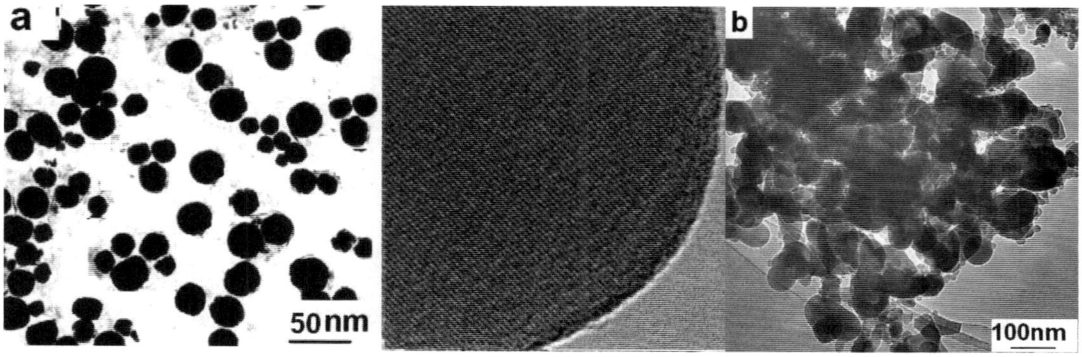

Figure 2. TEM morphology of the powder calcined for (a) 1h; (b)2h.

The micrographs clearly indicate that the crystallites are remarkably less agglomerated compared to those mentioned in other reports [20,21]. The high-resolution lattice image of the surface demonstrates that the crystals are structurally perfect, i.e., no defects are found the crystal structure. The lattice planes visible are the (420) planes of the cubic (a_0= 1.2002nm) phase of YAG. When the holding time is prolonged to 2h or more, the YAG grains grow to reveal enhanced agglomeration, as shown Fig 2b. It is well known that as the grains grow, the attractive force between them is increased, and aggregation among the particles occurs. Recent studies have suggested that an un-aggregated spherical powder with a narrow size distribution is the most desirable for high sinter-ability and high packing densities [22,23]. This can be explained from the standpoint of packing uniformity of particles in the green form. Heavily agglomerated materials exhibit a higher friction force during compaction, making it more difficult to achieve uniform compaction and a high final density. This result thus indicates that a holding time of 1h is suitable for the synthesis of YAG powder with good sinterability.

Fig.3 shows the relative density of the compacted body sintered from 1300℃ to 1750℃ for 5h under vacuum. The relative density of YAG sintered at 1650℃ is 99.1%, and the density of the specimen sintered at 1750℃ approaches the theoretical value. Our results also suggest that a vacuum-sintering temperature of 1750℃ remains the optimum sintering temperature for uniform grain growth to up 3-5μm. The sintered ceramic has a relative density of about 99.98% of the theoretical value (4.55g/cm^3).

By comparison, Ikesue et al[6] reported that transparent YAG ceramic was fabricated at 1850℃ for 5h using high purity Al_2O_3 and Y_2O_3 powders as starting materials. It is apparent that the lower sintering temperature (1750℃) in the present study results from good dispersion of the synthesized YAG powder. Fig.5a-c show the microstructure evolution during different sintering stages. From the micrographs some difference in grain size and porosity during the three stages of sintering can be observed. It also can be seen that the sinterability and the grain growth phenomena in YAG are enhanced with increasing temperature.

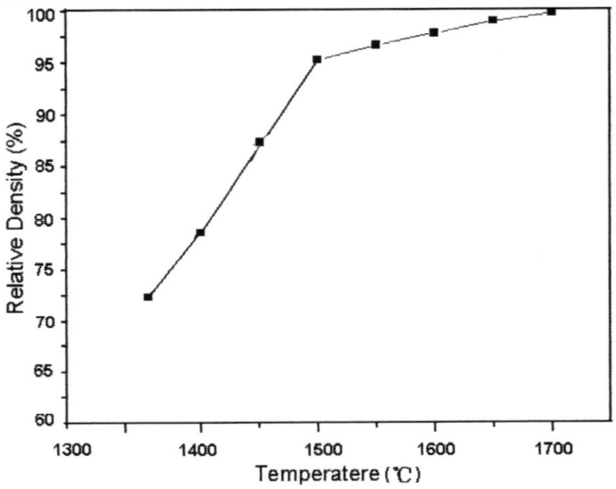

Figure 3. Relationship between relative density and sintering temperature

During the initial stage of sintering (Fig4a), it is observed that the spherical particles seen at lower sintering temperatures begin to coalesce through the phenomenon of neck formation between adjacent particles. Grain boundaries between the particles in the compacted sintered material develop at the small cross-section of the contact regions and the pores are continuous. At the intermediate stage of the sintering process (as shown in Fig4b), the pores form a continuous network that is attached to the grain boundaries. The system of continuous pores breaks down and more or less equilibrium-shaped isolated pores remain at the grain boundaries. Also shown is the microstructure that develops as the pores migrate with the grain boundaries and finally disappear. This sequence is essentially different from the case of conventionally prepared YAG powder, where many pores remain within the grains. The present technique for synthesizing the powder yields low agglomeration and thus the material can be compacted homogeneously. In the final stage of sintering (as shown in Fig 4c), the pores that remain on the grain boundaries can be eliminated and may be accompanied by grain growth. A dense and nearly pore-free microstructure is observed at 1700℃and above. The trend also shows that the porosity decrease with the increase in the sintering temperature is consistent with the increase of the densification of the specimens.

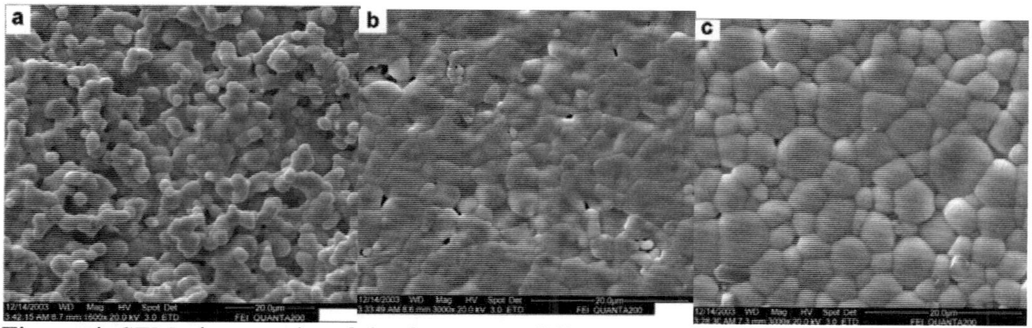

Figure 4. SEM photographs of the fracture at different sintering stages (a) initial stage; (b) intermediate stage(c) final stage.

Fig 5 shows an SEM photograph of a specimen that was sintered at 1750℃ for 5h, polished and then thermally etched. The sintered body had a homogeneous microstructure with grains of about 2-3μm in diameter. The grain boundaries are found to be fairly sharp. The sintered sample is highly translucent, suggesting that an increase in sintering temperature would result in a transparent ceramic. Because the sintering proceeded homogeneously and the sintering temperature was kept relatively low, abnormal grain growth was not observed.

Figure 5. SEM photograph of the microstructure of polished surface of specimen sintered at 1750℃.

Fig.6 shows the optical transmission spectrum of Nd:YAG ceramic sintered at 1750℃ over the wavelength range from 400nm to 1000nm. The single crystal data is also shown for comparison. The optical transmittance is 55% in the visible range, and reaches almost 70% in the infrared, but the transmittance is slightly lower than that of the single crystal. This can be attributed to structural differences between the crystalline and polycrystalline ceramics. The optical transmittance of YAG ceramic is mainly affected by pores, inclusions and grain boundaries [24]. This indicates that YAG ceramic contains internal inhomogeneities that can serve as scattering sites that reduce the transparency to translucency or even opacity. Another concern in the ceramic samples is the scattering loss caused by grain boundaries. If the grain boundary width is too large or impurities coexist with it, the transparency can be reduced because of the energy loss at the grain boundaries due to reflection and scattering.

Nano-crystalline YAG powder with spherical particle shape was synthesized by the solvo-thermal method under modest reaction conditions. The precursor is converted directly to the pure YAG phase at about 300℃. The resulting YAG powder shows good dispersion and high sinter-ability. Using the synthesized nano-crystalline YAG powder, fully transparent YAG ceramic was fabricated by sintering under vacuum at 1750℃ for 5h. The sintered transparent ceramic exhibits a homogeneous microstructure with no abnormal grain growth. The average grain size obtained is about 3-5μm. The transmittance of the YAG ceramic is 55% in the visible range and 70% in the infrared range. Through the present work, we have demonstrated that using a low-agglomeration powder is important for obtaining a transparent ceramic.

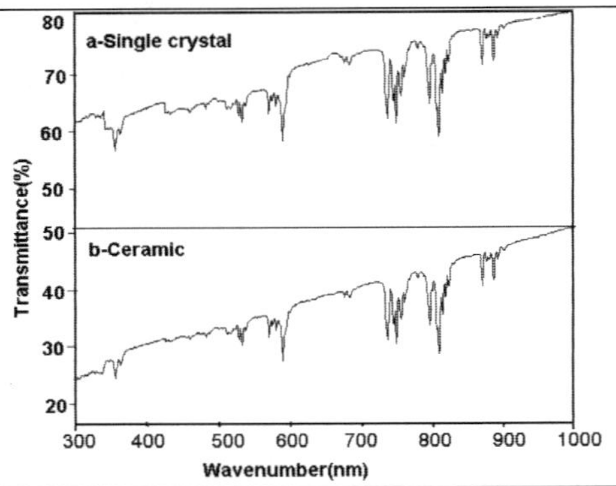

Figure 6. Optical transmission spectra of YAG single crystal and polycrystalline YAG (a) single crystal; (b) polycrystalline ceramic sintered at 1750℃ .

Acknowledgement
This work was supported by the Natural Science Foundation of China (Grant No.50772051).

References

[1] Geusic J E,, Marcos H M and Van Uitert L G 1964 *Appl. Phys. Lett.* **4**182
[2] Motscher G N, Chen K C and Mazdiyasni K S 1994 *Ceram. Eng. Sci. Pro.* **15**181
[3] Corman G S 1991 *Ceram.Engi.Sci. Proc.* **12**1745
[4] Lu J R, Prabhu M and Song J 2000 *Appl. Phys. Lett*, **71**469
[5] Vrolijk J W G A, Willems J W M M and Metsellar R 1990 *J.Eur.Ceram. Soc.* **6**47
[6] Ikesue A, Furusato I and Kamata K 1995 *J. Am. Ceram. Soc.* **78**225
[7] Ikesue A, Kinshita T and Kamata K 1995 *J. Am. Ceram. Soc.* **78**1033
[8] Bates J L and Garnier J E 1981 *J.Am.Ceram.Soc.* **64**138
[9] Warshaw I and Roy R 1959 *J.Am. Ceram. Soc.*, **42**434
[10] DeWith G and Dijk H van 1984 *Mater. Res.Bul.* **19**1669
[11] Veith M, Mathur S, Kareiva A, Jilavi M, Zimmer M and Huch V 1999 *J. Mater. Chem.* **9**3069
[12] Vaqueiro P and Lopez-Quintela M A 1998 *J. Chem. Mater.***8**161
[13] Li J G, Ikegami T, Lee J H, Mori T and Yajima Y 2000 *J. Euro. Ceram. Soc.* **2**2395
[14] Mastsushita N, Tsuchiya N and Nakatsuka K 1999 *J. Am. Ceram. Soc.* **82**1977
[15] Pillai K T, Kamat R V, Vaigya V N and Sood D D 1996 *Mater. Chem. Phys.* **44**255
[16] Hess N J, Maupin G D, Chick L A, Sunbery D S, Mecreedy D E and Armstrong T R1994 *J. Mater. Sci.* **29**873
[17] Kang Y C, Roh H S and Park S B 2000 *Adv. Mater.* **12** 451
[18] Kang Y C, Park S B, Lenggoro I W and Okuyama K 1999 *J. Electrochem. Soc.* **146**1227
[19] Lu Q, Dong W, Wang H and Wang X 1997 *J.Am.Ceram.Soc.* **85**490
[20] Hakuta Y, Seino K, Ura H, Adschiri T, Takizawa H and Arai K 1999 *J.Mater.Chem.* **9**2671
[21] GuoX Z, Devi P S, Ravi B G, Parise J B, Sampath S and Hanson J C 2004 *J.Mater.Chem.*, **14**1288
[22] Lange F F 1984 *J.Am.Ceram.Soc*, **67**83
[23] Lange F F 1989 *J.Am.Ceram. Soc.* **72**3
[24] Ikesue A, Yoshida K, Yamamoto T and Yamaga I 1997 *J.Am.Ceram.Soc.* **80**1517

Journal of Physics: Conference Series 152 (2009) 012080 doi:10.1088/1742-6596/152/1/012080

Experimental research on pulse forming based on high-temperature SMES applied in pulsed power

Yusheng ZHOU [1, 2, a], Jianghua KUANG [1, b], Yuejin TANG [2], Meng SONG [1, 2], Bin WEI[2], Shijie CHENG[2] and Yuan PAN[2]

[1]College of Electrical Engineering and Information, Changsha University of Science and Technology, Changsha 410076, China

[2]R& D Center of Applied Superconducting, Huazhong University of Science and Technology, Wuhan 430074, China

[a]E-mail: zysdl0193@sina.com, [b]E-email: zhouysdl@vip.sina.com

Abstract. To research the key problems of storage energy and pulse forming in pulsed power, a pulse magnet made of Bi2223/Ag high-temperature superconducting (HTS) tapes applied in pulsed power experiment was developed. After determining electromagnetic characteristics of the magnet, a pulse forming network was designed. HTS magnet was immersed in liquid nitrogen bath, experiments were carried out about discharging pulse current to resistance load based on HTS magnet energy storage (SMES). The results show that pulse current waves were obtained through adjusted circuit construction and magnet parameters by acting delay of switches in the pulse forming network. The technical schemes about pulse forming based on SMES were presented.

1. Introduction

The key problems of high pulsed power technology are energy storage systems which can support high density of energy storage and power, discharge pulse current wave well controllable and repetitive, satisfy different load required, and require simple composing of system. Thus improving energy storage density, repeated frequency, making the devices light, miniaturization and practicality, these are the developing direction in the future. Superconducting magnet energy storage (SMES) provides advantages of high-efficiency storage, high density storage, fast discharge energy and easy control [1].

Low-temperature superconducting technology was applied to produce some strong magnetic-field and storage energy [2], high-temperature superconducting (HTS) tape of Bi series can work at 5-10T under 30 K temperature, more lower the temperature, more stronger the working magnetic-field [3], the stronger magnetic-field can be obtained at higher temperature range by Y series high-temperature superconductor. The sufferance energy density of high-temperature superconductor are more several amount levels than low-temperature superconductor's [4], high-temperature superconducting magnet has better resisting ability to disturb than low-temperature superconducting magnet [5], such as disturbing in normal run, magnet charging and discharging [6]. The capability of high-temperature magnet can be improved by augmenting thermal capacity of superconducting coil under a well cooling condition [7]. Some researches on high-temperature SMES applied in pulsed power have been reported [8, 9]. The authors' former investigation has discovered the high-temperature superconductor a good characteristic for pulse over-current [10, 11]. Therefore the application of HTS for future pulsed power would be very promising.

To discuss methods on pulse forming based on high-temperature SMES, a pulse magnet made of Bi2223/Ag high-temperature superconducting tapes applied in pulsed power experiment was developed. After determining electromagnetic characteristics of the magnet, a pulse forming network was designed. HTS magnet was immersed in liquid nitrogen bath, experiments were carried out about discharging pulse current to resistance load based on HTS magnet energy storage. The results show that pulse current waves were obtained through adjusting circuit construction and magnet parameters, the technical schemes about pulse forming based on SMES were put forward.

2. Experimental HTS pulse magnet

2.1. Magnet Structure

Hollow single-solenoid HTS pulse magnet is shown in figure 1, HTS double-pancake coil is shown in figure 2, and the specifications are shown in table 1. Double-pancake coil was formed framework with copper-coil, the insulated Bi2223/Ag HTS tapes were enlaced on the framework, and double-pancake coils were piled up to be a single-solenoid coil. To avoid short circuit effect at transformer subaltern side, the cooper-coil is cracking little, and the brim is protruding at the top and concave at the below along inside copper-coil, buttoning one another along coaxial, so that improve the whole magnet resisting machine stress effect. The magnet exterior-diameter is 272 mm, the inner-diameter 142 mm, height 100 mm. Use 1 mm epoxy board to insulate and exchange heat. The magnet is linked upper and nether flange and 6 heat exchange pieces of epoxy between two coils, and by using 8 radixes of steeliness heat exchange poles. The length of heat exchange pole is 130 mm, its diameter is 12 mm, the passing aperture are all 13 mm, the gaps is crammed with indium to insure the heat exchange well.

Figure 1. Experimental HTS pulse magnet

Figure 2. HTS double-pancake coil

To consider the requirement of research on pulse forming method based on high-temperature SMES and electromagnetic characteristics of single double-pancake coil, there are 5 tie-ins in magnet, each double-pancake coil was designed alone current lead, each current lead was fetch out by copper down-lead (Using two voluble copper leads with 150 mm length and 3 mm diameter) to provide experimental connection. The current lead was made of copper bar, the same width with the tape, which is fixed on the epoxy bracket, and the bracket board is fixed on magnet flange. Two superconducting taps of each double-pancake coil and the jointing length of current down-lead are all 5 cm, jointing well. There are two voltage detection taps in each double-pancake, and the distance between voltage detection tap and the jointing point is 8 cm, so that it avoids detection error because of jointing superconducting tape over temperature to damage.

Table 1. Specifications of HTS double-pancake coil

Specification	Parameter
Tape manufacturer	Beijing Innao ST,Ltd.
Tape	Bi-2223/Ag tape
Width	4.2 mm
Thickness	0.23 mm
Tape critical current	68.7 A (77 K， self-field)
Length of double-pancake coil tape	231.25 m
Coil inner diameter	150 mm
Coil exterior diameter	272 mm
Coil(double-pancake) circle number	176×2 circles
Critical current of coil	28.7 A (77 K， self-field)
Insulation between pancakes	0.5 mm epoxy
Diameter of inside heat exchange canister	145 mm
Thickness of inside heat exchange canister	3 mm
Outboard insulation of inside heat exchange canister	2 mm solidify apoxy glass fabric
Double-pancake coil self-induction	123 mH

2.2. Magnet Parameter

The HTS double-pancake coils are 5, the height of the superconducting magnet is 56 mm, the sum of coil circles: $352 \times 5 = 1760$, the length of HTS Bi2223/Ag tape in magnet: $231.25 \times 5 = 1156.25\ m$, the quench voltage of magnet under DC current: 115.6mV (criterion: 1 μV/cm).

Magnetic field result: under 77 K, current 20 A, radial magnetic field $7.616 \times 10^{-2}\ T$, axis magnetic field 0.3174 T, inductance 0.615 H, energy storage 123.0 J.

Immersed in liquid nitrogen bath, and self-field, the magnet critical current 19.4 A. The magnet resistance in room temperature 20.87 Ω (13 °C)

3. The DC characteristic of magnet

As a first step, to make sure electric characteristic of HTS magnet, immersed in liquid nitrogen bath and self-field, DC characteristic experiment of HTS magnet was carried out, the critical current of the magnet was determined.

The DC characteristic of magnet is shown in figure 3, use 1 μV/cm as criterion of quench, the total length of HTS tapes applied for magnet is 1156.25 m, quench voltage is $115\ mV + 230\ mV$ (the resistance voltage of current copper lead between HTS double-pancake coils)$=345\ mV$, corresponding current is 19.4 A. As is shown in figure 3, when the magnet doesn't quench, the $I-V$ curve of DC current flatness, the obtained voltage is resistance voltage of current lead (copper lead) between 5 double-pancake coils. When the magnet quench, the normal region enlarge rapidly, the normal resistance augment incessantly, the normal voltage augment rapidly, the $I-V$ curve become cragged, therefore it shows that the critical current of magnet is 19.4 A under 77 K and self-field.

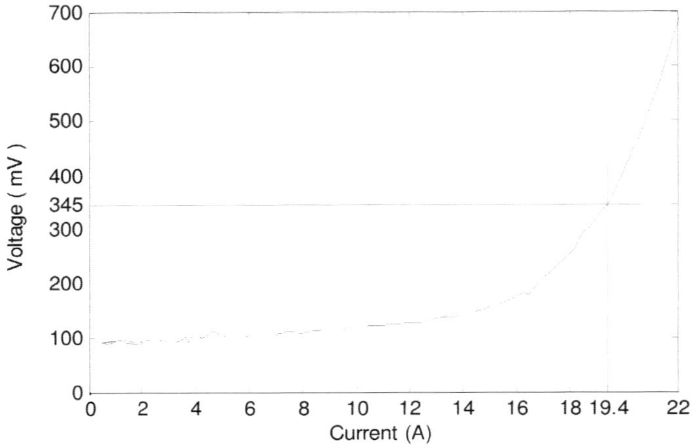

Figure 3. DC characteristics of HTS magnet(77 K, self-field)

4. Pulse forming experiment

4.1. Experiment circuit.

As is shown in figure 4, the cooling condition of magnet is immersed in liquid nitrogen bath, Ls_1, Ls_2 are two superconducting coils of HTS magnet, I_s is constant current source adjustable, D_1、 D_2、 D_3、 D_4 are diodes, R_1 is energy consumption resistance of magnet protection (1.6 Ω), R_2 is load resistance (0.86 Ω), F_1 is DC diffluent implement (200A/75mV), F_2 is coaxial diffluent implement, the resistance 3.1058 mΩ. There are three processes in the experiment, magnet charge, energy storage, and magnet discharge. The charge current and discharging load current are detected with oscillograph through F_1, F_2, respectively. The magnet voltage and the load voltage are detected with oscillograph also.

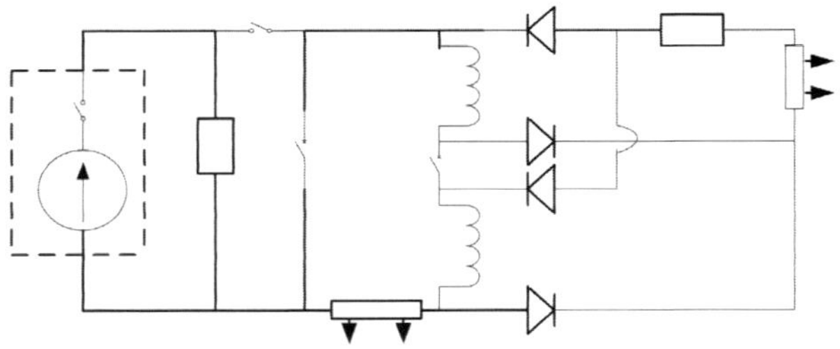

Figure 4. Experiment circuit of pulse forming based on SMES

4.2. Magnet in Series Discharge

K_2 is cut off, K_3 and K_1 are closed, superconducting coil L_{S1} and L_{S2} are charged in series. Close K_2, cut off K_1, L_{S1}, L_{S2} storage energy. As discharge experiment, cut off K_2, the two superconducting coils in series discharge to load. Figure 5 shows the obtained representative pulse current wave, the pulse current amplitude of load is 14.1 A, the attenuation time of the pulse current is 230 ms, the export maximal pulse power is 128.3 W.

4.3. Mixed Discharge of Magnet in Series and parallel

After two superconducting coils in series charged and storage energy, cut off K_2 and K_3 at the same time, the two superconducting coils in parallel discharge, or adjust the acting times of two switches, change discharge circuit connection, the HTS magnet in series or parallel discharge to obtained load pulse current with different waves.

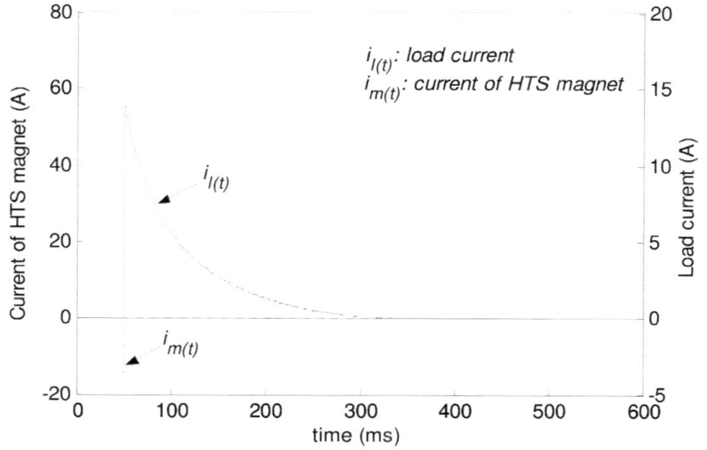

Figure 5. Pulse current wave as two HTS coils in series discharge

Figure 6 shows the experimental results that K_3 delay cutting off K_2 31.163 ms, the two load current peak value are 12.8 A and 13.1 A respectively, the current attenuation time is 101.6 ms, the two peak value of export pulse power are 107.8W and 106.4W. It shows that a good load pulse current wave can be obtained through adjusting switch action time to change the circuit or parameter of superconducting magnet discharging to load.

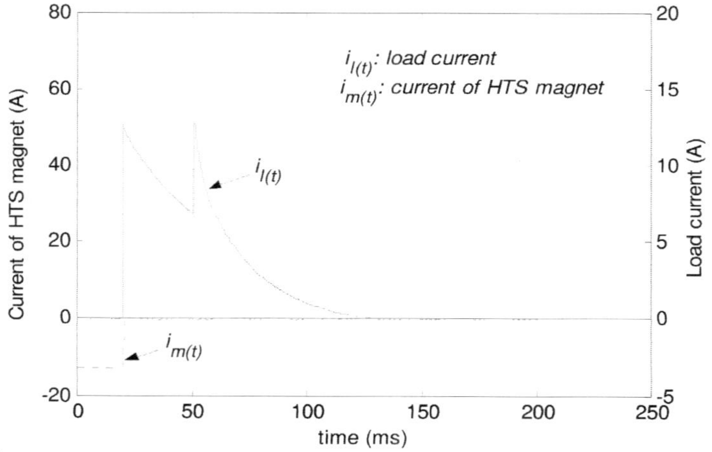

Figure 6. Pulse current wave as two HTS coils in series and parallel discharge

5. Conclusions

A experimental HTS pulse magnet was developed, under the condition of immersed nitrogen bath and self-field, DC characteristic experiment on magnet was carried out. After determining the magnet critical current, the pulse forming circuit was designed. To discuss pulse forming method based on high-temperature SMES, pulse forming discharge experiment to 0.86 Ω resistance load was carried out after the magnet charged. The experimental results show that the HTS magnet has a good pulse discharge characteristic, pulse current waves can be generated through adjusting circuit construction and magnet parameters by acting delay of switches in the pulse forming network. The technical schemes about pulse forming based on SMES would be very promising.

Acknowledgments
The work was supported by National High Technology Research and Development Program of China (863 Program) under Contract No.2007AA03Z243, National Natural Science Foundation of China under Contract No.50647026, Hunan Provincial Natural Science Foundation of China under Contract No.07JJ3101 and 06JJ20044, and by Hunan Provincial Science and Technology Program of China under Contract No.2007FJ3008.

References

[1] Barnes P N, Rhoads G L, Tolliver J C , Sumption M D and Schmaeman K W 2005 *IEEE Trans. on Magn.* **41** 268
[2] Liu X and Wang F 2004 *Proceedings of the CSEE* **24** 172
[3] Sneary A B, Friend C M, Vallier J C and Hampshire D P 1999 *IEEE Trans. on Appl. Supercond.* **9** 2585
[4] Lue J W, Dresner L, Schwenterly S W, Aized D, Campbell J M and Schwall R E 1995 *IEEE Trans. on Appl. Supercond.* **5** 230
[5] Wu C, Bai Z and Wang J 2005 *Proceedings of the CSEE* **25** 109
[6] Iwasa Y 1994 *Case studies in superconducting magnet* (New York: Plenum Press)
[7] Hisashi I, Benjamin J H and Yukikazu I 2001 *IEEE Trans. on Appl. Supercond.* **11** 1852
[8] Oberly C E, Kozlowski G, Gooden C E, Lenard R X, Sarkar A K, Maartense I and Ho J C 1991 *IEEE Trans. on Magn.* **27** 509
[9] Juengst K P and Salbert H 1996 *IEEE Trans. on Magn.* **32** 2272
[10] Zhou Y, Tang Y, Li J, Wei B and Shi J 2005 *Physica C* **433** 37
[11] Zhou Y, Tang Y, Li J, Song Q, Wei B,Yao T and Shi J *IEEE Trans. on Appl. Supercond.* **15** 3906

Energy transfer in Dy^{3+} and Al^{3+} co-doping $ZnNb_2O_6$

Jing GOU[1], Yuhua WANG[1,2], Huizhang GUO[1]

1 Department of Material Science, Lanzhou University, Lanzhou 730000, P. R. China

2 Key Laboratory for Magnetism and Magnetic Materials of the Ministry of Education, Lanzhou University, Lanzhou 730000, P. R. China

E-mail: wyh@lzu.edu.cn

Abstract. Dy^{3+}, Al^{3+} codoping $ZnNb_2O_6$ phosphors were synthesized by solid state reaction. The luminescence intensity of $ZnNb_2O_6{:}xDy^{3+}(0.02 \leqq x \leqq 0.13)$ can be enhanced by codoping with Al^{3+} under 365 nm excitation, but weakened under 254 nm excitation, because energy can be tranferred from CTS to the excitation states of Dy^{3+}. The results indicate that under 365 nm excitation the relative intensities of the peaks at 490 nm and 575 nm were varied with the increase of Al^{3+} content, this phenomenon causes the movement of chromaticity coordinates from the white to the yellow side, and could be helpful for developing new LEDs phosphors.

1. Introduction

Nowadays, the versatility of luminescent materials have attracted researchers' attention, because of their application in various fields, e.g., fluorescent lamps, display devices, detector systems, immunoassays, and scintillates of phosphor marking [1-4]. In these devices, luminescent materials absorb the energy and then convert it to visible light.

As interesting photoactive host materials, the luminescent properties of niobates have been studied extensively by many researchers, e.g., $LiNbO_3$ [5-7], $KNbO_3$ [8,9], $LnNbO_4$ (Ln= Pr, La, Ga, and Y) [10,11], $KLaNb_2O_7$ [12], $Sr_xBa_{1-x}Nb_2O_6$ [13,14], and $Ba_5Nb_4O_{15}$ [15]. Zinc niobate ($ZnNb_2O_6$, ZN) with a columbite crystal structure has been known as one of binary niobate compounds with excellent dielectric properties at microwave frequencies: $Q \times f$=87300 GHz, ε_r =25 and $\tau_f = -56$ ppm/℃ [16-19]. It has very low loss and high dielectric constant, thus is a promising candidate for applications in microwave devices [20-22].

Dy^{3+} ion is a good activator for preparation of electron-trapping luminescence materials. Defects and traps could be produced by substitution of the divalent alkaline-earth ions in the matrix with trivalent Dy^{3+} ion or by high-energy radiation. Electrons or holes could be trapped by these defects, then the electrons and holes recombine, and light will emit by radiation released from traps with different wavelength.

Dy^{3+} has two dominant emission bands in the emission spectra in many host matrixes such as $Ba_3Gd(PO_4)_3$, $Mg_2Gd_8(SiO_4)_6O_2$ and SrB_4O_7 [23-26]. The band located at 574 nm (yellow) attributes to the hypersensitive transition $^4F_{9/2}$-$^6H_{13/2}$ ($\triangle L = 2$, $\triangle J = 2$), and another band located at 487 nm (blue) corresponds to the transition $^4F_{9/2}$-$^6H_{15/2}$. The intensity of the former varies along with the environment more obviously than that of the latter. In the CIE 1931 chromaticity diagram, the line linking the yellow with blue wavelength crosses the white light region; therefore, by adjusting a suitable yellow-to-blue intensity ratio (Y/B), the chromaticity coordinates of the phosphor doped with Dy^{3+} ions could be controlled in the white region. There are several reports about the phosphors doping with Dy^{3+} ions, which were studied for direct white light production [23-27]. Furthermore, the codoping Al^{3+} may be beneficial in enhancing the luminescent intensity of phosphor [28,29]. In this work, the photoluminescence (PL) of $ZnNb_2O_6$:Dy^{3+} codoped with Al^{3+} evaluated under the excitation of ultraviolet (UV) was investigated.

2. Experimental:

Powder samples $ZnNb_2O_6$:$x$$Dy^{3+}$ ($0.02 \leqq x \leqq 0.13$) and $ZnNb_2O_6$:*0.09*Dy^{3+},$y$$Al^{3+}$($0.02 \leqq y \leqq 0.13$) were prepared by solid state reactions. The purities of ZnO and Nb_2O_5 are analytical reagent (AR), and that of Dy_2O_3 and Al_2O_3 is better than 99.99%. The doping amount x of Dy_2O_3 is from 0 to 13%, and the same to that of Al_2O_3. Stoichiometric amounts of the starting reagents ZnO, Nb_2O_5, Dy_2O_3, and Al_2O_3 were thoroughly mixed and ground together. The mixture was heated at 1000°C for 4 h and cooled to room temperature subsequently.

The powder X-ray diffraction (XRD) patterns of all samples were carefully collected in the 2θ range 10–80° (Rigaku D/MAX-2400 X-ray diffractometer with Ni-filtered Cu Kα radiation). The luminescence spectra were recorded on a FLS-920T fluorescence spectrophotometer with Xe 900 (450 W xenon arc lamp) as the light source. All the spectra were recorded at room temperature.

3. Results and discussion:

The XRD patterns of the samples $ZnNb_2O_6$:$x$$Dy^{3+}$,$y$$Al^{3+}$ are shown in Fig. 1 (a) and (b), through indexed by orthorhombic structure, those were proved as single phase, in line with the JCPDS card 76-1827.

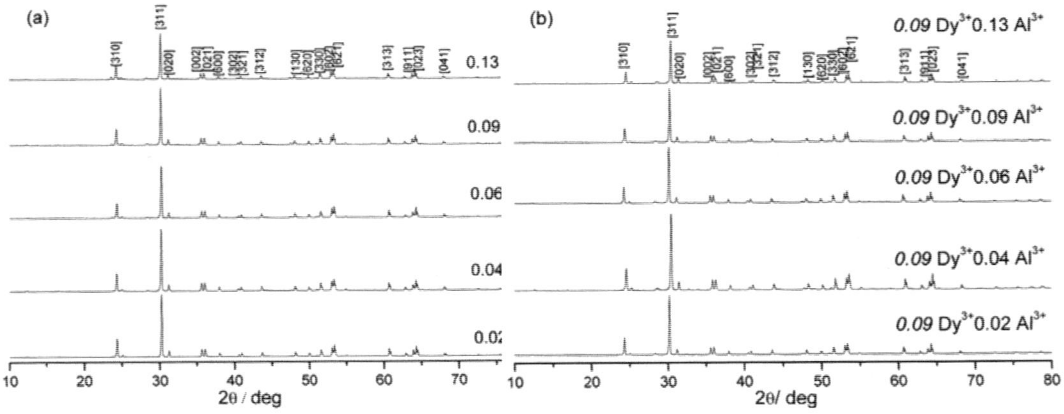

Figure 1. (a). The X-ray powder diffraction patterns of $ZnNb_2O_6$:$x$$Dy^{3+}$; (b). The X-ray powder diffraction patterns of $ZnNb_2O_6$:*0.09*Dy^{3+},$x$$Al^{3+}$

It is well known that the host lattice $ZnNb_2O_6$ crystallizes in the orthorhombic symmetry with space group Pbcn (60), through analyzing XRD ,it can be calculated that a=14.2527Å, b=5.7170Å and c=5.0378Å. Fig.2 shows the crystal structure of $ZnNb_2O_6$. Along the a-axis, the layers of Zn^{2+} and Nb^{5+} are separated by planes of O^{2-}. The ZnO_6 and NbO_6 octahedra are connected by common corners, and each NbO_6 octahedron is linked with six ZnO_6 octahedron, each ZnO_6 octahedron is linked with three ZnO_6 octahedron and three NbO_6 octahedron, the NbO_6 and ZnO_6 are edge-shared groups.

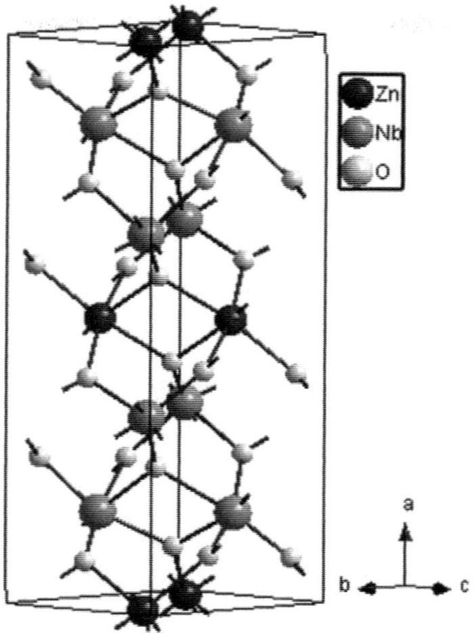

Figure 2. The crystal structure of $ZnNb_2O_6$

Figure 3 shows several emission spectra of $ZnNb_2O_6$:$x$$Dy^{3+}$ under 254 nm excitation. From Fig. 3, the peaks at 490 nm and 575 nm are observed and could be attributed to $^4F_{9/2}$-$^6H_{15/2}$, $^4F_{9/2}$-$^6H_{13/2}$ transitions of Dy^{3+} ion. It's obvious that the intensitys of the peaks at 490 nm and 575 nm are gradually enhanced along with x increasing and become dominant when x=0.09, then weakened when x>0.09.

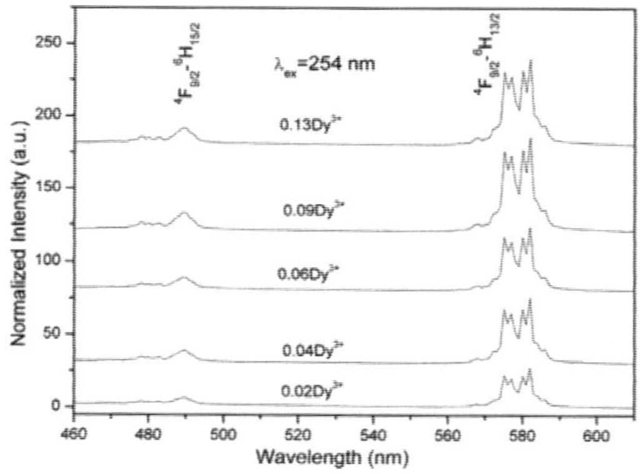

Figure 3. The emission spectra of $ZnNb_2O_6$:$x$$Dy^{3+}$($0.02 \leqq x \leqq 0.13$) ($\lambda_{ex}$=254 nm)

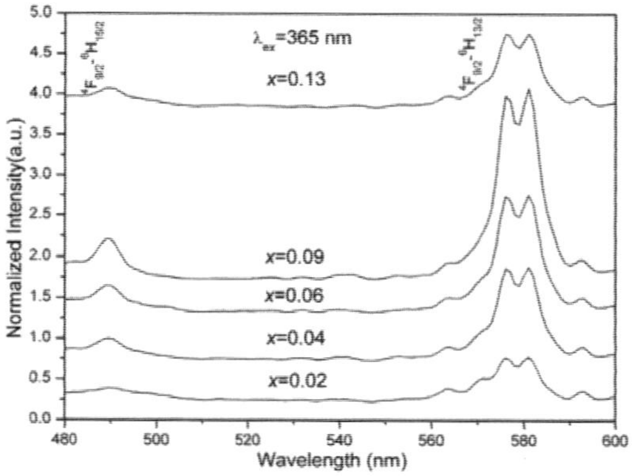

Figure 4. The emission spectra of $ZnNb_2O_6$:$x$$Dy^{3+}$($0.02 \leqq x \leqq 0.13$) ($\lambda_{ex}$=365 nm)

Several emission spectra of $ZnNb_2O_6$:$x$$Dy^{3+}$ and $ZnNb_2O_6$:*0.09*Dy^{3+} $x$$Al^{3+}$ under 365 nm excitation are shown in Figure 4 and 5, respectively. The emission peaks at 490 nm and 575 nm are observed, which could be also attributed to $^4F_{9/2}$-$^6H_{15/2}$, $^4F_{9/2}$-$^6H_{13/2}$ transitions of Dy^{3+} ion. It's obvious that the intensitys of the peaks at 490 nm and 575 nm are gradually enhanced along with x increasing and become dominant when x=0.09, then weakened when x>0.09 in the two figures, however, doping Al^{3+} ions can enhance the intensity of luminescence effectively and the intensity of dominant emission peak of $ZnNb_2O_6$:*0.09*Dy^{3+} *0.09*Al^{3+} could be enhanced as 3.15 times stronger than that of

$ZnNb_2O_6$:*0.09*Dy^{3+}, so that it's a potensitial LED phosphors. As we know, the increase in the Dy^{3+} or Al^{3+} concentration too much would cause a cross-relaxation process between neighboring Dy^{3+} ions, leading to quenching of the emission. Y. H. Wang et al, have reported that the luminescence intensity could be enhanced by doping with less Al^{3+} ions [28,29]. But in Figure 6, the intensity of dominant emission peak of $ZnNb_2O_6$:*0.09*Dy^{3+} *0.09*Al^{3+} is much weaker than that of $ZnNb_2O_6$:*0.09*Dy^{3+} under 254 nm excitaion, that is opposite to Figure 5.

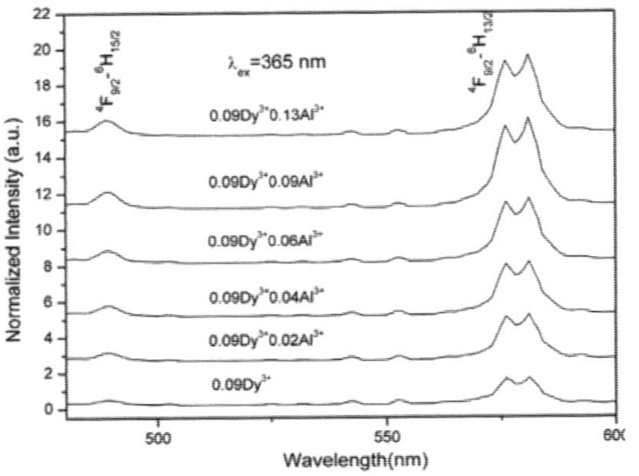

Figure 5. The emission spectra of $ZnNb_2O_6$:*0.09*Dy^{3+} xAl^{3+} ($0.02 \leqq x \leqq 0.13$) (λ_{ex}=365 nm)

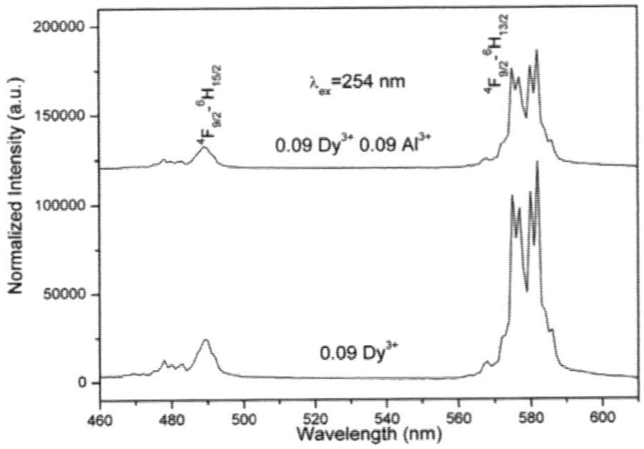

Figure 6. The emission spectra of $ZnNb_2O_6$:*0.09*Dy^{3+} *0.09*Al^{3+} and $ZnNb_2O_6$:*0.09*Dy^{3+} (λ_{ex}=254 nm)

Figure 7 shows the excitation spectra of $ZnNb_2O_6$:*0.09*Dy^{3+} *0.09*Al^{3+} and $ZnNb_2O_6$:*0.09*Dy^{3+} with the emission at 575 nm. In the excitation spectrum, it's obvious observed that there is a stronge absorption band near 270 nm, that can be attributed to the charge transfer band (CTS) between O^{2-} and Dy^{3+} ions. It is obviously exhibited that the CTS became littler narrower than another and had a little bathochromic shift when codoping with Al^{3+}. In the range of 320 nm-470 nm, the absorption is enhanced by coding with Al^{3+}, the series of line spectra in this range are ascribed to the transitions from the ground state to excitation states in the $4f^9$ configuration of Dy^{3+} ions. The peaks at 327 nm, 354 nm, 367 nm, 384 nm, 426 nm and 454 nm are attributed to $^6H_{15/2}$-$^4K_{15/2}$, $^6H_{15/2}$-$^4M_{15/2}$, $^6H_{15/2}$-$^4P_{3/2}$, $^6H_{15/2}$-$^4M_{21/2}$, $^6H_{15/2}$-$^4G_{11/2}$ and $^6H_{15/2}$-$^4I_{15/2}$ transitions of Dy^{3+} respectively. The radii of Zn^{2+}, Nb^{5+} and Al^{3+} ions are 0.74 Å, 0.64 Å and 0.51 Å, and the radius of Al^{3+} ions is more close to that of Nb^{5+} ions, thus sites of Nb^{5+} could be more probably occupied by Al^{3+} ions. In addition, the electronegativity of Al^{3+} ions is little bigger than that of Nb^{5+} ions. It's obviously showed in figure 2 that Nb^{5+} and Zn^{2+} ions are connected with O^{2-} ions, when codoping with Al^{3+} ions in Nb^{5+} sites, the covalence of Dy^{3+} ions at Zn^{2+} sites and O^{2-} ions increases, thus the energy difference for electrons transition between the two ions become smaller, CTS between the two ions moves to lower energy location. Thus when doping with a few Al^{3+} ions, it's more easily transfering energy from CTS to the excitation states of Dy^{3+} by relaxation effect.

Figure 8 presents the chromaticity coordinates (x, y) of $ZnNb_2O_6$:*0.09*Dy^{3+} xAl^{3+} excited under 365 nm in the CIE 1931 chromaticity diagram. The blackbody radiation locus is indicated by the solid curve. It is obvious that the chromaticity coordination of $ZnNb_2O_6$:*0.09*Dy^{3+} xAl^{3+} gradually moves to the warm-white side and approaches the yellow region with increasing x,then x>0.09 move to the opposite direction.

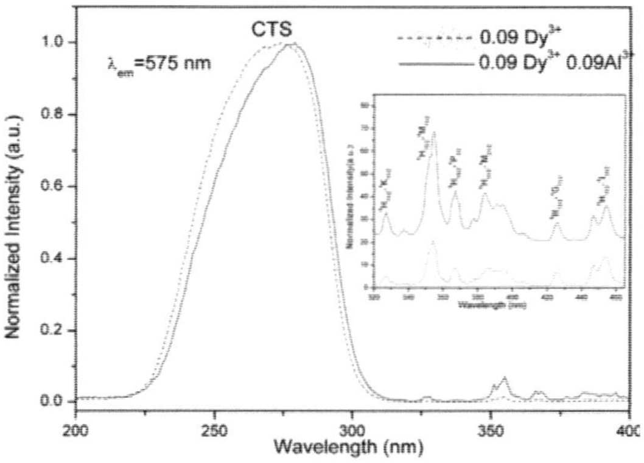

Figure 7. The excitation spectra of $ZnNb_2O_6$:*0.09*Dy^{3+} *0.09*Al^{3+} and $ZnNb_2O_6$:*0.09*Dy^{3+} (λ_{em}=575 nm)

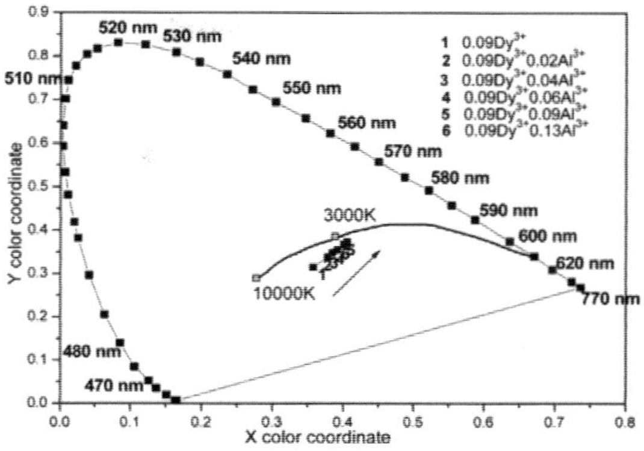

Figure 8. the chromaticity coordinates (*x*, *y*) of $ZnNb_2O_6$:*0.09*Dy^{3+} xAl^{3+} excited under 365 nm in the CIE 1931 chromaticity diagram. The blackbody radiation locus is indicated by the solid curve.

4. Conclusions

The luminescence property of $ZnNb_2O_6$:xDy^{3+},yAl^{3+} phosphor was investigated under UV excitation. Through an analysis of the luminescence spectra, it is obvious that energy can be transferred from CTS to excitation states of Dy^{3+} ions by codoping with Al^{3+} ions, the peaks located in the range 320 nm-470 nm is accordingly enhanced in the excitation spectra. Thus the emission peaks can be enhanced under 365 nm excitation and the relative intensity variety of main emission peaks at 490 and 575 nm make the results in the movement of the chromaticity coordinates from the cold-white to the warm-white side by codoping with Al^{3+} ions. In other words, through varying the amount of *x*, the chromaticity coordinates of $ZnNb_2O_6$:*0.09*Dy^{3+} xAl^{3+} can be expediently adjusted, thus $ZnNb_2O_6$:*0.09*Dy^{3+} *0.09*Al^{3+} is a potential LED phosphors.

Acknowledgments

This work was supported by the Project of the combination of Industry and Research by the Ministry of Education and Guangdong Province (Grant No.0712226100023) and Key Science and Technology Project of Gansu Province (Grant no. 2GS064-A52-036-03).

References

[1] Lei B, Liu Y, Liu J, Ye Z and Shi C 2004 J. *Solid State Chem.* **177** 1333.
[2] Matsuzawa T, Aoki Y, Takeuchi N and Murayama Y 1996 *J. Eletrochem. Soc.* **143** 2670.
[3] Gordon W O, Carter J A and Tissue B M 2004 *J. Lumin.* **108** 339.
[4] Bertrand-Chadeyron G, El-Ghozzi M, Boyer D, Mahiou R and Cousseins J C 2001 *J. Alloys Compds.* **317** 183.
[5] Hreniak D, Speghini A, Bettinelli M and Strek W 2006 *J. Lumin.* **219** 119.
[6] Yao J H, Yan B X, Chen Y H, Xu J J, Zhang G Y, Ruan Y F and Li W R 2007 *Opt. Mater* **27** 373.
[7] Koepke C, Wisniewski K, Dyl D, Grinberg M and Malinowski M 2006 *Opt. Mater.* **28** 137.
[8] Wen C H, Chu S Y and Wen C K 2004 *J. Cryst. Growth* **269** 479.
[9] Fang T H, Hsiao Y J, Chang Y S and Chang Y H 2006 *Mater. Chem.Phys.* **100** 418.
[10] Misra S K and Andronenko S I 2004 *J. Phys. Chem. B* **108** 9397.
[11] Xiao X Z and Yan B 2006 *J. Alloys Compd.* **421** 252.
[12] Akihiko K 1997 *Chem. Mater.* **9** 664.
[13] Han T P J, Jaque F, Jaque D and Ivleva L 2006 *Chem. Phys. Lett.* **417** 196.

[14] Ramirez M O, Bausa´ L E, Speghini A, Bettinelli M, Ivleva L, Garı´ and Sole´ 2006 *J. Phys. ReV. B* **73** 35119.
[15] Srivastava A M, Ackerman J F and Beers W W 1997 *J. Solid State Chem.* **134** 187.
[16] Pullar R C, Breeze J D and Alford N M 2005 *J. Am. Ceram. Soc.* **88** 2466.
[17] Lee H J, Kim S J and Kim I T 1997 *Mater. Res. Bull.* **32** 847.
[18] Lee H J, Kim I T and Hong K S 1997 *Jpn. J. Appl. Phys.*, **36** 1318.
[19] Kim D W, Ko K H and Hong K S 2001 *J. Am. Ceram. Soc.* **84** 1286.
[20] Zhang Y C, Yue Z X, Qi X, Li B, Gui Z L and Li L T 2004 *Mater. Lett.* **58** 1392.
[21] Zhang Y C, Li L T, Yue Z X and Gui Z L 2003 *Mater. Sci. Eng., B, Solid-State Mater. Adv. Technol.* **99** 282.
[22] Simoes A Z, Gonzalez A H M, Cavalheiro A A, Zaghete M A, Stojanovic B D and Varela J A 2002 *Ceram. Int.* **28** 265.
[23] Su Q, Liang H B, Li C Y, He H, Lu Y H, Li J and Tao Y 2007 *J. Lumin.*, **122-123** 927.
[24] Su Q, Pei Z, Lin J and F. Xue 1995 *J. Alloys Compd.*, **225** 103 .
[25] Gou J, Wang Y H and Li F 2008 *J. Lumin.*, **128** 728.
[26] Gou J, Wang Y H and Li F 2007 *J. Eletrochem. Soc.* **10** F31.
[27] Su Q, Pei Z, Zeng Q, Chi L, Liu S, Lin J, Wang S and Lu Y 1999 *Mater. Sci. Forum* **228** 315.
[28] Wang Y H and Wang L L 2006 *Mater. Lett.*, **60** 2645.
[29] Ryan T G and Jackson S D 2007 *Opt. Commun.* **273** 159.

Improved Performance of Strontium Aluminate Luminous Coating on the Ceramic Surface

Fang GAO[1], Zhaoxian XIONG[2], Hao XUE[2], Yongxi LIU[2]

[1]College of Chemistry and Chemical Engineering,

[2]College of Materials, Xiamen University, Xiamen 361005, P.R. China

Email: zxxiong@xmu.edu.cn

Abstract. Phosphor of strontium aluminate co-actived by Eu^{2+} and Dy^{3+} is one kind of important afterglow luminescent materials. In this paper, the phosphors were used with transparent glaze for an inorganic luminous coating on the ceramic surface, which was stable even at high temperature. The chemical structure and microstructure of the luminous coating were identified with X-ray diffraction (XRD) and observed with scanning electron microscopy (SEM), respectively. The photoluminescence of the coating was measured by a HITACHI F-4500 fluorescence spectrophotometer. The afterglow property was recorded by a ST-86LA-3 brightness meter. The samples behaved good performances such as high lighting brightness and long after-glowing time.

1. Introduction

Photoluminescent material with long afterglow is a kind of energy storage material, which can absorb both ultraviolet (UV) and visible lights from the sunlight and gradually release the energy in the darkness at a certain wavelength. This kind of long afterglow materials had been widely studied by many researchers [1,2].

In recent years, $Sr_4Al_{14}O_{25}$: Eu^{2+}, Dy^{3+}, as one of the persistent luminescent materials, has been extensively focused because it provides typical emission bands in blue and green regions [3]. Compared with sulfide phosphorescent phosphors, $Sr_4Al_{14}O_{25}$: Eu^{2+}, Dy^{3+} phosphor possesses safe, chemically stable, very bright and long-lasting photoluminescence in addition to no radiation [4,5], which results in wide field of applications, such as luminous paints in highway, airport, building and ceramics, as well as in textile, dial plate of glowing watch, warning signs and escaping routeway [6].

In this paper, luminous phosphor of $Sr_4Al_{14}O_{25}$: Eu^{2+}, Dy^{3+} was mixed with transparent glaze, coated onto ceramic surface and then heat-treated in different conditions. Microstructure, photoluminescence and afterglow properties were characterized in details.

2. Experimental

The powders were uni-axially pressed at 5 MPa into green body disks of 13 mm in diameter and 3.5 mm in thickness. The luminous glazes were prepared by mixed strontium aluminate phosphors with transparent glazes in anhydrous ethanol, which were transferred to a ultrasonic cleaning bath to sonicate for 10 min, then were vigorously grinded for l0 min in a mortar. The contents of strontium aluminate phosphors in the luminous glazes were 10%~30% in weight. The luminous glazes were then coated onto the surface of ceramics.

Moreover, the samples were removed into a crucible with a lid, which were then introduced into a muffle furnace and heated at a rate of 5~10 °C/min and maintained at 920 °C for 10~60 min in air or an active carbon atmosphere, and at last the sample was cooled to ambient temperature in the furnace.

The crystal structures were identified by an X-ray powder diffractometer with CuKα radiation generated at 40 kV/30mA. The morphologies of the coating of the samples were observed with scanning electron microscopy. The excitation and emission spectra of photoluminescence were measured using a fluorescence spectrophotometer (HITACHI, F-4500). The decay curves of the luminous coatings were recorded by a brightness meter (Peking Normal University optical instrument plant, China, ST-86LA-3), after the samples were exposed to irradiation from a 8 W conventional tricolor fluorescent lamp (about 1000 lux) for 30 min.

3. Results and discussion

XRD patterns of the samples were shown in Figure. 1, which were heat-treated at a rate of 5 °C/min and soaked at 920 °C for 15 min in an active carbon atmosphere. The diffraction peaks became stronger with the increasing weight ratios of strontium aluminate powders with the luminous glazes.

The SEM image of the coating of sample was also displayed in Figure. 2. It could be seen that the strontium aluminate powders were randomly distributed into the transparent glazes.

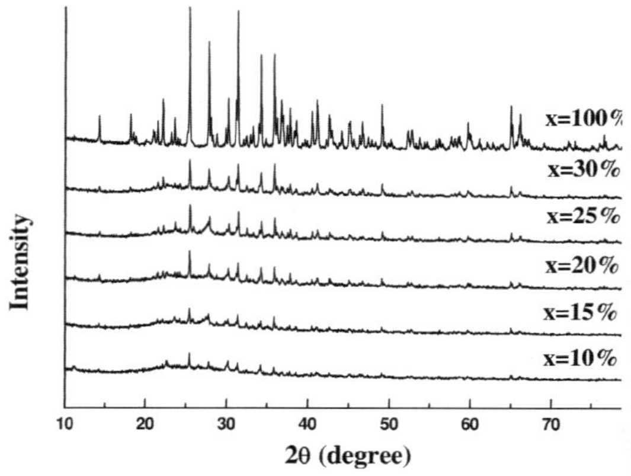

Figure 1. XRD patterns of the samples with different weight ratios of phosphors.

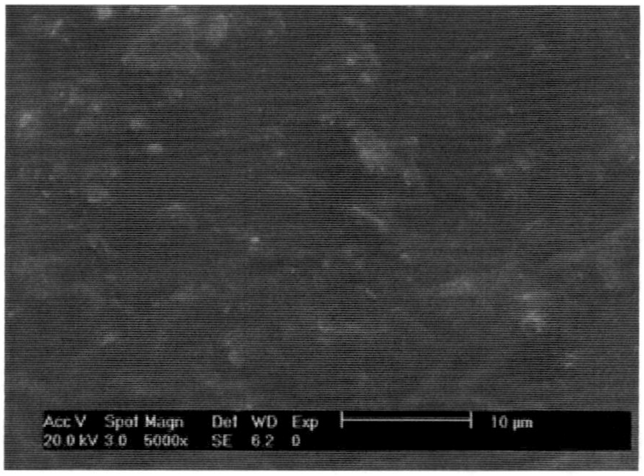

Figure 2. SEM image of the surface coated with 20% in weight of phosphors.

The excitation and emission spectra of the samples were present in Figure. 3 in different weight ratios of strontium aluminate phosphors. The results indicated that main excitation peak was at 360 nm and main emission peak at 489 nm. The $Sr_4Al_{14}O_{25}$: Eu^{2+}, Dy^{3+} luminous phosphor yielded high brightness of blue green luminescence with only one band at ambient temperature. The samples, heat-treated at 920 °C for 15 min, generated a emission at 489 nm, which corresponded to the emission of Eu^{2+} in $Sr_4Al_{14}O_{25}$ lattice [7]. The intensity of the coatings was increased with the increasing weight of the strontium aluminate phosphors, as shown in Figure. 3. When the samples were heat-treated in an active carbon atmosphere, the intensity of the samples were higher than that in air, as displayed in Figure. 4. Compared with the samples at heating rate of 5 °C/min and 8 °C /min, those at a heating rate of 10 °C/min and soaked at 920 °C for 15 min possessed the stronger intensity, as shown in Figure. 5. Meanwhile when the sample was heat-treated at 920 °C for 10 min, the intensity of the coating was stronger than that for the other holding time, as present in Figure. 6.

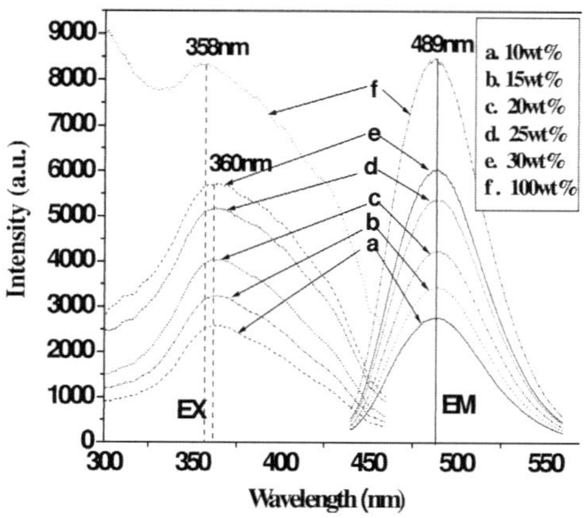

Figure 3. Excitation and emission spectra of different weight of phosphors.

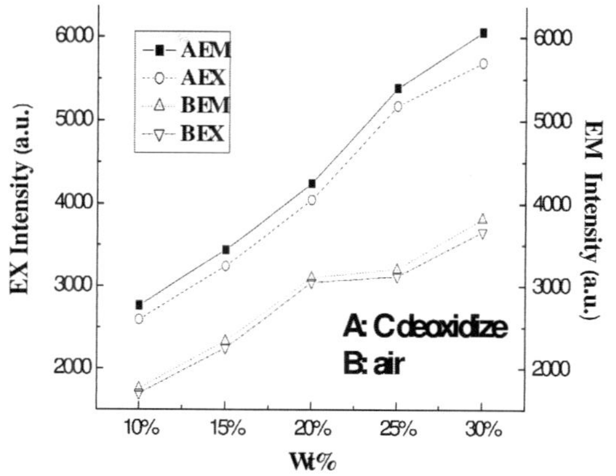

Figure 4. Excitation and emission spectra of the sample heat-treated in different atmosphere.

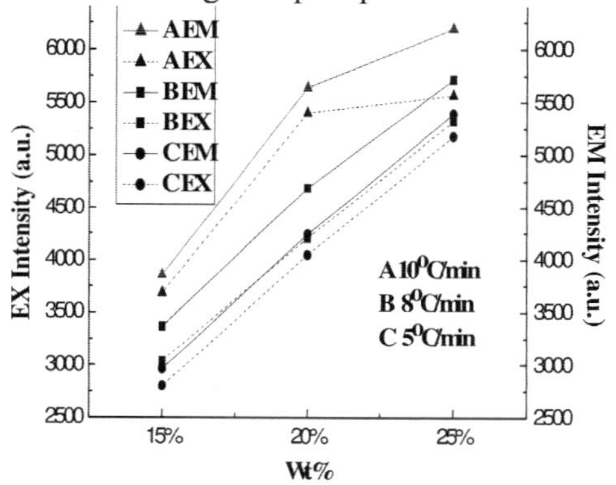

Figure 5. Excitation and emission spectra of the sample treated at different heating rate.

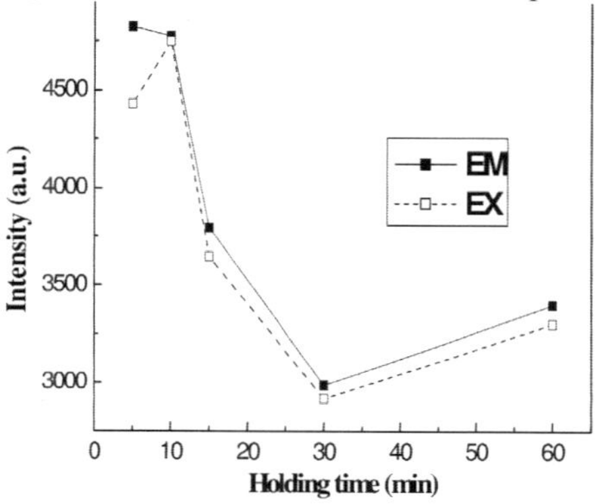

Figure 6. Spectra of the sample treated at 920°C for different holding time.

Afterglow curve of the coating was displayed in Figure. 7, which was heat-treated at a rate of 10 °C/min and soaked at 920 °C for 10 min in an active carbon atmosphere. It could be seen that the coating showed long afterglow property when the coating was efficiently excited by a conventional tricolor fluorescent lamp (8 W, about 1000 lux). When the source lamp was switched off, the intensity

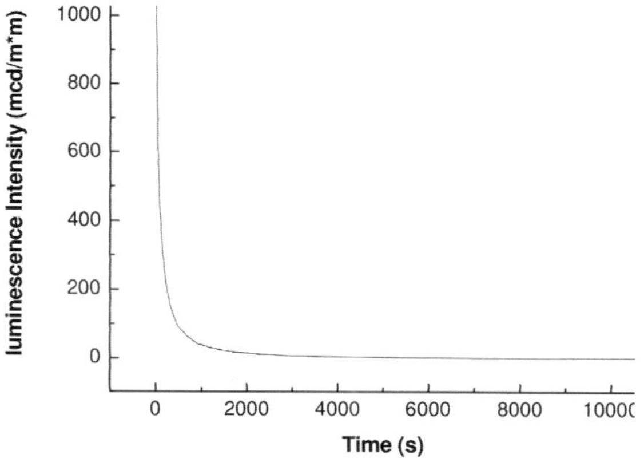

Figure 7. Afterglow decay curve of the coating after exciting for 30 min.

of the afterglow firstly decreased rapidly and finally formed a stable long persistent emission. It was clear that the changing of the intensity of the sample with time was in good accordance with the double-curve equation, $I = ct^{-n}$ [8] , which revealed that the afterglow decay of the phosphor was a complex process. After the excited source was cut off, the afterglow of luminous ceramics allowed to be visually recognized ($\geq$0.32 mcd/m^2) and lasted for over 20 h.

The luminous glazes were coated onto the ceramics in size of 50*50 mm, the digital photo of the luminous ceramics in dark room was shown in Figure 8. The luminous glazes were coated onto on a ceramic disk with 13 mm in diameter and 3.5 mm in thickness, which were also imaged in dark (Figure. 9). All of the samples were heat-treated at a rate of 10 °C/min and soaked at 920 °C for 10 min in an active carbon atmosphere.

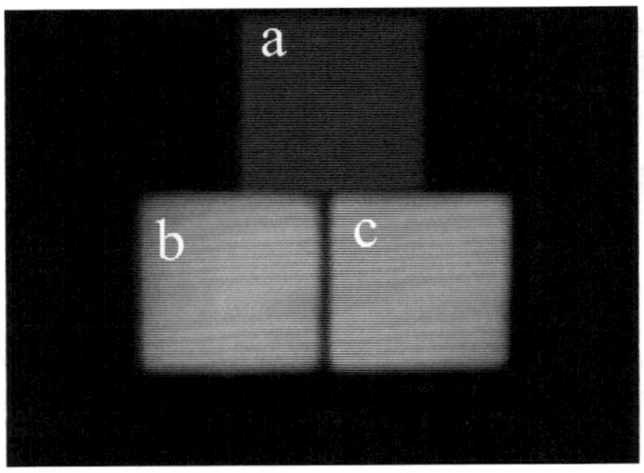

Figure 8. Photo of luminous coating with (a) blank, (b) and (c) 30% in weight.

Figure 9. Photo of luminous coating on a ceramic disk in darkness.

4. Conclusions

In this paper, $Sr_4Al_{14}O_{25}$: Eu^{2+}, Dy^{3+} luminous ceramic coatings were prepared, which showed excellent performance of long afterglow. The main excitation peak of the samples was at 360 nm and the main emission peak at 490 nm, respectively. The coating heat-treated at a rate of 10 °C/min and soaked at 920 °C for 10 min in an active carbon atmosphere possessed the best performance of luminescence and long afterglow. The afterglow time allowed to be visually recognized for over 20 h at darkness after removal of the excited light source.

References

[1] Blasse G and Bril A 1968 *Philips Res. Rep.* **23** 201
[2] Matsuzawa T, Aoki Y, Takeuchi N and Murayama Y 1996 *J. Electrochem. Soc.* **143** 2670
[3] Zhao C L, Chen D H, Yuan Y H and Wu M 2006 *Materials Science and Engineering B.* **133** 200-204
[4] Lin Y H, Tang Z L, Zhang Z T and Nan C W 2002 *Appl. Phys. Lett.* **81** 996
[5] Lin Y H, Tang Z L and Zhang Z T 2001 *Mater. Lett.* **51** 14
[6] Murayama Y, Takeuchi N, Aoki Y and Matsuzawa T 1995 *Phosphorescent phosphor, US Patent* 5,424,006
[7] Chang C K and Mao D L 2004 *Thin Solid Films.* **460** 48–52
[8] Peng T, Liu H, Yang H and Yan C 2004 *Mater. Chem. Phys.* **85** 68–72

Journal of Physics: Conference Series **152** (2009) 012083 doi:10.1088/1742-6596/152/1/012083

Preparation and luminescence properties of Eu^{2+}-doped $CaSi_2O_{2-\delta}N_{2+2/3\delta}$ phosphors

Yunxin GU[1], Qinghong ZHANG[1], Yaogang LI[2], Hongzhi WANG[1]

[1]State Key Laboratory for Modification of Chemical Fibers and Polymer Materials,

[2]College of Materials Science and Engineering, Donghua University,Shanghai 201620, People's Republic of China

E-mail: wanghz@dhu.edu.cn, yaogang_li@dhu.edu.cn.

Abstract. Eu^{2+}-doped $CaSi_2O_{2-\delta}N_{2+2/3\delta}$ phosphors for white LED lamps were prepared by solid-state reaction, and the effects of heat-treatment conditions and the overall composition of host lattice on the optical properties have been discussed. Eu^{2+}-doped $CaSi_2O_{2-\delta}N_{2+2/3\delta}$ displayed a single broad emission band peak at 540 nm, which could be assigned to the allowed transition of Eu^{2+} from the lowest crystal field component of $4f^65d$ to $4f^7$ ground-state level. The excitation band of samples, extending from UV to blue, is extremely wide, so the phosphors are suitable for white LED lamps in combination with a UV or blue LED dies. The highest PL intensity is found for the sample sintered at 1400 ℃. Moreover, the emission intensity decreases when N partially replaces O. A red shift of emission wavelength did not occur with increasing of the N content.

1. Introduction

Rare-earth (RE)-activated oxynitride or nitride luminescent materials have attracted much attention in recent years due to their nontoxicity, interesting luminescence properties, and potential applications as phosphors and pigments. Examples include $M_2Si_5N_8:Eu^{2+}$ (M=Ca, Sr, Ba) [1], $Sr_2Si_5N_8:Eu^{2+}$ [2], Ca-α-Sialon:Eu^{2+} [3-5], Li-α-sialon:Eu^{2+}[6], Y-Si-O-N:Ce^{3+}[7], and $MSi_2O_2N_2:Eu^{2+},Ce^{3+}$(M=Ca, Sr, Ba) [8,9]. Among these (oxy)nitride luminescence materials, Eu^{2+}-doped $CaSi_2O_2N_2$ represents a new class of yellow phosphors with high efficiency, which has strong absorption in the UV-visible spectral region and exhibits a broad emission band peaking at 560 nm [8]. Therefore, this novel yellow $CaSi_2O_2N_2:Eu^{2+}$ phosphor is expected to have applications in white LEDs when a GaN-based blue chip is coupled.

Comparing to the procedure of other (oxy)nitride phosphors mentioned above, the preparation condition of $CaSi_2O_2N_2$ is easier to achieve. As Li et al [8] reported, $CaSi_2O_2N_2$:Eu^{2+} phosphor can be obtained at a low temperature of 1400℃ and no high pressure was needed. To widely apply this phosphor, low sintering temperature is needed. On this aspect, the influence of the temperature on the structure and the optical property is necessarily investigated. In this paper, we attempt to reveal the structural evolution and the optical property variation when calcining temperatures in the range of 1100~1400℃ were applied. Based on these experimental results, optimal processing conditions were obtained.

Besides, a wide range of the O/N ratio enables Eu^{2+}-doped sialon materials to tune their emission spectra by compositional modification [3]. Here, the influence of O/N ratio on the photoluminescence properties of $CaSi_2O_{2-\delta}N_{2+2/3\delta}$ is also investigated.

2. Experimental Section

All powder samples of Eu^{2+}-doped $CaSi_2O_{2-\delta}N_{2+2/3\delta}$ were synthesized by a high-temperature solid-state reaction. The starting materials were $CaCO_3$, SiO_2, Si_3N_4, and Eu_2O_3. The Eu^{2+} mole fractions with respect to the Ca ion is 5% (x=0.05). The starting materials were weighed out in various amounts and subsequently homogeneously wet-mixed by a planetary ball mill for 4-5 h in ethanol. After mixing, the slurry was dried and ground in an agate mortar. Subsequently, the mixtures were fired in alumina crucibles at 1100~1400℃for 10 h in N_2 atmosphere. After that, the samples were cooled to room temperature in N_2 atmosphere and were ground again with an agate mortar.

Phase identification was done by the powder X-ray diffraction pattern (XRD, D/max 2550V, Rigaku, Japan) with Cu-K_α irradiation (λ=1.5406 nm). The luminescence spectra were measured at room temperature using a fluorescent spectrophotometer (JASCO，FP-6600, Japan).

3. Results and Discussion

Figure 1 shows the typical excitation and emission spectra of the Eu^{2+}-doped $CaSi_2O_2N_2$ prepared at 1400℃. The excitation spectra are characterized by a broad band, which corresponds to $4f^7 \rightarrow 4f^65d$ absorption of Eu^{2+} ion. The broad emission band, characteristic of 5d-4f transition in Eu^{2+}, is located in the range of 500~650 nm resulting in a green-yellow emission. The characteristic luminescence of Eu^{3+} is not observed, which exhibits sharp lines between 580 and 650 nm. This suggests that the europium ion in $CaSi_2O_2N_2$ is in the divalent state.

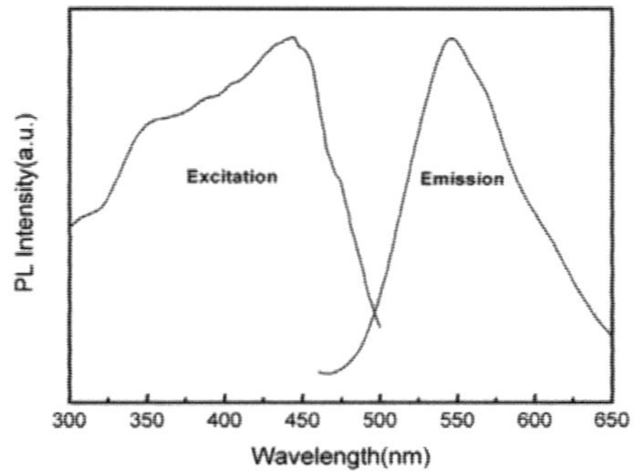

Figure 1. Typical excitation and emission spectra for $CaSi_2O_2N_2$: Eu^{2+} sintered at 1400℃. (λ_{em}= 540nm, λ_{ex}=445nm)

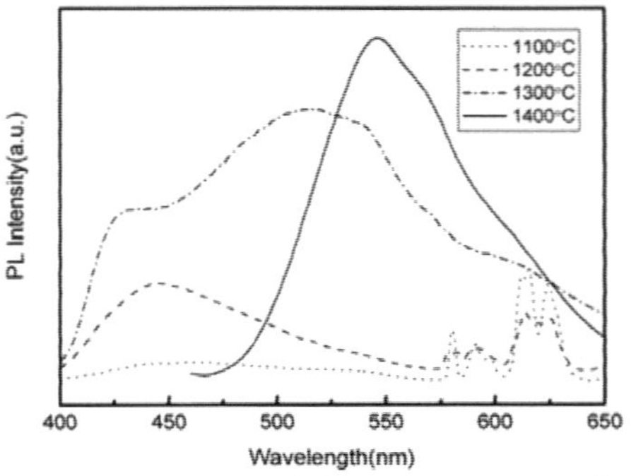

Figure 2. Emission spectra for $CaSi_2O_2N_2$: Eu^{2+} sintered at temperatures in the range of 1100~1400℃.

The Stokes shift of the Eu^{2+}-doped $CaSi_2O_2N_2$, roughly calculated from the spectral data is about 4200 cm^{-1}, which is a typical value for the fd emission from Eu^{2+}. The full width at half-maximum (FWHM) of the emission band is 83 nm, slightly smaller than that of common Eu^{2+} fd emission [10]. The position of the emission band is at longer wavelengths than for Eu^{2+} in oxosilicates showing Eu^{2+} emission typically in the blue spectral region [11]. The observed unusual long-wavelength excitation and emission bands may be mainly because of the nitrogen-rich surrounding of Eu^{2+} ions in $CaSi_2O_2N_2$. Both the higher formal charge of N^{3-} compared with O^{2-} and the lower electronegativity of nitrogen (3.04) compared with oxygen (3.44) (Nephelauxetic effect) would lead to larger ligand-field splitting of the 5d levels and the center of gravity of the 5d states at lower energy. It is also suggested that the preferential orientation of a d-orbital or low-lying state of the conduction band would account for the long emission wavelength [12].

The emission spectra for samples prepared at temperatures in the range of 1100~1400℃ are depicted in Figure 2. At 1400℃, it is the typical emission spectra of the Eu^{2+}-doped $CaSi_2O_2N_2$, and the emission intensity is the highest. When the sintering conducted at lower temperature, the intensity and wavelength of the emission band are obviously changed. At 1300℃, it is observed that a wider band with a maximum at 515 nm and a shoulder at 430 nm. The difference becomes more evident at the temperature of 1200℃, at which the emission band changed to 445 nm. Besides, a sharp line is found at 615 nm, indicating that there is still a part of Eu^{3+} ion unreduced. The emission spectrum at 1100℃ is similar to the one at 1200℃, except that the emission band at 445 nm is almost disappeared, and the peak at 615 nm is much higher.

Figure 3 shows the XRD patterns of the samples synthesized at different temperatures. At 1200℃, it is observed that there are mainly two crystal phases, $CaSiO_3$ and Si_3N_4. Poort et al.[13] has reported that the emission of $CaSiO_3$:Eu^{2+} synthesized at 1200℃ exhibits a broad band centered at 460 nm. Therefore, the blue emission band around 450nm in the emission spectra may be originated from $CaSiO_3$ phase. When the temperature increased to 1300℃, the phase assemblages of the sample was different, with a new crystalline phase detected being $CaSi_2O_2N_2$. Thus, the Eu_{2+} ion incorporated in

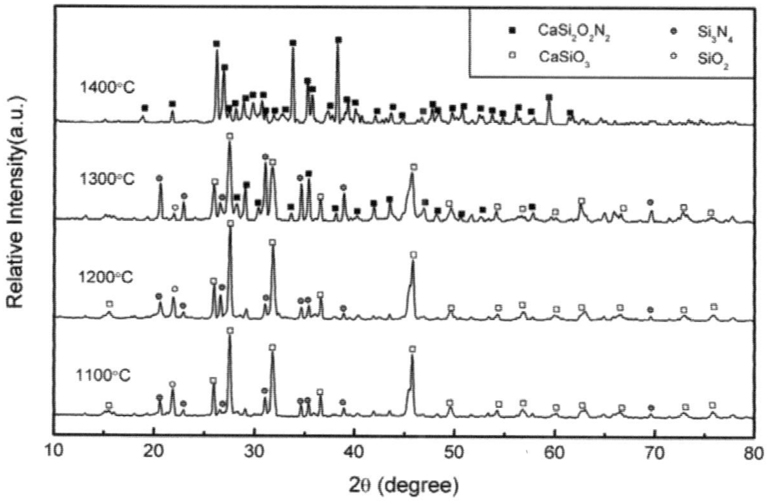

Figure 3. XRD patterns of the $CaSi_2O_2N_2$: Eu^{2+} synthesized at different temperatures.

the two phases at this temperature could exhibit different emission bands, including one shoulder at 430 nm and a maximum at 515 nm. Up to 1400℃, almost pure $CaSi_2O_2N_2$ is observed. The fine crystallized phase leads to high intensity in the emission spectra.

This result indicates that for the completion of the reaction, the temperature of 1400℃ is needed. When a lower temperature was applied, impurity phases would be detected, and the emission intensity would decrease.

The emission spectra of the samples with various δ values sintered at 1400℃ are shown in Figure 4. We can observe that the emission intensities of $CaSi_2O_{2-\delta}N_{2+2/3\delta}$ decrease with the increasing of the δ value, while the wavelength of peaks remain constant. According to the reports, in Eu^{2+}-doped Sialon system [3], as N content increased, the emission peak shifted to a longer wavelength, which may be due to less rigidity of the α-Sialon lattice caused by partial replacement of Si-N bonds. However, in $CaSi_2O_{2-\delta}N_{2+2/3\delta}$, the red shift of emission wavelength did not occur with increasing of the N content as our expectation. This may indicate that the replacement of O ion by N ion had little influence on the crystal field or the Eu^{2+} ion is not very sensitive to the changes of the local structure.

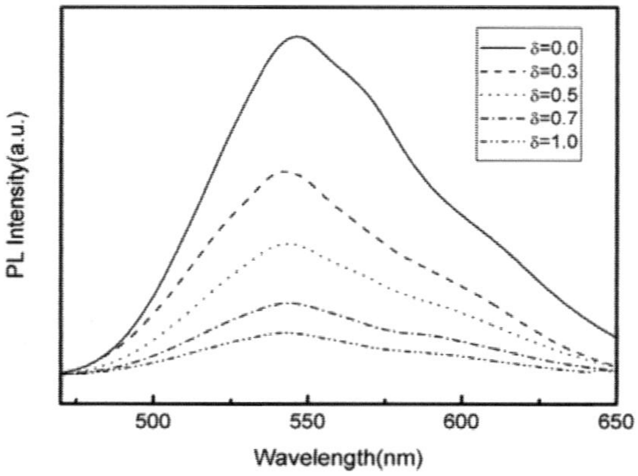

Figure 4. Emission spectra of $CaSi_2O_{2-\delta}N_{2+2/3\delta}:Eu^{2+}$ with various δ values sintered at 1400℃.

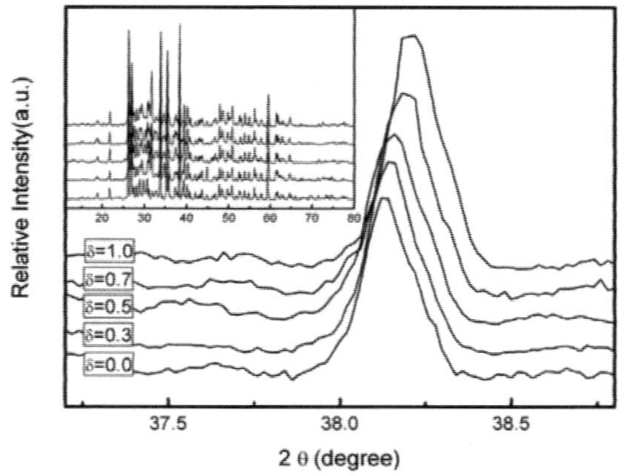

Figure 5. XRD patterns of $CaSi_2O_{2-\delta}N_{2+2/3\delta}:Eu^{2+}$ with various δ values sintered at 1400℃.

Owing to their very similar scattering factors, the direct experimental differentiation between O and N in the $[Si_2O_2N_2]^{2-}$ framework is difficult with X-ray methods. In the inset of Figure 5, it is found that the XRD patterns with different O/N ratio are almost the same, except for small differences in the location of the strongest peaks. As seen in the Figure 5, it is observed that for the oxygen-rich $CaSi_2O_{2-\delta}N_{2+2/3\delta}$ (δ=0) samples, the strongest peak is at about 2θ=38.12° . In reverse, for nitrogen -rich $CaSi_2O_{2-\delta}N_{2+2/3\delta}$ (δ=1), the strongest peak is located at 2θ=38.22° . This detail reflects that as the O atom replaced by N atom, the peaks shift to larger degree, and the unit cell volume decrease.

As we known, Eu^{2+} (1.17 Å, CN= 6) is significantly larger (about 15%) in size than Ca^{2+} (1.00 Å, CN= 6) [14]. When Eu2+ was incorporated in $CaSi_2O_{2-\delta}N_{2+2/3\delta}$, Ca^{2+} ion was replaced by the larger ion, and the unit cell would expand. In reverse, the shrinkage of the unit cell may indicate that solubility of Eu^{2+} in $CaSi_2O_{2-\delta}N_{2+2/3\delta}$ decreased. The decrease of active ion concentration may be one of reasons for the decline in the emission intensity mentioned above.

4. Conclusion

Eu^{2+}-doped $CaSi_2O_{2-\delta}N_{2+2/3\delta}$ materials were obtained by a solid-state reaction at 1100℃~1400℃. To obtain pure $CaSi_2O_2N_2$, the temperature of 1400℃ is necessary. When a lower sintering temperature was applied, $CaSiO_3$ was detected and gradually became the main phase in the product. Under the influence of $CaSiO_3$, the emission intensity decreased and the peak wavelength changed from 540 nm to 445 nm.

Moreover, the luminescence properties of Eu^{2+}-doped $CaSi_2O_{2-\delta}N_{2+2/3\delta}$ with various δ values were also investigated. It is found that the emission intensity decreased as O replaced by N, while the wavelength remained constant. This may be ascribed to the decrease of Eu^{2+} content in the host matrix.

Acknowledgements

We gratefully acknowledge the financial supports by Ministry of Education of the People's Republic of China (No. NECT-05-0419), Shanghai Municipal Education Commission (No.07SG37), National Natural Science Foundation of China (NO. 50772022, 50772127 and 50572116), and Shanghai Leading Academic Discipline Project (B603), and the Program of Introducing Talents of Discipline to Universities (No.111-2-04).

References

[1] Li Y Q, van Steen J E J, van Krevel J W H, Botty G, Delsing A C A, DiSalvo F J, de With J and Hintzen H T 2006 *J. Alloys Compd* **417** 273
[2] Xie R -J, Hirosaki N, Suehiro T, Xu F -F and Mitomo M 2006 *Chem. Mater* **18** 5578
[3] Xie R -J, Hirosaki N, Mitomo M, Yamamoto Y and Suehiro T 2004 *J. Phys. Chem.* B **108** 12027
[4] Suehiro T, Hirosaki N, Xie R -J and Mitomo M 2005 *Chem. Mater* **17** 308
[5] Xie R -J, Mitomo M, Uheda K, Xu F -F and Akimune Y *J. Am. Ceram. Soc* **85** 1229
[6] Xie R -J, Hirosaki N, Mitomo M, Sakuma K and Kimura N 2006 *Appl. Phys. Lett* **89** 241103
[7] van Krevel J W H, Hintzen H T, Metselaar R and Meijerink A 1998 *J. Alloys Compd* **268** 272
[8] Li Y Q, Delsing A C A, de With G and Hintzen H T 2005 *Chem. Mater* **17** 3242
[9] Li Y Q, de With G and Hintzen H T 2005 *J. Mater. Chem* **15** 4492
[10] Dorenbos P 2003 *J. Lumin* **104** 239
[11] Poort S H M, Meyerink A and Blasse G 1997 *J. Phys. Chem. Solids* **58** 1451
[12] Höppe H A, Lutz H, Morys P, Schnick W and Seilmeier A 2000 *J. Phys. Chem. Solids* **61** 2001
[13] Poot S H M, Reijnhoudt H M, van der Kuip H O T and Blasse G 1996 *J. Alloys Compd* **241** 75
[14] Shannon R D 1976 *Acta Cryst* **A32**751

Microstructure and photoluminescence properties of Nd-doped (Ba,Sr)TiO$_3$ thin films

Jingyang WANG[1,2], Tianjin ZHANG[1], Juan JIANG[1], Ruikun PAN[1], Fen ZHAN[1]

[1] Faculty of Materials Science & Engineering, Hubei University, Wuhan 430062 , China

[2] Physics & Electronics Information Technology Departmert, Xiangfan University, Xiangfan 441053, PR China

E-mail:zhangtj@hubu.edu.cn

Abstract. $Ba_{0.8}Sr_{0.2}TiO_3$(BST) thin films with different Nd-doped concentrations have been frabracated on silicon substrates by modified Sol-gel process. All the samples were annealed at 700℃ for 60 min.The structural and morphological properties of the films were examined by X-ray diffraction (XRD) and Atomic Force Microscopy (AFM). The XRD studies indicated that both pure and doped BST thin films show crystallized perovskite structures, the particle size is about 20 nm. The AFM images displayed that the films have high quality and fine-grained polycrystalline structure. The photoluminescence spectra(PL) were carried out using the 808 nm Ar^+ laser line as the excitation source at room temperature. The Nd-doped BST thin films show three NIR luminescence peaks corresponding to $^4F_{3/2}$–$^4I_{9/2}$ transition at 880 nm; $^4F_{3/2}$–$^4I_{11/2}$ transition at 1066 nm and $^4F_{3/2}$–$^4I_{13/2}$ transition at 1345 nm; respectively. The influence of Nd^{3+} concentration on the microstructure and the luminescence properties were discussed. All the results showed that the Nd-doped BST films have great potential uses for novel integrated optic devices.

1. Introduction

In recent years, the attention of reseachers has been focused on the fabrication of rare-earth-doped materials for novel optoelectronics devices and next generation flat panel displays[1-4]. Rare-earth ions exhibit a characteristic intra-4f shell luminescence which is both nearly host and temperature independent. According to the previous investigation, many Nd-doped materials have been extensively studied because the laser actions have been realized in a variety of glasses and fluoride crystals[5-7].

Comparing with the previously mentioned materials, $Ba_{1-x}Sr_xTiO_3$(BST) have drawn a good deal of attention due to their excellent dielectric and electro-optic properties as well as high dielectric constant, low dielectric loss, large electric-optic coefficient and low absorption coefficient. Therefore, it is a good candidate for a rare-earth doped host material.

In this work, Nd-doped $Ba_{0.8}Sr_{0.2}TiO_3$ thin films were fabricated on Si substrate by Sol-gel process. The influence of Nd doping on the microstucture and photoluminescence properties were investigated.

2. Experimental

The precursor solutions for both pure and Nd-doped $Ba_{0.8}Sr_{0.2}TiO_3$ were prepared by the sol-gel process using barium acctate ($Ba(CH_3COO)_2$), strontium acetate ($Sr(CH_3COO)_2$), neodymium acetate

($Nd(CH_3COO)_3$), titanium (IV) isopropoxide ($Ti(C_4H_9O)_4$) as starting materials. The solid-state barium acetate, strontium acetate, and neodymium acetate (as a dopant precursor with concentrations ranging from 0 to 5mol%) were initially dissolved in acetic acid and then mixed to obtain a Nd-doped (Ba, Sr) stock solution. Titanium (IV) isopropoxide was dissolved in methoxyethanol. Finally, both starting solutions were mixed to prepare the stoichiometric, transparent, and stable precursor for making Nd-doped BST films.

Both the undoped and doped BST thin films were deposited on silicon substrate by a multilayer spin-coating technique at 4000 rpm for 30 s. After the spin-coating procedure, the samples were kept at 450℃ for 30 min in air to remove the organic radicals. Thicker films can be obtained by repeating the spin-coating process. Then, the films were annealed at 700℃ for 1 h in oxygen atmosphere for crystallization.

The crystal structure of BST films was examined using X-ray diffraction (XRD,Rigaku D/Max-III C) with Cu-K_α radiation. The surface morphology of BST films was analyzed by an atomic force microscope (AFM, Digital Instrument Nanoscope IIIa) with a tapping mode. The photoluminescence spectra (PL) were carried out using Edinburgh Analytical Instruments F900 type UV-vis spectrophotometer, the 808 nm Ar^+ laser line was used as the excitation source at room temperature.

3. Results and discussion

The XRD patterns of the pure and Nd-doped BST thin films on Si substrate was shown in Figure 1. The results indicated that both undoped and Nd-doped BST thin films show crystallized perovskite polycrystalline structures and without any crystalline orientation. No additional peaks indicating the presence of rare earth oxides are observed. It indicated that the rare earth ions can completely solid dissolve in the BST lattice. In order to investigate the influence of Nd-doped on the microstructure of the BST thin films, the (200) peaks of the samples were chosen for further discussion. It can be seen from Figure. 2 that when the Nd concentration increased from 0% to 3 mol%, the (200) peak of the sample shifted toward higher diffraction angle, but when the Nd concentration increased from 3 mol% to 5 mol%, the peak shifted toward low diffraction angle slightly. It shows that the lattice parameter of the BST thin films decreased as Nd concentration increased from 0% to 3 mol%, but slightly increased when Nd-doping ratio increased from 3 mol% up to 5 mol%. It can be explained by the Nd ions may substitute A site and/or B site ions in ABO_3 perovskite structure. The ionic radius of Nd, Ba, Sr and Ti are 0.995 Å, 1.35 Å, 1.12 Å and 0.605 Å, respectively. When the Nd content below 3 mol%, Nd ions mainly substiute A site (Ba, Sr) ions in the lattice. It result in decrease of the lattice parameter. But when the Nd concentration increased from 3 mol% up to 5 mol%, Nd ions can substitute both A site (Ba, Sr) ions and B site (Ti) ions in the BST structure. The total effect is the slightly increase of the lattice parameter compared to the sample doped with 3 mol% Nd. In our former study, the similar phenomenon was seen in the Ho-doped BST thin films[8]. This result is also corresponding to the report by S.Munrakami et al. [9],they assumed that Eu^{3+} ions substitute La and (Zr, Ti) sites in PLZT.

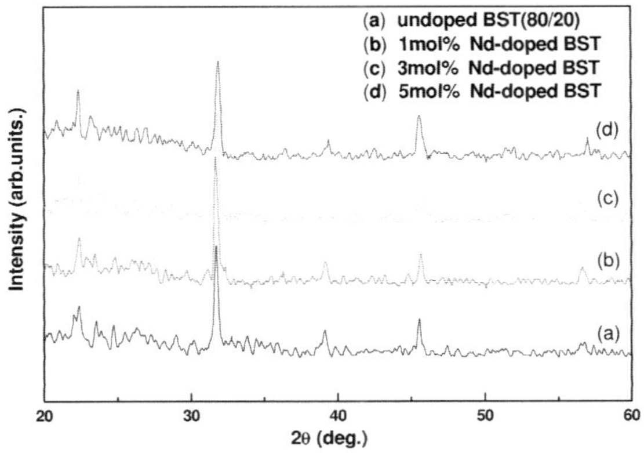

Figure 1. X-ray diffraction patterns of undoped and and Nd-doped BST thin films.

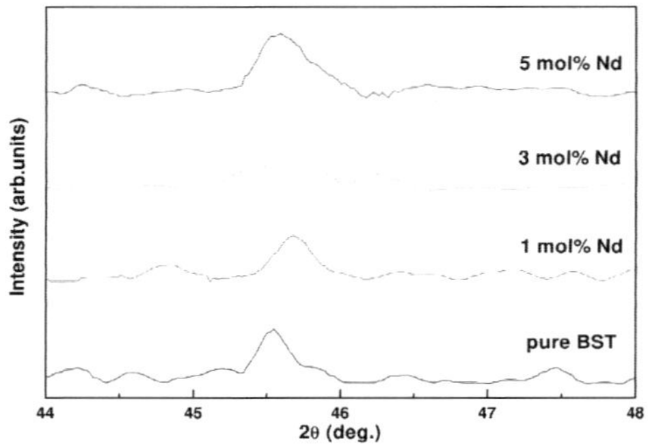

Figure 2. The (200) peaks of undoped and Nd-doped BST thin films.

The average grain size and the surface roughness of the undoped and doped BST thin films were analyzed with AFM in tapping mode with amplitude modulation. The AFM images were obtained using an area of 2μm×2μm. The surface AFM micrographs for the undoped and Nd doped BST thin films are shown in Figure. 3. All the AFM images show smooth surface morphology and the absence of both cracks and pinholes. A trend of decreasing grain size with increasing Nd content is observed and the average grain size of the samples was about 20–30 nm. Figure. 4 shows that the root mean square roughness (RMS) of the films decreases as the Nd concentration increases from 0 mol% to 3 mol%, but it increases when the Nd doping concentration exceed 3 mol%.

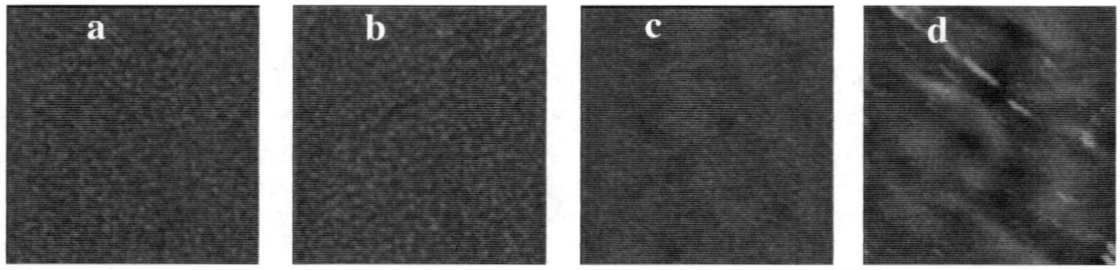

Figure 3. Two-dimensional AFM images of the surface of the BST:Nd thin films(a. Undope, b. 1 mol% Nd-doped, c. 3 mol% Nd-doped, d. 5 mol% Nd-doped).

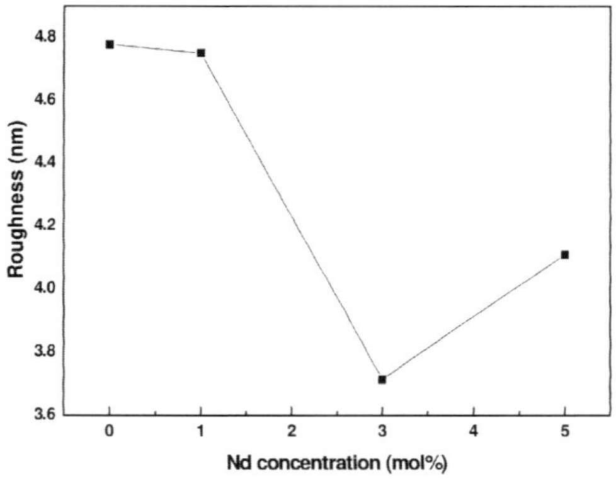

Figure 4. The roughness of BST thin films as a function of Nd content.

The mechanism of emission of Nd-contained materials has been well established in the literatures [10-13] and the energy diagram is schematically shown in Figure. 5. The PL spectra of undoped and

Nd-doped BST thin films are shown in Figure. 6. It can be explained that the peak in the dashed area was caused by one and half time double-frequency of the 808 nm pump laser. In this case, the host material has tetragonal rather than cubic structure and the Nd-related emission can be observed. The near infrared (NIR) luminescence peaks at 880, 1066 and 1345 nm are attributed to the Nd^{3+} inner shell $^4F_{3/2}$ to the $^4I_{9/2}$, $^4I_{11/2}$ and $^4I_{13/2}$ levels,respectively. It can be seen that the 1066 nm emission is more stronger than the others. The characteristic emission peaks in these films were due to the Stark splitting of the degenerate 4f levels under the crystalline field and homogeneous and inhomogeneous broadening caused by the films texture structures and multi-domain structures [14]. As shown in Fig. 6, we found the luminescence intensities reached maximum at 1 mol% Nd ions dopant concentration, meanwhile, the PL intensity decreased as Nd doping concentration exceed 1 mol%. The reason for this phenomenon may be due to the cross-relaxtion process between the two closely placed Nd ions.From charge compensation mechanism, the Nd^{3+}- Nd^{3+} distance is too small in BST lattice and cross relaxation is greatly enhanced, resulting in decrease in luminescence intensity. Very efficient cross relaxation can occur when two or more rare earth ions sit to one another to form a cluster that result in almost immediate interaction between the ions.

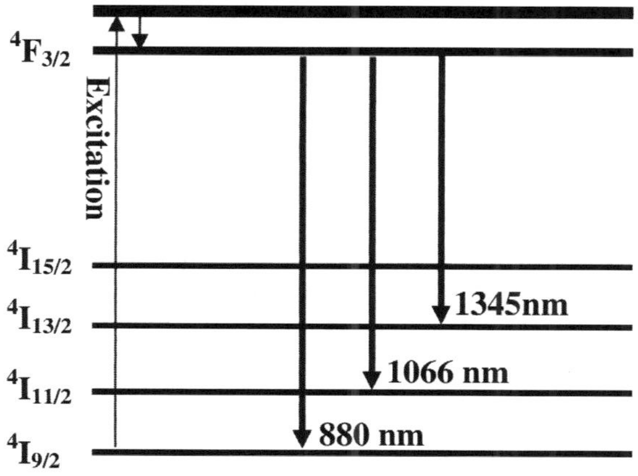

Figure 5. Relevant energy levels in the PL of BST:Nd.

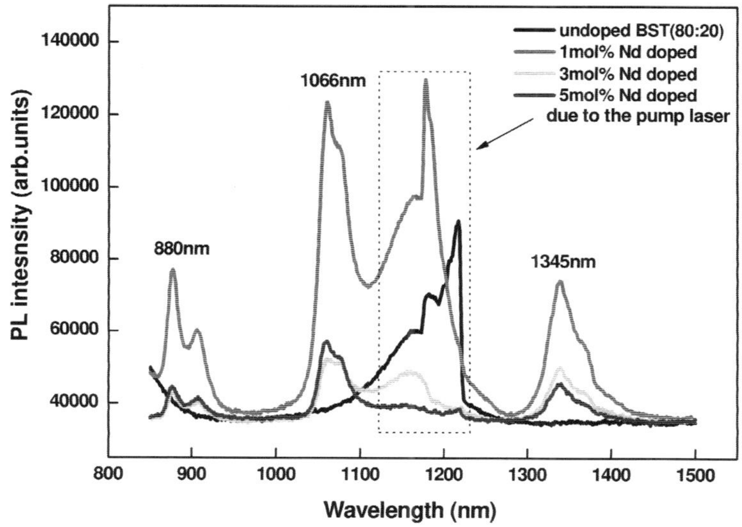

Figure 6. PL spectra from BST:Nd thin films. excitation wavelength 808 nm.

4. Conclusion

The BST:Nd thin films were prepared by sol-gel process and sintered at 700℃, and studied by XRD, AFM and PL. We found that the Nd doped BST thin films belong to polycrystalline perovskite structure. The surface morphologics of the films is smooth and crack-free. The 3 mol% Nd-doped BST thin film had the minimum root-mean-square roughness. The Nd-doped BST thin films show three NIR luminescence peaks, the 1066 nm emission was the most distinct luminescence. The emission intensities reached maximum at 1 mol% Nd ions dopant concentration and decreased when the Nd doping concentration exceed 1 mol%. The emission efficiency of BST:Nd was proposed to be influenced by the doping concentration of Nd^{3+} and mean distance between Nd^{3+} ions as well. Good performance on both microstructure and luminescence properties of BST:Nd materials may show potential applications in 1066 nm laser and novel photonic devices.

Acknowledgements
This work has been supported by the National Natural Science Foundation of China through Grant No. 50672022 and State Key Laboratory of Ferro and Piezoelectric Materials and Devices of Hubei Province.

References

[1] D.Y. Wang, K.P. Lor, K.K. Chung, H.P. Chan, K.S. Chiang, H.L.W. Chan, C.L. Choy 2006 Optical rib waveguide based on epitaxial $Ba_{0.7}Sr_{0.3}TiO_3$ thin film grown on MgO. *Thin Solid Films*. 510, 329-333.

[2] Tianjin Zhang, Haoshuang Gu, Jianghua Liu 2002 Structural and optical properties of BST thin films prepared by the sol–gel process. *Microelectronic Engineering*. 66, 860-864.

[3] Dal-Young Kim, Seung Eon Moon, Eun-Kyung Kim, Su-Jae Lee, Jong-Jin Choi, and Hyoun-Ee Kim 2003 Electro-optic characteristics of (001)-oriented $Ba_{0.6}Sr_{0.4}TiO_3$ thin films. *Appl. Phys. Lett*. 82,1455.

[4] S. B. Rai 2002 Optical properties of Ho^{3+} doped oxyfluoroborate glass. *Spectrochimica Acta Part A: Molecular and Biomolecular Spectroscopy*. 58, 1559.

[5] I. Iparraguirre, J. Azkargorta, R. Balda, J. Fernández 2005 Laser dynamics and upconversion processes in Nd3+-doped yttrofluorite crystals.*Optical Materials*. 27,1697.

[6] J. H. Campbell, T. I. Suratwala 2000 Nd-doped phosphate glasses for high-energy/high-peak-power lasers. Journal of Non-Crystalline Solids.263-264,318-341.

[7] Lilia Coronato Courrol, Edison Puig Maldonado, Laércio Gomes, Nilson Dias Vieira Jr., Izilda Marcia Ranieri, Spero Penha Morato 2000 Diode pumping Nd-laser efficiency limitations due to up-conversion processes in Nd:YLF and Nd:GLF. *Optical Materials*. 14, 81-90.

[8] Tianjin Zhang, Jun Wang, Juan Jiang, Ruikun Pan, Baishun Zhang 2007 Microstructure and photoluminescence properties of Ho-doped $(Ba,Sr)TiO_3$ thin films. Thin Solid Films. 515, 7721-7725.

[9] S. Murakami, M. Morita, M. Herren, T.Sakurai, D. Rau 2000 Near-infrared luminescence and spectral anomaly in PLZT ceramics doped with Nd^{3+}, Er^{3+}, Yb^{3+} and Cr^{5+} ions at low temperatures. *Journal of Luminescence*. 87-89, 694-696.

[10] J. Pisarska, W. Ryba-Romanowski, G. Dominiak-Dzik, T. Goryczka, W.A. Pisarski 2008 Nd-doped oxyfluoroborate glasses and glass-ceramics for NIR laser applications. *Journal of Alloys and Compounds*. 451, 223-225.

[11] W. Strek, A. Bednarkiewicz, D. Hreniak, P. Mazur, W. Lojkowski 2007 Fabrication and optical properties of transparent Nd^{3+}:YAG nanoceramics. *Journal of Luminescence* 122-123,70-73.

[12] A.F.H. Librantz, L. Gomes, S.L. Baldochi, I.M. Ranieri, G.E. Brito 2006 Luminescence study of the $^4f_2 5d$ configuration of Nd^{3+} in $LiYF_4$, $LiLuF_4$ and BaY_2F_8 crystals. *Journal of Luminescence* 121,137-148.

[13] E. Macho, R. Balda, M. J. Elejalde, J. Fernández, J. L. Adam 1993 Luminescence thermal quenching of Nd^{3+} doped fluoride BIGaZYTZr glass. *Journal of Non-Crystalline Solids*. 161,

Journal of Physics: Conference Series **152** (2009) 012084 doi:10.1088/1742-6596/152/1/012084

245-248.

[14] Cai Shen, Qian Liu, Qing-Feng Liu 2004 Photoluminescence properties of Er^{3+}-doped $Ba_{0.5}Sr_{0.5}TiO_3$ prepared by sol–gel synthesis. *Materials Science and Engineering: B*. 111, 31-35.

Composition, Microstructures and Properties of Al_2O_3/ZrO_2 (Y_2O_3) Self-growing Ceramic Composites Prepared by Combustion Synthesis under High Gravity

Z.M. ZHAO[1], L. ZHANG[1], Y.G. SONG[2], W.G. WANG[2] and H.B. LIU[2]

[1]Institute of Advanced Materials, Shijiazhuang Mechanical Engineering College, Shijiazhuang, 050003, China

[2]Science Research Department, Shijiazhuang Mechanical Engineering College, Shijiazhuang, 050003, China

E-mail: zhao_zhongmin@163.net

Abstract. By introducing ZrO_2 (4Y) powder into the thermit, Al_2O_3/ZrO_2 (4Y) composite ceramics were prepared through combustion synthesis under high gravity, and the correlations of composition, microstructures and mechanical properties of composite ceramics were investigated. The results of XRD, SEM and EDS showed that Al_2O_3/33%ZrO_2 (4Y) were composed of random-orientated rod-shaped colonies consisting of a triangular dispersion of orderly submicron-nanometer t-ZrO_2 fibers, surrounded by inter-colony regions consisting of spherically-shaped micronmeter t-ZrO_2 grains; Al_2O_3/45%ZrO_2 (4Y) were comprised of spherically-shaped micron-meter t-ZrO_2 grains, surround by irregularly-shaped α-Al_2O_3 grains and a few colonies. Compared to the directionally solidified Al_2O_3/ZrO_2 (Y_2O_3), the increase in hardness and flexural strength of Al_2O_3/33%ZrO_2 (4Y) in the experiment was due to high densification, small-size defect and high fracture toughness induced by compressive residual stress toughening and transformation toughening mechanisms; meanwhile, in despite of the moderate decrease in hardness, high flexural strength of Al_2O_3/45%ZrO_2 (4Y) was considered to be a result of small-size defect in spherically-shaped micronmeter t-ZrO_2 grain matrix and high fracture toughness induced by transformation toughening and micro-crack toughening mechanisms.

1. Introduction

It is well known that phase distribution, size and shape as well as interface characteristics play a crucial role in determining the behavior of multiphase materials, and the significant improvements in the properties are often associated with the development of novel microstructures. In this regard, the self-growing ceramic composites grown from the melts constitute a family of in-situ composites with a very fine microstructure and excellent bonding between the phases. Hence, in recent years, the oxide/oxide composite ceramics prepared by directional solidification procedure have attracted considerable interests because the small phase size together with the absence of the amorphous phase hinders the presence of defects and significantly increases mechanical properties of brittle ceramics [1-8]; meanwhile, the phases in the self-growing composites are thermodynamically compatible up to the temperature approaching the melting point of the self-growing composites, and also chemically stable in oxidizing environments even at very high temperature, which opens great possibilities for high-temperature application.

Among the different self-growing ceramic oxides, the Al_2O_3/ZrO_2 (Y_2O_3) composite is one of the most interesting materials because of its high mechanical properties, excellent strength retention at high temperature and surprising fracture toughness at both room temperature and elevated temperature. Up to now, micro-pulling down (μ-PD) and laser heated floating zone (LHFZ), two advanced technologies, have been used to prepare Al_2O_3/ZrO_2 (Y_2O_3) self-growing composites [1-8]. However, the μ-PD method seems to be more suitable for preparing high-performance fibers with sub-millimeter diameters [8]. In contrast, the LHFZ method can be used to obtain cylindrical geometries several millimeters in diameter [1] and large plates up to 30 cm^2 [2]; however, because of the high absorption coefficient of the molten oxides to laser radiation and low thermal diffusion in solid oxides, the size of the melt, as well as the diameter or thickness of the subsequently solidified rod or plate, are limited to several millimeters [1, 2, 4]. Hence, the problem of how to produce thick plates of Al_2O_3/ZrO_2 (Y_2O_3) self-growing composites on a large-scale is still a key for application of the materials in practical engineering.

Recently, we have worked out a new way to prepare Al_2O_3/ZrO_2 (Y_2O_3) self-growing composites by combustion synthesis [9, 10]. Subsequently, in order to obtain large-scale thick plates of Al_2O_3/ZrO_2 (Y_2O_3) self-growing composites, we also introduce high gravity into combustion synthesis, and low-cost, large-scale thick plates of the self-growing composites are achieved [11]. Hence, in order to obtain high-performance thick plates of the materials, we renew the experiment to prepare self-growing ceramic composites of different composition, microstructures and properties, and the correlations of composition, microstructures and mechanical properties of the self-growing composite ceramics are investigated.

2. Experimental procedure

Raw materials 0.074–0.147 mm in size were prepared from high purity (>99.9%) CrO_3, Fe_2O_3, Al, Al_2O_3, ZrO_2 and Y_2O_3, and two combustion systems were in succession prepared as follows: one was labeled as A1, which contained 10.9 mol.% CrO_3, 10.9 mol.% Fe_2O_3, 43.7 mol.% Al, 12.7 mol.% Al_2O_3, 20.9 mol.% ZrO_2 and 0.9 mol.% Y_2O_3; the other was labeled as A2, which contained 10.6 mol% CrO_3, 10.6 mol.% Fe_2O_3, 42.5 mol.% Al, 7.2 mol.% Al_2O_3, 27.9 mol.% ZrO_2 and 1.2 mol.% Y_2O_3. After mixing the powders of two above systems respectively, the graphite rooms were at first filled with A1 reactive mixture at constant pressure and inserted into two combustion chambers at the end of the rotating arms of the centrifugal machine, as shown in Figure 1. An optical pyrometer was mounted firmly on one rotating arm of the centrifugal machine and centered on the window of one combustion chamber to measure the synthesis temperature. The combustion systems were ignited with the W wire (diameter of 0.25 mm) after the centrifugal machine had provided a centrifugal force in the combustion chambers greater than 200g. As the combustion chambers were cooled to ambient temperature, the centrifugal machine was stopped and the graphite rooms were taken out of the combustion chambers. Finally, the ceramic plates of 100 mm diameter and 20 mm thickness were obtained after the products were taken out from the graphite rooms and the metallic layers at the bottom of the products were removed, as shown in Figure 2. Subsequently, by filling A2 reactive mixture in the graphite rooms and repeating the above experimental procedure, the ceramic plates of same dimension and different composition were also obtained.

The samples were cut and ground into two rectangular bars measuring 3mm (width)×4mm (height)×36mm (length) and 2.5mm (width)×5mm (height)×25mm (length), respectively, to be used to determine flexural strength and fracture toughness. The flexural strength was measured by the three-point bending method with a cross-head speed of 0.5 mm min^{-1} and a span of 30 mm. The fracture toughness was measured by the single-edge-notched beam method with a cross-head speed 0.05 mm min^{-1}; the specimens were pre-notched and the depth of the notch was 0.4–0.5 of the height of the specimen, the width-to-height ratio was 0.5 and the span-to-height ratio was 4. The indentation test was conducted for a loading time of 30s and a load scale of 294N. The phase constitution and microstructures of the composites were determined by X-ray diffractometry (XRD) and scanning electron microscopy (SEM), respectively, and electron probe microanalysis (EPMA) was conducted by energy dispersive spectrometry (EDS). The volume fraction of monoclinic ZrO_2 was determined according to the following equation:

$$V_m = \frac{I_{(\bar{1}11)m} + I_{(111)m}}{I_{(\bar{1}11)m} + I_{(111)m} + I_{(011)t}} \quad (1)$$

where I is the integral intensity, and the subscripts m and t refer to the monoclinic phase and tetragonal phase, respectively. The intensities were corrected for the background.

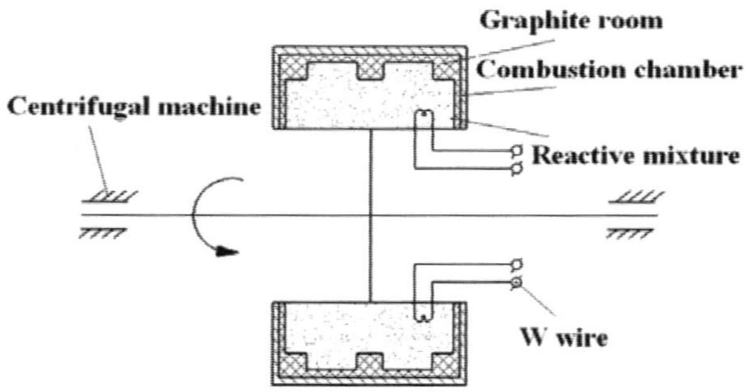

Figure 1. The sketch of preparing Al_2O_3/ZrO_2 (Y_2O_3) self-growing composite.

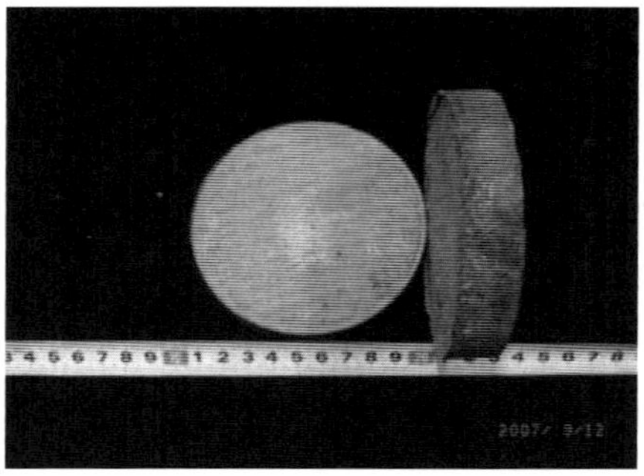

Figure 2. Plates of Al_2O_3/ZrO_2 (Y_2O_3) self-growing composites.

3. Results and discussion

3.1. Phase constitution of Al_2O_3/ZrO_2 (Y_2O_3) self-growing ceramic composites

XRD pattern showed that there were mainly α-Al_2O_3, t-ZrO_2 and m-ZrO_2 phases in the ceramic composites of A1 and A2, and the main diffraction crystal planes of α-Al_2O_3, t-ZrO_2 and m-ZrO_2 phases were (104), (011) and ($\bar{1}$11), respectively, as shown in Figure 3. By adding Si powder of 20% mass fraction into the sample powder and using K value method to determine the content of each phase in the ceramics, it was detected that there were 67.1% α-Al_2O_3, 20.1% t-ZrO_2 and 12.8% m-ZrO_2 in A1 sample (% stood for volume fraction), whereas there were 55.5% α-Al_2O_3, 22.7% t-ZrO_2 and 21.8% m-ZrO_2 in A2 sample. Hence, it was considered that the actual composition of A1 and A2 samples were Al_2O_3/33%ZrO_2 (4Y) and Al_2O_3/45%ZrO_2 (4Y).

3.2. Microstructures and properties of Al_2O_3/33%ZrO_2 (4Y)

SEM images of Al_2O_3/33%ZrO_2 (4Y) showed that the composite ceramics were mainly composed of randomly orientated colonies, as shown in Figure 4, and EDS images indicated that the colony consisted of Al, O and Zr, as shown in Figure 5. Combining with the XRD pattern, there was no segregation of Y_2O_3 into Al_2O_3 or formation of yttrium–aluminum garnet in the ceramics, so it was determined that Y_2O_3 had been completely dissolved into the ZrO_2 phases, and the ZrO_2 phases in the colony were in the form of tetragonal structures. Meanwhile, the colonies were surrounded by boundary regions containing spherically shaped micrometer size ZrO_2 grains, as shown in Figure 4, and the elements of the amorphous phase were yet not detected by EDS in the inter-colony regions, as shown in Figure 5. Meanwhile, a triangular dispersion of orderly t-ZrO_2 fibers in the colony was also observed, as shown in Figure 6, similar to the microstructures of directionally solidified self-growing ceramic composites. By combining with the literature [1, 4], the phenomenon was considered a result of the faceted eutectic growth of two Al_2O_3/t-ZrO_2 phases, in other words, Al_2O_3 crystal grew along its

c-axis but in its three rhombohedral R-planes $(10\bar{1}4)$, $(\bar{1}104)$ and $(0\bar{1}14)$ while the ZrO_2 fibers grew perpendicular to three rhombohedral R-planes of Al_2O_3.

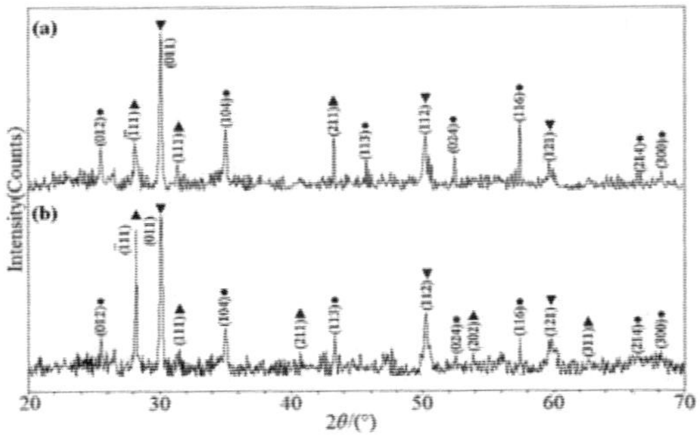

Figure 3. XRD pattern of A1 and A2 samples A1, Al_2O_3/33% ZrO_2 (4Y) (a) A2, Al_2O_3/45% ZrO_2 (4Y) (b).

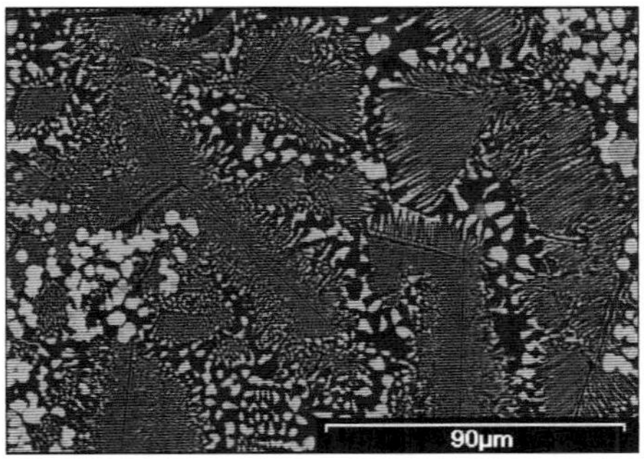

Figure 4. SEM Microstructures of Al_2O_3/33%ZrO_2 (4Y).

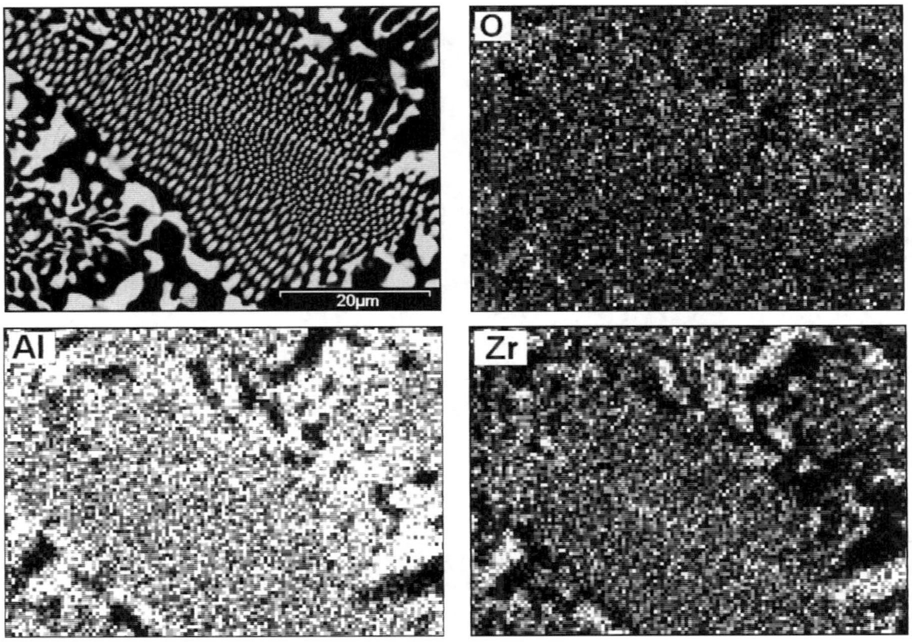

Figure 5. EDS patterns of Al_2O_3/33%ZrO_2 (4Y).

The relative density of Al_2O_3/33%ZrO_2 (4Y) and Al_2O_3/45%ZrO_2 (4Y) measured 98.4% and 98.8%, respectively, which indicated the actual densities of the materials were very close to their theoretical density.

The hardness, flexural strength and fracture toughness of Al_2O_3/33%ZrO_2 (4Y) measured 18.3GPa, 1550MPa and 13.8MP·m$^{1/2}$, respectively. Compared to the directionally solidified Al_2O_3/ZrO_2 (Y_2O_3) (hardness, flextural strength and fracture toughness were 18.2-18.5GPa, 1580MPa and 4-5 MP·m$^{1/2}$[1]), the hardness and flexural strength of the materials were almost same as ones of directionally solidified Al_2O_3/ZrO_2 (Y_2O_3), whereas the materials was enhanced 176%-245% in fracture toughness.

According to the literature [1], the flexural strength of directionally solidified self-growing composites is dependent on colony size and the thickness of inter-colony regions. Increasing the solidification rate reduces the colony size but also increases the thickness of the intercolony regions, so the flexural strength, which is actually dictated by the inter-colony regions containing pores and microcracks, is reduced greatly with increasing thickness of the inter-colony regions. Hence, Pastor et al. [12] found that the highest strength (close to 1.6 GPa) was measured in the composites obtained at low growth rate (about 0.005 mm^{-1}), which were made up of a homogeneous dispersion of t-ZrO_2 lamellae within the Al_2O_3 matrix.

As for the Al_2O_3/33%ZrO_2 (4Y) obtained in the experiment, which has the same composition as the one of the directionally solidified Al_2O_3/ZrO_2 (Y_2O_3), it is considered that the introduction of high gravity greater than 200g brings about a considerable increase in the cooling velocity and solidification rate of the melts, resulting in the formation of a colony surrounded by a thick inter-colony region. However, in the presence of high gravity above 200g, the gas in the melts of the ceramics is accelerated enough to escape, while the melt flow is enhanced greatly; thus, there is enough ceramic liquid forced to flow into the inter-colony regions, where it can solidify, resulting in a considerable decrease in volume fraction and size of pores and shrinkage cavities. Hence, it is considered reasonable that the introduction of high gravity above 200g is the unique reason for the high densification and high hardness of the ceramic composites, and finally high flexural strength of the materials, which is very close to the highest one of the directionally solidified Al_2O_3/ZrO_2 (Y_2O_3), is obtained as a result of the remarkable decrease of the actual defect size in the ceramic composite.

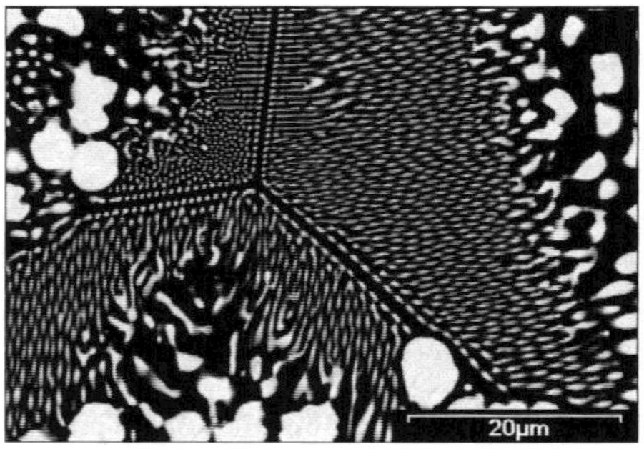

Figure 6. SEM image of triagonal dispersion of ZrO_2 (Y_2O_3) fibers.

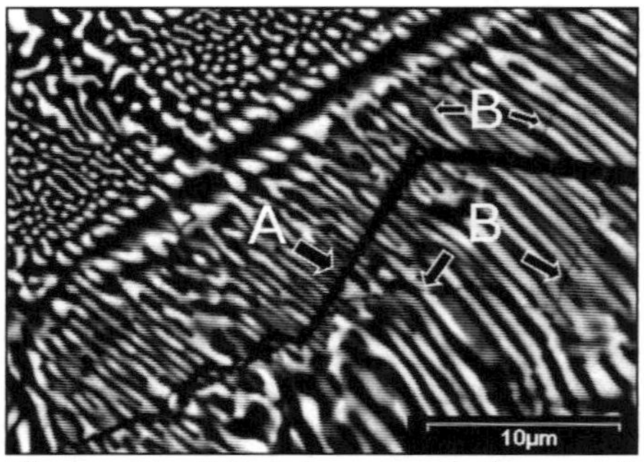

Figure 7. SEM image of the crack propagation in the colonies.

With respect to high fracture toughness of the materials, an indentation test showed that there were two kinds of fracture modes in the ceramics: one is a crack propagated in the colonies, shown by arrow A in Figure 7, and the other is a crack propagated in the inter-colony regions, as shown in Figure 8. As discussed above, because the ceramics grow directly from the high-temperature melts, the significant residual stresses caused by thermal expansion mismatch within Al_2O_3/ZrO_2 build up when the sample cools from the solidification temperature. In terms of the literature [1, 13], the compressive residual stresses in the Al_2O_3 matrix of the colony might be close to 1 GPa because the average diameter of t-ZrO_2 fibers and inter-phases spacing measured 300nm. Hence, as the crack is met by the colonies, as a result of the interaction between the compressive residual stress field and the crack tip, a shielding effect is induced, leading to an increase in the applied stress intensity factor; moreover, as the crack propagates in the colony along the growth direction of the colony, the toughening effect contributed by the compressive residual stress is so intense that the main crack is arrested and some microcracks develop in the colony, as shown by arrow B in Figure 7, so it is considered that under the strong compressive residual stresses in the colony, the crack is usually forced to deviate from the colony and propagates in inter-colony regions. By combining the XRD patterns of polished samples with those of the ceramic powder and using the equation (1) to calculate the volume fraction of m-ZrO_2 in the ZrO_2 phases, it is found that the volume fraction of m-ZrO_2 phases increased from 26.7% to 38.9% as the sample was broken up to be powder; in other words, 12.2% t-ZrO_2 was transformable in the ceramics. Hence, it is considered reasonable that as the crack inevitably meets the spherically shaped t-ZrO_2 micron-meter grains in the inter-colony regions, the transformation of t-ZrO_2 → m-ZrO_2 is sure to happen and stress-induced transformation toughening is initiated, leading to the cracking of spherically shaped ZrO_2 grains, as shown in Figure 8. Hence, this procedure for toughening the ceramics is controlled by the cooperative action of the compressive residual stresses in the colony and stress-induced transformation of spherically shaped t-ZrO_2 grains in inter-colony regions.

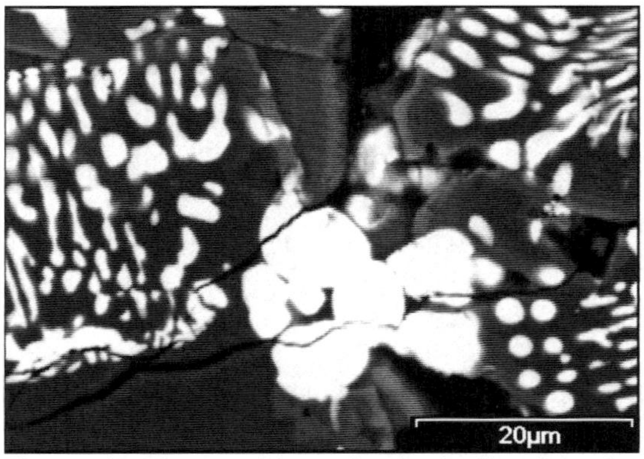

Figure 8. The cracking of spherically shaped t-ZrO_2 grains in inter-colony regions.

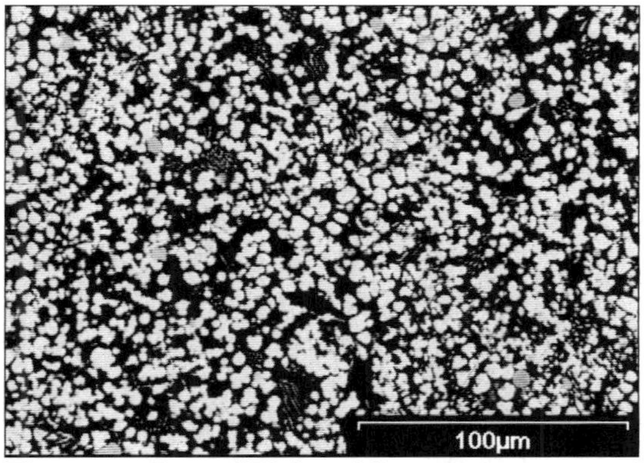

Figure 9. SEM Microstructures of Al_2O_3/45%ZrO_2 (4Y).

3.3. Microstructures and properties of Al_2O_3/45%ZrO_2 (4Y)

SEM images of Al_2O_3/45%ZrO_2 (4Y) showed that the ceramic matrix were comprised of spherically shaped t-ZrO_2 micron-meter grains, surround by irregularly-shaped α-Al_2O_3 grains and a few colonies, as shown in Figure 9. The elements of the amorphous phase between the Al_2O_3 and ZrO_2 phases were yet not detected by EDS, as shown in Figure 10, and it was indicated no amorphous phase is formed at the interface between the Al_2O_3 and ZrO_2 phases, so it was possible that two phases were bound together strongly at this interface. Meanwhile, the size distribution of spherically shaped t-ZrO_2 grains ranged mainly from 2.5μm to 10μm, and their average diameter measured 4.0μm. Hence, it was considered that the size, distribution and morphology of t-ZrO_2 primary solids were predominated by high gravity; in other words, the introduction of high gravity greater than 200g brought an intensive convection in the ceramic melts, resulting in the presence of a high shear force on t-ZrO_2 primary solids, so the t-ZrO_2 dendrites were inclined to be broken up, and t-ZrO_2 primary solids gradually grew in the form of the spheres.

The hardness, flexural strength and fracture toughness of Al_2O_3/45%ZrO_2 (4Y) measured 15.2GPa, 1650MPa and 18.8 $MP{\cdot}m^{1/2}$, respectively, and compared to the directionally solidified Al_2O_3/ZrO_2 (Y_2O_3) (hardness, flexural strength and fracture toughness were 18.2-18.5GPa, 1580MPa and 4-5 $MP{\cdot}m^{1/2}$[1]), the materials was enhanced 4.4% and 276-370%, respectively, in flexural strength and fracture toughness, while the hardness of the materials decreased by 20.3%.

It is well known that the hardness of the ceramics at ambient temperature is dependent on the elastic modulus of the ceramics, and the elastic modulus of the ceramics is also a function of the porosity of the ceramics, the elastic modulus and the respective volume fraction of each phase in the ceramics. Because the elastic modulus of ZrO_2 (160−241GPa) is lower than one of Al_2O_3 (390GPa), it is obvious that the decrease in hardness of Al_2O_3/45%ZrO_2 (4 Y_2O_3) is a result of high content of ZrO_2 in the ceramics, compared to the directionally solidified Al_2O_3/ZrO_2 (Y_2O_3) and Al_2O_3/33%ZrO_2 (4Y) obtained in the experiment.

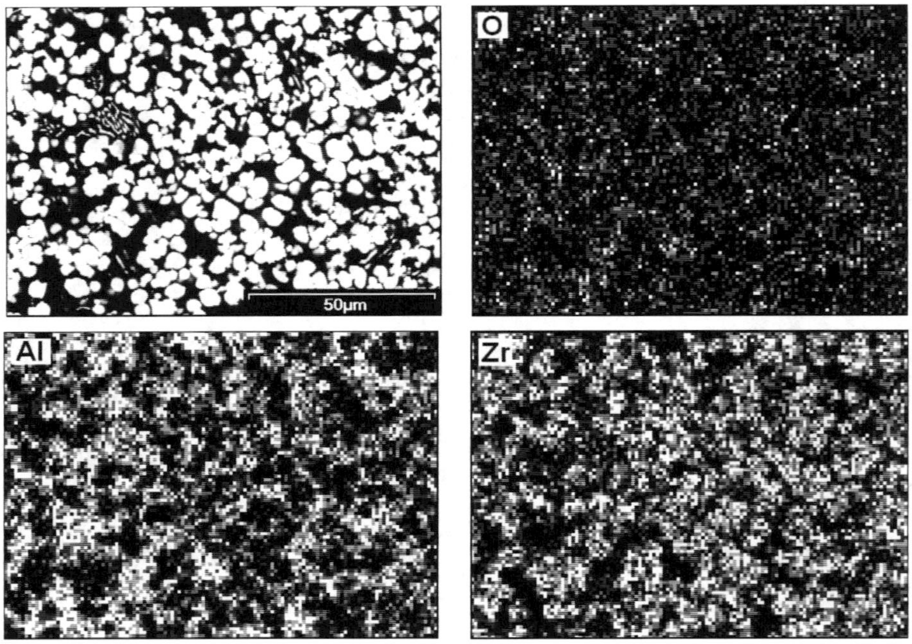

Figure 10. EDS patterns of Al_2O_3/45%ZrO_2 (4Y).

Meanwhile, SEM images of crack propagation paths showed that trans-crystalline fracture of spherically shaped ZrO_2 micron-meter grains is usually observed, as shown in Figure 11 and Figure 12. By combining the XRD patterns of polished samples with those of the ceramic powder and using Eq. (1) to calculate volume fraction of m-ZrO_2 in the ZrO_2 phases, it is found that 18.4% t-ZrO_2 was transformable in the ceramics. Hence, it is considered that as the crack propagate in Al_2O_3/45%ZrO_2 (4Y), stress-induced transformation toughening mechanism is initiated, followed by the occurrence of the ZrO_2 grain cracking and the microcracks nearby the main crack, respectively, as shown by the arrows A and B in Figure 11; meanwhile, because of new micro-cracking surfaces nearby the main crack, the stress concentration at the main crack tip has to be relaxed and the driving force of the main crack has to be reduced correspondingly, as shown by the arrows in Figure 12, resulting in the initiation of transformation-induced microcrack toughening mechanism. Hence, high fracture toughness of Al_2O_3/45%ZrO_2 is attributed to the coupled mechanisms of stress-induced transformation toughening and transformation-induced microcrack toughening.

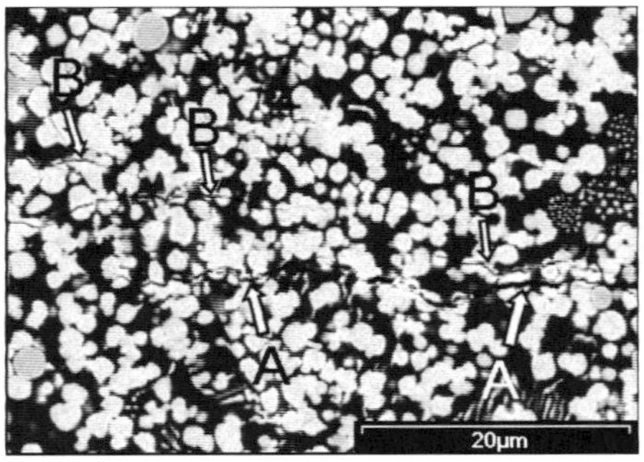

Figure 11. Crack propagation paths in Al_2O_3/45%ZrO_2 (4Y).

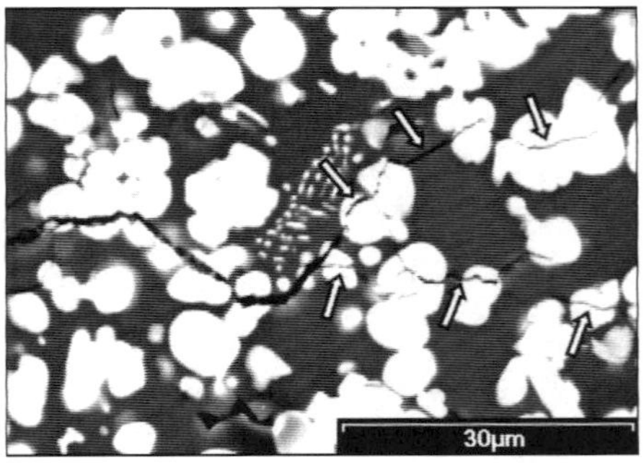

Figure 12. Micro-cracking around the main crack tip.

With respect to the strength of the materials, according to the linear elastic fracture mechanics for I-mode crack [14], the strength of the materials is controlled by flaw size and fracture toughness of the materials. In terms of mechanical behavior of materials [14], providing there is high relative density in the ceramics, or the density of the ceramics approaches to its theoretical density, the initial flaw size in the ceramics is similar to the grain size. As discussed above, the relative density of Al_2O_3/45%ZrO_2 (4Y) measured 98.8%, so it is considered that the initial flaw size is similar to the average size (4.0μm) of spherically shaped ZrO_2 grains. Therefore, because of small initial flaw size and high fracture toughness, the high strength of the Al_2O_3/45%ZrO_2 (4Y) is reached.

It is worthy to be noted that the most conventional ceramic composites are prepared by the powder sintering method, so they always contain many impurities, and there are many amorphous phases forming at grain boundaries. In many cases, these amorphous phases and grain boundaries effectively act to increase fracture toughness and strength at room temperature, but there are microstructural factors though to have a deleterious effect on high-temperature strength and creep characteristics, such as grain growth and grain-boundary sliding at high temperature. However, in the Al_2O_3/33%ZrO_2 (4Y) and Al_2O_3/45%ZrO_2 (4Y) obtained in the experiment, there are no amorphous phases and grain boundaries but phase interfaces, similar to the directionally solidified Al_2O_3/ZrO_2 (Y_2O_3); moreover, because the ceramics grow directly from the melts, the interfacial structures between two Al_2O_3/ZrO_2 phases are also similar to ones of the directionally solidified Al_2O_3/ZrO_2 (Y_2O_3). As just discussed by Sayir and Farmer [13] and Mazerolles et al. [15], the interfaces between two Al_2O_3/ZrO_2 (Y_2O_3) phases in self-growing ceramic composite have homopolar surfaces where they share a common oxygen plane, which makes the two phases bond strongly even at high temperature. Hence, compared to the conventional Al_2O_3/ZrO_2 ceramic composite prepared by the powder sintering method, the Al_2O_3/33%ZrO_2 (4Y) and Al_2O_3/45%ZrO_2 (4Y) obtained in the experiment are promised to have excellent strength retention and high creep resistance at high temperature, in addition to their high hardness, high flexural strength and high fracture toughness at ambient temperature.

4. Conclusions

By introducing ZrO_2 (4Y) powder into the thermit, Al_2O_3/33%ZrO_2 (4Y) and 45%ZrO_2 (4Y) composite ceramics with different microstructures were prepared through combustion synthesis under high gravity. XRD, SEM and EDS results indicated that Al_2O_3/33%ZrO_2 (4Y) were mainly composed of random-orientated rod-shaped colonies consisting of a triangular dispersion of orderly submicron-nanometer t-ZrO_2 fibers, and Al_2O_3/45%ZrO_2 (4Y) were mainly comprised of spherically-shaped micron-meter t-ZrO_2 grains. As a result of the introduction of high gravity during the preparation of the materials, the relative density of Al_2O_3/33%ZrO_2 (4Y) and 45%ZrO_2 (4Y) measured 98.4% and 98.8%. The hardness, flexural strength and fracture toughness of Al_2O_3/33%ZrO_2 (4Y) measured 18.3GPa, 1550MPa and 13.8MP·$m^{1/2}$, respectively. High flexural strength of Al_2O_3/33%ZrO_2 (4Y) is not only a result of small-size flaw in the ceramics, but also a result of high fracture toughness contributed by the cooperative action of the compressive residual stresses in the colony and stress-induced transformation of spherically shaped micron-meter t-ZrO_2 grains in inter-colony regions. The hardness, flexural strength and fracture toughness of Al_2O_3/45%ZrO_2 (4Y) measured 15.2GPa, 1650MPa and 18.8 MP·$m^{1/2}$, respectively. High fracture toughness of Al_2O_3/45%ZrO_2 (4Y) is attributed to the presence of a coupled mechanism of stress-induced transformation toughening and transformation-induced microcrack toughening, and high flexural strength of the ceramics is a coupled result of small-size flaw and high fracture toughness. Because of no amorphous phases and grain boundaries but clean and stable phase interfaces in the ceramic composite, the Al_2O_3/33%ZrO_2 (4Y) and Al_2O_3/45%ZrO_2 (4Y) obtained in the experiment are promised to have excellent strength retention and high creep resistance at high temperature, compared to ones prepared by powder sintering method.

Acknowledgements

This work is supported by National Natural Science Foundation of China (Grant No. 50672131).

References

[1] Llorca J and Orera V M 2006 *Prog. Mater. Sci.* **51** 711
[2] Larrea A, Orera V M, Merino R I and Pena J I 2005 *J. Eur. Ceram. Soc.* **25** 1419
[3] Larrea A, Fuente G F and Merino R I 2002 *J. Eur. Ceram. Soc.* **22** 191
[4] Pena J I, Merino R I, Harlan N R, Larrea A, de la Fuente G F and Orera V M 2002 *J. Eur. Ceram. Soc.* **22** 2595
[5] Oliete P B, Pena J I, Larrea A, Orera V M, Llorca J and Pastor J Y 2007 *Adv. Mater.* **19** 2313
[6] Waku Y 1998 *Adv. Mater.* **10** 615
[7] Pastor J Y, Poza P and Llorca J 2001 *Mater. Sci. Eng.* **A 308** 241
[8] Lee J H, Yoshikawa A, Kaiden H, Lebbon K, Fuhuda T, Yoon D and Waku Y 2001 *J. Cryst. Growth* **231** 179
[9] Zhao Z, Zhang L, Zheng J and Bai H 2005 *Scripta Mater.* **53** 995
[10] Zhao Z, Zhang L, Song Y and Wang W 2006 *Scripta Mater.* **55** 819
[11] Zhao Z, Zhang L, Song Y and Wang W 2008 *Scripta Mater.* **58** 207
[12] Pastor J Y, Llorca J, Poza P, de Francisco I, Merino R I and Pena J I 2005 *J. Eur. Ceram. Soc.* **25** 1215
[13] Sayir A and Farmer S C 2000 *Acta Mater.* **48** 4661
[14] Courtney T H 2000 *Mechanical Behavior of Materials* (New York: the McGraw-Hill Company) 404-450
[15] Mazerolles L, Michel D and Hytch M J 2005 *J. Eur. Ceram. Soc.* **25** 1389

Journal of Physics: Conference Series **152** (2009) 012086 doi:10.1088/1742-6596/152/1/012086

Influence of grain boundary energy on the grain size evolution in nanocrystalline materials

Zheng CHEN, Feng LIU, Gencang YANG, Yaohe ZHOU

State Key Laboratory of Solidification Processing, Northwestern Polytechnical University, Xi'an, Shaanxi 710072, People's Republic of China

E-Mail address: liufeng@nwpu.edu.cn

Abstract. A study for the effect of grain boundary (GB) energy on the evolution of grain size in nanocrystalline (NC) materials was given. Incorporating an empirical relation for GB energy with a generalized parabolic law, numerically, an evolution of grain size was investigated. Subsequently, a combination of grain-size dependent GB energy with the generalized parabolic growth law allows an analytic description for the average grain size as a function of annealing time. Satisfactory agreement of the model predictions with the experimental results obtained from grain growth occurring in NC Ni-P and Ni-W alloys and oxygen-doped nickel was achieved.

1. Introduction

The thermal stability of nanocrystalline (NC) materials is an important property for industrial applications. So far, two strategies have been devised to improve the thermal stability of NC materials, i.e., a kinetic one, in which the grain boundary (GB) migration is controlled by a retarding force from solute or precipitates [1-3], or a thermodynamic one [4-6], in which the GB energy σ_b as the driving force for the GB migration, is suppressed.

As for GB migration driven by the GB energy, the so-called parabolic growth law has been proposed and applied extensively [1-3]. Recently, Michels et al. [2] introduced a grain-size-dependent drag term to stabilize the grain size of the NC materials. Noticeably, an important assumption (i.e., constant σ_b) prevails in the above kinetic models. According to Gibbs adsorption theorem, σ_b reduces with addition of solute atoms [7]. Kirchheim and co-workers [5, 6] derived additional analytical models recently. Above all, GB energy cannot be considered as constant covering throughout the grain growth for nano-sized alloy. Liu et al. [8] attempted to introduce variable activation energy Q and σ_b into the parabolic growth law. Unfortunately, from the thermodynamic aspect, a physically sounded description for the effect of GB energy on the evolution of grain size with time is not available yet.

In this work, we will incorporate the GB energy, as a function of solute segregation and grain size, into the generalized parabolic growth law, in order to analyze the effect of GB energy on the grain size evolution with time in NC materials.

2. Incorporation of the grain-size dependent GB energy into the parabolic growth law

A simple power law from Burke and Turnbull [1] of the type applicable to grain growth in highly pure and coarse-grained polycrystalline materials gives,

$$dD/dt = M\sigma_b/D \Rightarrow D^2 - D_0^2 = 2M\sigma_b t \quad (1)$$

with M as the GB mobility and D_0 the initial mean grain size at annealing time t=0.

For nanosized alloy with a large fraction of interfacial volume, the decrease of GB energy caused by solute segregation can not be neglected. According to a thermodynamic treatment of Weissmüller [4] and Kirchheim [5], σ_b reduces with solute segregation,

$$\sigma_b = \sigma_0 - \Gamma_{b0}\left(RT\ln(x_0 - 3\Gamma_b V_m/D) + \Delta H_{seg}\right) \quad (2)$$

where σ_0 is the GB energy for pure solvent, x_0 the average concentration, Γ_b the solute excess, Γ_{b0} the saturated solute excess, V_m the molar volume of the alloy and ΔH_{seg} the segregation enthalpy. Γ_b in Eq.(2) is the solute excess at GBs, which can be expressed as [9, 10]

$$\Gamma_b = \delta\left(\rho_b x_B^{GB} - \rho_\alpha x_B^{\alpha}\right) \approx \delta\rho x_B^{GB} \quad (3)$$

with ρ_b and ρ_α as the density, and x_B^{GB} and x_B^{α} as the concentration, of the GB and the matrix, and δ (≈0.8 nm) as the GB thickness. Generally, a relation between the solute concentration in the matrix and in the bulk is given for a binary system as [9, 10]

$$x_0 = x_B^{GB} f + x_B^{\alpha}(1 - f) \quad (4)$$

Suppose that the volume fraction of GBs follows,

$$f = \left((D+\delta)^3 - D^3\right)/D^3 \approx 3\delta/D \quad (5)$$

The equilibrium GB segregation of solute can be described using Mclean's equation [11]

$$\frac{x_B^{GB}}{x_B^{GB0} - x_B^{GB}} = \frac{x_B^{\alpha}}{1 - x_B^{\alpha}}\exp\left(\frac{\Delta H_{seg}}{RT}\right) \quad (6)$$

with x_B^{GB0} as the saturation level for x_B^{GB} i.e., the fraction of GB monolayer available for segregated atoms at saturation. Combining Eqs. (3-6), the grain size dependent solute excess Γ_b can be rewritten as,

$$\frac{\Gamma_b}{\Gamma_{b0} - \Gamma_b} = \frac{x_0 - \left((D+\delta)^3 - D^3\right)/D^3 \times \Gamma_b/\delta\rho}{1 - x_0 + \left((D+\delta)^3 - D^3\right)/D^3 \times (\Gamma_b/\delta\rho - 1)}\exp\left(\frac{\Delta H_{seg}}{RT}\right) \quad (7)$$

with $\Gamma_{b0}=\delta\rho x_B^{GB0}$ as the saturated solute excess. Γ_b infinitely approaches Γ_{b0} with grain growth.

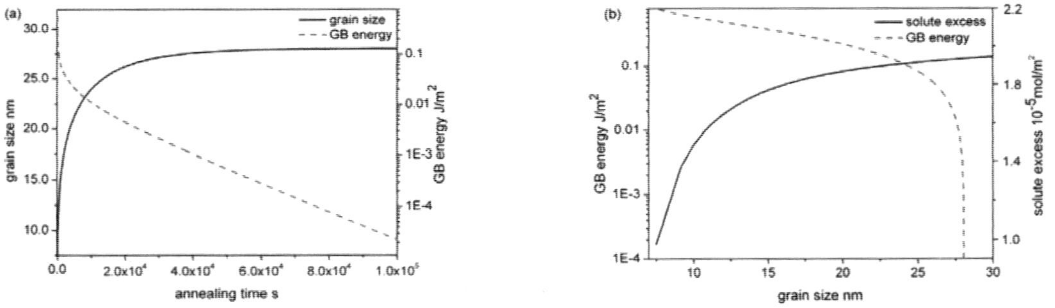

Figure 1. (a) grain size and GB energy *vs.* annealing time; (b) GB energy and excess *vs.* grain size using Eqs.(1,2,7) and typical values of σ_0=0.8 J/m^2, ΔH_{seg}=50 kJ/m^2, Γ_{b0}=2.0×10^{-5} mol/m^2, M=10^{-19} m^4/Js.

Up to now, combining Eqs. (1), (2) and (7) gives an evolution of the average grain size with the annealing time (Fig.1) [10]. As shown in Fig. 1a, the grain size is increased rapidly (i.e., within a relative short time duration) and is then stabilized. This phenomenon has been observed in a lot of experiments [12].

As is well known, grain growth in a polycrystal with solute segregation is accompanied with a competitive process between the energy lost due to a decrease of GB area and the energy gain due to a transfer of solute atoms into the lattice caused by the decreased segregation sites. That is to say, Γ_b increases up to its saturated value Γ_{b0} with grain growth (Fig.1b, the solid line), whereas, the GB energy reduces from the value of pure solvent σ_0 (the dashed lines in Fig.1a). As solute segregation is strengthened upon grain growth, a metastable equilibrium will be established, corresponding to a thermodynamic zero σ_b.

3. Analytic model for GB energy and grain size evolution

From Eqs. (1-7), the evolution of the average grain size can be given only by an iterated interpolation method. Furthermore, four adjusted parameters (i.e., σ_0, ΔH_{seg}, Γ_{b0} and M) will alleviate the validity of the procedure. If possible, a simple analytic relationship is preferred.

A reduction of GB energy can be illustrated by a simple model proposed by Krill [13]. Thus a useful expression for σ_b is derived as [13],

$$\sigma_b = \sigma_0 - \Delta H_{seg}\Gamma_b \tag{8}$$

Inserting typical values of Γ_b=2.0×10^{-5}mol/m^2 and ΔH_{seg}=50kJ/m^2 into Eq.(8), we obtain $\sigma_b \approx 0$ when σ_0=1 J/m^2. σ_b reduces at a linear rate with a slope equal to the enthalpy change ΔH_{seg} (Fig.2). Using the typical values in Section 2, σ_b could also be calculated by Eqs. (2) and (7), as shown in Fig.2. It is shown that the solid line is almost overlapped with the dashed one within a Γ_b range marked as 2, although there are relatively large differences within regions marked by 1 and 3. This could be attributed to the ignored entropic contributions in the system [13]. As for a physically realistic system, approximately, σ_b following Eq. (8) could be considered as a substitution of Eq. (2).

Assuming the GB concentration as $x_B^{GB}=\xi x_B^{\alpha}$ with ξ as the distribution coefficient, for NC systems with strong segregation tendency, i.e., ξ>>1 [2], combining Eqs. (3-5) gives

$$x_B^{GB} - x_B^{\alpha} = x_B^{GB}\,\xi - 1/\xi = x_0(\xi-1)/[1+(\xi-1)f/(1+f)] \tag{9}$$

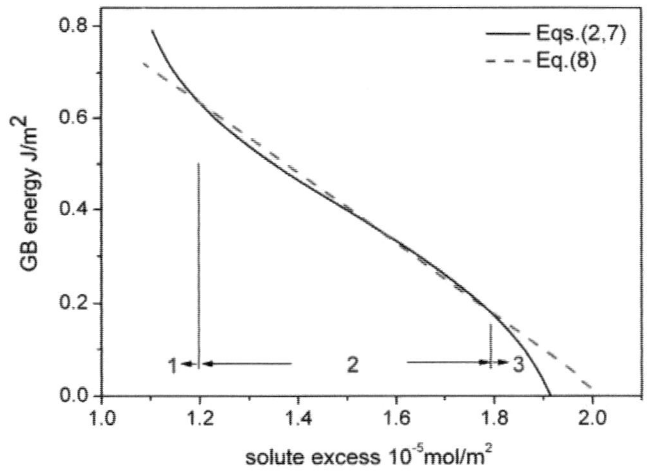

Figure 2. GB energy *vs.* solute excess. The solid and dashed lines are calculated by Eqs.(2,7) and Eq.(8), respectively. It is noted that the two lines are almost overlapped within 2.

Further combining Eq. (3) reduces to

$$\Gamma_b = \delta\rho\left(x_B^{GB} - x_B^{\alpha}\right) = x_0\rho\, D/3(1+3\delta/D) \tag{10}$$

Integrating Eqs. (8) and (10) gives,

$$\sigma_b = \sigma_0 - \lfloor \Delta H_{seg} x_0 \rho (D+3\delta)\rfloor/3 \tag{11}$$

Until now, a simplified model for the grain size dependent GB energy is obtained. Substituting Eq. (11) into (1) leads to,

$$\frac{\sigma_1}{\sigma_2^2}\ln\left(\frac{\sigma_1-\sigma_2 D}{\sigma_1-\sigma_2 D_0}\right)+\frac{D-D_0}{\sigma_2} = -Mt \tag{12}$$

where $\sigma_1=\sigma_0-\Delta H_{seg}x_0\delta\rho$ and $\sigma_2=\Delta H_{seg}x_0\rho/3$. Eq. (12) shows that the grain growth kinetics is correlated with thermodynamics. It can be inferred from Eq. (12) that the grain growth is an integrative one coupling the effects due to thermodynamics and kinetics.

4. Comparison with experimental results

4.1. Grain growth in electrodeposited NC Ni-P and Ni-W alloys

The stability of Ni-P alloys with P-content from 3 to 13 at.% [9] at a annealing temperature of 573 K has been shown in Fig.3 [9]. As grain growth proceeds, GB segregation leads to a metastable thermodynamic equilibrium corresponding to $\sigma_b=0$ [5, 6]. From Eq. (11),

$$D^* = 3\sigma_0/(\Delta H_{seg}\rho x_0) - 3\delta \tag{13}$$

Simultaneous fits of Eq. (13) to the average grain sizes are shown in Fig.3 (the solid line). Note that σ_0 is estimated as 0.87 J/m^2 [14]. The segregation enthalpy ΔH_{seg} is fitted to be 36 kJ/mol, which is close to an estimated value using $\Delta H_{seg}\approx RT\ln x_{op}$-(10±6) kJ/mol with x_{op} as the terminal solute solubility (x_{op}=0.32 at.%P in Ni at 1143 K) [5]. As shown in Fig.3, a decrease of grain size is predicted by Eq. (13) as the increase of P content, which is analogous to the calculation of Liu [6] using Eq. (2). However, less fitting parameters (only ΔH_{seg}) are employed, the validity of this fitting are enhanced.

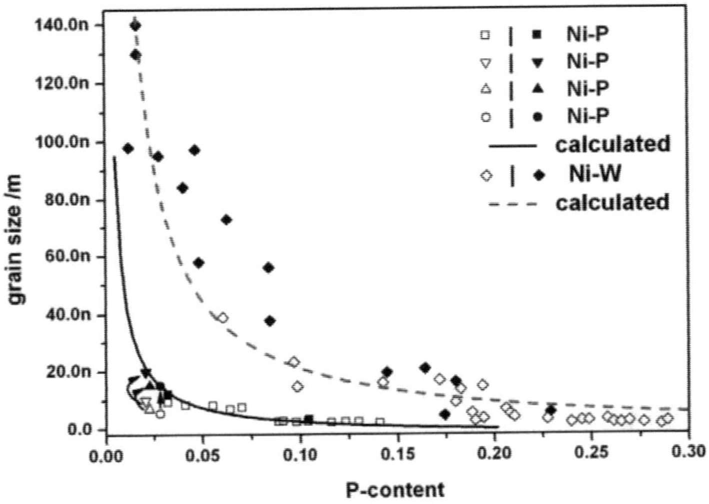

Figure 3. Relationship between grain size and solute content of Ni-P [5] and Ni-W [14] alloys.

Different from Ni-P system, the electrodeposited Ni–W system exhibits significant solute solubility in the crystalline lattice [14]. On this basis, a weak tendency for GB segregation is expected in Ni–W. Suppose a threshold value of C below which grain growth is effectively suppressed, the stop of grain growth in Ni-W thus corresponds not to $\sigma_b=0$, but $\sigma_b=C\times\sigma_0$. Then Eq. (11) is rewritten as

$$D^* = 3(1-C)\sigma_0/(\Delta H_{seg}\rho x_0) - 3\delta \tag{14}$$

The fits of Eq. (14) to the average grain sizes are shown in Fig.3 (the dashed line). The parameters C=0.4 and ΔH_{seg}=5.6kJ/mol are close to the values from computer simulations [14]. Note that a reduction of about 60% from 0.87 J/m^2 (for pure Ni) [14] is shown. This reduction in GB energy is significant and should substantially stabilize the NC structure. However, the thermodynamical ideal stability (i.e., $\sigma_b=0$) has apparently not been achieved because of the low ΔH_{seg} (5.6kJ/mol).

This further evidences that the so-called thermodynamical conclusion that the suppressing growth corresponds to zero GB energy can only occur in alloys with strong segregation tendency.

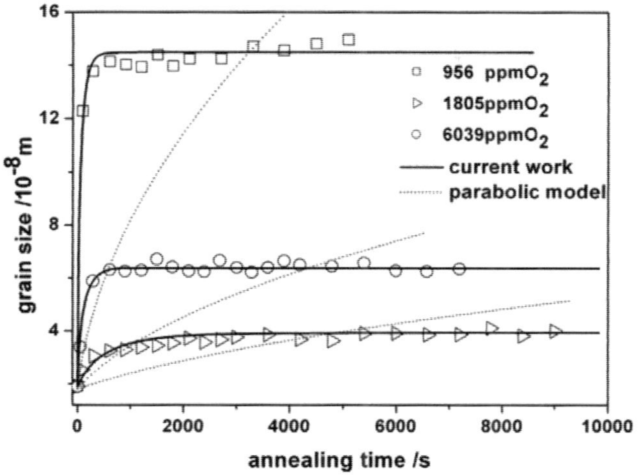

Figure 4. Average grain size *vs.* annealing time of the n-nickel samples at 673 K [12]. The dotted lines are the results of fits with the parabolic law (Eq. (1)); The solid lines show the fit results using Eq. (12).

4.2. The evolution of grain size in NC nickel doped nickeloxide

According to Ref. [12], at a temperature of 673 K, the effect of GB nickel oxide on thermal stability of NC-Nickel was studied using samples with oxygen contents between 956 and 6039 wt. ppm. The current work could give good fits to the experimental data (Fig.4). However, the generalized parabolic growth model cannot be brought into agreement. The parameters are listed in Table 1. As shown in Fig.4 and Table 1, a decrease of ΔH_{seg} from 66 and 76 to 37 kJ/mol has to be performed to describe well the data point with oxygen content from 956 and 1805 to 6039 ppm. Since ΔH_{seg} can be defined as a difference in energy between a solute atom in the grain interior and in the GBs [11], the high value (66 and 76 kJ/mol) is expected in the dilute limit which is very close to the calculated value of Ni-P with the similar systems chemically and the low value (37 kJ/mol) corresponds to the substantially increased solute content (6039 ppm) [14].

The above analysis does show that grain growth is quite influenced by solute segregation to GBs.

5. Discussion

Either Eqs.(1, 2 and 7) or Eq.(12) could predict a stabilized grain size after a fast increase within a relatively short time duration. As is evidence by a detailed comparison between Ni-P and Ni-W data, thermodynamically, the solute segregation to GBs stabilizes the NC structure. This phenomenon strongly suggests the GB energy reduction due to GB segregation is a reasonable mechanism.

In order to analysis the effect of GB energy on the grain growth, differentiation of Eq. (11) over t, then combining Eqs.(1) and (11) gives,

$$-\sigma_1 \ln \sigma_b/\sigma_0 + \sigma_b - \sigma_0 = \sigma_2^{\,2} Mt \tag{15}$$

Table 1 Parameters of Eq. (12) to the experimental data with different O contents [12].

Parameter /wt.%		956ppm	1805ppm	6039ppm
Mobility (m^4/J s)	M	2.2×10^{-16}	3×10^{-17}	1.4×10^{-18}
GB energy (J/m^2)	σ_0	0.87	0.87	0.87
Enthalpy change (KJ/mol)	ΔH_{seg}	66	76	37
Error	%	2.3	2.19	3.7

Using the parameters listed in Table 1, the calculation of Eq. (15) is shown in Fig.5. The GB energy decreases with the grain growth, but with annealing time, it can only approaches infinitely zero (Fig.5). Thermodynamically, ''zero'' value should be considered as an ideally limitation for σ_b.

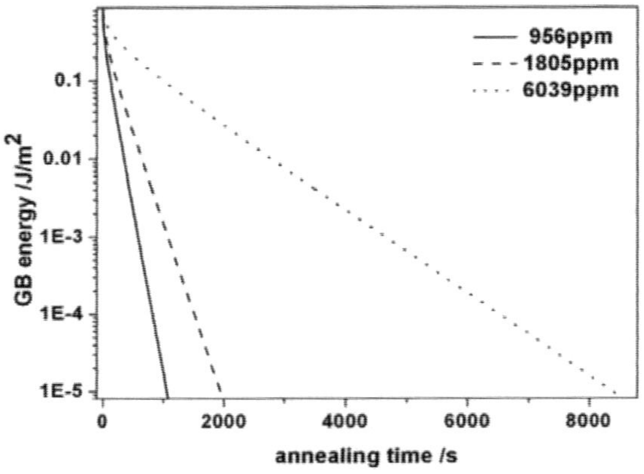

Figure 5. The GB energy *vs*. annealing time with the O-contents of 956, 1805 and 6039 wt. It is shown that GB energy infinitely approaches zero with annealing time.

In connection with Fig.5 and Table 1, the stagnated time of grain growth increases with the increase of solute concentration, meanwhile, the GB mobility M reduces. It is assumed that M has an Arrhenius relation with temperature [8, 10]

$$M = M_0 \exp(-Q/(RT)) \tag{16}$$

with R as the gas constant and M_0 a constant. From Fig.5, it can be implied that increasing solute concentration reduces M i.e., enhances Q. Since the consumed annealing time in the whole kinetic process of grain growth should be determined by GB mobility, which is described through Q of atom diffusion [10], longer annealing is needed for the stagnation of grain growth (Fig.5).

6. Conclusions

The effect of GB energy on the evolution of grain size in NC materials was studied detailedly. Incorporating an empirical relation for GB energy with a generalized parabolic law, numerically, an evolution of grain size was investigated. A combination of grain-size dependent GB energy with the generalized parabolic law allows an analytic model for the evolution of the average grain size as a function of annealing time. Satisfactory agreement of the model prediction with the experimental results obtained from grain growth occurring in NC Ni-P and Ni-W alloys and oxygen-doped nickel was achieved. The inhibition of grain growth can be attributed to the elimination of driving force for solute segregation, whereas, the consumed annealing time before the stop of grain growth should be determined by GB mobility.

Acknowledgements

The authors are grateful to the financial support of New Century Excellent Person Supporting Project (NCET-05-870), the Scientific and Technological Creative Foundation of Youth in Northwestern Polytechnical University, the Natural Science Foundation of China (Grant no. 50501020, 50771084, 50395103), the Huo Yingdong Young Teacher Fund (111052), the Fundamental Research Project of National Defense of China (A2720060295), the Fundamental Research Fund of Northwestern Polytechnical University and the Project Sponsored by the Scientific Research Foundation for the Returned Overseas Chinese Scholars, The Ministry of Education of the People's Republic of China (N6CJ0002).

References

[1] Burke J E, Turnbull D 1952 Recrystallization and grain growth. Prog. Met. Phys. 3 220.
[2] Michels A, Krill C E, Ehrhardt H, Birringer R, Wu D T 1999 Modelling the influence of grain-size-dependent solute drag on the kinetics of grain growth in nanocrystalline materials. Acta Mater. 47 2143.
[3] Rabkin E 2000 On the grain size dependent solute and particle drag. Scripta Mater. 42 1199.
[4] Weissmüller J 1993 Alloy effects in nanostructures. Nanostructured Mater. 3 261.
[5] Kirchheim R 2002 Grain coarsening inhibited by solute segregation. Acta Mater. 50 413.
[6] Liu F, Kirchheim R 2004 Nano-scale grain growth inhibited by reducing grain boundary energy through solute segregation. J. Cryst. Growth. 264 385.
[7] Gibbs J W Trans Conn Acad III 108 (1876); J. W. Gibbs: Trans Conn Acad III 343 (1878); JW Gibbs: The Collected Works of J. W. Gibbs vol. 1, Longmans, Green and Co, New York (1928) pp. 55–354.
[8] Liu F, Kirchheim R 2004 Comparison between kinetic and thermodynamic effects on grain growth. Thin Solid Films. 466 108.
[9] Färber B, Cadel E, Menand A, Schmitz G, Kirchheim R 2000 Phosphorus segregation in nanocrystalline Ni–3.6 at.% P alloy investigated with the tomographic atom probe (TAP). Acta Mater. 48 789.
[10] Chen Z, Liu F, Yang W, Wang H F, Yang G C, Zhou Y H Influence of grain boundary energy on the grain size evolution in nanocrystalline materials. J all. Compd.: 10. 1016 / j. jallcom. 2008. 08. 040.
[11] D. Mclean 1957 Grain Boundaries in Metals, Oxford University Press, Oxford.
[12] Natter H, Löffler M S, Krill C E, Hempelmann R 2001 Crystallite growth of nanocrystalline transition metals studied in situ by high temperature synchrotron X-ray diffraction. Scr. Mater. 44 2321.
[13] Krill C E, Ehrhardt H, Birringer R 2005 Thermodynamic stabilization of nanocrystallinity. Z. Metallkd. 96 1134.
[14] Detor A J, Schuh C A 2007 Grain boundary segregation, chemical ordering and stability of nanocrystalline alloys: Atomistic computer simulations in the Ni–W system. Acta Mater. 55 4221.

A comparative study of substrate effects on the orientation-selective growth of epitaxial $Mg_{0.4}Zn_{0.6}O$ thin films by pulsed laser deposition

L. ZHUANG[1, 2, a)], K. H. WONG[2]

[1] Institute for Solar Energy Systems, State Key Laboratory of Optoelectronic Materials and Technologies, Sun Yat-sen University, Guangzhou, 510006 China

[2] Department of Applied Physics and Materials Research Centre, The Hong Kong Polytechnic University, Kowloon, Hong Kong, PR China

E-mail: stszhl@mail.sysu.edu.cn

Abstract. Selective growth of single-oriented epitaxial $Mg_{0.4}Zn_{0.6}O$ (MZO) films on $LaAlO_3$ (LAO) (001) and Al_2O_3 (0001) substrates were obtained by the pulsed laser deposition method. It was found that the orientation of the MZO films was determined at the initial deposition stage by the substrates orientation only. Highly (001)-oriented MZO films could be stabilized on the LAO (001) substrates, while (111)-oriented MZO films were shown on the Al_2O_3 (0001) substrates. X-ray diffractometer and transmission electron microscope characterizations exhibit excellent epitaxial relationships of (001) and (111) orientation of films on LAO and Al_2O_3 substrates, respectively. Under the present deposition conditions, the surfaces of MZO films with different single orientation were smooth and featureless.

1. Introduction

Oxide semiconductors have attracted considerable scientific and technological attention in last few years. The rich choice of oxides and their dopants provides a wealth of potentially useful electrical properties [1-3]. ZnO is a wide bandgap semiconductor and an ideal material for realizing room temperature excitonic devices [4, 5]. Recently, as an alternative to the group III nitride material system, the wide bandgap wurtzite ZnO (E_{gap} ~3.37 eV) [6, 7], cubic MgO (E_{gap} ~7.50 eV) [8] and their ternary alloy $Mg_xZn_{1-x}O$ are of substantial interest. The success in the wide band gap engineering of $Mg_xZn_{1-x}O$ to short wavelengths paves the way for heterostructure device applications, which may eventually compete with the group III nitrides [9-11]. Pulsed laser deposition method has revealed that pure ZnO wurtzite phase can be maintained at high MgO content of 33% [12]. For MgO content above 33% cubic phase segregation occurs and both cubic MgO and hexagonal ZnO coexist. At higher MgO content, only cubic phase is observed. Ohtomo et al [13] and others [14, 15] have grown high-quality, single-phase $Mg_xZn_{1-x}O$ thin films with Mg concentrations up to 33 at.% using PLD. All the above-mentioned $Mg_xZn_{1-x}O$ thin films possess hexagonal structure. Choopun et al [12] and Narayan et al [16] reported the growth of cubic $Mg_xZn_{1-x}O$ thin films grown by PLD with Mg composition x in the range of 0.50-0.86 and 0.82-1.00, respectively. However, the cubic MgO and hexagonal ZnO show phase separation after rapid thermal annealing.

Journal of Physics: Conference Series **152** (2009) 012087 doi:10.1088/1742-6596/152/1/012087

In this study, we report the different orientation growth of $Mg_{0.4}Zn_{0.6}O$ (MZO) films on cubic $LaAlO_3$ (LAO) (001) substrates and hexagonal Al_2O_3 (0001) substrates. MZO films grown from ablating MZO target have mixed phases in this stage. Its formation of (001) or (111) orientation could be induced by using LAO (001) or Al_2O_3 (0001) substrates with different crystal orientation. To the author's knowledge, $Mg_xZn_{1-x}O$ mixed phase is reported for the first time by the real experiment technique. This provides templates to grow ZnO related compounds on different structure substrates with different orientations. It is particularly useful if one wants to study the phase transition in $Mg_xZn_{1-x}O$ series of films. The ZnO doping compounds may lead to devices with performance that supersedes those offered by group-III nitrides.

2. Experimental detail

Deposition of MZO films was carried out by PLD (KrF excimer laser, λ=248 nm, Lambda Physik, Complex 205, duration=10 Hz, laser energy=250 mJ) of a MZO ceramic target in a stainless steel chamber with base pressure of about 2×10^{-5} torr. The target was prepared by sintering a pellet of ZnO (99.95% purity) and MgO (98% purity) powders with the conventional solid state reaction. Grinding, pressing, and sintering were repeated three times during the target preparation and the sintering temperature was progressively increased from 1100°C to 1400°C. Cubic LAO (001) substrates and hexagonal Al_2O_3 (0001) substrates cleaned by acetone were used and positioned parallel to and at a distance of about 45 mm from the target. The temperature was kept at 650°C during the depositions. After deposition, the films were annealed and then naturally cooled to room temperature. The deposition time for all samples was 30 min, resulting in films of about 350-500 nm thickness. The orientation and crystalline quality of films were measured by the θ-2θ and ω scans of X-ray diffraction (XRD) with Cu-Kα radiation and transmission electron microscope (TEM), which is carried out using a JEOL JEM-2011 transmission electron microscope operated at 200 kV. The morphology of the film surface was examined by scanning electron microscopy (SEM).

3. Results and discussion

The MZO target used for film deposition in this study is mixed-phase compound, as shown of its XRD pattern in figure 1. Apart from reflections that could be indexed according to the hexagonal ZnO phase, some reflections from the cubic MgO phase appeared. To create a reduced atmosphere for the growth of stoichiometric MZO compound, we used a multiphase target and the vacuum atmosphere (2×10^{-5} torr) to deposit MZO films by the PLD technique.

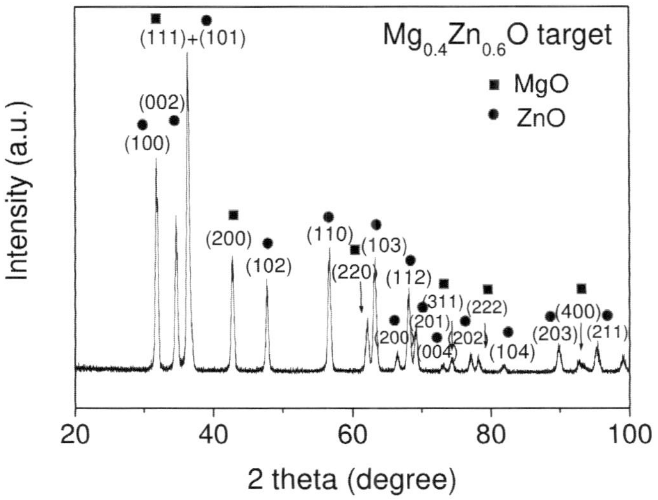

Figure 1. XRD pattern of the air-sintered target from the mixture of stoichiometrical MZO composition.

Figure 2, 3 and 4 show XRD patterns including θ-2θ, ω and ϕ scans of a typical MZO film deposited at 650°C on LAO (001) and Al_2O_3 (0001) substrates. No impurity phases were recorded for the as-deposited films on different substrates although the target is not single phase. In figure 2(a) with 2θ range of 20° - 100°, we observed the peaks corresponding to (002) and (004) planes of MZO, (001), (002) and (003) planes of LAO. The 2θ peak of MZO (002) plane which equals to 42.6° with a full width at half maximum (FWHM) of 1.116° is coincided with the LAO substrate (002) $CuK_β$ peak with the FWHM equals to 0.297°. Some weak peaks due to the $CuK_β$ radiation were also indexed. This suggests that MZO film grows epitaxially on LAO substrate with preferred (001) orientation. Lattice parameter for MZO is determined to be 4.24Å. The inset shows close up view at 2 theta near 94°. The MZO (004) and LAO (004) $CuK_β$ diffractions are clearly discernable. In figure 2(b), strong reflections from MZO (111) and (222) diffraction planes appeared, we also observed the peaks corresponding to (0006) and (00012) planes of sapphire. It is shown that MZO grows epitaxially on sapphire with preferred (111) orientation. The Al_2O_3 substrate (0006) $CuK_β$ peak and (00012) $CuK_β$ peak were also indexed.

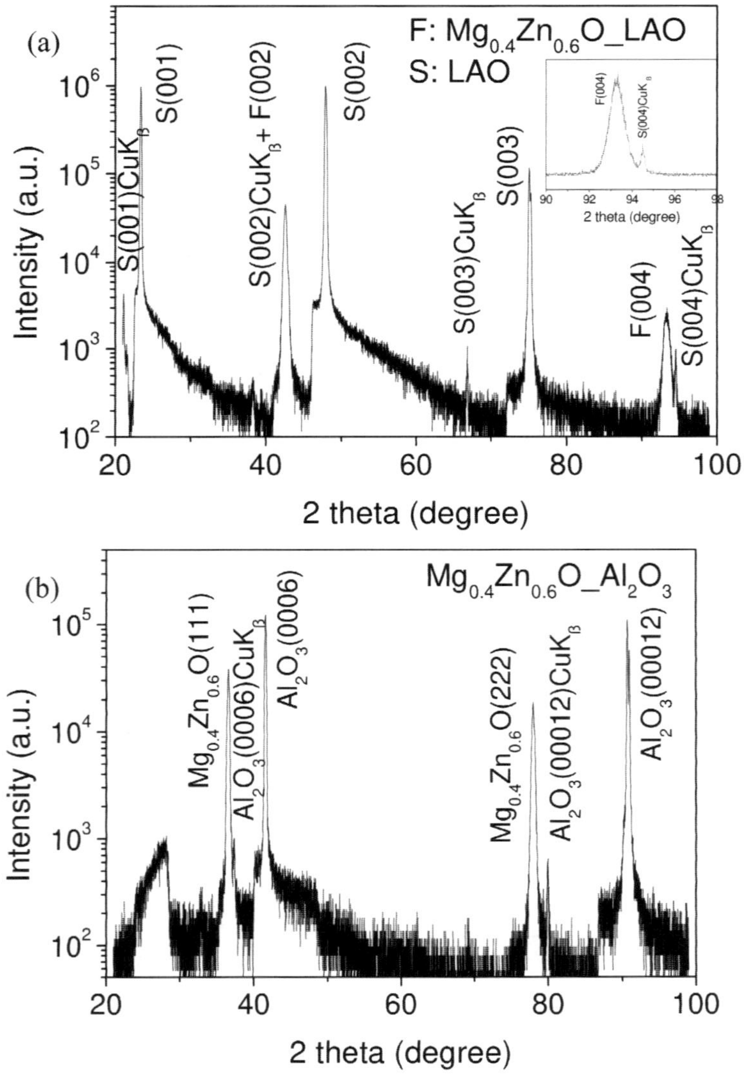

Figure 2. XRD 2-theta scans of the (a) MZO/LAO(001) film (pressure: 3 × 10 − 5 torr; temperature: 650 ° C). The inset shows the close up view of 2 theta at near 94° and (b) MZO/Al2O3(0001) film.

Figure 3 show the XRD rocking curves on the MZO(004)/LAO(001) reflection and MZO (222)/Al_2O_3 (0001) reflection. The full width at half maximum (FWHM) of the curves are 1.0488° and 0.5441°, respectively, indicating good crystalline quality of the films. Note that the FWHM of the rocking curves on the LAO (003) reflection and Al_2O_3 (00012) reflection are 0.297° and 0.207°, respectively.

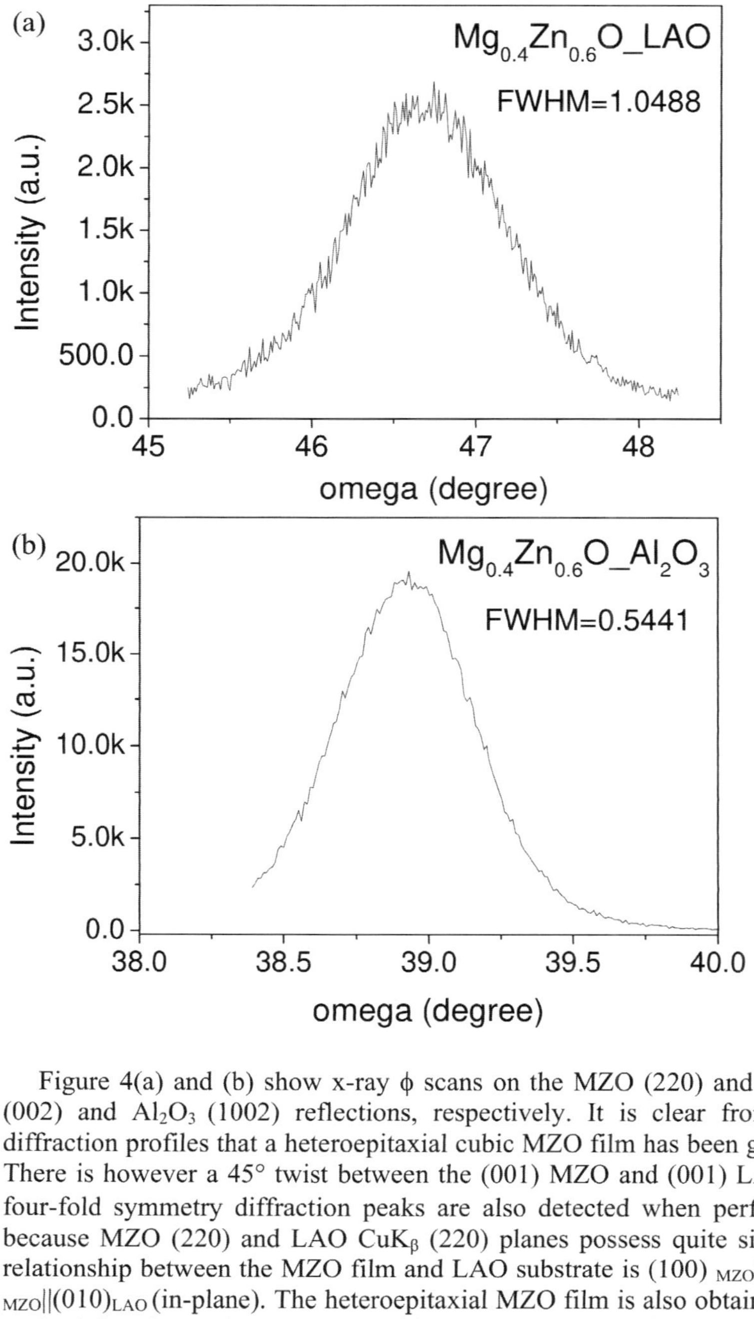

Figure 3. Rocking curves on the (a) MZO(004)/LAO(001) diffraction plane and (b) MZO(222)/A12O3(0001) diffraction plane.

Figure 4(a) and (b) show x-ray ϕ scans on the MZO (220) and LAO (220) reflections and MZO (002) and Al_2O_3 (1002) reflections, respectively. It is clear from the clean four-fold symmetry diffraction profiles that a heteroepitaxial cubic MZO film has been grown on the LAO (001) substrate. There is however a 45° twist between the (001) MZO and (001) LAO planes. The LAO CuK_β (220) four-fold symmetry diffraction peaks are also detected when performing the ϕ scan of MZO film because MZO (220) and LAO CuK_β (220) planes possess quite similar 2 theta values. The epitaxy relationship between the MZO film and LAO substrate is $(100)_{MZO}||(100)_{LAO}$ (out-of-plane) and $(011)_{MZO}||(010)_{LAO}$ (in-plane). The heteroepitaxial MZO film is also obtained on Al_2O_3 (0001) substrate. It is noticed that the peaks are separated by 60° due to a six-fold symmetry in the diffraction profiles of MZO (002), while the peaks are separated by 120° due to a three-fold symmetry in the diffraction profiles of Al_2O_3 (1002). Also there is a 30° twist between the (002) MZO and (0001) Al_2O_3 planes. The epitaxy relationship is concluded to be $(111)_{MZO}||(0001)_{Al2O3}$ (out-of-plane) and $(002)_{MZO}||(1002)_{Al2O3}$(in-plane). Detailed epitaxy relationship should be studied further with TEM images.

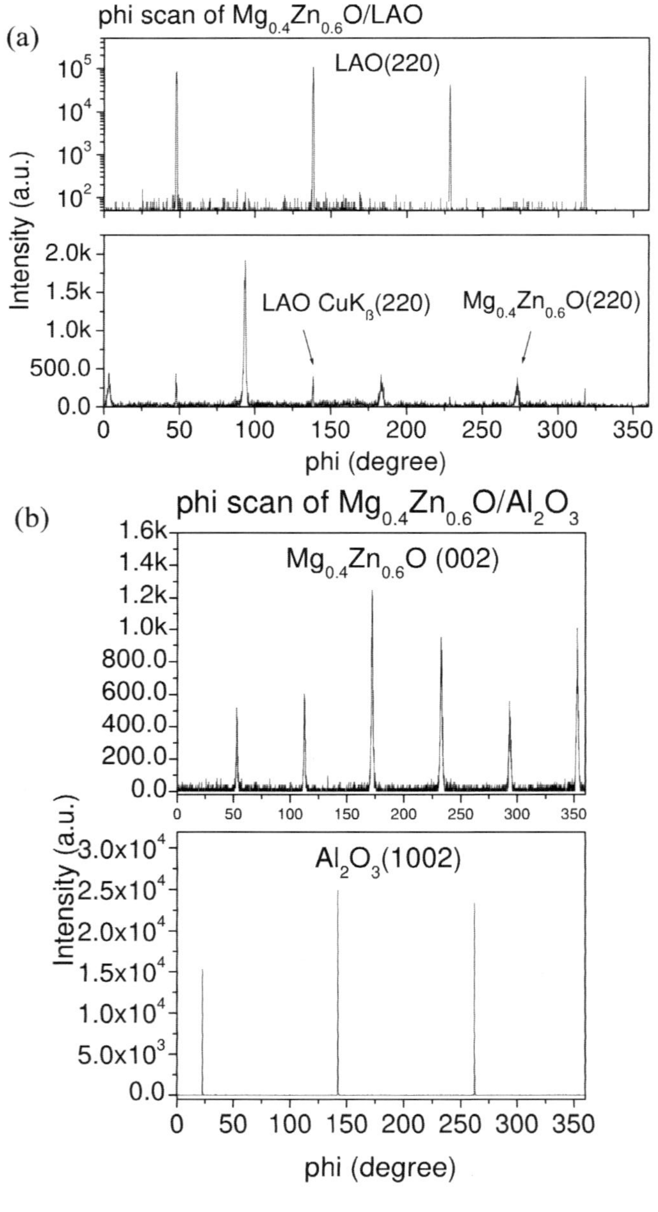

Figure 4. XRD scans on the (a) MZO (220) and LAO (220) reflections and (b) MZO (002) and Al2O3(1002) reflections.

In order to get more precise information about the nature of the MZO/LAO and MZO/Al_2O_3 interfaces and the crystalline quality of the MZO films on both substrates we preformed transmission electron microscopic investigation on this heterostructure. The bright-field cross-section TEM micrographs from the MZO/LAO(001) film and MZO/Al_2O_3(0001) film are shown in figure 5(a) and (c). For the film deposited on LAO substrate, crystal grains are alloy-like grown. Apparently the film at near the interface has higher density of defects than the film at near the surface. It is also seen that there are regular dark patches in the LAO substrate along the interface boundary. All these are clear evidences of lattice mismatch induced straining on the film and substrate. For the film deposited on sapphire, typical columnar crystal grains fabricated by PLD vertical to the substrate could be obtained and the top of the column is very flat. The corresponding selected area electron diffraction (SAED) patterns of these two films at region near the film/substrate interface are shown in figure 5(b) and (d). These patterns contain diffraction spots corresponding to the epitaxial MZO films with LAO and sapphire substrates, respectively. The results confirm the epitaxial relationships obtained from X-ray diffraction.

Figure 5. Cross-section TEM micrographs of (a) MZO/LAO(001) film and (c) MZO/Al2O3(0001) film and corresponding SAED patterns at region near the film/substrate interface of (b) MZO/LAO(001) film and (d) MZO/Al2O3(0001) film.

SEM photographs of cross section of these two kinds of oriented films on LAO and sapphire substrates are shown in figure 6. The surfaces of the MZO films grown at the vacuum atmosphere are reasonably smooth, and the surfaces of these orientated films remained featureless at magnifications up to 50000. In particular, there were no large droplets or particulates that are common in PLD deposited films. The cross-section morphology reveals sharp boundary and layer thickness is about 350-500 nm. Also, from the pictures, we know that the MZO/LAO film has a larger average roughness than that of the MZO/Al_2O_3 film. Substrate factor is the only reason to interpret this phenomenon.

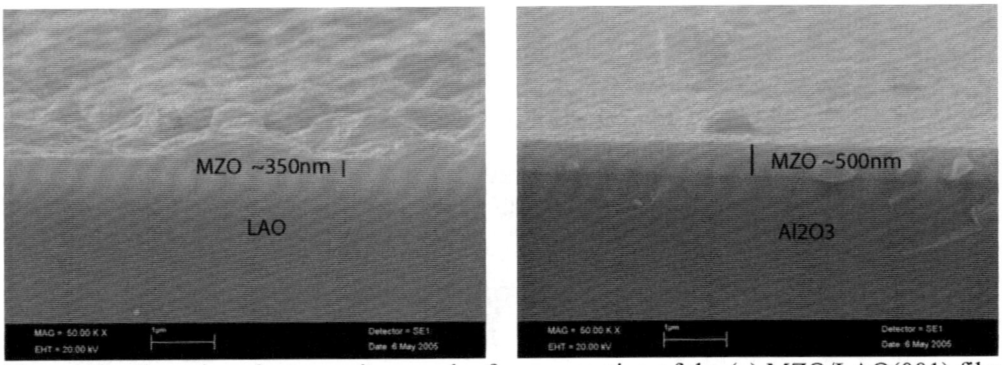

Figure 6. Scanning electron micrograph of cross section of the (a) MZO/LAO(001) film and (b) MZO/Al_2O_3(0001) film.

4. Conclusion

Single-oriented epitaxial cubic MZO films were prepared on LAO (001) and Al_2O_3 (0001) substrates by the PLD technique. All films were smooth and no large of droplets. The film on the LAO substrate has a larger average roughness than that of the film on the Al_2O_3 substrate. It was found that the orientation selection is controlled at the initial stage of deposition and the growth orientation is solely determined by the substrates orientation. Heteroepitaxial relationship of $(100)_{MZO}||(100)_{LAO}$ (out-of-plane) and $(011)_{MZO}||(010)_{LAO}$ (in-plane) can be realized on LAO substrate. $(111)_{MZO}||(0001)_{Al2O3}$ (out-of-plane) and $(002)_{MZO}||(1002)_{Al2O3}$(in-plane) relationship could be induced by using of Al_2O_3 (0001) substrate. The cross-section morphology reveals layer thickness of about 350-500 nm.

References

[1] Kudo A, Yanagi H, Ueda K, Hosono H, Kawazoe H and Yano Y 1999 *Appl. Phys. Lett.* **75** 2851
[2] Watanabe Y 1995 *Appl. Phys. Lett.* **66** 1770
[3] Mathews S, Ramesh R, Venkatesan T and Benedetto J 1997 *Science* **276** 238
[4] Vispute R D, Talyansky V, Trajanovic Z, Choopun S, Downes M, Sharma R P and Venkatesan T 1997 *Appl. Phys. Lett.* **70** 2735
[5] Ashrafi A B M A, Ueta A, Avramescu A, Kumano H and Suemune I 2000 *Appl. Phys. Lett.* **76** 550
[6] Kaldis E 1981 *Current Topics in Materials Science* vol 7 (Amesterdam: North-Holland) p 143
[7] Jaffe J E and Hess A C 1993 *Phys. Rev. B* **48** 7903
[8] Roessler D M and Walker W C 1966 *Phys. Rev. Lett.* **17** 310
[9] Bundesmann C, Schubert M, Spemann D, Butz T, Lorenz M, Kaidashev E M and Grundmann M 2002 *Appl. Phys. Lett.* **81** 2376
[10] Chen J and Shen W Z 2003 *Appl. Phys. Lett.* **83** 2154
[11] Takeuchi I, Yang W, Chang K S, Aronova M A and Venkatesan T 2003 *J. Appl. Phys.* **94** 7336
[12] Choopun S, Vispute R D, Yang W, Sharma R P and Venkatesan T 2002 *Appl. Phys. Lett.* **80** 1529
[13] Ohtomo A, Kawasaki M, Koida T, Masubuchi K, Koinuma H, Sakurai Y, Yoshida Y, Yasuda T, Segawa Y 1998 *Appl. Phys. Lett.* **72** 2466
[14] Sharma A K, Narayan J, Muth J F, Teng C W, Jin C, Kvit A, Kolbas R M, Holland O W 1999 *Appl. Phys. Lett.* **75** 3327
[15] Teng C W, Muth J F, Ozgur U, Bergmann M J, Everitt H O, Sharma A K, Jin C and Narayan J 2000 *Appl. Phys. Lett.* **76** 979
[16] Narayan J, Sharma A K, Kvit A, Jin C, Muth J F and Holland O W 2002 *Solid State Commun.* **121** 9

Journal of Physics: Conference Series **152** (2009) 012088 doi:10.1088/1742-6596/152/1/012088

The Polarization Fatigue Behavior in $Pb(Mg_{1/3}Nb_{2/3})O_3$-$0.32PbTiO_3$ Single Crystals

Dabin LIN, Zhenrong LI*, Zhuo XU and Xi YAO

Electronic Materials Research Laboratory, Key Laboratory of Education Ministry, Xi'an Jiaotong University, Xi'an 710049, People's Republic of China

E-mail: zhrli@mail.xjtu.edu.cn

Abstract. Polarization hysteresis (P-E) change was investigated as a function of the strength, frequency and cycles of electric field alternating for $[001]_c$, $[110]_c$, $[111]_c$ oriented $Pb(Mg_{1/3}Nb_{1/3})O_3$-$0.32PbTiO_3$ (PMN-0.32PT) single crystals. Under 1.5kV/cm (< E_c), the fatigue rates after 10^5 cycles were 13%, 16% and 34% for $[001]_c$, $[110]_c$ and $[111]_c$ orientations, respectively. Under 7.5kV/cm (>E_c), obvious polarization fatigue was presented for $[001]_c$, $[110]_c$, $[111]_c$ orientations and the biggest fatigue rate was around ~70% in $[110]_c$ orientations after 10^5 cycles. The fatigue rate decreased with the increasing of electric field frequency. The influences of crystallographic orientation and electric field strength and frequency on polarization fatigue behavior in PMN-0.32PT single crystals was discussed based on space charge, domain reverse and phase transition process.

1. Introduction

Relaxor-based ferroelectric single crystals $Pb(Mg_{1/3}Nb_{2/3})O_3$-$PbTiO_3$(PMN-PT) and $Pb(Zn_{1/3}Nb_{2/3})O_3$-$PbTiO_3$(PZN-PT) with morphtropic phase boundary (MPB) composition along $[001]_c$ orientation exhibit extraordinary large electromechanical coupling coefficient k_{33}>90% and piezoelectric coefficient d_{33}>2000pC/N[1]. The intermediate phase monoclinic (M) and orthorhombic (O) phases between rhombohedra (R) and tetragonal (T) phases were observed in PMN-PT single crystals with MPB composition based on the results of high-resolution synchrotron X-ray diffraction, polarized light microscopy, and dielectric properties[2-6]. It is reasonable that intermediate phases induced by electric field play an important role for enhanced piezoelectric properties.[7] Ferroelectric materials exhibit polarization fatigue under applied cyclic electric field, meanwhile, electric field could induce phase transition in PMN-PT single crystals. Much attention has been focused on the physical origin of great piezoelectric properties in relaxor-based single crystal. However less study was concerned with the effects of complex phase transition under electric field on ferroelectric fatigue.

In earlier studies, fatigue in ferroelectrics is thought from internal stress, switching mechanisms, domain configuration and space charge[8-10] etc. The formation of space charges is usually from oxygen vacancies[11-13] or electrons-hole pairs[14]. It is expected to display the intrinsic origin of ferroelectric fatigue based on ferroelectric single crystals[15-17]. The fatigue behavior of PZN-0.045PT single crystals concerned with the crystallographic orientation[18]. The rhombohedra PZN-0.045PT single crystal have no fatigue along $[001]_c$ orientation after 10^5 cycles under 30kV/cm and fatigue in $[111]_c$ direction[19]. $[001]_c$ oriented PZN-0.10PT crystal around MPB was serious fatigue after 10^3 cycles at low temperatures[18]. The fatigue rate is severe along $[001]_c$ orientation for tetragonal PZN-0.12PT single crystal [20]. Metin Ozgul et al. reported the polarization fatigue of PZN-PT with respect to composition, orientation, temperature and electric field strength. The ferroelectric fatigue behavior of PMN-PT single crystal has been less reported until now. In this paper, the effects of orientation, electric field strength (E<Ec and E>Ec) and frequency on the polarization fatigue behavior of PMN-0.32PT single crystal was studied. The ferroelectric fatigue mechanism in PMN-0.32PT single crystals was discussed.

2. Experimental Procedure

PMN-0.32PT single crystals were grown by a modified Bridgman method in our laboratory [21]. The pseudo-cubic $[001]_c$, $[110]_c$ and $[111]_c$ directions of samples were determined by X-ray diffraction (XRD). The final dimensions of samples for ferroelectric fatigue measurement were nominally $4 \times 4 \times 0.6$ mm^3. Silver electrodes were painted onto both surfaces. Hysteresis loops were measured using a modified Sawyer-Tower circuit (TF ANALYZER 2000 and MTI 2000 FOTONIC SENSOR). The remnant polarizations (P_r) and the coercive fields (E_c) were computed from the recorded hysteresis loops. The electric field was applied with a triangular bipolar wave form for fatigue experiments. The applied ac electric field strength were 1.5kV/cm ($<E_c$), 7.5kV/cm ($>E_c$) and the frequencies were 1Hz, 10Hz, 100Hz, respectively. The hysteresis loops after 10^1, 10^2, 10^3, 10^4, 10^5 switching cycles at different electric fields were measured during fatigue experiments. For hysteresis loop measurement of before and after fatigue in PMN0.32PT single crystal, the applied electric field was generally 7.0kV/cm at 1Hz with a sine wave form.

3. Results and Discussion

Figure 1(a), (b) and (c) show unfatigue (1cycle) P-E loops and loops after 10^5 cycles under E=1.5kV/cm ($<E_c$) ac electric field in PMN-0.32PT single crystals along with $[001]_c$, $[110]_c$ and $[111]_c$ orientations at room temperature, respectively. The remnant polarization (P_r) obviously decreased and coercive field strength increased along $[111]_c$ orientation after 10^5 cycles, comparing with $[001]_c$ and $[110]_c$ orientations. The fatigue rate is defined as the remnant polarization change, $\Delta P_r/P_r$, where the remnant polarization at 1 cycle is denoted as P_r and the difference of remnant polarization before and after fatigue is denoted as ΔP_r. $\Delta P_r/P_r$ as a function of crystallographic orientation and switching cycles is shown in Figure 2. The remnant polarization decreases obviously after 10^2 cycles for both $[001]_c$ and $[110]_c$ orientations and decreases quickly after 10 cycles for $[111]_c$ orientation. After 10^5 cycles, the fatigue rates are 13%, 16% and 34% for $[001]_c$, $[110]_c$ and $[111]_c$ orientations, respectively. The ferroelectric fatigue rate along $[111]_c$ direction is the biggest among three orientations in PMN-0.32PT single crystals.

The domain reversion is not easy during switching cycles if the applied electric field is below E_c. So ferroelectric fatigue in PMN0.32PT single crystal isn't directly related to the domain reversion when E=1.5kV/cm ($<E_c$). For $[001]_c$ and $[110]_c$ orientations, it is maybe assumed that domain wall pinning induced by the accumulating of space charge which restricts the polarization. PMN-0.32PT single crystal is considered to be R phase at room temperature, in which the spontaneous polarization is along $[111]_c$ direction. The polarization vector is paralleled to the electric field and the switching is presumably dominated by 180° domain walls. Ferroelectric fatigue is mostly induced by 180° domain reversion for [111] orientation. Furthermore the accumulating of space charge intensifies the local electric field and accelerates the deflection of spontaneous polarization. As a result, the strong fatigue is presented for [111] orientation PMN0.32PT single crystal.

Figure 3(a), (b) and (c) show unfatigue (1cycle) P-E loops and loops after 10^5 cycles under E=7.5kV/cm ($>E_c$) ac electric field in PMN-0.32PT single crystals along with $[001]_c$, $[110]_c$ and $[111]_c$ orientations at room temperature, respectively. As shown in Figure3, ferroelectric fatigue is obvious for all samples after 10^5 cycles. Meanwhile, the coercive field E_c increases in $[110]_c$ and $[111]_c$ orientation and E_c is almost unchanged in $[001]_c$ orientation. The fatigue rate as a function of crystallographic orientation and switching cycles is shown in Figure 4. The polarization decreases quickly after 10 cycles for all samples. After 10^5 cycles, the fatigue rates are about 38%, 71% and 55%

for $[001]_c$, $[110]_c$ and $[111]_c$ orientations, respectively. Comparing Figure 2 and Figure 4, the fatigue under 7.5kV/cm (>Ec) is stronger than that under 1.5 kV/cm (<Ec).

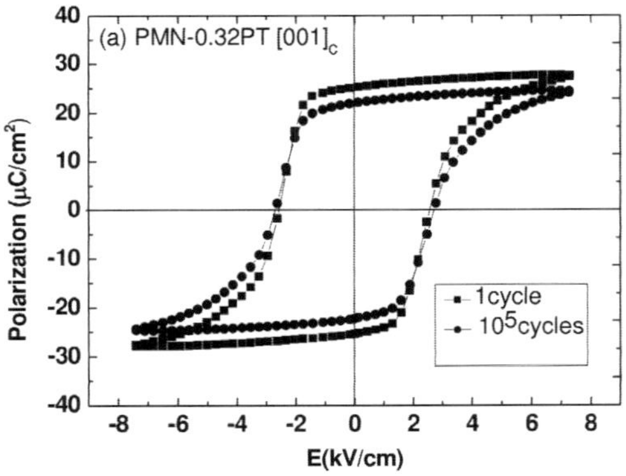

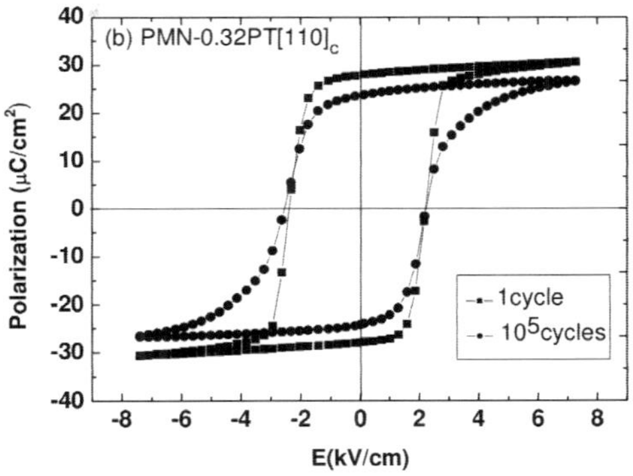

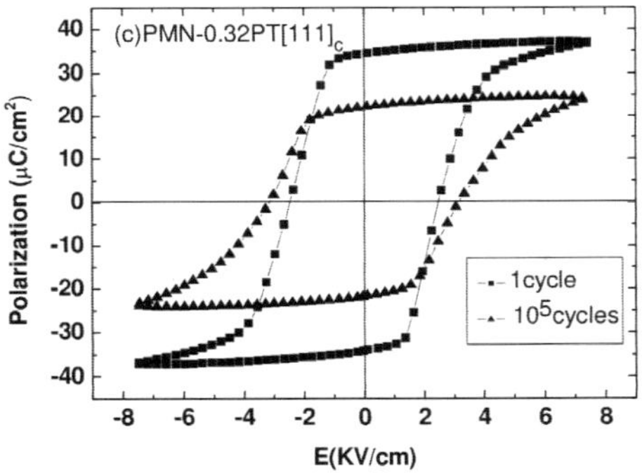

Figure 1. Hysteresis loop change after 10^5 cycles under 1.5kV/cm in PMN-0.32PT single crystal along with (a) $[001]_c$, (b) $[110]_c$ and (c) $[111]_c$ orientation

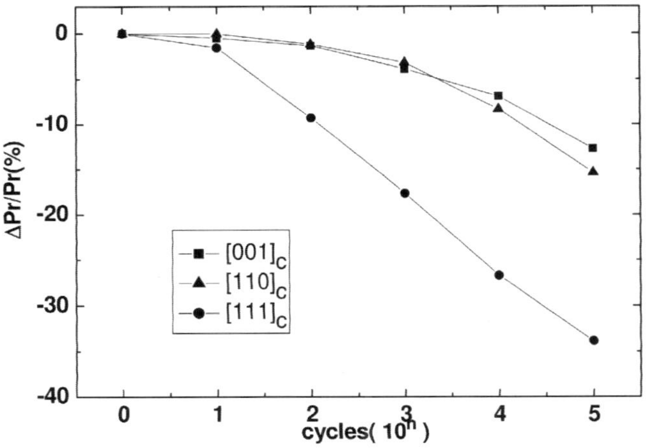

Figure 2. Polarization fatigue behavior of PMN-0.32PT single crystals under 1.5kV/cm ac field at 10Hz.

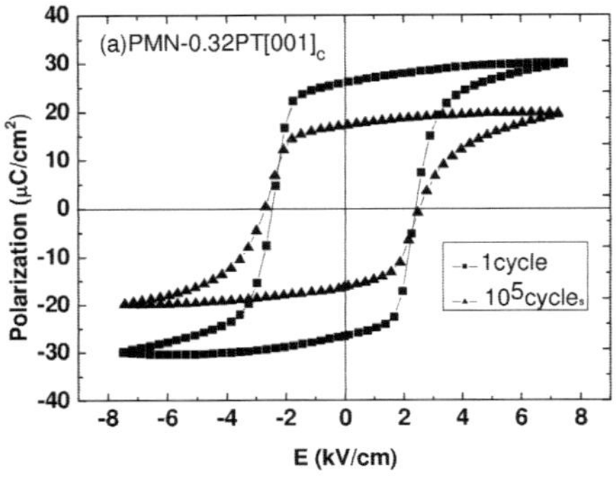

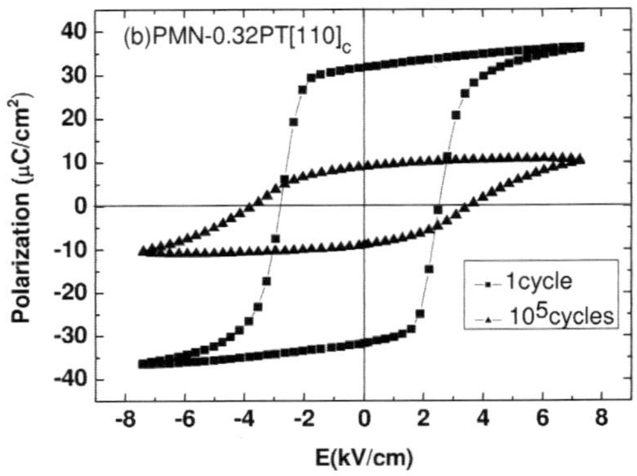

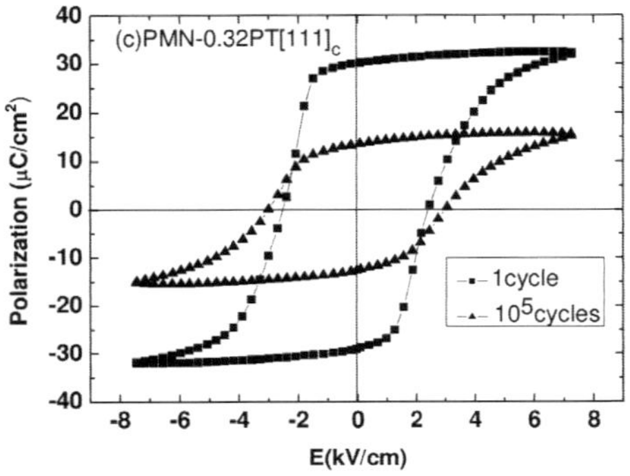

Figure 3. Hysteresis loop change after 10^5 cycles under 7.5kV/cm in PMN-0.32PT single crystal along with (a) $[001]_c$, (b) $[110]_c$ and (c) $[111]_c$ orientation

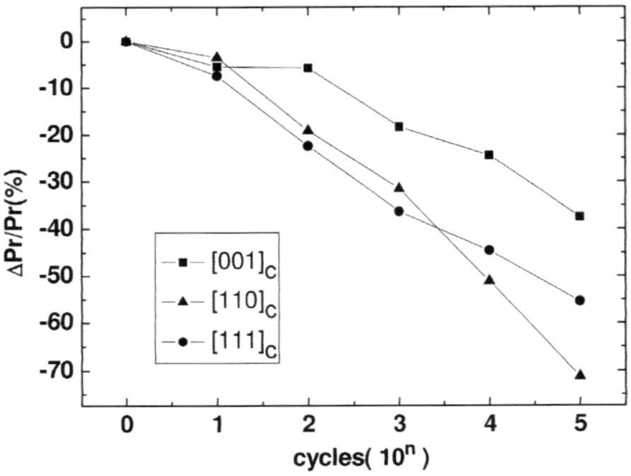

Figure 4. Polarization fatigue behavior of PMN-0.32PT single crystals under 7.5kV/cm ac field at 10Hz.

Vanderbilt and Cohen[22] conceived a thermodynamic model to describe three possible monoclinic phases (M_A, M_B and M_C) from Devonshire theory. The intermediate phases (monoclinic and orthorhombic phases) between rhombohedra and tetragonal phase have been found in PMN-PT single crystal around MPB [23]. The phase transformation sequence in MPB PMN-PT crystals is strongly dependent on the amplitude and the direction of applied electric field. For example, the phase transition sequence of $[001]_c$, $[110]_c$ and $[111]_c$ oriented are $R \rightarrow M_A \rightarrow M_C \rightarrow T$[24], $R \rightarrow M_B \rightarrow O$[25] and

Journal of Physics: Conference Series **152** (2009) 012088 doi:10.1088/1742-6596/152/1/012088

$R \rightarrow M_A \rightarrow T \rightarrow M_A \rightarrow R_{111}$[26], respectively. According to our dielectric measurement results, the different phase transition process induced by electric field for different orientations in PMN-0.32PT single crystals were exhibited[27]. When the applied electric field was 7.5kV/cm, it was presented that M_C phase induced from R phase through M_A phase for $[001]_c$ orientation, mixed phase between R and O phase including M_B phase for $[110]_c$ and only R phase for $[111]_c$ orientation in PMN-0.32PT single crystals, respectively[27]. It is supposed that the fatigue is controlled by the polarization reverse in Mc phase, the mixed phases and R phase for $[001]_c$, $[110]_c$ and $[111]_c$ orientation under 7.5kV/cm, respectively. From Figure 3 and Figure 4, the biggest remnant polarization change is shown in $[110]_c$ orientation after 10^5 cycles. The spontaneous polarization in O-phase is along with $[110]_c$ direction. So the direction of polarization is paralleled to applied electric field for $[110]_c$ orientation. As a result, the switching is presumably dominated by 180° domain walls. At the same time, besides O phase, R and M phases coexist. The phase transition among O, R and M phases induced by ac electric field would contribute to the ferroelectric fatigue. And R phase only exists for $[111]_c$ orientation sample under electric field 7.5kV/cm. The phase transition process in $[110]_c$ orientation is more complex than that in $[001]_c$ and $[111]_c$ orientations. This may be the main reason that the fatigue rate in $[110]_c$ orientation is the biggest among $[100]_c$, $[110]_c$ and $[111]_c$ orientations.

In PMN-0.32PT single crystal, the polarization vectors in ferroelectrics phase is paralleled to the electric field direction induced by high alternating electric fields. This implies that switching mechanisms become more dominant when the polarization direction paralleled to the electric field. The complex domain switching mechanism induced by the existence of mixed ferroelectric phases may lead to fatigue isotropy. Domain switching anisotropy may be one of the origins of the fatigue anisotropy. It presumably shows that the ferroelectric fatigue is concerned with the phase transition process. Therefore, the existence of mixed phase under high electric field and the phase transition among the mixed phase under cycling electric field will enhance the ferroelectric fatigue.

The frequency dependence of the fatigue rate for $[110]_c$ orientation sample under 1.5kV/cm and 7.5kV/cm are shown in Figure 5(a) and 5(b), respectively. In Figure 5(a), $\Delta P_r/P_r$ at low frequency is bigger than that at high frequency. The long interval of electric field at low frequency is favored for accumulating space charge at domain boundaries. The domain wall motion is restricted during polarization switching due to space charges. It implies that the space charges play a dominant role under 1.5kV/cm. Comparing Figure 5(a) with 5(b), the effect of frequencies on the fatigue rate under 7.5kV/cm obviously decreases. That is to say, effect of space charges on ferroelectric fatigue is reduced under high electric field. It is assumed that the fatigues are dominated by the space charges under below E_c and the complex domain switching mechanism under above E_c in PMN-0.32PT single crystals. Actually, PMN-PT single crystals have complex phase relation under electric field. The fatigue mechanism will be studied further.

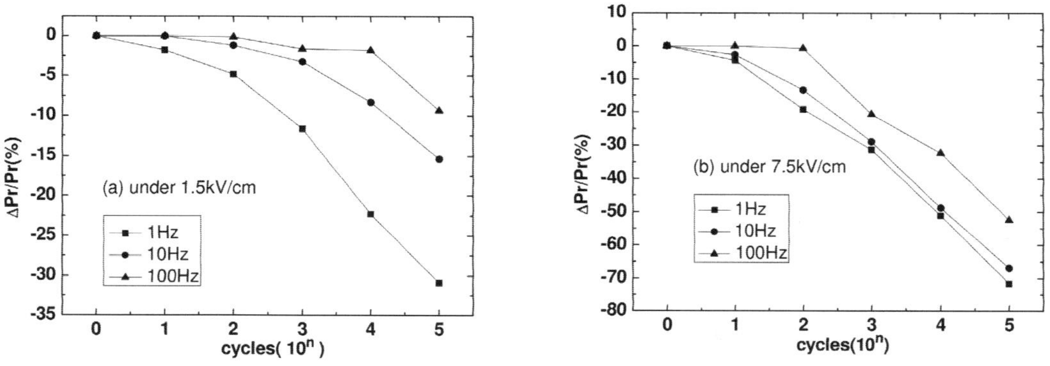

Figure 5. Comparison of the fatigue rate in $[110]_c$ oriented PMN-0.32PT single crystal at different frequency under (a) 1.5kV/cm, and (b) 7.5kV/cm

4. Conclusions

The ferroelectric fatigue was investigated as a function of the strength, frequency and cycles of electric field alternating for $[001]_c$, $[110]_c$, $[111]_c$ orientations in PMN-0.32PT single crystals. Under 1.5kV/cm ($< E_c$), the fatigue rate after 10^5 cycles for $[111]_c$ orientation was the biggest among three orientations. The ferroelectric fatigue is mostly induced by 180° domain reversion for [111] orientation. Under 7.5kV/cm ($>E_c$), obvious fatigue was presented for $[001]_c$, $[110]_c$, $[111]_c$ orientations and the biggest fatigue rate is around ~70% in $[110]_c$ orientations after 10^5 cycles. It was related the mixed phases between R and O phase including M_B phase for $[110]_c$ orientation. The fatigue rate decreased with electric field frequency increasing. The domain switching mechanism can be expected to control the fatigue rate of single crystal with single phase structure. The fatigue was concerned with the relation between the polarization direction and electric field direction in single crystal with mixed ferroelectric phases.

Acknowledgements

This work was supported by National Natural Science Foundation of China No.90205030 and New Century Excellent Talent (2007) of the Ministry of Education

Reference

[1] Shrout T R, Chang Z P and Klm N 1990 Markgraf S: Ferroelectric. Lett. **12** 63
[2] Lu Y, Jeong D.-Y, Cheng Z.-Y, Zhang Q M, Luo H S, Yin Z W and Viehland D 2001 Appl. Phys. Lett. **78** 3109
[3] Ye Z.-G, Noheda B, Dong M, Cox D and Shirane G 2001 Phys. Rev. B **64** 184114
[4] Xu G, Luo H, Xu H and Yin Z W 2001 Phys. Rev. B **64** 020101
[5] Noheda B, Cox D E, Shirane G, Gao J and Ye Z.-G 2002 Phys. Rev. B **66** 054104
[6] Tu C S, Hung L.-W, Chien R R and Schmidt V H 2004 J. Appl. Phys **96** 4411
[7] Viehland D, Amin A and Li J F 2001 Appl. Phys. Lett. **79** 1006
[8] Arlt G and Pertsev N A 1991 J. Appl. Phys. **70** 2283
[9] Duiker H M, Beale P D, Scott J F, Paz de Araujo C A, Melnick B M, Cuchiaro J D and McMillan L D 1990 J. Appl. Phys. **68** 5783
[10] Lohkamper R, Neumann H and Arlt G 1990 J.Appl. Phys. **68** 4220
[11] Scott F, Araujo C, Melnick B M, McMillan L D and Zuleeg R 1991 J. Appl. Phys. **70** 382
[12] Chen, Harmer M P and Smyth D M 1994 J. Appl. Phys. **76** 5394
[13] Duiker H M and Beale P D 1990 Phys. Rev.B **41** 490
[14] Warren W L, Dimos D, Tuttle B A, Nasby R D and Pike G E 1994 Appl. Phys. Lett. **65** 1018
[15] Nakamura T, Nakao Y, Kanisawa A and Takasu H 1994 Appl. Phys. Lett. **65** 152
[16] Ramesh R, Girchlist H, Sands T, Keramidas V G, Haakenaasen R and Fork D K 1993 Appl. Phys. Lett. **63** 3592
[17] Dat R, Lee J K, Auciello O and Kingon A I 1995 Appl. Phys. Lett. **67** 572
[18] Ozgul M, Takemura K, Susan Trolier McKinstry and Clive A. Randall 2001 J. Appl. Phys. **89** 5100
[19] Metin Ozgul, Susan Trolier-McKinstry and Clive A. Randall 2004 J. App. Phys. **V95** 4296-4302.
[20] Koichi Takemura, Metin Ozgul, Veronique Bornand, Susan Trolier-McKinstry and Clive A. Randall 2000 J. Appl. Phys. **88** 7272
[21] Li Z R, Xi Z Z, Xu Z and Yao X 2002 J. Mater. Sci. Lett. **21** 1325
[22] Vanderbilt David and Cohen Morrel H 2001 Phys. Rev. B **63** 214112
[23] Noheda B 2002 Curr. Opin. Solid State Mater. Sci. **6** 27
[24] Feiming Bai, Naigang Wang, Jiefang Li and Viehland D 2004 J. Appl. Phys. **96** 1620
[25] Viehland V and Li J F 2002 J. Appl. Phys. **92** 7690
[26] Chi-shun Tu, Hugo Schmidt V, Shih I C and Chien R 2003 Phys. Rev. B. **67** 020102
[27] Li Z R, Xu Z, Yao X and Cheng Z.-Y 2008 J. Appl. Phys. **104** 024112

Symposia D, E and F from MRS International Materials Research Conference IOP Publishing
Journal of Physics: Conference Series **152** (2009) 012089 doi:10.1088/1742-6596/152/1/012089

Novel red phosphors for light emitting diodes: $Sr_{2-y}Ce_{1-x}Ti_xO_4$:yEu^{3+}

Haiyan JIAO[1], Yuhua WANG[1,2,*], Jiachi ZHANG[1]

[1]Department of Materials Science, Lanzhou University, Lanzhou 730000, P. R. China

[2]Key Laboratory for Magnetism and Magnetic Materials of the Ministry of Education, Lanzhou University, Lanzhou 730000, P. R. China

E-mail address: wyh@lzu.edu.cn

Abstract. A new luminescent inorganic oxide with the composition $Sr_{2-y}Ce_{1-x}Ti_xO_4$:yEu^{3+} ($0 \leq x \leq 0.03$, $0.005 \leq y \leq 0.10$), and as new near ultraviolet excited phosphors were synthesized using conventional solid-state reaction method and their luminescence properties under 365 nm excitation were investigated in detail. It indicated that, by doping small amount of Ti^{4+} into Ce^{4+} sites, the excitation intensity of charge transfer (CT) band of Ce^{4+}-O^{2-}(330-400 nm) was significantly broaden and enhanced. This can be explained by the special crystal structure of Sr_2CeO_4, which contained two kinds of Ce-O bonds: Ce-O-Ce and Ce-O-Sr, the incorporation of Ti^{4+} would change one of the Ce-O bond and result in the change of charge transfer band of Ce^{4+}-O^{2-}. As a result, the emission intensity of phosphor $Sr_2Ce_{1-x}Ti_xO4$ have improved about 85% by doping 0.01 mol Ti^{4+}. $Sr_{2-y}Ce_{0.99}Ti_{0.01}O_4$:$yEu^{3+}$ shows white emission at $y \leq 0.01$ because of the dominant intensities of $^5D_2 - ^7F_{0-3}$ transitions in blue-green region whereas the intensities of $^5D_0 - ^7F_{0-3}$ transitions in orange-red regions dominate at concentrations $y \geq 0.03$ and give red emission. Comparing with the commercial red phosphor Y_2O_2S:Eu^{3+}, the $Sr_{1.95}Ce_{0.99}Ti_{0.01}O_4$:$0.05Eu^{3+}$ phosphor exhibited excellent color purity and much higher brightness. The results showed that $Sr_{1.95}Ce_{0.99}Ti_{0.01}O_4$:$0.05Eu^{3+}$ phosphors could be considered as a promising red phosphor for application in LEDs.

1. Introduction

White light-emitting diodes (LEDs), the so-called next generation solid-state lighting (SSL), with the advantages of long lifetime, saving energy consumption, and environmental-friendly characteristics, are attracting more attentions. In view of application, white LEDs is expected to open up a great number of exciting new application fields: white light sources to replace traditional incandescent and fluorescent lamps, backlights for portable electronics, medical, and architecture lightings, etc.[1]. Generally, there are two ways to gain white light[2-4]: one is combination of a blue LED (460 nm) with a yellow phosphor (YAG:Ce), however, the spectral composition of the light produced by this approach is different from that of natural white light due to the poor color rendering property caused by the lack of red component; and the second approach to achieve white light is combination of a near-UV(NUV) LED (350-420 nm) with red, green, and blue phosphors. Presently, the emission bands of LEDs are shifted to NUV range about 350-420 nm, because this wavelength can offer a higher efficiency solid-state lighting [5]. The current tricolor phosphors for NUV GaN-based LEDs are Y_2O_2S:Eu^{3+} for red, ZnS: (Cu^+, Al^{3+}) for green, and $BaMgAl_{10}O_{17}$:Eu^{2+} for blue [6]. However, the

$Y_2O_2S:Eu^{3+}$ red phosphor cannot absorb efficiently in NUV region and its brightness is about eight times less than that of the green and blue phosphors. In addition, the lifetime of the $Y_2O_2S:Eu^{3+}$ is inadequate under NUV irradiation for its instability [7, 8]. Therefore, it is urgent to search for novel and highly efficient red phosphors that can be excited by NUV (350 -420 nm).

A suitable red-emitting NUV-LED phosphor should meet the following necessary conditions in general: the host is stable, the phosphor exhibits strong and broad absorption around 350 - 420 nm (LED emission wavelength), and the phosphor shows strong emission under NUV excitation and with the chromaticity coordinates near the NTSC (National Television Standard Committee) Standard values. In order to find novel efficient red-emitting LED phosphors, in this paper, Eu^{3+} activated $Sr_{2-y}Ce_{1-x}Ti_xO_4:yEu^{3+}$ were prepared in terms of the following considerations. It is well know that Sr_2CeO_4 had been intensively studied for many years [9-12] for their application in UV light, cathode rays or X-rays excitation because of its unique one-dimensional chains structure[13], which consists of linear chains of edge-sharing CeO_6 octahedra, with four equatorial μ^2-O atoms and two terminal Ce-O bonds per octahedron, surrounded by inter-chain Sr^{2+} cations, and the terminal Ce-O bonds (Ce-O-Sr) are shorter than the equatorial bonds (Ce-O-Ce) by ~0.1 Å. As a result, Sr_2CeO_4 shows not only desirable absorption in UV region but also excellent thermal and chemical stability. At present, most of the research for Sr_2CeO_4 is concentrated in two areas: introduction of rare earth ions as a luminous centres [14, 15] and the improvement of synthesis methods [16-19]. Recently, Abanti Nag et al. [20] have investigated the phosphors of $Sr_{2-y}CeO_4:yEu^{3+}$, when increasing the Eu^{3+} to higher concentrations, which emitted red light under 282 nm and 336 nm excitation, but the emission intensity of the phosphors were not strong enough to be used in LED, because of the absorption of Sr_2CeO_4 excited at NUV regions is not strong and inadequate for application. In order to enhance the emission intensity of $Sr_{2-y}CeO_4:yEu^{3+}$ excited at 365 nm further, one of the important approaches could be introducing some ions in the phosphor to broaden and enhance the excited band of Sr_2CeO_4 at NUV regions. We considered that Sr_2CeO_4 shows a very effective absorption in UV region because of its unique one-dimensional structure, which is induced by a stable quadrivalent cerium cation with CeO_6 octahedron. It is well known that there are many other elements can form a stable quadrivalence such as Zr, Pr, Ti, Dy and so on. Therefore, it can be predict that by doping small amount of quadrivalent cation (Zr^{4+}, Pr^{4+}, Ti^{4+}, Dy^{4+}) into Ce^{4+} sites, may be broaden and enhance the excitation intensity of charge transfer (CT) band of Ce^{4+}-O^{2-}(330-400nm). According to reference [21], zhang lei et.al have already reported the luminescence property of Sr_2ZrO_4. Moreover, we synthesized the $Sr_2(Ce, Pr)O_4$ and $Sr_2(Ce, Dy)O_4$ phosphors using a solid-state reaction, but didn't obtain their single phase. Accordingly, in this paper, the $Sr_{2-y}Ce_{1-x}Ti_xO_4:yEu^{3+}$ phosphors are synthesized and their luminescence properties under 365 nm excitation were investigated in detail.

2. Experimental

Samples $Sr_{2-y}Ce_{1-x}Ti_xO_4:yEu^{3+}$ ($0\leq x\leq 0.03$, $0.005\leq y\leq 0.10$) were obtained by solid-state reaction in air as follows: the starting materials $SrCO_3$ (99%), CeO_2 (99.99%), TiO_2 (99%) and Eu_2O_3 (99.99%) were weighed according to appropriate stoichiometric ratio. After this, powders were blended in an agate mortar, the homogeneous mixture obtained was put into an alumina crucible and calcined in a muffle furnace at the temperature of 1200°C for 6 h, and then we obtained the intentional samples. The samples with x=0, 0.01, 0.03; y=0 and x=0.01; y=0.01, 0.05 are denoted as SCT series with SCT1, SCT2, SCT3, SCT4, and SCT5, respectively.

The phase of the phosphor powder was checked by X-ray powder diffraction (XRD) pattern using Rigaku D/Max-2400 X-ray diffractometer with Ni-filtered Cu Kα radiation. The luminescence spectra were measured using an Edinburgh Instruments FLS920 T combined Fluorescence Lifetime and steady-state spectrophotometer with Xe900 (450W xenon arc lamp) as excitation source. All spectra were recorded at room temperature.

3. Results and discussion

3.1. XRD analysis

The XRD patterns show that all samples are single phase and their color is white. As examples, the powder XRD patterns of $Sr_{2-y}Ce_{1-x}Ti_xO_4$:yEu^{3+} ($0 \le x \le 0.03$, $0.005 \le y \le 0.10$) are shown in Figure 1. According to the JCPDS No. 22-1422, all the samples are single orthorhombic Sr_2CeO_4 phase. No second phase is observed. The relationships between the crystal cell parameter a, c and the Ti^{4+} concentration in the system $Sr_2Ce_{1-x}Ti_xO_4$ ($0 \le x \le 0.03$) is exhibited in Figure 2. It is found that with the increase of Ti^{4+} concentration, the crystal cell parameter has gradually decreased. Because the ionic radius of Ce^{4+} (0.0970 nm) is larger than that of Ti^{4+} (0.0750 nm), it can be suggested that Ti^{4+} occupies the Ce^{4+} site in this phosphor.

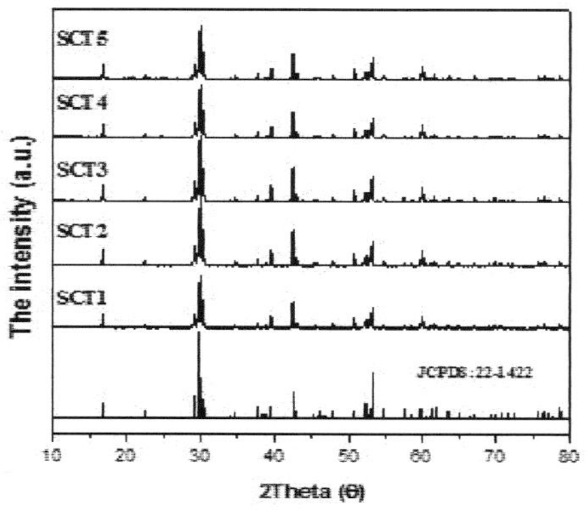

Figure 1. XRD patterns of $Sr_{2-y}Ce_{1-x}Ti_xO_4$:yEu^{3+} ($0 \le x \le 0.03$, $0.005 \le y \le 0.10$).

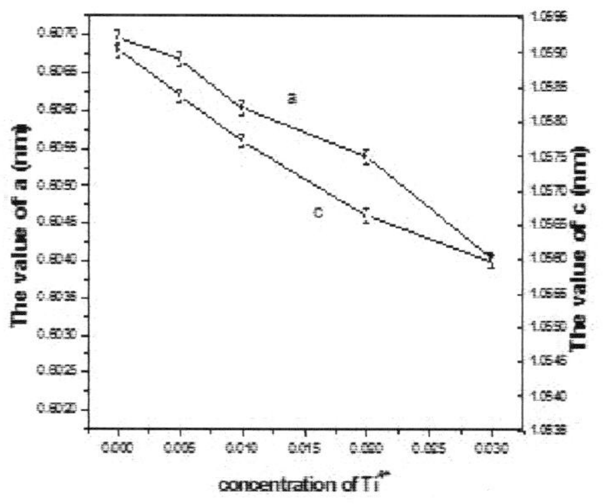

Figure 2. The relationships between the crystal cell parameter a, c and the Ti^{4+} concentration in the system $Sr_2Ce_{1-x}Ti_xO_4$ ($0 \le x \le 0.03$).

3.2. Luminescence properties

The excitation and emission spectra of $Sr_2Ce_{1-x}Ti_xO_4$ ($0 \le x \le 0.03$) are similar broad bands and those for Sr_2CeO_4 and $Sr_2Ce_{0.99}Ti_{0.01}O_4$ are shown as representative in Figure 3 and Figure 4, respectively. The excitation and emission spectra show broad bands peaked around 260 and 350 nm for excitation and 470 nm for emission. These peaks in the excitation and emission spectra can be assigned to Ce^{4+} - O^{2-} charge-transfer transition[9, 14].In Sr_2CeO_4, two kinds of Ce^{4+} exist, that is, there are two different Ce^{4+}- O^{2-} bond lengths in the lattice[9].The two excitation peaks observed in the figure are thus attributed to the different charge-transfer transitions.

As shown in Figure3, by doping small amount of Ti^{4+} into Ce^{4+} sites, the CT band of Ce^{4+}-O^{2-} (330-400 nm) was significantly broaden and enhanced. It can be considered that there are two different Ce^{4+}-O^{2-} bond lengths in the Sr_2CeO_4 lattice, the introduction of Ti^{4+} could change one of the Ce-O bond and result in the change of CT band of Ce^{4+}-O^{2-}. So, In the case of $Sr_2Ce_{1-x}Ti_xO_4$, the excited band (350 nm) was broaden and enhanced with doping of Ti^{4+}, these wavelength make the excitation intensity during NUV relatively strong and make the phosphors suitable to be used for 365 nm excitation. As shown in the inset of Figure4, doping of appropriate Ti^{4+} is helpful to improve the NUV–PL intensity of $Sr_2Ce_{1-x}Ti_xO_4$ ($0 \le x \le 0.03$). The PL intensity has been enhanced by 85% when Ti^{4+} doped concentration arrived 0.01mol.

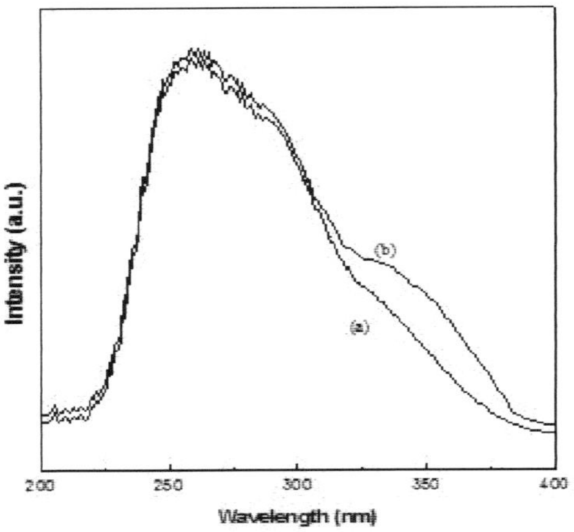

Figure 3. Excitation spectrum of (a) Sr_2CeO_4 and (b) $Sr_2Ce_{0.99}Ti_{0.01}O_4$ (λ_{em}=470 nm).

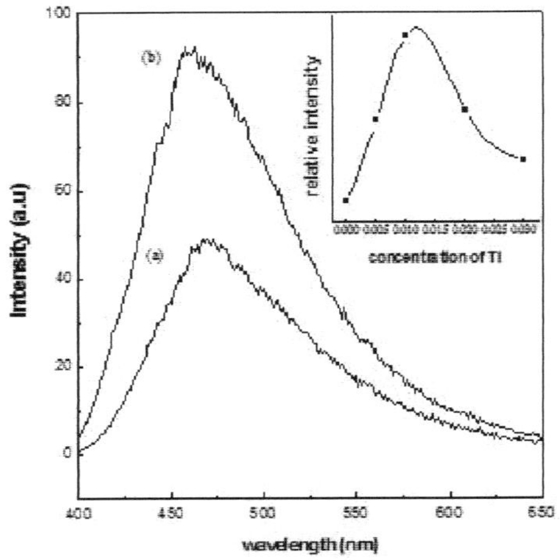

Figure 4. Emission spectrum of (a) Sr_2CeO_4 and (b) $Sr_2Ce_{0.99}Ti_{0.01}O_4$ (λ_{ex}=365 nm).

The emission spectra of $Sr_{2-y}Ce_{0.99}Ti_{0.01}O_4:yEu^{3+}$ ($0.005\leq y\leq 0.10$) recorded at λ_{ex}=365 nm are shown in Figure 5. We could see that the profiles of emission spectra of $Sr_{2-y}Ce_{0.99}Ti_{0.01}O_4:yEu^{3+}$ phosphors are similar to $Sr_{2-y}CeO_4:yEu^{3+}$ phosphors which have already reported by Abanti Nag and T. R. Narayanan Kutty [20].The emission spectrum of $Sr_{1.995}Ce_{0.99}Ti_{0.01}O_4:0.005Eu^{3+}$ [Figure 5(a)] is composed of a broad band of lower intensity in the green-blue spectral region (which is already observed in the undoped $Sr_2Ce_{0.99}Ti_{0.01}O_4$[Figure 4(b)]) that overlaps with a set of Eu^{3+} intra-$4f^6$ sharp lines in the blue, green and red regions. As shown in Figure5, they are assigned to the transitions between the $^5D_0$, $^5D_1$ and $^5D_2$ excited states and $^7F_{0\text{–}3}$ ground states. When the Eu^{3+} concentration y=0.01, $Sr_{1.99}Ce_{0.99}Ti_{0.01}O_4:0.01Eu^{3+}$ shows white emission because of the intensities of the Ce^{4+}–O^{2-} CT emission band and $^5D_2$–$^7F_0$ transitions of Eu^{3+} in blue–green region are comparable the $^5D_0$–$^7F_2$ transitions of Eu^{3+} in orange–red regions[Figure 5(b)]. It is found that the intensity of the Ce^{4+}– O^{2-} CT emission decreases whereas the Eu^{3+} emission lines ($^5D_0$–$^7F_2$) enhances gradually with increasing of Eu^{3+}-doped concentration. By increasing the Eu^{3+} concentration to y = 0.03 ($Sr_{1.97}Ce_{0.99}Ti_{0.01}O_4:0.03Eu^{3+}$), the intensity of the Ce^{4+}–O^{2-} CT emission band becomes negligible when compared to those of the Eu^{3+} emission lines [Figure 5(c)]. The Ce^{4+}–O^{2-} CT emission band completely disappears and the $^5D_0$–$^7F_2$ line in the red region becomes the most intense line for $Sr_{1.95}Ce_{0.99}Ti_{0.01}O_4:0.05Eu^{3+}$, so the sample shows red emission [Figure 5(d)]. However, on going to higher concentrations of Eu^{3+} ($x\geq 0.1$), the $^5D_0$–$^7F_2$ emission line decreases gradually [Figure 5(e)]. Among the emission peaks of $^5D_0$–^7F_j (j =1, 2, 3) transition of Eu^{3+}, the 614 nm peak of $^5D_0$–$^7F_2$ transition is dominant. It indicates that the Eu^{3+} in $Sr_{2-y}Ce_{0.99}Ti_{0.01}O_4:yEu^{3+}$ ($0.005\leq y\leq 0.10$) samples occupy the lattice sites without inversion symmetry, and it also implies that our samples would have good color purity. Figure 6 exhibited the luminescence intensity of $^5D_0$–$^7F_2$ transition of Eu^{3+} (614nm) as a function of Eu^{3+}-doped concentration in the system $Sr_{2-y}Ce_{0.99}Ti_{0.01}O_4:yEu^{3+}$ ($0.005\leq y\leq 0.10$). The results indicate that the emission intensity of 614 nm ($^5D_0$–$^7F_2$) enhances gradually with increasing of Eu^{3+}-doped concentration, and the quenching concentration is 0.05mol. The CIE chromaticity coordinates of $Sr_{1.95}Eu_{0.05}Ce_{0.99}Ti_{0.01}O_4$ sample calculated by "Zolix Color Convert" program is (0.665, 0.330).

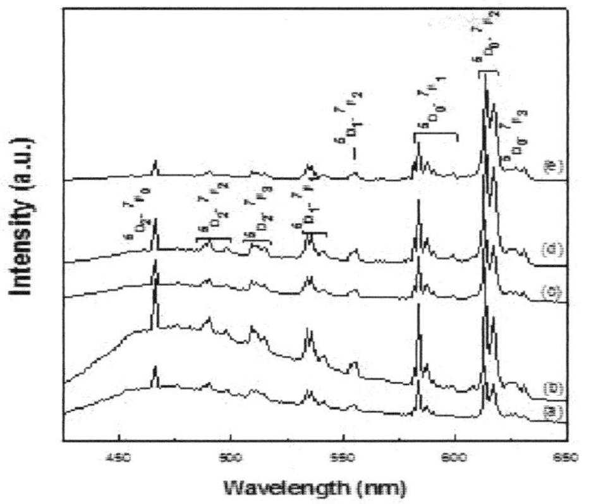

Figure 5. Emission spectra of $Sr_{2-y}Ce_{0.99}Ti_{0.01}O_4$: yEu^{3+} ($0.005 \leq y \leq 0.10$) samples excited at 365 nm: (a) y=0.005, (b) y=0.01, (c) y=0.03, (d) y=0.05, (e) y=0.10

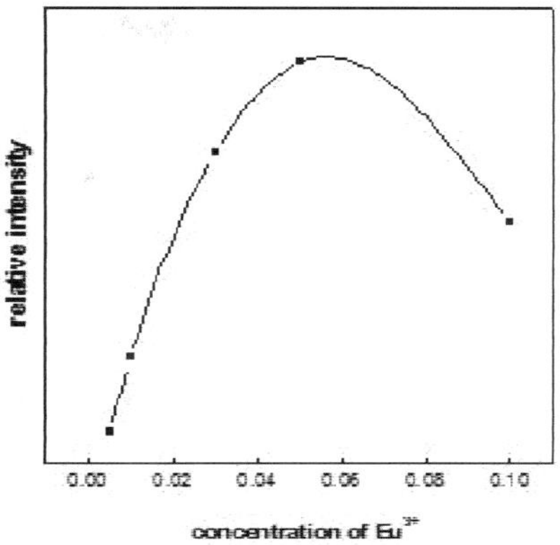

Figure 6. The luminescence intensity of $^5D_0$–$^7F_2$ transition of Eu^{3+} (614 nm) as a function of Eu^{3+}-doped concentration in the system $Sr_{2-y}Ce_{0.99}Ti_{0.01}O_4$:$yEu^{3+}$ ($0.005 \leq y \leq 0.10$).

The emission spectra of $Sr_{1.95}Ce_{1-x}Ti_xO_4$:$0.05Eu^{3+}$ ($0 \leq x \leq 0.03$) samples under 365 nm excitation are not given, because of the profiles of emission spectra of all samples are similar to each other and do not show obvious differences from the $Sr_{1.95}Ce_{0.99}Ti_{0.01}O_4$:$0.05Eu^{3+}$ sample (see in Figure 5(d)).The relative emission intensities and CIE chromaticity coordinates of $Sr_{1.95}Ce_{1-x}Ti_xO_4$:$0.05Eu^{3+}$ ($0 \leq x \leq 0.03$) and commercial phosphor Y_2O_2S:$0.05Eu^{3+}$ are listed in Table 1. Among all the samples we obtained, the $Sr_{1.95}Ce_{0.99}Ti_{0.01}O_4$:$0.05Eu^{3+}$ sample is the best one with the emission intensity 1.5 times higher than that of the $Sr_{1.95}CeO_4$:$0.05Eu^{3+}$sample. We can find that $Sr_{1.95}Ce_{0.99}Ti_{0.01}O_4$:$0.05Eu^{3+}$ phosphor shows not only strong emission intensity but also good color purity. Compared with the commercial red phosphor Y_2O_2S:$0.05Eu^{3+}$, the emission intensity of $Sr_{1.95}Ce_{0.99}Ti_{0.01}O_4$:$0.05Eu^{3+}$ is about 3.0 times higher than that of Y_2O_2S:$0.05Eu^{3+}$ under 365 nm excitation which can also be found from their emission spectra (see in Figure7). The CIE chromaticity coordinates of $Sr_{1.95}Ce_{0.99}Ti_{0.01}O_4$:$0.05Eu^{3+}$ are also closer to the NTSC standard values than that of Y_2O_2S:$0.05Eu^{3+}$.These results showed that the $Sr_{1.95}Ce_{0.99}Ti_{0.01}O_4$:$0.05Eu^{3+}$ phosphor could be considered as a potential red phosphor for application in LEDs.

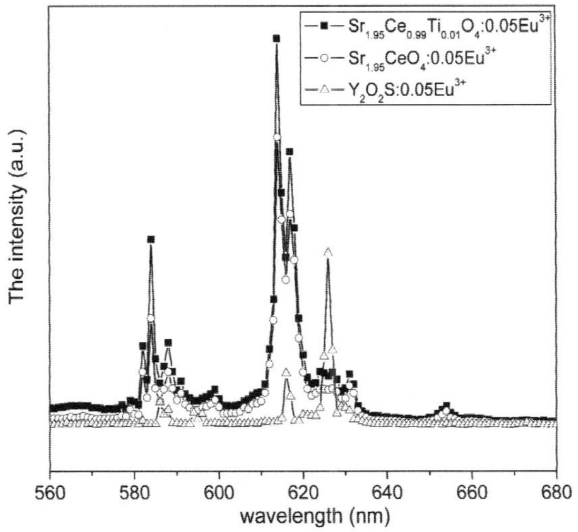

Figure 7. Emission spectra of $Sr_{1.95}Ce_{1-x}Ti_xO_4$:$0.05Eu^{3+}$ (x=0, 0.01) and Y_2O_2S:$0.05Eu^{3+}$ (λ_{ex}=365 nm).

Table1 The CIE chromaticity coordinates and relative intensities of $Sr_{1.95}Ce_{1-x}Ti_xO_4$: $0.05Eu^{3+}$ ($0 \le x \le 0.03$) and Y_2O_2S:$0.05Eu^{3+}$.

Phosphor	CIE chromaticity coordinates[a]		Relative intensity of $^5D_0$–$^7F_2$ (100%)[b]
	X	Y	
$Sr_{1.95}CeO_4$:$0.05Eu^{3+}$	0.685	0.358	100
$Sr_{1.95}Ce_{0.995}Ti_{0.005}O_4$:$0.05Eu^{3+}$	0.672	0.335	134
$Sr_{1.95}Ce_{0.99}Ti_{0.01}O_4$:$0.05Eu^{3+}$	0.665	0.330	152
$Sr_{1.95}Ce_{0.98}Ti_{0.02}O_4$:$0.05Eu^{3+}$	0.668	0.330	126
$Sr_{1.95}Ce_{0.97}Ti_{0.03}O_4$:$0.05Eu^{3+}$	0.660	0.328	112
Y_2O_2S:$0.005Eu^{3+}$	0.631	0.350	48

a The NTSC standard values x=0.67, y=0.33.
b The intensity of $Sr_{1.95}CeO_4$:$0.05Eu^{3+}$ is regarded as 100%.

4. Conclusions

A new series of red-emitting phosphors $Sr_{2-y}Ce_{1-x}Ti_xO_4$:yEu^{3+} ($0 \le x \le 0.03$, $0.005 \le y \le 0.10$) were successfully synthesized using a solid-state reaction at 1200°C for 6h. The photoluminescence investigation showed that the profiles of emission spectra of $Sr_{2-y}Ce_{1-x}Ti_xO_4$:yEu^{3+} are similar to $Sr_{2-y}CeO_4$:yEu^{3+} and the luminescence intensity of $Sr_{2-y}CeO_4$:yEu^{3+} excited at 365 nm could be enhanced largely by substituting Ce^{4+} with Ti^{4+}. In particular, the emission color can be changed by tuning activator's concentration in $Sr_{2-y}Ce_{1-x}Ti_xO_4$:yEu^{3+} phosphors: the doped compound emits strong white light when the concentration of the doped Eu^{3+} is low ($y \le 0.01$), when the Eu^{3+} concentration is increased ($y \ge 0.03$), it emits red light. An optimal phosphor $Sr_{1.95}Ce_{0.99}Ti_{0.01}O_4$:$0.05Eu^{3+}$ showed much stronger emission intensity and better color purity than the commercial red phosphor Y_2O_2S:Eu^{3+}. As a conclusion, this novel red phosphor could be considered as a promising red phosphor for application in LEDs.

Acknowledgments

This study was supported by Key Science and Technology Project of Gansu Province (Grant No. 2GS064-A52-036-03), and the Project of the combination of Industry and Research by the Ministry of Education and Guangdong Province (Grant No.0712226100023).

References

[1] Hirosaki N, Xie R J and Kimoto K 2005 *Appl. Phys. Lett.* **86** 211905
[2] Rack P D, Naman A, Holloway P H, Sun S S and Tuenge R T 1996 *MRS Bull.* **21** 49
[3] Nakamura S 1997 *MRS Bull.* **22** 29
[4] Zhao X, Wang X, Chen B, Meng Q, Di W, Ren G and Yang Y 2007 *J. Alloys Compd.* **433** 352
[5] Steigerwald D A, Bhat J C, Collins D, Fletcher R M, Holcomb M O, Ludowise M J, Martin P S and Rudaz S L 2002 *IEEE J. Sel. Top. Quantum Electron.* **8** 310
[6] Shinoya M and Yen W M 1999 *Phosphor HandBook* CRC Press Boca Raton FL
[7] Neeraj S, Kijima N and Cheetham A K 2004 *Chem. Phys. Lett.* **387** 2
[8] Kutty T R N and Nag A 2003 *J. Mater. Chem.* **13** 2271
[9] Danielson E, Devenney M, Giaquinta D M, Golden J H, Haushalter R C, McFarland E W, Poojary D M, Reaves C M, Weinberg W H and Wu X D 1998 *Science* **279** 837
[10] Pieterson V L, Soverna S and Meijerink A 2000 *J. Electrochem. Soc.* **147** 4688
[11] Chiga T, Masui T, Imanaka N and Adachi G Y 2001 *Chem. Lett.* **12** 1246
[12] Masui T, Chiga T, Imanaka N and Adachi G Y 2003 *Mater. Res. Bull.* **38** 17
[13] Hinatsu Y and Wakeshima M 1999 *J. Solid. State. Chem.* **144** 20
[14] Sankar R and Subba Rao G V 2000 *J. Electrochem. Soc.* **147** 2773

[15] Hirai T and Kawamura Y 2004 *J. Phys. Chem. B* **108** 12763
[16] Park C, Kim C, Pyun C and Choy J 2000 *J. Lumin.***87–89** 1062
[17] Jiang Y D, Zhang F, Summers C J and Wang Z L 1999 *Appl.Phys.Lett.***74** 1677
[18] Serra O A, Severino V P, Calefi P S and Cicillini S A 2001 *J. Alloys Compd.* **323–324** 667
[19] Tang Y X, Guo H P and Qin Q Z 2002 *Solid State Commun.* **121** 351
[20] Nag A and Narayanan Kutty T R 2003 *J. Mater.Chem.***13** 370
[21] Zhang L, Hong G Y and Sun X L 1999 *Chin.chem lett* **10** 799

Journal of Physics: Conference Series **152** (2009) 012090 doi:10.1088/1742-6596/152/1/012090

Enhancement of long-persistence by Ce^{3+} co-doping in $Sr_2MgSi_2O_7:Eu^{2+}$, Dy^{3+} blue phosphor

Yanqin LI[1], Yuhua WANG[1, 2,*]

[1] Department of Materials Science, School of Physical Science and Technology, Lanzhou University, Lanzhou, 730000, PR China

[2] Key Laboratory for Magnetism and Magnetic Materials of the Ministry of Education, Lanzhou University, Lanzhou 730000, P. R. China

E-mail address: wyh@lzu.edu.cn

Abstract. $Sr_2MgSi_2O_7:Ce^{3+}$, $Sr_2MgSi_2O_7:Eu^{2+}$, $Sr_2MgSi_2O_7:Eu^{2+}$, Ce^{3+}, $Sr_2MgSi_2O_7:Eu^{2+}$,Dy^{3+} and $Sr_2MgSi_2O_7$: Eu^{2+}, Dy^{3+}, Ce^{3+} were prepared under a weak reducing atmosphere by a solid state reaction. Ce^{3+} singly doped sample had no afterglow. $Sr_2MgSi_2O_7:Eu^{2+}$, Ce^{3+}showed some afterglow with a short persistenence. By incorporation of Ce^{3+}, an efficient energy transfer from Ce^{3+} to Eu^{2+} was found and emission intensity of Eu^{2+} was enhanced. The triply doped phosphor $Sr_2MgSi_2O_7$: Eu^{2+}, Dy^{3+}, Ce^{3+} has higher brightness and longer lasting time than $Sr_2MgSi_2O_7:Eu^{2+}$, Dy^{3+}.

1. Introduction

The blue-luminescent $Sr_2MgSi_2O_7:Eu^{2+}$, Dy^{3+} phosphor has drawn many research interests since it was first developed by Zhiguo Xiao et al [1]. Comparing with previously developed aluminates materials, the silicate phosphor has more advantages on chemical stability, heat stability, lower cost and excellent weather resistance [2]. However, its long-lasting phosphorescence property was inferior to that of aluminates materials. So it is necessary to further improve its long-lasting phosphorescence properties. So far, $Sr_2MgSi_2O_7:Eu^{2+}$, Dy^{3+} phosphor has been extensively studied on the synthesis techniques [3~6], and little research has been done on improving its long-lasting phosphorescence properties.

It is known that Co-doping is one of the most commonly used methods to make long lasting phosphors. Usually with proper codopants, the persistent time can be increased by a factor of ten, for example, Dy^{3+} in $SrAl_2O_4:Eu^{2+}$, Dy^{3+} [7]. Recently, Huajie Song et al [8, 9] reported by doping the $SrAl_2O_4:Eu^{2+}$, Dy^{3+} lattice with other rare earth ions like Tb^{3+} and Gd^{3+}, the phosphor has higher brightness and longer lasting time. Ren Chen et al [10] obtained a similar result by doping the $SrAl_2O_4:Eu^{2+}$, Dy^{3+} lattice with Yb^{3+}.

Ce^{3+} ion activated long lasting phosphors have been reported by Dongdong Jia et al [11, 12] and Kodama et al [13]. Ce^{3+} is well known as an efficient sensitizer and Ce^{3+} to Eu^{2+} energy transfer is possible [14, 15]. Dongdong Jia et al[14]found that through energy transfer from Ce^{3+} to Eu^{2+}, $CaS:Eu^{2+}$,Tm^{3+},Ce^{3+} has higher brightness and longer lasting time than $CaS:Eu^{2+}$,Tm^{3+}.

In this work, Ce^{3+} was used to enhance the long-lasting afterglow of $Sr_2MgSi_2O_7:Eu^{2+}$, Dy^{3+} through energy transfer from Ce^{3+} to Eu^{2+}.Combining the facts that extra electron traps were created due to nonequivalent substitution by the incorporation of Ce^{3+}, we expected to have a brighter afterglow and longer persistence time.

2. Experimental

Samples were prepared by solid-state reaction $SrCO_3$ (AR), H_2SiO_3 (AR), Mg $(NO_3)_2{\cdot}6H_2O$ (AR), Eu_2O_3 (99.99%), Dy_2O_3 (99.99%) and CeO_2 (99.99%) were used as starting materials. A stoichiometric amount of the starting materials were weighed and thoroughly mixed in an agate mortar, subsequently fired at 1230°C for 5 h in a reducing atmosphere.

All the samples were characterized by powder X-ray diffraction (XRD) using a Rigaku diffractometer with Ni-filtered Cu Kα radiation. The photoluminescence spectra were measured by a FLS-920T fluorescence spectrophotometer with Xe 900 (450 W xenon arc lamp) as the light source. The decay curves were measured by a PR305 Phosphorophotometer. The thermoluminescence (TL) curves were measured by a FJ-427A1 meter with a heating rate of 1 K/s. Before the measurement, the samples were irradiated by UV light (365 nm) for 10 min, and then were kept in a darkroom for 3 h. All the measurements are performed at room temperature except TL curves.

3. Results and discussion

3.1. Crystal structure

Figure 1 shows XRD patterns of select compositions. All the peaks can be indexed to the phase of $Sr_2MgSi_2O_7$ (JCPDS 75-1736) and all the compositions show single phase formation. No impurity phase has been observed in any of the compositions, clearly implying that the little amount of doped rare earth ions have almost no effect on the $Sr_2MgSi_2O_7$ phase composition. All of the other samples were also characterized to be single phase by XRD.

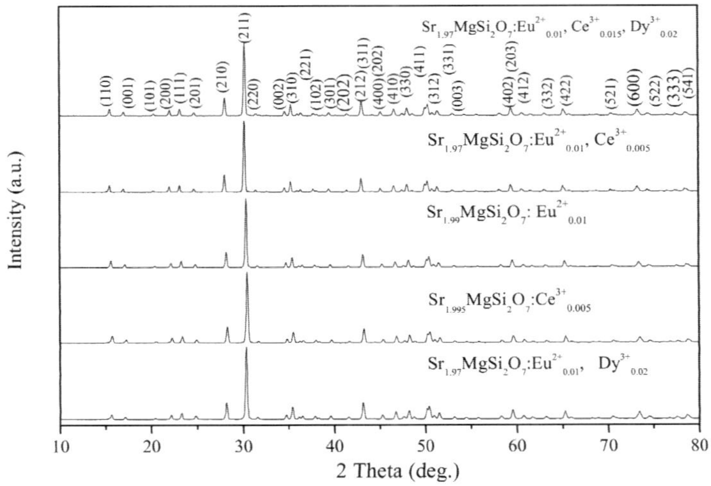

Figure 1. Powder XRD patterns of $Sr_{1.97}MgSi_2O_7$:$Eu^{2+}_{0.01}$, $Dy^{3+}_{0.02}$(a), $Sr_{1.995}MgSi_2O_7$:$Ce^{3+}_{0.005}$(b), $Sr_{1.99}MgSi_2O_7$: $Eu^{2+}_{0.01}$(c), $Sr_{1.97}MgSi_2O_7$:$Eu^{2+}_{0.01}$, $Ce^{3+}_{0.005}$(d), $Sr_{1.97}MgSi_2O_7$:$Eu^{2+}_{0.01}$, $Ce^{3+}_{0.015}$, $Dy^{3+}_{0.02}$(e).

3.2. Photo-luminescence properties

3.2.1. $Sr_2MgSi_2O_7$:Ce^{3+} and $Sr_2MgSi_2O_7$:Eu^{2+}.

The excitation and emission spectra of the samples are show in Figure 2. The excitation and emission spectra of $Sr_2MgSi_2O_7$:Ce^{3+} is shown in Figure 2, the excitation spectra mainly consists of two broad bands centered about 260 nm and 325 nm, which is due to $4f^1 - 4f^0 5d^1$ transition. The emission peaks at 350 and 370nm are due to the transition from 5d to $^2F_{7/2}$ and $^2F_{5/2}$ "ground state", respectively. Figure 3 shows the excitation and emission spectra of $Sr_2MgSi_2O_7:Eu^{2+}$. The excitation spectra shows a broad band from 230 to 450 nm .The emission spectra is also a broad band centered at 466 nm ,which is attributed to the typical $4f^6 5d^1$-$4f^7$ transition of Eu^{2+}. The results are in agreement with the previous literature reports [1, 2, 5].

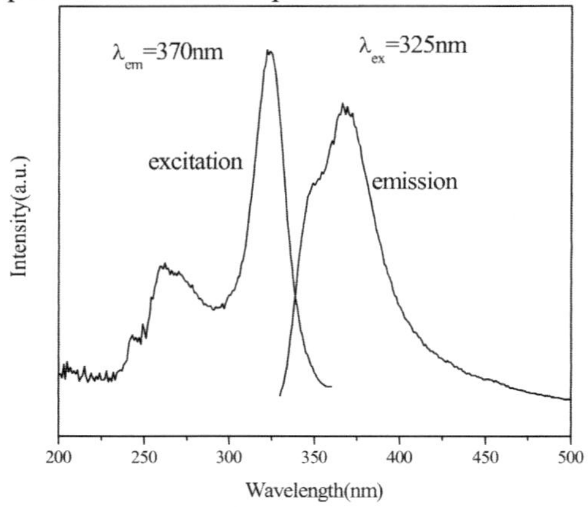

Figure 2. Excitation and emission spectra of $Sr_{1.999}MgSi_2O_7$: $Ce^{3+}_{0.001}$

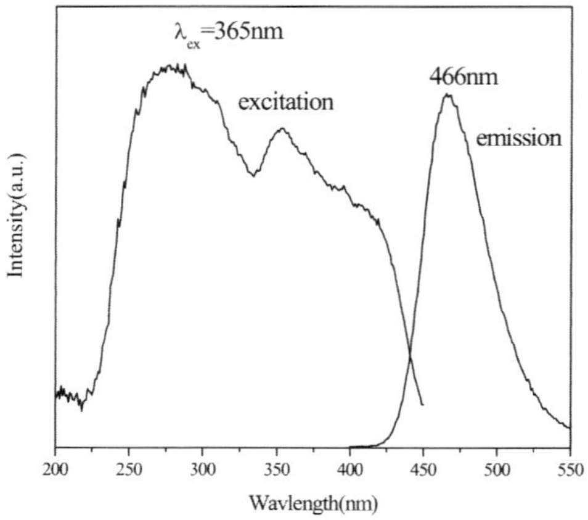

Figure 3. Excitation and emission spectra of $Sr_{1.99}MgSi_2O_7$: $Eu^{2+}_{0.01}$

3.2.2. $Sr_2MgSi_2O_7$:Ce^{3+}, Eu^{2+}.

The excitation and emission spectra of $Sr_2MgSi_2O_7:Ce^{3+}$, Eu^{2+} is shown in Figure 4. The excitation wavelength of the emission spectra was at 325 nm (Ce^{3+} excitation peak), even though there is no Ce^{3+} emission found; Both the position and the shape of the excitation and emission spectra are the same with the single doped Eu^{2+} sample (Figure 3), but the intensity is greatly enhanced, which implies that Ce^{3+} can efficiently transfer the absorbed energy to Eu^{2+}.

Generally energy transfer occurs only when the emission band of sensitizer overlaps spectrally with the absorption band of activator. Spectral overlap has been observed in Ce^{3+} emission and Eu^{2+} excitation in $Sr_2MgSi_2O_7$, which is shown in Figure 5. Energy transfer usually took place at high doping concentration [16]. But for $Sr_{1.985}MgSi_2O_7:Eu^{2+}_{0.01}$, $Ce^{3+}_{0.001}$, the overlap of Ce^{3+} emission and

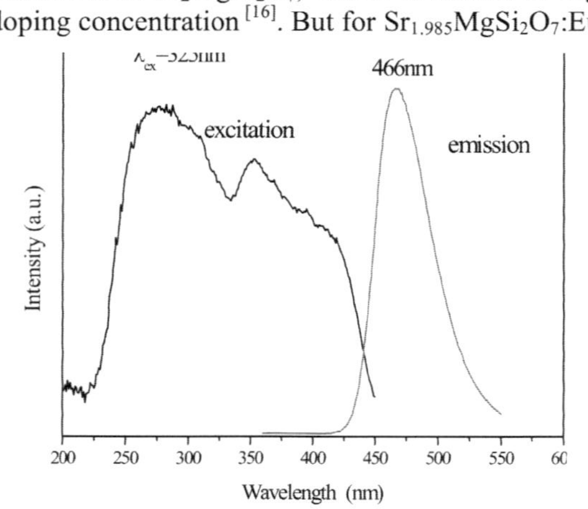

Figure 4. Excitation and emission spectra of $Sr_{1.989}MgSi_2O_7:Eu^{2+}_{0.01}$, $Ce^{3+}_{0.001}$.

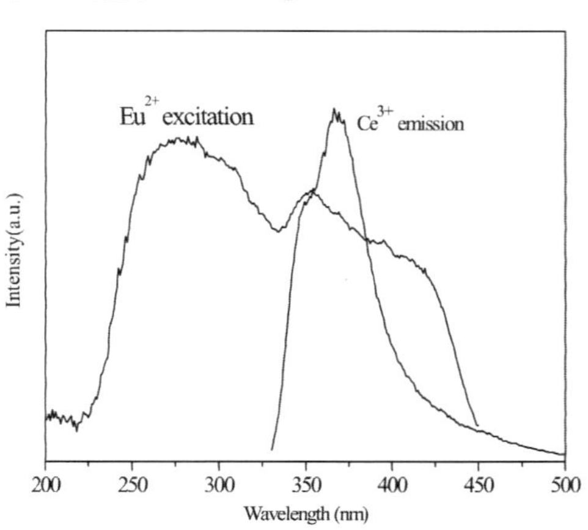

Figure 5. Spectral overlaps between Eu^{2+} excitation and Ce^{3+} emission.

Eu^{2+} absorption was almost 100% (Figure 5).Therefore, energy transfer from Ce^{3+} to Eu^{2+} was strong and Ce^{3+} emission was totally quenched by Ce^{3+}-Eu^{2+} energy transfer, the similar phenomenon was reported by Jia et al [14] in CaS:Eu^{2+},Ce^{3+} system.

3.2.3. *$Sr_2MgSi_2O_7$:Ce^{3+}, Eu^{2+}, Dy^{3+}*

The emission spectrum of $Sr_{1.965}MgSi_2O_7$:$Ce^{3+}{}_{0.005}$, $Eu^{2+}{}_{0.01}$, $Dy^{3+}{}_{0.02}$ under 325nm excitation is shown in Figure 6.The emission band centered at 466 nm is attributed to the typical $4f^65d^1$- $4f^7$ transition of Eu^{2+}. Weak Ce^{3+} emission centered at 370nm appears in the emission spectrum and the emission is very weak compared to the Eu^{2+} emission suggesting efficient energy transfer from Ce^{3+} to Eu^{2+}. In our work ,the special Dy^{3+} emission are not present, indicating that Dy^{3+} ions maybe serve as the hole or electron traps and energy transporting media[17].

3.3. Afterglow properties

Figure 7 shows decay curves of the phosphors $Sr_{1.99}MgSi_2O_7$:$Eu^{2+}{}_{0.01}$ and $Sr_{1.989}MgSi_2O_7$:$Eu^{2+}{}_{0.01}$, $Ce^{3+}{}_{0.001}$. It is found that $Sr_2MgSi_2O_7$:Eu^{2+} shows some afterglow with a short persistenence and its afterglow can be enhanced by co-doping Ce^{3+}. However, $Sr_2MgSi_2O_7$:Ce^{3+} sample shows no afterglow.

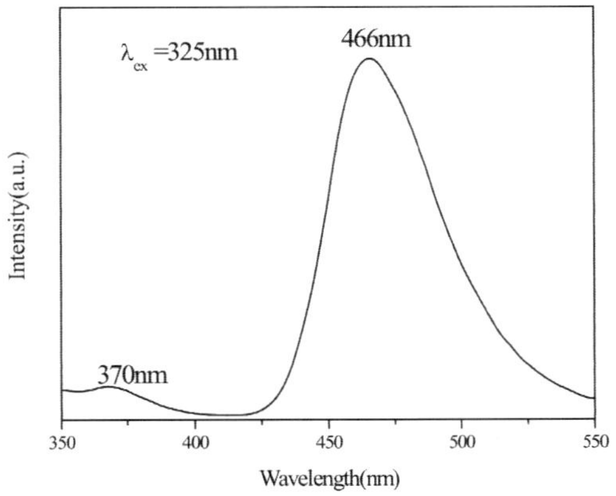

Figure 6. Emission spectra of $Sr_{1.965}MgSi_2O_7$:$Ce^{3+}{}_{0.005}$, $Eu2^{+}{}_{0.01}$, $Dy^{3+}{}_{0.02}$.

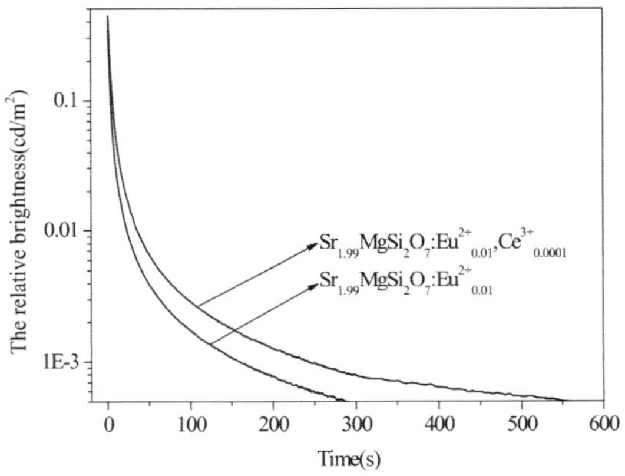

Figure 7. Decay curves of the phosphors $Sr_{1.99}MgSi_2O_7$:$Eu^{2+}{}_{0.01}$ and $Sr_2MgSi_2O_7$:$Eu^{2+}{}_{0.01}$, $Ce^{3+}{}_{0.001}$.

The decay curves of $Sr_2MgSi_2O_7$: Eu^{2+}, Dy^{3+} phosphors with different molar ratio doping of Ce^{3+} ions is shown in Figure 8. All of them show both a rapid decay and subsequential long-lasting phosphorescence. It is clear from the curves that the initial intensity of the phosphors rise dramatically at first and then decreased when the molar ratio of Ce^{3+} ions doping is over 0.005.When the molar ratio of Ce^{3+} ions doping is 0.005, the initial brightness and afterglow duration of the phosphor is greatly enhanced (inset).

3.4. Thermoluminescence

TL curves of $Sr_2MgSi_2O_7$:Eu^{2+}, $Dy^{3+}$and $Sr_2MgSi_2O_7$:Eu^{2+}, Dy^{3+}, Ce^{3+} samples are shown in Figure 9. Only one peak at 352 K is observed above room temperature. It is important to note that the TL intensity of the $Sr_2MgSi_2O_7$:Ce^{3+}, Eu^{2+}, Dy^{3+} phosphor is much stronger than that of $Sr_2MgSi_2O_7$: Eu^{2+}, Dy^{3+}.

The thermal activation energy E_t (associated with the trap depth) can be calculated from the glow-peak parameters by the following equation [18].

$$E_t = [2.52 + 10.2(\mu_g - 0.42)]\left(\frac{k_B T_m^2}{\omega}\right) - 2k_B T_m \qquad (1)$$

where ω, the FWHM, is known as the shape parameter and defined as $\omega = \delta + \tau$, with δ being the high-temperature half width and τ the low-temperature half width. The asymmetric glow-peak shape is defined by the asymmetry parameter $\mu_g = \delta/\omega$. Tm is the temperature of the glow peaks.

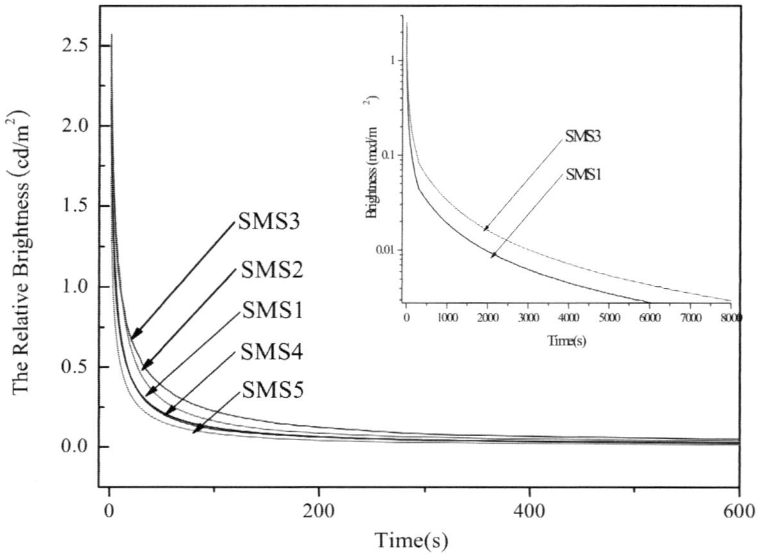

Figure 8. Decay curves of the phosphors $Sr_{1.97-x}MgSi_2O_7:Eu^{2+}_{0.01}$, $Dy^{3+}_{0.02}$, Ce^{3+}_{x}, x is 0.00, 0.001, 0.005, 0.01, 0.015, respectively, corresponding to SMS1, SMS2, SMS3, SMS4, SMS5

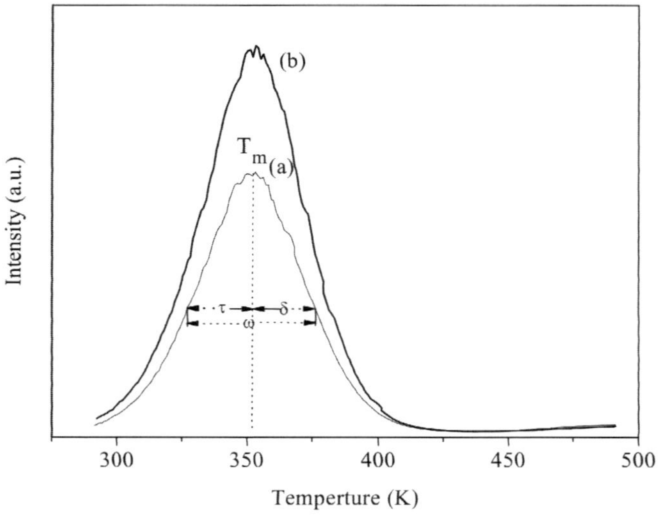

Figure 9.TL curves of (a) $Sr_{1.97}MgSi_2O_7:Eu^{2+}_{0.01}$, $Dy^{3+}_{0.02}$ (b) $Sr_{1.965}MgSi_2O_7:Eu^{2+}_{0.01}$, $Dy^{3+}_{0.02}$,$Ce^{3+}_{0.005}$

Table 1 gives the TL parameters and calculated results of the trap depth. It can be seen from Table.1 that the trap depth of $Sr_{1.97}MgSi_2O_7:Eu^{2+}_{0.01}$, $Dy^{3+}_{0.02}$ is about 0.65 eV, slightly shallower than that of $Sr_{1.965}MgSi_2O_7$:$Eu^{2+}_{0.01}$, $Dy^{3+}_{0.02}$, $Ce^{3+}_{0.005}$ (0.71ev). It is reported that a suitable trap depth (0.65~0.75eV) is essential for phosphors to show long persistence [19], therefore, the trap density of both the above phosphors is suitable for long afterglow. So, we conclude that the difference in afterglow property comes mainly from the difference in trap density. It is well known that a higher trap density normally lead to a higher initial afterglow and longer persistence, and the intensity of glow peaks is proportional to the trap density. From Figure 9, we can clearly see the TL intensity of the $Sr_{1.965}MgSi_2O_7$:$Ce^{3+}_{0.005}$, $Eu^{2+}_{0.01}$, $Dy^{3+}_{0.02}$ is much stronger than that of $Sr_{1.97}MgSi_2O_7$: $Eu^{2+}_{0.01}$, $Dy^{3+}_{0.02}$, indicating the former phosphor shows a better afterglow than the latter. This result is in good agreement with the afterglow curve in Figure 8.

Table 1.TL parameters of $Sr_{1.97}MgSi_2O_7$:$Eu^{2+}_{0.01}$, $Dy^{3+}_{0.02}$ (SMS-EN) and $Sr_{1.965}MgSi_2O_7$:$Eu^{2+}_{0.01}$, $Dy^{3+}_{0.02}$, $Ce^{3+}_{0.005}$ (SMS-ENC)

Composition	T_m(K)	Glow peak shape parameters				TL parameters
		τ	δ	ω	μ_g	E(eV)
SMS-EN	352	24	23	47	0.49	0.65
SMS-ENC	352	23	23	46	0.50	0.71

4. Conclusion

$Sr_2MgSi_2O_7$ doped with Ce^{3+}, or Eu^{2+}, or Ce^{3+} and Eu^{2+}, or Eu^{2+} and Dy^{3+}, or Ce^{3+}, Eu^{2+} and Dy^{3+} were synthesized by solid state reaction. $Sr_2MgSi_2O_7$:Ce^{3+} showed an intense intensive UV emission around 370nm and it had no afterglow. An efficient energy transfer from Ce^{3+} to Eu^{2+} was found in $Sr_2MgSi_2O_7$. Co-doping with Ce^{3+}, both $Sr_2MgSi_2O_7$:Eu^{2+} and $Sr_2MgSi_2O_7$:Eu^{2+}, Dy^{3+} exhibited longer persistent afterglow and enhanced afterglow brightness.

Acknowledgment
This work is supported by Key Science and Technology Project of Gansu Province (Grant no. 2GS064-A52-036-03).

References

[1] Z.G. Xiao, Z.Q. Xiao Chinese patent No. 98105078.6(1998)
[2] X.X. Luo, J.X Duan, G.X Lin, J. Xu,Y. Yang and Z.G. Xiao 2003 *Chin.J. Lumin.* **24**(2)165-170
[3] H.M Ji, G.J Xie and Y. Lv 2007 *J. Sol-gel. Sci. Technol.* **44** 133-137
[4] S. S., T. K, I. Y 2005 *J. Ceram. Soc. Jpn.* **113**(1319) 484-487
[5] F.L. Song, D.H. Chen and Y.H Yuan 2008 *J. Alloys. Compd.* **458** 564-568
[6] Xu, Y. and D. Chen, In Press, Corrected Proof
[7] Matsuzawa, Y. Aoki, N. Takeuchi and Y. Murayama 1996 *J. Electrochem. Soc.* **143**(8) 2670-2673
[8] H.J. Song, D.H. Chen and W.J. Tang 2008 *Displays* **29**(1) 41-44
[9] H.J. Song and D.H. Chen 2007 *Luminescence* **22**(6) 554-558
[10] C. Ren, Y.H. Wang, Y.H Hu, Z.F. Hu and C. Liu 2008 *J. Lumin.* **128**(7) 1180-1184
[11] D. Jia, X. J. Wang, E.van der Kolk and W.M. Yen 2002 *Opt. Mater.* **204** 247-251
[12] D. Jia and W. M. Yen. 2003 *J. Electrochem. Soc.* **150**(3): 61-65
[13] Kodama, N., Y Tanii and M Yamaga 2000 *J. Lumin.* **87**(9): 1076-1078
[14] D.Jia 2006 *J. Electrochem. Soc.* **153**(11): 198-201
[15] Sivakumar, V. and U. V. Varadaraju 2007 *J. Electrochem. Soc.* **154**(5) 167-171
[16] D.Jia, W. Wang,W.Jia and W.M.Yen 2003 *J. Appl. Phys.* **93** 148-152
[17] Y. Lin, Z. Zhang, Z. Tang, X. Wang, J. Zhang and Z. Zheng 2001 *J. Euro. Ceram. Soc.* **21** 683
[18] R. Chen 1969 *J. Appl. Phys.* **40** 570
[19] T. Zhang and Q. Su 2000 *J. Soc. Inf. Disp.* **8**(1) 27-30

$Ca_{1-x}Mo_{1-y}Nb_yO_4:Eu_x^{3+}$: A novel Red Phosphor for White Light Emitting Diodes

Zhipeng CI[1], Yuhua WANG[1,2]*, Jiachi ZHANG[1]

[1]Department of Materials Science, Lanzhou University, Lanzhou, 730000, P. R. China.

[2]Key Laboratory for Magnetism and Magnetic Materials of the Ministry of Education, Lanzhou University, Lanzhou 730000, P. R. China

E-mail: wyh@lzu.edu.cn

Abstract. Single phase of $Ca_{1-x}Mo_{1-y}Nb_yO_4:Eu_x^{3+}$ ($0.18 \leqq x \leqq 0.26, 0 \leqq y \leqq 0.07$) was synthesized by solid state method. The photoluminescence investigation indicated that $Ca_{1-x}MoO_4:Eu_x^{3+}$ ($0.18 \leqq x \leqq 0.26$) could be effectively excited by 393nm and 464nm, and it exhibited an intense red emission at 615nm. The incorporation of Nb^{5+} strongly enhanced the emission intensity of $^5D_0$-$^7F_2$ transition of Eu^{3+} under 393nm and 464nm excitation and the obtained products showed excellent colour purity of which the best chromaticity coordination was x=0.2, y=0.03. The optimal emission intensity excited by 393nm was about 4.2 times higher than that of $Y_2O_2S:0.05Eu^{3+}$ phosphor. The phosphor could be suitable for the UV LED chips application.

1. Introduction

Phosphor-converted light-emitting diodes (LEDs) technique is an important solid-state illumination [1-4]. In comparison with conventional incandescent and fluorescent lamps, the white LEDs show many advantages such as high luminescent efficiency, long lifetime, compactness, environment friendly and designable features. Generally, the white LED is realized by a combination of an GaN-based blue-LED chip (emitting blue light at 465nm) with a yellow phosphor (i.e. $YAG:Ce^{3+}$ based materials) [5,6]. This method is relatively easy to be fulfilled and the device has been commercialized. However, the scheme of blue LED + yellow phosphor also appears many serious drawbacks such as colour shift, low reproducibility and low colour rendering index (CRI) mainly due to two colour mixing [7,8]. In order to overcome these drawbacks, the white LED has been fabricated by employing blue, green and red emitting multiphase phosphors excited by a near-ultraviolet (UV) InGaN chip recently. The current tri-colour phosphors for near-UV InGaN-LEDs chips are $Y_2O_2S:Eu^{3+}$ for red, $ZnS:(Cu^+,Al^{3+})$ for green, and $BaMgAl_{10}O_{17}:Eu^{2+}$ for blue[9]. Unfortunately, the $Y_2O_2S:Eu^{3+}$ red phosphor can't absorb efficiently in near-UV region and its brightness is about 8 times less than that of the green and blue phosphors. As a result, it is necessary to use a phosphor mixture containing 80% red, 10% green and 10% blue in order to obtain good colour rendering. In addition, the lifetime of the $Y_2O_2S:Eu^{3+}$ is inadequate under extended UV irradiation for its instability. It is the bottleneck for the advancement of white LEDs.

$CaMoO_4$ belongs to the scheetlite-like ($CaWO_4$) iso-structure and the central Mo^{6+} is coordinated by four O^{2-} in tetrahedral symmetry (Td) [10]. As reported by Hu et al, $CaMoO_4:Eu^{3+}$ shows not only desirable absorption in near-UV region but also excellent thermal and chemical stability [11]. Thus, $CaMoO_4:Eu^{3+}$ is considered as a promising red candidate that may substitute sulfide phosphors in white LEDs. However, nowadays, the brightness of $CaMoO_4:Eu^{3+}$ is still inadequate for application in white LEDs. Wang et al co-doped Sm^{3+}, Bi^{3+}, Eu^{3+} in molybdate and obtained an enhanced emission intensity of Eu^{3+} [12,14]. Up to now, there are few investigations considering the substitution for the sites of Mo^{6+}. According to Ropp [13], the Eu^{3+} in the $EuNbO_4$ exhibited intense excitation and emission in UV and near-UV region because of the strong coupling between Eu^{3+} and NbO_4 groups [13,15]. Therefore, in this paper, the $Ca_{1-x}Mo_{1-y}Nb_yO_4:Eu_x^{3+}$ ($0.18 \leqq x \leqq 0.26, 0 \leqq y \leqq 0.07$) phosphors are synthesized and the properties of luminescence are investigated in detail.

2. Experimental

The $Ca_{1-x}Mo_{1-y}Nb_yO_4:Eu_x^{3+}$ ($0.18 \leqq x \leqq 0.26$, $0 \leqq y \leqq 0.07$) samples were prepared by solid-state reaction in air as follows: The starting materials MoO_3(AR), CaO(AR), Nb_2O_5(AR) and Eu_2O_3(4N) were weighed with the stoichiometric ratios. These powders were blended in an agate mortar. Then, the homogeneous mixture obtained was put into an alumina crucible and calcined in a muffle furnace at 750°C for 4h.

The samples were characterized by powder X-ray diffraction (XRD) using a Rigaku diffractometer with Ni filtered Cu-Kα radiation at room temperature. The photoluminescence spectra were measured by a Xe lamp (FLS-920T) at room temperature.

3. Results and discussion

The X-ray powder diffraction patterns show that all samples are single phase, and their colors are white. As an example, the powder XRD patterns of $Ca_{0.8}Mo_{1-x}Nb_xO_4:Eu_{0.2}^{3+}$ ($0 \leqq x \leqq 0.07$) are showed in Figure.1, according to the JCPDS card 29-0351, all peaks can be indexed with space group of I41/a (88). No second phases are observed.

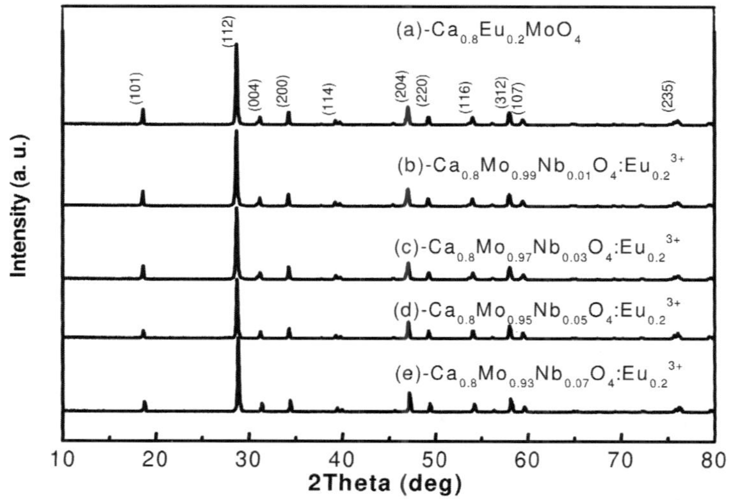

Figure 1. XRD patterns of the samples $Ca_{0.8}Mo_{1-x}Nb_xO_4:Eu_{0.2}^{3+}$ (x_a=0, x_b=0.01, x_c=0.03, x_d=0.05, x_e=0.07).

Figure.2 shows the excitation spectrum of $Ca_{0.8}MoO_4:Eu_{0.2}^{3+}$ phosphor (λ_{em}=615nm). The excitation spectrum is divided into two regions: the board absorption band centered at 270nm is assigned to the combination of the charge transfer(CT) transitions of Eu-O and MoO_4^{2-} group and the contribution of the CT transition of MoO_4^{2-} group is dominant in this broad band [12,14]. The weak and narrow peaks above 350nm are assigned to the f-f transitions of Eu^{3+}. Among them, the dominant sharp peaks at 393nm and 464nm are due to $^7F_0$-$^5L_6$ and $^7F_0$-$^5D_2$ transitions of Eu^{3+}, respectively [12,14,15]. Fig.3 shows the emission spectrum of $Ca_{0.8}MoO_4$:$Eu_{0.2}^{3+}$ and the variation of the luminescent intensities of the phosphors $Ca_{1-x}MoO_4$:Eu_x^{3+} ($0.18 \leqq x \leqq 0.26$) with Eu^{3+} concentration under 393nm excitation. The two red peaks at 615nm and 612nm are assigned to the Eu3+ electric dipole transition of $^5D_0$-$^7F_2$, induced by the lattice sites occupied by Eu^{3+} without inversion symmetry and the break of parity selection rules [11]. The orange peak at 590nm is weak and ascribed to the Eu^{3+} magnetic dipole transition $^5D_0$-$^7F_1$, which is insensitive to the site symmetry [11]. It can be found that the emission intensity at 615nm reaches maximum at 20% in Figure.3. So the Eu^{3+}-doped concentration of $Ca_{1-x}Mo_{1-y}Nb_yO_4$:Eu_x^{3+} ($0.18 \leqq x \leqq 0.26$, $0 \leqq y \leqq 0.07$) samples in the following work is fixed to 20%.

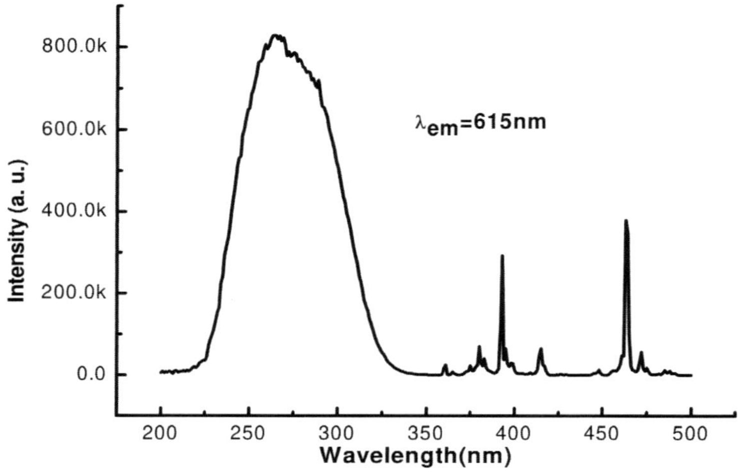

Figure 2. the excitation spectrum of $Ca_{0.8}MoO_4$:$Eu_{0.2}^{3+}$ phosphor

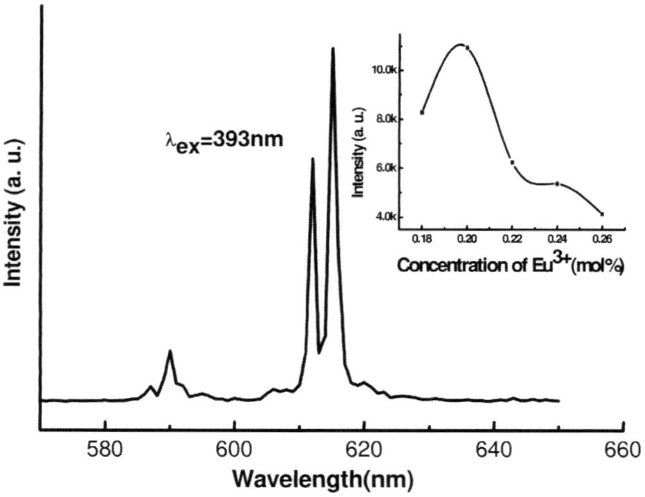

Figure 3. the emission spectrum of $Ca_{0.8}MoO_4$:$Eu_{0.2}^{3+}$ and the variation of the luminescent intensity with Eu^{3+} concentration in the system $Ca_{1-x}MoO_4$:Eu_x^{3+} ($0.18 \leqq x \leqq 0.26$).

The excitation spectra of the $Ca_{0.8}Mo_{1-x}Nb_xO_4$:$Eu_{0.2}^{3+}$ ($0 \leqq x \leqq 0.07$) samples monitored at 615nm are exhibited in Figure.4. These sharp peaks at about 393nm and 464nm are enhanced obviously by increasing of Nb^{5+}. When the Nb^{5+}-doped concentration is up to 3%, the intensity of sharp peak at

393nm is the highest and is nearly higher by 75% than that of the original $Ca_{0.8}MoO_4:Eu_{0.2}^{3+}$ sample (a) while the improvement percent of intensity at 464nm is about 90%. According to Ropp [13], there existed strong coupling between NbO_4^{3-} and Eu^{3+} in some host leading to improvement of emission intensity of Eu^{3+} [16,17,18]. This result indicates that the coupling between NbO_4^{3-} and Eu^{3+} in $Ca(Mo,Nb)O_4$ is strong and the role of Nb^{5+} on increasing the luminescence of $CaMoO_4:Eu^{3+}$ is significant. When the concentration of Nb^{5+} is larger than 7%, the excitation intensities of f-f transitions of Eu^{3+} at 393nm and 464nm are gradually decreased due to the concentration quenching effect. As representatives, the emission spectra of $Ca_{0.8}Mo_{1-x}Nb_xO_4:Eu_{0.2}^{3+}$ ($0 \leqq x \leqq 0.07$) samples excited at 393nm are exhibited in Figure.5. It is observed that the profiles of these emission lines of all samples are similar to each others. The emission intensity is strongest when Nb^{5+} doped content reaches 3% and the optimal emission intensity improves 75% than the original $Ca_{0.8}MoO_4:Eu_{0.2}^{3+}$ sample (a). This observation is coincident with the results derived from excitation spectra.

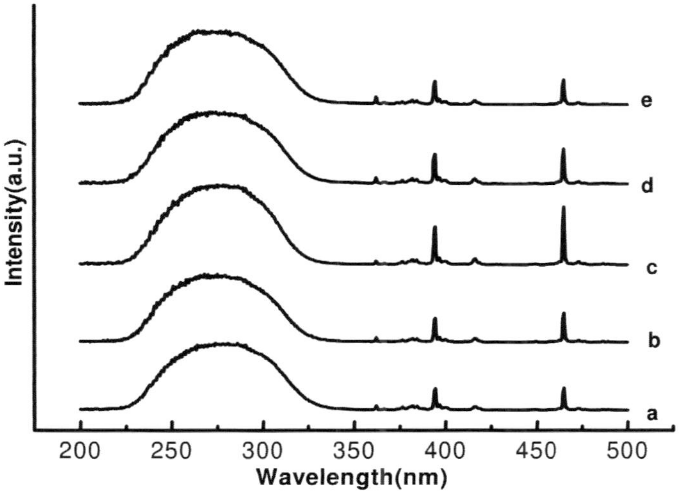

Figure 4. The excitation spectra of $Ca_{0.8}Mo_{1-x}Nb_xO_4:Eu_{0.2}^{3+}$ (x_a=0, x_b=0.01, x_c=0.03, x_d=0.05, x_e=0.07, λ_{em}=615nm).

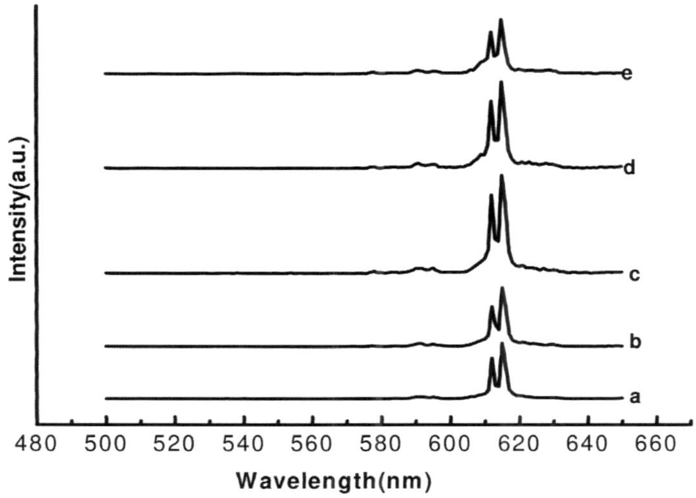

Figure 5. The emission spectra of $Ca_{0.8}Mo_{1-x}Nb_xO_4:Eu_{0.2}^{3+}$ (x_a=0, x_b=0.01, x_c=0.03, x_d=0.05, x_e=0.07, λ_{ex}=393nm).

The relative emission intensities and CIE chromaticity coordinates of $Ca_{0.8}Mo_{1-x}Nb_xO_4:Eu_{0.2}^{3+}$ ($0 \leqq x \leqq 0.07$) and $Y_2O_2S:0.05Eu^{3+}$ are listed in Table I. We can found that not only emission intensity but also chromaticity coordinate of $Ca_{0.8}Mo_{0.97}Nb_{0.03}O_4:Eu_{0.2}^{3+}$ is the optimum. Compared with the red phosphor $Y_2O_2S:0.05Eu^{3+}$, the emission intensity of $Ca_{0.8}Mo_{0.97}Nb_{0.03}O_4:Eu_{0.2}^{3+}$ is about 4.2 times higher than that of $Y_2O_2S:0.05Eu^{3+}$ under 393nm excitation which can also be found from their emission spectra (see Figure.6). The CIE chromaticity coordinates of $Ca_{0.8}Mo_{0.97}Nb_{0.03}O_4:Eu_{0.2}^{3+}$ is closer to the NTSC standard values than that of $Y_2O_2S:0.05Eu^{3+}$. These results imply that the luminescent properties of $Ca_{0.8}Mo_{0.97}Nb_{0.03}O_4:Eu_{0.2}^{3+}$ may be better than that of $Y_2O_2S:0.05Eu^{3+}$ when they are applied in UV LED.

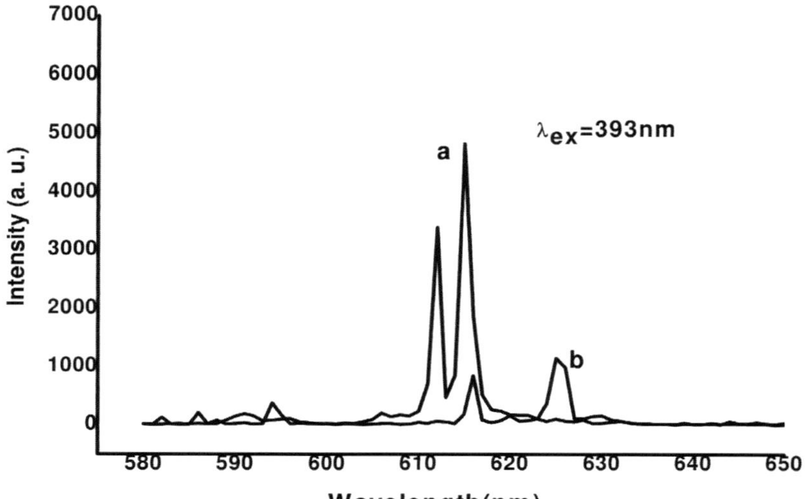

Figure 6. The emission spectra of $Ca_{0.8}Mo_{0.97}Nb_{0.03}O_4:Eu_{0.2}^{3+}$ (a) and Y_2O_2S: $0.05Eu^{3+}$ (b)

Table I. The CIE chromaticity coordinates and relative intensities of $Ca_{0.8}Mo_{1-x}Nb_xO_4:Eu_{0.2}^{3+}$

Phosphor	CIE chromaticity coordinates*		Relative intensities
	x	y	of $^5D_0$-$^7F_2$(100%)[a]
$Ca_{0.8}MoO_4:Eu_{0.2}^{3+}$	0.694	0.366	100
$Ca_{0.8}Mo_{0.99}Nb_{0.01}O_4:Eu_{0.2}^{3+}$	0.665	0.334	115
$Ca_{0.8}Mo_{0.97}Nb_{0.03}O_4:Eu_{0.2}^{3+}$	0.662	0.330	175
$Ca_{0.8}Mo_{0.95}Nb_{0.05}O_4:Eu_{0.2}^{3+}$	0.661	0.330	155
$Ca_{0.8}Mo_{0.93}Nb_{0.07}O_4:Eu_{0.2}^{3+}$	0.662	0.330	120
$Y_2O_2S:0.05Eu^{3+}$	0.631	0.350	41

* The NTSC standard values x=0.67, y=0.33, [a]the intensity of $Ca_{0.8}MoO_4:Eu_{0.2}^{3+}$ is regarded as 100%.

4. Conclusion

A novel red phosphor $Ca_{1-x}Mo_{1-y}Nb_yO_4:Eu_x^{3+}$ ($0.18 \leqq x \leqq 0.26, 0 \leqq y \leqq 0.07$) has been successfully obtained by solid-state reaction at 750℃ for 4h. The x-ray measurements showed that all samples were single phase. The samples $Ca_{1-x}MoO_4:Eu_x^{3+}$ ($0.18 \leqq x \leqq 0.26$) can be excited efficiently by UV(393nm) and blue(464nm) light and emit the red light at 615nm with line spectra. With the introduction of Nb^{5+}, the intensity of emission can be improved and the CIE chromaticity coordinate (x=0.662, y=0.33) was closed to NTSC standard values. Compared with the sulfides phosphors, this novel phosphor has much stronger emission intensity and better colour purity under the excitation of 393nm UV light. Most importantly, the novel phosphor is much more stable than sulfide phosphors. All the results indicated that this red phosphor was more suitable for commercial use in near UV InGaN based LEDs.

Acknowledgements
This study was supported by the Project of the combination of Industry and Research by the Ministry of Education and Guangdong Province (Grant No.0712226100023).

References

[1] Mueller-Mach R, Mueller G O, Krames M R and Trottier T 2002 *IEEE J. Select. Top. Quant. Electron.* **8** 339
[2] Yum J H, Seo S Y, Lee S and Sung Y E 2003 *J. Electron. Soc.* **150** H47
[3] Kim J S, Jeon P E, Park Y H, Chol J C, Park H L, Kim G C and Kim T W 2004 *Appl. Phys. Lett.* **85** 3696
[4] Park J K, Kim C H, Park S H, Park H D and Chol S Y 2004 *Appl. Phys. Lett.* **84** 1647
[5] Schlotter P, Baur J, Hielscher Ch, Kunzer M, Obloh H, Schmidt R and Schneider J 1999 *Mater. Sci. Eng. B* **59** 390
[6] Nakamura S and Fasol G 1997 Springer.Berlin pp.216–221
[7] Narukawa Y 2004 *Opt. Photonics News* **4** 25
[8] Yamada M, Naitou T, Izuno K, Tamaki H, Murazaki Y, Kameshima M and Mukai T 2003 *Jpn. J. Appl. Phys.* **42** L20
[9] Neeraj S, Kijima N and Cheetham A K 2004 *Chem. Phys. Lett.* **387** 2
[10] Barbosa L B, Reyes Ardila D, Cusatis C and Andreeta J P 2002 *Journal of Crystal Growth* **235** 327–332
[11] Hu Yunsheng, Zhuang Weidong, Ye Hongqi, Wang Donghui, Zhang Shusheng and Huang Xiaowei 2005 *Journal of Alloys and Compounds* **390** 226–229
[12] Wang Zhengliang, Liang Hongbin, Gong Menglian and Su Qiang 2006 *Optical Materials* **29** 896-900
[13] Ropp R C 1991 Luminescence and the Solid State. Studies in Inorganic Chemistry Vol. 12 (Elsevier Science.Amsterdam)
[14] Ci Zhipeng, Wang Yuhua, Zhang Jiachi and Sun Yunkui 2008 *Physica B* **403** 670 - 674
[15] Fang Te-Hua, Hsiao Yu-Jen, Chang Yee-Shin and Yen-Hwei Chang 2006 *Materials Chemistry and Physics* **100** 418–422
[16] Blasse G, Bleijenberg K C and Powell R C 1980 Luminescence and Energy Transfer (Springer Heidelberg)
[17] Sang H S,Duk Y J, Kyung S S 2001 *J. Appl. Phys.* **90** 5986
[18] Massabni A M G, Montandon G J M and Couto dos Santos M A 1998 *Mater. Res.* **1** 1

AUTHOR INDEX

AUTHOR INDEX

AUTHOR INDEX